AEROSOLS HANDBOOK

Measurement, Dosimetry, and Health Effects

Second Edition

AEROSOLS HANDBOOK

Measurement, Dosimetry, and Health Effects

Second Edition

Edited by
Lev S. Ruzer and Naomi H. Harley

CRC Press
Taylor & Francis Group
Boca Raton London New York

CRC Press is an imprint of the
Taylor & Francis Group, an **informa** business

CRC Press
Taylor & Francis Group
6000 Broken Sound Parkway NW, Suite 300
Boca Raton, FL 33487-2742

First issued in paperback 2019

© 2013 by Taylor & Francis Group, LLC
CRC Press is an imprint of Taylor & Francis Group, an Informa business

No claim to original U.S. Government works

ISBN-13: 978-1-4398-5510-2 (hbk)
ISBN-13: 978-0-367-86611-2 (pbk)

Visit the Taylor & Francis Web site at
http://www.taylorandfrancis.com

and the CRC Press Web site at
http://www.crcpress.com

Contents

Preface to the Second Edition

Seven years have passed since the publication of the first edition of this book. During this time, considerable research has been accomplished. Still more is underway, particularly regarding nanoparticles and their extensive applications in evolving landscapes. There has been a dramatic increase in the number of publications plus media attention regarding airborne particles and their impact on human health in local and global situations. This has raised people's awareness about what is in the air they breathe or in products they use.

During this period, nanotechnology industries have grown rapidly with federal and state initiatives, and these promise substantial economic benefits. These initiatives will have a significant global impact on scientific aspects applicable to engineering, electronic devices, and sensors. Medically, they were first used in a treatment for AIDS-related Kaposi's sarcoma. They are now injected into patients' bodies as cancer-fighting medicines and imaging agents for disease.

Nanoparticle synthesis, especially for biomedical use, is exquisitely sensitive to reaction conditions, purity of precursors or surfactants used, temperature, etc. At present, there is concern over nanomaterial purity, especially with respect to their biomedical use.

Aerosols are airborne suspensions of biologic, organic, and inorganic origin. They can affect both the global climate and, when inhaled, human health. Climate change alone can affect health. Aerosols consist mainly of carbon- or mineral-based materials and have a broad range of sizes, from nanometers to hundreds of micrometers (four to five orders of magnitude). Their behavior in the global atmosphere, outdoors, indoors, and especially in the lungs is brought up to date in this second edition.

The USEPA assessment of benefits to human health by the Clean Air Act attributed 90% of the estimated benefits to reductions in particulate matter during 1990–2010.

The human health effects associated with aerosols depend on the deposition of the particles in the bronchial and pulmonary regions of the lung. Radioactive aerosols are of interest because their deposition in the lung can be determined and the energy deposited in the known target cells calculated to estimate radiation dose and subsequent health effects.

New chapters are included in the second edition that deal with important practical problems: nanoparticle cell penetration, high aspect ratio nanomaterials, aerosols and climate change, health effects of metals in air, and unattached fraction of radon progeny as a tool for nanoparticles.

Other chapters in this edition update the areas of aerosol physics, medical and pharmaceutical aerosols, bioaerosols, health effects of ultrafine particles, radon epidemiology, long-lived radionuclides in the atmosphere, and lung deposition mechanisms. Gaps in knowledge are identified where further research is needed.

The objective of this second edition remains the same, to provide useful information to understand the science concerning stable and radioactive aerosol particles in the environment with emphasis on human health effects.

Preface to the First Edition

Aerosols consist of particles in the very broad range of sizes from nanometers to hundreds of micrometers (4 to 5 orders of magnitude). Therefore, their behavior is complicated in the atmosphere, indoors, and especially in the lung.

Health effects associated with aerosols depend on the physical parameter that we call "dose." Dose depends on the quantity of aerosols in target cells. With the exception of some radioactive aerosols, it is practically impossible to measure dose directly. In practice, assessment of the dose is provided by measuring air concentration and calculating some known parameters.

According to the U.S. EPA, "in epidemiological studies, an index of exposure from personal or stationary monitors of selected pollutants is analyzed for associations with health outcomes, such as morbidity or mortality. However, it is a basic tenet of toxicology that the dose delivered to the target site, not the external exposure, is a proximal cause of a response. Therefore, there is increased emphasis on understanding the exposure–dose–response relationship. Exposure is what gets measured in the typical study and what gets regulated; dose is the causative factor."

In this book, we present a general, up-to-date overview of all aspects of aerosols, from their properties to the health outcomes. First, current issues related to aerosol measurement are detailed: standardization of measurements for different types of aerosols (indoor, medical and pharmaceutical, industrial, bioactive, and radioactive), with a special emphasis on breathing zone measurements. The handbook also discusses the problems of aerosol dosimetry, such as the definitions of aerosol exposure and aerosol dose, including the issue of nanometer particles, the mechanism of aerosol deposition in the lung, and modeling deposition with an emphasis on the corresponding uncertainty in risk assessment.

A separate part on radioactive aerosols includes aspects such as radon; natural and artificial aerosols; radioactive aerosols and the Chernobyl accident; dosimetry and epidemiology in miners, including direct dose measurement in the lungs; radon and thoron; and long-lived radionuclides in the environment.

It is especially important that the handbook includes an overview of nonradioactive and radioactive aerosols together, because behavior of radioactive aerosols in the lungs, including deposition and biokinetic processes, depends not on their activity, but on particle size distribution and breathing parameters. On the other hand, radioactivity of aerosols is the most useful tool for the study of their behavior in the lungs.

The handbook concludes with overviews of different aspects related to the health effects of diesel aerosols, health risks from ultrafine particles, and epidemiology to molecular biology.

Editors

Lev S. Ruzer enrolled at the recently founded Department of Nuclear Physics at Moscow University after being demobilized from the Soviet Army. After graduation, he could not work as a scientist for political reasons. He had relatives in the United States, and his father was executed during Joseph Stalin's regime. For 8 years, he worked as a teacher in schools at Moscow.

After Stalin's death, Ruzer found a job as a scientist in a medical institute. The research included the assessment of the dose to animals exposed to radon and its decay products. Based on this theoretical and experimental work, he defended his degree as a candidate of physico-mathematical sciences (equivalent to PhD) in 1961.

From 1961 to 1979, he worked as the founder and chairman of Aerosol Laboratory at the Institute of Physico-Technical and Radiotechnical Measurements in Moscow. Under his supervision, the set of installations for generating and measuring different types of aerosols was certified as a State Standard of Aerosols in the USSR. This work does not have analogy in the world.

In 1968, Ruzer published a book on radioactive aerosols. In 1970, he became a doctor of technical sciences, and in 1977 he became a professor. He also served as a scientific supervisor to eight candidates of sciences. In 1979, he was discharged from his position for political reasons. His children were involved in dissident activity. He spent the following eight years without work and unable to obtain permission to emigrate. He finally arrived to the United States in 1987 and published his second book in English.

In 1989, Ruzer was invited as a visiting researcher to Lawrence Berkeley National Laboratory, where he currently works. During this period, he published papers in the new field of dosimetry of nanoparticles. He also published a book on radioactive aerosols (2001), in Russian, and served as the editor of *Aerosol Handbook: Measurement, Dosimetry, and Health Effects* (Taylor & Francis Group, 2004). He has now completed the second edition.

Ruzer has more than 130 publications and 3 patents to his credit.

Naomi H. Harley received her BS in electrical engineering from The Cooper Union, her ME in nuclear engineering, and her PhD in radiological physics from the New York University Graduate School of Science. She obtained an APC in management from the New York University Graduate Business School. Dr. Harley was elected a council member to the National Council on Radiation Protection and Measurements (NCRP) in 1982 and was made an honorary member in 2000. She is an advisor to the United Nations Scientific Committee on the Effects of Atomic Radiation (UNSCEAR).

Dr. Harley's major research interests are the measurement of inhaled or ingested radionuclides, the measurement of environmental radioactivity, the modeling of their fate within the human body, the calculation of the detailed radiation dose to cells specifically implicated in carcinogenesis, and risk assessment from exposure to internal radioactivity.

Dr. Harley has authored over 150 journal publications, 8 book chapters, and has 5 patents for radiation detection instrumentation. The most recent patent, issued in 2009, was for an integrating miniature particle size sampler. Dr. Harley is a researcher and professor at the New York University School of Medicine.

Contributors

Michael G. Apte
Indoor Environment Department
Environmental Energy Technologies Division
Ernest O. Lawrence Berkeley National
    Laboratory
Berkeley, California

A.M. Baklanov
Institute of Chemical Kinetics and Combustion
Russian Academy of Sciences
Novosibirsk, Russia

Michelle L. Bell
School of Forestry and Environmental Studies
Yale University
New Haven, Connecticut

V.V. Boldyrev
Scientific and Education Centre "Molecular
    Design and Ecologically Safe Technologies"
Novosibirsk State University
and
Institute of Solid State Chemistry and
    Mechanochemistry
Russian Academy of Sciences
Novosibirsk, Russia

A.O. Bryzgalov
Institute of Organic Chemistry
Russian Academy of Sciences
Novosibirsk, Russia

A.K. Budyka
Karpov Institute of Physical Chemistry
Moscow, Russia

Judith C. Chow
Desert Research Institute
Reno, Nevada

Beverly S. Cohen
Nelson Institute of Environmental Medicine
New York University School of Medicine
New York

Daniel J. Cooney
Novartis Pharmaceuticals Corporation
San Carlos, California

Hugo Destaillats
Indoor Environment Department
Environmental Energy Technologies Division
Ernest O. Lawrence Berkeley National
    Laboratory
Berkeley, California

Jeroen Douwes
Centre for Public Health Research
Massey University
Wellington, New Zealand

D.E. Fertman
Joint Stock Company Specialized Scientific
    Research Institute for Instrumentation
    Engineering
Moscow, Russia

Isabel M. Fisenne (retired)
Environmental Measurements Laboratory
United States Department of Energy
New York, New York

V.M. Fomin
Institute of Theoretical and Applied Mechanics
Russian Academy of Sciences
Novosibirsk, Russia

Lucila Garcia-Contreras
Oklahoma University
Oklahoma City, Oklahoma

Robert J. Garmise
Bristol-Myers Squibb
New Brunswick, New Jersey

Lara A. Gundel
Indoor Environment Department
Environmental Energy Technologies Division
Ernest O. Lawrence Berkeley National
    Laboratory
Berkeley, California

Steven M. Hankin
SAFENANO, Institute of Occupational
    Medicine
Edinburgh, United Kingdom

Naomi H. Harley
School of Medicine
New York University
New York, New York

Anthony J. Hickey
Research Triangle Institute
Research Triangle Park, North Carolina

Kristin K. Isaacs
United States Environmental Protection
    Agency
National Exposure Research Laboratory
Research Triangle Park, North Carolina

Latarsha D. Jones
School of Pharmacy
University of North Carolina at Chapel Hill
Chapel Hill, North Carolina

V.V. Karasev
Institute of Chemical Kinetics and Combustion
Russian Academy of Sciences
Novosibirsk, Russia

M.V. Khvostov
Novosibirsk Vorozhtsov Institute of Organic
    Chemistry of the Siberian Branch of Russian
    Academy of Sciences
Novosibirsk, Russia

A.A. Kirsch
Karpov Institute of Physical Chemistry
and
Russian Scientific Centre Kurchatov Institute
Moscow, Russia

V.L. Kustova (retired)
Joint Stock Company Specialized Scientific
    Research Institute for Instrumentation
    Engineering
Moscow, Russia

Yu.V. Kuznetzov (retired)
All-Russian Scientific Institute to Physico-
    Technical and Radiotechnical Measurements
    (VNIIFTRI)
Moscow, Russia

Morton Lippmann
Department of Environmental Medicine
 School of Medicine
New York University
Tuxedo, New York

Janet M. Macher
Environmental Health Laboratory
California Department of Public Health
Richmond, California

Stephen R. Marek
College of Pharmacy
University of Texas at Austin
Austin, Texas

Ted B. Martonen
Cyberlung, Inc.
Laguna Beach, California

B.I. Ogorodnikov
Karpov Institute of Physical Chemistry
Moscow, Russia

A.A. Onischuk
Institute of Chemical Kinetics and Combustion
Russian Academy of Sciences
Novosibirsk, Russia

Craig A. Poland
SAFENANO, Institute of Occupational
    Medicine
Edinburgh, United Kingdom

Brad Prezant
Centre for Public Health Research
Massey University
Wellington, New Zealand

P.A. Purtov
Institute of Chemical Kinetics and Combustion
Russian Academy of Sciences
Novosibirsk, Russia

Tiina Reponen
Department of Environmental Health
University of Cincinnati
Cincinnati, Ohio

A.I. Rizin
Joint Stock Company Specialized Scientific
    Research Institute for Instrumentation
    Engineering
Moscow, Russia

Charles E. Rodes
Center for Aerosol Technology
Research Triangle Institute International
Research Triangle Park, North Carolina

Jacky A. Rosati
United States Environmental Protection
    Agency
National Homeland Security Research Center
Research Triangle Park, North Carolina

Lev S. Ruzer
Ernest O. Lawrence Berkeley National
    Laboratory
Berkeley, California

Richard G. Sextro
Indoor Environment Department
Environmental Energy Technologies Division
Ernest O. Lawrence Berkeley National
    Laboratory
Berkeley, California

Hugh D.C. Smyth
College of Pharmacy
University of Texas at Austin
Austin, Texas

I.V. Sorokina
Institute of Organic Chemistry
Russian Academy of Sciences
Novosibirsk, Russia

Ira B. Tager
School of Public Health
Division of Epidemiology
University of California at Berkeley
Berkeley, California

Jonathan W. Thornburg
Center for Aerosol and Nanomaterials
    Engineering
Research Triangle Institute International
Research Triangle Park, North Carolina

George D. Thurston
Department of Environmental Medicine
School of Medicine
New York University
Tuxedo, New York

G.A. Tolstikov
Novosibirsk Vorozhtsov Institute of Organic
    Chemistry of the Siberian Branch of Russian
    Academy of Sciences
Novosibirsk, Russia

T.G. Tolstikova
Novosibirsk Vorozhtsov Institute of Organic
    Chemistry of the Siberian Branch of Russian
    Academy of Sciences
Novosibirsk, Russia

S.V. Vosel
Institute of Chemical Kinetics and Combustion
Russian Academy of Sciences
Novosibirsk, Russia

John G. Watson
Desert Research Institute
Reno, Nevada

1 Medical and Pharmaceutical Aerosols

Stephen R. Marek, Hugh D.C. Smyth, Lucila Garcia-Contreras, Daniel J. Cooney, Robert J. Garmise, Latarsha D. Jones, and Anthony J. Hickey

CONTENTS

1.1   INTRODUCTION

1.1.1   HISTORICAL PERSPECTIVE

The use of aerosol therapy for the treatment of pulmonary disorders can be traced to India over 5000 years ago.[1] These therapies were either palliative, in the form of smokes and mists, or therapeutic, containing pharmacologically active agents such as stramonium alkaloids.[2] There have been periodic improvements in our understanding of diseases, which give relevance to aerosol approaches.

Modern aerosol therapy with pure drugs was initiated in the 1950s with the development of the pressurized metered dose inhaler (MDI) for the delivery of β-adrenergic agonists to facilitate bronchodilatation to relieve the bronchoconstriction occurring in asthmatic patients.[3] In the past 50 years, aerosol therapy has become a central element of asthma management and its potential for the treatment of other pulmonary and systemic diseases has been explored.

1.1.2   FUTURE PROSPECTS

The future of aerosol therapy seems assured. New developments occur at frequent intervals in the areas of drug discovery, formulation, and device development.[4,5] The major areas in which new therapies can be anticipated are treatment of lung diseases such as chronic obstructive pulmonary disease (COPD),[6] emphysema,[7] lung cancer,[8] and systemic diseases such as diabetes[9] and prostate cancer.[10]

New drugs continue to be developed for the treatment of asthma and COPD such as the anticholinergic agent, tiotropium.[11] Undoubtedly, new compounds will be developed as the incidence and severity of these diseases justifies biomedical research into their underlying causes and mechanisms of pathogenesis.

Formulation strategies have focused on methods of particle manufacture[12] and desirable physicochemical characteristics.[13,14] These particles can then be placed in one of three general categories of device: pressurized MDIs, dry powder inhalers (DPIs), or nebulizers.[15] There continue to be new and exciting developments in each of the categories of device.[5]

1.2   THERAPEUTIC AGENTS

1.2.1   LOCALLY ACTING MEDICAL AND PHARMACEUTICAL AEROSOLS

1.2.1.1   β-Adrenergic Agonists

Figure 1.1a shows the structure of a β-adrenergic agonist. This category of compounds was the first to be commercially available for the treatment of the symptom of bronchoconstriction in asthmatic individuals. The mechanism of action of β_2-adrenergic agonists is to act on the sympathetic system to cause muscle relaxation and, therefore, bronchodilation in the lungs. Structurally these

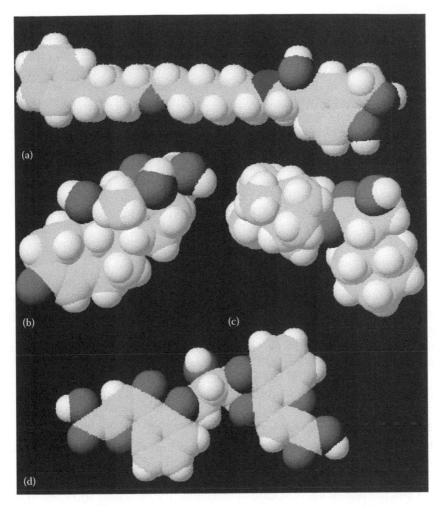

FIGURE 1.1 Examples of structures of common pharmacological agents used in medical and pharmaceutical aerosols for delivery to the lungs (space-filling models of (a) salmeterol, (b) fluticasone, (c) ipratropium, and (d) cromolyn).

agents fell into three groups: catechols, resorcinols, and saligenins.[16] Each compound was a structural analog to epinephrine (adrenaline). Indeed, epinephrine appeared in an early product, which is now available over the counter at pharmacies (Primatene, Whitehall-Robins, Richmond, VA). After some early toxicity issues with nonspecific β-adrenergic agonists, the specific β_2-adrenergic agonists were produced and introduced with great success to manage the symptoms of asthma. The first pharmacologically specific product, albuterol (GSK, RTP, NC), was a short-acting agent introduced in the 1960s. It was followed by a number of variants including fenoterol (BI, Ingelheim, Germany) and terbutaline (Astra-Zeneca, Lund, Sweden). In the late 1980s and early 1990s, these were replaced for routine maintenance therapy by long-acting β_2-adrenergic agonists, notably formoterol (BI, Ingelheim, Germany) and salmeterol (GSK, RTP, NC). The short-acting agents were retained as rescue medications for patients experiencing acute exacerbations. The newest treatment options being investigated allow for a single daily dose and fall under the class of ultra-long-acting β_2-adrenergic agonists (ultra-LABAs). Indacaterol has been approved for usage in Europe (Onbrez, Novartis, Basel, Switzerland), and the other products currently in the most advanced stages of development are carmoterol and GSK-642444 (GSK, RTP, NC).[17]

1.2.1.2 Corticosteroids

Figure 1.1b shows the structure of a glucocorticosteroid molecule. Inhaled corticosteroids (ICS) were introduced to treat the underlying cause of asthma, inflammation, and are rapidly becoming the preferred treatment for asthma.[18] Corticosteroids had been used as an oral therapy for severe, life-threatening conditions in the form of prednisone and prednisolone. The development of agents with some lung specificity gave rise to the concept of inhaled steroid therapy. The first agent to be developed was beclomethasone (GSK, RTP, NC and Schering Plough, White Plains, NJ), followed closely by triamcinolone (Aventis, Collegeville, PA); both were used in pressurized MDIs. In the late 1980s and early 1990s, two steroids with a high degree of lung specificity, budesonide and fluticasone, were developed as DPI products. These molecules were shown to induce lower side effects, notably cortisol suppression, than their predecessors. Recently, more drugs have arrived on the market such as mometasone furoate (MF) as a DPI (Asmanex Twisthaler, Schering Corp., Kenilworth, NJ)[19] and ciclesonide as a pressurized MDI (Alvesco, Nycomed GmbH, Zurich, Switzerland).[20] MF may also be prescribed for COPD.[21]

1.2.1.3 Anticholinergics

Figure 1.1c shows the structure of an anticholinergic agent. Anticholinergic agents are known parasympathetic antagonists that operate on the opposing arm, balancing bronchomotor tone to the sympathomimetic agents described in Section 1.2.1.1. The first anticholergic agent was ipratropium, which was followed by oxitropium. Cholinergic receptors are known to be centrally located in the airways and, consequently, these agents are effective in dilating the bronchioles in this region. Tiotropium is a recent addition to this category of drugs, which has been shown to be very effective in COPD, which requires not only bronchodilatation but also the loosening of mucus. All of these agents are produced by Boehringer Ingelheim (Ingelheim, Germany).

1.2.1.4 Anti-Inflammatory Agents

Figure 1.1d shows the structure of a unique anti-inflammatory agent. Disodium cromoglycate (cromolyn sodium) (Aventis, Collegeville, PA) was developed in the 1960s and 1970s and has been shown to have a number of effects in the lungs. The most prominent effect is mast cell stabilization and the prevention of release of inflammatory mediators. However, it is not clear what the dominant action is that renders this molecule therapeutically effective. This molecule has the distinction of being the first in modern times to be administered as a dry powder aerosol. A follow-up molecule was developed by the same company in the 1980s, nedocromil sodium. Contained in a pressurized MDI (Tilade, King Pharmaceuticals, Cary, NC), it was a treatment for asthma until the FDA ban on CFCs in inhalers took effect in 2008 and U.S. production was terminated.[22]

1.2.1.5 Antimicrobials

A number of antimicrobial agents have been delivered to the lungs to treat different diseases. Pentamidine (Fujisawa, North Chicago, IL and Aventis, Collegeville, PA) was delivered for the treatment of *Pneumocystis carinii* pneumonia (PCP). This organism was originally considered to be a parasite but has recently been redesignated taxonomically as a fungus. Fungal therapy had already been attempted with amphotericin B for the treatment of aspergillosis. The occurrence of *Pseudomonas aeruginosa* infections as a corollary to cystic fibrosis has engendered considerable interest in antibiotic aerosol therapy. Initial work on amikacin products has resulted in the development of the tobramycin aerosol treatment (Novartis), which was the first aerosol antibiotic approved in the United States.[23] A monobactam antibiotic, aztreonam, is now available in a nebulizer formulation as Cayston and was approved by the FDA in February, 2010. It was reformulated as the lysine salt to reduce post-inhalation inflammation. Sputum taken after nebulization showed that it retained its antimicrobial properties.[23–25]

1.2.1.6 Biotechnological Agents, Genes, and DNA Aerosols

The most sustained effort in the area of delivery of biotechnological agents has been for the treatment of cystic fibrosis. In the early 1990s, rDNase aerosols (Genetech, South San Francisco, CA) were delivered to cleave leukocyte DNA, which was contributing to the viscosity of mucus in cystic fibrotic lungs. Delivery of this aerosol allowed the patient to expectorate readily and clear their lungs of mucin blockages. In combination with tobramycin to treat the bacterial infection, this appears to have been a successful approach. However, it has long been the objective of those involved in gene therapy to challenge their technology by expressing the gene for cystic fibrosis transport receptor (CFTR) in the epithelium of airways cells, thereby correcting the underlying chloride ion imbalance, which gives rise to thickened mucus and poor mucociliary clearance. As yet this approach has met with limited success, but it remains the goal of a number of researchers.[26–29] Oligonucleotides have also been used to target the adenosine A(1) receptor, a G-protein-coupled receptor (GPCR) that plays an important role in the etiology of asthma.[30] Targeting of the Akt pathway for the suppression of lung tumorigenesis has been achieved using shRNA.[31,32] Pulmonary delivery of the siRNA ALN-RSV01 may prove an effective antiviral treatment for respiratory syncytial virus (RSV).[33,34]

1.2.2 Systemically Acting Agents

The lungs have been considered a route of administration for systemically acting agents for decades. The first systemically acting agent delivered as an aerosol product in the twentieth century was ergotamine tartrate for the treatment of migraine headaches.[35]

More recently, the focus has been on proteins and peptides. Among these, the notable candidates to date have been insulin,[9] for the treatment of diabetes, and leuprolide acetate, a luteinizing hormone releasing hormone analog[10] for the treatment of prostate cancer.

Other agents have been evaluated, including calcitonin, human growth hormone,[36] parathyroid hormone,[37] interferons, erythropoietin,[38] and granulocyte–monocyte colony-stimulating factor (GM-CSF),[39] but these have progressed more slowly than the candidates mentioned in the previous paragraph. Alexza Pharmaceuticals has shown success with their Staccato® aerosol delivery device on a wide array of drugs. Loxapine for schizophrenia and bipolar disorders is currently awaiting FDA approval after successful clinical trials.[40,41] Alexza also has alprazolam and prochlorperazine in phase II clinical trials.

Recent reviews of these systemic treatments and more were written by Siekmeier and Scheuch.[42,43] There are several challenges that systemic aerosols must overcome to be suitable treatment options, including the avoidance of the mucus layer and the prevention of protease and peptidase degradation. Typically, the larger the molecular weight of a therapeutic agent, the more difficult these barriers tend to be. Yet, the benefits, such as high blood perfusion and the lack of first-pass metabolism, make pulmonary delivery a highly desirable delivery route.[42]

Inhaled vaccines have also been explored recently. Reports indicate that the administration of a vaccine topically to the airways has several benefits: enhanced mucosal immunity, induction of systemic immunity and immunity at distant mucosal sites, noninvasive administration, reduction in risk of infection spread through needle use/misuse, improved stability, and the potential of elimination of the cold-chain requirements. As such, several diseases have been the focus of inhaled vaccines including measles,[44,45] tuberculosis,[46] HIV,[47–49] influenza,[50–52] and several others.

1.3 CLASSIFICATION

1.3.1 Metered Dose Inhalers

Many devices used to deliver drugs to the respiratory tract do so by producing a metered dose of aerosolized drug that is inhaled by the patient. However, MDIs are specifically recognized as those devices that contain a pressurized formulation that is aerosolized through an atomization nozzle.

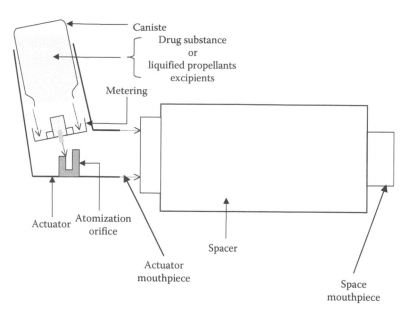

FIGURE 1.2 Basic components of a pMDI system.

More correctly, these devices should be called pressurized metered dose inhalers (pMDIs) to avoid confusion with MDIs that incorporate and use dry powders and aqueous-based systems (discussed in the following).

The first MDI was commercialized in the mid-1950s to compete with glass nebulizers for the delivery of asthma medications. The acceptance and utility of these delivery systems was quickly established and they have now become the most common system for drug delivery to the respiratory tract. The basic design of pMDIs (Figure 1.2) has not changed greatly since their inception and they typically contain three basic components: the active substance, the propellant system, and other stabilizing excipients. These components are enclosed within an aerosol container with a metering valve that connects to an actuator or aerosolization nozzle. An adaptor mouthpiece, which may include one of a variety of holding chambers, allows the patient to draw the aerosol into the lungs. The composition and design of pMDIs affects the performance of the drug delivery system.[53] Propellants serve as a source of energy for atomization of the liquid formulation as it exits the nozzle. They also function as a liquid phase for the dispersion of drug and other excipients that often are present in suspension. Surfactants aid dispersion or dissolution in addition to providing lubrication for valve components.[53–55] Solution formulations are typically attained by incorporation of cosolvents such as ethanol.[53,56] Currently marketed products and those in development can be divided into three classes based upon the propellant system used.

1.3.1.1 CFC Systems

Chlorofluorocarbon (CFC) propellants have been the most common propellant type used in pMDIs. These propellants include CFC 12 (dichlorodifluoromethane), CFC 11 (trichlorofluoromethane), and CFC 114 (1,2-dichloro-1,1,2,2-tetrafluoroethane). The widespread use of these propellants was a consequence of their low pulmonary toxicity, high chemical stability, purity, and compatibility.[53] Also, mixtures of these propellants can be formulated to yield desirable vapor pressures, densities, and solvency properties for successful formulation of a variety of drug substances. However, pMDIs containing CFCs have been phased out and are being replaced by other propellant systems. This is a result of the linking of CFCs to the depletion of stratospheric ozone and the signing of the Montreal Protocol on substances that deplete the ozone layer.[57,58] On April 14, 2010, the U.S. Food and Drug Administration (FDA) issued a "final rule" that confirms the fate of the last seven remaining pMDI products containing CFC propellants.[59] In a phased approach, these remaining products will be removed from the market completely by December 31, 2013. Other countries may still allow marketing of CFC products while alternatives are being developed and while confirmation of reliable supplies of the alternatives is established.

1.3.1.2 HFC Systems

Several hydrofluorocarbon (HFC) compounds were identified as possible CFC propellant alternatives because of their non-ozone-depleting capacity and similar desirable characteristics in common with CFC pMDI propellants (nonflammability, chemical stability, and similar vapor pressures). These propellants currently include HFC 134a (1,1,1,2-tetrafluoroethane) and HFC 227ea (1,1,1,2,3,3,3-heptafluoropropane).[60] Extensive toxicological and safety testing demonstrated that these propellants were at least as safe as the CFC propellants.[60–62] Subsequently, inhalers were approved by regulatory agencies for medical use (Table 1.1). Despite apparent similarities with CFCs, the HFC alternatives have required different formulation strategies and device designs.[57,63,64] In general, these challenges included overcoming different solvency properties,[64–66] designing different materials and coatings for compatibility issues,[67] and identifying different atomization

TABLE 1.1

Common Marketed pMDIs and Their General Composition

Therapeutic Group	Drug	Surfactants/ Excipients	Propellant System	Formulation Type	Particle Size Estimate (Microns)	References
Bronchodilators						
Atrovent HFA	Ipratropium bromide	Citric acid, water	HFA 134a, ethanol	Suspension		
Combivent inhalation aerosol	Ipratropium bromide and albuterol sulfate	Soya lecithin	CFC 11, CFC 12, CFC 114	Suspension	3, 2.4	[196]
Dulera	Mometasone furoate and formoterol fumarate dihydrate	Oleic acid	HFA 227, anhydrous alcohol	Suspension		[197]
Maxair autohalor	Pirbuterol acetate	Sorbitan trioleate	CFC 11, CFC 12	Suspension	3.1	[198]
Proventil HFA	Albuterol sulfate	Oleic acid	HFA 134a, ethanol	Suspension	1.96 2.21	[198]
Ventolin HFA	Albuterol sulfate	None	HFA 134a	Suspension	2.5	[199]
Xopenex HFA	Levalbuterol tartarate	Oleic acid	HFA 134a, ethanol	Suspension		
Corticosteroids						
Azmacort	Triamcinolone acetonide		CFC 12, 1% w/w ethanol	Suspension	4.33	[200]
Flovent HFA	Fluticasone propionate		HFA 134a	Suspension		
QVAR autohaler	Beclomethasone dipropionate		HFA 134a, ethanol	Solution	1.0	[201]
Symbicort	Budesonide and formoterol fumarate dihydrate	Povidone K25 USP, polyethylene glycol 1000 NF	HFA 227	Suspension		

behaviors.[68–70] An example of a reformulated product is an HFA (hydrofluoroalkane) inhaler containing beclomethasone dipropionate, which is now marketed as a solution formulation instead of a suspension formulation. Regional lung deposition was shown to be different from the existing CFC product due to a smaller droplet size and throat deposition characteristics.[71] Thus, in addition to pharmaceutical issues, regulatory issues are also important in the transition to more "environmentally friendly" propellant systems. These issues have meant that in 2001, 14 years since the signing of the Montreal Protocol, over half of the world's pMDIs were still CFC-based systems.[60]

1.3.1.3 Alternative Propellants

This third category of propellant systems includes propellants such as dimethylether and low-molecular-weight hydrocarbons such as butane and propane.[72,73] Compressed gases have also been considered for incorporation into medical aerosol systems.[53,73] Although under development, the commercialization of these propellant systems has been restricted by formulation issues and/or safety issues. One concern has been the potential flammability of propellants such as butane, propane, and dimethylether.[53] Handheld aqueous systems are also under development and have similar appearance and operation to pMDIs. However, these systems are more closely related to aqueous-based nebulizers and are discussed in the following.

1.3.2 Dry Powder Inhalers

DPIs have been available since the 1970s when the Rotahaler® (GlaxoSmithKline) and the Spinhaler® (Aventis) products were introduced.[74] DPIs provide an alternative to MDIs that offers the advantage of not using environmentally unfriendly propellants.[75] However, without a propellant to create an aerosol cloud, other dispersion methods must be employed. Most DPIs employ the energy from patient inhalation to disperse the powder. The airflow is directed over or through the static powder bed in order to fluidize it and allow it to be entrained into the inspiratory airflow. This technique offers the advantage of automatic coordination with patient inhalation, a problem with MDIs, but does bring up a number of challenges. The energy from the patient inhalation must be applied in a manner that is able to overcome the cohesive and adhesive forces in the powder to allow dispersion to the lungs.[76] Some patients may not be able to generate a high enough flow to disperse particles and varying flow rates can lead to varying delivered doses.[77] These characteristics, along with hydration of the powders and surface electrical properties, are determined by a combination of the drug formulation and the dispersion device. For this reason, the dispersion problem must be viewed from two different angles: formulation and device design.

The aim of formulation optimization is to improve powder dispersion properties. The efficiency of an inhaler device is often measured by the amount of particles that enter the periphery of the lungs during inhalation. These particles need to have an aerodynamic diameter smaller than 5–7 μm, and the portion of the aerosol falling into this category is generally referred to as the fine particle fraction (FPF).[78] The interparticulate adhesive forces become dominant in particles in this size range and they tend to form agglomerates or aggregates.[5] The adhesive forces are influenced not only by particle size, but also by shape, crystallinity, surface morphology, and surface chemistry.[76] A number of approaches have been taken to reduce these forces and increase the flowability and dispersibility of the powders.[79,80] The manufacture of small particles is accomplished by several means. Breaking methods such as milling are employed, but those particles tend to have irregular size, shape, and surface characteristics, and high cohesiveness.[5] Constructive methods like spray drying, evaporation, extraction, and condensation have also been used and tend to produce more uniform and lower-energy particles.[12] Even uniform particles in the 5 μm size range tend to adhere to one another and have poor dispersion and flow properties. For this reason excipients are often included in dry powder formulations.

Carrier particles such as lactose are blended with the active compound to control the dispersion.[76] The carrier particles are designed to keep the small drug particles from forming agglomerates by allowing adhesion to the carrier until inhalation energy frees the particles. Similar to drug particles alone, the size, shape, and surface of the carrier particles affect the dispersion. Ternary components such as L-Leucine and fine particle lactose have been added to blended formulation systems to decrease binding during inhalation, thus increasing FPF.[81,82] Another approach to increasing FPF of formulation is to produce large, low-density particles.[83] These particles have aerodynamic diameters in the appropriate range, but the adhesive forces are smaller in ratio because of the large size.

Recent work has evaluated the possibility of delivering liposomes via DPIs.[84,85] β-glucuronidase was used in one study as a model protein inside dimyristoyl phosphatylcholine and cholesterol liposomes, which were lyophilized, micronized, and then aerosolized. These showed promise, but only 15% of the aerosolized particles were below 6.4 μm, and, thus, these formulations require more optimization.[86] A second study indicated that budesonide could be entrapped inside egg phosphatidyl choline and cholesterol liposomes, freeze dried, and then aerosolized. Long shelf life was demonstrated, along with similar respirable fractions (RFs).[84]

1.3.2.1 Passive

Formulation is only one side of the DPI. Even the best formulation needs a vehicle for dispersion, the inhaler device. The first DPIs developed used the patient's own inhalation airflow directed through or across a capsule that is broken prior to inhalation to aerosolize the powder.[5] Because the only energy source used to deaggregate the particles is the inhalation, these devices are deemed passive inhalers. To increase the deaggregation energy, baffles or deflected airflow are used to create turbulence around the powder bed during inhalation.[76] The first passive inhalers, like the Spinhaler and Rotahaler, required loading of a new capsule for each use as single-dose inhalers. Since then multi-single-dose (device holds multiple capsules or blisters) and true multidose inhalers (dose taken from reservoir for each inhalation) have become available. Table 1.2 contains a list of some inhalers and their properties.

1.3.2.2 Active

Using the inspiratory force of the patient is no longer the only way that powder in dispersed in inhaler design. There has been a recent push to develop active rather than passive inhalers. The inhalation flow rate and force developed vary from breath to breath and certainly from person to person. This makes it very difficult to both insure that a proper dose is delivered and that the inhaler is operating at maximum efficiency. Active DPIs use a source of energy other than inhalation to disperse the powder. Compressed air from a user-operated pump is used as the dispersion energy in the Inhance™ (Inhale, San Carlos, CA) inhaler. Other patents have been filed for active DPI designs using energy sources such as vacuum pressure, an impaction hammer, and vibration.[5,87–90]

1.3.3 Nebulizers

Nebulization is probably the oldest means of administering drugs to the lungs as aerosols. Because coordination between breathing and aerosol generation is not necessary, nebulization of therapeutic agents is mostly used by children, elderly patients, in hospital settings, and for the treatment of lung diseases such as cystic fibrosis and asthma.[91–93] Adaptive aerosol delivery (AAD) systems, such as the I-neb, are able to deliver aerosols at different time points during inhalation, such as the tidal breathing mode (TBM) or the target inhalation mode (TIM), which may provide better delivery compared to conventional nebulizers.[94–96]

Nebulizers produce small polydisperse droplets capable of delivering therapeutic agents to the deep lung in large doses and can deliver a dose 10 times larger than that of DPIs or MDIs.[91] Mass median diameters of droplets generated by nebulizers normally range from 2 to 5 μm and have been used to deliver solutions and suspensions of a great variety of therapeutic agents including macromolecules and biotechnology products.[97] Miniaturization of the hardware and introduction

TABLE 1.2
Selected DPIs and Some of Their Properties

Inhaler	Company	Energy Source	Carrier	Powder Supply	Dosing	Doses
Rotahaler®	GSK	Passive	Lactose	Capsule	Single dose	1
Spinhaler®	Fission/Aventis	Passive	None	Capsule	Single dose	1
Handihaler™	Boehringer Ingelheim	Passive	Lactose	Capsule	Single dose	1
Aerolizer™	Novartis	Passive	Lactose	Capsule	Single dose	1
Inhalator®	Boehringer Ingelheim	Passive	Glucose	Capsule	Multiple unit dose	6
Diskus®/ Accuhaler®	GlaxoSmithKline	Passive	Lactose	Blister	Multiple unit dose	60
Aerohaler®	Boehringer Ingelheim	Passive	Lactose	Capsule	Multiple unit dose	6
Diskhaler®	GlaxoSmithKline	Passive	Lactose	Blister	Multiple unit dose	4, 8
Taifun®	LAB Pharma	Passive	Lactose	Reservoir	Multiple unit dose	200
Easyhaler®	Orion	Passive	Lactose	Reservoir	Multidose	200
Airmax™	IVAX	Passive	Lactose	Reservoir	Multidose	
Novolizer®	Sofotec	Passive	Lactose	Reservoir	Multidose	200
Twisthaler®	Schering-Plough	Passive		Reservoir	Multidose	60
Turbuhaler®	AstraZeneca	Passive	None	Reservoir	Multidose	200
Spiros®	Elan Pharmaceuticals	Impeller	N/A	Blister, Cassette	Single and Multiple unit dose	1, 16, or 30
Inhance™	Inhale	Compressed gas	Lactose	Blister	Single dose	1
Dynamic Powder Disperser™	Pfeiffer	Compressed gas	Lactose	Cartridge	Multiple unit dose	12
Jethaler®	RatioPharm GmbH	Mechanical	Lactose	Ring tablet	Single dose	1
Aspirair®	Vectura	Compressed air	Mannitol	Blister	Single dose	1

Source: Adapted from Dunbar, C. et al., *KONA*, 16, 7, 1998; Newman, S.P. and Busse, W.W., *Respir. Med.*, 96, 293, 2002; Son, Y. and McConville, J.T., *Drug Dev. Ind. Pharm.*, 34, 948, 2008; Islam, N. and Gladki, E., *Int. J. Pharm.*, 360, 1, 2008.

of high-output nebulizers that shorten dosing times and increase drug delivery to the patient have increased the usage of nebulizers.[98,99]

However, some disadvantages related to their use include their lack of portability, use of other accessories such as tubing, mouthpiece or facemasks, obstructive dosing, and cleaning requirements.[100] Others, depending on the nebulizer type, may include inter-device variability, greater costs of drug delivery as a result of the need for extensive assistance from health-care personnel, and requirement of high doses to achieve therapeutic effect.[91]

Based on the mechanism of aerosol production, nebulizers can be classified as air-jet or ultrasonic. In jet nebulizers, the aerosol is produced by applying a high-velocity air stream from a pressurized source at the end of a capillary tube; liquid can be drawn up the tube from a reservoir in which it is immersed. When the liquid reaches the end of the capillary, it is drawn into the airstream and forms droplets that disperse to become an aerosol.[16] In ultrasonic nebulizers, the solution is aerosolized by the vibration of a piezoelectric crystal to induce waves in a reservoir of solution. Interference of these waves at the reservoir surface leads to the production of droplets in the atmosphere above the reservoir. An airstream is passed through this atmosphere to transport the droplets as an aerosol.[101]

Ultrasonic nebulizers require an electrical power source, which can either be AC (wall plug) or DC (battery). The movement of the transducing crystal causes frictional heating, which increases the temperature of the nebulized solution and may be detrimental for proteins and thermolabile drugs. Ultrasonic nebulizers are less popular than jet nebulizers because they are more expensive and not disposable. A common problem with ultrasonic nebulizers is that continuous atomization does not occur if the volume in the chamber falls below 10 mL.[91] Detailed information on these mechanisms of aerosolization has been described elsewhere.[102,103] A newer nebulizer design uses a vibrating mesh or plate. Several advantages include faster nebulization (shorter treatment times), increased efficiency (less drug needed), lower fill volume, and less heating of the nebulized solution.[104,105]

Some factors that influence nebulizer performance and droplet size are density and velocity of the atomizing air, surface tension and viscosity of liquid, concentration, temperature, and nebulizer design.[102] Efficiency of nebulizers is often expressed in terms of energy use, function, or output. Output can be measured by simply weighing nebulizer before and after operation. Variation in total output, particle size, and overall efficiency has been reported between different nebulizers. Several studies have evaluated the differences between nebulizers. Chan et al. examined the aerosol characteristics of five handheld nebulizers in terms of aerosol output and droplet size.[106] Smith et al. tested the variability among 23 different nebulizer/compressor combinations in terms of MMD,[107] while Weber et al. optimized the nebulizer conditions of several nebulizers used to administer antibiotics for the treatment of cystic fibrosis.[108]

1.3.3.1 Solutions

Nebulizers are commonly used with solutions of bronchodilators, such as albuterol or terbutaline and other drugs like sodium cromoglycate, corticosteroids, and pentamidine.[109,110] Cystic fibrosis treatment with antimicrobials has become common as well.[94,111–113] Combinations of drugs could be administered at the same time; however, the stability and possible interactions of the components should be evaluated before administration.[114] A possible outcome can be an insoluble complex in certain conditions. Interactions during nebulization of amiloride hydrochloride and nucleotide UTP have been documented.[115] Other factors that should be accounted for are the effects that some excipients included in the solution to be nebulized may have on the patient. Inclusion of preservatives in the solutions of bronchodilators has been reported to cause bronchoconstriction instead and the addition of osmotic agents may also cause side effects.[16,100,116–118] The effect of other additives, ionic strength, and contamination could also influence the effect of the nebulized solution.[119]

1.3.3.2 Suspensions

Insoluble or inert particles can be suspended in a solution and delivered by nebulization provided that the particle size is smaller than the droplet size and the density of the particle is relatively small. Most steroids are not soluble in water and, therefore, have to be administered as suspensions. It is important to note that when nebulized, suspensions of drugs behave differently than solutions. Cameron et al.[120] compared the performance of five different nebulizers with an amino-phylline aerosol solution and a suspension of budesonide. They found that even with the same nebulizer, different aerosol characteristics were obtained after nebulization of the solution and the suspension. Tiano and Dalby determined how the differences in aerosolization mechanism (jet vs. ultrasonic) affected droplet and insoluble particle deposition of a nebulized model respiratory suspension. Both nebulizers produced droplets large enough to incorporate <6 μm insoluble latex spheres. However, droplets generated by the jet nebulizer contained spheres of all sizes, while with ultrasonic nebulizer, 99% of the spheres were not aerosolized and recovered from the nebulizer.[121]

Liposomal formulations have also been delivered by nebulization.[122–126] Effects of air pressure, temperature, buffer, osmotic strength, and pH on the nebulized liposome dispersions were studied. Changes in air pressure produced large changes in the percentage of release of the encapsulated substance; increasing air pressure increased the percentage of release. The leakage of liposomes was increased in hypotonic solution but decreased in hypertonic solution. At low pH, the leakage was increased compared to higher pH. Stability of liposomes was affected by the operating and environmental conditions of aerosolization, with air pressure having the greatest effect.[122] Droplet size was shown to not significantly influence the drug content in Arikace™ amikacin liposomes after dispersion in an ACI.[127]

Niosomes, which are considered to be more stable than liposomes, have been used to encapsulate all-trans retinoic acid and have been effectively delivered by the PARI-LC STAR nebulizer with droplet in the respiratory size and good entrapment.[128]

1.3.3.3 Macromolecules

Nebulization of macromolecules has been a delicate issue, since some factors inherent to nebulization such as shear stresses and volume have an effect on the stability and integrity of macromolecules.[102] Furthermore, the characteristics of macromolecule solutions are often different from those of regular solutions or suspensions, which in turn may affect nebulizer performance. Biotechnology products frequently form viscous solutions, with modified interfacial and surface tension.[119] The effects of pH, additive, and ionic strength on the delivery of recombinant consensus alpha interferon have been published.[129] It was found that interferon molecule was destabilized by air-jet nebulizer aggregation of the plasmid. This was influenced by pH; the smaller the pH, the bigger the aggregation. Ionic strength of the solution did not influence aggregation. Ultrasonic nebulization of the solution of plasmid also resulted in aggregation but denaturation was dependent upon the type of nebulizer used and related to the heating of nebulizer solutions.

Contradictory statements are made regarding the nebulization of proteins by ultrasonic nebulizers. Some authors state that denaturation of proteins is unavoidable by ultrasonic nebulizers, while others have successfully aerosolized protein solutions with this type of nebulizers.[130–132] It has been observed that by preventing heating of the nebulizer fluid during operation, denaturation of the proteins was altered. In addition, by including 0.01% w/v Tween 80 or 1% w/v PEG 8000 almost all activity of the proteins was retained. Therefore, cooling of the solution in conjunction with the addition of different surfactants[133] is one approach that could be used to stabilize proteins to ultrasonic nebulization. However, cooling may also reduce solute output from the nebulizer.[134,135]

Stribling et al. were among the first to use a jet nebulizer to aerosolize plasmid/liposome vectors to mice for the treatment of cystic fibrosis, with encouraging results.[26] Cipolla et al.[136] have also used this approach efficiently in monkeys. A review has been published by Niven on the delivery of biotherapeutics by inhalation aerosol.[97] It outlines the advantages of delivery of these molecules by inhalation, stability problems when they are aerosolized, and the problems with aqueous protein formulations, and highlights a variety of biotherapeutics given by aerosols.

1.3.4 HANDHELD AQUEOUS SYSTEMS

A major disadvantage of nebulizer systems has been their lack of portability.[137] The next-generation aqueous systems include several smaller battery-powered devices that address this issue. Some examples of these are summarized in Table 1.3. These devices typically deliver small volume doses equivalent to those emitted from pMDIs (10–50 μL).

1.3.5 TOPICAL DRUG DELIVERY SPRAYS

Therapeutic topical aerosols have been used for centuries for various skin diseases. Topical sprays are used due to the ability of a spray to coat evenly the target surface with a minimum of excess drug that may lead to undesirable effects.[138] Modern topical sprays include anti-infectives (e.g., tolnaftate), local anesthetics (e.g., lidocaine, benzocaine), antiseptics (phenol), scabicides (piperonyl butoxide, pyrethrins), and sunscreens (octyl methoxycinnamate). These agents generally act locally or topically. Subsequently, the generation of the aerosol is generally not required to have the same stringent quality controls on particle size or dose as those aerosols used for delivery to the lungs. However, droplet sizes are generally larger so that the probability of inertial impaction on the skin

TABLE 1.3

Specific Challenges of Particle Size Measurement of Pharmaceutical Aerosols

Device Manufacturer	Trade Name	Description of Atomization Mechanism	References
Aerogen	Aerodose®	Electrically induced vibrations to a concave surface generates an aerosol from aqueous solution or suspension formulation	[203]
Aradigm	AERx®	Computer-controlled device using a laser-machined nozzle through which the formulation is mechanically extruded to produce an aerosol	[204] [205]
Battelle	Mystic™	Electrohydrodynamic atomization	[8]
Boehringer Ingelheim	Respimat	Mechanical propulsion	[137]
ODEM	TouchSpray™	Vibrating perforated membrane	[206]

Source: Hickey, A.J. and Swift, D., Characterization of pharmaceutical and diagnostic aerosols, in: Baron, P.A. and Willeke, K. eds., *Aerosol Measurement: Principles, Techniques and Applications*, Wiley, New York, pp. 1031–1052, 2001.

is increased, while respirable particles are minimized.[138] In addition, systemically acting sprays have been developed (NitroLingual™). This sublingual spray is used to deliver nitroglycerin for the treatment of angina. A metered spray device is used. Topical sprays are also being investigated for other systemically acting agents such as hormones.[139,140] Sinomenine hydrochloride has been delivered locally, for the treatment of rheumatoid arthritis, using a topical spray.[141] Related technologies also include high-speed powder transdermal delivery (Powderject®). This involves the acceleration of powder particles of a specific size range, density, and strength so that they penetrate the skin without the need for injections. Applications of this technology include traditional small-molecule pharmaceutical agents, peptides, proteins, and vaccines.[142,143]

1.4 SPECIFIC MEASUREMENT TECHNIQUES AND CALIBRATION

As with other aerosols, pharmaceutical and medical aerosols are described using properties such as particle size, electrostatics, hygroscopicity, and uniformity of drug dispersion.[138] Measurement is performed for product development and also for quality and regulatory purposes. Measurement techniques focus on characterizing the efficiency and reproducibility by which aerosols are generated and delivered to the respiratory tract of patients. *In vitro* and *in vivo* techniques are used and are summarized in the following.

1.4.1 PARTICLE SIZE MEASUREMENT

Particle size analysis is extremely important in the characterization of medical and pharmaceutical aerosols as particle or droplet size and distribution are the most important physicochemical properties influencing lung deposition.[144] Hence, the measurement of particle size is important in several aspects of pharmaceutical aerosols, including particle manufacture, formulation optimization and stability, and quality control. Medical and pharmaceutical aerosols are generally sized using two general classes of particle size analysis: inertial methods and optical methods.

Particle size analysis in medical and pharmaceutical aerosols has unique issues relating to aerosol sampling. There are spatial restrictions between the generation site (device) and the inhalation target site (lungs). The presence of inhalation flow rates and cyclical breathing patterns *in vivo* also complicates the interpretation of particle size observations. Measurement can be particularly challenging given the often unstable nature of generated particles due to evaporation, condensation, temperature, and humidity changes.

1.4.1.1   Inertial Methods

1.4.1.1.1   Cascade Impactor

Cascade impactors are commonly used to characterize environmental materials such as pollutants; however, this chapter will focus on the use of the instrument for characterizing particles designed for delivery to the lung. The cascade impactor has been a tool to determine the particle size of pharmaceutical aerosol dispersions for over 30 years. The impactor applies the principles of inertia discussed earlier in this text to separate particles by size. Particles are dispersed and travel through multiple stages with sequentially decreasing jets. Each stage plate should be thinly coated in oil to avoid particle bounce and blow-off.[145]

There are multiple impactor systems currently on the market and their size specifications can be seen in Table 1.4.[76] The Next Generation Impactor (MSP, Inc., Niwot, CO) was developed to make particle size testing more efficient. It has low internal particle loss and requires minimal washing between tests.[146–148]

There are some differences in the setup of the apparatus when sampling for pulmonary delivery as opposed to environmental testing. First, a throat is added to the impactor to model the anatomy of the human body so as to catch the larger aerodynamic particles.[149] Second, the flow rate of air through the impactor differs from that of environmental sampling.[150] When performing impactions of aerosols intended for pulmonary delivery, airflow rates of 28.3–90 L/min are typically employed.

1.4.1.2   Optical Methods

Optical particle size analysis techniques are commonly used in pharmaceutical aerosol development and testing. Due to the unstable and transient nature of emitted aerosols, these methods frequently perform dynamic measurements. Particle size characterization of pharmaceutical aerosols by static measurements (i.e., microscopy) is also essential during formulation development.[151–153]

Phase Doppler anemometry (PDA) is an open laser system and can sample the aerosol from the point of generation to the extremities.[154,155] Accordingly, PDA has been most often used to characterize the development of nebulizers and MDI aerosols.[154] Time-of-flight (TOF) particle size analyzers are real-time optically based and are used frequently in medical and pharmaceutical aerosols.[156,157] Particles are separated on the basis of their inertia by accelerating the particles through a well-defined flow field and measuring the particle TOF across a split laser beam. TOF analyzers measure single particles and collect distributions by collecting size data on a statistically valid number of particles. Powder dispersion, DPIs, MDIs, and nebulizers have all been investigated using this type of instrument.[157,158]

Laser diffraction as a method of particle size analysis for pharmaceutical aerosols has also been reviewed.[159–161] Laser diffraction techniques are widely used in the characterization of pharmaceutical aerosols due to the rapid nature of data collection. The laser light scattering patterns of an aerosol passing through the laser region allow instantaneous evaluation of multiple particles. Original systems used the principle of Fraunhofer diffraction, although newer laser diffraction equipment relies on Mie scattering theory that more accurately allows the sizing of particles near the same size as the wavelength of the laser.[162] Like PDA systems, various regions and time points of an aerosol produced from an inhaler can be sampled.[159]

A new system currently being investigated for pharmaceutical aerosol measurements is the optical particle counter (OPC). A white light aerosol spectrometer can measure single particle sizes on the order of 0.3–40 µm, although typically it requires very dilute sample conditions. A welas (Palas GmbH) sensor, however, in conjunction with an appropriate dilution mechanism, can be used to measure particle sizes of pMDIs comparable to laser diffraction or cascade impactor values. The aerosol amount can also be accurately measured with such a system, possibly saving some time during the characterization of pMDI aerosols.[163]

TABLE 1.4

Aerodynamic Cutoff Diameters (μm) for Instruments That Determine Particle Size by Inertial Impaction as Reported by the Manufacturer

Apparatus Stage	Twin-Stage Liquid Impinger	USP B	Multistage Impinger				Andersen Viable Impactor	Andersen Nonviable Impactor			Next Generation Impactor			Marple–Miller Impactor			Delron
Flow Rate (L/min)	60	60	30	60	80	100	28.3	28.3	60	90	30	60	100	30	60	90	12.5
Pre-separator	—	—	—	—	—	—	—	—	—	—	15.0	13.0	10.0	—	—	—	—
−1	—	9.8	—	—	—	—	—	—	8.6	8.0	—	—	—	—	—	—	—
0	6.4	—	—	—	—	—	7.1	9.0	6.5	6.5	—	—	—	10.0	10.0	8.0	11.2
1	—	—	16.9	13.3	11.8	10.4	4.7	5.8	4.4	5.2	11.0	7.8	6.0	5.0	5.0	4.0	5.5
2	—	—	9.3	6.7	5.6	4.9	3.3	4.7	3.3	3.5	6.6	4.6	3.6	2.5	2.5	2.0	3.3
3	—	—	4.5	3.2	2.7	2.4	2.1	3.3	2.0	2.6	3.9	2.7	2.1	1.25	1.25	1.0	2.0
4	—	—	2.5	1.7	1.4	1.2	1.1	2.1	1.1	1.7	2.3	1.6	1.2	0.63	0.63	0.6	0.9
5	—	—	—	—	—	—	0.65	1.1	0.54	1.0	1.4	0.96	0.72	—	—	—	0.5
6	—	—	—	—	—	—	—	0.7	0.25	0.43	0.84	0.57	0.42	—	—	—	—
7	—	—	—	—	—	—	—	0.4	—	—	0.51	0.33	0.23	—	—	—	—
MOC	—	—	—	—	—	—	—	—	—	—	0.31	0.13	0.055	—	—	—	—

Source: Dunbar, C. et al., *KONA*, 16, 7, 1998.

MOC, micro-orifice collector.

Currently, optical particle sizing techniques are useful tools in evaluating pharmaceutical aerosols. In general, these techniques are not accepted as pharmacopeial methods because, from a regulatory point of view, they suffer from representative sampling issues and an absence of chemical analysis.[164]

1.4.2 OTHER *IN VITRO* MEASUREMENT TECHNIQUES COMMONLY EMPLOYED FOR PHARMACEUTICAL AEROSOLS

From a development and regulatory viewpoint, particle size is a dominant analytical method used to characterize medical and pharmaceutical aerosols. However, regulatory agencies also recommend that other characteristics of medical aerosols be measured.[75] These include dose content uniformity of the emitted aerosol,[53] spray pattern and geometry measurements,[165] and the effect of storage conditions (temperature, relative humidity, duration), flow rates, and stability challenges.[53]

1.4.2.1 Angle of Repose

The angle of repose of a powder employed in a DPI is useful in determining the relative flowability of the sample. Highly cohesive powders tend to have larger angles of repose, whereas minimally cohesive powders have a very small angle of repose. Several methods are used in determining this value and most methods employ forming a cone of powder and measuring the angle formed between the horizontal surface and the side of the cone. This property relates to the now defunct USP <1174> standard.

A fairly new piece of equipment, the rotating drum[166] (e.g., the Revolution Powder Analyzer by Mercury Scientific Inc.), allows the measurement of the dynamic angle of repose, as well as many other properties. A drum with clear sides is partially loaded with powder, and the drum is rotated at a desired rate while a camera captures the avalanching of the powder. These images are then used in determining the average time to avalanche, the avalanche energy, the dynamic angle of repose, and many other useful properties which can be related to flow.[87] The fluidization, packing, and granulation behavior of these powders can also be analyzed with the Revolution Powder Analyzer.

1.4.2.2 Atomic Force Microscopy

Atomic force microscopy (AFM) is a relatively new characterization technique, invented in 1986, which uses a microscopic cantilever to determine such features as surface topology and adhesion forces. Force measurements have been used to probe the interaction between pMDI surfactants and solvents[167] or drugs and pMDI components.[168] The cohesive–adhesive balance (CAB) approach has shown that the ratio between the cohesion of the drug and the adhesion of the drug to carrier can influence the performance of DPIs.[169,170] These measurements are performed by attaching a probe, such as the drug of interest, to an AFM cantilever and then measuring the forces between this probe and different surfaces. The CAB approach can potentially help during the formulation process to optimize drug–drug and drug–carrier interactions.

1.4.2.3 Electrostatics

Electrostatics can play an important role in several aspects of aerosol formulation and delivery, and, thus, measuring these forces can provide some insight into how they will affect the final product. In particular, triboelectric charging due to contact between particles may increase the lung deposition due to charge attraction between the particles and the lung tissue.[171] The electrostatic charge can be measured by a Faraday pail or Faraday cup, which essentially consists of an electromagnetically shielded and insulated metal collection pail connected to either an electrometer or a series of amplifiers to measure the total charge delivered from the powder to the pail. The powder is either manually added to the pail, or it impinges on the pail from the aerosol device itself.[172–174] A newer method includes coupling particle sizing with charge measurement, by either modifying a Berner multi-jet low-pressure impactor (BLPI) to create an electrical low-pressure impactor (ELPI),[175] or modifying a Next Generation Impactor (NGI) to create an Electrical Next Generation Impactor (eNGI).[176,177] These apparatus allow the elucidation of triboelectrification as a function of particle size, and these data can be used both in the design of an inhaler or the formulation of the drug.

1.5 RESPIRATORY DEPOSITION, RETENTION, AND DOSIMETRY

1.5.1 DEPOSITION

1.5.1.1 Gamma Scintigraphy, PET, and SPECT

The assessment of lung deposition of pharmaceutical and medical drug delivery systems is a key step in the evaluation of pharmaceutical aerosol device performance. It has also been shown that with inhaled anti-asthma drugs, lung deposition data can act as a surrogate for clinical response. Thus, lung deposition studies can facilitate optimized drug delivery to the lungs during product development and also can be used in therapeutic equivalency studies. Gamma scintigraphy has been widely used in quantifying regional lung deposition of aerosolized drugs.[178] Gamma scintigraphy is a planar imaging technique where a gamma-ray-emitting nuclide of appropriate half-life (typically 99 m Technetium for pulmonary systems) is included in the pharmaceutical formulation such that its aerodynamic particle size and deposition parallels that of the drug substance. The radionuclide distribution in the lungs is then analyzed using a gamma camera and can be quantified using computer-based software. Single-photon emission computed tomography (SPECT) is similar to gamma scintigraphy in that it involves similar radio-labeling procedures and measurement of deposition using a gamma camera.[179] However, in SPECT the emitted gamma rays are detected using a rotating gamma camera from various angles in order to obtain several views of lung deposition. Thus, it may be possible to define drug deposition more closely with respect to the three-dimensional anatomy of the lung. Nebulizers and MDIs have been investigated using this technique, but some significant drawbacks are that larger amounts of radionuclides are required for satisfactory scintigraphic data collection and that longer imaging periods result in drug clearance from the lung.[179] Positron emission tomography (PET) is a relatively new technique in analyzing lung deposition resulting from pharmaceutical systems.[179,180] The basis of the method is the use of positron emitting nuclides (typically 11C, 15O, 13N, or 18F) that are detected after positron collision with electrons. This interaction results in two photons being emitted in opposite directions that are detected using coincident counting of photons on ring-shaped array of detector elements. Advantages of the technique include the small radiation doses resulting from the relatively short half-lives of the radionuclides used, scanning times are very short, picomolar concentrations of nuclides are detectable, and nuclides are often chemically incorporated into the drug molecule rather than physical attachment.[179] However, these are balanced against practical disadvantages of handling radionuclides with such short half-lives: transportation, formulation and labeling steps, and dosing volunteers. Apart from time limitations the cost of PET can also be significant due to the need for a nearby cyclotron and detector equipment. These factors need to be weighed against the PET images that yield more precise relation of deposition data with three-dimensional structural features of the respiratory tract.[179]

1.5.2 RETENTION

The retention of aerosols in the lungs is related to the physicochemical properties of the material, particularly those linked to dissolution rate,[181,182] and the mechanisms of clearance from the lungs, that is, absorption, mucociliary transport, and cell-mediated transport.[183]

Rapidly dissolving particles are subject to clearance from lungs by absorption and mucociliary clearance depending on the site of deposition. Slowly dissolving particles are cleared predominantly by mucociliary clearance in the upper airways and macrophage uptake in the periphery of the lungs.

Delaying dissolution has been used as a strategy for controlling drug delivery by increasing lung retention time.[184,185] However, there may be some concerns about the safety of such an approach.[181] Another approach is to make inhalable hydrogel particles that swell to sizes larger than the macrophage uptake limit. These particles can then release the drug, allowing longer delivery times than traditional drug particles.[186]

1.5.3 Dosimetry

Dosimetry is generally linked to the physicochemical properties of a drug; however, other factors that may influence the decision to formulate or administer a drug by a determined device or devices are technological implications, the health condition of the patient, target site in the airways, and the therapeutic dose. Doses are drug specific and they are influenced by the potency of the drug itself. In addition, the throughput of a device and the RF need to be considered. Throughput is used to describe the amount of drug deliverable from an aerosol device, while RF describes the fraction of aerosolized dose surviving filtration and impaction mechanisms of the nasopharynx.[187]

Although it was generally believed that dosimetry from MDIs was more efficient, convenient, and reproducible than DPIs and nebulizers, emerging technologies have changed this perspective. Gamma scintigraphy and pharmacokinetic (charcoal-block) methods have been used to quantify the amount and pattern of deposition in human lungs.[188–190] The influence of pulmonary physiology and pressure gradients on the dosimetry of inhaler devices has been studied.[191,192] Martonen et al. have developed models to study human lung dosimetry and deposition.[193] These models can simulate effects of aerosol polydispersity and hygroscopicity, lung morphology, and patient ventilation, age, and airway disease. Another model of mechanistic dosimetry developed by Lazardis et al. describes the dynamics of respirable particles. Model predictions of deposition and transport of aerosols are based on equations that describe changes in particle size and mass distribution as a consequence of nucleation, condensation, coagulation, and deposition processes.

1.5.3.1 Bolus Delivery

Table 1.5 shows some common drugs administered by inhalation device, the doses dispensed with each inhalation, and the therapeutic doses required per day.[194] In general, DPIs and MDIs are used for single delivery of medications on a single breath; doses range from 4 to 500 µg and require administration two to six times a day. MDIs are capable of delivering very accurate and reproducible doses. The valves in the device can deliver volumes of the formulation between 25 and 100 µL. Characteristics that can influence the amount of drug to be formulated and, therefore, the amount of drug that can be dosed are drug characteristics such as solubility and concentration, and the addition of excipients such as surfactants (to minimize particle aggregation which improves physical stability and dose uniformity), solvents (to aid in drug solubility, carrier properties), and propellants. The use of accessories such as spacers with MDIs improves inhalant technique and drug delivery.[16] Some factors that influence the size of doses delivered by DPIs are drug characteristics, the method used to produce fine particles, blend uniformity, and factors related to the device itself such as fine particle fraction and emitted dose of the formulation.

1.5.3.2 Continuous Delivery

As mentioned in previous sections, the use of nebulizers is preferred in hospital settings, acute conditions, and treatments in elderly patients and children. This is mainly because required doses of the prescribed drug are very large and have to be administered over long periods of time or because the health condition of the patient has limited his/her ability to breath adequately to use other inhalation device. In most cases, once the clinical condition of an adult patient improves, the prescribed

TABLE 1.5
Drugs Delivered by the Inhalation Route

Product	Active	Dose/Inhalation	Dose/Day
MDI[a]			
Atrovent HFA	Ipratropium bromide HFA	17 µg	2 inhalations 4 times, max 12 times (136–204 µg)
Azmacort	Triamcinolone acetonide	75 µg	2 inhalations 3–4 times or 4 inhalations bid, max 16 times (0.45–1.2 mg)
Combivent inhalation aerosol	Ipratropium bromide and albuterol sulfate	18/103 µg	2 inhalations 4–6 times (0.14/0.82–0.216/1.24 mg)
Dulera	Mometasone furoate and formoterol fumarate dihydrate	100/5 µg 200/10 µg	2 inhalations bid (400/20–800/40 µg)[197]
Flovent HFA	Fluticasone propionate	44 µg 110 µg 220 µg	2 inhalations bid (0.176–0.880 mg)
Maxair autohaler	Pirbuterol acetate	200 µg	2 inhalations 4–6 times (1.6–2.4 mg)
Proventil HFA	Albuterol sulfate	108 µg	1–2 inhalations 4–6 times (0.432–1.296 mg)
Qvar	Beclomethasone dipropionate	40 µg 80 µg	2 inhalations bid (0.16–0.320 mg)
Symbicort	Budesonide and formoterol fumarate dihydrate	80/4.5 µg 160/4.5 µg	2 inhalations bid (320/18–640/18 µg)
Xopenex HFA	Levalbuterol tartrate	45 µg	1–2 inhalations 4–6 times (0.18–0.540 mg)
DPI[a]			
Advair Diskus	Fluticasone propionate and salmeterol xinafoate	93/45 µg 165/45 µg 233/45 µg	1 inhalation bid (0.186/0.090–0.330/ 0.090 mg)
Asmanex Twisthaler	Mometasone furoate	100 µg 200 µg	1–2 inhalations 1–2 times (0.1–0.4 mg)
Foradil Aerolizer	Formoterol fumarate	10 µg	1 inhalation bid (0.02 mg)
Pulmicort flexhaler	Budesonide	90 µg 180 µg	1–4 inhalations bid (0.18–1.44 mg)
Relenza	Zanamivir	4 mg	2 inhalations bid (16 mg)
Serevent diskus	Salmeterol xinofoate	50 µg	1 inhalation bid (0.1 mg)
Spiriva Handihaler	Tiotropium bromide	18 µg	1 inhalation bid (36 µg)
Nebulizer[b]			
Accuneb	Albuterol sulfate	1.25 mg/3 mL 0.63 mg/3 mL	1 nebulization 3–4 times as needed (5 mg)
Atrovent solution	Ipratropium bromide	0.02% 500 µg/2.5 mL	1 nebulization 3–4 times (2 mg)
Duoneb solution	Ipratropium bromide/Albuterol sulfate	0.5/3.0 mg/3 mL 42%/46%/3 mL	1 nebulization 3 times (1.5/9.0 mg)
Perforomist	Fomoterol fumarate	20 µg/2 mL	1 nebulization bid for a maximum of 40 µg

(*continued*)

TABLE 1.5 (continued)
Drugs Delivered by the Inhalation Route

Product	Active	Dose/Inhalation	Dose/Day
Pulmicort respules	Budesonide Suspension	0.25 mg/2 mL 0.5 mg/2 mL	1 nebulization once or twice for a maximum of 0.5 mg
Pulmozyme solution	Dornase alfa	1.0 mg/mL (2.5 mL)	1 nebulization once (1 mg)

Source: Murray, L. and Kelly, G.L., *Physicians' Desk Reference*, Medical Economics Company, Inc., Montvale, NJ, 2010.

[a] Doses in MDIs and DPIs are described from the mouthpiece.

[b] Nebulizations usually take between 15 and 30 min in average if used with the nebulizer and compressor recommended by manufacturer observing the recommended parameters.

medications are administered by either MDI or DPI. Table 1.5 shows some commercial preparations to be delivered by nebulization. Doses range from a few milligrams, to be used at home, to several hundred milligrams or even grams when they are used in hospital settings. Nebulizers are more effective generators of small particles than MDIs and DPIs.[100] Recent nebulizer designs provide increased efficiency of deliver by gauging the patient's breathing and delivering aerosol only during inhalation. Other systems achieve increased delivery efficiency by collecting the aerosol generated during exhalation in a reservoir. Nebulizer reservoirs increase the delivery rate of nebulized medications by conserving the aerosol generated during exhalation and making it available to the patient during the next inhalation.[195] The development of medications for nebulization should include the specification and testing of a nebulizer and the compression source that is fixed by prescription along with the medication. Alterations of nebulizer/compression source must be carefully considered at the physician level as such alterations could significantly change the dose received by the patient even if the same nebulizer charge dose is used.[114]

1.6 CONCLUDING REMARKS

Aerosols have a long history in the medical and pharmaceutical fields, and research in this area will continue to push the boundaries of drug delivery for several decades to come. While most aerosol therapeutics currently on the market are locally acting, systemic drug delivery via the lungs has great potential. Current research, other than improving present designs for higher efficiency, includes protein delivery in order to bypass the oral and intravenous routes, avoiding first-pass metabolism of the therapeutic and allowing higher patient compliance and comfort.

The three main types of aerosol delivery systems are pMDIs, DPIs, and nebulizers, each of which has its own pros and cons. While nebulizers are typically used in a hospital setting, pMDIs and DPIs are portable devices that are able to keep up with the active lifestyle of the patient. Research into portable nebulizers is still being conducted due to their increased efficacy. However, there is no one universal method for pulmonary delivery, and each drug must be investigated to determine the appropriate delivery method by balancing cost, performance, and patient compliance. Before a formulation can successfully be implemented via one of these devices, it has to be characterized in numerous ways to determine particulate size, cohesion, adhesion, dispersion, deposition, and dosimetry. The overall efficiency of DPIs and pMDIs is still fairly low; future devices will need to have much higher RFs in order to deliver expensive drugs in a cost-effective manner.

REFERENCES

1. Sciarra, J.J. and Cutie, A.J. Pharmaceutical aerosols. *Drugs and the Pharmaceutical Sciences* **40**, 605–634 (1990).
2. Duke, J.A. *Handbook of Edible Weeds*. CRC: Boca Raton, FL (1992).
3. Thiel, C.G. From Susie's question to CFC free: An inventor's perspective on forty years of MDI development and regulation. In: Dalby, R.N., Byron, P.R., and Farr, S.J., eds. *Respiratory Drug Delivery V*. Interpharm Press: Buffalo Groove, IL, pp. 115–123 (1996).
4. Dunbar, C.A. and Hickey, A.J. A new millenium for inhaler technology. *Pharmaceutical Technology* **21**, 116–125 (1997).
5. Crowder, T.M., Louey, M.D., Sethuraman, V.V., Smyth, H.D.C., and Hickey, A.J. 2001: An odyssey in inhaler formulation and design. *Pharmaceutical Technology* **25**, 99–113 (2001).
6. Barnes, P.J. New concepts in chronic obstructive pulmonary disease. *Annual Review of Medicine* **54**, 113–129 (2003).
7. Hillerdal, G. New principles for the treatment of diffuse pulmonary emphysema. *Journal of Internal Medicine* **242**, 441–448 (1997).
8. Ding, J.Y. et al. Delivery of a chemotherapy agent for cancer treatment via nebulization. In: Dalby, R.N. et al. eds. *Respiratory Drug Delivery VIII*. DHI: Tucson, AZ, pp. 359–362 (2002).
9. Patton, J.S., Bukar, J., and Nagarajan, S. Inhaled insulin. *Advanced Drug Delivery Reviews* **35**, 235–247 (1999).
10. Adjei, A. and Garren, J. Pulmonary delivery of peptide drugs: Effect of particle size on bioavailability of leuprolide acetate in healthy male volunteers. *Pharmaceutical Research* **7**, 565–569 (1990).
11. Shukla, V.K. Tiotropium: A potential replacement for ipratropium in patients with COPD. *Issues in Emerging Health Technologies* **35**, 1–4 (2002).
12. Sacchetti, M. and Van Oort, M.M. Spray-drying and supercritical fluid particle generation techniques. *Inhalation Aerosols: Physical and Biological Basis for Therapy* **94**, 337–384 (1996).
13. Crowder, T.M., Rosati, J.A., Schroeter, J.D., Hickey, A.J., and Martonen, T.B. Fundamental effects of particle morphology on lung delivery: Predictions of Stokes' Law and the particular relevance to dry powder inhaler formulation and development. *Pharmaceutical Research* **19**, 239–245 (2002).
14. Edwards, D.A. et al. Large porous particles for pulmonary drug delivery. *Science* **276**, 1868–1872 (1997).
15. Hickey, A.J. Delivery of drugs by the pulmonary route. *Modern Pharmaceutics* **121**, 479–500 (2002).
16. Hickey, A.J. Summary of common approaches to pharmaceutical aerosol administration. In: Hickey, A.J., ed. *Pharmaceutical Inhalation Aerosol Technology*. Marcel Dekker: New York, Vol. 54, pp. 255–288 (1992).
17. Cazzola, M., Segreti, A., and Matera, M.G. Novel bronchodilators in asthma. *Current Opinion in Pulmonary Medicine* **16**, 6–12 (2010).
18. Tan, R.A. and Corren, J. Mometasone furoate in the management of asthma: A review. *Therapeutics and Clinical Risk Management* **4**, 1201–1208 (2008).
19. Bousquet, J. Mometasone furoate: An effective anti-inflammatory with a well-defined safety and tolerability profile in the treatment of asthma. *International Journal of Clinical Practice* **63**, 806–819 (2009).
20. Deeks, E. and Perry, C. Ciclesonide—A review of its use in the management of asthma. *Drugs* **68**, 1741–1770 (2008).
21. Calverley, P. et al. One-year treatment with mometasone furoate in chronic obstructive pulmonary disease. *Respiratory Research* **9**, 73 (2008).
22. Carter, E. Letter from King Pharmaceuticals to Healthcare Professionals. (2009). At http://www.fda.gov/downloads/Drugs/DrugSafety/DrugShortages/ucm089433.pdf
23. Geller, D.E. Aerosol antibiotics in cystic fibrosis. *Respiratory Care* **54**, 658–670 (2009).
24. O'Sullivan, B.P., Yasothan, U., and Kirkpatrick, P. Inhaled aztreonam. *Nature Reviews Drug Discovery* **9**, 357–358 (2010).
25. Elborn, J. and Henig, N. Optimal airway antimicrobial therapy for cystic fibrosis: The role of inhaled aztreonam lysine. *Expert Opinion on Pharmacotherapy* **11**, 1373–1385 (2010).
26. Stribling, R., Brunette, E., Liggitt, D., Gaensler, K., and Debs, R. Aerosol gene delivery in vivo. *Proceedings of the National Academy of Sciences of the United States of America* **89**, 11277–11281 (1992).

27. Rochat, T. and Morris, M.A. Gene therapy for cystic fibrosis by means of aerosol. *Journal of Aerosol Medicine* **15**, 229–235 (2002).

28. Flotte, T. and Laube, B. Gene therapy in cystic fibrosis. *Chest* **120**, 124S–131S (2001).

29. Moss, R. et al. Repeated adeno-associated virus serotype 2 aerosol-mediated cystic fibrosis transmembrane regulator gene transfer to the lungs of patients with cystic fibrosis—A multicenter, double-blind, placebo-controlled trial. *Chest* **125**, 509–521 (2004).

30. Tang, L. et al. RASONs: A novel antisense oligonucleotide therapeutic approach for asthma. *Expert Opinion on Biological Therapy* **1**, 979–983 (2001).

31. Jiang, H. et al. The suppression of lung tumorigenesis by aerosol-delivered folate-chitosan-graft-polyethylenimine/Akt1 shRNA complexes through the Akt signaling pathway. *Biomaterials* **30**, 5844–5852 (2009).

32. Xu, C. et al. Poly(ester amine)-mediated, aerosol-delivered Akt1 small interfering RNA suppresses lung tumorigenesis. *American Journal of Respiratory and Critical Care Medicine* **178**, 60–73 (2008).

33. Devuni, D. and Wu, G. ALN-RSV01 siRNA targeting RSV nucleocapsid N gene treatment of RSV infection. *Drugs of the Future* **34**, 781–783 (2009).

34. DeVincenzo, J. RNA interference strategies as therapy for respiratory viral infections. *Pediatric Infectious Disease Journal* **27**, S118–S122 (2008).

35. BPC British Pharmaceutical Codex, ed., The Pharmaceutical Press, London, (1973).

36. Lechuga-Ballesteros, D. et al. Trileucine improves aerosol performance and stability of spray-dried powders for inhalation. *Journal of Pharmaceutical Sciences* **97**, 287–302 (2008).

37. Byron, P.R. and Patton, J.S. Drug delivery via the respiratory tract. *Journal of Aerosol Medicine* **7**, 49–75 (1994).

38. Bitonti, A.J. et al. Pulmonary delivery of an erythropoietin Fc fusion protein in non-human primates through an immunoglobulin transport pathway. *Proceedings of the National Academy of Sciences of the United States of America* **101**, 9763–9768 (2004).

39. Wylam, M.E. et al. Aerosol granulocyte-macrophage colony-stimulating factor for pulmonary alveolar proteinosis. *European Respiratory Journal* **27**, 585–593 (2006).

40. Spyker, D., Munzar, P., and Cassella, J. Pharmacokinetics of loxapine following inhalation of a thermally generated aerosol in healthy volunteers. *Journal of Clinical Pharmacology* **50**, 169–179 (2010).

41. Anon. Staccato goes to FDA for review. *Journal of Psychosocial Nursing and Mental Health Services* **48**, 47–47 (2010).

42. Siekmeier, R. and Scheuch, G. Systemic treatment by inhalation of macromolecules—Principles, problems, and examples. *Journal of Physiology and Pharmacology* **59**, 53–79 (2008).

43. Siekmeier, R. and Scheuch, G. Treatment of systemic diseases by inhalation of biomolecule aerosols. *Journal of Physiology and Pharmacology* **60**, 15–26 (2009).

44. Omer, S., Hiremath, G., and Halsey, N. Respiratory administration of measles vaccine. *Lancet* **375**, 706–708 (2010).

45. Burger, J. et al. Stabilizing formulations for inhalable powders of live-attenuated measles virus vaccine. *Journal of Aerosol Medicine and Pulmonary Drug Delivery* **21**, 25–34 (2008).

46. Shi, S. and Hickey, A. PLGA microparticles in respirable sizes enhance an in vitro T cell response to recombinant *Mycobacterium tuberculosis* antigen TB10.4-Ag85B. *Pharmaceutical Research* **27**, 350–360 (2010).

47. Corbett, M. et al. Aerosol immunization with NYVAC and MVA vectored vaccines is safe, simple, and immunogenic. *Proceedings of the National Academy of Sciences of the United States of America* **105**, 2046–2051 (2008).

48. Hunter, Z., Smyth, H.D., Durfee, P., and Chackerian, B. Induction of mucosal and systemic antibody responses against the HIV coreceptor CCR5 upon intramuscular immunization and aerosol delivery of a virus-like particle based vaccine. *Vaccine* **28**, 403–414 (2009).

49. Im, E. et al. Vaccine platform for prevention of tuberculosis and mother-to-child transmission of human immunodeficiency virus type 1 through breastfeeding. *Journal of Virology* **81**, 9408–9418 (2007).

50. DiNapoli, J. et al. Newcastle disease virus-vectored vaccines expressing the hemagglutinin or neuraminidase protein of H5N1 highly pathogenic avian influenza virus protect against virus challenge in monkeys. *Journal of Virology* **84**, 1489–1503 (2010).

51. Orson, F. et al. Protection against influenza infection by cytokine-enhanced aerosol genetic immunization. *Journal of Gene Medicine* **8**, 488–497 (2006).

52. Giudice, E. and Campbell, J. Needle-free vaccine delivery. *Advanced Drug Delivery Reviews* **58**, 68–89 (2006).

53. Purewal, T.S. Formulation of metered dose inhalers. In: Purewal, T.S. and Grant, D.J.W., eds. *Metered Dose Inhaler Technology.* Interpharm Press: Buffalo Groove, IL (1998).

54. Clarke, J.G., Wicks, S.R., and Farr, S.J. Surfactant mediated effects in pressurized metered dose inhalers formulated as suspensions. I. Drug/surfactant interactions in a model propellant system. *International Journal of Pharmaceutics* **93**, 221–231 (1993).

55. Blondino, F.E. and Byron, P.R. Surfactant dissolution and water solubilization in chlorine-free liquified gas propellants. *Drug Development and Industrial Pharmacy* **24**, 935–945 (1998).

56. Harnor, K. et al. Effect of vapour pressure on the deposition pattern from solution phase metered dose inhalers. *International Journal of Pharmaceutics* **95**, 111–116 (1993).

57. McDonald, K.J. and Martin, G.P. Transition to CFC-free metered dose inhalers—Into the new millennium. *International Journal of Pharmaceutics* **201**, 89–107 (2000).

58. Forte, J. and Dibble, C. The role of international environmental agreements in metered-dose inhaler technology changes. *Journal of Allergy and Clinical Immunology* **104**, s217–s220 (1999).

59. Kux, L. Use of ozone-depleting substances; removal of essential-use designation (flunisolide, etc.). *Federal Register* **75**, 19213–19241 (2010).

60. Noakes, T. Medical aerosol propellants. *Journal of Fluorine Chemistry* **118**, 35–45 (2002).

61. Alexander, D. and Libretto, S. An overview of the toxicology of HFA-134a (1,1,1,2-tetrafluoroethane). *Human and Experimental Toxicology* **14**, 715–720 (1995).

62. Emmen, H.H. et al. Human safety and pharmacokinetics of the CFC alternative propellants HFC 134a (1,1,1,2-tetrafluoroethane) and HFC 227 (1,1,1,2,3,3,3-heptafluoropropane) following whole-body exposure. *Regulatory Toxicology and Pharmacology* **32**, 22–35 (2000).

63. Keller, M. Innovations and perspectives of metered dose inhalers in pulmonary drug delivery. *International Journal of Pharmaceutics* **186**, 81–90 (1999).

64. Vervaet, C. and Byron, P.R. Drug-surfactant-propellant interactions in HFA-formulations. *International Journal of Pharmaceutics* **186**, 13–30 (1999).

65. Byron, P.R., Miller, N.C., Blondino, F.E., Visich, J.E., and Ward, G.H. Some aspects of alternative propellant solvency. In: Byron, P., Dalby, R., and Farr, S., eds. *Proceedings of Respiratory Drug Delivery IV.* Interpharm Press: Buffalo Groove, IL, pp. 231–242 (1994).

66. Dickinson, P., Seville, P., Mchale, H., Perkins, N., and Taylor, G. An investigation of the solubility of various compounds in the hydrofluoroalkane propellants and possible model liquid propellants. *Journal of Aerosol Medicine* **13**, 179–186 (2000).

67. Tiwari, D., Goldman, D., Dixit, S., Malick, W.A., and Madan, P.L. Compatibility evaluation of metered-dose inhaler valve elastomers with tetrafluoroethane (P134a), a non-CFC propellant. *Drug Development and Industrial Pharmacy* **24**, 345–352 (1998).

68. Brambilla, G. et al. Modulation of aerosol clouds produced by pressurised inhalation aerosols. *International Journal of Pharmaceutics* **186**, 53–61 (1999).

69. Dunbar, C.A., Watkins, A.P., and Miller, J.F. Theoretical investigation of the spray from a pressurized metered-dose inhaler. *Atomization and Sprays* **7**, 417–436 (1997).

70. Dunbar, C., Watkins, A., and Miller, J. An experimental investigation of the spray issued from a pMDI using laser diagnostic techniques. *Journal of Aerosol Medicine* **10**, 351–368 (1997).

71. Leach, C. Effect of formulation parameters on hydrofluoroalkane-beclomethasone dipropionate drug deposition in humans. *Journal of Allergy and Clinical Immunology* **104**, s250–s252 (1999).

72. Dalby, R.N. Possible replacements for CFC-propelled metered-dose inhalers. *Medical Device Technology* **2**, 21–25 (1991).

73. Sommerville, M.L., Cain, J.B., Johnson Jr, C.S., and Hickey, A.J. Lecithin inverse microemulsions for the pulmonary delivery of polar compounds utilizing dimethylether and propane as propellants (2000). At http://informahealthcare.com/doi/abs/10.1081/PDT-100100537

74. Clark, A.R. Medical aerosol inhalers: Past, present, and future. *Aerosol Science and Technology* **22**, 374 (1995).

75. Draft Guidance for Industry: Metered Dose Inhaler (MDI) and Dry Powder Inhaler (DPI) Drug Products. Department of Health and Human Services, Food and Drug Administration, Center for Drug Evaluation and Research, (1998). At http://www.fda.gov/downloads/Drugs/GuidanceComplianceRegulatory Information/Guidances/UCM070573.pdf

76. Dunbar, C., Hickey, A.J., and Holzner, P. Dispersion and characterization of pharmaceutical dry powder aerosols. *KONA* **16**, 7–45 (1998).

77. Hindle, M. and Byron, P.R. Dose emissions from marketed dry powder inhalers. *International Journal of Pharmaceutics* **116**, 169–177 (1995).

78. Hickey, A.J., Martonen, T.B., and Yang, Y. Theoretical relationship of lung deposition to the fine particle fraction of inhalation aerosols. *Pharmaceutica Acta Helvetiae* **71**, 185–190 (1996).

79. Hickey, A.J., Concessio, N.M., Van Oort, M.M., and Platz, R.M. Factors influencing the dispersion of dry powders as aerosols. *Pharmaceutical Technology* **18**, 58–58 (1994).

80. Calvert, G., Ghadiri, M., and Tweedie, R. Aerodynamic dispersion of cohesive powders: A review of understanding and technology. *Advanced Powder Technology* **20**, 4–16 (2009).

81. Lucas, P., Anderson, K., Potter, U.J., and Staniforth, J.N. Enhancement of small particle size dry powder aerosol formulations using an ultra low density additive. *Pharmaceutical Research* **16**, 1643–1647 (1999).

82. Lucas, P., Anderson, K., and Staniforth, J.N. Protein deposition from dry powder inhalers: Fine particle multiplets as performance modifiers. *Pharmaceutical Research* **15**, 562–569 (1998).

83. Musante, C.J. et al. Factors affecting the deposition of inhaled porous drug particles. *Journal of Pharmaceutical Sciences* **91**, 1580–1590 (2002).

84. Joshi, M. and Misra, A. Liposomal budesonide for dry powder inhaler: Preparation and stabilization. *AAPS PharmSciTech* **02**, e25 (2001).

85. Bi, R. and Zhang, N. Liposomes as a carrier for pulmonary delivery of peptides and proteins. *Journal of Biomedical Nanotechnology* **3**, 332–341 (2007).

86. Lu, D. and Hickey, A.J. Liposomal dry powders as aerosols for pulmonary delivery of proteins. *AAPS PharmSciTech* **6**, E641–E648 (2005).

87. Crowder, T. and Hickey, A. Powder specific active dispersion for generation of pharmaceutical aerosols. *International Journal of Pharmaceutics* **327**, 65–72 (2006).

88. Newman, S.P. and Busse, W.W. Evolution of dry powder inhaler design, formulation, and performance. *Respiratory Medicine* **96**, 293–304 (2002).

89. Son, Y. and McConville, J.T. Advancements in dry powder delivery to the lung. *Drug Development and Industrial Pharmacy* **34**, 948–959 (2008).

90. Peart, J. and Clarke, M.J. New developments in dry powder inhaler technology. *American Pharmaceutical Review* **14**, 37–45 (2001).

91. Smith, S.J. and Bernstein, J.A. Therapeutic uses of lung aerosols. *Inhalation Aerosols: Physical and Biological Basis for Therapy* **94**, 233–269 (1996).

92. Taburet, A. and Schmit, B. Pharmacokinetic optimisation of asthma treatment. *Clinical Pharmacokinetics* **26**, 396–418 (1994).

93. Ilowite, J. Techniques of aerosol therapy. *Emergency Medicine* 23(7), 68–76 (1991).

94. Denyer, J. et al. Evaluation of the target inhalation mode (TIM) breathing maneuver in simulated nebulizer therapy in patients with cystic fibrosis. *Journal of Aerosol Medicine and Pulmonary Drug Delivery* **23**, S29–S36 (2010).

95. Nikander, K., Prince, I., Coughlin, S., Warren, S., and Taylor, G. Mode of breathing-tidal or slow and deep-through the I-neb adaptive aerosol delivery (AAD) system affects lung deposition of Tc-99 m-DTPA. *Journal of Aerosol Medicine and Pulmonary Drug Delivery* **23**, S37–S43 (2010).

96. Hardaker, L. and Hatley, R. In vitro characterization of the I-neb adaptive aerosol delivery (AAD) system. *Journal of Aerosol Medicine and Pulmonary Drug Delivery* **23**, S11–S20 (2010).

97. Niven, R.W. Delivery of biotherapeutics by inhalation aerosol. *Critical Reviews in Therapeutic Drug Carrier Systems* **12**, 151–231 (1995).

98. Newnham, D.M. and Lipworth, B.J. Nebuliser performance, pharmacokinetics, airways and systemic effects of salbutamol given via a novel nebuliser delivery system ("Ventstream"). *Thorax* **49**, 762–770 (1994).

99. Baker, P. and Stimpson, P. Electronically controlled drug delivery systems based on the piezoelectric crystal. In: Byron, P., Dalby, R.N., and Farr, S.J., eds. *Proceedings of Respiratory Drug Delivery IV.* Interpharm Press: Buffalo Grove, IL, pp. 273–285 (1994).

100. Dalby, R.N., Tiano, S.L., and Hickey, A.J. Medical devices for the delivery of therapeutic aerosols to the lungs. *Inhalation Aerosols: Physical and Biological Basis for Therapy* **94**, 441–473 (1996).

101. Boucher, R.M. and Kreuter, J. The fundamentals of the ultrasonic atomization of medicated solutions. *Annals of Allergy* **26**, 591–600 (1968).

102. Niven, R.W. Atomization and nebulizers. In: Hickey, A.J., ed. *Inhalation Aerosols: Physical and Biological Basis for Therapy*. Marcel Dekker, Inc.: New York, pp. 273–312 (1996).

103. Greenspan, B.J. Ultrasonic and electrohydrodynamic methods for aerosol generation. *Inhalation Aerosols: Physical and Biological Basis for Therapy* **94**, 313–335 (1996).

104. Johnson, J., Waldrep, J., Guo, J., and Dhand, R. Aerosol delivery of recombinant human DNase I: In vitro comparison of a vibrating-mesh nebulizer with a jet nebulizer. *Respiratory Care* **53**, 1703–1708 (2008).

105. Dhand, R. Nebulizers that use a vibrating mesh or plate with multiple apertures to generate aerosol. *Respiratory Care* **47**, 1406 (2002).
106. Chan, K.N., Clay, M.M., and Silverman, M. Output characteristics of DeVilbiss No. 40 hand-held jet nebulizers. *European Respiratory Journal* **3**, 1197–1201 (1990).
107. Smith, E.C., Denyer, J., and Kendrick, A.H. Comparison of twenty three nebulizer/compressor combinations for domiciliary use. *European Respiratory Journal* **8**, 1214–1221 (1995).
108. Weber, A. et al. Effect of nebulizer type and antibiotic concentration on device performance. *Pediatric Pulmonology* **23**, 249–260 (1997).
109. Montgomery, A.B. Aerosolized pentamidine for treatment and prophylaxis of *Pneumocystis carinii* pneumonia in patients with acquired immunodeficiency syndrome. In: *Pharmaceutical Inhalation Aerosol Technology*, 2nd Edn., Hickey, A.J., Ed. Marcel Dekker: New York, pp. 459–472 (2004).
110. Niven, R.W. Modulated drug therapy with inhalation aerosols. In: Hickey, A.J., ed. *Pharmaceutical Inhalation Aerosol Technology*. Marcel Dekker: New York, Vol. 54, pp. 321–359 (1992).
111. Collins, N. Nebulizer therapy in cystic fibrosis: An overview. *Journal of the Royal Society of Medicine* **102**, S11–S17 (2009).
112. Hubert, D. et al. Pharmacokinetics and safety of tobramycin administered by the PARI eFlow® rapid nebulizer in cystic fibrosis. *Journal of Cystic Fibrosis* **8**, 332–337 (2009).
113. Falagas, M.E., Michalopoulos, A., and Metaxas, E.I. Pulmonary drug delivery systems for antimicrobial agents: Facts and myths. *International Journal of Antimicrobial Agents* **35**, 101–106 (2010).
114. Garcia-Contreras, L. and Hickey, A.J. Pharmaceutical and biotechnological aerosols for cystic fibrosis therapy. *Advanced Drug Delivery Reviews* **54**, 1491–1504 (2002).
115. Jones, L.D., McGlynn, P., Bovet, L., and Hickey, A.J. Analysis and stability of pharmaceutical aerosols. *Pharmaceutical Technology* **24**, 40–54 (2000).
116. Summers, Q.A., Nesbit, M.R., Levin, R., and Holgate, S.T. A non-bronchoconstrictor, bacteriostatic preservative for nebuliser solutions. *British Journal of Clinical Pharmacology* **31**, 204–206 (1991).
117. Rocchiccioli, K. and Pickering, C. Air-flow obstruction induced by ultrasonically nebulized water—The underlying mechanism. *Thorax* **39**, 710–710 (1984).
118. Borland, C., Chamberlain, A., Barber, B., and Higenbottam, T. The effect of ultrasonically nebulized distilled water on air-flow obstruction, regional ventilation and lung epithelial permeability in asthma. *Thorax* **39**, 240–240 (1984).
119. Atkins, P., Barker, P.N., and Matiesen, D. The design and development of inhalation drug delivery systems. In: *Pharmaceutical Inhalation Aerosol Technology*, 2nd Edn., Hickey, A.J., Ed., Marcel Dekker: New York, pp. 258–285 (1992).
120. Cameron, D., Clay, M., and Silverman, M. Evaluation of nebulizers for use in neonatal ventilator circuits. *Critical Care Medicine* **18**, 866–870 (1990).
121. Tiano, S.L. and Dalby, R.N. Comparison of a respiratory suspension aerosolized by an air-jet and an ultrasonic nebulizer. *Pharmaceutical Development and Technology* **1**, 261–268 (1996).
122. Niven, R.W., Carvajal, T.M., and Schreier, H. Nebulization of liposomes. III. The effects of operating conditions and local environment. *Pharmaceutical Research* **9**, 515–520 (1992).
123. Niven, R.W. and Schreier, H. Nebulization of liposomes. I. Effects of lipid composition. *Pharmaceutical Research* **7**, 1127–1133 (1990).
124. Niven, R.W., Speer, M., and Schreier, H. Nebulization of liposomes. II. The effects of size and modeling of solute release profiles. *Pharmaceutical Research* **8**, 217–221 (1991).
125. Lange, C.F., Hancock, R.E.W., Samuel, J., and Finlay, W.H. *In vitro* aerosol delivery and regional airway surface liquid concentration of a liposomal cationic peptide. *Journal of Pharmaceutical Sciences* **90**, 1647–1657 (2001).
126. Zaru, M. et al. Rifampicin-loaded liposomes for the passive targeting to alveolar macrophages: In vitro and in vivo evaluation. *Journal of Liposome Research* **19**, 68–76 (2009).
127. Li, Z. et al. Characterization of nebulized liposomal amikacin (Arikace™) as a function of droplet size. *Journal of Aerosol Medicine and Pulmonary Drug Delivery* **21**, 245–253 (2008).
128. Desai, T.R. and Finlay, W.H. Nebulization of niosomal all-trans-retinoic acid: An inexpensive alternative to conventional liposomes. *International Journal of Pharmaceutics* **241**, 311–317 (2002).
129. Ip, A.Y., Arakawa, T., Silvers, H., Ransone, C.M., and Niven, R.W. Stability of recombinant consensus interferon to air-jet and ultrasonic nebulization. *Journal of Pharmaceutical Sciences* **84**, 1210–1214 (1995).
130. Gale, A. Drug degeneration during ultrasonic nebulization. *Journal of Aerosol Science* **16**, 265 (1985).
131. Kosugi, T., Nakamura, M., Noguchi, S., and Huang, G.W. Effect of ultrasonic nebulization of miraclid® on the proteolytic activity in tracheobronchial secretions of rats. *The Laryngoscope* **99**, 1281–1285 (1989).

132. Niven, R.W., Ip, A.Y., Mittelman, S., Prestrelski, S.J., and Arakawa, T. Some factors associated with the ultrasonic nebulization of proteins. *Pharmaceutical Research* **12**, 53–59 (1995).

133. Steckel, H., Eskandar, F., and Witthohn, K. Effect of excipients on the stability and aerosol performance of nebulized aviscumine. *Journal of Aerosol Medicine-Deposition Clearance and Effects in the Lung* **16**, 417–432 (2003).

134. Niven, R.W., Whitcomb, K.L., Woodward, M., Liu, J., and Jornacion, C. Systemic absorption and activity of recombinant consensus interferons after intratracheal instillation and aerosol administration. *Pharmaceutical Research* **12**, 1889–1895 (1995).

135. Niven, R.W., Whitcomb, K.L., Shaner, L., Ip, A.Y., and Kinstler, O.B. The pulmonary absorption of aerosolized and intratracheally instilled rhG-CSF and monoPEGylated rhG-CSF. *Pharmaceutical Research* **12**, 1343–1349 (1995).

136. Cipolla, D.C. et al. Coarse spray delivery to a localized region of the pulmonary airways for gene therapy. *Human Gene Therapy* **11**, 361–371 (2000).

137. Dolovich, M. New propellant-free technologies under investigation. *Journal of Aerosol Medicine* **12(Suppl. 1)**, S9–S17 (1999).

138. Hickey, A.J. and Swift, D. Characterization of pharmaceutical and diagnostic aerosols. In: Baron, P.A. and Willeke, K. eds. *Aerosol Measurement: Principles, Techniques and Applications*. Wiley: New York, pp. 1031–1052 (2001).

139. Finnin, B. and Morgan, T. Transdermal penetration enhancers: Applications, limitations, and potential. *Journal of Pharmaceutical Sciences* **88**, 955–958 (1999).

140. Chik, Z. et al. Pharmacokinetics of a new testosterone transdermal delivery system, TDS (R)-testosterone in healthy males. *British Journal of Clinical Pharmacology* **61**, 275–279 (2006).

141. Li, X. et al. Development of patch and spray formulations for enhancing topical delivery of sinomenine hydrochloride. *Journal of Pharmaceutical Sciences* **99**, 1790–1799 (2010).

142. Lahm, K. and Lee, G. Penetration of crystalline powder particles into excised human skin membranes and model gels from a supersonic powder injector. *Journal of Pharmaceutical Sciences* **95**, 1511–1526 (2006).

143. Dorey, E. PowderMed to trial three vaccines. *Chemistry and Industry* 21, 5 (2004).

144. Gonda, I. Targeting by deposition. In: Hickey, A.J., ed. *Pharmaceutical Inhalation Aerosol Technology*. Marcel Dekker: New York, pp. 61–82 (1992).

145. Esmen, N.A. and Lee, T.C. Distortion of cascade impactor measured size distribution due to bounce and blow-off. *American Industrial Hygiene Association Journal* **41**, 410–419 (1980).

146. Marple, V. et al. Next generation pharmaceutical impactor: A new impactor for pharmaceutical inhaler testing. Part I: Design. *Journal of Aerosol Medicine-Deposition Clearance and Effects in the Lung* **16**, 283–299 (2003).

147. Marple, V. et al. Next generation pharmaceutical impactor: A new impactor for pharmaceutical inhaler testing. Part II: Archival calibration. *Journal of Aerosol Medicine-Deposition Clearance and Effects in the Lung* **16**, 301–324 (2003).

148. Marple, V. et al. Next generation pharmaceutical impactor: A new impactor for pharmaceutical inhaler testing. Part III. Extension of archival calibration to 15 L/min. *Journal of Aerosol Medicine-Deposition Clearance and Effects in the Lung* **17**, 335–343 (2004).

149. Hallworth, G.W. and Andrews, U.G. Size analysis of suspension inhalation aerosols by inertial separation methods. *Journal of Pharmacy and Pharmacology* **28**, 898–907 (1976).

150. Rubow, K.L., Marple, V.A., Olin, J., and Mccawley, M.A. A personal cascade impactor: Design, evaluation and calibration. *American Industrial Hygiene Association Journal* **48**, 532–538 (1987).

151. Williams, R.O.3. and Hu, C. Moisture uptake and its influence on pressurized metered-dose inhalers. (2000). At http://informahealthcare.com/doi/abs/10.1081/PDT-100100530

152. Clarke, M.J., Tobyn, M.J., and Staniforth, J.N. The formulation of powder inhalation systems containing a high mass of nedocromil sodium trihydrate. *Journal of Pharmaceutical Sciences* **90**, 213–223 (2001).

153. Evans, R. Determination of drug particle size and morphology using optical microscopy. *Pharmaceutical Technology* **17**, 146–146 (1993).

154. Ranucci, J.A. and Chen, F.C. Phase Doppler anemometry: A technique for determining aerosol plume-particle size and velocity. *Pharmaceutical Technology* **17**, 62–62 (1993).

155. Dunbar, C.A. and Hickey, A.J. Selected parameters affecting characterization of nebulized aqueous solutions by inertial impaction and comparison with phase-Doppler analysis. *European Journal of Pharmaceutics and Biopharmaceutics* **48**, 171–177 (1999).

156. Mitchell, J.P., Nagel, M.W., and Cheng, Y.S. Use of the aerosizer® aerodynamic particle size analyzer to characterize aerosols from pressurized metered-dose inhalers (pMDIs) for medication delivery. *Journal of Aerosol Science* **30**, 467–477 (1999).

157. Mitchell, J.P. and Nagel, M.W. Time-of-flight aerodynamic particle size analyzers: Their use and limitations for the evaluation of medical aerosols. *Journal of Aerosol Medicine* **12**, 217–240 (1999).

158. Mitchell, J.P., Nagel, M.W., and Archer, A.D. Size analysis of a pressurized metered dose inhaler–delivered suspension formulation by the API aerosizer time-of-flight aerodynamic particle size analyzer. *Journal of Aerosol Medicine* **12**, 255–264 (1999).

159. Ranucci, J. Dynamic plume-particle size analysis using laser diffraction. *Pharmaceutical Technology* **16**, 108–108 (1992).

160. Annapragada, A. and Adjei, A. An analysis of the fraunhofer diffraction method for particle size distribution analysis and its application to aerosolized sprays. *International Journal of Pharmaceutics* **127**, 219–227 (1996).

161. de Boer, A.H., Gjaltema, D., Hagedoorn, P., and Frijlink, H.W. Characterization of inhalation aerosols: A critical evaluation of cascade impactor analysis and laser diffraction technique. *International Journal of Pharmaceutics* **249**, 219–231 (2002).

162. de Boer, G.B., de Weerd, C., Thoenes, D., and Goossens, H.W. Laser diffraction spectrometry: Fraunhofer diffraction versus Mie scattering. *Particle and Particle Systems Characterization* **4**, 14–19 (1987).

163. Kuhli, M., Weiss, M., and Steckel, H. A new approach to characterise pharmaceutical aerosols: Measurement of aerosol from a single dose aqueous inhaler with an optical particle counter. *European Journal of Pharmaceutical Sciences* **39**, 45–52 (2010).

164. Hickey, A.J. and Jones, L.D. Particle-size analysis of pharmaceutical aerosols. *Pharmaceutical Technology* **24**, 48–58 (2000).

165. Barry, P.W. and O'Callaghan, C. Video analysis of the aerosol cloud produced by metered-dose inhalers. *Pharmaceutical Sciences* **1**, 119–121 (1995).

166. Evesque, P. Analysis of the statistics of sandpile avalanches using soil-mechanics results and concepts. *Physical Review A* **43**, 2720 (1991).

167. Wu, L., Peguin, R.P.S., and da Rocha, S.R.P. Understanding solvation in hydrofluoroalkanes: Ab initio calculations and chemical force microscopy. *The Journal of Physical Chemistry B* **111**, 8096–8104 (2007).

168. James, J., Davies, M., Toon, R., Jinks, P., and Roberts, C.J. Particulate drug interactions with polymeric and elastomeric valve components in suspension formulations for metered dose inhalers. *International Journal of Pharmaceutics* **366**, 124–132 (2009).

169. Hooton, J., Jones, M., Harris, H., Shur, J., and Price, R. The influence of crystal habit on the prediction of dry powder inhalation formulation performance using the cohesive-adhesive force balance approach. *Drug Development and Industrial Pharmacy* **34**, 974–983 (2008).

170. Hooton, J., Jones, M., and Price, R. Predicting the behavior of novel sugar carriers for dry powder inhaler formulations via the use of a cohesive-adhesive force balance approach. *Journal of Pharmaceutical Sciences* **95**, 1288–1297 (2006).

171. Saini, D., Gunamgari, J., Zulaloglu, C., Sims, R.A., and Mazumder, M.K. Effect of electrostatic charge and size distributions on respirable aerosol deposition in lung model. *Industry Applications Conference, 2004. 39th IAS Annual Meeting. Conference Record of the 2004 IEEE*, Vol. **2**, pp. 948–952 (October 3–7, 2004).

172. Zhao, H., Castle, G.S.P., and Inculet, I.I. The measurement of bipolar charge in polydisperse powders using a vertical array of Faraday pail sensors. *Journal of Electrostatics* **55**, 261–278 (2002).

173. Murtomaa, M., Pekkala, P., Kalliohaka, T., and Paasi, J. A device for aerosol charge measurement and sampling. *Journal of Electrostatics* **63**, 571–575 (2005).

174. Chow, K., Zhu, K., Tan, R., and Heng, P. Investigation of electrostatic behavior of a lactose carrier for dry powder inhalers. *Pharmaceutical Research* **25**, 2822–2834 (2008).

175. Keskinen, J., Pietarinen, K., and Lehtimäki, M. Electrical low pressure impactor. *Journal of Aerosol Science* **23**, 353–360 (1992).

176. Hoe, S., Traini, D., Chan, H., and Young, P. Measuring charge and mass distributions in dry powder inhalers using the electrical next generation impactor (eNGI). *European Journal of Pharmaceutical sciences* **38**, 88–94 (2009).

177. Hoe, S., Young, P., Chan, H., and Traini, D. Introduction of the electrical next generation impactor (eNGI) and investigation of its capabilities for the study of pressurized metered dose inhalers. *Pharmaceutical Research* **26**, 431–437 (2009).

178. Newman, S.P. and Wilding, I.R. Gamma scintigraphy: An in vivo technique for assessing the equivalence of inhaled products. *International Journal of Pharmaceutics* **170**, 1–9 (1998).
179. Dolovich, M. Measuring total and regional lung deposition using inhaled radiotracers. *Journal of Aerosol Medicine* **14**, 35–44 (2001).
180. Lee, Z., Berridge, M.S., Finlay, W.H., and Heald, D.L. Mapping PET-measured triamcinolone acetonide (TAA) aerosol distribution into deposition by airway generation. *International Journal of Pharmaceutics* **199**, 7–16 (2000).
181. Gonda, I. Drugs administered directly into the respiratory tract: Modeling of the duration of effective drug levels. *Journal of Pharmaceutical Sciences* **77**, 340–346 (1988).
182. Byron, P.R. Prediction of drug residence times in regions of the human respiratory tract following aerosol inhalation. *Journal of Pharmaceutical Sciences* **75**, 433–438 (1986).
183. Hickey, A.J. and Thompson, D.C. Physiology of the airways. In: Hickey, A.J., ed. *Pharmaceutical Inhalation Aerosol Technology*. Marcel Dekker: New York, pp. 1–27 (1992).
184. Pillai, R., Yeates, D., Miller, I., and Hickey, A. Controlled-release from condensation coated respirable aerosol-particles. *Journal of Aerosol Science* **25**, 461–477 (1994).
185. Pillai, R.S., Yeates, D.B., Miller, I.F., and Hickey, A.J. Controlled dissolution from wax-coated aerosol particles in canine lungs. *Journal of Applied Physiology* **84**, 717–725 (1998).
186. El-Sherbiny, I.M., McGill, S., and Smyth, H.D. Swellable microparticles as carriers for sustained pulmonary drug delivery. *Journal of Pharmaceutical Sciences* **99(5)**, 2343–2356 (2010).
187. Adjei, A.L., Qiu, Y., and Gupta, P.K. Bioavailability and pharmacokinetics of inhaled drugs. *Inhalation Aerosols: Physical and Biological Basis for Therapy* **94**, 197–231 (1996).
188. Borgström, L., Newman, S., Weisz, A., and Morén, F. Pulmonary deposition of inhaled terbutaline: Comparison of scanning gamma camera and urinary excretion methods. *Journal of Pharmaceutical Sciences* **81**, 753–755 (1992).
189. Newman, S., Steed, K., Hooper, G., Källén, A., and Borgström, L. Comparison of gamma scintigraphy and a pharmacokinetic technique for assessing pulmonary deposition of terbutaline sulphate delivered by pressurized metered dose inhaler. *Pharmaceutical Research* **12**, 231–236 (1995).
190. Newman, S., Pitcairn, G., Steed, K., Harrison, A., and Nagel, J. Deposition of fenoterol from pressurized metered dose inhalers containing hydrofluoroalkanes. *Journal of Allergy and Clinical Immunology* **104**, s253–s257 (1999).
191. Altiere, R.J. and Thompson, D.C. Physiology and pharmacology of the airways. *Inhalation Aerosols: Physical and Biological Basis for Therapy* **94**, 85–137 (1996).
192. Clark, A.R. and Hollingworth, A.M. The relationship between powder inhaler resistance and peak inspiratory conditions in healthy volunteers—Implications for in vitro testing. *Journal of Aerosol Medicine* **6**, 99–110 (1993).
193. Martonen, T.B. et al. Lung models: Strengths and limitations. *Respiratory Care* **45**, 712–736 (2000).
194. Murray, L. and Kelly, G.L. *Physicians' Desk Reference*. Medical Economics Company, Inc.: Montvale, NJ (2010).
195. Corcoran, T., Dauber, J., Chigier, N., and Iacono, A. Improving drug delivery from medical nebulizers: The effects of increased nebulizer flow rates and reservoirs. *Journal of Aerosol Medicine* **15**, 271–282 (2002).
196. Guo, C., Gillespie, S.R., Kauffman, J., and Doub, W.H. Comparison of delivery characteristics from a combination metered-dose inhaler using the Andersen cascade impactor and the next generation pharmaceutical impactor. *Journal of Pharmaceutical Sciences* **97**, 3321–3334 (2008).
197. Dulera full prescribing information. (2010). At http://www.spfiles.com/pidulera.pdf
198. Tzou, T. Aerodynamic particle size of metered-dose inhalers determined by the quartz crystal microbalance and the Andersen cascade impactor. *International Journal of Pharmaceutics* **186**, 71–79 (1999).
199. Wu, L., Al-Haydari, M., and da Rocha, S.R. Novel propellant-driven inhalation formulations: Engineering polar drug particles with surface-trapped hydrofluoroalkane-philes. *European Journal of Pharmaceutical Sciences* **33**, 146–158 (2008).
200. Warren, S.J. and Farr, S.J. Formulation of solution metered dose inhalers and comparison with aerosols emitted from conventional suspension systems. *International Journal of Pharmaceutics* **124**, 195–203 (1995).
201. Stein, S.W. Size distribution measurements of metered dose inhalers using Andersen Mark II cascade impactors. *International Journal of Pharmaceutics* **186**, 43–52 (1999).

202. Islam, N. and Gladki, E. Dry powder inhalers (DPIs)—A review of device reliability and innovation. *International Journal of Pharmaceutics* **360**, 1–11 (2008).

203. Fink, J., McCall, A., Simon, M., and Uster, P. Enabling aerosol delivery technology for critical care. In: Dalby, R.N., Byron, P.R., and Farr, S.J., eds. *Respiratory Drug Delivery VIII.* Interpharm Press: Phoenix, AZ, pp. 323–325 (2002).

204. Deshpande, D. et al. Aerosolization of lipoplexes using AERx® pulmonary delivery system. *American Association of Pharmaceutical Scientists* **4**, 12–21 (2002).

205. Farr, S., Schuster, J., and Nicholas, C. Expanding applications for precision pulmonary delivery. *Drug Delivery Technology* **2**, 42 (2002).

206. Smart, J. et al. TouchSpray technology: Comparison of the droplet size measured with cascade impaction and laser diffraction. In: Dalby, R.N., Byron, P.R., and Farr, S.J., eds. *Respiratory Drug Delivery VIII.* Interpharm Press: Phoenix, AZ, pp. 525–527 (2002).

2 Breathing Zone Exposure Assessment

Charles E. Rodes and Jonathan W. Thornburg

CONTENTS

2.1 WHY BREATHING ZONE EXPOSURE ASSESSMENT?

Concentrations measured at fixed locations can differ substantially from those made by personal exposure assessments made in the very same microenvironments (Breslin et al., 1967; Rodes et al., 1991). Only the most ubiquitous and evenly distributed contaminants provide comparable data when monitored either way. The concept that simplistic measurements of concentration do not necessarily reflect actual human exposures can be illustrated by the classical risk paradigm (Rodes and Wiener, 2001):

$$\text{Sources} \rightarrow \text{Emissions} \rightarrow \text{Concentrations} \rightarrow \textit{Exposures} \rightarrow \text{Doses} \rightarrow \text{Effects}$$

where sources produce emissions that result in microenvironmental concentrations via some dispersion/transformation process that can lead to exposures—*if* the individual encounters the stressors across both time and space domains. Exposures are dependent on many influencing factors, and especially time-weighted proximity to the emissions from both distant and localized sources. The encountered exposure scenarios can produce significant body doses, dependent upon influencing uptake factors for that exposure route, such as the ventilation (inhalation) rate for air exposures.

31

The most appropriate location to monitor personal-level exposures is in the breathing zone (BZ) in relatively close proximity to the nose and mouth to minimize the influence of strong gradients that may exist near localized sources and near the body.

This chapter explores questions such as the following: (a) What information content is added by conducting BZ exposures instead of simplistic, fixed location monitors? (b) does the monitoring location on the body matter? (c) how are exposure misclassification bias and confounding issues best handled during BZ exposure assessments? (d) how do burden issues from BZ exposure monitors impact the representativeness (and bias) of the resultant data? (e) does the application of BZ exposures rather than applying exposure surrogate methods strengthen the epidemiologic associations between elevated toxicant levels and adverse health outcomes? and (f) can potential doses also be estimated from BZ exposure concentrations?

While concentrations are produced by emissions from sources, exposures only occur if an individual is close enough (proximal) for a sufficiently extended period to result in a significant exposure level. Fixed-location monitors do not incorporate the element of proximity, nor do they account for periods of time the person is actually nearby. The proximity to point, area, and line contaminant sources produces concentration gradients that can be substantial. Personal activity sources such as particle resuspension while walking (Oberoi et al., 2010), cooking (Koistinen et al., 2004), and workplace operations (Maynard and Jensen, 2001) often comprise the single strongest source category for many exposures. Since proximity changes as a person moves, personal exposure assessment provides the most complete and integrated picture of exposure. The process of capturing personal exposure data is more time consuming and expensive, and can be very burdensome to the individuals being studied, compared with less representative surrogate monitoring locations. Thus, apply BZ exposures involves trade-offs balancing the values of enhanced accuracy and representativeness of the collected data against the cost, burden, and complexity benefits of simpler surrogate estimating methods and locations. The importance of conducting BZ exposure studies for children is addressed in a review paper by Ashmore and Dimitroulopoulou (2009) for both gases and particles. They stress the importance of a holistic approach to managing aspects of indoor air pollution by utilizing personal exposure methods to robustly define the spatial and temporal ranges of their exposure distributions and their impacts on children's health.

While BZ exposures are routinely collected for both gas- and particle-phase contaminants, the inertial, diffusional, and transformational characteristics of airborne particles tend to add an additional level of complexity to the issues inherent in conducting accurate and representative assessments. This chapter intentionally focuses on aerosol sampling issues, given their greater complexity, often larger and more burdensome monitors, and potential for exhibiting the greatest biases between locations imposed by either differences in monitoring methods or strong spatial and temporal gradients encountered by individuals.

2.2 MINIMIZING EXPOSURE MISCLASSIFICATION BIASES

2.2.1 BZ Exposures versus Fixed-Location Concentrations

2.2.1.1 Rationale

The strategy of sampling of contaminants in or near the human BZ as most representative of inhalation exposure has been acknowledged as the most accurate approach (e.g., Cohen et al., 1984; NRC, 2004; Sarnat et al., 2007). These researchers recognized that relatively simple, fixed-location metrics are still needed to estimate cohort exposures in a simple and cost-effective manner over long study periods. BZ exposure metrics are inherently more complex and costly to administer. But they also recognized that fixed-location monitors (and exposure models) are only surrogates for true—"gold standard"—BZ exposures and cannot possibly estimate every nuance of the inherent spatial and temporal gradients individuals in a cohort experience daily. The key is defining the bias levels inherent in the surrogate method and conducting careful trade-off analyses to define the costs and benefits (Sarnat et al., 2007). Rodes et al. (2010) observed that accurate assessments for the most exposed required BZ methodologies (compared with fixed-location surrogates) in a general population study in metro Detroit, MI. Clearly, if understanding the levels defining the most exposed is an important goal for the exposure study, BZ level metrics must be given serious consideration.

An early study by Sherwood (1966) of radioactive aerosol concentrations in a nuclear materials laboratory showed that collecting data from workers using personal exposure monitors (PEMs) in the BZ provided levels of beta activity that exceeded single-location, room-average, microenvironmental* exposure measurement (MEM) values by a mean ratio of 7.7 (mean BZ-to-room ratio). The primary explanation for the substantially elevated personal levels was the workers' closer proximity to the beta sources than the fixed-location, room monitor. In a similar study for both alpha and beta activity, Stevens (1969) similarly reported much higher BZ PEM levels compared with concurrent MEM levels, ranging from ratios of 2 to 3 when workers were near scattered, multiple point sources in the same room, and ratios from 5 to 15 when they were in close proximity to a single-room source. Again the rationale was suggested to be driven primarily by the composite point source-to-worker proximity during the 8 h exposure interval. Parker et al. (1990) studied the release of polydisperse 0.5 μm particles into a test room and reported that personal exposure measurements at the lapel of a manikin were 5–10 times higher 0.5 m from a point source than room average samplers a few meters away. They also reported that real-time measurements at the lapel and at the mouth, separated by only 0.3 m, showed very poor correlation for the concentration fluctuations. The potential for point sources to provide nonuniform concentrations in occupational and nonoccupational indoor settings (respectively) was discussed by Nicas (1996) and Furtaw et al. (1996), who provided models to estimate the influence of the room ventilation system on the room concentrations. Rodes et al. (1991) showed that the ratios of BZ to fixed-location indoor measurements for residential settings differed significantly from those in workplace environments, attributed primarily to the stronger workplace localized sources. However, they reported that the residential 90th percentile BZ exposures were still typically as much as four times higher than fixed metrics at the cohort median levels.

The limited ability of some particle sizes to readily penetrate to the indoor environment (e.g., Thornburg et al., 2001) can play a key role in defining the representativeness of outdoor sampling methods to estimate exposures. Without utilizing personal-level exposure characterization, the times spent outdoors and indoors can only be estimated from outdoor monitoring, or exposure models that combine the two. Burke et al. (2001) developed the U.S. EPA CHAD exposure model to establish the exposure distributions for $PM_{2.5}$ from ambient monitoring data for a cohort in Philadelphia, PA, and observed that while the model reasonably estimated the interquartile range (IQR), the uncertainties became substantial and excess at the 90th percentile. Accurate definition of the most exposed would require BZ exposure assessments. Koistinen et al. (2004) reported that data from ambient monitoring locations significantly overestimate the contributions of outdoor traffic and long-range transport aerosol compared with personal exposures, and underestimate the contributions from indoor sources.

2.2.1.2 Characterizing the "Most Exposed"

Rodes et al. (1991) observed that activity pattern information during the integration period, in addition to source proximity, was critical to understanding nonoccupational personal exposures for aerosols. They suggested that the PEM-to-MEM ratios for residential exposure settings can be significantly different than occupational, due to typically weaker and more dispersed residential point sources and significant periods that include minimal or no sources. They reported median PEM-to-MEM ratios for nonoccupational aerosol exposure studies ranging from approximately 1.5 to 2.0 (much lower than typical occupational ratios), but still reported that the 90th or 95th percentile—"most exposed"—portions of the residential study populations exhibited ratios exceeding 4.0. Wallace et al. (2006), using a personal real-time aerosol nephelometer, identified a wide variety of sources

---

* Microenvironment is defined here to mean a localized, contained volume that generally defines the concentration—most often approximately bounded by the perimeter of the room when indoors.

that contribute to daily nonoccupational personal exposures, in both outdoor locations and indoor microenvironments. They also reported that a significant portion of the personal aerosol exposures occurred in locations where the participants spent only 4%–13% of their time. Ott (2007) examined the dispersion of a gas tracer indoors in a private residence and noted a strong proximity effect (concentration decrease with distance) at distances up to 2 m from the source. Room background levels were typically measured within experimental error beyond 2 m from the source. They also reported that the room air exchange rate had little effect on the proximity effect within 0.5 m of the source. These data suggest that the room average MEM would need to be significantly closer to the PEM than 2 m in order to reflect the elevated BZ levels in close proximity to a source. Rodes et al. (1995) also reported that the velocity profile near the body decreases sharply, defining a low-velocity boundary layer adjacent to the body only a few centimeters in thickness. Since the typical BZ inlet typically extends only a few centimeters from the body, particles will require a minimum size and inertia to penetrate across the flow streamlines to the inlet to be sampled. This bias is discussed subsequently in more detail in Section 2.2.3.

It has become increasingly clear (e.g., Rodes et al., 2010) that the strengths of epidemiologic investigations are strongly influenced by the ability to identify and characterize the most exposed in either a residential or occupational cohort. The exposure data from this segment of the cohort define the range of the exposure distributions and are heavily responsible for obtaining the strongest possible epidemiologic (cause and effect) associations, as well as defining the strongest relative risks when comparing exposed and unexposed populations. Additionally, it is now evident (Edwards and Jantunen, 2009) that the chemical compositions of exposure samples for the most exposed can be dramatically different from those defining the median or IQR. Rodes et al. (2010) note that minimally biased and confounded BZ personal exposures are the most representative of the true levels that form the "gold standard" against which to achieve the most representative epidemiologic findings. But technologies to conduct exposure studies applying minimally burdensome and cost approaches have historically not been available (Weis et al., 2005). The National Institute for Environmental Health Sciences at NIH initiated the Genes and Environment Initiative to develop better personalized exposure methodologies to fill this gap (Schmidt, 2006). These new approaches will tap into new sensor technologies that can be applied at the personal level to provide less biased and confounded exposure data for both the IQR and the most exposed from which to provide the strongest predictions of adverse biological and health responses.

2.2.2 Exposure Biases versus Confounding

2.2.2.1 Biases, Including Representativeness

The NRC (2004) recognized that the single most important misclassification bias to understand and minimize when attempting to characterize human exposures was that imposed when surrogate, fixed-location data were used to estimate BZ exposures. This aspect was discussed and reviewed by Sarnat et al. (2007) examining the impacts of traditional precision and accuracy type parameters, as well as biases from interpreting total particulate exposures, when the subset of exposures attributed to a single source category such as ambient aerosol was more appropriate. They observed that "…the use of (BZ) personal exposure measurements can lead to different interpretations than those derived from the use of ambient (fixed-location) concentrations.…" They strongly recommended that where possible, the contributions of source categories (e.g., aerosol of ambient origin) be taken into account to help minimize the biases between surrogate and BZ exposure assessments. Rabinovitch et al. (2005) noted that personal BZ exposures to the known inflammatory agent endotoxin contained in residential particulate matter were found to be statistically associated with severity levels in asthmatic children, while concurrent indoor and outdoor levels showed no associations. They attributed the finding to the stronger representativeness of personal exposures conducted in the BZ during personal activities when the children resuspended dust.

2.2.2.2 Confounding from Secondhand Smoke

The single largest confounder to personal level contaminant exposures during nonoccupational studies is likely to be secondhand smoke (SHS) from tobacco. Paoletti et al. (2006) reported the substantial influences of SHS on PM_{10} in indoor air (and personal exposures), noting the huge range of gas and particle-phase chemicals found in this smoke than can confound exposure studies. Slezakova et al. (2009) similarly reported on the wide range of metallic elements from SHS and their potential to confound $PM_{2.5}$ exposure measurements. Rodes et al. (2010) reported significant SHS levels in a general population personal exposure study of nonsmoking households that resulted from both study protocol violations and passive smoking exposures when away from home. These levels were quantified by applying the optical absorbance signature methodology of Lawless et al. (2004). Data were presented on the number of personal exposure samples that exceeded a $1.5\,\mu g/m^3$ level, defined as excessively confounding the target mass concentrations from all other sources. For these participants in presumed nonsmoking households, tobacco smoke contributes to the mean 24 h $PM_{2.5}$ mass concentrations that typically exceeded $3.0\,\mu g/m^3$ for 7.9%–30.3% of the time. Since the average personal $PM_{2.5}$ levels across the study averaged only $20.3\,\mu g/m^3$, this was a substantial contribution. Without applying some adjustment method for SHS confounding, these contributions could seriously bias the resultant data interpretations.

2.2.2.3 "Bluff Body" Biases

Studies of aerosol trajectories around objects have long suggested that the presence or absence of the human body shape can affect particle capture methodologies (e.g., Wood and Birkett, 1979; Mark and Vincent, 1986). They describe test methodologies that evaluate the performance of BZ samplers incorporating the presence of a manikin behind the inlet to simulate the human body. Trajectories of particles approaching the body (and the PEM inlet) must navigate through the flow streamlines decelerating and diverging around the body shape. A mathematical study by Ingham and Yan (1994) attempted to estimate the level of sampling bias using a particle trajectory model and suggested that the presence of the human "bluff body" behind an aspirated personal sampling inlet could bias the aerosol collection by as much as factors of 2 or more. The models showed that important influencing variables included factors such as (a) the particle Stokes number (particle size), (b) the level of particle bounce from inlet and clothing surfaces (and subsequent total capture by the inlet), (c) the distance the sampling inlet projects away from the body, (d) the sampling inlet velocity relative to the local air velocity, and (e) the width of the inlet relative to that of the human body. The relative sampling efficiency data in Table 2.1 illustrate the predicted influences of particle size, airspeed, and presence or absence of a bluff body behind the BZ inlet (compared with an inlet alone in the

TABLE 2.1
Predicted Fractional Overall Sampling Efficiencies, Illustrating the Influence of Airspeed, Particle Size, and Scenario, with and without a Bluff Body Present

Particle Size (µm)	Indoor (15 cm/s)		Walking (100 cm/s)		Outdoor (400 cm/s)	
	With Body	Without Body	With Body	Without Body	With Body	Without Body
0.5	1.0	1.0	1.2	1.1	2.0	1.5
1.0	1.5	1.5	2.0	2.0	3.5	3.5
2.5	1.6	2.5	2.2	3.0	4.0	6.0
10	1.7	3.7	2.5	4.2	4.2	7.5

Source: Ingham, D.B. and Yan, B., *J. Aerosol Sci.*, 25(3), 535, 1994.

Rainshield

PEM inlets

FIGURE 2.1 Use of an elliptical (aluminum here) body on which to mount a personal exposure (BZ) inlet used as a fixed location in an outdoor location 2 m above the ground. Location indoors at a 1.5 m elevation to represent the BZ of a seated adult is recommended.

free stream). Note that larger particles (e.g., 10 μm) are expected to be more impacted by bluff body effects than are accumulation mode, fine particles (e.g., 2.5 μm).

These data bolstered the notion that fixed-location MEM sampling without a bluff body simulating the human shape would potentially bias relationships with personal exposure sampling—even if the contaminant was uniformly distributed. The model also suggested that differences between personal sampling systems (size, shape, flow rate, etc.) could result in between-PEM-type biases. In order to minimize biases when intercomparing personal and fixed-location monitors, an elliptically shaped body such as that shown in Figure 2.1 can be used behind the inlet. This approach is not thought to require a physiologically correct manikin, but simply an approximate body shape to produce comparable flow streamlines around the body. This aluminum bluff body is 40.6 cm tall, with a 2-to-1 elliptical cross section 40.6 cm wide by 20.3 cm deep.

Only limited empirical data have become available to validate these model projections for either between-PEM biases or biases between MEM with and without a bluff body shape for nonoccupational settings. Heist et al. (2003) defined the flow streamline vectors around a heated, child-size manikin in a simulated residential setting and provided data on scaling between body sizes. Data by Rodes and Wiener (2001) examining the influence of the presence or absence of a bluff body for fixed-location sampling in a controlled wind tunnel setting simulating private residences, and utilizing both fine Arizona Test Dust challenges, found that the measured biases were less than 20% in all cases typically found in nonoccupational sampling. The data showed that for this aerosol type, the samplers with a bluff body provided lower concentrations than those without a bluff body. The empirical data indicated that while the direction of the biases for selected parameters appears to be correctly predicted by the model, the degree of particle bounce and subsequent capture by the PEM inlet was much smaller than was predicted by the total aspiration efficiency model of Ingham and Yan (1994). Further empirical studies are still needed to confirm these findings for the smallest and largest particle sizes in residential microenvironments, given the potential for adding biases to BZ assessments.

2.2.3 Defining Protocol Wearing Compliance

As noted previously, the burden imposed on a study participant by BZ sampling can result in serious protocol compliance problems in nonoccupational studies where participation is voluntary.

Occupational compliance is rarely of concern, since carrying a BZ monitor periodically is usually defined as a requirement of the job. Technologies are currently available to monitor wearing compliance using activity sensors (e.g., Lawless, 2003). Laboratory tests with controlled movements of the BZ sampler are used to set a threshold movement level below which the monitor can reasonably be assumed to be stationary and not being worn. Rodes et al. (2010) found that approximately 50% of the adult participants in a general population residential study wore the monitor less than 60% of the protocol-defined sampling time. Two of the 35 participants wore the monitor less than 25% of the time. For these participants, the levels of exposure misclassification bias would be extremely large, dependent upon the representativeness of the location where the monitors were left. Burden levels (weight, inconvenience) are typically the largest contributing factors to poor compliance. Rodes et al. (2010) reported similar results (1/3 of the participants wore the monitor <50% of the time) for a 3 year, general population study in metro Detroit, MI. However, a comparable fraction of the cohort was essentially fully compliant, suggesting that person-to-person wearing compliance, especially for general population cohorts is difficult to quantify without applying some level of wearing compliance monitoring.

While it is difficult to extrapolate broadly from these data, the findings strongly suggest that compliance can definitely be an important factor in personal exposure studies. When the monitor is not being worn as prescribed, it becomes a fixed-location monitor, resulting in data that are not representative of personal exposures. Some likely locations for leaving the BZ monitors (e.g., on the seat or in the trunk of a car) could seriously bias the contaminant levels, when only the monitor was being exposed. In settings with strong point sources, the distinction between personal and fixed-location monitors can be dramatically different. It is strongly recommended that efforts be made in any BZ study where burden might be construed to be a compliance factor, to assess the degree of compliance. While this may be as simple as asking follow-up questions, this approach may be inadequate for some participants.

2.3 DEFINING THE BREATHING ZONE

2.3.1 RATIONALE

A relevant consideration here is the physiological definition of the "breathing zone." Occupational exposure studies almost always require that the inlet be placed in the BZ to capture Bees but the rationale and specifics for that recommendation are not always provided. Baron (2003) recommended that personal exposure inlets sample air that would "most nearly represent that inhaled by the employee" for occupational assessments. Fortunately, most participants in occupational studies (a) are healthy adults, (b) are required to participate as part of the job, and (c) only need to wear the monitor for 8 h periods. This simplifies the complexity design requirements for monitoring technologies applied from the perspective of the total burden (weight, size, obtrusiveness). Nonoccupational studies, on the other hand, often include unhealthy, elderly adults and children, utilize voluntary participants, and often require 24 h assessment periods. This places a premium on the burden imposed on the study participants. Ideally, miniature passive badge samplers that can unobtrusively be attached to the lapel are most effective. These have been developed for a number of gases, and most recently efforts have been made to address aerosol collections passively (Brown and Monteith, 2001; Wagner and Leith, 2001). The size and weight of active (pumped) sampling systems for personal exposure pose substantial burden concerns for these classes of participants. The most common locations for the inlets in or near the BZ are clipped to a coat, vest lapel, or shoulder strap, or using a vest pocket.

Simplistically, the air parcel immediately in front of the nose and mouth from which the inhalation volume is withdrawn might could be construed as the BZ. Defining the boundaries of this parcel is not simple and varies with many factors including the inhalation/exhalation rate, microenvironmental air velocity and turbulence, nose versus mouth breathing, walking/running speed, etc. Often the location for obtaining BZ measures is defined simply by indicating that the inlet be attached to a "lapel" (e.g., NRC, 2003). Shapiro (1990) indicates that radiation personal monitors should be worn between the waist and the shoulders, also noting that BZ sampling should be conducted at the lapel location. In an effort to standardize measurement methodologies, while allowing a greater range of inlet types (many are simply too heavy to clip to a lapel), spatial definitions are defined for the BZ, such as "...the hemisphere of 300 mm (0.3 m) radius extending in front of the face and measured from the midpoint of an imaginary line joining the ears" (NOHSC, 2003). Rock (2001) similarly defines a 300 mm distance from the nose, and observes that such sampling is designated "breathing zone," even when a respirator is worn. While such a rigidly defined space is useful for uniformity, limited data are available to define the concentration uniformity throughout such a volume, especially for aerosols. Martinelli et al. (1983) documented biases for selected elements and suggested that resuspension of relatively large particle clothing dust was a contributing factor. Bradley et al. (1994) reported that sampling efficiencies for respirable aerosol at various locations across the chest and back varied from 0.45 to 0.61. Part of the chest bias was surmised to result from the inlet being in the downwash of cleaner exhaled breath from the nostrils. This observation lends credence to placement of a BZ inlet to the left or right rather than directly under the head. Bull et al. (1987) generated 0.26 μm particles in 60 s releases at point locations into a chamber and found that integrated samples collected in front of the nose and at the chest correlated to the same level (R = 0.84) and nose versus the waist. The data suggested that within experimental error, extending the BZ across the chest and as far down as the waist, produced essentially equivalent data for these relatively small particles.

A continuous tracer gas plume from a point source 1 m from a manikin-mounted PEM in a simulated indoor setting (equivalent air velocity and turbulence levels) was reported by Rodes et al. (1995) to be only 15 cm wide at the chest, suggesting that strong, nearby sources may produce significant differences between BZ and other personal sampling methods, if sufficiently long exposures occur. This scenario is most common in occupational settings. Similarly, Coker (1981) reported that workers handedness (left vs. right) during daily activities (in this case, paint spraying) affected concentrations by as much as 50% across the chest—the right lapel concentrations being much higher than the left for right-handed individuals. Rodes et al. (2001) reported that body dander could be a significant contributor to the collected mass nonoccupational BZ aerosol. Importantly, this source and resuspended clothing dust are associated with activities and character of the person being monitored, rather than external sources. The electrostatic charging of both a manikin (simulating the body) and the PEM were found to affect aerosol collection performance in a study by Smith and Bartley (2003). The efficiency was found to increase by ~10% for 7 μm particles when the charge could be effectively neutralized. Since participant clothing (e.g., sweaters) are surmised to periodically be highly charged, this may also influence BZ uniformity for aerosols.

The listing of spatial bias scenarios provided in the previous paragraph suggest that BZ assessments can certainly be substantially different from personal exposure assessment measurements made at more distant body locations. Table 2.2 provides typical separation distances when the methodologies listed are used for personal exposure sampling. Placing the inlet within 0.3 m of the oral/nasal plane (spherical distance) is recommended to provide measurements well within 20% of the true BZ values for gases and total mass of fine particles (<2.5) and coarse particles as large as 10 μm. Separation distances greater than 0.3 m may produce significant biases.

2.3.2 MONITORING ZONE LOCATIONS

The most applicable method of obtaining personal air exposures is through sampling directly in the BZ, and is specifically prescribed for most occupational scenarios to demonstrate compliance with permissible exposure limits (PELs) (e.g., DOE, 2001; MSHA, 2003; NRC, 2003; OSHA, 2003). There are no U.S. regulations requiring personal exposure monitoring for nonoccupational exposures. Personal exposure studies that integrate both occupational and nonoccupational periods

TABLE 2.2
Advantages and Disadvantages of the BZ Approach

Active Sampling Approaches	Advantage	Disadvantage	Relevant Citations
BZ	"Gold standard"; truest representation of inhalation	Likely to be the most burdensome	Cohen et al. (1984), Rodes et al. (2010), Delfino et al. (2008), Williams et al. (2003)
Waist pack	Removes inlet burden from lapel location; eliminates need for shoulder strap	At least 0.5 m from BZ; potentially biased by "ground cloud" for coarse particles	Pellizzari et al. (1999)
Backpack	Integral straps easy to wear and train participant	At least 0.5 m from BZ unless lapel inlet used; potentially biased by "ground cloud" for coarse particles	Rodes and Wiener (2001), Eisner et al. (2002), Choi et al. (2008)
Pull cart	Pull cart handle	Pull cart platform	Evans et al. (2000)
Surrogate person	No physical burden	Minimal personal cloud representativeness; constant presence of surrogate person	Stevens et al. (2003)
Predictive exposure models	No field monitoring required; low cost; validated for predicting the IQR	Models to predict the most exposed not yet developed or validated	Burke et al. (2001), Ott et al. (2000)

should utilize BZ sampling approaches to allow the greatest flexibility in utilizing existing databases and models. Only limited bias data are available to define the representativeness that more distant estimates of BZ exposures provide.

Personal exposure assessments can place an excessive burden upon study participants if the technology is not sufficiently miniaturized. Minimal participant burdens are strongly encouraged by governmental and private institutional review boards (IRBs) in dealing with human subjects in nonoccupational settings (e.g., NIH, 2003). Specifically, obtaining BZs can be the most burdensome approach to assessing personal exposures, but alternative approaches are available when this becomes an issue. Miniaturized PEMs typically have low collection rates, and often poor sensitivities (minimum detection limits) that counterproductively may limit the comparability of the data with more robust, fixed-location technologies. Detection limit compromises will be required if sufficiently low-burden personal monitors are simply not available.

2.3.3 MONITORING COMPROMISES

A number of strategies have been utilized to minimize the burden of conducting personal exposure assessments. In most scenarios, only the inlet system is placed in the BZ, with the pump located at the waist or in a backpack. The inlet can be clipped to the lapel, attached to special tabs on a vest, or attached to a shoulder strap adjacent to the lapel. Examples are shown in Figure 2.2. The swinging mass of a heavy inlet system can be annoying, and may affect the long-term comfort of the participant. This type of burden may result in two undesirable occurrences: (a) early dropout from the study (nonoccupational only) and/or (b) not wearing the sampler at all times. Moving the inlet away from the BZ can reduce the perceived burden, but at the potential expense of reduced representativeness to BZ. Such relocations were reported by Pellizzari et al. (1999) to have resulted in acceptable biases for fine particles ($PM_{2.5}$). This would suggest that personal monitors for gas phase contaminates (e.g., passive badges) could similarly be relocated with minimal bias. Burden

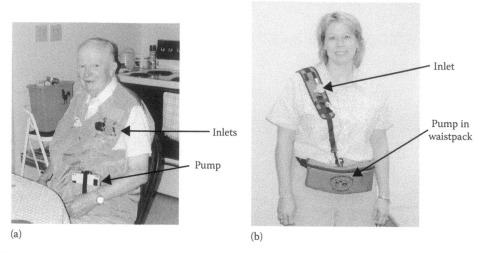

FIGURE 2.2 BZ inlets fastened to (a) a low-lint vest that can be worn over clothing and (b) a shoulder strap located within the BZ rather than on the more distant waist pack.

reduction approaches include utilizing a waist pack with the inlet mounted on the front, a back-pack with the inlet mounted on the side, or for those participants who have physical disabilities, a luggage-cart-mounted personal exposure system. Examples of these are shown in Figure 2.3.

Validated exposure models are the most obvious approach to developing exposure databases with minimal cost and no burden applied to the target cohorts. Such models for sized particles in residential settings (e.g., Ott et al., 2000; Burke et al., 2001) are becoming more widely applied and have been shown to be acceptable predictors of at least the IQR for cohort exposure distributions. Rodes et al. (2010) reviewed the distributions of personal exposures to ambient fixed-site $PM_{2.5}$ data and noted that the personal exposures at times were two or more times higher than the ambient data from a combination of SHS (see Section 2.2.2.2) and personal activity cloud sources.

2.4 APPLICATIONS

2.4.1 GENERAL

The most common application of BZ personal exposures assessments is to determine the distributions of exposure as related to a defined PEL. This requires that the methodologies be as accurate and representative of the true BZ concentrations. If the inlet cannot physically be placed continuously in the subject's BZ (e.g., small children), additional information is needed to subsequently allow any needed adjustment induced by the separation distance. As noted previously, gases and fine particles ($<\sim2.5\,\mu m$) may be uniformly distributed in the sampled air. In this case, personal measurements made by all the approaches given in Table 2.3 may provide statistically equivalent results. Particles larger than $\sim2.5\,\mu m$, however, may exhibit significant microenvironmental concentration gradients that would result in detectable biases between inlet locations. Resuspended clothing dust and body dander inadvertently collected by the inlet may provide samples that are not representative of the air in the BZ. Utilizing a (low-lint) sampling vest can reduce clothing resuspension problems.

Not all personal exposure studies are conducted to estimate actual exposure levels. While occupational BZ measurements are most often collected to make assessments relative to PELs, nonoccupational studies are often conducted to determine how well personal exposures associate with a dose or adverse health effect. In the case of associative studies, the absolute values are not nearly as

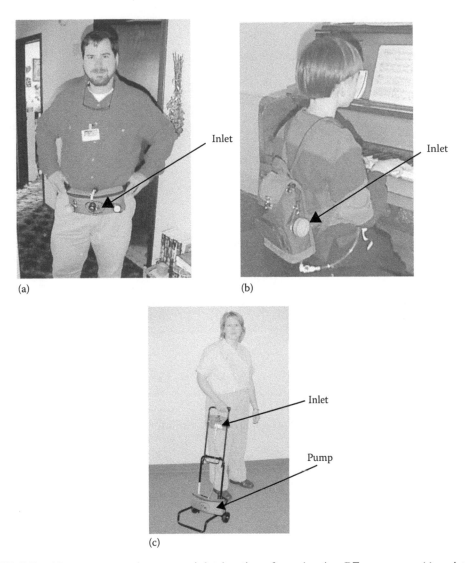

FIGURE 2.3 Alternate personal exposure inlet locations for estimating BZ exposures: (a) waist pack, (b) backpack, and (c) cart pulled by participant.

important as the consistency and representativeness of the data. The lower filter face velocities of typical personal exposures samplers possibly result in significantly higher collected aerosol mass concentrations in warmer environments compared to high-flow-rate fixed-locations samplers, due to reduced loss of volatile species. However, a Monte Carlo simulation analysis showed that either low-flow PEM or high-flow MEM mass concentrations would likely have produced statistically equivalent associations with concurrent adverse health effects.

2.4.2 SAMPLING BY CONTAMINANT TYPE AND INTEGRATION INTERVAL

The application of specific contaminant categories (e.g., coarse aerosols) can accentuate specific performance areas such as accuracy, sampling artifacts, and representativeness. While this level of detail is beyond the scope of this chapter, it is appropriate to identify selected sources for additional information on BZ exposure assessment. Sources of information describing specific sampling

TABLE 2.3

Active (Pumped) Personal Exposure Sampling Approaches

Active Sampling Approaches	PEM inlet Location	Pump Location	Participant Burden Level	Target Cohort	Notes
BZ	Within 0.3 m; lapel, vest pocket, shoulder strap	Integrated BZ package, waist pack, backpack	Medium to high	Healthy adults only if high burden; compromised adults and older children if minimally intrusive	"Gold standard" approach
Waist pack	Front of pack, waist level	Inside waist pack	Medium	Healthy adults, compromised adults and children	Okay for fine particles; coarse particles may be biased by "ground cloud"
Backpack	Side of backpack, or connected across shoulder to lapel	Inside backpack	Medium	Younger children or for excessively heavy monitors	May interfere with children's book bags; excessive burden may result in low wearing compliance
Pull cart	Pull cart handle	Pull cart platform	Low to medium	Compromised adults; elderly	May be biased by "ground cloud"
Surrogate person	On trailing/ nearby person	On trailing/ nearby person	Minimal	Adults or children that are burden averse	Surrogate labor time and cost; not representative of individual ground cloud
Predictive exposure models	None required	None required	None	Selected cohorts; metro populations	Utilizes ambient monitoring and estimates of times spent indoors and outdoors; poorly estimates the most exposed

methodologies for gases, aerosols, and radiation are summarized in Table 2.3. The collection of personal exposure samples for gases and vapors, including those directly from the BZ, is described by Brown and Monteith (2001), including an array of active (pumped) and passive techniques. The active techniques require the air sample to be drawn through miniature absorber tubes clipped to the lapel containing specific substrates (e.g., charcoal, XAD-2) by a controlled flow rate pump. The pump is often attached to the waist. Low-burden passive badge samplers are also described that attach to the lapel for a wide range of contaminants including organic vapor, amines, aldehydes, ozone, nitrogen dioxide, and mercury.

Particle phase contaminants are almost universally sampled in an active mode, but typically require an aerodynamic sizing step that significantly increases the size and complexity of the inlet. Often the resulting device is simply too cumbersome to attach directly to a lapel. The air stream is first sampled through a size-classifying device (e.g., cyclone, impactor) and the aerosol then collected on either a filter or media (e.g., impactor plate, agar [viable], or liquid [bubbler]). Hering

(2001) describes many of the commonly used aerosol samplers for both fixed-location and personal exposures. Rodes and Wiener (2001) also summarize personal aerosol monitors, including two miniature nephelometers.

Personal-level BZ exposures can now be considered on both acute (real-time sensing) and chronic (integrated filter collection) bases for sized aerosol. Advances in the miniaturization of nephelometric particle sensors now allow real-time exposure data at the personal level (e.g., Delfino et al., 2008; Adams et al., 2009). These BZ systems are currently fairly burdensome, weighing at least several kilograms with batteries, but may become much less burdensome if the research program described by Schmidt (2006) is successful. Without such burden level (weight, size, and noise level) reductions, the likelihood of widespread use of these approaches in residential settings is limited.

2.4.3 Estimation of Potential Exposures

McNabola et al. (2008) suggest that estimated potential doses of fine particles from BZ exposures to commuters is likely to present a different picture than that built solely upon concentration data. It is certainly reasonable that concentration adjustments for the levels of toxicants reaching into the respiratory system should provide stronger associations with biological uptake and any resultant adverse health outcomes. While McNabola et al. used models to estimate the potential doses, technologies are now available that should allow such estimates to be made by the monitoring system. Personal exposure characterizations provide data in concentration units (e.g., for sized particles, $\mu g/m^3$) over the integration period of interest. Personal, real-time monitoring provides either instantaneous concentrations or data integrated over short times (minutes), but still in concentration units. The tiny size of current technology accelerometer chips have now allowed them to be worn as personal-level sensors for prediction of metabolic functions (e.g., Rosenberger et al., 2008). This provides the potential for collecting estimates of ventilation (breathing) rate that can be combined with the exposure data to estimate the potential dose reaching target respiratory system locations. This would be accomplished by first relating the accelerometric response to estimated ventilation rate in a hopefully linear regression manner (see Equation 2.1) and then computing the potential dose from the measured concentrations (see Equation 2.2) by multiplying the integrated exposure concentrations over time t, by the estimated ventilation rate during the same time interval, and then dividing this quantity by the weight of the person being monitored.

$$VR = m_x \times AC + b_y \tag{2.1}$$

where
    VR is the ventilation rate
    m_x is the regression slope
    b_y is the regression intercept

The estimated ventilation rate can then be multiplied by the concentration and normalized by the body weight:

$$D = C \times \frac{VR}{BW} \tag{2.2}$$

where D [mass/time/body weight] = {C [aerosol mass/volume × VR [volume/time]}/BW [body mass]

Equation 2.2 can alternatively be stated as follows: dose [$\mu g/min/kg$] = {concentration [$\mu g/m^3$] × ventilation rate [l/min]}/body weight [kg]. At rest, the dose can be computed, utilizing a fixed ventilation rate, which in this scenario is the intercept (b_y) term from the regression. For both at-rest and active participants, a full regression would be needed.

Adding a tiny onboard accelerometer (as per Rosenberger et al., 2008) that is sufficiently sensitive can potentially enable development of predictive regressions between the composite motion signals and ventilation rate in real time by conducting scripted activities over a range of adults and/or children. While this technology is still evolving and may not be applicable for all subjects, it will set the stage for the next generation of personal-level BZ monitors where both BZ concentrations and estimated concurrent potential dose from Equations 2.1 and 2.2 are applied. The added value from this approach is that the burdensome ventilation measurements are conducted only during the scripted exercises, while the subsequent application of the same accelerometers can be applied transparently to the participants being monitored for exposure levels.

REFERENCES

Adams, C., Riggs, P., and J. Volckens (2009) Development of a method for personal, spatiotemporal exposure assessment, *J. Environ. Monit.*, 11:1331–1339.

Ashmore, M. and C. Dimitroulopoulou (2009) Personal exposure of children to air pollution, *Atmos. Environ.*, 43:128–141.

Baron, P. A. (April 6, 2003) NIOSH, Cincinnati, OH, personal communication to C. E. Rodes, RTI International, Research Triangle Park, NC.

Bradley, D. R., Johnson, A. E., Kenny, L. C., Lyons, C. P., Mark, D., and S. L. Upton (1994) The use of a manikin for testing personal aerosol samplers, *J. Aerosol Sci.*, 25 (Suppl. 1):S155–S156.

Breslin, A., Ong, L., Glauberman, H., George, A., and P. LeClare (1967) The accuracy of dust exposure estimates obtained from conventional air sampling, *AIHAJ*, 28 (1):56–61.

Brown, R. H. and L. E. Monteith (2001) Gas and vapor sample collectors, Ch. 16, pp. 415–457, in *Air Sampling Instrumentation of Atmospheric Contaminants*, 9th edn., B. Cohen and C. McCammon, eds., ACGIH: Cincinnati, OH.

Bull, R., Stevens, D., and M. Marshall (1987) Studies of aerosol distributions in a small laboratory and around a humanoid phantom, *J. Aerosol Sci.*, 18 (3):321–335.

Burke, J., Zufall, M., and H. Ozkaynak (2001) A population exposure model for particulate matter: Case study results for $PM_{2.5}$ in Philadelphia, PA, *J. Expo. Anal. Environ. Epidemiol.*, 11:470–489.

Choi, H., Perera, F., Pac, A., Wang, L., Flak, E., Mroz, E., Jacek, R. et al. (2008) Estimating individual-level exposure to airborne polycyclic aromatic hydrocarbons throughout the gestational period based on personal, indoor, and outdoor monitoring, *Environ. Health Perspect.*, 116 (11):1509–1518.

Cohen, B. S., Harley, N., and M. Lippmann (1984) Bias in air sampling techniques used to measure inhalation exposure, *AIHAJ*, 45 (3):187–192.

Coker, P. (1981) Recent developments in personal monitoring for exposure to organic vapors, *Int. Environ. Saf.*, 12:13–15.

Delfino, R., Staimer, N., Tjoa, T., Gillen, D., Kleinman, M., Sioutas, C., and D. Cooper (2008) Personal and ambient air pollution exposures and lung function decrements in children with asthma, *Environ. Health Perspect.*, 115 (4):550–558.

Department of Energy (DOE) (2001) DOE Occupational Radiation Exposure, 2001 Report, report DOE/EH-0660, U.S. Department of Energy, Office of Safety and Health, Washington, DC.

Edwards, R. and M. Jantunen (2009) Subgroups exposed to systematically different elemental composition of $PM_{2.5}$, *Atmos. Environ.*, 43:3571–3578.

Eisner, A. D., Heist, D. K., Drake, Z. E., Mitchell, W. J., and R. W. Wiener (2002) On the impact of the human (child) microclimate on passive aerosol monitor performance, *Aerosol Sci. Technol.*, 36:803–813.

Evans, G. F., Highsmith, R. V., Sheldon, L. S., Suggs, J. C., Williams, R. W., Zweidinger, R. B., Creason, J. P., Walsh, D., Rodes, C. E., and P. A. Lawless (2000) The 1999 Fresno particulate matter exposure studies: Comparison of community, outdoor, and residential PM mass measurements, *J. Air Waste Manage. Assoc.*, 50:1887–1896.

Furtaw, E. J., Pandian, M. D., Nelson, D. R., and J. V. Behar (1996) Modeling indoor air concentrations near emission sources in imperfectly mixed rooms, *J. Air Waste Manage. Assoc.*, 46:861–868.

Heist, D. K., Eisner, A. D., Mitchell, W., and R. Wiener (2003) Airflow around a child-size manikin in a low-speed wind environment, *Aerosol Sci. Technol.*, 37:303–314.

Hering, S. (2001) Impactors, cyclones, and other particle collectors, Ch. 2, pp. 316–376, in *Air Sampling Instrumentation of Atmospheric Contaminants*, 9th edn., B. Cohen and C. McCammon, eds., ACGIH: Cincinnati, OH.

Ingham, D. B. and B. Yan (1994) The effect of a cylindrical backing body on the sampling efficiency of a cylindrical sampler, *J. Aerosol Sci.*, 25 (3):535–541.

Koistinen, K., Edwards, R., Mathys, P., Urrskanen, J., Kunzli, N., and M. Jantunen (2004) Sources of fine particulate matter in personal exposures and residential indoor, residential outdoor, and workplace microenvironments in the Helsinki phase of the EXPOLIS study, *Scand. J. Work Environ. Health*, 30 (Suppl. 2):36–46.

Lawless, P. A. (2003) Portable air sampling apparatus including non-intrusive activity monitor and methods of using same, U.S. Patent 6,502,469B2, awarded January 7, 2003, U.S. Patent Office.

Lawless, P. A., Rodes, C. E., and D. S. Ensor (2004) Multiwavelength absorbance of filter deposits for determination of environmental tobacco smoke and black carbon, *Atmos. Environ.*, 38:3373–3383.

Mark, D. and J. H. Vincent (1986) A new personal sampler for airborne total dust in workplaces, *Ann. Occup. Hyg.*, 30 (1):89–102.

Martinelli, C. A., Harley, N. H., Lippmann, M., and B. S. Cohen (1983) Monitoring real-time aerosol distributions in the breathing zone, *AIHAJ*, 44:280–285.

Maynard, A. and P. Jensen (2001), Aerosol measurement in the workplace, Ch. 25, in *Aerosol Measurement: Principles, Techniques and Applications*, Wiley Interscience, New York.

McNabola, A., Broderick, B., and L. Gill (2008) Relative exposure to fine particulate matter and VOC's between transport microenvironments in Dublin: Personal exposure and uptake, *Atmos. Environ.*, 42:6496–6512.

Mine Safety and Health Administration (MSHA) (March 6, 2003) Verification of underground coal mine operators' dust control plans and compliance sampling for respirable dust; proposed rule, *Federal Register*, 30 CFR Parts 70, 75, and 90, 68 (44):10784–10872.

National Occupational Health & Safety Commission (NOHSC) (2003), Commonwealth of Australia, Glossary of Terms, web site: www.nohsc.gov.au/OHSInformation/NOHSCPublications/fulltext/docs/h4/607.htm

National Research Council (NRC) (2004) Research priorities for air borne particulate matter: IV. Continuing research progress, pp. 65–70, *Research Priorities for Airborne Particulate Matter, Research*, NAS: Washington, DC.

Nicas, M. (1996) Estimating exposure intensity in an imperfectly mixed room, *AIHAJ*, 57:542–550.

NIH (2003) Guidance on reporting adverse events to institutional review boards for NIH-supported multicenter clinical trials, web site: http://grants.nih.gov/grants/guide/notice-files/not99-107.html

Nuclear Regulatory Commission (NRC) (January 1, 2003) Standards of protection against radiation, *Federal Register*, 10 CFR Part 20:321–424.

Oberoi, R., Choi, J., Edwards, J., Rosati, J., Thornburg, J., and C. Rodes (2010) Human-induced particle re-suspension in a room, *Aerosol Sci. Technol.*, 44:216–229.

Occupational Safety and Health Administration (OSHA) (2003) Permissible exposure limits, web site: http://www.osha-slc.gov/SLTC/pel/

Ott, W. (2007) Exposure analysis: A receptor-oriented science, Ch. 1, in *Exposure Analysis*, Ott, W.R., Steinemann, A.C., Wallace, L.A., Eds., CRC Press, New York.

Ott, W., Wallace, L., and D. Mage (2000) Predicting particulate (PM_{10}) personal exposure distributions using a random component superposition statistical model, *J. Air Waste Manage. Assoc.*, 50:1390–1406.

Paoletti, L., Berardis, B., Arrizza, L., and V. Granato (2006) Influence of tobacco smoke on indoor PM_{10} particulate matter characteristics, *Atmos. Environ.*, 40:3269–3280.

Parker, R. C., Bull, R. K., Stevens, D. C., and M. Marshall (1990) Studies of aerosol distributions in a small laboratory containing a heated phantom, *Ann. Occup. Hyg.*, 1:35–44.

Pellizzari, E. D., Clayton, A. C., Rodes, C. E., Mason, R. E., Piper, L. L., Fort, B., Pfeifer, G., and D. Lynam (1999) Particulate matter and manganese exposures in Toronto, Canada., *Atmos. Environ.*, 33:721–734.

Rabinovitch, N., Liu, A., Zhang, L., Rodes, C., Foarde, K., Dutton, S., Murphy, J., and E. Gelfand (2005) Importance of the personal endotoxin cloud in school-age children with asthma, *J. Allergy Clin. Immunol.*, 116:1053–1057.

Rock, J. C. (2001) Occupational air sampling strategies, Ch. 2, pp. 20–50, in *Air Sampling Instrumentation of Atmospheric Contaminants*, 9th edn., B. Cohen and C. McCammon, eds., ACGIH: Cincinnati, OH.

Rodes, C. E., Kamens, R. M., and R. W. Wiener (1991) The significance and characteristics of the personal activity cloud on exposure assessment measurements for indoor contaminants, *Indoor Air*, 2:123–145.

Rodes, C. E., Kamens, R. M., and R. W. Wiener (1995) Experimental considerations for the study of contaminant dispersion near the body, *AIHAJ*, 56:535–545.

Rodes, C. E., Lawless, P. A., Evans, G. F., Sheldon, L. S., Williams, R. W., Vette, A. F., Creason, J. P., and D. Walsh (2001) The relationships between personal PM exposures for elderly populations and indoor and outdoor concentration for three retirement center scenarios, *J. Expo. Anal. Environ. Epidemiol.*, 11:1–13.

Rodes, C., Lawless, P., Thornburg, J., Williams, R., and C. Croghan (2010) DEARS particulation relationships for personal, indoor, outdoor, and central site settings for a general population, *Atmos. Environ.*, 44:1386–1399.

Rodes, C. E. and R. W. Wiener (2001) Indoor aerosols and exposure assessment, in *Aerosol Measurement*, p. 860, P. A. Baron and K. Willeke, eds., Wiley Interscience: New York.

Rosenberger, M. E., Skrinar, G., Haskell, W. L., Intille, S. S., and E. Munguia Tapia (2008) Multiple wireless accelerometers and heart rate accurately predict energy expenditure during level walking, *Med. Sci. Sports Exerc.*, 40 (5):S62–S63.

Sarnat, J., Wilson, W., Strand, M., Brook, J., Wyzga, R., and T. Lumley (2007) Panel discussion review: Session 1—Exposure assessment and related errors in air pollution epidemiologic studies, *J. Exposure Sci. Environ. Epidemiol.*, 17:S75–S82.

Schmidt, C. (2006) Monitoring environmental exposures, now it's personal, *Environ. Health Perspect.*, 114 (9):A529–A534.

Shapiro, J. (1990) *Radiation Protection—A Guide for Scientists and Physicians*, 3rd edn., Harvard University Press, Cambridge, MA.

Sherwood, R. J. (1966) On the interpretation of air sampling for radioactive particles, *Am. Ind. Hyg. Assoc. J.*, 27:98–109.

Slezakova, K., Pereira, M., and M. Alvim-Ferraz (2009) Influence of tobacco smoke on the elemental composition of indoor particles of different sizes, *Atmos. Environ.*, 43:486–493.

Smith, J. and D. Bartley (2003) Effect of sampler and manikin conductivity on the sampling efficiency of manikin-mounted personal samplers, *Aerosol Sci. Technol.*, 37:79–81.

Stevens, D. C. (1969) The particle size and mean concentration of radioactive aerosols measured by personal and static air samples, *Ann. Occup. Hyg.*, 12:33–40.

Stevens, C., Williams, R., Leovic, K., Chen, F., Vette, A., Seila, R., Amos, E., Rodes, C., and J. Thornburg (2003) Preliminary exposure assessment findings from the Tampa Asthmatic Children's Pilot Study (TACS), abstract prepared for the *2003 National Meeting of the AAAR*, Anaheim, CA, May, 2003.

Thornburg, J. W., Ensor, D. S., Rodes, C. E., Lawless, P. A., Sparks, L. E., and R. B. Mosley (2001) Penetration of particles into buildings and associated physical factors, Part I: Model development and computer simulations, *Aerosol Sci. Technol.*, 34:284–296.

Wagner, J. and D. Leith (2001) Passive aerosol sampler: 1. Principle of operation, *Aerosol Sci. Technol.*, 34:186–192.

Wallace, L., Williams, R., Rea, A., and C. Croghan (2006) Continuous weeklong measurements of personal exposures and indoor concentrations of fine particles for 37 health-impaired NC residents for up to four seasons, *Atmos. Environ.*, 40:399–414.

Weis, B., Balshaw, D., Barr, J., Brown, D., Ellisman, M., Lioy, P., Omenn, G. et al. (2005) Personalized exposure assessment: Promising approaches for human environmental health research, *Environ. Health Perspect.*, 113 (7):840–848.

Williams, R., Suggs, J., Zweidinger, R., Rea, A., Sheldon, L., Rodes, C., and J. Thornburg (2003) The Research Triangle Park Particulate Matter Panel Study: Modeling ambient source contribution to personal and residential PM mass concentrations, *Atmos. Environ.*, 37:5365–5378.

Wood, J. D. and J. L. Birkett (1979) External airflow effects on personal sampling, *Ann. Occup. Hyg.*, 22:299–310.

3 Mechanisms of Particle Deposition*

Jacky A. Rosati, Kristin K. Isaacs, and Ted B. Martonen

CONTENTS

---

* This chapter has been subject to an administrative review but does not necessarily reflect the views of the EPA. No official endorsement should be inferred. EPA does not endorse the purchase or sale of any commercial products or services.

3.1 INTRODUCTION

Particle deposition is an important topic of concern to diverse areas of aerosol science and technology, ranging from the design and manufacture of equipment for aerosol generation and characterization to the study of human health effects in aerosol therapy and inhalation toxicology. This chapter shall focus on the application of aerosol science to particle deposition in the human respiratory system. Particle deposition in human airways is governed by multiple mechanisms. Inertial impaction, sedimentation, and diffusion are often considered the primary mechanisms of deposition, while interception, charging, and cloud motion may be important in some situations. These respective mechanisms are described and formulated in this chapter. The deposition efficiency of each mechanism is dependent upon interactions among aerosol characteristics, ventilatory parameters, and respiratory system morphologies.

In this chapter, we discuss factors governing both the motion and deposition of particles in the respiratory system. We review the kinematic behavior of particles immersed in a fluid medium. We then consider the motion of particles and fluids within tubular structures (e.g., airways), and we discuss specific mechanisms of particle deposition in branching networks. We discuss particle inhalability and the current state of knowledge. We also briefly discuss deposition in the extrathoracic airways, but refer the reader to the peer-reviewed literature for more information on this evolving field.

3.2 FUNDAMENTALS OF INHALED AEROSOLS

An aerosol is a suspension of particulate matter in a gaseous carrying medium. Several textbooks have been written that address the general field of aerosol science and technology. We refer the interested reader to the works of Fuchs,[1] Mercer,[2] Reist,[3] Hinds,[4] and Friedlander.[5]

Particle size, density, and shape are important factors in the prediction of particle kinetics and aerosol deposition in human airways. In this section, we discuss particle characteristics, their effects on Stokes's Law and terminal settling velocity, and their role in particle deposition.

3.2.1 STOKES'S LAW

The interaction between an aerosol particle and the carrying gas is quantified by Stokes's Law. Stokes's Law provides a basis for the study of aerosol particle motion. The drag force on a particle moving through a fluid may be expressed as

$$F_D = 3\pi\mu\, d_p V_p \tag{3.1}$$

It is important to recognize that the application of Equation 3.1 requires the following assumptions: (1) the particle is a rigid sphere, (2) the carrying gas is incompressible, (3) the inertia of the particle is negligible compared to drag force, (4) there are no hydrodynamic interactions or boundary effects affecting the particle, (5) particle motion is constant, and (6) the fluid velocity at the particle surface is zero. In practice, these conditions must be checked for validity.

3.2.2 TERMINAL SETTLING VELOCITY AND RELAXATION TIME

An aerosol particle will reach its terminal settling velocity when the drag force, F_D, on it is equal and opposite to the force of gravity, F_g:

$$F_D = F_G \tag{3.2}$$

By recognizing that $F_g = mg$ and substituting in Equation 3.1, this relationship may also be expressed as

$$3\pi\mu\, d_p V_p = mg \tag{3.3}$$

The equation for terminal settling velocity, V_{TS}, is derived by expressing the particle mass in terms of ρ_p and solving for V_p:

$$V_{TS} = \frac{\rho_p d_p^2 g}{18\mu} \tag{3.4}$$

The time it takes for a particle starting at rest to reach its terminal settling velocity is the relaxation time, τ, expressed as

$$\tau = \frac{d_p^2 \rho_p}{18\mu} = \frac{d_{ae}^2 \rho_o}{18\mu} \tag{3.5}$$

Note that τ is dependent upon both particle properties (e.g., particle size and density) and fluid viscosity. When we substitute Equation 3.5 into Equation 3.4, we obtain V_{TS} expressed in terms of τ:

$$V_{TS} = \tau g \tag{3.6}$$

3.2.3 AERODYNAMIC DIAMETER

The aerodynamic diameter of a particle (d_{ae}) is an important parameter used to relate particles of differing shapes and densities. The d_{ae} of a particle is defined as the diameter of a unit density sphere that has the same V_{TS} as the particle in question. The d_{ae} is commonly used to characterize the kinetic behavior of larger (>1 μm) aerosol particles, and can be calculated as

$$d_{ae} = d_g \sqrt{\rho_p} \tag{3.7}$$

where d_g is the geometric diameter of the particle.

3.2.4 MODIFICATIONS TO THE AERODYNAMIC EQUATIONS

Modifications may be made to adapt Stokes's Law for use in nonidealized situations. For example, Equation 3.1 assumes that a particle is spherical in shape, an assumption that may not always be appropriate (e.g., fibers, spheroids, cubes). Without appropriate correction, calculations of aerodynamic properties and deposition probabilities for inhaled particles may be inaccurate.[6]

3.2.4.1 Correction for Nonspherical Particles

When particles are nonspherical, a dynamic shape correction factor, Π, may be applied to Stokes's Law. This quantity is defined as the ratio of the drag force of the nonspherical particle to the drag force of a sphere having the same volume and velocity. Stokes's Law, corrected for shape, may then be expressed as

$$F_D = 3\pi\mu\, d_p V \chi \tag{3.8}$$

TABLE 3.1
Dynamic Shape Factors for Common Particle Shapes and Types

Type of Particle	Dynamic Shape Factor, Π	Reference
Particles of regular shape		
Sphere	1.00	12
Cube	1.08	12
Cylinder (vertical axis)	1.01	12
Cylinder (horizontal axis)	1.14	12
Cube octahedron	1.03	2
Octahedron	1.06	2
Tetrahedron	1.17	2
Other particles		
Bituminous coal	1.05–1.11	12
Sand	1.57	12
Talc	1.88–2.04	12
Quartz (0.65–1.85 μm)	1.84	2
Quartz (>4 μm)	1.23	2

The correction factor χ can also be used to modify d_{ae} for shape irregularity[7]:

$$d_{ae} = d_g \sqrt{\frac{\rho_p}{\chi}} \tag{3.9}$$

The terminal settling velocity, corrected for shape, is given by

$$V_{TS} = \frac{\rho_p d_p^2 g}{18 \mu \chi} \tag{3.10}$$

Moss,[8] Stober,[9,10] and Leith[11] have investigated the effects of shape on the aerodynamic properties of particles. Table 3.1 provides dynamic shape factors for common particle shapes and types.

3.2.4.2 Correction for Slip

Stokes's Law assumes that the fluid velocity at the particle surface is zero. This no-slip assumption is not valid for small particles; thus, the Cunningham slip correction factor, C_c, should be applied when particles smaller than 10 μm are being considered. C_c may be expressed as

$$C_c = 1 + Kn \left[A_1 + A_2 \exp - \left(\frac{A_3}{Kn} \right) \right] \tag{3.11}$$

where
   Kn is the dimensionless Knudsen number (defined in the following)
   A_1, A_2, and A_3 are constants derived from experimental measurements

While the values of these constants vary in the literature,[1,13–15] the most commonly accepted values, as determined by Davies,[13] are $A_1 = 1.257$, $A_2 = 0.4$, and $A_3 = 1.1$.

Kn is the ratio of the mean free path of the gas molecules (λ) to the particle diameter. It may be expressed as

$$Kn = \frac{2\lambda}{d_p} \qquad (3.12)$$

The mean free path may be calculated using

$$\lambda = \frac{1}{\sqrt{2} n \pi d_m^2} \qquad (3.13)$$

If the Kn is very small ($Kn \ll 1$), a particle will decelerate due to gas molecule collisions upon the particle's surface. Conversely, if the Kn is large ($Kn \gg 1$), a particle will not be affected by gas molecules.

When the Cunningham slip correction factor is included, the Stokes drag force on a particle becomes

$$F_D = \frac{3\pi\mu V d_p \chi}{C_c} \qquad (3.14)$$

The terminal settling velocity, corrected for shape and slip, is given by

$$V_{TS} = \frac{\rho_p d_p^2 g C_c}{18\mu\chi} \qquad (3.15)$$

Table 3.2 demonstrates the effect of slip correction on V_{TS} for particles of various sizes. Note that there are significant differences between the corrected and uncorrected V_{TS} values for submicron particles.

TABLE 3.2
Terminal Settling Velocity Values Calculated Using the Cunningham Slip Correction Factor

Particle Diameter (μm)	C_c	V_{TS}[a], w/o C_c (cm/s)	V_{TS}[a], with C_c (cm/s)
0.001	228.234	3.02E–09	6.61E–07
0.01	23.3443	3.02E–07	6.77E–06
0.1	2.97393	3.02E–05	8.71E–05
0.5	1.34743	7.54E–04	1.005E–03
1	1.17273	3.02E–03	3.516E–03
3	1.05757	2.71E–02	2.864E–02
5	1.03454	7.54E–02	7.790E–02
10	1.01727	3.02E–01	3.066E–01

Source: Crowder, T.M. et al., *Pharm. Res.*, 19, 239, 2002.
[a] Values given are calculated for a unit density sphere of a given size at 20°C and 101 kPa using Davies's constants.[13]

3.2.5 HYGROSCOPICITY

In human airways, particle deposition may depend on the physicochemical (hygroscopic) properties of inhaled substances. Hygroscopicity may be defined as a particle's propensity to absorb water from a warm, humid environment, thereby changing its diameter and density. The degree and rate of hygroscopic growth are influenced by many factors, including chemical composition, temperature, relative humidity, initial particle size, and duration of exposure.[17]

As the relative humidity within the lungs may approach saturation (99.5%),[18] hygroscopic effects can substantially affect inhaled aerosols. As particle size and density are important factors in determining deposition, hygroscopic growth is a critical concern in the risk assessment of inhaled aerosols (inhalation toxicology) and the development of inhaled pharmaceutics (aerosol therapy). Hygroscopic growth has been shown to occur in many aerosols, including a number of environmental pollutants and pharmacological agents.[19,20]

Theoretical models are useful in characterizing properties (i.e., growth rate, equilibrium size, aerodynamic diameter) of hygroscopic particles under different *in vivo* environmental conditions. Various theoretical models of particle hygroscopicity have been developed for cigarette smoke,[21] aqueous droplets,[22] and soluble particles.[23]

Martonen[24] derived equations for the density and aerodynamic diameter of a particle in different lung airway generations, provided the particle growth rate, r_g, is known. The density ρ_i of a particle entering airway generation i can be calculated as

$$\rho_i = \left[\frac{d_0}{d_i}\right]^3 \left[\rho_0 - \rho_{H_2O}\right] + \rho_{H_2O} \tag{3.16}$$

where
ρ_0 is the initial particle density
d_0 is the initial particle diameter
d_i is the particle diameter at the entrance to airway generation i, given by

$$d_i = d_0 + r_g \left(\sum_{j=0}^{j=i-1} \frac{L_j}{V_j}\right) \tag{3.17}$$

where
L_j is the length of airway j
V_j is the mean velocity of the particle in airway j

The aerodynamic diameter of the particle in airway i is then given by

$$d_{ae,i} = \sqrt{\rho_i C_c}\, d_i \tag{3.18}$$

A particle's size and density may change while traveling through the respiratory system (Figure 3.1). It may not be prudent, therefore, to expect the deposition pattern of inhaled aerosols to be determined by the pre-inspired sizes and densities of its constituent particles. The dynamic process of hygroscopic growth must be accounted for during a breathing cycle. It is also interesting to conjecture that evaporation may occur when aerosols from the warm, moist, deep lung are cooled during expiration.

Both experimental and computational studies have been performed to determine deposition patterns for inhaled hygroscopic particles in human airway networks. Experimental studies have

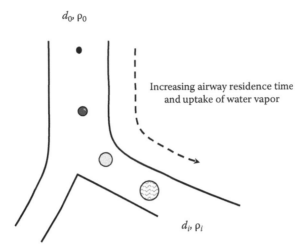

d_0, ρ_0

Increasing airway residence time
and uptake of water vapor

d_i, ρ_i

FIGURE 3.1   Variation in particle properties due to hygroscopic growth in a lung airway (not to scale).

been performed for NaCl[25,26] and cigarette smoke.[27] Computational models of the deposition of hygroscopic particles have been developed for sulfate aerosols,[28,29] cigarette smoke,[30,31] saline,[32–35] phosphoric acid aerosols,[36] atmospheric salts,[37] and several inhaled pharmaceutics.[38]

3.3   DOMAINS OF PARTICLE DYNAMICS

The theory employed to quantify particle dynamics in a transporting medium will depend on the properties of both the particles and gases involved. Particle dynamics can be classified by size into three regimes of behavior. In the free-molecule regime, the behavior of very small particles is influenced by the motion of individual gas molecules. In the continuum regime, the surrounding gas acts on large particles as a continuum or viscous fluid. In the slip-flow regime, a correction for fluid slip at the surface of the particle is employed. A fourth (transition) regime, situated between the slip-flow and free-molecule regimes, is sometimes considered,[38] although the behavior of particles within this regime is not independently defined. In some texts, the entire region between the free-molecule and continuum regimes is called the transition regime.[4]

The regimes are defined by Kn values, or (at constant temperature and pressure) by particle sizes. However, the actual values of the threshold Kn or d_p parameters for each regime vary in the literature. The regimes, their corresponding particle characteristics, and their associated dynamic theories are discussed in the following.

3.3.1   Free-Molecule Regime

The free-molecule regime has been defined to include particles having $Kn > 10$,[38] or $Kn > 20$.[7] In air at STP, these values correspond to particle diameters of 0.01 and 0.02 µm, respectively.

In the free-molecular regime, the motion of particles can be influenced by interactions with individual gas molecules. Particles and gas molecules in Brownian motion collide randomly, and after many collisions, the force exerted by the gas molecules will affect the direction of motion of the much more massive particles. In this case, particle motion must be quantified using gas dynamic theory.

Particle diffusion becomes a very important consideration in the study of particle dynamics in the free-molecule regime. In this regime, the diffusional velocity of the particle due to random Brownian motion is several orders of magnitude greater than the terminal settling velocity[38] and, thus, in this regime the effects of gravitational forces may be neglected.

3.3.2 Continuum Regime

In the continuum regime, the gas in which the particles are suspended can be assumed to act as a continuous, viscous fluid. The continuum regime may also be called the Stokes regime.[38] This regime is alternately defined as $Kn < 0.1$,[38] $Kn < 0.4$,[7] and $d_p < \lambda$ $(Kn < 2)$.[4] A value of $Kn < 0.1$ corresponds to particle diameters of greater than 1.3 μm in air at STP.

In the continuum regime, particle motion within the gas is governed by the momentum (Navier–Stokes) equations for a viscous fluid, and the particle drag force and terminal settling velocity can be calculated from Stokes's Law. In this region, no-slip conditions are assumed to exist at the surface of the particle.

3.3.3 Slip-Flow Regime

The slip-flow regime falls between the free-molecule and continuum regimes and has been variously defined as including particles having a value of Kn of 0.1–0.3 (with a transitional regime defined having Kn of 0.3–10),[38] or 0.4–20.[4]

In the slip-flow regime, the assumption of no-slip conditions at the surface of the particle is not applicable. In this case, there exists a quantifiable velocity of the gas relative to the particle at its surface. In the slip-flow regime, the drag force exerted on a particle is overestimated (and the terminal settling velocity is underestimated) by Stokes's Law. Therefore, within this regime Stokes's Law must be corrected by the Cunningham slip correction factor (Equation 3.5).

3.4 INHALABILITY

To understand the deposition of particles in the respiratory system, one must first understand the inhalability of a particle, or the ability of the particle to enter the nose or mouth during breathing. Many experimental studies have used "breathing" mannequins to determine the inhalability of various particle sizes and types. These mannequins act as "samplers" and the inhalability is actually the aspiration efficiency of the mannequin breathing system.[39,40] Mathematical models such as the ICRP 66[41] and ACGIH[42] have been used to predict inhalability of particles through the nose.[43] Computational fluid dynamics (CFD) has also been used to estimate the inhalability of particles.[44] Breysse and Swift studied nasal inhalability from still air by four human subjects; the only human study of inhalability from still air in the literature at this time.[45] A comprehensive overview of inhalability work is provided by Millage et al.[43]

Inhalability (I), or the ratio of the concentration of particles of a particular aerodynamic diameter that enter the nose or mouth to the ratio of the concentration of particles of the same aerodynamic diameter in the inhaled ambient air,[41] may be basically defined by Equation 3.19[43]:

$$I(d_{ae}) = \left[\frac{Cinhaled(d_{ae})}{Cambient(d_{ae})} \right] \tag{3.19}$$

Inhalability is dependent on wind speed with different inhalability fractions resulting from still air than from low or moderate wind speeds.[41,43,46] Inhalability is also affected by face orientation to the wind and facial feature dimensions,[43,46–48] as well as breathing mode (nasal versus oral) and breathing rate.[43] Because of the dependence of inhalability fraction on each of these factors, various inhalability efficiency equations have been developed primarily for the Industrial Hygiene community.[41–43,46,49]

The overarching message from all of this work is that there is no single equation that describes all situations when it comes to inhalability; there are so many variables (e.g., wind speed, particle size, and breathing rate) that different equations must be used for different situations. Finally, it is important to note that all of this work deals with the assumption of spherical or semi-spherical particles (aerodynamic diameters), with no work being performed on fiber inhalability. Fiber research is needed to better understand inhalability, particularly in the wake of the World Trade Center collapse that released large quantities of fibers into lower Manhattan (man-made vitreous fibers [MMVFs], asbestos, carbon nanotubules, etc.),[50,51] also because of concerns about asbestos exposure and inhalation during remediations,[52,53] and because of the emerging research and development area of nanoparticles/nanotubules.[54,55]

3.5 DEPOSITION IN THE EXTRATHORACIC REGION

Particle deposition in the extrathoracic region of the human respiratory system has been studied far less than deposition in the human lung due to the extreme complexities and irregular geometries of the extrathoracic region (nose, mouth, throat, and larynx). The area of the extrathoracic region that has been studied the most is the nose. Experimental work in nasal casts, nasal replicas, and even resin models has been performed, with the most recent work on the nasal replicas of infants and children.[56–58] In their human subject study, Bennett et al.[58] explored the contribution of the nose to fine particle deposition in children. They found that the child's nose was less efficient at filtering larger (2 µm) particles and under higher flow conditions than the adult nose. Cheng, in his 2003 paper,[59] reviews the literature on the deposition of particles in the extrathoracic region of the human respiratory system and performs an analysis combining data from casts as well as human subjects. This analysis resulted in regional deposition efficiency equations for the oral and nasal cavities based on the diffusion and impaction mechanisms of deposition discussed in Section 3.7. Recent work by Smith et al.[59] has also explored the clearance of inhaled particles from the human nose.

Modeling work, both mathematical and CFD in nature, has been performed to predict particle deposition in the extrathoracic airways.[60–69] One computational particle fluid dynamics (CPFD) model in particular,[70] combines both the extrathoracic and the airways for a complete three-dimensional model of the human respiratory system. This model was developed using a nose and mouth constructed from imaging data from the U.S. National Library of Medicine Visible Human Project[71] and a five-lobe lung airway model constructed using the morphological data of Yeh and Schum.[72] For additional information on particle deposition in the extrathoracic region, we refer the reader to Chapter 5, as well as the emerging literature on this subject.

3.6 FLUID DYNAMICS IN AIRWAYS

Both motion and deposition of aerosols in the respiratory system are dependent upon airflow conditions within the airways. Particles will be affected by the nature of the velocity and pressure fields in which they are carried. Particles are entrained in the airflow, and are transported with both the bulk convective motion of the flow and any other secondary flow patterns initiated by airway geometries. Therefore, the fluid dynamic conditions (e.g., laminar versus turbulent motion) of air in the bronchial tree are important considerations in both inhalation toxicology and aerosol therapy.

The dynamic behavior of air in the respiratory passages is governed by both morphological and respiratory parameters. Morphological considerations include airway dimensions (airway diameters and lengths), bifurcation angles, spatial arrangement of the branching network, and airway surface characteristics. Respiratory parameters describe the mechanics of ventilation, and include respiratory rates and tidal volumes. In this section, different lung airway morphologies and their corresponding airflow characteristics will be discussed. Both idealized and anatomically realistic morphologies will be introduced. A more advanced discussion of fluid flow modeling and its incorporation into aerosol deposition modeling will be provided in Chapter V, Modeling Inhaled Aerosols.

3.6.1   FUNDAMENTALS OF FLOW

A discussion of the dynamics of airflow in the respiratory airways requires the introduction of some basic flow considerations.

3.6.1.1   Steady versus Unsteady Flow

If the fluid properties (e.g., density, velocity) at each point in a prescribed flow field are time invariant, then the flow is considered steady. In contrast, in unsteady flow the fluid properties at each point in the flow field vary with time. In some circumstances, when the time-dependent changes in fluid properties are very small, the flow may be assumed to be quasi-steady. During normal respiration, the flow in the lung airways can be considered quasi-steady. However, under high-frequency breathing or forced expiration, flow in the airways is unsteady.[73] It is important for investigators to recognize the potential significance of transient, time-dependent, phenomena when analyzing airflow patterns in their particular studies.

3.6.1.2   Laminar versus Turbulent Flow

Flow in airways may be either laminar or turbulent. Simply stated, in laminar flow, the fluid motion occurs in smooth layers (laminae), and there is no fluid mixing between adjacent layers. In the turbulent regime, the flow structure is random and characterized by the chaotic motion of fluid particles within the larger structure of the mean fluid flow. In this case, mixing of fluid between layers may be initiated locally and propagated to downstream regions.

Flows in an idealized cylindrical passage can be characterized as either laminar or turbulent by calculating the value of dimensionless Reynolds number:

$$Re = \frac{\rho_a d V_a}{\mu} \tag{3.20}$$

In general, the flow in idealized tubes will be laminar for $Re < 2300$.[74] Idealized tubes (i.e., those being straight and having smooth, rigid walls, circular cross sections, and constant diameters) are rarely observed *in vivo*.

Table 3.3 provides estimated Reynolds numbers in 24 different airway generations for two different respiratory rates. We note that the Re values predict laminar flow in every airway at a ventilation rate of 15 L/min, but predict turbulent flow in the larger airways at a rate of 60 L/min. However, it has been determined that due to disturbances in the inflow of air to the lungs caused by the larynx and the cartilaginous rings, turbulent flow is usually observed in the large airways,[73] and thus, care must be exercised when attempting to relate Re values with *in vivo* flow conditions.[75]

3.6.2   FLOW IN IDEALIZED TUBES

A straightforward way of simulating airflow within respiratory passageways is to consider airways as idealized tubes, as defined earlier. Flow in such an airway under both laminar and turbulent conditions is shown in Figure 3.2. As flow enters, the velocity of the fluid V_0 is uniform within the cross section of the tube. This type of velocity pattern is called plug flow. As the flow proceeds, the velocity is retarded by the shear force that the boundary surface of the tube imparts on the airflow. The no-slip condition exists at the tube surface, so the velocity of flow at the wall is zero. The axial distribution of flow velocity is called the velocity profile. The region of the flow where $V < V_0$ is called the boundary layer. As the fluid moves along the tube, pressure and shear forces within the fluid will equilibrate, and a fully developed velocity profile will be attained. The distance from the entrance of the tube to the point where the laminar velocity profile is fully developed is

TABLE 3.3
Reynolds Numbers by Airway Generation (Calculated at 1 atm and 37°C)

Airway Generation	Airway Cross Section[76] (cm²)	Airway Diameter[76] (cm)	Velocity[a] (cm/s)	Velocity[b] (cm/s)	Re[a]	Re[b]
0	2.54	1.800	98.43	393.70	1077.42	4309.68
1	2.33	1.220	107.30	429.18	796.07	3184.27
2	2.13	0.830	117.37	469.48	592.44	2369.76
3	2.00	0.560	125.00	500.00	425.70	1702.80
4	2.48	0.450	100.81	403.23	275.87	1103.49
5	3.11	0.350	80.39	321.54	171.10	684.41
6	3.96	0.280	63.13	252.53	107.50	430.00
7	5.10	0.230	49.02	196.08	68.57	274.26
8	6.95	0.186	35.97	143.88	40.69	162.76
9	9.56	0.154	26.15	104.60	24.49	97.97
10	13.40	0.130	18.66	74.63	14.75	58.99
11	19.60	0.109	12.76	51.02	8.46	33.82
12	28.80	0.095	8.68	34.72	5.02	20.06
13	44.50	0.082	5.62	22.47	2.80	11.21
14	69.40	0.074	3.60	14.41	1.62	6.49
15	117.00	0.050	2.14	8.55	0.65	2.60
16	225.00	0.049	1.11	4.44	0.33	1.32
17	300.00	0.040	0.83	3.33	0.20	0.81
18	543.00	0.038	0.46	1.84	0.11	0.43
19	978.00	0.036	0.26	1.02	0.06	0.22
20	1743.00	0.034	0.14	0.57	0.03	0.12
21	2733.00	0.031	0.09	0.37	0.02	0.07
22	5070.00	0.029	0.05	0.20	0.01	0.04
23	7530.00	0.025	0.03	0.13	0.01	0.02

[a] 15 L/min.
[b] 60 L/min.

called the entrance length, $L_{e,l}$. Under laminar conditions, the final velocity profile is parabolic, with the velocity increasing from zero at the tube walls to the free stream velocity V_0 at the center of the tube. The actual shape of the final parabolic profile is determined by both the absolute fluid viscosity and the pressure gradient driving the flow. In turbulent flow, the boundary layer grows more rapidly, the entrance length ($L_{e,t}$) is shorter, and the fully developed velocity profile is flatter.[74] In turbulent flow, a laminar sublayer will exist near the wall, and a transitional layer will lie between this laminar layer and the entirely turbulent flow that exists in the tube's center. These important details are shown in Figure 3.2.

Since inhaled particles are transported via airstreams, knowledge of both flow development characteristics and velocity profiles are fundamental to predicting particle deposition patterns. However, the airflow patterns in human lungs are much more complex than the idealized situation described earlier. The human bronchial tree is a branching network of bifurcating, irregular tubes. This network also contains morphological surface features that affect flow patterns. A vast body of work has been generated in attempting to characterize airflow in anatomically realistic human lung models. In this chapter, the discussion will be limited to a short overview of flow in curved tubes, bifurcations, and branching networks.

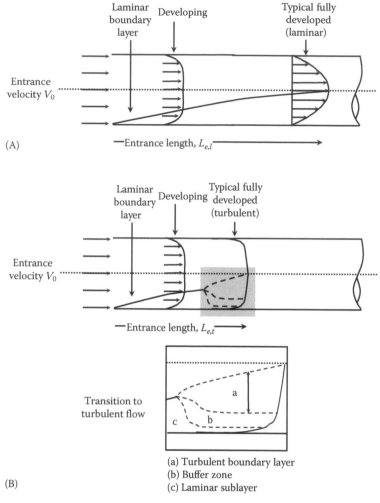

FIGURE 3.2 Developing flows and velocity profiles in an idealized airway for laminar and turbulent flow (not to scale).

3.6.3 FLOW IN CURVED TUBES

As a laminar, steady flow moves though a curved tube, secondary velocity patterns develop in the transverse plane of the tube.[73] These secondary patterns (Figure 3.3) are a result of the fast-moving fluid in the center of the tube being pushed toward the outside of the bend by inertia. Any distortion of the velocity profile can be predicted by the Dean number:

$$De = Re\left(\frac{d}{2R}\right)^{1/2} \tag{3.21}$$

where
   R is the radius of curvature of the bend
   d is the tube diameter

The velocity profile will be affected for $De > 25$, and large distortions will occur for $De > 300$.[77] In curved tubes, velocity profiles are no longer radially symmetric. Instead, in laminar flow the profile becomes M-shaped in the transverse plane and skewed toward the outside wall in the plane of the bend.[78]

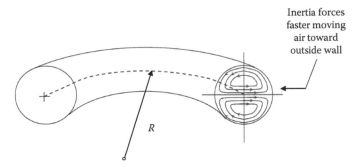

FIGURE 3.3   Secondary motion in a curved tube with radius of curvature R.

Flow in curved tubes has been studied experimentally,[79] theoretically,[77,80] and numerically.[81–83] Experimental[84,85] and computational[86,87] studies of flow in curved airway configurations have also been performed.

3.6.4   FLOW IN BIFURCATIONS AND BRANCHING NETWORKS

Because the lungs are a system of branching airways, much work has been done to characterize the airflow patterns within both single bifurcations and branching networks.[84,85,88–91] Different types of bifurcations have been considered, incorporating two-dimensional, three-dimensional, symmetric, and asymmetric geometries. In general, experimental studies have indicated that a parabolic velocity profile will be skewed as it passes through a bifurcation, in a manner similar as in a curved tube.[92] Figure 3.4 provides a qualitative example of these skewed velocity profiles.

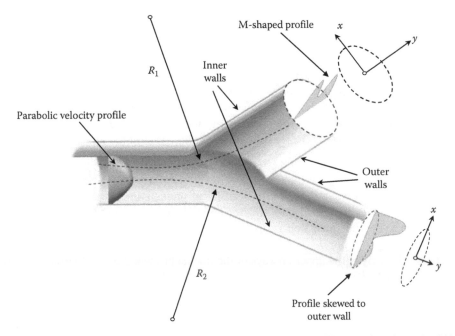

FIGURE 3.4   Qualitative description of velocity profiles in a bifurcation. Flow passing through a bifurcation behaves in a manner similar to flow passing through a curved tube. The radii of curvature R_1 and R_2 for the two daughter branches are shown. In the x plane (the plane of the bifurcation), the velocity profile becomes skewed toward the outer wall. In the y plane (which is perpendicular to the plane of the bifurcation), an M-shaped profile is observed.

Flows in physical models of more complex branching networks (e.g., branching models derived from casts of human or animal lungs) have been studied experimentally.[84,85,90] However, the characterization of features such as secondary currents, particularly in downstream daughter branches, is difficult in such studies. Recently, CFD has been used to model airflow characteristics in a variety of branching models.[92–97] Such models, and their use in predicting particle deposition, will be reviewed in detail in Chapter V, Modeling Inhaled Aerosols.

3.7 PARTICLE MOTION

The quantitative assessment of factors affecting inhaled particles was pioneered by Findeisen.[98] The subsequent work of Beeckmans[99] and Landahl[100] refined the analysis of particle deposition. The importance of these early efforts should be recognized.

Particle deposition within airways is governed by three primary mechanisms, (inertial impaction, sedimentation, and diffusion), and several secondary mechanisms (interception, electric charge, and cloud motion). The respective deposition efficiencies of these mechanisms are dependent upon interactions among aerosol characteristics, ventilatory parameters, and lung morphologies.

3.7.1 Primary Deposition Mechanisms

Particle deposition efficiencies for inertial impaction, sedimentation, and diffusion are dependent upon fluid dynamics, airway geometries, and particle characteristics. It is, therefore, necessary to formulate expressions for deposition efficiencies that specifically consider these factors. Deposition efficiency is defined as the ratio of the number of particles deposited within a given respiratory system region or airway to the total number of entering (or inhaled) particles.

Many individual equations have been developed to determine particle deposition efficiencies within different regions of the respiratory system. The equations most commonly used include those formulated by Martonen,[24,101] Beeckmans,[99] Landahl,[100] and Landahl and Herrman,[102,103] and Ingham.[104–106] The respective equations offered by the authors were derived using differing assumptions (e.g., turbulent versus laminar conditions) and velocity fields (e.g., uniform versus parabolic velocity profiles). Some investigators have selected equations for use from the aforementioned authors without recognizing their incompatibilities. For instance, authors have used a sedimentation equation for a uniform velocity field while simultaneously using an impaction equation for a parabolic velocity field. This has led to confusion in the literature. To address this problem, Martonen[24,101] presented a consistent, compatible set of formulae, being derived from well-defined conditions applicable to airways. These expressions are discussed in the following.

3.7.1.1 Inertial Impaction

Inertial impaction occurs when particles have sufficient momentum to deviate from fluid streamlines and strike boundary airway surfaces (Figure 3.5A). Because momentum is the product of mass and velocity (mV), inertial impaction is an important deposition mechanism for large particles, usually greater than $1\,\mu m$ in size. Inertial impaction is increasingly effective at higher velocities; thus, it occurs primarily in the upper airways of the tracheobronchial tree.[75] Inertial impaction as a mechanism of particle deposition has been widely studied.[107–111]

3.7.1.1.1 Laminar Conditions

Martonen[24,101] expressed particle deposition efficiency from inertial impaction under laminar plug flow conditions as

$$P(I) = \frac{2}{\pi}\left[\beta(1-\beta^2)^{1/2} + \sin^{-1}(\beta)\right] \qquad (3.22)$$

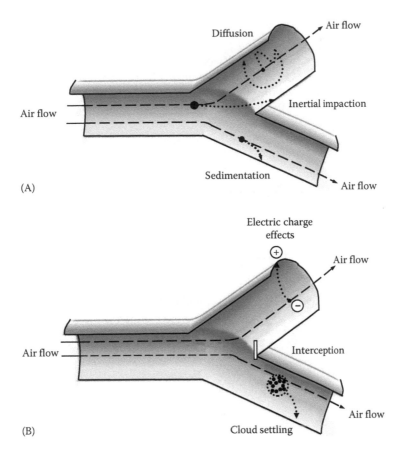

FIGURE 3.5 Particle deposition mechanisms in a bifurcating airway (A). Inertial impaction, sedimentation, and diffusion (B). Interception, electrostatic charge, and cloud motion. (Adapted from Martonen, T.B. et al., *Respir. Care*, 45, 712, 2000. With permission.)

where

$$\beta = \frac{\tau V_a \theta_b}{d_a} \tag{3.23}$$

(Note that deposition efficiency is dependent upon τ, as defined in Section 3.2.2.) These equations can be used to simulate particle deposition in the peripheral airways of the tracheobronchial tree. An equation applicable to fully developed (parabolic) velocity profile, appropriate for more distal airways, has been presented by Landahl.[100] However, airstream velocities are typically quite small in such areas, so the impaction deposition efficiency *per se* is negligible.

3.7.1.1.2 Turbulent Conditions
Deposition efficiency for inertial impaction for turbulent flow conditions is[24,101]

$$P(I) = 1 - \exp\left[\frac{-4\tau V_a \theta_b}{\pi d}\right] \tag{3.24}$$

Turbulent flows, such as those requiring the use of Equation 3.24, may occur in the upper tracheobronchial airways of the human lung.

3.7.1.2   Sedimentation

Reduced flow velocities lead to increased particle residence times in airways. When this occurs (e.g., when the peripheral airways of the tracheobronchial tree or alveolar region are being considered), particles may be deposited by gravitational forces (Figure 3.5A). This is called sedimentation, and while particles as small as 0.1 μm may be affected, this mechanism increases in importance with increasing particle size.[112,113] Particle deposition from sedimentation, like inertial impaction, is directly dependent upon τ, the relaxation time of the aerosol particle. Sedimentation of particles has been studied by Yu and Thiagarajan[114] and Gebhardt and Heyder.[115]

3.7.1.2.1   Laminar Conditions

Particle deposition efficiency from sedimentation, $P(S)$, under laminar plug flow conditions can be expressed by[24,101]

$$P(S) = \frac{2}{\pi}\left[\beta(1-\beta)^{2\frac{1}{2}} + \sin^{-1}(\beta)\right] \tag{3.25}$$

The orientation of airways with respect to gravity must be considered, and, therefore, for upward flow,

$$\beta = \frac{L_a V_{TS}\cos(\theta_b)}{d_a\left[V_a + V_{TS}\sin(\theta_b)\right]} \tag{3.26}$$

and for downward flow,

$$\beta = \frac{L_a V_{TS}\cos(\theta_b)}{d_a\left[V_a - V_{TS}\sin(\theta_b)\right]} \tag{3.27}$$

These equations are suitable for calculating deposition in the peripheral airways of the tracheobronchial network. For downstream regions where flow is fully developed, and parabolic velocity profiles exist, the formulation of Beeckmans[99] may be employed.

3.7.1.2.2   Turbulent Conditions

Under turbulent flow conditions, deposition efficiency from sedimentation can be expressed as[24,99,101]

$$P(S) = 1 - \exp\left[\frac{-4\beta}{\pi}\right] \tag{3.28}$$

where β is given by either Equation 3.26 or 3.27.

3.7.1.3   Diffusion

Deposition via diffusion (Figure 3.5A) occurs when particles exhibiting random Brownian motion collide with the airway surface. While deposition of particles between 0.1 and 1 μm in size occurs by both diffusion and sedimentation,[112,113] diffusion is the primary deposition mechanism for particles less than 0.1 μm in size,[116] since diffusion is governed by geometric diameter rather than aerodynamic diameter. Many studies of varying emphases have been performed to quantify the diffusion of aerosol particles.[104–106,117 124]

3.7.1.3.1 *Laminar Conditions*

Under laminar plug flow conditions, deposition by diffusion may be described by[101]

$$P(D) = 4\sqrt{\frac{K}{\pi}} - K \tag{3.29}$$

where

$$K = \frac{DL_a}{V_a d_a^2} \tag{3.30}$$

and the diffusion coefficient, D is given by

$$D = \frac{C_c kT}{3\pi\mu d_g} \tag{3.31}$$

Under parabolic laminar flow conditions, deposition by diffusion, $P(D)$, can be calculated from the expression derived by Ingham:[104]

$$P(D) = 1 - 0.819e^{-3.66K} + 0.0976e^{-22.3K} + 0.0325e^{-57K} + 0.0509e^{-49.9K^{2/3}} \tag{3.32}$$

The formulations given earlier have two major limitations: they are only valid for smooth-walled tubes and they ignore entrance effects. Regarding application to airways, the limitations may be quite serious because airways have natural surface features (e.g., cartilaginous rings).[121] In addition, due to the serially branching quality of the respiratory network, velocity profiles may be developing after each airway bifurcation. To address these issues, studies analyzed entrance effects in smooth-walled and rough-walled tubes.[117,118,122] They determined that the deposition of ultrafine (~0.01 μm) particles would be underestimated by ~35% if airway surface structures were ignored.

3.7.1.3.2 *Turbulent Conditions*

Particle deposition by diffusion under turbulent conditions can be calculated as[21,101]

$$P(D) = 1 - \exp\left[-\frac{8L_a \nu^{1/4} D^{3/4} Re^{7/8}}{57 V_a d_a^2} \right] \tag{3.33}$$

which in air at STP reduces to

$$P(D) = 1 - \exp\left[-\frac{0.088 L_a D^{3/4} Re^{7/8}}{V_a d_a^2} \right] \tag{3.34}$$

3.7.2 Secondary Deposition Mechanisms

3.7.2.1 Interception

Deposition by interception occurs when a particle contacts an airway surface while passing it a distance less than or equal to its radius, without deviating from the flow streamline (Figure 3.5B). Interception is an important deposition mechanism for fibers because as length increases, the likelihood of that fiber touching an airway surface increases.[125,126] Fiber orientation also plays a critical factor in deposition by interception. While fibers oriented with the flow streamlines deposit similarly to particles of their equivalent diameter, fibers oriented away from the streamline, or fibers that are "tumbling," have an increased likelihood of deposition by interception.[127,128] Interception may also be important for non-fibrous particles having low density and large geometric diameters, because the larger geometric diameter increases the likelihood of a particle coming into contact with an airway surface.[2] It has been suggested that such "porous" particles may have important applications to aerosol therapy.[127,128] However, recent findings have indicated that certain issues remain unresolved concerning the use of porous particles as vehicles for drug delivery.[129]

Several studies have experimentally characterized fiber movement and deposition in human lung casts or bifurcated tubes.[130,131] Other studies have modeled fiber movement and deposition in the human lung,[126,132,133,159] while still others have modeled interception of both spherical particles and fibers.[134] The motion of particles, including fibers, in airway bifurcations has been systematically studied by Balásházy et al.[135–137] Harris and Fraser[132] generated simplified equations to model the deposition of fibers by interception under laminar and turbulent conditions. They determined that deposition by interception increases as fiber length increases and airway diameter decreases. Because of the potential importance of fibers in inhalation toxicology and aerosol therapy, their expressions for interception deposition efficiencies are presented in the following.

3.7.2.1.1 Laminar Conditions

Harris and Fraser formulated expressions for interception of fibers in laminar parabolic flows. For fibers that are tumbling in the flow, the deposition efficiency may be given by[132]

$$P(\text{Int}) = \frac{f_t}{\pi r_a^4}\left[r_a^2 \theta_s \left(L_f^2 + \bar{L}_f^2\right) - \frac{\theta_s}{8}\left(L_f^4 - \bar{L}_f^4\right) - \bar{L}_f^2\left(\tan\theta_s\right)\left(r_a^2 + \frac{\bar{L}_f^2}{8}\right) + \frac{\bar{L}_f^4}{24}\tan^3\theta_s\right] \quad (3.35)$$

where the fraction of fibers tumbling is given by

$$f_t = \left(\frac{2}{\pi}\right)\tan^{-1}\left(\frac{\tan\varphi}{\beta_e}\right) \quad (3.36)$$

where φ has a value of 80°, and

$$\theta_s = \frac{\pi}{2} - \sin^{-1}\left(\frac{\bar{L}_f}{L_f}\right) \quad (3.37)$$

$$\bar{L}_f = L_f \cos\left[\tan^{-1}\left(0.34\beta_e + \frac{0.68L_a}{r_a}\right)\right] \quad (3.38)$$

and[126]

$$\beta_e = 1.07\left(\frac{L_f}{d_f}\right)^{0.087} \quad (3.39)$$

Harris and Fraser also present an equation for interception of fibers that are oriented with the laminar flow streamlines.[132]

3.7.2.1.2 Turbulent Conditions

Under turbulent flow conditions, interception of fibers in the lung may be expressed as[132]

$$P\left(\text{Int}\right) = \frac{2}{\pi r_a^2}\left(\frac{1}{4}\sqrt{r_a^2 - \frac{L_f^2}{16}} + r_a^2 \sin^{-1}\frac{L_f}{4r_a}\right) \qquad (3.40)$$

Asgharian and Yu[133] developed a more complex set of equations to characterize deposition of large fibers by interception. They found that their equations agreed with the results of Harris and Fraser (at all aspect ratios and fiber sizes tested) in airways from the trachea to the 15th airway generation. However, they suggested that beyond the 15th generation, their equations are required to adequately quantify interception, while recognizing that the equations formulated by Harris and Fraser may still be valid for small fibers having $L_f/d_f \le 10$ and $d_f < 1\,\mu m$.

Interception can be an important deposition mechanism, particularly for fibers. However, its significance is often underemphasized. The link between the inhalation of asbestos fibers and lung disease has been well established.[138,139] Martonen and Schroeter[140] have described how the mechanism of interception, and its function in fiber deposition in the human lung, may play a significant role in respiratory health and disease.

Due to their high-deposition and low-clearance rates (which will be discussed in detail in Chapter V), hollow fibers (microtubules) may be an effective and efficient vehicle for the delivery of inhaled pharmacological drugs. Techniques to generate liposomes as microtubules have been reported by Johnson et al.[141] The hollow fibers may be filled with pharmaceuticals of choice, and serve to transport the airborne drugs. The aerodynamic size of a particle may be estimated as[10,142]

$$d_{ae} = 2.19d_f\left(\frac{L_f}{d_f}\right)^{0.116} \qquad (3.41)$$

We note that the d_{ae} of a fiber is relatively independent of length; therefore, the microtubule can be packed with a large quantity of drugs without adversely affecting its aerodynamic properties.

Equations have been developed for the aerodynamic properties and behavior of tumbling fibers.[142,143] The integration of these formulae into the aerosol delivery program[75] will produce an inhalation therapy model for the administration of aerosolized liposomes as microtubules.

3.7.2.2 Electrostatic Charge

When, during aerosol generation, a friction or shear force is applied to the substance to be aerosolized, an electrostatic charge is often imparted to the particles.[3,144,145] In addition, aerosol particles may pick up charges as a result of collisions with atmospheric aerosols. These electrostatic charges will affect how particles, especially ultrafine particles, are deposited within human lungs.[146]

Two types of charging are important for particle deposition in the human lung: image charge and space charge. Image charge occurs when a charge-carrying particle induces the opposite charge in an airway surface, thus creating an attraction of the particle to the surface. Image charge occurs most often with highly charged particles. Space charge occurs when two particles of the same charge simultaneously repel, resulting in a particle colliding with the airway wall. Space charge occurs most often with high concentration, unipolar charged aerosols.[4] A number of studies have dealt with the various aspects of electric force effects.[147–154]

Experimental studies in human airway casts have been used to investigate how charge affects the deposition of particles in the tracheobronchial tree. Chan et al.[147] found that highly charged particles, between 2 and 7 μm in size, deposited significantly more than uncharged particles of the same size. Cohen et al.[148] found that charged ultrafine aerosols deposit at least three times more than charge-neutralized aerosols. Other experimental work has been carried out in human subjects. Scheuch et al.[150] found that negatively charged particles deposit more effectively than uncharged particles, with the greatest effects of charge occurring for submicron particles. Melandri et al.[153] found that unipolar charge on monodisperse aerosols resulted in an increase in deposition efficiency in the human lung. Several models of the effect of charge on particle deposition in the respiratory system have been developed.[151,152,154]

The published data from experimental and modeling studies, albeit limited, indicate that electric forces may have highly relevant effects on the deposition of inhaled particles. Therefore, the role of electric forces should not be ignored. However, their significance may be more of a concern for aerosol therapy (in which larger mass concentrations may be inhaled) than for inhalation toxicology.

3.7.2.3 Cloud Motion

Under certain circumstances, an array of particles may behave as an entity rather than as individual constituent units. This type of behavior is called cloud motion (Figure 3.6). The motion of such an entity in a gas differs from that predicted by theory for individual particles. Cloud motion may play a role in the deposition of cigarette smoke in the human lung.[155]

Cloud motion occurs when the terminal settling velocity of a cloud of particles is much greater than the settling velocity for a single particle. The terminal settling velocity of a particulate cloud having a characteristic diameter, d_c, and drag coefficient, C_D, is given by[156]

$$V_c = \left(\frac{4\rho_c d_c g}{3 C_D \rho_g} \right)^{1/2}$$

(3.42)

The cloud drag coefficient C_D is a function of cloud Reynolds number:

$$Re_c = \frac{\rho_a V_c d_c}{\mu}$$

(3.43)

Using the relationship between C_D and Re recommended by Klyachko,[2] the following expression was derived for V_c (Equation 3.41) for Re_c between 3 and 400[156]:

$$V_c = \frac{\rho_c g d_c^2}{3\mu(6 + Re_c^{2/3})}$$

(3.44)

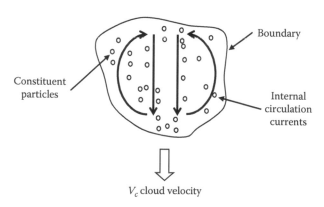

FIGURE 3.6 Cloud motion of concentrated aerosols. The settling velocity of the cloud is superimposed upon the motion of the constituent particles.

The requirement of $V_c \gg V_p$ led directly to the following criteria for the existence of cloud motion:

$$\rho_c \gg \frac{1}{6} \rho_p C_c \left[\frac{d_p}{d_c}\right]^2 (6 + Re_c^{2/3}) \qquad (3.45)$$

Therefore, cloud motion may be a factor in airway deposition when the mass concentration of inhaled particles is very high.

3.8   CONCLUSIONS

The deposition of inhaled particles is a function of aerosol characteristics, ventilatory parameters, and respiratory system morphologies. Deposition in human airways is especially linked to air motion and fluid dynamics. We have provided the fundamental theory behind many of the factors that affect deposition, and have presented equations for the deposition efficiencies of several deposition mechanisms under different flow conditions. However, we stress that the many of the basic formulae presented herein were derived for idealized or simplified conditions. The formulation of more advanced, simulation-based methods of predicting particle deposition in the respiratory system are currently being developed. We will present an overview of these methods in Chapter V, Modeling Particle Deposition.

NOMENCLATURE

Particle Quantities

d_p	particle diameter (µm)
d_g	geometric diameter (µm)
d_{ae}	aerodynamic diameter (µm)
V_p	particle velocity (cm·s^{-1})
V_{TS}	particle terminal settling velocity (cm·s^{-1})
ρ_p	particle mass density (g·cm^{-2})
ρ_u	unit density (g·cm^{-2})
m	particle mass (g)
Π	dynamic shape correction factor (dimensionless)
F_D	drag force on a particle (g·cm·s^{-2})
τ	relaxation time (s)
r_g	particle hygroscopic growth rate (µm·s^{-1})
C_c	Cunningham slip correction factor (dimensionless)
D	diffusion coefficient (cm^2·s^{-1})
n	molecular density (molecules·cm^{-3})
d_m	molecular diameter (µm)
C_D	particle drag coefficient (dimensionless)
I	inhalability ratio (dimensionless)

Air Characteristics

STP	standard temperature and pressure: $T = 273$ K (0°C) $P = 1$ atm $= 1.01 \times 10^6$ g·cm^{-1}·s^{-2}
µ	absolute air viscosity (g·cm^{-1}·s^{-1}) at STP, µ $= 1.82 \times 10^{-4}$ g·cm^{-1}·s^{-1}
ρ_a	air density (g·cm^{-3}) at STP, $\rho_g = 1.2 \times 10^{-3}$ g·cm^{-3}

ν $\qquad \dfrac{\mu}{\rho_g}$ = kinematic air viscosity (cm^2·s^{-1}) at STP, $\nu = 1.52 \times 10^{-1}$ cm^2·s^{-1}

λ $\qquad$ mean free path of air (cm) at STP, $\lambda = 6.5 \times 10^{-6}$ cm

V_a $\qquad$ mean velocity of air in an airway (cm·s^{-1})

Deposition Efficiencies

$P(I)$ $\qquad$ deposition efficiency for inertial impaction fraction
$P(S)$ $\qquad$ deposition efficiency for sedimentation fraction
$P(D)$ $\qquad$ deposition efficiency for diffusion fraction
$P(Int)$ $\qquad$ deposition efficiency for interception fraction

Dimensionless Parameters

Re $\qquad \dfrac{\rho\, dV}{\mu}$ = Reynolds number

Kn $\qquad \dfrac{2\lambda}{d_p}$ = Knudsen number

De $\qquad Re\left(\dfrac{d}{2R}\right)^{1/2}$ = Dean number

Fiber Characteristics

L_f $\qquad$ fiber length (μm)
$\bar{L}_f$ $\qquad$ mean length of fiber projection into the plane normal to the airway axis (μm)
d_f $\qquad$ fiber diameter (μm)

Particle Cloud Characteristics

d_c $\qquad$ cloud terminal diameter (cm)
V_c $\qquad$ cloud terminal settling velocity (cm·s^{-1})
Re_c $\qquad \dfrac{\rho_a V_c d_c}{\mu}$ = cloud Reynolds number (dimensionless)
ρ_c $\qquad$ mass concentration of particles in a cloud (g·cm^{-3})

Airway Properties

L_a $\qquad$ airway length (cm)
d_a $\qquad$ airway diameter (cm)
r_a $\qquad$ airway radius (cm)
θ_b $\qquad$ airway branching angle (°)

Other Quantities

k $\qquad$ Boltzmann's constant
$\qquad k = 1.38 \times 10^{-16}$ g·cm^2·s^{-2}·molecule^{-1}·K^{-1}
T $\qquad$ absolute temperature (Kelvin)
g $\qquad$ gravitational acceleration rate
$\qquad g = 9.8 \times 10^2$ cm·s^{-2}

ρ_{H_2O} density of water at STP
          $\rho_{H_2O} = 1\,\mathrm{g\,cm}^{-3}$

d diameter of an idealized tube (cm)

R radius of curvature of an idealized tube (cm)

REFERENCES

1. Fuchs, N.A., *The Mechanics of Aerosols*, Pergamon Press, New York, 1964.
2. Mercer, T.T., *Aerosol Technology in Hazard Evaluation*, Academic Press, New York, 1973.
3. Reist, P.C., *Aerosol Science and Technology*, McGraw-Hill, New York, 1993.
4. Hinds, W.C., *Aerosol Technology: Properties, Behavior and Measurement of Airborne Particles*, Wiley, New York, 1999.
5. Friedlander, S., *Smoke, Dust and Haze: Fundamentals of Aerosol Dynamic*, 2nd Edn., Oxford University Press, New York, 2000.
6. Hofmann, W., Morawska, L., Winkler-Heil, R., and Moustafa, M., Deposition of combustion aerosols in the human respiratory tract: Comparison of theoretical predictions with experimental data considering nonspherical shape. *Inhal. Toxicol.*, 21, 1154, 2009.
7. Baron, P.A. and Willeke, K., Gas and particle motion, in *Aerosol Measurement: Principles, Techniques, and Applications*, 2nd edn., Baron, P. and Willeke, K. (Eds.), Van Nostrand Reinhold, New York, 2001, Chap. 4.
8. Moss, O., Shape factors for airborne particles, *Am. Ind. Hyg. Assoc. J.*, 32, 221, 1971.
9. Stober, W.A., A note on the aerodynamic diameter and the mobility of non-spherical aerosol particles, *J. Aerosol Sci.*, 2, 453, 1971.
10. Stober, W., Dynamic shape factors of nonspherical aerosol particles, in *Assessment of Airborne Particles*, Mercer, T., Morrow, P., and Stober, W. (Eds.), Charles C. Thomas, Springfield, IL, 1972, Chap. 14.
11. Leith, D., Drag on nonspherical object, *Aerosol Sci. Technol.*, 6, 153, 1987.
12. Davies, C.N., Particle fluid interaction, *J. Aerosol Sci.*, 10, 477, 1979.
13. Davies, C.N., Definitive equations for the fluid resistance of spheres, *Proc. Phys. Soc.*, 57, 259, 1945.
14. Allen, M.D. and Raabe, O.G., Re-evaluation of Milikan's oil drop data for the motion of small particles in air, *J. Aerosol Sci.*, 6, 537, 1982.
15. Jennings, S.G., The mean free path in air, *J. Aerosol Sci.*, 19, 159, 1988.
16. Crowder, T.M., Rosati, J.A., Schroeter, J.D., Hickey, A.J., and Martonen, T.B., Fundamental effects of particle morphology on lung delivery: Predictions of Stokes's law and the particular relevance to dry powder inhaler formulation and development, *Pharm. Res.*, 19, 239, 2002.
17. Dennis, W.L., The growth of hygroscopic drops in a humid air stream, in *The Physical Chemistry of Aerosols*, The Faraday Society, Aberdeen University Press, Aberdeen, Scotland, 1961, Chap. 2.
18. Ferron, G.A., Haider, B., and Kreyling, W.G., Inhalation of salt aerosol particles. I. Estimation of the temperature and relative humidity of the air in the human upper airways, *J. Aerosol Sci.*, 19, 343, 1988.
19. Morrow, P.E., Factors determining hygroscopic aerosol deposition in airways, *Physiol. Rev.*, 66, 330, 1986.
20. Hiller, F.C., Health implications of hygroscopic particle growth in the human respiratory tract, *J. Aerosol Med.*, 4, 1, 1991.
21. Robinson, R.J. and Yu, C.P., Theoretical analysis of hygroscopic growth rate of mainstream and side-stream cigarette smoke particles in the human respiratory tract, *Aerosol Sci. Technol.*, 28, 21, 1998.
22. Finlay, W.H., Estimating the type of hygroscopic behavior exhibited by aqueous droplets, *J. Aerosol Med.*, 11, 221, 1998.
23. Ferron, G.A., The size of soluble aerosol particles as a function of the humidity of the air: Application to the human respiratory tract, *J. Aerosol Sci.*, 8, 251, 1977.
24. Martonen, T.B., Analytical model of hygroscopic particle behavior in human airways, *Bull. Math. Biol.*, 44, 425, 1982.
25. Blanchard, J.D. and Willeke, K., Total deposition of ultrafine sodium chloride particles in human lungs, *J. Appl. Physiol.*, 57, 1850, 1984.
26. Chan, H.K., Eberl, S., Daviskas, E., Constable, C., and Young, I., Changes in lung deposition of aerosols due to hygroscopic growth: A fast SPECT study, *J. Aerosol Med.*, 15, 307, 2002.
27. Hofmann, W., Morawska, L., and Bergmann, R., Environmental tobacco smoke deposition in the human respiratory tract: Differences between experimental and theoretical approaches, *J. Aerosol Med.*, 14, 317, 2001.

28. Martonen, T.B. and Patel, M., Computation of ammonium bisulfate aerosol deposition in conducting airways, *J. Toxicol. Environ. Health*, 8, 1001, 1981.
29. Martonen, T.B., Barnett, A.E., and Miller, F.J., Ambient sulfate aerosol deposition in man: Modeling the influence of hygroscopicity, *Environ. Health Perspect.*, 63, 11, 1985.
30. Muller, W.J., Hess, G.D., and Scherer, P.W., A model of cigarette smoke particle deposition, *Am. Ind. Hyg. Assoc. J.*, 51, 245, 1990.
31. Schroeter, J.D., Musante, C.J., Hwang, D., Burton, R., Guilmette, R., and Martonen, T.B., Hygroscopic growth and deposition of inhaled secondary cigarette smoke in human nasal pathways, *Aerosol Sci. Technol.*, 34, 1, 2001.
32. Ferron, G.A., Kreyling, W.G., and Haider, B., Inhalation of salt aerosol particles. II. Growth and deposition in the human respiratory tract, *J. Aerosol Sci.*, 19, 611, 1988.
33. Ferron, G.A. and Soderholm, S.C., Estimation of times for evaporation of pure water droplets and for stabilization of salt solution particles, *J. Aerosol Sci.*, 21, 415, 1990.
34. Finlay, W.H. and Stapleton, K.W., The effect on regional lung deposition of coupled heat and mass transfer between hygroscopic droplets and their surrounding phase, *J. Aerosol Sci.*, 26, 655, 1995.
35. Eberl, S., Chan, H.K., Daviskas, E., Constable, C., and Young, I., Aerosol deposition and clearance measurement: a novel technique using dynamic SPET, *Eur. J. Nucl. Med.*, 28, 1365, 2001.
36. Martonen, T.B. and Clark, M.L., The deposition of hygroscopic phosphoric acid aerosols in ciliated airways of man, *Fundam. Appl. Toxicol.*, 3, 10, 1983.
37. Broday, D.M. and Georgopoulos, P.G., Growth and deposition of hygroscopic particulate matter in the human lungs, *Aerosol Sci. Technol.*, 34, 144, 2001.
38. Ferron, G.A., Oberdorster, G., and Hennenberg, R., Estimation of the deposition of aerosolised drugs in the human respiratory tract due to hygroscopic growth, *J. Aerosol Med.*, 2, 271, 1989; Hesketh, H.E., *Fine Particles in Gaseous Media*, Lewis Publishers, Inc., Chelsea, MA, 1986.
39. Schmees, D.K., Wu, Y.H., and Vincent, J.H., Visualization of the airflow around a life-sized, heated, breathing mannequin at ultralow windspeeds, *Ann. Occup. Hyg.*, 52, 351, 2008.
40. Sleeth, D.K. and Vincent, J.H., Proposed modification to the inhalable aerosol convention applicable to realistic workplace wind speeds, *Ann. Occup. Hyg.*, 55(5), 476–484, 2011.
41. International Commission on Radiological Protection (ICRP)., ICRP Publication 66. *Annals of the ICRP Human Respiratory Tract Model for Radiological Protection*, Elsevier, Tarrytown, NY, 1994.
42. American Conference of Governmental Industrial Hygienists (ACGIH)., *Air Sampling Procedures: Particle Size-Selective Sampling in the Workplace*, Cincinnati, OH.
43. Millage, K.K., Bergman, J., Asgharian, B., and McClellan, G., A review of inhalability fraction models: Discussion and recommendations, *Inhal. Toxicol.*, 22, 151, 2010.
44. Se, C.M., Inthavong, K., and Tu, J., Inhalability of micron particles through the nose and mouth, *Inhal. Toxicol.*, 22, 287, 2010.
45. Breysse, P.N. and Swift, D.L., Inhalability of large particles into the human nasal passage—In vivo studies in still air, *Aerosol Sci. Technol.*, 13, 459, 1990.
46. Brown, J.S., Particle inhalability at low wind speeds, *Inhal. Toxicol.*, 17, 831, 2005.
47. Anthony, T.R., Contribution of facial feature dimensions and velocity parameters on particle inhalability, *Ann. Occup. Hyg.*, 54, 710, 2010.
48. Vincent, J.H., Mark, D., Miller, B.G., Armbruster, L., and Ogden, T.L., Aerosol inhalability at higher windspeeds, *J. Aerosol Sci.*, 21, 577, 1990.
49. Menache, M.G., Miller, F.J., and Raabe, O.G., Particle inhalability curves for humans and small laboratory animals, *Ann. Occup. Hyg.*, 39, 317, 1995.
50. Rosati, J.A., Bern, A.M., Willis, R.D., Blanchard, F.T., Conner, T.L., Kahn, H.D., and Friedman, D., Multi-laboratory testing of a screening method for world trade center (WTC) collapse dust, *Sci. Total Environ.*, 390, 514, 2008.
51. Wu, M., Gordon, R.E., Herbert, R., Padilla, M., Moline, J., Mendelson, D., Litle, V., Travis, W.D., and Gil, J., Case report: Lung disease in World Trade Center responders exposed to dust and smoke: Carbon nanotubes found in the lungs of World Trade Center patients and dust samples, *Environ. Health Perspect.*, 118, 499, 2010.
52. Dufresne, A., Dion, C., Frielaender, A., Audet, E., and Perrault, G., Personal and static sample measurements of asbestos fibres during two abatement projects, *Bull. Environ. Contam. Toxicol.*, 82, 440, 2009.
53. Lange, J.H., Sites, S.L., Mastrangelo, G., and Thomulka, K.W., Exposure to airborne asbestos during abatement of ceiling material, window caulking, floor tile and roofing material, *Bull. Environ. Contam. Toxicol.*, 80, 10, 2008.

54. Becker, H., Herzberg, F., Schulte, A., and Kolossa-Gehring, M., The carcinogenic potential of nanomaterials, their release from products and options for regulating them, *Int. J. Hyg. Environ. Health*, 214(3), 231–238, 2011.

55. Aschberger, K., Johnston, H.J., Stone, V., Aitken, R.J., Hankin, S.M., Peters, S.A., Tran, C.L., and Christensen, F.M., Review of carbon nanotubes toxicity and exposure—Appraisal of human health risk assessment based on open literature, *Crit. Rev. Toxicol.*, 40, 759, 2010.

56. Golshahi, L., Finlay, W.H., Olfert, J.S., Thompson, R.B., and Noga, M.L., Deposition of inhaled ultrafine aerosols in replicas of nasal airways of infants, *Aerosol Sci. Technol.*, 44, 741, 2010.

57. Minocchieri, S., Burren, J.M., Bachmann, M.A., Stern, G., Wildhaber, J., Buob, S., Schindel, R., Kraemer, R., Frey, U.P., and Nelle, M., Development of the premature infant nose throat-model (PrINT-Model)—An upper airway replica of a premature neonate for the study of aerosol delivery, *Pediatr. Res.*, 64, 141, 2008.

58. Janssens, H.M., de Jongste, J.C., Fokkens, W.J., Robben, S.G., Wouters, K., and Tiddens, H.A., The Sophia Anatomical Infant Nose-Throat (Saint) model: A valuable tool to study aerosol deposition in infants, *J. Aerosol Med.*, 14, 433, 2001.

59. Cheng, Y.S., Aerosol deposition in the extrathoracic region, *Aerosol Sci. Technol.*, 37, 659, 2003.

60. Xi, J.X. and Longest, P.W., Transport and deposition of micro-aerosols in realistic and simplified models of the oral airway, *Ann. Biomed. Eng.*, 35, 560, 2007.

61. Zhang, Z., Kleinstreuer, C., and Kim, C.S., Micro-particle transport and deposition in a human oral airway model, *J. Aerosol Sci.*, 33, 1635, 2002.

62. Xi, J.X. and Longest, P.W., Effects of oral airway geometry characteristics on the diffusional deposition of inhaled nanoparticles, *J. Biomech. Eng.*, 130, 2008.

63. Inthavong, K., Wen, H., Tian, Z.F., and Tu, J.Y., Numerical study of fibre deposition in a human nasal cavity, *J. Aerosol Sci.*, 39, 253, 2008.

64. Tian, Z.F., Inthavong, K., and Tu, J.Y., Deposition of inhaled wood dust in the nasal cavity, *Inhal. Toxicol.*, 19, 1155, 2007.

65. Shi, H., Kleinstreuer, C., and Zhang, Z., Laminar airflow and nanoparticle or vapor deposition in a human nasal cavity model, *J. Biomech. Eng.*, 128, 697, 2006.

66. Schroeter, J.D., Kimbell, J.S., and Asgharian, B., Analysis of particle deposition in the turbinate and olfactory regions using a human nasal computational fluid dynamics model, *J. Aerosol Med.*, 19, 301, 2006.

67. Wang, S.M., Inthavong, K., Wen, J., Tu, J.Y., and Xue, C.L., Comparison of micron- and nanoparticle deposition patterns in a realistic human nasal cavity, *Respir. Physiol. Neurobiol.*, 166, 142, 2009.

68. Heenan, A.F., Matida, E., Pollard, A., and Finlay, W.H., Experimental measurements and computational modeling of the flow field in an idealized human oropharynx, *Exp. Fluids*, 35, 70, 2003.

69. Chen, X.B., Lee, H.P., Chong, V.F., and Wang de, Y., A computational fluid dynamics model for drug delivery in a nasal cavity with inferior turbinate hypertrophy, *J. Aerosol Med. Pulm. Drug Deliv.*, 23, 329, 2010.

70. Rosati, J., Burton, R., McCauley, R., and McGregor, G., Three dimensional modeling of the human respiratory system, presented at *29th Annual Meeting of the American Association for Aerosol Research*, Portland, OR, October 25–29, 2010.

71. National Institutes of Health (NIH). National Library of Medicine Visible Human Project. http://www.nlm.nih.gov/research/visible/visible_human.html. Accessed January 31, 2011.

72. Yeh, H.C. and Schum, G.M., Models of human lung airways and their application to inhaled particle deposition, *Bull. Math. Biol.*, 42, 461, 1980.

73. Pedley, T.J. and Kamm, R.D., Dynamics of gas flow and pressure-flow relationships, in *The Lung: Scientific Foundations, Volume I*, Crystal, R.G. and West, J.B. (Eds.), Raven Press, New York, 1991.

74. Fox, R.W. and McDonald, A.T., *Introduction to Fluid Mechanics*, 3rd Edn., Wiley & Sons, New York, 1985.

75. Martonen, T.B., Musante, C.J., Segal, R.A., Schroeter, J.D., Hwang, D., Dolovich, M.A., Burton, R., Spencer, R.M., and Fleming, J.S., Lung models: Strengths and limitations, *Respir. Care*, 45, 712, 2000.

76. Weibel, E., Design of airways and blood vessels considered as branching trees, in *The Lung: Scientific Foundations, Volume I*, Crystal, R.G. and West, J.B. (Eds.), Raven Press, New York, 1991.

77. Dean, W.R., The streamline of motion of a curved pipe, *Philos. Mag.*, 5, 623, 1928.

78. Pedley, T.J. and Drazen, J.M., Aerodynamics theory, in *Handbook of Physiology, Section 3: The Respiratory System, Volume 3: Mechanics of Breathing*, Mackelm, P.T. and Mead, J. (Eds.), Williams & Wilkins, Baltimore, MD, 1986.

79. Taylor, G.I., The criterion for turbulence in curved pipes, *Proc. R. Soc. London Ser. A*, 124, 243, 1929.

80. Yao, L.S. and Berger, S.A., Entry flow in a curved pipe, *J. Fluid Mech.*, 67, 177, 1975.

81. Greenspan, D., Secondary flow in a curved tube, *J. Fluid Mech.*, 57, 167, 1973.
82. McConalogue, D.J. and Srivastava, R.S., Motion of a fluid in a curved tube, *Proc. R. Soc. London Ser. A.*, 307, 37, 1968.
83. Humphrey, J.A.C., Some numerical experiments on developing laminar flow in circular sectioned bends, *J. Fluid Mech.*, 154, 357, 1985.
84. Isabey, D. and Chang, H.K., A model study of flow dynamics in human central airways. Part II: Secondary flow velocities, *Respir. Physiol.*, 49, 97, 1982.
85. Chang, H.K. and El Masry, O.A., A model study of flow dynamics in human central airways. Part I: Axial velocity profiles, *Respir. Physiol.*, 49, 75, 1982.
86. Guan, X. and Martonen, T.B., Simulations of flow in curved tubes, *Aerosol Sci. Technol.* 26, 485, 1997.
87. Kleinstreuer, C. and Zhang, Z., Laminar-to-turbulent fluid-particle flows in a human airway model, *Int. J. Multiphase Flow*, 29, 271–289, February 2003.
88. Schroter, R.C. and Sudlow, M.F., Flow patterns in models of the human bronchial airways, *Respir. Physiol.*, 7, 341, 1969.
89. Brech, R. and Bellhouse, B.J., Flow in branching vessels, *Cardiovasc. Res.*, 7, 593, 1973.
90. Pedley, T.J., Schroter, R.C., and Sudlow, M.F., Flow and pressure drop in systems of repeatedly branching tubes, *J. Fluid Mech.*, 46, 365, 1971.
91. Guan, X. and Martonen, T.B., Flow transition in bends and applications to airways. *J. Aerosol Sci.*, 31, 833, 2000.
92. Martonen, T.B., Yang, Y., Xue, Z.Q., and Zhang, Z., Motion of air within the tracheobronchial tree, *Part. Sci. Technol.*, 12, 175, 1994.
93. Martonen, T.B., Guan, X., and Schrek, R.M., Fluid dynamics in airway bifurcations: I. Primary flows, *Inhal. Toxicol.*, 13, 261, 2001.
94. Martonen, T.B., Guan, X., and Schrek, R.M., Fluid dynamics in airway bifurcations: II. Secondary currents, *Inhal. Toxicol.*, 13, 281, 2001.
95. Martonen, T.B., Guan, X., and Schrek, R.M., Fluid dynamics in airway bifurcations: III. Localized flow conditions, *Inhal. Toxicol.*, 13, 291, 2001.
96. Yu, G., Zhang, G., and Lessman, R., Fluid flow and particle diffusion in the human upper respiratory system, *Aerosol Sci. Technol.*, 28, 146, 1998.
97. Calay, R.K., Kurujareon, J., and Hóldo, A.E., Numerical simulation of respiratory flow patterns within human lung, *Respir. Physiol. Neurol.*, 130, 201, 2002.
98. Findeisen, W., Über die von der molecularkinetischen theorie der wärmer geforderte bewegun von in ruhenden flüssigkeiten suspendierten teilchen, *Ann. Phys.*, 17, 549, 1935.
99. Beekmans, J.M., The deposition of aerosols in the respiratory tract. I. Mathematical analysis and comparison with experimental data, *Can. J. Physiol. Pharmacol.*, 43, 157, 1965.
100. Landahl, H.D., On the removal of air-borne droplets by the human respiratory tract: I. The lung, *Bull. Math. Biophys.*, 12, 43, 1950.
101. Martonen, T.B., Mathematical model for the selective deposition of inhaled pharmaceuticals, *J. Pharm. Sci.*, 82, 1191, 1993.
102. Landahl, H.D. and Herrmann, R.G., On the retention of airborne particulates in the human lung, *J. Ind. Hyg. Toxicol.*, 30, 181, 1948.
103. Landahl, H.D. and Herrmann, R.G., Sampling of liquid aerosols by wires, cylinders, and slides, and the efficiency of impaction of the droplets, *J. Colloid Sci.*, 4, 103, 1949.
104. Ingham, D.B., Diffusion of aerosols from a stream flowing through a cylindrical tube, *J. Aerosol Sci.*, 6, 125, 1975.
105. Ingham, D.B., Diffusion of aerosols from a stream flowing through a short cylindrical tube, *J. Aerosol Sci.*, 15, 637, 1984.
106. Ingham, D.B., Diffusion of aerosols in the entrance region of a smooth cylindrical pipe, *J. Aerosol Sci.*, 22, 253, 1991.
107. May, K.R. and Clifford, R., The impaction of aerosol particles in cylinders, spheres, ribbons, and discs, *Ann. Occup. Hyg.*, 10, 83, 1967.
108. Starr, J.R., Inertial impaction of particulates upon bodies of simple geometry, *Ann. Occup. Hyg.*, 10, 349, 1967.
109. Cheng, Y.-S. and Wang, C.-S., Inertial deposition of particles in a bend, *J. Aerosol Sci.*, 6, 139, 1975.
110. Reeks, M.W. and Skyrme, G., The dependence of particle deposition velocity on particle inertia in turbulent pipe flow, *J. Aerosol Sci.*, 7, 485, 1976.
111. Crane, R.I. and Evans, R.L., Inertial impaction of particles in a bent pipe, *J. Aerosol Sci.*, 8, 161, 1977.

112. Stahlhofen, W., Rudolf, G., and James, A.C., Intercomparison of experimental regional aerosol deposition data, *J. Aerosol Med.*, 2, 285, 1989.
113. Heyder, J., Gebhart, J., Rudolph, G., Schiller, C.F., and Stahlhofen, W., Deposition of particles in the size range 0.005–15 μm, *J. Aerosol Sci.*, 17, 811, 1986.
114. Yu, C.P. and Thiagarajan, V., Sedimentation of aerosols in closed finite tubes in random orientation, *J. Aerosol Sci.*, 9, 315, 1978.
115. Heyder, J. and Gebhart, J., Gravitational deposition of particles from laminar aerosol flow though inclined circular tubes, *J. Aerosol Sci.*, 8, 289, 1977.
116. Foster, W.M., Deposition and clearance of inhaled particles, in *Air Pollution and Health*, Holgate, S.T., Koren, H.S., Maynard, R.L., and Samet, J.M. (Eds.), Academic Press, New York, 1999, Chap. 14.
117. Martonen, T.B., Zhang, Z., and Yang, Y., Particle diffusion from developing flows in rough-walled tubes, *Aerosol Sci. Technol.*, 26, 1, 1997.
118. Martonen, T.B., Zhang, Z., Yang, Y., and Bottei, G., Airway surface irregularities promote particle diffusion in the human lung, *Radiat. Prot. Dosimetry*, 59, 5, 1995.
119. Shaw, D.T. and Rajendran, N., Diffusional deposition of airborne particles in curved bronchial airways, *J. Aerosol Sci.*, 8, 191, 1977.
120. Scheuch, G. and Heyder, J., Dynamic shape factor of nonspherical aerosol particles in the diffusion regime, *Aerosol Sci. Technol.*, 12, 270, 1990.
121. Russo, J., Robinson, R., and Oldham, M. J., Effects of cartilage rings on airflow and particle deposition in the trachea and main bronchi, *Med. Eng. Phys.*, 30, 581, 2008.
122. Martonen, T.B., Zhang, Z., and Yang, Y., Particle diffusion with entrance effects in a smooth-walled cylinder, *J. Aerosol Sci.*, 27, 139, 1996.
123. Hamill, P., Particle deposition due to turbulent diffusion in the upper respiratory system, *Health Phys.*, 36, 355, 1979.
124. Martin, D. and Jacobi, W., Diffusion deposition of small-sized particles in the brochial tree, *Health Phys.*, 23, 23, 1972.
125. Martonen, T. and Yang, Y., Deposition mechanisms of pharmaceutical particles in human airways, in *Inhalation Aerosols: Physical and Biological Basis for Therapy*, Hickey, A.J. (Ed.), Marcel Dekker, New York, 1996.
126. Sussman, R.G., Cohen, B.S., and Lippman, M., Asbestos fiber deposition in a human tracheobronchial cast. I. Empirical model, *Inhal. Toxicol.*, 3, 161, 1991.
127. Edwards, D.A., Hanes, J., Caponetti, G., Hrkach, J., Ben-Jebria, A., Eskew, M.L., Mintzes, J., Deaver, D., Lotan, N., and Langer, R., Large porous particles for pulmonary drug delivery, *Science*, 276, 1868, 1997.
128. Edwards, D.A., Ben-Jebria, A., and Langer, R., Recent advances in pulmonary drug delivery using large, porous, inhaled particles *J. Appl. Physiol.*, 85, 379, 1998.
129. Musante, C.J., Schroeter, J.D., Rosati, J.A., Crowder, T.M., Hickey, A.J., and Martonen, T.B., Factors affecting the deposition of inhaled porous drug particles, *J. Pharm. Sci.*, 91, 1590, 2002.
130. Myojo, T., Deposition of fibrous aerosols in model bifurcations, *J. Aerosol Sci.*, 18, 337, 1987.
131. Myojo, T. and Takaya, M., Estimation of fibrous aerosol deposition in upper bronchi based on experimental data with model bifurcation, *Ind. Health*, 39, 141, 2001.
132. Harris, R.L. and Fraser, D.A., A model for deposition of fibers in the human respiratory system, *Am. Ind. Hyg. Assoc. J.*, 37, 73, 1976.
133. Asgharian, B. and Yu, C.P., A simplified model of interceptional deposition of fibers at airway bifurcations, *Aerosol Sci. Technol.*, 11, 80, 1989.
134. Cai, F.S. and Yu, C.P., Inertial and interceptional deposition of spherical particles and fibers in a bifurcating airway, *J. Aerosol Sci.*, 19, 679, 1988.
135. Balásházy, I., Martonen, T.B., and Hofmann, W., Fiber deposition in airway bifurcations, *J. Aerosol Med.*, 3, 243, 1990.
136. Balásházy, I., Martonen, T.B., and Hofmann, W., Inertial impaction and gravitational deposition of aerosols in curved tubes and airway bifurcations, *Aerosol Sci. Technol.*, 13, 308, 1990.
137. Balásházy, I., Martonen, T.B., and Hofmann, W., Simultaneous sedimentation and impaction of aerosols in two-dimensional channel bends, *Aerosol Sci. Technol.*, 13, 20, 1990.
138. Becklake, M.R., Asbestos-related diseases of the lung and other organs: Their epidemiology and implications for clinical practice, *Am. Rev. Respir. Dis.*, 114, 187, 1976.
139. Timbrell, V., Inhalation and biological effects of asbestos, in *Assessment of Airborne Particles*, Mercer, T.T., Morrow, P.E., and Stober, W. (Eds.), Charles C. Thomas, Springfield, IL, 1972, Chap. 22.

140. Martonen, T.B. and Schroeter, J., Deposition of inhaled particles within human lungs, in *The Asbestos Legacy, Volume 23 of the Sourcebook on Asbestos Diseases: Medical, Legal, and Engineering Aspects*, Peters, G.A. and Peters, B.J. (Eds.), Matthew Bender and Company, Newark, NJ, 2001, Chap. 3.

141. Johnson, D.L., Polikandritou-Lambros, M., and Martonen, T.B., Drug encapsulation and aerodynamic behavior of a lipid microtubule aerosol, *Drug Deliv.*, 3, 9, 1996.

142. Johnson, D.L. and Martonen, T.B., Behavior of inhaled fibers: Potential applications to medicinal aerosols, *Part. Sci. Technol.*, 12, 161, 1994.

143. Johnson, D.L. and Martonen, T.B., Fiber deposition along airway walls: Effects of fiber cross section on rotational interception, *J. Aerosol Sci.*, 24, 525, 1993.

144. Rosati, J.A., Leith, D., and Kim, C., Monodisperse and polydisperse aerosol deposition in a packed bed, *Aerosol Sci. Technol.*, 37, 1, 2003.

145. Kousaka, Y., Okuyama, K., Adachi, M., and Ebie, K., Measurement of electrical charge of aerosol particles generated by various methods, *J. Chem. Eng. Jpn*, 14, 54, 1981.

146. Wilson, L.B., The deposition of charged particles in tubes with reference to the retention of therapeutic aerosols in the human lung, *J. Colloid Sci.*, 2, 271, 1947.

147. Chan, T.L., Lippman, M., Cohen, V.R., and Schlesinger, R.B., Effect of electrostatic charges on particle deposition in a hollow cast of the human larynx-tracheobronchial tree, *J. Aerosol Sci.*, 9, 463, 1978.

148. Cohen, B.S., Xiong, J.Q., Fang, C., and Li, W., Deposition of charged particles in lung airways, *Health Phys.*, 74, 554, 1998.

149. Grover, S.N. and Pruppacher, H.R., A numerical determination of the efficiency with which spherical aerosol particles collide with spherical water drops due to inertial impaction and phoretic and electrical forces, *J. Atmos. Sci.*, 34, 1655, 1977.

150. Scheuch, G., Gebhart, J., and Roth, C., Uptake of electrical charges in the human respiratory tract during exposure to air loaded with negative ions, *J. Aerosol Sci.*, 21, S439, 1990.

151. Chan, T.L. and Yu, C.P., Charge effects on particle deposition in the human tracheobronchial tree, *Ann. Occup. Hyg.*, 26, 65, 1982.

152. Hashish, A.H., Bailey A.G., and Williams, T.J., Modelling the effect of charge on selective deposition of particles in a diseased lung using aerosol boli, *Phys. Med. Biol.*, 39, 2247, 1994.

153. Melandri, C., Tarroni, G., Prodi, V., De Zaiacomo, T., Formignani, M., and Lombardi, C.C., Deposition of charged particles in the human airways, *J. Aerosol Sci.*, 14, 657, 1983.

154. Yu, C.P. and Chandra, K., Precipitation of submicron charged particles in the human lung airways, *Bull. Math. Biol.*, 49, 471, 1977.

155. Martonen, T.B. and Musante, C.J., Importance of cloud motion on cigarette smoke deposition in lung airways, *Inhal. Toxicol.*, 12, 261, 2000.

156. Martonen, T.B., The behavior of cigarette smoke in human airways, *Am. Ind. Hyg. Assoc. J.*, 53, 6, 1992.

157. Bennett, W.D., Zeman, K.L., and Jarabek, A.M., Nasal contribution to breathing and fine particle deposition in children versus adults, *J. Toxicol. Environ. Health A*, 71, 227, 2008.

158. Smith, J.R., Bailey, M.R., Etherington, G., Shutt, A.L., and Youngman, M.J., An experimental study of clearance of inhaled particles from the human nose, *Exp. Lung Res.*, 37(2):109–129, 2011.

159. Zhou, Y., Su, W.C., Cheng, Y.S., Fiber deposition in the tracheobronchial region: Deposition equations. *Inhal. Toxicol.*, 20(13), 1191–1198, 2008.

4 Aerosol Dose

Lev S. Ruzer, Michael G. Apte, and Richard G. Sextro

CONTENTS

> All substances are poisons,
> there is none that is not a poison.
> The right dose differentiates
> a poison and a remedy.
>
> **Paraselsus (1493–1541)**

4.1 INTRODUCTION

A well-defined pharmacokinetic relationship exists between chemicals administered orally, dermally, or intravenously and the quantity of substance delivered to the specific target site. Certainly, some variability in dose exists due to interindividual differences in metabolism and transport kinetics, but it is relatively easy to define the final delivery of an agent to the target site. In the case of aerosols, the understanding of this relationship is complicated for several reasons:

1. Aerosols are a complex medium consisting of particles in a very wide size range, with diameters ranging from nanometers to tens of micrometers (4–5 orders of magnitude).
2. Diversity in particle size results in different mechanisms of interaction inside the three main branches of the lung: the extrathoracic, tracheobronchial, and alveolar regions.
3. The particle size distribution is altered inside the lung due to changes in humidity and temperature.
4. In each branch of the lung, the particle size distribution differs due to selective filtration.
5. Individual breathing characteristics and lung morphology vary widely.
6. The variability in clearance, translocation, and other biokinetic processes taking place after aerosol delivery to the site of the lung must be considered.

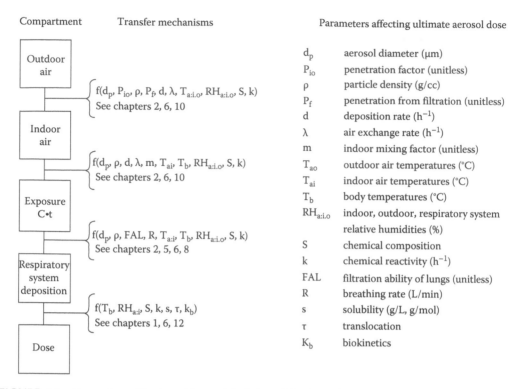

Compartment   Transfer mechanisms   Parameters affecting ultimate aerosol dose

Outdoor air

$f(d_p, P_{io}, \rho, P_f, d, \lambda, T_{a:i.o}, RH_{a:i.o}, S, k)$
See chapters 2, 6, 10

Indoor air

$f(d_p, \rho, d, \lambda, m, T_{ai}, T_b, RH_{a:i.o}, S, k)$
See chapters 2, 6, 10

Exposure C•t

$f(d_p, \rho, FAL, R, T_{a:i}, T_b, RH_{a:i.o}, S, k)$
See chapters 2, 5, 6, 8

Respiratory system deposition

$f(T_b, RH_{a:i}, S, k, s, \tau, k_b)$
See chapters 1, 6, 12

Dose

d_p	aerosol diameter (μm)
P_{io}	penetration factor (unitless)
ρ	particle density (g/cc)
P_f	penetration from filtration (unitless)
d	deposition rate (h^{-1})
λ	air exchange rate (h^{-1})
m	indoor mixing factor (unitless)
T_{ao}	outdoor air temperatures (°C)
T_{ai}	indoor air temperatures (°C)
T_b	body temperatures (°C)
$RH_{a:i.o}$	indoor, outdoor, respiratory system relative humidities (%)
S	chemical composition
k	chemical reactivity (h^{-1})
FAL	filtration ability of lungs (unitless)
R	breathing rate (L/min)
s	solubility (g/L, g/mol)
τ	translocation
K_b	biokinetics

FIGURE 4.1 Parameters affecting ultimate inhaled particle dose, showing physical transfer mechanisms across to microenvironmental compartments. The chapters in this book relating to the various transfer mechanisms are listed.

In the study of the pathway of aerosol exposure to dose to effect, three important conditions should be taken into account:

1. Dose, the physical parameter responsible for the health effect, should be defined correctly.
2. If, as in most practical cases, the dose is not measured directly, the correlation between the value of the measurable surrogate and the dose should be established.
3. The uncertainty related to the value of the dose surrogate should be assessed correctly.

The last condition is especially important in the case of aerosols where uncertainties are very high due to spatial and temporal variability, as are uncertainties associated with the interaction between aerosols and humans.

Figure 4.1 describes the particle dose pathway, indicating the complex set of physical parameters and mechanisms at play. The parameters and mechanisms shown in the figure are intended to be illustrative rather than comprehensive. Further details may be found in the chapters of this book as indicated in the figure.

4.2   ENVIRONMENTAL DOSIMETRY

Assessing the aerosol dose resulting from an environmental exposure has a particular set of issues. These problems vary depending upon environmental conditions, particle sources within environments, and the activities of the occupants and particle sources within them.

The issues may vary greatly, depending upon whether the exposed individuals are outdoors, indoors, and whether they occur in an occupational, residential, or another setting.

Regardless of the environmental conditions, in order to accurately determine the burden of aerosol exposure on an individual, dose must be measured or estimated. Typically, surrogates for actual dose such as aerosol concentration or exposure (time-weighted average concentration* duration of exposure) of particular aerosol species, or lung tissue dose estimates using inhalation models are used, since true dose measurements are usually not possible.

The use of any of these quantities in either epidemiological studies or control involves, explicitly or implicitly, assumptions about their relationships with one another. However, although it is the dose that is the cause of the biological effect, it is typically the airborne concentration that is measured, and it is the corresponding estimates of exposure that are used as correlates to health effects or predictors of risk.

It is the dose or intake to lung tissue that is studied to understand the actual insult leading to lung cancer or other diseases. If exposure is used as an index of dose, for example, in epidemiological studies, assumptions are made, typically implicitly, about a comparability or constancy of breathing rates, and of retention of aerosols in the lung.

It should be noted that, especially in industrial environments, the inhalation rates depend on the groups involved and on working conditions, specifically the physical load, which can vary substantially. Retention depends on the properties of the aerosols and also on the load of work and is typically not known accurately.

In Ruzer (2001) and Ruzer et al. (1995), a correlation was established between measured aerosol concentration in the breathing zone and gamma activity (dose) in the lungs of miners. The transition coefficient called filtration ability of lungs (FAL) was defined and measured. FAL is a combination of a breathing rate and deposition coefficient. It was shown that FAL depends on physical activity and is different for different groups of miners.

The measured concentration is not complete enough to directly provide accurate exposure characterization, because the concentrations to which individuals are exposed vary substantially in time and space. For example, a variation of the ventilation rate even for a short period of time can lead to a substantial change in concentrations. Moreover, the concentrations and especially the particle size distribution of aerosols in the breathing zone (which is directly responsible for local deposition) may differ substantially from those measured by standard instruments (Domanski et al., 1989; Scherwood and Greenhalgh, 1960; Schulte, 1967).

Finally, in the occupation setting, the very concept of "workplace"-associated concentration and workload are indefinite and job dependent, because workers are typically at a number of places during their working shift, with variable concentrations and nature of work. A similar situation exists with indoor exposure. Since the measurements of concentration in many environments may be performed only infrequently or even constructed, we cannot expect reliable correspondence between exposures estimated from spatially and temporally spare measurements and actual personal exposures. Thus, the use of measurements of airborne concentration as a basis for estimating or comparing dose can lead to substantial errors. In aggregate, these errors may constitute as much as an order of magnitude or more and, therefore, make risk assessment data unreliable.

Unfortunately, even after the use of measured concentration for radioactive and nonradioactive aerosols was established as a measure of the dose to the lung more than 40 years ago, little discussion and experimental study were conducted to determine the degree of correlation between the measured concentration and actual dose for individuals, both workers and for the general population. Experimental data on the correlation between the measured concentrations according to standard procedures (site measurement) and breathing zone measurements, not to mention dose, are very scarce. The only systematic study for radioactive aerosols provided in Polish mines (Domanski et al., 1989) showed poor correlation between standard area measurement and breathing zone sampling.

Using direct measurement of naturally occurring radioactive radon decay products in the lungs of miners, it was shown that the ratio of the true dose and that based on measured concentration could be large: 8 (Ruzer et al., 2004).

4.3   EXPOSURE AND DOSE DEFINITION

The term "exposure" came to the aerosol field from the study of the effect of gases. Aerosol science, including epidemiological studies of the health effect of aerosol exposure, was developed mainly in the second half of the twentieth century.

In 1924, German chemist Fritz Haber proposed the following definition of exposure:

> A simple and practical measure for toxicity can be obtained that suffices for all practical purposes. For each war gas, the amount (c) present in one cubic meter of air expressed in milligrams and multiplied by the time (t) in minutes necessary for the experimental animal inhaling this air to obtain a lethal effect. The bigger this products (c*t), the greater is the toxicity of the war gas.

This was the original formulation of "Haber's law." The product of concentration and time was simply a measure of acute lethality in cats, nothing else. To expect a constant response to the same (**c*t**) product had been postulated in 1921 by experimental toxicologist and biochemist Flury. Haber's law did not take into account the difference between acute and chronic exposure.

The definition of exposure in the case of aerosols is much more complicated. The main difference in dosimetry between gases and aerosols is that aerosol distribution, spatially and temporally, and deposited aerosols in the lungs are very nonuniform. The reason for this is that respirable fractions of aerosols consist of particles with diameters ranging from nanometers to 5–10 μm. Therefore, their particle size distribution and their deposition inside the lungs are very uneven.

According to the 1991 National Academy of Science report, the definition of exposure is as follows:

An event that occurs when there is contact at a boundary between humans and the environment with a contaminant of a specific concentration for an interval of time; the units are concentration multiplied by time.

A statistical definition of exposure has been proposed (Ott, 1966):
An exposure at some instant of time is a joint occurrence of two events:

1. The pollutant of concentration C is present at a particular location in space at a particular time.
2. The person is present at the same time and location in space.

A later definition (Duan and Ott, 1990) addresses the notion that the target remains important, and also that different parts of the target can receive different exposures at the same time.

The last definition is more adequate for aerosols, because it takes into account the specific aerosol problem of nonuniformity.

In all these definitions, the key word is contact, which means that in the case of aerosols only breathing zone measurement should be used for the exposure and particle size measurement. If concentration (and particle size characterization) is provided at a distance from the breathing zone, the correlation should be established between breathing zone and sampling site measurement.

Based upon this discussion, we can formulate the definition of dose as a physical value responsible for a biological effect:

Dose is the specific quantity of aerosols delivered to a target site that is directly responsible for a biological effect.

The term "quantity" is defined as follows:

1. In the case of radioactive aerosols, deposited energy per unit mass for alpha, beta, or gamma radiation is expressed in units of J/kg (Gray) or rads (100 erg/g) or the equivalent.
2. In the case of nonradioactive aerosols, quantity is the deposited number of particles, surface area, or mass of a discrete particle size.
3. The term "directly" means that dose is a quantity of the deposited amount of aerosol particles after the completion of all biokinetic processes.

According to the Environment Protection Agency of the United States,

In epidemiological studies, an index of exposure from personal or stationary monitors of selected pollutants is analyzed for associations with health outcomes, such as morbidity or mortality. However, it is a basic tenet of toxicology that the dose delivered to the target site, not the external exposure, is the proximal cause of a response. Therefore, there is increased emphasis on understanding the exposure–dose–response relationship. Exposure is what gets measured in the typical study and what gets regulated; dose is the causative factor.

Despite the fact that the importance of dosimetry is well understood, some uncertainties in interpretations exist even in terms of aerosol dose.

4.4 UNCERTAINTY IN THE DOSE ASSESSMENT

It has been shown (Ruzer et al., 1995; Ruzer, 2001) that in the case of natural radioactive aerosols, that is, the decay products of radon (in mines), it is possible to measure dose due to alpha radiation directly by measuring gamma activity from the lungs.

Such opportunities do not exist for nonradioactive aerosols; hence, the main method for nonradioactive aerosol dose assessment is through modeling. It should be noted, however, that calculated dose is very sensitive to input parameters (Phalen et al., 1990); in other words, the dose assessment based only on modeling can be unreliable. One of the main sources of uncertainty in lung dosimetry is in the assessment of particle size distribution in the breathing zone, the location where contact between aerosols and the organism occurs.

Reliable determination of the dose includes the measurement of aerosol particle size distribution (concentration) in the breathing zone, a knowledge of the parameters of transformation of particle size distribution due to humidity, temperature, and other factors inside the respiratory tract, calculating or measuring directly (e.g., by using radioactive markers) aerosol deposition, and, finally, a knowledge or direct measurement of parameters of biochemical processes inside the lungs: translocation, clearance, absorption, etc.

Most current information on the clearance of aerosol from the respiratory tract comes from radioactive aerosol inhalation studies; much less is known about respiratory biokinetics of inhaled nonradioactive particles.

In terms of biokinetics, the main difference between radioactive and nonradioactive aerosols is that mass concentrations of radioactive aerosols are typically low in comparison with nonradioactive aerosols, which results in differences in clearance and translocation (EPA, 1996).

4.5 AEROSOL CONCENTRATION STANDARDS

In 1997, the EPA established a new standard for particulate matter as an indicator of mass concentration: particle mass with aerodynamic diameters of 10 and 2.5 μm or less (PM10 and PM 2.5, respectively).

In the light of the theoretical and experimental results of the last 10 years, it has become clear that quantifying aerosol mass deliveries to the lung is not always adequate for biological effects. One such very important example was mentioned in Valberg and Watson (1998). According to the EPA (1996), for a mass concentration of $50\,\mu g/m^3$ of ambient particulate aerosol, the daily deposition in the alveolar region was close to $50\,\mu g$.

It is easy to calculate that such a mass will produce, at the alveolar surface, only 1 particle per day per $1.5\,mm^2$ and 1 particle per $44\,mm^2$ for 1 and 3 nm particles, respectively. Such amounts will cover less than a millionth part of the lung surface. No known chemical constituents of ambient particulate matter have such a threshold of toxicity at this daily level.

However, the corresponding numbers of deposited particles per mm² and part of the covered surface of alveoli will be orders of magnitude higher for particles at the level of tens of nanometers. For the same mass, the number of particles is inversely proportional to the cube of the diameter. Therefore, in case of substantial nanometer particle concentration, even exposure to nontoxic chemicals can result, as experiments on animals have shown, in acute biological effect. Thus, in terms of dosimetry, the concentration of nanometer particles should be measured correctly. It should be mentioned that the measurement of particles in this size presents some difficulties. In Banse et al. (2001), it was shown that detection efficiency in this range of sizes decreases dramatically.

4.6 NONUNIFORMITY OF DEPOSITION AND ULTRAFINE/ NANOMETER-SIZED PARTICLES

Many reasons can explain the discrepancy between dosimetric factors and health effects in the case of aerosols, the first being biological factors. However, from the point of view of dosimetry, two main factors can be used for the explanation:

1. Nonuniformity in the distribution of aerosols both in breathing space and inside the lungs, making the use of average numbers inappropriate
2. The special role of ultrafine/nanometer aerosols in health effects

Ultrafine particles contribute little mass to the fine fraction; however, they dominate particle surface area and particle number concentration.

The special role of nanometer-sized, ultrafine particles, and nonuniformity in aerosol dosimetry was demonstrated by Báláshazy and Hoffman (2000). Present lung dosimetry models for radon decay products are based on deposition efficiencies for straight cylindrical airways, which are equivalent to the commonly accepted assumption that inhaled particles are uniformity deposited in these airways.

Because aerosol deposition depends only on particle size and not on radioactivity, we can assume that the same approach will be correct for nonactive aerosols. In the present models, depth–dose distributions on bronchial epithelium are obtained by integrating the surface activities over the surface of the cylindrical airways within the range of alpha particles, thereby assuming again that alpha particle sources are uniformly distributed on airway surfaces. The assumption of uniformity is further supported by the theory that Brownian motion in straight cylindrical tubes *a priori* produces uniform deposition patterns.

In contrast, experimental studies about molecular-sized, ultrafine, and submicron particle deposition in single-pathway tracheobronchial models, in airway casts of the human tracheobronchial tree, and in single bronchial bifurcation models have demonstrated that diffusion-dominated particle deposition patterns are highly nonuniform.

Experimental evidence exists to show that the main features of the deposition patterns within airways exist at bifurcation points in the lung.
Research showed that

1. Deposition is enhanced at airway branching zones relative to cylindrical airway portions.
2. Deposition within a bifurcation is highest at the dividing spur.
3. Deposition is also enhanced at the inner sides of the daughter airways.

Computed enhancement factors indicate that the cells located at carinal ridges may receive localized doses that are 20–40 times (1 nm) and 50–115 times higher (10–200 nm), respectively, than the corresponding average doses.

In current dosimetric models, the implicit assumption was made that epithelial cells on bronchial airway surfaces will receive the same average dose. In contrast, Báláshazy and Hoffman (2000) proposed that the target should be divided into two fractions:

1. The fraction of the surface without any particles being deposited there (i.e., the dose is zero)
2. The remaining fraction, which may have many more particles deposited than indicated by the average deposition density

Since it has been recently suggested that the number of multiple cellular hits may play a crucial role in the extrapolation of lung cancer from occupational to domestic environments, maximum enhancement factors may serve as a measure of the probability of multiple hits.

In spite of the fact that the nature of hits to lung cells from alpha particles and nanometer-sized and ultrafine particles is completely different, it is interesting to compare these numbers for a given concentration (we can call "hits" for alpha particles and "touchdowns" for nanometer particles). The number of hits from alpha particles of radon decay products is given in Tables 2–14 of BEIR VI 1999.

4.7   CONCLUSION

The main physical (dosimetric) characteristics related to the biological effect of aerosols are as follows:

For radioactive aerosols (attached versus unattached),

1. Concentration (Bq/m^3)
2. Exposure (Bq/m^3) s
3. Dose (absorbed dose) (J/kg)

For nonradioactive aerosols,

1. Particle concentration (particle size distribution, PSD) (m^{-3})
2. Surface area concentration (m^{-1})
3. Mass concentration (g/m^3)
4. Exposure (concentration multiplied by time)
5. Dose (number of particles, surface area, mass)

The concepts of dose for radioactive and nonradioactive aerosols are different. In the case of alpha, beta, or gamma aerosols, we are talking about absorbed energy in the lung tissue at the target cells.

In the case of nonradioactive aerosols, dose represents the number of particles per mm^2 of the lung surface, surface of the aerosol particle, or mass of the particles, depending on the range of sizes. Special attention should be given to nanometer-sized particles and high deposition areas in the lung. The main sources of uncertainty in the dose assessment are nonradioactive characteristics: space and time distribution of aerosol concentration and PSD, especially differences in the breathing and sampling site, changes in humidity and temperature inside the lungs, and unknown characteristics of biokinetic processes.

REFERENCES

Báláshazy, I. and Hoffman, W., Quantification of local deposition pattern of inhaled radon decay products in human bronchial airway bifurcations, *Health Phys.*, 78, 147–158, 2000.
Banse, D.F. et al., Particle counting efficiency of TSI CPC 3762, *J. Aerosol Sci.*, 32, 157–161, 2001.

Deposition, retention and dosimetry of inhaled radioactive substances, NCRP Report No. 125, 1997.

Domanski, T., Kluszynski, D., Olszewski, J., and Chrusciewski, W., Field monitoring vs individual miner dosimetry of radon daughter products in mines, *Pol. J. Occup. Med.*, 2, 147–160, 1989.

Duan, N. and Ott, W.R., Comprehensive definitions of exposure and dose to environment pollution, in *Proceedings of the EPA/A&WMA Specialty Conference on Total Exposure Assessment Methodology*, November 1989, Las Vegas, NV, Air and Waste Management Association, Pittsburg, PA, 1990.

EPA, Review of the National Ambient Air Quality Standards for Particular Matter: Policy Assessment of Scientific and Technical Information, OAQPS Staff Paper, EPA/452/R-96-013, Office of Air Quality Planning and Standards, Research Triangle Park, NC, 1996.

Phalen, R.F., Schum, G.M., and Oldham, M.J., The sensitivity of an inhaled aerosol tracheobronchial deposition model to input parameters, *J. Aerosol Med.*, 3, 271–282, 1990.

Ruzer, L.S. and Harley, N.H., eds., Aerosol Handbook: Measurment, Dosimetry, and Health Effects, CRC Press, New York, 2004.

Ruzer, L.S., Radioaktivnie Aerozoli, Energoatomizdat, Moscow, Russia, 2001 (in Russian).

Ruzer, L.S., Nero, A.V., and Harley, N.H., Assessment of lung deposition and breathing rate of underground miners in Tadjikistan, *Radiat. Prot. Dosim.* 58, 261–268, 1995.

Schulte, H.F., Personal air sampling and multiple stage sampling, Los Alamos Scientific Laboratory, Los Alamos, New Mexico, U.S.A., *Symposium on Radiation Dose Measurements*, Paris, France, 1967.

Sherwood, R.J. and Greenhalgh, D.M.S., Personal air sampler, *Ann. Occup. Hyg.*, 2, 127–132, 1960.

Valberg, P.A. and Watson, A.Y., Alternative hypotheses linking outdoor particulate matter with daily morbidity and mortality, *Inhal. Toxicol.*, 10, 641–662, 1998.

5 Modeling Deposition of Inhaled Particles*

Kristin K. Isaacs, Jacky A. Rosati, and Ted B. Martonen

CONTENTS

* The U.S. Environmental Protection Agency through its Office of Research and Development funded and managed the research described here. It has been subjected to the Agency's administrative review and approved for publication.

5.1   INTRODUCTION

The mathematical modeling of the deposition and distribution of inhaled aerosols within human lungs is an invaluable tool in predicting both the health risks associated with inhaled environmental aerosols and the therapeutic dose delivered by inhaled pharmacological drugs. However, mathematical modeling of aerosol deposition requires knowledge of the intricate geometry of the respiratory network and the resulting complex motion of air and particles within the airways. In this chapter, an overview of the basic engineering theory and respiratory morphology required for deposition modeling is covered. Furthermore, current deposition modeling approaches are reviewed, and many factors affecting deposition are discussed. Experimental methods for measuring lung deposition are presented, albeit briefly, and the comparison between experimental results and modeling predictions is examined for a selection of modeling efforts.

5.2   FLUID DYNAMICS IN AIRWAYS

The deposition patterns associated with inhaled particulate matter are intrinsically linked to the airflow patterns within the respiratory system. Therefore, any effort to realistically model particle deposition requires an understanding of the fundamental fluid dynamics theory behind the motion of air in the extrathoracic and lung airways.

5.2.1 Fundamental Equations

Fluid motion is governed by the conservation of mass (continuity) equation and conservation of momentum (Navier–Stokes) equation. The flow of air in the respiratory airways is usually assumed to be incompressible.[1] For incompressible flow, the continuity equation is given by

$$\nabla \cdot V = 0 \qquad (5.1)$$

and the Navier–Stokes equation is

$$\left[\frac{\partial V}{\partial t} + (V \cdot \nabla)V \right] = \rho f - \nabla p + \mu \nabla^2 V \qquad (5.2)$$

where
  $\nabla \cdot$ and ∇^2 are the gradient and Laplacian operators, respectively (defined in the following)
  ρ is the fluid density
  μ is the absolute fluid viscosity
  p is the hydrodynamic pressure

The parameter f is any externally applied volumetric force, such as gravity.

In studying fluid flow in airways, it is convenient to use the cylindrical coordinate representation of the motion equations. Noting that the gradient operator ∇ in cylindrical coordinates is

$$\frac{\partial}{\partial r} + \frac{1}{r}\frac{\partial}{\partial \theta_\theta} + \frac{\partial}{\partial z} \qquad (5.3a)$$

continuity equation in cylindrical coordinates becomes

$$\frac{1}{r}\frac{\partial}{\partial r}\left(rV_r\right) + \frac{1}{r}\frac{\partial}{\partial \theta}V_\theta + \frac{\partial}{\partial z}V_z = 0 \qquad (5.3b)$$

where V_r, V_θ, and V_z are the components of the fluid velocity in the radial (r), circumferential (θ), and axial (z) directions, respectively (Figure 5.1). The corresponding equations for momentum in the r, θ, and z directions become

$$\frac{\partial V_r}{\partial t} + (V \cdot \nabla)V_r - \frac{1}{r}V_\theta^2 = -\frac{1}{\rho}\frac{\partial p}{\partial r} + f_r + \frac{\mu}{\rho}\left(\nabla^2 V_r - \frac{V_r}{r^2} - \frac{2}{r^2}\frac{\partial V_\theta}{\partial \theta}\right) \qquad (5.4)$$

$$\frac{\partial V_\theta}{\partial t} + (V \cdot \nabla)V_\theta + \frac{V_r V_\theta}{r} = -\frac{1}{\rho r}\frac{\partial p}{\partial \theta} + f_\theta + \frac{\mu}{\rho}\left(\nabla^2 V_\theta - \frac{V_\theta}{r^2} + \frac{2}{r^2}\frac{\partial V_r}{\partial \theta}\right) \qquad (5.5)$$

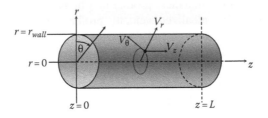

FIGURE 5.1 Cylindrical coordinate system for an arbitrary airway and corresponding velocity components at an arbitrary point.

$$\frac{\partial V_z}{\partial t} + (V \cdot \nabla) V_z = -\frac{1}{\rho}\frac{\partial p}{\partial z} + f_z + \frac{\mu}{\rho}\nabla^2 V_z \tag{5.6}$$

where

$$V \cdot \nabla = V_r \frac{\partial}{\partial r} + \frac{1}{r} V_\theta \frac{\partial}{\partial \theta} + V_z \frac{\partial}{\partial z} \tag{5.7}$$

and the Laplacian operator in cylindrical coordinates is defined as

$$\nabla^2 = \frac{1}{r}\frac{\partial}{\partial r}\left(r\frac{\partial}{\partial r}\right) + \frac{1}{r^2}\frac{\partial^2}{\partial \theta^2} + \frac{\partial^2}{\partial z^2} \tag{5.8}$$

If the flow is steady, then the time derivatives in Equations 5.4 through 5.6 can be ignored.

5.2.2 BOUNDARY CONDITIONS

The aforementioned system of four nonlinear partial differential equations can only be solved analytically when a number of assumptions are made for very simple flow geometries. Therefore, in most cases, numerical methods are required to determine a solution for the velocity and pressure fields. Numerical solution of the equations of motion requires knowledge about the velocity or pressure at some or all boundaries of the flow geometry. The nature of the flow characteristics and geometry determines which boundary conditions need defining.

As flow has been studied in a vast variety of geometric lung models, it is difficult to recommend boundary conditions that are appropriate in every circumstance. In almost all cases, however, it is assumed that the no-slip condition (V at the wall $= 0$) exists along the walls of any airway. The definition of any inlet and outlet boundary conditions is not as straightforward. For instance, the flow velocity at the inlet may be designated steady (time-invariant) or unsteady and may be uniform over the diameter of the inlet or vary with radial direction. In addition, in different cases, the velocity, the velocity gradient, the pressure, or the pressure gradient may be defined. In airway flow modeling, the velocity profile at the exit of an airway is often assumed parabolic, and the conducting airway is assumed sufficiently long enough for the flow to be fully developed.[2]

In some simplified cases, only the inlet conditions may need to be specified. For example, if one assumes that the pressure in the fluid varies only in the axial direction, then the mathematical character of the equations changes and only the inlet velocity profile is needed to obtain a solution.[3] In this case the outlet velocity profile is determined as part of the model solution.

5.2.3 IDEALIZED VELOCITY PROFILES

The velocity profiles present in real lung airways are determined by a great many factors including ventilatory conditions, lung morphology, and the airway generation(s) being considered. Experimental studies[4–6] have shown that these profiles may be skewed and, in general, geometrically complex. However, by assuming idealized velocity profiles in individual airway generations, fundamental equations of fluid motion can be reduced and the development of simplified expressions for deposition via different mechanisms can be obtained. Idealized velocity profile types include laminar fully developed (parabolic), laminar undeveloped (plug), and turbulent. The type of velocity profile selected determines the appropriate expressions for particle deposition (e.g., via sedimentation, diffusion, or inertial impaction) in a given airway. Examples of such expressions were presented in detail in Chapter 3.

Martonen[7] presented recommendations for velocity profiles in different generations of lung airways, based on previously published experimental measurements.[8] At the entrance to the trachea, it was assumed that the velocity profile was determined by the action of the laryngeal jet. In the upper tracheobronchial (TB) airways, the flow was considered turbulent, while in the lower TB airways the flow was assumed laminar undeveloped. In the pulmonary region, both laminar undeveloped and developed profiles were considered to serve as limiting cases, since the actual velocity profiles in this region are likely to lie in the transitional region between plug and parabolic flow. These flow conditions have been used in several particle deposition model studies.[2] A more recent modeling study[9] provided further evidence that the flow in the TB airways is not fully developed. Specifically, it was determined that an undeveloped flow model more accurately predicted total deposition of ultrafine particles than did a fully developed flow model.

5.2.4 Computational Fluid Dynamics

The field of study concerning the numerical solution of the equations of motion for fluids is known as *computational fluid dynamics* (CFD). CFD is an invaluable tool in the study of fluid and particle motion under circumstances that are difficult to simulate with a physical experiment or a simplified theoretical model. CFD involves segmenting a flow area (e.g., a system of airways and bifurcations) into many discrete elements or volumes, collectively called a *mesh*. In each element, the partial differential equations that describe the fluid motion are converted into algebraic equations by relating the fluid properties of the elements to the properties in adjacent elements. There exist several schemes for performing this *discretization*, including finite difference, finite element, and finite volume methodologies. In the finite difference method, the differential equations are approximated by Taylor series expansions. The theories behind finite element and finite volume discretizations are beyond the scope of this chapter.

Many commercial packages are available for performing CFD, including the finite-volume-based programs CFX-F3D and FLUENT (both now provided by Ansys, Inc.). FIDAP, a finite-element-based package that has been used to study respiratory system airflow in studies by Martonen et al.[10–15] and Kimbell et al.,[16] is now mostly unavailable. CFX-F3D has been used by Yu et al.[17] and Martonen et al.[18]

5.3 AEROSOL DEPOSITION MODELS

5.3.1 Classes of Models

The modeling of aerosol deposition in the lungs has been approached from several different conceptual directions. However, the vast majority of aerosol deposition models can be categorized as empirical, deterministic, or stochastic in nature. Empirical models are based on fitting numerical relationships to experimental data. In deterministic modeling, both the physical nature of the lung and the fluid and particle dynamics associated with respiration are quantified by simplified expressions and the resulting particle motion is calculated. In stochastic modeling, the geometry of the airways is varied randomly to account for inter- and intrasubject variation.

Empirical and deterministic models were the first efforts designed to simulate and predict aerosol deposition in airways, and stochastic models were later developed to intrinsically incorporate biological variability into such simulations. However, as computer resources became more powerful, computational fluid particle dynamics (CFPD) models of particle motion and deposition (based on CFD analyses) began to be developed. While CFPD models do not comprise a distinct class by themselves (i.e., they may be either deterministic or stochastic in nature), they are also discussed in the following.

5.3.1.1 Empirical Models

In empirical deposition models, regional or total aerosol deposition is described by equations derived by fitting algebraic relationships to experimental data. Stalhofen et al.[19] developed a semi-empirical model for total and regional deposition, deriving equations for aerosol deposition in the extrathoracic (nasal, oral, and laryngeal), "fast-cleared" thoracic (ciliated airways), and "slow cleared" thoracic (peripheral) regions as a function of respiratory parameters and aerodynamic diameter. The model, which was based on previous empirical models,[20] considered data from a wide variety of deposition experiments; theoretical relationships were used for particle sizes for which no data were available. The empirical ICRP respiratory tract dosimetry model[21] was developed by the International Commission on Radiological Protection. This model can be used to estimate regional deposition in the lungs as a function of particle characteristics and ventilatory conditions via a number of algebraic equations based on the work of Rudolf et al.[20,22–24] Other empirical deposition models have been developed for the head,[25,26] TB,[25–27] and alveolar[28] regions, and the entire lung.[29]

Empirical models have been combined with pharmacokinetic models,[30] and modeling predictions derived from empirical relationships have been used to validate and confirm results from mechanistic deposition models.[9,31]

5.3.1.2 Deterministic Models

Deterministic models are developed using an engineering approach to the simulation of air and particle motion. In deterministic models, simplifying assumptions about airway geometries and airflow conditions are made in order to derive expressions for particle trajectories from particle momentum equations. Such models vary in complexity; deterministic modeling efforts may range from simple analytical expressions that can be solved algebraically to systems of nonlinear ordinary or partial differential equations. In addition, deterministic models may describe particle deposition in a single airway, a bifurcation, or a complete branching network of respiratory airways.

Martonen[7,32] has developed deterministic models of particle deposition in human lungs. These models were formulated by modeling the airways of the lung as either straight or curved smooth-walled tubes, and assuming a fixed (laminar or turbulent) velocity profile in each airway generation. Other deterministic models of particle deposition in the respiratory system have been developed by Gradón and Orlicki,[33] Yu et al.,[34,35] Egan and Nixon,[36] Anjilvel and Asgharian,[37] Asgharian et al.,[38] Phalen et al.,[39,40] and Choi and Kim.[41]

In deterministic models, the simulated particle deposition patterns are determined solely by the input parameters to the model. Therefore, for any set of model input parameters, the same deposition pattern is found. This is not necessarily true of the next class of models, stochastic models.

5.3.1.3 Stochastic Models

Models of particle deposition are categorized as stochastic if the morphological description of the lung is considered to vary in a random manner, within prescribed limits. The concept of stochastic deposition modeling, first introduced by Koblinger and Hofmann,[42] has been used by Hofmann and coworkers to simulate aerosol particle deposition in both human[42–45] and rat[46–48] lungs. Radon progeny,[49] cigarette smoke,[50] and diesel exhaust[51] deposition have been specifically addressed.

In stochastic deposition modeling, the morphometric parameters describing the geometry of the lung are not given constant values, but are described instead by statistical (e.g., lognormal) distributions, which are in turn based on experimental measurements. These morphometric parameters may include airway diameters, lengths, and branching angles. Other, less obvious, parameters (such as ratios of parent airway cross-sectional area to the sum of daughter airway cross-sectional areas) may also be stochastically defined. Then, as each modeled particle enters the lungs, its pathway is determined by a random selection of values for each of the required morphometric parameters within their corresponding lognormal distribution. For example, the properties of the daughter airways are randomly assigned for each bifurcation.[52] The average resulting deposition in each airway is then calculated from the behavior of the entire ensemble of particles.

5.3.1.4 Computational Fluid-Particle Dynamics

Computational fluid-particle dynamics (CFPD) refers to the study of the motion of particles as determined by CFD simulations. In CFPD studies of particle deposition, CFD solutions of fluid velocities are coupled with the solution of particle trajectory equations developed from Newton's Second Law.[53] Particles are deposited when their trajectory intersects with an airway wall.

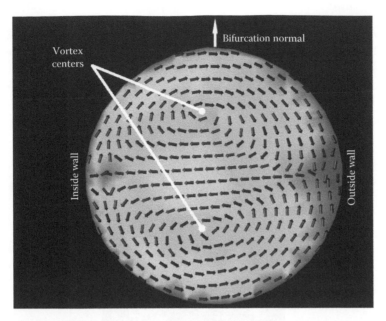

FIGURE 5.2   Image of a CFD simulation of secondary velocity currents in the eighth generation of a bifurcating airway, using FLUENT 6.0.

CFPD methods have been used to examine the effects of complex flow patterns on particle motion and deposition in the respiratory system. An example of such flow patterns in a model airway is shown in Figure 5.2. Martonen and Guan[54,55] performed linked fluid dynamics and particle motion studies that examined the effects of tumors on the deposition of particles in an idealized two-dimensional (2D) airway bifurcation. In recent years, as desktop computing capabilities have increased, many three-dimensional (3D) simulation studies have been performed. These studies can be categorized into three groups: (1) studies in idealized bifurcations, (2) studies in physiologically realistic (data-driven) bifurcating lung airways, and (3) studies in models of extrathoracic airways.

5.3.1.4.1   CFPD Studies in Idealized Bifurcations
In these types of studies, CFPD is used to examine airflow and particle deposition in smooth-walled, tubular airway systems having regular (usually symmetric) branching angles. An example of a CFPD simulation in a multiply bifurcating idealized airway (performed in FLUENT) is shown in Figure 5.3. Zhang et al.,[56-59] Comer et al.,[53,60] Longest et al.,[61,62] and Martonen et al.[18,63] have all performed CFPD studies aimed at predicting particle deposition in idealized bifurcating airways or networks. Zhang et al.[64] have also considered idealized representations of tumors within the airways.

5.3.1.4.2   CFPD Studies in Physiologically Realistic Bifurcating Airways
As available computing resources have grown, it has become easier to generate 3D morphological models of the lung airways from medical images. As this capability has grown, more and more CFD studies are being performed to study the airflow and particle deposition in bifurcating networks developed from images of real human lungs. Studies have been performed in the trachea and upper airways[65] and in more complicated networks involving several airway generations,[66-68] and in models representative of asthmatic lungs.[69]

5.3.1.4.3   CFPD Studies in the Extrathoracic Airways
Much of the recent work being done in CFPD is in simulations of particle deposition in the extrathoracic airways. The extrathoracic airways, such as the mouth, throat, and larynx, have very irregular geometries, and, thus, generalizing flow and deposition patterns in these areas is more difficult than

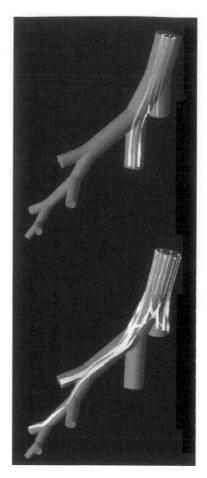

FIGURE 5.3   CFPD simulations of particle trajectories in a bifurcating airway using FLUENT 6.0. The flow rate is 120 L/min, and the two panels show trajectories for two different sets of initial particle locations.

in lung airways. Morphological models of these airways are discussed later in this chapter. CFPD studies have been performed for the oral cavity,[70–72] nose,[73–77] and pharynx (throat).[78]

5.3.2   Merits and Limitations of Deposition Models

Each deposition model class has both scientific strengths and drawbacks. In selecting a suitable deposition model one must consider many factors, including the scientific foundations of the models, the desired level of biological realism, and the available data and computing resources.

5.3.2.1   Scientific Foundations

The scientific foundations of the different classes of models must be considered when selecting a model for the interpretation of experimental data or the prediction of particle deposition pattern. Empirical models may be valid for the physical system for which they were derived, but application to other systems may result in spurious conclusions. In addition, deterministic models are only as valid as the assumptions made in their derivation. One must be aware of these assumptions and use the models accordingly. Stochastic models are derived from observations of morphological variability; however, such formulations can be generated only if an appropriate amount of experimental data is available. One must be mindful of the limitations of stochastic models derived from limited experimental data. CFPD modeling is based on well-established methods that have been in use for years in a wide variety of scientific fields, especially in mechanical and aerospace engineering. However, the scientific validity of any CFPD simulation is based on the appropriate application of these methods to the problem at hand. For example, the geometry of realistic respiratory system passages is more irregular than that in many industrial or engineering applications. Appropriate discretization of the complex geometry (for solution of the Navier–Stokes equations) may be very difficult, and an invalid mesh may result in specious computational results. However, more robust discretization (meshing) algorithms are continuously being developed and tested.

5.3.2.2 Biological Realism

Particle deposition models vary greatly in their level of biological realism. For example, empirical models contain no information about particle motion or the physiology or anatomy of the respiratory system (i.e., the respiratory system is treated as a "black box"), yet they may be useful in interpreting data from experiments in which subjects breath well-characterized aerosols in a known manner. Many deterministic models take into account both respiratory system morphology and the motion of individual particles. They are, therefore, able to model deposition in different respiratory system regions (extrathoracic, TB, or pulmonary), in individual lung airway generations or within single airways.

Stochastic models may present a limited "anatomically realistic" model *per se* (i.e., surface features not considered); however, they have the important advantage of being able to model the realistic biological variability that is present among each lung pathway in a single subject and between individual subjects.[79] Stochastic models by their construction provide estimations of intra- and intersubject variability in deposition.[79]

CFPD models are capable of simulating deposition in realistic airway configurations, considering flow conditions that arise from airway surface features. Therefore, CFPD models can predict *local* deposition (i.e., at cells, bifurcations, carinal rings) caused by secondary flow currents, which cannot be predicted from simplified analytical models. CFPD can be used to predict deposition in complex anatomical geometries where the assumption of a smooth-walled cylinder should not be made, for example, in the larynx, mouth, or nasal passages. A particular advantage of CFPD models is that they provide the potential for coupling imaging studies with deposition modeling. Irregular respiratory system morphologies can be extracted from CT or MRI images, and these morphologies can provide a basis for CFPD studies.

5.3.2.3 Hardware and Software Issues

Deposition models of different classes can require vastly different hardware and software resources. For example, empirical modeling requires no specialized computer programs, as the models are simple algebraic relationships. Deterministic models (those derived from particle motion and flow equations using various simplifying assumptions) vary greatly in computational efficiency, based on the nature of the assumptions made and the complexity of the airway geometry considered. However, CFPD modeling is typically computationally intensive, requiring high-performance hardware with adequate processing and memory resources. In general, CFPD also requires either expensive third-party software or a large amount of complicated, challenging, in-house programming. As an example, Zhang and Kleinstreuer[80] used the CFD package CFX 4.3 (AEA Technology, Inc., Carlsbad, CA) to perform their studies of secondary flow patterns in an airway branching network. The software ran on a multiprocessor Silicon Graphics workstation, and a typical run time for a fluid flow and particle transport simulation for a single breathing cycle was approximately 72 h. However, the use of CFPD models is becoming more frequent as the computing power of desktop personal computers and workstations rapidly increases.

5.3.2.4 Advantages of Modeling and Simulation

So far we have focused on the merits and limitations of particular types of deposition models; we would also like to comment on the advantages and challenges of using models in the study of inhaled particle deposition. Modeling can be a powerful research tool, as it can be used to predict behaviors, phenomena, or physiological parameters that cannot be measured. In addition, modeling has the potential to help maximize both financial and animal resources, by aiding in the design of appropriate experiments for a given scientific hypothesis. As noted by Martonen,[81] modeling studies should be integrated in a complementary manner with human inhalation exposure studies.

One of the main challenges in simulation is that its valid and rigorous use may require uniquely trained scientific personnel. For example, performing CFPD requires appropriately trained interdisciplinary scientists who are capable of understanding the computational, mathematical, and physiological nuances of modeling complex biological systems. Therefore, rigorous modeling studies may require collaboration among physicians, toxicologists, and engineers.

5.4   FACTORS INFLUENCING AEROSOL DEPOSITION PATTERNS

In Section 5.3, a brief overview of the different classes of lung deposition models was presented. Now we will discuss some of the morphological, ventilatory, and situational factors that affect deposition patterns in the human respiratory system. These are factors that may be considered in both deterministic and stochastic models of particle deposition and distribution.

5.4.1   AEROSOL PROPERTIES

As described in detail in Chapter 6, the primary mechanisms by which particles deposit in the respiratory tract are inertial impaction, sedimentation, and diffusion. The influence of each of these mechanisms is dependent on various particle characteristics. Therefore, the vast majority of deposition models will include input parameters describing the size and density of the particles being studied. Some models only consider a single size of particles, while some may allow the user to describe a polydisperse aerosol having multiple sizes. Recently, the study of deposition of particles of nanometer or ultrafine scale (nanoparticles) has increased, both in inhalation toxicology[82,83] and aerosol medicine.[84,85] In addition to size, models also may consider the influence of particle shape (e.g., spherical, tubular, fibroid) on deposition.

5.4.2   INHALABILITY

Inhalability is the ability of particles to be inhaled from the ambient environment into the mouth or nose. For small particles (those less than 5–10 μm in diameter) inhalability is essentially equal to 1. For larger particles, inhalability becomes a concern and, thus, should be considered when modeling respiratory tract deposition. Particle aerodynamic diameter, breathing conditions, and ambient conditions (such as wind speed) may all have an influence on the fraction of particles that can be inhaled. Typically, inhalability curves are empirical models that have been developed using experiments.[86,87] Millage et al.[88] provide a review of several algebraic models of inhalability. In addition, CFPD has recently been used to examine particle inhalability,[89] including investigation of the effect of facial morphology.[90]

5.4.3   RESPIRATORY SYSTEM MORPHOLOGY

Any modeling of the deposition of aerosols in the human respiratory system requires a description of the morphology of the airway(s) being studied. Both the overall branching structure of the airway tree and dimensions (e.g., diameters and lengths) of individual airways must be considered. Both idealized morphology models and models based on specific experimental observations have been used in particle deposition modeling.

5.4.3.1   Idealized Models

Many morphology models of the respiratory system have been derived from experimentally obtained morphometric data. Early morphometric models were simplified to provide idealized representations of branching network of the human lung. The most widely used idealized model is the symmetric model of Weibel.[91,92] In his model, the lung is characterized as a symmetric and dichotomously branching tree of tubular airways. Idealized asymmetrical models of the airway tree have also been developed. These include Weibel's asymmetric "B" model[92] and the models of Horsfield[93] and Horsfield and Cumming.[94,95]

5.4.3.2 Data-Driven Models

Other more biologically realistic morphological models also have been derived from experimentally obtained morphometric data. Soong et al.[96] presented a stochastic model of the human TB tree, using Weibel's symmetric model as the underlying average model, and incorporating probability distributions of lengths and airways based on several experimental studies (e.g., Jesseph and Merendino[97] and Parker et al.[98]). Horsfield et al.[99] developed a theoretical model of the bronchial tree based on measurements from human casts. Kitaoka et al.[100] also presented a model of the 3D branching structure of the lung based on morphometric studies. Their model was derived using an algorithm guided by several morphology- or physiology-based "rules," such as prescribed relationships between airway length and diameter and between airway diameter and flow rate.

5.4.3.3 Three-Dimensional Morphological Models for Interpretation of Lung Images

3D morphological models have been built for use in the interpretation of images in experimental deposition studies. These models consist of a branching network of line segments, oriented in 3D space via idealized or physiologically realistic branching and gravity angles. Martonen et al.[101,102] have developed models for image interpretation based on idealized branching networks. An example of such a model is shown in Figure 5.4. More recently, Montesantos et al.[103] have developed hybrid 3D models of branching networks for image analysis. These methods construct the first five airway generations of the branching network high-resolution computed tomography data, potentially allowing the network to be personalized to a particular subject within a lung deposition study. The use of branching network models in gamma scintigraphy is discussed further in the section on experimental deposition measurements.

5.4.3.4 Three-Dimensional Morphological Models for CFPD

With the increase in use of CFPD as a research tool, there has been a drive to develop 3D models of respiratory system morphologies for use in CFPD studies. These models, which may be developed

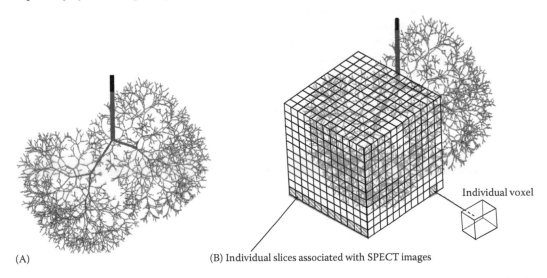

(A)

(B) Individual slices associated with SPECT images

Individual voxel

FIGURE 5.4 Branching lung network model (A) and an associated network of voxels (B, not to scale) for analysis of SPECT images of aerosol distribution.

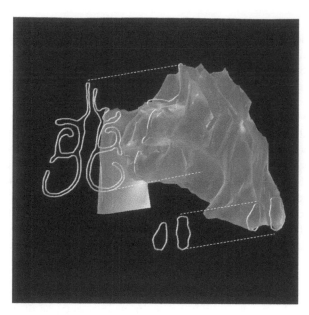

FIGURE 5.5 Morphological model of the human nasal passages derived from MRI imaging data. Note the complex cross section of the nasal passage.

directly from *in vivo* medical images or from the imaging of casts, provide a physiologically realistic foundation for studying the influence of local airway features on the motion and deposition of particles. Spencer et al.[104] have derived a morphological model of the lung airways using data-driven surface modeling techniques. In this model, anatomical data were used to define airway lengths, diameters, and orientation angles. The surfaces of the resulting airway network were then realized using advanced graphics rendering techniques. Specifically, nonuniform rational B-splines (NURBS) were used to model smooth airway connections and realistic lung surface features.

Models derived from imaging or cast data are especially useful in developing representations of the extrathoracic airways, as these passages are not easily modeled as idealized tubes. Figure 5.5 depicts a geometric model of the morphology of the human nasal passages derived from MRI images of an adult male. The irregular, tortuous shape of the nasal passages (as seen in the cross sections) will result in a distinct particle deposition pattern that cannot be predicted using simplified geometries. Morphological models have been developed for the nose[105] and the oral cavity.[106] Recently, Rosati et al.[107] have presented a combined extrathoracic-lung model for use in particle deposition studies. This model uses a nose and mouth constructed from imaging data from the U.S. National Library of Medicine Visible Human Project.[108] The extrathoracic model is combined with a five-lobe lung airway model constructed using the morphological data of Yeh and Schum.[109] The combined system is depicted in Figure 5.6, which shows modeled deposition in a typical lobar path within the lung.

5.4.3.5 Surface Features

There is much evidence to suggest that surface features in the lungs should be considered when modeling particle deposition. Both the cartilaginous rings (which are a pronounced anatomical feature of the TB airways), and the carinal ridges (which are situated at airway bifurcations) have been studied. Bronchoscopy images depicting these surface features are shown in Figure 5.7.

Cartilaginous rings affect airflow patterns in the large airways.[12,105] Using CFD modeling, Musante and Martonen[110] demonstrated that small eddies, produced between the rings, may increase localized particle deposition. They also predicted that flow instabilities produced by the rings could affect deposition in locations downstream from the rings themselves. In addition, errors in large airway deposition of up to 35% were possible if the rings were ignored.[111]

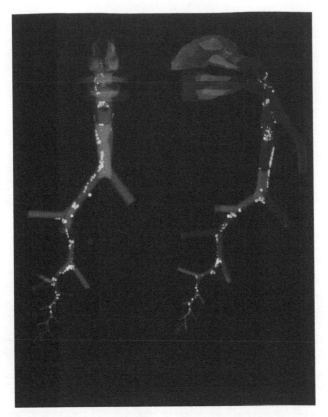

FIGURE 5.6 Combined extrathoracic and lung airway model of Rosati et al.[105] Deposition of 1 μm particles in a typical lobar path is shown.

Martonen et al.[10] demonstrated via CFD modeling that localized deposition of particles at the carinal ridges could be explained by localized flow instabilities arising from the ridge geometry. They also predicted that the ridges could initiate flow effects that propagate to later generations, especially at high inspiratory flow rates. Using a numerical model, Balásházy and Hofmann[112] quantified deposition at carinal ridge sites for radon progeny of different sizes. They predicted that cells located at the ridge sites experienced deposition 20–115 times greater (depending on particle size) than the average airway dose.

5.4.4 VENTILATORY CONDITIONS

Ventilatory conditions have a distinct effect on aerosol deposition and distribution. Both the mode of respiration and breathing pattern must be considered when modeling particle deposition and clearance in the human lung.

5.4.4.1 Mode of Respiration

Humans have the ability to breathe either nasally or orally. In contrast, rodents such as mice and rats are obligate nasal breathers. Thus, extrapolation of particle deposition data from these animals to human beings is difficult.

When sedentary, humans breathe through their nose, efficiently heating and humidifying the inhaled air. However, during exertion such as exercise, they switch over to oronasal breathing, or breathing through both mouth and nose. This switch is thought to occur when the breathing rate

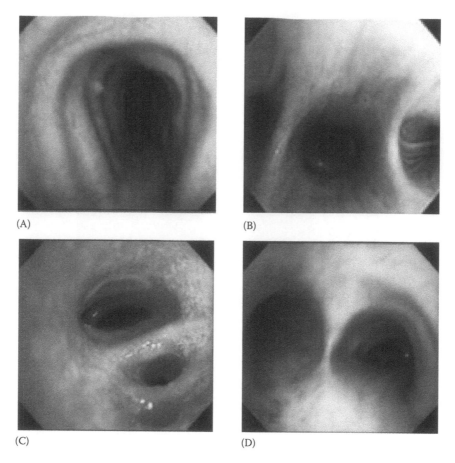

(A)   (B)

(C)   (D)

FIGURE 5.7 Airways and surface features photographed with videobronchoscopy. (A) The trachea and the main bronchi with the cartilaginous rings clearly visible. (B) A quadruple bifurcation, indicative of the complex branching pattern of the lung. (C) Blunt carinal ridges. (D) Sharp carinal ridges. (Reprinted with permission from Virtual Hospital, University of Iowa, Iowa city, IA, 1992–2003, www.vh.org.)

becomes so great that overcoming the relatively high pressure drop and resistance in the nasal passages is no longer an efficient means of respiration. In healthy adults, the switching point between nasal and oronasal breathing is thought to be approximately 35 L/min.[113,114] In children, however, this switching point is more variable.[115] Gender does not seem to play a role in switching point determination.[113]

5.4.4.1.1 Effect of Oral or Nasal Breathing on Particle Delivery to Lungs

The route of breathing influences the quantity of inhaled contaminants or therapeutics delivered to human lungs. Particle penetration to the human lung is lower during nasal breathing (versus oral breathing)[116] due to higher deposition efficiency in the nasal region, and there is thus more effective filtering of inhaled particles.[117,118] The higher deposition efficiency of nasal breathing is due to increased particulate matter removal by nasal hairs, impaction on pathway walls, and diffusion. Thus, it is less efficient to administer therapeutic aerosols to the human lung via the nose as opposed to the mouth.

5.4.4.1.2 Effect of Oral or Nasal Breathing on Total Particle
Deposition within the Respiratory Tract

Due to efficient particle removal by the nose,[119–121] total respiratory tract deposition is higher for nasal breathing than for oral breathing in nonsmokers.[50,122] As stated, the higher deposition efficiency in the nose versus the mouth is due to increased particle removal by nasal hairs, and inertial impaction. However, particulate matter removal in the nasal passages (and thus total deposition during nasal breathing) is highly dependent on particle size and inhalatory volume and flow,[118,120,121,123] as well as nasal passage morphology and development.[119,124,125]

5.4.4.2 Breathing Pattern

Breathing pattern is an important factor in the respiratory deposition of therapeutic particles, as it may affect treatment efficiency.[126] Both spontaneous and regulated breathing patterns (Figure 5.8) have been used in human inhalation studies of particle deposition and clearance. The type of breathing pattern will affect the amount and pattern of particle deposition in the human lung. Studies comparing deposition data for spontaneous breathing patterns have been performed.[19,126–129]

Breathing pattern is typically described in terms of tidal volume (volume of air inhaled) and flow rate. In general, larger tidal volumes result in higher particle deposition in the human lung as particle-laden air penetrates deeper into the lung. Lower flow rates also result in higher particle deposition in the peripheral lung as velocities are slower and particles have more time to deposit by sedimentation or diffusion (see Chapter 6 for discussion on deposition mechanisms).

5.4.4.2.1 Spontaneous Breathing

Spontaneous breathing is unprescribed, with subjects breathing at their own pace, with only approximate tidal volume or flow requirements. Figure 5.8A illustrates the relationship between tidal volume and time for a typical spontaneous breathing pattern. Chan and Lippmann[26] and Brown et al.[130] have used spontaneous breathing patterns in the study of deposition of inhaled particles. Other studies have shown that spontaneous breathing of human subjects resulted in large intersubject variability in the fractional deposition of inhaled particles.[127,131,132] Furthermore, Bennett and Smaldone[128] determined that it was differences in spontaneous breathing pattern (and not peripheral air space size or morphology) that influenced intersubject variation in peripheral deposition.

5.4.4.2.2 Regulated Breathing

Regulated or academic breathing occurs when human subjects are required to follow a specified breathing pattern that varies from their own natural breathing rate. In this pattern, illustrated in Figure 5.8B, a constant flow rate is maintained. Regulated breathing has been used to attempt to mitigate the intersubject variability introduced by spontaneous breathing.[128] Studies

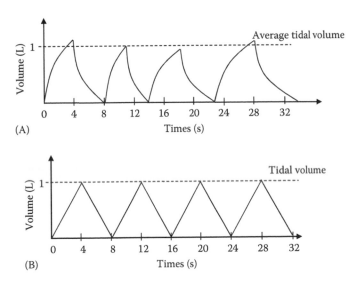

FIGURE 5.8 Spontaneous (A) and controlled (B) breathing patterns. Note that in spontaneous breathing, rate and tidal volume may vary, while these are constant in controlled breathing.

by Heyder et al.,[133] Stahlhofen et al.,[134] and Svartengren et al.[135] utilized regulated breathing in the determination of particle deposition in the human respiratory tract.

5.4.5 Respiratory System Environment

Temperature and relative humidity (RH) in the human respiratory system varies with mode of respiration and anatomical location. Table 5.1 provides temperature and humidity data[136–145] for different anatomical locations for both oral and nasal respiration. In general, a temperature of 37°C and an RH of 99.5% may be assumed for nasal respiration.[146] For oral respiration, 37°C and 90% RH may be assumed for air entering the trachea, with RH increasing by 1% per airway generation until reaching 99.5% at the tenth generation.[146]

RH and temperature affect the growth of hygroscopic particles in the human lung. Hygroscopic growth occurs when the absorption of water from a humid environment causes changes in particle diameter and density. Since RH and temperature vary throughout the human lung, a particle's size and density may change while traveling through the respiratory system. Thus, a measured size distribution for hygroscopic particles is not likely to reflect actual particle sizes *in vivo*. A more extensive discussion of hygroscopicity and its effect on particle deposition in the human lungs may be found in Chapter 6.

Hygroscopic growth has been widely observed in many environmental and pharmaceutical aerosols[147–158] (Table 5.2). Therefore, it is often desirable to account for hygroscopic growth in particle deposition modeling. Deposition modeling studies[7,32] have accounted for hygroscopicity by incorporating the growth rate of aerosol particles as a function of residence time in the lung. Such experimental growth rate measurements for different aerosols have been presented.[159,160] The growth of aerosol particles has also been computationally predicted.[161–164]

5.4.6 Clearance

Inhaled foreign material is continually cleared from the respiratory tract. From a simplified perspective, inhaled insoluble particles are cleared from the human lung in two phases, mucociliary (or fast phase) clearance and phagocytosis (or slow phase) clearance. In addition, free particles may translocate out of the alveolar region of the lung into the lymphatic system or the lung interstitium. Depending on their lipophilicity, hydrophilicity, and/or size, soluble particles may be dissolved prior to physical clearance. Other particles, such as asbestos fibers and other biopersistent fibrous minerals, are often unable to be cleared from the lung; their retention may result in inflammation, tissue damage, and eventual disease. Oberdörster[165] provides a review of the clearance of both soluble and insoluble particles. Because the toxicity of a substance may be related to its time of residence in the respiratory system, clearance is an importance consideration in the risk assessment of inhaled particles. Thus, we will present a brief overview of its mediation by drugs, inhaled contaminants, age, and activity.

5.4.6.1 Mucociliary Clearance

Mucociliary (or fast phase) clearance occurs in the TB airways of the lung. Mucus is secreted by mucous glands in the bronchial walls and by goblet cells in the bronchial epithelium. This mucus is propelled by millions of cilia (collectively referred to as the mucociliary escalator) toward the pharynx, in the process transporting particles out of the conducting airways. At the pharynx, the mucus and particles are swallowed. The velocity of the mucus varies from a rate of 1 mm/min in the smaller airways to 2 cm/min in the trachea.[166] Mucociliary clearance of deposited particles generally occurs within 24 h after deposition in healthy individuals.[166–169] A comprehensive review of mucociliary clearance is presented by Yeates et al.[170]

5.4.6.1.1 *Effect of Drugs and Inhaled Contaminants on Mucociliary Clearance*

Numerous studies have shown that drugs can have a significant effect on mucociliary clearance. Beta-andrenergics, histamines, and amiloride have all been shown to increase mucociliary clearance

TABLE 5.1

Temperatures and RH in the Human Respiratory Tract, for Both Inspiration and Expiration

Anatomical Location	Nasal				Oral				References
	T_{insp} (°C)	T_{exp} (°C)	RH_{insp}	RH_{exp}	T_{insp} (°C)	T_{exp} (°C)	RH_{insp}	RH_{exp}	
Nasal Passages									
Distance from nostril									
1.5 cm	28.9 ± 2.3	—	69 ± 6.5	—	—	—	—	—	Keck et al.[134]
2.5 cm	30.3 ± 1.6	—	78.7 ± 7.2	—	—	—	—	—	Keck et al.[134]
6.0 cm	32.6 ± 1.5	—	90.3 ± 5.3	—	—	—	—	—	Keck et al.[134]
Laryngeal cavity	32.3 ± 0.8	36.4 ± 0.2	98–99	98–99	30.6 ± 0.8	36.2 ± 0.3	90	99	Ingelstedt[135]
Airway generation, i									
$i = 0$ (trachea)	34–35	36–37	—	—	33–34	36–37	—	—	Cole[136]
$i = 0$ (trachea)	35.3	35.7	98	99.9	—	—	—	—	Perwitzschky[137]
$i = 0$ (trachea)	35.4	36.2	—	—	34.5	35.8	—	—	Verzar et al.[138]
$i = 0$ (trachea)	32.6	35.3	—	—	32.9	34.4	—	—	Herlitzka[139]
$i = 0$ (trachea)	—	—	—	—	26.7	—	82.7	—	Dery et al.[140]
$i = 0$ (trachea)	—	—	—	—	31.4–31.9	33.2–33.4	73.6–80.4	—	Dery[141]
$i = 0$ (trachea)	—	—	—	—	31.2	32.6	—	—	McFadden et al.[142]
$i = 0$ (trachea)	—	—	—	—	32	33	—	—	McFadden et al.[142]
$i = 0$ (trachea)	—	—	—	—	32.2	33.4	—	—	Dery et al.[140]
$i = 0$ (trachea)	36	36.2	98.3	99.2	—	—	—	—	McRae et al.[143]
$i = 1$ (main)	—	—	—	—	30.6	—	85.8	—	Dery[141]
$i = 1$ (main)	—	—	—	—	32.2	33.7	87	—	McFadden et al.[142]
$i = 2$ (lobar)	—	—	—	—	33	34	—	—	McFadden et al.[142]
$i = 3$ (segmental)	—	—	—	—	33.1	—	91.3	—	Dery et al.[140]
$i = 4$ (subsegmental)	—	—	—	—	33.9	—	94.6	—	McFadden et al.[142]
$i = 4 – 5$	—	—	—	—	33.9	35	—	—	Dery[141]
$i = 10 – 11$	—	—	—	—	34.6	36	—	—	McFadden et al.[142]

Source: Adapted from *Extrapolation of Dosimetric Relationships for Inhaled Particles and Gases*, Martonen, T.B., Hoffman, W., Eisner, A.D., and Ménache, M.G., The role of particle hygroscopicity in aerosol therapy and inhalation toxicology, pp. 303–316, Copyright 1989, with permission from Elsevier.

TABLE 5.2

Hygroscopic Growth of Various Environmental and Pharmacological Aerosols

Substance	Conditions T (°C)	Conditions RH (%)	Diameter Increase (%)	Source
Mainstream cigarette smoke				
0.44 μm		100	65	Kousaka et al.[145]
~0.3 μm		99.5	~60	Li and Hopke[146]
~0.2 μm		99.5	~45	Li and Hopke[146]
—		100	~70	Hicks et al.[147]
NaCl				
0.1 μm	20	90	129	Gysel et al.[148]
0.3–0.5 μm	25	75	~90	Tang et al.[149]
0.3–0.5 μm	25	85	~110	Tang et al.[149]
0.3–0.5 μm	25	98	~280	Tang et al.[149]
$(NH_4)_2SO_4$				
0.05 μm	20	90	66	Gysel et al.[148]
0.1 μm	20	90	68	Gysel et al.[148]
$NaNO_3$				
0.05 μm	20	90	86	Gysel et al.[148]
0.1 μm	20	90	91	Gysel et al.[148]
NH_4HSO_4		98	~220	Tang and Munkelwitz[150]
Cromolyn sodium		98	31	Smith et al.[151]
Metaproterenol sulfate		98	29	Hiller et al.[152]
Isoproterenol sulfate		98	13	Hiller et al.[152]
Beclomethasone dipropionate		98	33	Hiller et al.[153]
Isoproterenol/phenynlephedrine		90	24	Kim et al.[152]
Epinephrine		90	11	Kim et al.[152]
Metaproterenol		90	10	Kim et al.[152]
Albuterol		90	8	Kim et al.[152]
Isoetharine/phenylephedrine		90	10	Kim et al.[152]
Triamcinolone		90	17	Kim et al.[152]
Aerodur		98	37	Seemann et al.[155]
Bricanyl		98	144	Seemann et al.[155]
Cromolind		98	48	Seemann et al.[155]
Intal powder		97.4	30	Seemann et al.[155]
Intal composite		97.4	30	Seemann et al.[155]
Atropine sulfate		99.5	160	Peng et al.[156]
Isoproterenol hydrochloride		99.5	186	Peng et al.[156]
Isoproterenol hemisulfate		99.5	142	Peng et al.[156]
Disodium cromoglycate		99.5	26	Peng et al.[156]

in the human lung.[165,171–175] In contrast, cholinergic antagonists, aspirin, and anesthetics have all been shown to markedly decrease mucociliary clearance.[165,170,176,177]

Inhaled chemical contaminants such as sulfur dioxide, sulfuric acid, ozone, and tobacco smoke affect mucociliary clearance. At low concentrations (<100 μg/m³), sulfuric acid increases mucociliary clearance while at high levels, it seems to impair clearance by paralyzing the cilia.[178,179] Sulfur dioxide, ozone, and acute cigarette smoke exposures all have been shown to increase clearance in humans, whereas chronic cigarette smoking decreases mucociliary clearance.[165,180–184]

5.4.6.1.2   Effect of Disease on Mucociliary Clearance

Mucociliary clearance is inhibited by numerous respiratory diseases. Acute respiratory infections such as pneumonia and influenza have been shown to impair clearance.[180,185,186] Chronic respiratory infections such as chronic bronchitis and bronchiectasis often result in an accumulation of mucus in the ciliary transport system, hindering mucociliary clearance.[187] While chronic obstructive pulmonary disease (COPD) patients generally have varied and erratic clearance rates,[169,170] asthma can result in reduced mucus transport rates and mucus plugging of the bronchi.[168,173,188,189] Patients with cystic fibrosis (CF) have been shown to have whole lung clearance impairment, as well as regional clearance impairment.[190] Small airway dysfunction or disease from chronic cigarette smoke also results in slowed mucociliary clearance.[182]

5.4.6.1.3   Effect of Age and Activity on Mucociliary Clearance

Mucociliary clearance has been shown to decrease with age, starting at the age of 20, with large differences between adults >54 years old and adults 21–37 years.[191–194] Ho et al.[195] showed that adults over the age of 40 have decreased ciliary beat frequency, thus leading to slowed mucociliary clearance.

Increased physical activity, particularly aerobic exercise, has been shown to increase mucociliary clearance in humans.[165,189,196,197] Normal activities that do not require significant exertion have no effect on mucociliary clearance, while sleep significantly slows mucociliary clearance.[165,189]

5.4.6.2   Macrophage Clearance or Phagocytosis

Macrophage (or slow phase) clearance occurs in the alveolar region of the lung. As alveoli do not have cilia, deposited particles are engulfed by large bodies called macrophages (phagocytosis). These alveolar macrophages then migrate to the cilia surface and are cleared by mucociliary clearance, or move into the lymphatic system or bloodstream for removal. This type of clearance may take months or years.[165]

5.4.6.2.1   Effect of Particle Size/Fiber Length and Shape on Macrophage Clearance

Particle size affects macrophage clearance rates, with clearance efficiency decreasing for particles smaller than 1 μm,[198] and optimal clearance occurring for particles between 1.5 and 3 μm.[165] Studies have shown that particle/fiber length affects alveolar clearance, with shorter fibers cleared more readily than longer fibers.[199–201] Particles such as asbestos and other nonsoluble fibers are often unable to be removed by phagocytosis because the macrophage is unable to engulf the entire fiber. This may result in incomplete phagocytosis, damage to the macrophage, and the release of the macrophage's digestive enzymes. These digestive enzymes can cause extensive tissue damage. Thus, the persistence of these fibers in the lung can result in inflammation and disease.[202,203]

In addition, toxic non-fiber particles such as silica, that persist in the alveoli, provoke reactions that can lead to lung disease (i.e., silicosis). When a macrophage engulfs silica particles, the macrophage may release enzymes that cause fibroblast proliferation. This release of enzymes results from the crystalline structure of silica, the crystalline structure having been linked to the particle's fibrogenic potential.[204]

5.4.6.2.2   Effect of Drugs and Inhaled Contaminants on Macrophage Clearance

Inhaled contaminants such as ozone, nitrogen dioxide, and cigarette smoke affect alveolar macrophage activity. Ozone and nitrogen dioxide reduce phagocytosis as well as the bactericidal activity of macrophages, making it more difficult to fight bacterial lung infections.[205] Acute cigarette smoke exposures, as well as high-level cigarette smoke exposures, have been shown to inhibit macrophage action while low levels of cigarette smoke exposure have been shown to increase macrophage action.[206,207] Exposure to nongaseous aerosolized contaminants such as nickel, cadmium, lead, manganese, chromium, and vanadium has been shown to damage alveolar macrophages.[204,205] Such exposures may result in disease (i.e., cadmium inhalation causing emphysema), potentially leaving the alveolar region unable to defend itself against or remove other inhaled contaminants, as well as causing the macrophages to release enzymes that damage the neighboring lung tissue. As discussed previously, toxic particulate contaminants such as silica and asbestos can cause macrophage impairment and damage, resulting in particle persistence and disease. The next section discusses inhaled particle overload in the alveolar region of the lung and resulting particle translocation.

5.4.6.3 Free Particle Uptake and Translocation to the Interstitium

Particles, particularly ultrafines, that are not rapidly cleared by macrophage action may persist in the alveoli or be taken up by epithelial cells and translocated from the alveolar region of the lung to the interstitial tissues and regional lymph nodes.[165,208–210] These particles may remain in the interstitium or regional lymph nodes for years, building up over time,[211] or may be removed by interstitial macrophages and/or penetrate into the post-nodal lymph circulation.[210–212] It has been suggested that impaired clearance or a significant burden of particles in the lung (particle overload) increases translocation of particles to the interstitium.[208,213] Studies in animals have found that increasing particle number and dose rate, and decreasing particle size enhances translocation of particles to the interstitium.[214–216] Enhanced interstitium translocation, particularly when toxic dusts (i.e., silica) are involved, has been linked with tissue damage, tumors, and fibrosis.[168,208,217]

5.4.6.4 Importance of Clearance in Particle Deposition Modeling

Several computational models of clearance mechanisms have been developed.[218–220] As clearance will affect both the residence time and local distribution of inhaled particles, the implementation of such models into particle deposition simulations would be desirable.

Hofmann et al.[221] developed a model of particle clearance in which different clearance rates (derived from experimental studies) were associated with different generations of TB airways. Furthermore, Martonen and Hofmann[222] described a model of clearance as a function of spatial location within airway branching sites. Specifically, they incorporated distinct clearance rates for tubular airway segments, bifurcation zones, and carinal ridges.

As discussed in the previous sections, age, activity level, drugs, inhaled contaminants, and disease can affect the efficiency of particle clearance, in turn influencing the number and local concentrations of inhaled particles. Particle deposition models that consider these overlapping factors could be of great use in both inhalation toxicology and aerosol therapy. Although we have discussed its influence as it relates to clearance, we will now provide a more complete discussion of the effects of disease on particle deposition and distribution.

5.4.7 Disease

Airway disease has a dual effect on particle deposition, both influencing breathing pattern and physically changing airway morphology. In this section, we will discuss common respiratory diseases and how they affect airflow and airway morphology, thus affecting the deposition and distribution of inhaled aerosols.

5.4.7.1 Chronic Obstructive Pulmonary Disease

COPD is a term that is generally applied to patients with emphysema and/or chronic bronchitis. Both of these diseases modify the structure of the human lung,[168] resulting in either obstructed airways and/or degeneration of the alveolar structure.

Emphysema may affect the respiratory or terminal bronchioles or the peripheral alveoli. Cigarette smoke and air pollution are the likely causes of emphysema, resulting in the destruction of elastin in the alveolar wall. This elastin destruction leads to enlarged airspaces and loss of alveolar structure, often resulting in intrapulmonary bronchi collapse during expiration.[117,168] In addition, the loss of the alveolar structure results in the loss of capillary bed that transfers oxygen from the lungs to the blood.[168]

Chronic bronchitis is characterized by excessive mucus generation and alveolar wall thickening. The excessive mucus is a result of hypertrophied mucus glands, and causes the formation of mucus plugs that obstruct airways and may fully occlude small bronchi. Alveolar wall thickening results in reduced elasticity of alveolar walls, limiting regional ventilation.[117,168]

5.4.7.2 Asthma

Asthma is characterized by a reduction of the airway lumen due to constriction of the bronchial airways in response to a stimulus. This constriction may also in turn result in an increase in the mucus layer thickness. Chronic asthma can result in subepithelial fibrosis.[168] Asthma stimuli may include pollutants, allergens, or exercise. It has been noted that asthma has attained epidemic proportions on a global scale.[223]

5.4.7.3 Cystic Fibrosis

CF is a genetic disease that causes the lung's epithelial cells to produce abnormally thick, excessive mucus. This slowly cleared mucus narrows airways and obstructs airflow making tissue vulnerable to inflammation and recurrent infection. This inflammation and infection causes progressive respiratory disease including bronchiectasis and chronic airway obstruction.[168,224] Impairment typically begins in the small airways and progresses proximally, with ventilation increasingly shifting from obstructed regions to healthy regions of the lung.[225]

5.4.7.4 Effect of Obstructive Disease on Particle Deposition and Distribution

Several studies have investigated the effect of obstructive disease on particle deposition and distribution in the human lung. Exploring the effect on particle deposition, Kim and Kang[226] found a marked increase in deposition of 1 μm particles in patients with COPD and asthma compared to normal subjects. Anderson et al.[227,228] found that the deposition of fine and ultrafine particles was increased in patients with CF and obstructive disease. Brown et al.[130] found that COPD patients had a greater dose rate for ultrafine particles than healthy subjects. Also, the deposition of particles increased with severity of obstruction or decrease in lung function.[225,229–231] Reasons for increased deposition in patients with obstructive disease include (1) reduction of airway diameter by constriction or mucus buildup, thus increasing inertial impaction on airway walls, (2) increased residence time of particles in the alveolar region resulting from nonuniform ventilation distribution, (3) collapse of airways due to flow limitation, and (4) flow perturbations or induced turbulence at sites of obstruction.[225,226,228]

Investigating the effect of obstructive disease on particle distribution, Brown et al.[232] found that a significant number of coarse particles deposit in the poorly ventilated TB airways of CF patients, while these particles follow regional ventilation in healthy subjects. Other studies indicate that the deposition pattern of particles in patients with obstructive disease is heterogeneous, with an enhancement of deposition in various local regions.[233–235]

5.4.7.5 Modeling Disease

Several models of particle deposition have been developed that specifically address disease. Segal et al.[236] modeled particle deposition in patients with COPD, using a modified deterministic model.[7,32] This work investigated the dependence of deposition pattern on the severity of disease. In addition, Martonen et al. simulated particle deposition in CF. This study found a proximal shift in particle deposition with severity of obstruction.[237–239] More recently, Martonen et al.[223] have simulated the effect of asthma on particle deposition patterns, comparing their results with data from imaging studies of asthma patients.

The United States Environmental Protection Agency (U.S. EPA) has identified people suffering from respiratory disease as a sensitive subpopulation needing particular consideration in risk assessment of particulate matter and in the establishment of air pollution standards.[240] Therefore, more advanced models of particle deposition for a variety of diseases are needed.

5.4.8 Age

Age can be a significant factor influencing the deposition and distribution of particles in the human lung, likely due to the differences in airway geometry and ventilation between children and adults. Studies of airway geometry as a function of age are presented by Ménache and Graham,[241] Hofmann et al.,[242] and Martonen et al.[243] Age has also been shown to affect a percentage of nasal breathing,[244] thus affecting an amount of particulate matter that makes it to the human lung. Bennett et al.[245] showed that children have enhanced upper airway deposition of coarse particles when compared to adults, but that total deposition amounts are comparable.

Several studies have modeled aerosol deposition as a function of human subject age. Martonen et al.[242] found that modeled total deposition within the human lung decreased with increasing age from 7 months to 30 years. In addition, Isaacs and Martonen[246] compared modeled lung deposition results for children with available experimental data. Using the International Commission on Radiological Protection (ICRP) 66 model,[21] Harvey and Hamby[247] found that for 1 µm particles, deposition and regional distribution varied by age, with extrathoracic deposition increasing significantly in younger age groups (young children and infants) that have smaller respiratory airway sizes. Hofmann et al.[243] showed that particle deposition does indeed depend on lung morphology and that dose per surface area decreases from 7 months of age to adulthood.

The U.S. EPA has also identified children as a sensitive subpopulation requiring additional consideration in the establishment of air quality standards.[240] Therefore, the development of particle deposition models for children is of particular importance. A model of particle deposition in a developing human lung has also been developed.[248] It was predicted that children may receive a localized PM dose three times that of adults. Such models will be of great use in both inhalation toxicology and inhalation therapy.

5.5 THEORY AND EXPERIMENT

Many different types of experimental protocols have been developed to measure particle deposition in the respiratory system. An overview of several of these methods will be presented, and a discussion of how the resulting data can be compared with particle deposition will be presented.

5.5.1 Predictions of Particle Deposition

Simulation studies can be used to predict the deposition of inhaled particles on differing spatial scales of resolution. Models can be developed that predict total respiratory system deposition or deposition in each of the regional (i.e., extrathoracic, TB, or pulmonary) compartments. In addition, models can predict deposition efficiencies in each individual airway generation or simulate the dose to a specific anatomical location (e.g., a carinal ridge or airway wall) within a respiratory passage. The level of detail desired and the type of experimental data available for model validation should govern the selection or development of an appropriate model for a given research purpose.

5.5.2 Particle Deposition Measurements

Much work has been done in attempting to quantify the deposition of particles in respiratory airways, encompassing a wide range of approaches and techniques. We provide an overview of aerosol deposition measurements that have been performed in casts, animal models, and human subjects. Comprehensive reviews of experimental measurements of respiratory particle deposition have been presented by Martonen,[8] Sweeny and Brain,[249] and Kim.[250]

5.5.2.1 Casts and Models

Particle deposition measurements have been performed in models and replica casts of both human and animal lungs. Such studies provide a means of examining particle deposition airway by airway

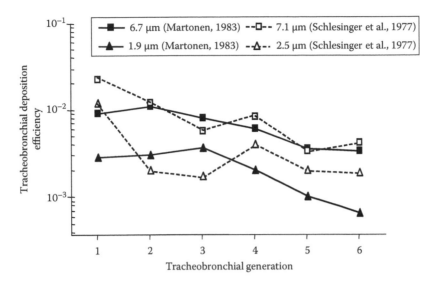

FIGURE 5.9 Deposition of particles in the TB region as measured by Martonen[254] using a silicone rubber cast, compared with the replica cast data of Schlesinger et al.,[256] for different particle sizes.

in realistic geometries. In addition, deposition studies in casts can be reproduced, and particles of different sizes and characteristics may be studied serially under the same conditions. Replica cast studies have been performed in the human TB region,[26,251–253] in canine lungs,[252] and in nasal passages.[254,255] Other studies have been performed using laryngeal casts combined with a silicone rubber model of the TB airways.[256,257] Deposition data obtained in this silicone model at a flow rate of 15 L/min are shown in Figure 5.9 for two particle sizes. Deposition in a TB replica cast[258] is also shown. In this figure, the ratio of the amount of aerosol deposited in a single generation to the amount entering the cast is plotted. Note that good agreement between the replica cast and the silicone model results were obtained. Table 5.3 summarizes localized (i.e., bifurcation) deposition from the two silicone model studies for several particle sizes at three different flow rates. Specifically, the table contains the bifurcation deposition ratio B_d, where

$$B_d = \left[\left(\text{aerosol mass deposited within a bifurcation} \right) / \left(\text{bifurcation surface area} \right) \right]$$

$$\div \left[\left(\text{total aerosol mass deposited within the two airways of a generation,} \right. \right.$$

$$\left. \left. \text{including the shared bifurcation} \right) / \left(\text{total airway and bifurcation surface area} \right) \right] \quad (5.9)$$

Note that in most cases, the deposition in the bifurcation zone is enhanced in relation to the adjacent airways. For example, at a flow rate of 60 L/min the deposition of 8.7 μm particles at the third generation bifurcation is three times greater than that in the adjacent airway segments. The values of B_d of less than 1 may be due to large regions of the bifurcations site having 0 deposition, as observed by Schlesinger et al.[251]

Recently, advances in stereolithography and "3D printing" techniques have allowed for the rapid development of plastic (resin) or polymer models from medical images for use in deposition studies. Such models include the Premature Infant Nose Throat (PrINT) model,[259] an MRI-based polymer model, and the Sophia Anatomical Infant Nose–Throat model.[260] In addition, Giesel et al.[261] presented a method of rapidly prototyping models of the upper (generations 0–5) TB airways using CT scanning and laser sintering.

TABLE 5.3
Bifurcation Deposition Ratio B_d at Different Flow Rates

Tracheobronchial Model Generation	Q=15 L/min				Q=30 L/min			Q=60 L/min			
	1.9 μm[b]	2.1 μm[a]	3.0 μm[b]	6.7 μm[a]	3.3 μm[a]	3.6 μm[a]	6.8 μm[b]	1.9 μm[a]	6.1 μm[a]	8.7 μm[b]	10.6 μm[b]
1	0.80	0.42	0.51	0.83	0.17	0.25	0.65	0.50	0.84	1.10	1.20
2	1.15	1.08	1.15	0.92	1.08	1.19	1.35	1.30	1.14	1.40	1.52
3	1.50	1.38	1.62	1.36	2.29	1.83	2.31	1.93	1.84	3.05	3.16
4	0.91	0.90	0.87	1.12	1.34	1.47	1.57	1.74	1.46	2.10	1.94
5	1.10	0.85	0.91	1.18	1.48	1.63	1.50	1.63	1.78	1.75	1.60

[a] Martonen.[254]
[b] Martonen and Lowe.[255]

5.5.2.2   Deposition Patterns Deduced from Clearance Studies

Experimental methods have been developed in which the regional deposition of particles is deduced from measurement of the time-course of clearance of particles from the thorax.[133] Specifically, radiolabeled particles are inhaled and a whole body counter is used to measure the amount of radioactive activity in the stomach, chest, and extrathoracic regions. Since the removal of particles (by mucociliary clearance) in the TB region occurs at a faster rate than removal (by macrophagic clearance) in the pulmonary region, deposition in the two regions can be deduced from the two-part slope of a normalized retention curve for the thorax.

5.5.2.3   Light-Scattering Methods

Traditionally, total deposition of particles in the respiratory tract has been quantified using light-scattering photometry to compare the concentration of particles in the inhaled and exhaled air.[133,250,262,263] When a monodisperse aerosol is used, and ventilation is simultaneously measured, the deposition fraction in the respiratory system can be calculated. However, photometry cannot distinguish between differences in inspiratory and expiratory aerosol concentration and changes in aerosol size distribution; therefore, these methods are inappropriate for polydisperse or hygroscopic aerosols.[250] Rosati et al.[264] have developed a light-scattering, particle-sizing system that may be the best option for determining total deposition of polydisperse aerosols in the respiratory tract. This system also has the potential to be applied to hygroscopic aerosols as it can determine particle sizes of inhaled and exhaled aerosols, and works well for varying-sized polydisperse aerosols.[264,265]

5.5.2.4   Imaging Studies

Radionuclide imaging has been widely used to measure both the concentration and spatial distribution of inhaled aerosols. In these studies, particles are tagged with a radioisotope (such as ^{99m}Tc) and then inhaled. 2D (planar) or 3D imaging modalities can then be used to measure the radioisotope emissions from specific locations within the body.

Planar gamma cameras can be used to obtain projections of the spatial distribution of inhaled radiolabeled aerosols. These images may be useful in predicting total deposition within the lung or extrathoracic passages, but the 2D nature of the images may obscure important deposition patterns.[266] Planar imaging studies have been performed for a variety of inhaled aerosols.[132,267,268] To assist in the interpretation of planar gamma camera data, Martonen et al.[267,268] have developed methods to associate regions of images (Figure 5.10A) with computer models of the human lung (Figure 5.10B). The computer model serves as a template to be superimposed on actual images, thus permitting the generational airway composition within the central (C), intermediate (I), and peripheral (P) zones of planar images to be predicted (Figure 5.10C). Tossici-Bolt et al.[269,270] have recently described methods of constructing 3D representations of aerosol deposition from planar scintigraphic images.

Recently, 3D tomographic imaging modalities have been applied to the study of particle deposition patterns. Both 3D single photon emission computed tomography (SPECT)[271–273] and positron emission tomography (PET)[274–276] have been employed in particle deposition studies. 3D methods provide a powerful means of associating particle deposition with distinct local regions within the respiratory system. In a series of papers, Martonen et al.[101,272,277–280] have recently presented computational methods for correlating the individual airways of a lung morphology model with the voxels of 3D SPECT images. These methods provide a means of validating 3D CFPD and deposition models using SPECT data while simultaneously providing a framework for predictive laboratory studies of targeted aerosol delivery. In practice, the computer model may be superimposed on the voxels of a SPECT image to allow a quantification of particle deposition. Figure 5.11 shows an example SPECT image and a detailed view of the airway composition of an associated voxel.

5.5.2.5   Microdosimetry

In the experimental methods described earlier, particle deposition may be measured by region (e.g., by clearance studies) or airway-by-airway (e.g., in casts). However, in the study of the health risks

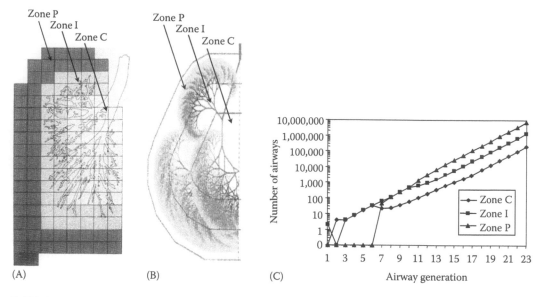

(A)   (B)   (C)   Airway generation

FIGURE 5.10 Association of the central, intermediate, and peripheral zones (A) of planar gamma camera images with a corresponding computerized lung branching network model (B). Panel (C) shows the resulting distribution of lung airways within each zone.

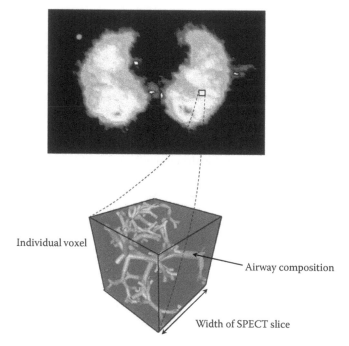

FIGURE 5.11 SPECT image of radiolabeled aerosol distribution in the human lung and an associated voxel of the branching computer model; the composition of airways within the voxel is shown.

associated with inhaled particles, it is desirable to obtain deposition measurements at ever-finer levels of spatial resolution, measuring the dose to individual airway structures or cells. Unfortunately, there is little such data available. Schlesinger et al.[258] performed local deposition studies of 8 μm particles in a cast of an airway bifurcation. After the aerosols were deposited, the airway cast (Figure 5.12) was cut open and a microscope was used to count the number of particles in 1 mm square regions

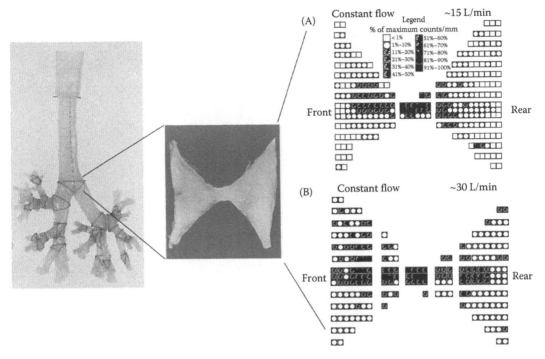

FIGURE 5.12 Determination of localized deposition by Schlesinger et al.[256] for flows of 15 (A) and 30 (B) L/min. The replica cast, the expanded bifurcation zones, and the resulting measured local deposition are shown. (Adapted from Schlesinger, R.B. et al., *Ann. Occup. Hyg.*, 26, 47, 1982. With permission from Pergamon Press.)

of the bifurcation surface. Panels A and B of Figure 5.12 show their results for constant flow rates of 15 and 60 L/min, respectively. There is a definite "hot spot" of deposition at the carinal ridge at 15 L/min, and this area becomes wider with increasing flow rate. In addition, Martonen[281] presented a qualitative description of the local concentrations of 6.7 μm ammonium fluorescein particles in several generations of a bifurcating cast at a constant flow rate. Following the deposition of the particles, the cast was cut open and the distribution of the particles was imaged (Figure 5.13). Note the "hot spots" of particles at the carinal ridges. Additional studies of the dosimetry or microdosimetry of inhaled particles would be of great use in the validation of CFPD studies.

5.5.3 COMPARISON OF MODELING AND DATA

Validation of deposition modeling results with experimental data is a crucial step in the modeling process. Studies comparing simulated particle deposition predictions with experimental results exist for a wide variety of models. In these studies, simulations of total and compartmental particle deposition have been examined, with many particle sizes being considered. Stahlhofen et al.[19] presented a comprehensive overview and summary of large amount of experimentally obtained particle deposition data, providing a resource for many subsequent modeling investigations.

In the study of the fate of inhaled particles, one must be aware that inherent uncertainties exist in both experimental data and in model simulations. For example, uncertainty and error may be imparted into experimental data by uncertainties in the measurement of flow rates, particle size distributions, or experimental deposition measurements, or by intersubject variability in these measurements.[19] Uncertainty in model simulations may depend on one or more of the following: (1) observational errors in any model input parameters, (2) natural (i.e., intersubject of intrasubject) variation of model input parameters, (3) the validity of the underlying theory of the model or any simplifying assumptions, (4) any approximation errors imparted by the computational numerical

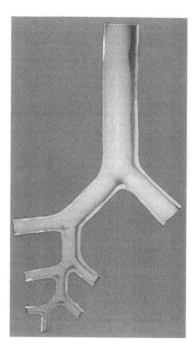

FIGURE 5.13 Enhanced deposition of 6.7 μm particles at different generations of bifurcations in a replica cast. (Adapted from Martonen, T.B., *Am. Ind. Hyg. Assoc. J.*, 53, 6, 1992. With permission from American Industrial Hygiene Association Press.)

methods, or (5) round-off errors imparted by limitations of the computer hardware or software being used. Therefore, any comparisons of model and data should be undertaken with the sources of uncertainty in both the experimental system and the modeling method in mind, and research aimed at explaining and/or controlling variability and uncertainty should be ongoing.

5.5.3.1 Simulations of Total Particle Deposition

Comparison of total deposition predictions with experimental data have been performed for both stochastic and deterministic models. Hofmann and Koblinger[282] presented a comparison between their estimated total deposition, as predicted by their stochastic model, and that obtained from a variety of experimental sources as a function of particle size. They also considered mouth versus oral breathing and breathing pattern in their comparisons. In general, they found good agreement between the model and experiment at all particle sizes.

Theoretical predictions of total particle deposition obtained using the deterministic model of Martonen et al.[7,32] are compared with the experimental data of Heyder et al.[283] in Figure 5.14. The deposition formulae presented in Chapter 6 form the foundation of this model. For the given ventilatory conditions, the predicted total deposition fractions are in relatively good agreement with the experimental data over a wide range of particle sizes. However, there are some systematic differences between the model predictions and the data, namely an overestimation of the total deposition fraction at larger particle sizes. Other comparisons of this deterministic model to human subject data have been performed by Segal et al.[284] More recently, Rosati et al.[107] used the total respiratory tract deposition of Kim and Hu[29] to assess total deposition results for a combined nose–lung CFPD model.

5.5.3.2 Simulations of Compartmental Particle Deposition

5.5.3.2.1 Extrathoracic

Due to the complex geometries present in the oral and nasal regions (e.g., the nasal geometry depicted in Figure 5.4), little theoretical modeling has been done in the extrathoracic compartment. Simple

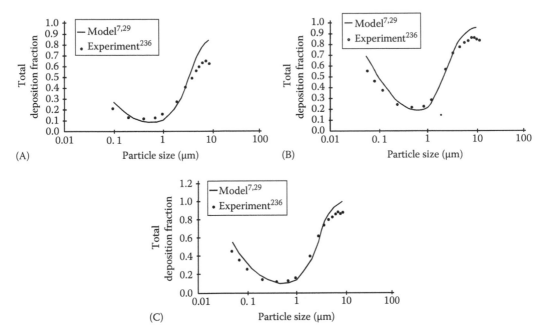

FIGURE 5.14 Model predictions of total deposition determined using the deterministic model of Martonen,[7,31] plotted against experimental data.[281] (A) Breathing frequency = 30 breaths/min, tidal volume = 500 mL, (B) breathing frequency = 15 breaths/min, tidal volume = 1000 mL, and (C) breathing frequency = 30 breaths/min, tidal volume = 1500 mL.

analytical models were developed early on for the nose,[285] but as the complexity of the nasal passages was recognized and more experimental data were published,[286] it became apparent that more complex models were needed. Experimental deposition data have also been reported for the oral cavity,[26,133,287] and analytical models have been developed.[288] Historically, most of the modeling that has been done for the nose[286,289,290] and mouth[289] region has been empirical in nature. Recently, CFPD has offered an alternative approach to the simulation of particle deposition in the extrathoracic passages, while at the same time advancements have been made in the manufacture of replica models of the tortuous extrathoracic airways for use in deposition experiments. Kelly et al. described the deposition of both coarse[291] and ultrafine[292] particles in stereolithography-based plastic nasal model. Their experimental results have been compared (with good agreement) to results obtained in CFPD models by Schroeter et al.[76] and Tian et al.[74] Similar comparisons have been done in the oral cavity as well. The mouth deposition efficiencies predicted by Xi et al.[70] in a realistic mouth geometry using CFPD compared well with the experiment results obtained by Cheng et al.[293] in a cast of the human mouth.

5.5.3.2.2 Tracheobronchial

Hofmann and Koblinger[282] compared their simulated TB deposition values with the experimental data of Heyder et al.[283] For particle sizes of 0.05–1 μm, their model predicted deposition in the TB region of 1%–11%, while zero measured deposition was reported by Heyder et al. It was hypothesized that these differences were due to inherent limitations of the definition of different regions in the model.

Theoretical predictions of TB deposition fraction for the model of Martonen[7,32] are plotted versus corresponding experimental data in Figure 5.15. Again, relatively good agreement between theory and average experimental data was observed over the range of particle sizes investigated, suggesting that the physics of the inhaled particles is being adequately simulated.

There have been many comparisons of CFPD models of the TB airways (either idealized or physiologically realistic models) with experimental data collected in physical models of branching networks. While these comparisons do not consider *in vivo* experimental data, they can be useful

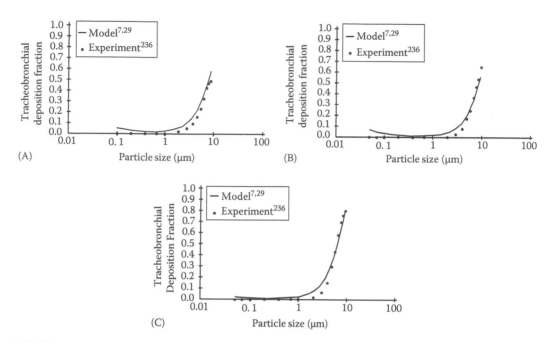

FIGURE 5.15   Model predictions of TB deposition determined using the deterministic model of Martonen,[7,31] plotted against experimental data.[281] (A) Breathing frequency = 30 breaths/min, tidal volume = 500 mL, (B) breathing frequency = 15 breaths/min, tidal volume = 1000 mL, and (C) breathing frequency = 30 breaths/min, tidal volume = 1500 mL.

in validating CFPD methods. For example, Kim and Fisher[294] described deposition in a series of physical tubes. Their results have been used to validate the CFPD models of Farkas et al.,[295] which were performed in geometrically similar configurations.

5.5.3.2.3   Pulmonary

Hofmann and Koblinger[282] also compared their simulated pulmonary deposition values with the experimental data of Heyder et al.[283]; their model very closely predicted measured values in the pulmonary region. Results from the deterministic model of Martonen[7,32] for the pulmonary region are plotted against the experimental data of Heyder et al. in Figure 5.16. While generally good agreement was seen between theory and experiment over the particle sizes simulated, there was a noticeable shift of the predicted deposition with respect to the experimental data. Such a systematic trend in the model bears further investigation.

5.5.3.3   Simulations of Particle Distribution Generation-by-Generation

Figure 5.17 depicts particle deposition by generation as predicted by the model of Martonen[7,32] for a variety of ventilatory conditions. At higher flow rates (Figure 5.17C), the model predicts enhanced deposition of large particles in the TB airways. Experimental measurements of deposition generation-by-generation are scarce, and additional accurate cast studies would be particularly useful in validating such simulations.

5.5.3.4   Simulations of Local Particle Deposition

Experimental data describing the local airway concentrations of inhaled particles are relatively scarce, and are mainly derived from observations of cadaver airways and cast studies. Therefore, it is challenging to validate CFPD studies of local particle deposition in all except very simple geometries, although predicted accumulations of particles at carinal ridges have been shown consistent with experimental observations.[53] Recent efforts by Martonen et al.[279,280] present a methodology for

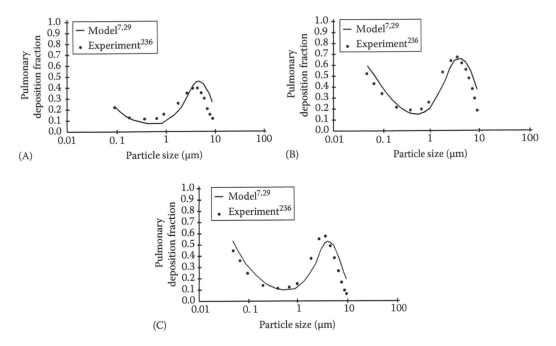

FIGURE 5.16 Model predictions of pulmonary deposition determined using the deterministic model of Martonen,[7,31] plotted against experimental data.[281] (A) Breathing frequency = 30 breaths/min, tidal volume = 500 mL, (B) breathing frequency = 15 breaths/min, tidal volume = 1000 mL, and (C) breathing frequency = 30 breaths/min, tidal volume = 1500 mL.

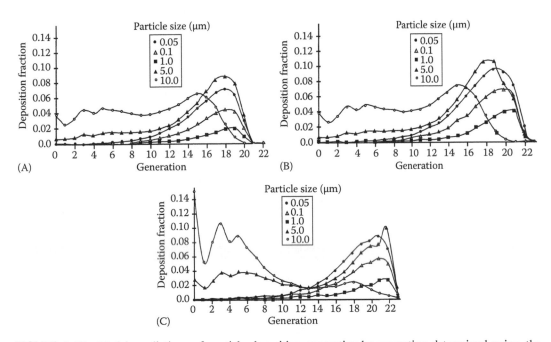

FIGURE 5.17 Model predictions of particle deposition generation-by-generation determined using the deterministic model of Martonen,[7,31] for a range of particle sizes. (A) Breathing frequency = 30 breaths/min, tidal volume = 500 mL, (B) breathing frequency = 15 breaths/min, tidal volume = 1000 mL, and (C) breathing frequency = 30 breaths/min, tidal volume = 1500 mL.

associating individual airways of a branching lung morphology model with specific voxels of a corresponding SPECT image, thereby allowing for the comparison of simulated deposition with actual deposition measurements.

Some CFPD studies have also compared local deposition patterns with experimental data. Isaacs et al.[63] created a CFPD model of a tracheal bifurcation, based on the geometry of the cast system of Schlesinger et al.[296] Modeled particle deposition results in the localized "hot spot" region of the bifurcation were favorably compared with the experimental results obtained in the cast.

5.6 APPLICATIONS OF DEPOSITION MODELS

5.6.1 Exposure and Risk Assessment

Aerosol particle deposition models can provide a quantitative estimate of the amount of material deposited in the lungs under certain conditions. The ability to provide a mechanistic link between human exposure to pollutants and intake dose received has a potential role in the assessment of risk associated with inhaled particulate matter. Several stochastic human exposure models developed and used by the U.S. EPA have recently been updated to incorporate aerosol dosimetry models. These models include the Air Pollutants Exposure (APEX) model[297,298] of the EPA's Office of Air Quality Planning and Standards and SHEDS-PM,[299] a stochastic exposure model developed by the EPA's Office of Research and Development (ORD).

In both APEX and SHEDS-PM, census and other input data are used to create a simulated population of the area being studied (usually a U.S. city, urban area, or state). A year-long time series of pollutant exposures (the PM concentration being encountered) are then estimated for each simulated person, based on ambient air quality data, housing characteristics, and human activity data. The exposure time series are combined with other physiological characteristics of the simulated person and particle composition information, and used as input to a modified version of the empirical ICRP dosimetry model[21] to predict population distributions of deposited particle doses. These dosimetry algorithms may be useful in future risk assessments of PM. An example of a population dose distribution curve is provided in Figure 5.18.

Other particle dosimetry models are also being used in other EPA efforts to associate human pollutant exposures and doses with health effects. The Exposure Model for Individuals (EMI)[300] is being developed by ORD for use in interpreting individual-level data (e.g., questionnaires) collected in a number of air pollution cohort health studies. Similar to APEX and SHEDS, EMI also contains an

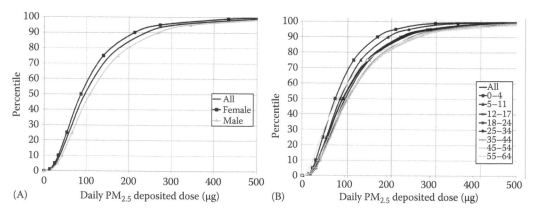

FIGURE 5.18 Distribution of daily deposited $PM_{2.5}$ dose in a year-long simulation of a population of 15,358 individuals in Philadelphia, PA, both by gender (A) and age (B). These doses were calculated using the SHEDS-PM exposure model,[299] incorporating an adapted version of the ICRP deposition algorithm (with no clearance). These simulations were based on $PM_{2.5}$ air quality data for 2008. (Courtesy of Janet Burke, EPA, Research Triangle Park, NC.)

exposure calculation module, but in this case the exposures are predicted not for a simulated population, but rather for a targeted group of real individuals. The exposure module will be coupled with the MPPD particle dosimetry model[301] and a particle clearance algorithm to estimate a time series of particle doses for the individuals being studied. Both the exposure and dose estimates produced by EMI will then be used in conjunction with epidemiological models to assess correlations with negative health outcomes.

5.6.2 Design of Inhaled Pharmaceuticals and Delivery Systems

One of the most promising applications of aerosol deposition modeling is in the design of optimized pharmaceutical aerosols and associated delivery systems. Computational models, especially CFPD models, have the potential to predict the ideal particle properties, inhalation patterns, and aerosol intake conditions (such as velocities or spatial distributions) for targeting the delivery of mass to the therapeutically relevant portions of the respiratory system. Typically, this means reducing deposition in the extrathoracic airways so that the drug can reach either the smooth muscle-lined airways of the lungs (for asthma therapies) or the pulmonary region (for systemic treatments), although in the case of some therapies (like nasal sprays) deposition in the extrathoracic region may be desirable.

In addition to their pure predictive power, deposition models can also be used to interpret observed deposition patterns produced by drug delivery systems, such as dry powder inhalers (DPIs), nebulizers, and pressured metered-dose inhalers (pMDIs). The use of models in conjunction with *in vitro* and *in vivo* experiments can illuminate mechanistic reasons for differences in inhaler performance.

Martonen et al.[106] provide a review of the theoretical modeling issues relevant to assessing different types of inhaled pharmaceuticals, and present a methodology for developing a physiologically based model of the entire upper respiratory system for use in CFPD studies of inhaler performance. Other CFPD models have recently been used to interpret human studies of dry powder inhalation,[302] compare deposition results among different pMDI formulations[303] and between an MDI and a DPI,[304] evaluate different inhaler mouthpiece configurations,[305–307] assess the influence of spray momentum on mouth and throat deposition,[308] and predict the deposition patterns of nasal sprays.[309]

Recently, hygroscopicity has been proposed as the basis for a new method of improving deposition of pharmaceutical aerosols. The method, called enhanced condensational growth (ECG), involves inhaling a submicron-sized aerosolized medication in combination with water vapor. Ideally, the initial small particle size would minimize extrathoracic deposition, while the hygroscopic growth of the particle as it moves into the lung would provide enhanced deposition. Longest and coworkers[310,311] have developed CFPD models to explore the potential of this method.

5.7 SUMMARY

The modeling of particle deposition is of great use in both inhalation toxicology and inhalation therapy. In particular, modeling provides a means of predicting total, regional, and local respiratory system concentrations of inhaled particles, and offers a foundation for the development of targeted delivery protocols. In addition, modeling aids in interpreting experimental measurements and advances the understanding of events and variables that cannot be experimentally quantified.

Particle deposition in the human respiratory system is an extremely complex phenomenon, governed by a wide variety of overlapping and interacting factors. Development and validation of increasingly sophisticated computational models that address particle deposition on local and regional scales, and consider both biological variability and realism, will be instrumental in improving the prediction of both the health effects of inhaled particles and the therapeutic value of inhaled pharmaceutics.

ACKNOWLEDGMENT

Kristin K. Isaacs was funded by the EPA/UNC DESE Cooperative Training Agreement CT827206, with the Department of Environmental Sciences and Engineering, University of North Carolina at Chapel Hill.

REFERENCES

1. Pedley, T. J. and Kamm, R. D., Dynamics of gas flow and pressure-flow relationships, *The Lung: Scientific Foundations*, Vol. I, Crystal, R. G., West, J. B., and Barnes, P. J., Eds., Raven Press, New York, 1991.
2. Martonen, T. B. and Yang, Y., Deposition mechanics of pharmaceutical particles in human airways, *Inhalation Aerosols: Physical and Biological Basis for Therapy*, Hickey, A. J., Ed., Marcel Dekker, Inc., New York, 1996.
3. White, F. M., *Viscous Fluid Flow*, McGraw-Hill, New York, 1991.
4. Chang, H. K. and El Masry, O. A., A model study of flow dynamics in human central airways. Part I: Axial velocity profiles, *Respir. Physiol.*, 49, 75, 1982.
5. Isabey, D. and Chang, H. K., A model study of flow dynamics in human central airways. Part II: Secondary flow velocities, *Respir. Physiol.*, 49, 97, 1982.
6. Schroter, R. C. and Sudlow, M. F., Flow patterns in the human bronchial airways, *Respir. Physiol.*, 7, 341, 1969.
7. Martonen, T. B., Mathematical model for the selective deposition of inhaled pharmaceuticals, *J. Pharm. Sci.*, 82, 1191, 1993.
8. Martonen, T. B., Surrogate experimental models for studying particle deposition in the human respiratory tract: An overview, *Aerosols: Research, Risk Assessment, and Control Strategies*, Lee, S. D., Schneider, T., Grant, L. D., and Verkerk, P. J., Eds., Lewis Publishers, Chelsea, MI, 1986.
9. Zhang, Z. and Martonen, T. B., Deposition of ultrafine aerosols in human tracheobronchial airways, *Inhal. Toxicol.*, 9, 99, 1997.
10. Martonen, T. B. Yang, Y., and Xue, Z. Q., Effects of carinal ridge shapes on lung airstreams, *Aerosol Sci. Technol.*, 21, 119, 1994.
11. Martonen, T. B., Yang, Y., Xue, Z. Q., and Zhang, Z., Motion of air within the human tracheobronchial tree, *Part. Sci. Technol.*, 12, 175, 1994.
12. Martonen, T. B., Yang, Y., and Xue, Z. Q., Influences of cartilaginous rings on tracheobronchial fluid dynamics, *Inhal. Toxicol.*, 6, 185, 1994.
13. Martonen, T. B., Guan, X., and Schrek, R. M., Fluid dynamics in airway bifurcations: I. Primary flows, *Inhal. Toxicol.*, 13, 261, 2001.
14. Martonen, T. B., Guan, X., and Schrek, R. M., Fluid dynamics in airway bifurcations: II. Secondary currents, *Inhal. Toxicol.*, 13, 281, 2001.
15. Martonen, T. B., Guan, X., and Schrek, R. M., Fluid dynamics in airway bifurcations: III. Localized flow conditions, *Inhal. Toxicol.*, 13, 291, 2001.
16. Kimbell, J. S., Gross, E. A., Joyner, D. R., Godo, N. M., and Morgan, K. T., Application of computational fluid dynamics to regional dosimetry of inhaled chemicals in the upper respiratory tract of the rat, *Toxicol. Appl. Pharmacol.*, 121, 253, 1993.
17. Yu, G., Zhang G., and Lessman, R., Fluid flow and particle diffusion in the human upper respiratory system, *Aerosol Sci. Technol.*, 28, 146, 1998.
18. Martonen, T. B., Zhang, Z., Yue, G., and Musante, C. J.,3-D particle transport within the human upper respiratory tract, *J. Aerosol Sci.*, 33, 1095, 2002.
19. Stahlhofen, W., Rudolph, G., and James, A. C., Intercomparison of experimental regional aerosol deposition data, *J. Aerosol Med.*, 2, 285, 1989.
20. Rudolf, G., Gebhart, J., Heyder, J., Schiller, C. F., and Stahlhofen, W., An empirical-formula describing aerosol deposition in man for any particle size, *J. Aerosol Sci.*, 17, 350, 1986.
21. Commission on Radiological Protection, ICRP Publication 66, Human respiratory tract model for radiological protection, *Ann. ICRP*, 24, 1–3, 1994.
22. Rudolf, G., Kobrich, R., and Stahlhofen, W., Modeling and algebraic formulation of regional aerosol deposition in man, *J. Aerosol Sci.*, 21, S403, 1990.
23. Rudolf, G., Gebhart, J., Heyder, J., Scheuch, G., and Stahlhofen, W., Modeling the deposition of aerosol-particles in the human respiratory-tract, *J. Aerosol Sci.*, 14, 188, 1983.
24. Rudolph, G., Gebhardt, J., Heyder, J., Schiller, C. F., and Stahlhofen, W., Mass deposition from inspired polydisperse aerosols, *Ann. Occup. Hyg.*, 32, 919, 1988.
25. Gonda, I., A semi-empirical model of aerosol deposition in the human respiratory tract for mouth inhalation, *J. Pharm. Pharmacol.*, 33, 692, 1981.
26. Chan, T. L. and Lippmann, M., Experimental measurements and empirical modelling of the regional deposition of inhaled particles in humans, *Am. Ind. Hyg. Assoc. J.*, 41, 399, 1980.
27. Cohen, B. S. and Asgharian, B., Deposition of ultrafine particles in the upper airways: An empirical analysis, *J. Aerosol Sci.*, 21, 789, 1990.

28. Asgharian, B., Wood, R., and Schlesinger, R. B., Empirical modeling of particle deposition in the alveo-lar region of the lungs: A basis for interspecies extrapolation, *Fund. Appl. Toxicol.*, 27, 232, 1995.
29. Kim, C. S. and Hu, S. C., Total respiratory tract deposition of fine micrometer-sized particles in healthy adults: Empirical equations for sex and breathing pattern, *J. Appl. Physiol.*, 101(2), 401–412, 2006.
30. Carpenter, R. L. and Kimmel, E. C., Aerosol deposition modeling using ACSL, *Drug Chem. Toxicol.*, 22, 73, 1999.
31. Zhang, Z. and Martonen, T. B., Comparison of theoretical and experimental particle diffusion data within human airway casts, *Cell Biochem. Biophys.*, 27, 97, 1995.
32. Martonen, T. B., Analytical model of hygroscopic particle behavior in human airways, *Bull. Math. Biol.*, 44, 425, 1982.
33. Gradón, L. and Orlicki, D., Deposition of inhaled aerosol particles in a generation of the tracheobronchial tree, *J. Aerosol Sci.*, 21, 3, 1990.
34. Yu, C. P., A two-component theory of aerosol deposition in lung airways, *Bull. Math. Biol.*, 40, 693, 1978.
35. Yu, C. P. and Diu, C. K., A comparative study of aerosol deposition in different lung models, *Am. Ind. Hyg. Assoc. J.*, 43, 54, 1982.
36. Egan, M. J. and Nixon, W., A model of aerosol deposition in the lung for use in inhalation dose assess-ments, *Radiat. Prot. Dosim.*, 11, 5, 1985.
37. Anjilvel, S. and Asgharian, B., A multiple-path model of particle deposition in the rat lung, *Fund. Appl. Toxicol.*, 28, 41, 1995.
38. Asgharian, B., Menache, M. G., and Miller, F. J., Modeling age-related particle deposition in humans, *J. Aerosol Med.*, 17, 213, 2004.
39. Phalen, R. F., Oldham, M. J., and Mautz, W. J., Aerosol deposition in the nose as a function of body size, *Health Phys.*, 57, 299, 1989.
40. Phalen, R. F. and Oldham, M. J., Methods for modeling particle deposition as a function of age, *Respir. Physiol.*, 128, 119, 2001.
41. Choi J. I. and Kim C. S., Mathematical analysis of particle deposition in human lungs: An improved single path transport model. *Inhal. Toxicol.*, 19, 925, 2007.
42. Koblinger, L. and Hofmann, W., Analysis of human lung morphometric data for stochastic aerosol depo-sition calculations, *Phys. Med. Biol.*, 30, 541, 1985.
43. Koblinger, L. and Hofmann, W., Monte Carlo modeling of aerosol deposition in human lungs: Part I: Simulation of particle transport in a stochastic lung structure, *J. Aerosol Sci.*, 21, 661, 1990.
44. Asgharian, B., Price, O. T., and Hofmann, W., Prediction of particle deposition in the human lung using realistic models of lung ventilation, *J. Aerosol Sci.*, 37, 1209, 2006.
45. Sturm, R. and Hofmann, W., 3D-visualization of particle deposition patterns in the human lung generated by Monte Carlo modeling: Methodology and applications, *Comput. Biol. Med.*, 35, 41, 2005.
46. Koblinger, L. and Hofmann, W., Stochastic morphological model of the rat lung, *Anat. Rec.*, 221, 533, 1988.
47. Hofmann, W., Koblinger, L., and Martonen, T. B., Structural differences between human and rat lungs: Implications for Monte Carlo modeling of aerosol deposition, *Health Phys.*, 57, 41, 1989.
48. Hofmann, W., Asgharian, B., Bergmann, R., Anjilvel, S., and Miller, F. J., The effect of heterogeneity of lung structure on particle deposition in the rat lung, *Toxicol. Sci.*, 53, 430, 2000.
49. Hofmann, W., Koblinger, L., and Mohamed, A., Incorporation of biological variability into lung dosim-etry by stochastic modeling techniques, *Environ. Int.*, 22, S995, 1996.
50. Hofmann, W., Morawska, L., and Bergmann, R., Environmental tobacco smoke deposition in the human respiratory tract: Differences between experimental and theoretical approaches, *J. Aerosol Med.*, 14, 317, 2001.
51. Alfoldy, B., Giechaskiel, B., Hofmann, W., and Drossinos, Y., Size-distribution dependent lung deposi-tion of diesel exhaust particles, *J. Aerosol Sci.*, 40, 652, 2009.
52. Hofmann, W. and Koblinger, L., Monte Carlo modeling of aerosol deposition in human lungs: Part II: Deposition fractions and their sensitivity to parameter variations, *J. Aerosol Sci.*, 21, 675, 1990.
53. Comer, J. K., Kleinstreuer, C., and Zhang, Z., Flow structures and particle deposition patterns in double-bifurcation airway models. Part 2. Aerosol transport and deposition, *J. Fluid Mech.*, 435, 55, 2001.
54. Martonen, T. B. and Guan, X., Effects of tumors on inhaled pharmacologic drugs. I. Flow patterns, *Cell Biochem. Biophys.*, 35, 233, 2001.
55. Martonen, T. B. and Guan, X., Effects of tumors on inhaled pharmacologic drugs. II. Particle motion, *Cell Biochem. Biophys.*, 35, 245, 2001.
56. Zhang, Z., Kleinstreuer, C., and Kim, C. S., Cyclic micron-sized particle inhalation and deposition in a triple bifurcation lung airway model, *J. Aerosol Sci.*, 33, 257, 2001.

57. Zhang, Z., Kleinstreuer, C., and Kim, C. S., Effects of curved inlet tubes on airflow and particle deposition in bifurcating lung models, *J. Biomech.*, 34, 659, 2001.
58. Zhang, Z., Kleinstreuer, C., and Kim, C. S., Flow structure and particle transport in a triple bifurcation airway model, *J. Fluids Eng.*, 123, 320, 2001.
59. Zhang, Z., Kleinstreuer, C., and Kim, C. S., Airflow and nanoparticle deposition in a 16-generation tracheobronchial airway model, *Ann. Biomed. Eng.*, 36, 2095, 2008.
60. Comer, J. K., Kleinstreuer, C., and Zhang, Z., Flow structures and particle deposition patterns in double-bifurcation airway models. Part 1. Air flow fields, *J. Fluid Mech.*, 435, 25, 2001.
61. Longest, P. W. and Oldham, M. J., Mutual enhancements of CFD modeling and experimental data: A case study of 1-mum particle deposition in a branching airway model, *Inhal. Toxicol.*, 18, 761, 2006.
62. Longest, P. W., Vinchurkar, S., and Martonen, T., Transport and deposition of respiratory aerosols in models of childhood asthma, *J. Aerosol Sci.*, 37, 1234, 2006.
63. Isaacs, K. K., Schlesinger, R. B., and Martonen, T. B., Three-dimensional computational fluid dynamics simulations of particle deposition in the tracheobronchial tree, *J. Aerosol Med.*, 19, 344, 2006.
64. Zhang, Z., Kleinstreuer, C., Kim, C. S., and Hickey, A. J., Aerosol transport and deposition in a triple bifurcation bronchial airway model with local tumors, *Inhal. Toxicol.*, 14, 1111, 2002.
65. Choi, L. T., Tu, J. Y., Li, H. F., and Thien, F., Flow and particle deposition patterns in a realistic human double bifurcation airway model, *Inhal. Toxicol.*, 19, 117, 2007.
66. Inthavong, K., Choi, L. T., Tu, J., Ding, S., and Thien, F., Micron particle deposition in a tracheobronchial airway model under different breathing conditions, *Med. Eng. Phys.*, 32, 1198–1212, 2010.
67. Ma, B. and Lutchen, K., CFD simulation of aerosol deposition in an anatomically based human large-medium airway model, *Ann. Biomed. Eng.*, 37, 271, 2009.
68. Gemci, T., Ponyavin, V., Chen, Y., Chen, H., and Collins, R., Computational model of airflow in upper 17 generations of human respiratory tract, *J. Biomech.*, 41, 2047, 2008.
69. Inthavong, K., Tu, J. Y., Ye, Y., Ding, S. L., Subic, A., and Thien, F., Effects of airway obstruction induced by asthma attack on particle deposition, *J. Aerosol Sci.*, 41, 587, 2010.
70. Xi, J. X. and Longest, P. W., Transport and deposition of micro-aerosols in realistic and simplified models of the oral airway, *Ann. Biomed. Eng.*, 35, 560, 2007.
71. Zhang, Z., Kleinstreuer, C., and Kim, C. S., Micro-particle transport and deposition in a human oral airway model, *J. Aerosol Sci.*, 33, 1635, 2002.
72. Xi, J. X. and Longest, P. W., Effects of oral airway geometry characteristics on the diffusional deposition of inhaled nanoparticles, *J. Biomech. Eng.*, 130, 011008, 2008.
73. Inthavong, K., Wen, H., Tian, Z. F., and Tu, J. Y., Numerical study of fibre deposition in a human nasal cavity, *J. Aerosol Sci.*, 39, 253, 2008.
74. Tian, Z. F., Inthavong, K., and Tu, J. Y., Deposition of inhaled wood dust in the nasal cavity, *Inhal. Toxicol.*, 19, 1155, 2007.
75. Shi, H., Kleinstreuer, C., and Zhang, Z., Laminar airflow and nanoparticle or vapor deposition in a human nasal cavity model, *J. Biomech. Eng.*, 128, 697, 2006.
76. Schroeter, J. D., Kimbell, J. S., and Asgharian, B., Analysis of particle deposition in the turbinate and olfactory regions using a human nasal computational fluid dynamics model, *J. Aerosol Med.*, 19, 301, 2006.
77. Wang, S. M., Inthavong, K., Wen, J., Tu, J. Y., and Xue, C. L., Comparison of micron- and nanoparticle deposition patterns in a realistic human nasal cavity, *Respir. Physiol. Neurobiol.*, 166, 142, 2009.
78. Heenan, A. F., Matida, E., Pollard, A., and Finlay, W. H., Experimental measurements and computational modeling of the flow field in an idealized human oropharynx, *Exp. Fluids*, 35, 70, 2003.
79. Hofmann, W., Modeling techniques for inhaled particle deposition: The state of the art, *J. Aerosol Med.*, 9, 369, 1996.
80. Zhang, Z. and Kleinstreuer, C., Transient airflow structures and particle transport in a sequentially branching lung airway model, *Phys. Fluids*, 14, 862, 2002.
81. Martonen, T. B. and Schroeter, J. D., Risk assessment dosimetry model for inhaled particulate matter: I. Human subjects, *Toxicol. Lett.*, 138, 119, 2003.
82. Asgharian, B. and Price, O. T., Deposition of ultrafine (nano) particles in the human lung, *Inhal. Toxicol.*, 19, 1045, 2007.
83. Geiser, M. and Kreyling, W. G., Deposition and biokinetics of inhaled nanoparticles, *Part. Fibre Toxicol.*, 7, 2, 2010.
84. Sung, J. C., Pulliam, B. L., and Edwards, D. A., Nanoparticles for drug delivery to the lungs, *Trends Biotechnol.*, 25, 563, 2007.

85. Sung, J. C., Padilla, D. J., Garcia-Contreras, L., Verberkmoes, J. L., Durbin, D., Peloquin, C. A., Elbert, K. J., Hickey, A. J., and Edwards, D. A., Formulation and pharmacokinetics of self-assembled rifampicin nanoparticle systems for pulmonary delivery, *Pharm. Res.*, 26, 1847, 2009.

86. Brown, J. S., Particle inhalability at low wind speeds, *Inhal. Toxicol.*, 17, 831, 2005.

87. Menache, M. G., Miller, F. J., and Raabe, O. G., Particle inhalability curves for humans and small laboratory animals, *Ann. Occup. Hyg.*, 39, 317, 1995.

88. Millage, K. K., Bergman, J., Asgharian, B., and McClellan, G., A review of inhalability fraction models: Discussion and recommendations, *Inhal. Toxicol.*, 22, 151, 2010.

89. Se, C. M. K., Inthavong, K., and Tu, J. Y., Inhalability of micron particles through the nose and mouth, *Inhal. Toxicol.*, 22, 287, 2010.

90. Anthony, T. R., Flynn, M. R., and Eisner, A., Evaluation of facial features on particle inhalation, *Ann. Occup. Hyg.*, 49, 179, 2005.

91. Weibel, E. R., *Morphometry of the Human Lung*, Academic Press, New York, 1963.

92. Weibel, E. R., Design of airways and blood vessels as branching trees, *The Lung: Scientific Foundations*, Vol. I, Crystal, R. G., West, J. B., Barnes, P. J., Cherniak, N. S., and Weibel E. R., Eds., Raven Press, New York, 1991.

93. Horsfield, K., Pulmonary airways and blood vessels considered as confluent trees, *The Lung: Scientific Foundations*, Vol. I, Crystal, R. G., West, J. B., Barnes, P. J., Cherniak, N. S., and Weibel E. R., Eds., Raven Press, New York, 1991.

94. Horsfield, K. and Cumming, G., Angles of branching and diameters of branches in the human bronchial tree, *Bull. Math. Biophys.*, 29, 245, 1967.

95. Horsfield, K. and Cumming, G., Morphology of the bronchial tree in man, *J. Appl. Physiol.*, 24, 373, 1968.

96. Soong, T. T., Nicolaides, P., Yu, C. P., and Soong, S. C., A statistical description of the human tracheobronchial tree geometry, *Respir. Physiol.*, 37, 161, 1979.

97. Jesseph, J. E. and Merendino, K. A., The dimensional interrelationships of the major components of the human tracheobronchial tree, *Surg. Gynecol. Obstet.*, 105, 210, 1957.

98. Parker, H., Horsfield, K., and Cumming, G., Morphology of the distal airways in the human lung, *J. Appl. Physiol.*, 31, 386, 1971.

99. Horsfield, K., Dart, G., Olson, D. E., Filley, G. F., and Cumming, G., Models of the human tracheobronchial tree, *J. Appl. Physiol.*, 31, 207, 1971.

100. Kitaoka, H., Takaki, R., and Suki, B., A three-dimensional model of the human airway tree, *J. Appl. Physiol.*, 87, 2207, 1999.

101. Martonen, T. B., Schroeter, J. D., and Fleming, J. S., 3D in silico modeling of the human respiratory system for inhaled drug delivery and imaging analysis, *J. Pharm. Sci.*, 96, 603, 2007.

102. Martonen, T. B., Schroeter, J. D., Hwang, D., Fleming, J. S., and Conway, J. H., Human lung morphology models for particle deposition studies, *Inhal. Toxicol.*, 12(Suppl. 4), 109, 2000.

103. Montesantos, S., Fleming, J. S., and Tossici-Bolt, L., A spatial model of the human airway tree: The Hybrid Conceptual Model, *J. Aerosol Med. Pulm. Drug Deliv.*, 23, 59, 2010.

104. Spencer, R. M., Schroeter, J. D., and Martonen, T. B., Computer simulations of lung airway structures using data-driven surface modeling techniques, *Comput. Biol. Med.*, 31, 499, 2001.

105. Inthavong, K., Wen, J., Tu, J. Y., and Tian, Z. F., From Ct scans to Cfd modelling—Fluid and heat transfer in a realistic human nasal cavity, *Eng. Appl. Comput. Fluid*, 3, 321, 2009.

106. Martonen, T. B., Smyth, H. D., Isaacs, K. K., and Burton, R. T., Issues in drug delivery: Concepts and practice, *Respir. Care*, 50, 1228, 2005.

107. Rosati, J., Burton, R., McCauley, R., and McGregor, G., Three dimensional modeling of the human respiratory system, presented at *29th Annual Meeting of the American Association for Aerosol Research*, Portland, Oregon, October 25–29, 2010, http://aaarabstracts.com/2010/viewabstract.php?paper-859.

108. National Institutes of Health, National Library of Medicine Visible Human Project. Available at: http://www.nlm.nih.gov/research/visible/visible_human.html

109. Yeh, H. C. and Schum, G. M., Models of human lung airways and their application to inhaled particle deposition, *Bull. Math Biol.*, 42, 461, 1980.

110. Musante, C. and Martonen, T. B., Computational fluid dynamics in human lungs I: Effects of natural airway features, *Medical Applications of Computer Modeling: The Respiratory System*, Martonen, T. B., Ed., WIT Press, Boston, MA, 2001.

111. Martonen, T. B., Zhang, Z., Yang, Y., and Bottei, G., Airway surface irregularities promote particle diffusion in the human lung, *Radiat. Prot. Dosim.*, 59, 5, 1995.

112. Balásházy, I. and Hofmann, W., Quantification of local deposition patterns of inhaled radon decay products in human bronchial airway bifurcations, *Health Phys.*, 78, 147, 2000.
113. Niinimaa, V., Cole, P., Mintz, S., and Shephard, R. J., The switching point from nasal to oronasal breathing, *Respir. Phys.*, 42, 61, 1980.
114. Niinimaa, V., Cole, P., Mintz, S., and Shephard, R. J., Oronasal distribution of respiratory airflow, *Respir. Phys.*, 43, 69, 1981.
115. James, D. S., Lambert, W. E., Mermier, C. M., Stidley, C. A., Chick, T. W., and Samet, J. M., Oronasal distribution of ventilation at different ages, *Arch. Environ. Health*, 52, 118, 1997.
116. Everard, M. L., Hardy, J. G., and Milner, A. D., Comparison of nebulised aerosol deposition in the lungs of healthy adults following oral and nasal inhalation, *Thorax*, 48, 1045, 1993.
117. Brain, J. D. and Sweeney, T. D., Effects of ventilatory patterns and pre-existing disease on deposition of inhaled particles in animals, Ch. 15, *Extrapolation of Dosimetric Relationships for Inhaled Particles and Gases*, Crapo, J. D., Miller, F. J., Smolko, E. D., Graham, J. A., and Hayes, A. W., Eds., Academic Press, Inc., New York, 1989.
118. Lennon, S., Shang, S., Lessmann, R., and Webster, S., Experiments on particle deposition in the human upper respiratory system, *Aerosol Sci. Technol.*, 28, 464, 1998.
119. Becquemin, M. H., Swift, D. L., Bouchikhi, A., Roy, M., and Teillac, A., Particle deposition and resistance in the noses of adults and children, *Eur. Respir. J.*, 4, 694, 1991.
120. Swift, D. L. and Strong, J. C., Nasal deposition of ultrafine 218Po aerosols in human subjects, *J. Aerosol Sci.*, 27, 1125, 1996.
121. Schwab, J. A. and Zenkel, M., Filtration of particles in the human nose, *Laryngoscope*, 108, 120, 1998.
122. Morawska, L., Barron., W., and Hitchins, J., Experimental deposition of environmental tobacco smoke submicrometer particulate matter in the human respiratory tract, *Am. Ind. Hyg. Assoc. J.*, 60, 334, 1999.
123. Anderson, I., Lundquist, G. R., Proctor, D. F., and Swift, D. L., Human response to controlled levels of inert dust, *Am. Rev. Respir. Dis.*, 119, 619, 1979.
124. Cheng, K., Cheng, Y., Yeh, H., Guilmette, R. A., Simpson, S. Q., Yang, Y., and Swift, D. L., In vivo measurements of nasal airway dimensions and ultrafine aerosol deposition in the human nasal and oral airways, *J. Aerosol Sci.*, 27, 785, 1996.
125. Kesavan, J., Bascom, R., Laube, B., and Swift, D. L., The relationship between particle deposition in the anterior nasal passage and nasal passage characteristics, *J. Aerosol Med.*, 13, 17, 2000.
126. Brand, P., Friemel, I., Meyer, T., Schulz, H., Heyder, J., and Haubinger, K., Total deposition of therapeutic particles during spontaneous and controlled inhalations, *J. Pharm. Sci.*, 89, 724, 2000.
127. Heyder, J., Gebhart, J., Stahlhofen, W., and Stuck, B., Biological variability of particle deposition in the human respiratory tract during controlled and spontaneous mouth-breathing, *Ann. Occup. Hyg.*, 26, 137, 1982.
128. Bennett, W. D. and Smaldone, G. C., Human variation in the peripheral air-space deposition of inhaled particles, *J. Appl. Physiol.*, 62, 1603, 1987.
129. Schiller-Scotland, C. F., Hlawa, R., and Gebhart, J., Experimental data for total deposition in the respiratory tract of children, *Toxicol. Lett.*, 72, 137, 1994.
130. Brown, J. S., Zeman, K. L., and Bennett, W. D., Ultrafine particle deposition and clearance in the healthy and obstructed lung, *Am. J. Respir. Crit. Care Med.*, 166, 1240, 2002.
131. Bennett, W. D., Messina, M., and Smaldone, G. C., Effect of exercise on deposition and subsequent retention of inhaled particles, *J. Appl. Physiol.*, 59, 1046, 1985.
132. Messina, M. A. and Smaldone, G. C., Evaluation of quantitative aerosol techniques for use in bronchoprovocation studies, *J. Allergy Clin. Immunol.*, 75, 252, 1985.
133. Stahlhofen, W., Gebhart, J., and Heyder, J., Experimental determination of the regional deposition of aerosol particles in the human respiratory tract, *Am. Ind. Hyg. Assoc. J.*, 41, 385, 1980.
134. Heyder, J., Armbruster, L., Gebhart, J., Grein, E., and Stahlhofen, W., Total deposition of aerosol particles in the human respiratory tract for nose and mouth breathing, *J. Aerosol Sci.*, 6, 311, 1975.
135. Svartengren, M., Svartengren, K., Aghaie, F., Philipson, K., and Camner, P., Lung deposition and extremely slow inhalations of particles. Limited effect of induced airway obstruction, *Exp. Lung Res.*, 25, 353, 1999.
136. Keck, T., Leiacker, R., Heinrich, A., Khunemann, S., and Rettinger, G., Humidity and temperature profile in the nasal cavity, *Rhinology*, 38, 167, 2000.
137. Ingelstedt, S., Studies on the conditioning of air in the respiratory tract, *Acta Otoalaryngol.*, 131, 1, 1956.
138. Cole, P., Recordings of respiratory air temperature, *J. Laryngol. Otol.*, 68, 295, 1954.
139. Perwitzschky, R., Die temperatur and feuchtigkeitsverhaltnisse der atemluft in den luftwegen.1, *Mitt. Arch. Ohren Nasen Kehlkopfh*, 117, 1, 1928.

140. Verzar, F., Keith, T., and Parchet, V., Temperatur and feuchtigkeit der lugt in den atemwegen, *Pflugers Arch. Ges. Physiol.*, 257, 400, 1953.
141. Herlitzka, A., Sur la temperature tracheale de l'air inspire et expire, *Arch. Int. Physiol.*, 18, 587, 1921.
142. Dery, R., Pelletier, J., Jaques, H., Clavet, M., and Houde, J. J., Humidity in anaesthesiology. III. Heat and moisture patterns in the respiratory tract during anaesthesia with the semi-closed system, *Can. Anaesth. Soc. J.*, 14, 287, 1967.
143. Dery, R., The evolution of heat and moisture in the respiratory tract during anaesthesia with a non-rebreathing system, *Can. Anaesth. Soc. J.*, 20, 296, 1973.
144. McFadden, E. R., Pichurko, B. M., Bowman, F. H., Ingenito, E., Burns, S., Dowling, N., and Solway, J., Thermal mapping of the airways in humans, *J. Appl. Physiol.*, 58, 564, 1985.
145. McRae, R. D. R., Jones, A. S., Young, P., and Hamilton, J., Resistance, humidity and temperature of the tracheal airway, *Clin. Otolaryngol.*, 20, 355, 1995.
146. Martonen, T. B., Hoffman, W., Eisner, A. D., and Ménache, M. G., The role of particle hygroscopicity in aerosol therapy and inhalation toxicology, *Extrapolation of Dosimetric Relationships for Inhaled Particles and Gases*, Crapo, J. D., Smolko, and E. D., Eds., Academic Press, Inc., New York, 1989.
147. Kousaka, Y., Okuyama, K. and Wang, C. S., Response of cigarette smoke particles to change in humidity, *J. Chem. Eng. Jpn.*, 15, 75, 1982.
148. Li, W. and Hopke, P. K., Initial size distributions and hygroscopicity of indoor combustion aerosol particles, *Aerosol Sci. Technol.*, 19, 305, 1993.
149. Hicks, J. F., Pritchard, J. N., Black, A., and Megaw, W. J., Experimental evaluation of aerosol growth in the human respiratory tract, *Aerosols: Formation and Reactivity*, Schikarski W., Fissan H. J., and Friedlander, S. K., Eds., Pergamon Press, Oxford, U.K., 1986, p. 243.
150. Gysel, M., Weingartner E., and Baltensperger, U., Hygroscopicity of aerosol particles at low temperatures. 2. Theoretical and experimental hygroscopic properties of laboratory generated aerosols, *Environ. Sci. Technol.*, 36(1), 63, 2002.
151. Tang, I. N., Munkelwitz, H. R., and Davis, J. G., Aerosol growth studies, II. Preparation and growth measurements of monodisperse salt aerosols, *J. Aerosol Sci.*, 8, 149, 1977.
152. Tang, I. N. and Munkelwitz, H. R., Aerosol growth studies, III. Ammonium bisulfate aerosols in a moist atmosphere, *J. Aerosol Sci.*, 8, 321, 1977.
153. Smith, G., Hiller, F. C., Mazumder, M. K., and Bone R. C., Aerodynamic size distribution of cromolyn sodium at ambient and airway humidity, *Am. Rev. Respir. Dis.*, 121, 513, 1980.
154. Hiller, F. C., Mazumder, M. K., Wilson, J. D., and Bone R. C., Effect of low and high relative humidity on metered-dose bronchodilator solution and powder aerosols, *J. Pharm. Sci.*, 69, 334, 1980.
155. Hiller, F. C., Mazumder, M. K., Wilson, J. D., and Bone R. C., Aerodynamic size distribution, hygroscopicity and deposition estimation of beclomethasone dipropionate aerosol, *J. Pharm. Pharmacol.*, 32, 605, 1980.
156. Kim, C. S., Trujillo, D., and Sackner, M. A., Size aspects of metered-dose inhaler aerosols, *Am. Rev. Respir. Dis.*, 132, 137, 1985.
157. Seemann, S., Busch B., Ferron, G. A., Silberg, A., and Heyder, J., Measurement of the hygroscopicity of pharmaceutical aerosols in situ. *J. Aerosol Sci.*, 26, 537, 1995.
158. Peng, C., Chow, A. H., and Chan, C. K., Study of the hygroscopic properties of selected pharmaceutical aerosols using single particle levitation, *Pharm. Res.*, 17, 1104, 2000.
159. Bell, K. A. and Ho, A. T., Growth rate measurements of hygroscopic aerosols under conditions simulating the respiratory tract, *J. Aerosol Sci.*, 12, 247, 1981.
160. Martonen, T. B., Bell, K. A., Phalen, R. F., Wilson, A. F., and Ho, A. T., Growth rate measurements and deposition modeling of hygroscopic aerosols in human tracheobronchial models, *Ann. Occup. Hyg.*, 26, 93, 1982.
161. Broday, D. M. and Georgopoulos, P. G, Growth and deposition of hygroscopic particulate matter in the human lungs, *Aerosol Sci. Technol.*, 34, 144, 2001.
162. Finlay, W. H., Estimating the type of hygroscopic behavior exhibited by aqueous droplets, *J. Aerosol Med.*, 11, 221, 1998.
163. Robinson, R. J. and Yu, C. P., Theoretical analysis of hygroscopic growth rate of mainstream and side-stream cigarette smoke particles in the human respiratory tract, *Aerosol Sci. Technol.*, 28, 21, 1998.
164. Ferron, G. A., The size of soluble aerosol particles as a function of the humidity of the air. Application to the human respiratory tract, *J. Aerosol Sci.*, 8, 251, 1977.
165. Oberdörster, G., Lung clearance of inhaled insoluble and soluble particles, *J. Aerosol Med.*, 1, 289, 1988.
166. West, J. B., *Pulmonary Pathophysiology, The Essentials*, 5th edn., Lippincott, Williams & Wilkins, Philadelphia, PA, 1998.

167. Gehr, P., Schurch, S., Im Hof, V., and Geiser, M., Inhaled particles deposited in the airways are displaced towards the epithelium, *Inhaled Particles VII, The Annals of Occupational Hygiene*, Walton, W. H., Critchlow, A., and Coppock, S. M., Eds., Pergamon Press, New York, 1994.

168. West, J. B., *Respiratory Physiology, The Essentials*, 6th edn., Lippincott, Williams & Wilkins, Philadelphia, PA, 2000.

169. Yeates, D. B., Gerrity, T. B., and Garrard, C. S., Characteristics of tracheobronchial deposition and clearance in man, *Inhaled Particles V, The Annals of Occupational Hygiene*, Walton, W. H., Critchlow, A., and Coppock, S. M., Eds., Pergamon Press, New York, 1982.

170. Yeates, D. B., Gerrity, T. B., and Garrard, C. S., Particle deposition and clearance in the bronchial tree, *Ann. Biomed. Eng.*, 9, 577, 1981.

171. Fazio, F. and Lafortuna, C., Effect of inhaled salbutamol on mucociliary clearance in patients with chronic bronchitis, *Chest*, 80, 827, 1981.

172. Foster, W. M., Langenback, E. G., and Bergofsky, E. H., Respiratory drugs influence mucociliary clearance in central and peripheral ciliated airways, *Chest*, 80, 877, 1981.

173. Foster, W. M., Langenback, E. G., and Bergofsky, E. H., Lung mucociliary function in man, *Ann. Occup. Hyg.*, 26, 227, 1982.

174. Mortensen, J., Lange, P., Nyboe, J., and Groth, S., Lung mucociliary clearance, *Eur. J. Nucl. Med.*, 21, 953, 1994.

175. Weiss, T., Dorrow, P., and Felix, R., Effects of a beta adrenergic drug and a secretolytic agent on regional mucociliary clearance in patients with COLD, *Chest*, 80, 881, 1981.

176. Gerrity, T. R., Cotormanes, E., Garrard, C. S., Yeates, D. B., and Lourenco, R. V., The effect of aspirin on lung mucociliary clearance, *N. Engl. J. Med.*, 308, 139, 1983.

177. Pavia, D., Sutton, P. P., Lopez-Vidriero, M. T., Agnew, J. E., and Clarke, S. W., Drug effects on mucociliary function, *Eur. J. Respir. Dis.*, 64, 304, 1983.

178. Leikauf, G., Yeates, D. B., Wales, K. A., Spektor, D., Albert, R. E., and Lippman, M., Effects of sulfuric acid aerosol on respiratory mechanics and mucociliary particle clearance in healthy non-smoking adults, *Am. Ind. Hyg. J.*, 42, 273, 1981.

179. Lippman, M., Schlesinger, R. B., Leikauf., G., Spektor, D., and Albert, R. E., Effects of sulfuric acid aerosols on respiratory tract airways, *Inhaled Particles V, The Annals of Occupational Hygiene*, Walton, W. H., Critchlow, A., and Coppock, S. M., Eds., Pergamon Press, New York, 1982.

180. Camner, P., Clearance of particles from the human tracheobronchial tree, *Clin. Sci.*, 59, 79, 1980.

181. Kenoyer, J. L., Phalen, R. F., and Davis, J. R., Particle clearance from the respiratory tract as a test of toxicity: Effect of ozone on short and long term clearance, *Exp. Lung Res.*, 2, 111, 1981.

182. Weiss, T., Dorrow, P., and Felix, R., Regional mucociliary removal of inhaled particles in smokers with small airways disease, *Respiration*, 44, 338, 1983.

183. Foster, W. M., Costa, D. L., and Langenback, E. G., Ozone exposure alters tracheobronchial mucociliary function in humans, *J. Appl. Physiol.*, 63, 996, 1987.

184. Vastag, E., Matthys, H., Zsamboki, G., Kohler, D., and Daileler, G., Mucociliary clearance in smokers, *Eur. J. Respir. Dis.*, 68, 107, 1986.

185. Jarstrand, C., Camner, P., and Philipson, K., *Mycoplasma pneumoniae* and tracheobronchial clearance, *Am. Rev. Respir. Dis.*, 110, 415, 1974.

186. Camner, P., Jarstrand, C., and Philipson, K., Tracheobronchial clearance in patients with influenza, *Am. Rev. Respir. Dis.*, 108, 131, 1973.

187. Pavia, D., Sutton, P. P., Agnew, J. E., Lopez-Vidriero, M. T., Newman, S. P., and Clarke, S. W., Measurement of bronchial mucociliary clearance, *Eur. J. Respir. Dis.*, 64, 41, 1983.

188. Bateman, J. R. M., Pavia, D., Sheahan, N. F., Agnew, J. E., and Clarke, S. W., Impaired tracheobronchial clearance in patients with mild stable asthma, *Thorax*, 38, 463, 1983.

189. Pavia, D., Lung mucociliary clearance, Ch. 6, *Aerosols and the Lung: Clinical and Experimental Aspects*, Clarke, S. W., and Pavia, D., Eds., Butterworths, Boston, MA, 1984.

190. Robinson, M., Everl, S., Tomlinson, C., Daviskas, E., Regnis, J. A., Bailey, D. L., Torzillo, P. J., Ménache, M., and Bye, P. T., Regional mucociliary clearance in patients with cystic fibrosis, *J. Aerosol Med.*, 13, 73, 2000.

191. Goodman, R. M., Yergin, B. M., Landa, J. F., Golinvaux, M. H., and Sackner, M. A., Relationship of smoking history and pulmonary function tests to tracheal mucous velocity in non-smokers, young smokers, ex-smokers, and patients with chronic bronchitis, *Am. Rev. Respir. Dis.*, 117, 205, 1978.

192. Puchelle, E., Sahm, J. M., and Bertrand, A., Influence of age on bronchial mucociliary transport, *Scand. J. Respir. Dis.*, 60, 307, 1979.

193. Vastag, E., Matthys, H., Kohler, D., Gronbeck, L., and Daileler, G., Mucociliary clearance and airway obstruction in smokers, ex-smokers and normal subjects who never smoked, *Eur. J. Respir. Dis.*, 139, 93, 1985.

194. Incalzi, R. A., Maini, C. L., Fuso, L., Giordano, A., Carbonin, P. U., and Galli, G., Effects of aging on mucociliary clearance, *Compr. Gerontol. A*, 3, 65, 1989.

195. Ho, J. C., Chan, K. N., Hu, W. H., Lam, W. K., Zheng, L., Tipoe, G. L., Sun, J., Leung, R., and Tsang, K. W., The effect of aging on nasal mucociliary clearance, beat, frequency, and ultrastructure of respiratory cilia, *Am. J. Respir. Crit. Care Med.*, 163, 983, 2001.

196. Wolff, R. K., Dolovich, M. B., Obminsky, G., and Newhouse, M. T., Effects of exercise and eucapnic hyperventilation on bronchial clearance in man, *J. Appl. Physiol.*, 43, 46, 1977.

197. Salzano, F. A., Manola, M., Tricarico, D., Precone, D., and Motta, G., Mucociliary clearance after aerobic exertion in athletes, *Acta Otorhinolaryngol. Ital.*, 20, 171, 2000.

198. Adamson, I. Y. and Bowden, D. H., Dose response of the pulmonary macrophagic system to various particulates and its relationship to transepithelial passage of free particles, *Exp. Lung Res.*, 2, 165, 1981.

199. Timbrell, V., Deposition and retention of fibres in the human lung, *Ann. Occup. Hyg.*, 26, 347, 1982.

200. Coin, P. G., Stevens, J. B., McJilton, C. M., Role of fiber length in the pulmonary clearance of amosite asbestos, *Am. Rev. Respir. Dis.*, 141, A521, 1990.

201. Coin, P. G., Roggli, V. L., and Brody, A. R., Persistence of long, thin chrysotile asbestos fibers in the lungs of rats, *Environ. Health Perspect.*, 102, 197, 1994.

202. Donaldson, K. and Tran, C. L., Inflammation caused by particles and fibers, *Inhal. Toxicol.*, 14, 5, 2002.

203. Oberdörster, G., Toxicokinetics and effects of fibrous and nonfibrous particles, *Inhal. Toxicol.*, 14, 29, 2002.

204. Witschi, H. R. and Last, J. A., Toxic responses of the respiratory system, Ch. 15, *Caserett and Doul's Toxicology, The Basic Science of Poisons*, 6th edn., Klaasen, C. D., Ed., McGraw-Hill, New York, 2001.

205. Hocking, W. G. and Golde, D. W., The pulmonary alveolar macrophage, *N. Engl. J. Med.*, 301, 580, 1979.

206. Holt, P. G. and Keast, D., Environmentally induced changes in immunologic function. Acute and chronic effects of inhalation of tobacco smoke and other atmospheric contaminants in man and experimental animals, *Bacteriol. Rev.*, 41, 205, 1977.

207. Thomas, W. R., Holt, P. G., and Keast, D., Cigarette smoke and phagocyte function: Effect of chronic exposure in vivo and acute exposure in vitro, *Infect. Immun.*, 20, 468, 1978.

208. Oberdörster, G., Lung particle overload: Implication for occupational exposures to particles, *Regul. Toxicol. Pharm.*, 27, 123, 1995.

209. Churg, A., The uptake of mineral particles by pulmonary epithelial cells, *Am. J. Respir. Crit. Care Med.*, 154, 1124–1140, 1996.

210. Churg, A., Wright, J. L., and Stevens, B., Exogenous mineral particles in the human bronchial mucosa and lung parenchyma. I. Nonsmokers in the general population, *Exp. Lung Res.*, 16, 159, 1990.

211. Dumortier, P., De Vuyst, P., and Yernault, J. C., Comparative analysis of inhaled particle contained in human bronchoalveolar fluids, lung parenchyma and lymph nodes, *Environ. Health Perspect.*, 102, 257, 1994.

212. Geiser, M., Morphological aspects of particle uptake by lung phagocytes, *Microsc. Res.*, 57, 512, 2002.

213. Morrow, P. E., Possible mechanisms to explain dust overloading of the lungs, *Fundam. Appl. Toxicol.*, 10, 369, 1988.

214. Ferin, J., Oberdörster, G., and Penney, D. P., Pulmonary retention of ultrafine and fine particles in rats, *Am. J. Respir. Cell Mol. Biol.*, 6, 535, 1992.

215. Ferin, J., Oberdörster, G., Soderhold, S. C., and Gelein, R., The rate of dose delivery affects pulmonary interstitialization of particles in rats, *Ann. Occup. Hyg.*, 38, 289, 1994.

216. Oberdörster, G., Finkelstein, J. N., Johnston, C., Gelein, R., Cox, C., Baggs, R., and Elder, A. C., Acute pulmonary effects of ultrafine particles in rats and mice, *Res. Rep. Health Eff. Inst.*, 96, 5, 2000.

217. Oberdörster, G., Ferin, J., and Lehnert, B. E., Correlation between particle size, in vivo, particle persistence and lung injury, *Environ. Health Perspect.*, 102, 173, 1994.

218. Sosnowski, T. R., Gradón, L., and Podgórski, A., Influence of insoluble aerosol depositions on the surface activity of the pulmonary surfactant: A possible mechanism of alveolar clearance retardation? *Aerosol Sci. Technol.*, 32, 52, 2000.

219. Gerrity, T. R., Garrard, C. S., and Yeates, D. B., A mathematical model of particle retention in the airspaces of the human lung, *Br. J. Ind. Med.*, 40, 121, 1983.

220. Sanchis, J., Dolovich, M., Chalmers, R., and Newhouse, M., Quantitation of regional aerosol clearance in the normal human lung, *J. Appl. Physiol.*, 33, 757, 1972.

221. Hofmann, W., Ménache, M. G., and Martonen, T. B., Age-dependent lung dosimetry of radon progeny, *Extrapolation of Dosimetric Relationships for Inhaled Particles and Gases*, Crapo, J. D., Smolko, E. D., Miller, F. J., Graham, J. A., and Hayes, A. W., Eds., Academic Press, San Diego, CA, 1989.

222. Martonen, T. B. and Hofmann, W., Dosimetry of localised accumulations of cigarette smoke and radon progeny at bifurcations, *Radiat. Prot. Dosim.*, 38, 81, 1991.

223. Martonen, T. B., Fleming, J., Schroeter, J., Conway, J., and Hwang, D., In silico modeling of asthma, *Adv. Drug Deliv. Rev.*, 55, 829, 2003.

224. Davis, P. B., Drumm, M., and Knostan, M. W., Cystic fibrosis, *Am. J. Respir. Crit. Care Med.*, 154, 1229, 1996.

225. Brown, J. S., Regional ventilation and particle deposition in the healthy and obstructed lung, UMI Dissertation Services, Ann Arbor, MI, 2000.

226. Kim, C. S. and Kang, T. C., Comparative measurement of lung deposition of inhaled fine particles in normal subjects and patients with obstructive airway disease, *Am. J. Respir. Crit. Care Med.*, 155, 3, 1997.

227. Anderson, P. J., Blanchard, J. D., Brain, J. D., Feldman, H. D., McNamara, J. J., and Heyder, J. Effect of cystic fibrosis on inhaled aerosol boluses. *Am. Rev. Respir. Dis.*, 140, 1317, 1989.

228. Anderson, P. J., Wilson, J. D., and Hiller, F. C., Respiratory tract deposition of ultrafine particles in subjects with obstructive or restrictive lung disease, *Chest*, 97, 115, 1990.

229. Love, R. G. and Muir, D. C. F., Aerosol deposition and airway obstruction, *Am. Rev. Respir. Dis.*, 114, 891, 1976.

230. Siekmeier, R., Schiller-Scotland, C. H. F., Gebhart, J., and Kronenberger, H., Pharmacon-induced airway obstruction in healthy subjects: Dose dependent changes of inspired aerosol boluses, *J. Aerosol Sci.*, 21, S423, 1990.

231. Anderson, P. J., Gann, L. P., Walls, R. C., Tennal, K. B., and Hiller, F. C., Utility of aerosol bolus behavior as a diagnostic index of asthma during bronchoprovocation, *Am. J. Respir. Crit. Care Med.*, 149, A1047, 1994.

232. Brown, J. S., Zeman, K. L., and Bennett, W. D., Regional deposition of coarse particles and ventilation distribution in healthy subjects and patients with cystic fibrosis, *J. Aerosol Med.*, 14, 443, 2001.

233. Ramana, L., Tashkin, D. P., Taplin, G. V., Elam, D., Detels, R., Coulson, A., and Rokaw, S. N., Lung imaging in chronic obstructive pulmonary disease, *Chest*, 68, 634, 1975.

234. Taplin, G. V., Tashkin, D. P., Chopri, S. K., Anselmi, O. E., Elam, D., Calvarese, B., Coulson, A., Detels, R., and Rokaw, S. N., Early detection of chronic obstructive pulmonary disease using radionuclide lung imaging procedures, *Chest*, 71, 567, 1977.

235. Ito, H., Ishii, Y., Maeda, H., Todo, G., Torizuka, K., and Smaldone, G. C., Clinical observations of aerosol deposition in patients with airway obstruction, *Chest*, 80, 837, 1981.

236. Segal, R. A., Martonen, T. B., Kim, C. S., and Shearer, M., Computer simulations of particle deposition in the lungs of chronic obstructive pulmonary disease patients, *Inhal. Toxicol.*, 14, 705, 2002.

237. Martonen, T. B., Katz, I., Hwang, D., and Yang, Y. Biomedical application of the supercomputer: Targeted delivery of inhaled pharmaceuticals in diseased lungs, *Computer Simulations in Biomedicine*, Power, H., Hart, R. T., Eds., Computational Mechanics Publications, Boston, MA, 1995.

238. Martonen, T. B., Katz, I., and Cress, W., Aerosol deposition as a function of airway disease: Cystic fibrosis, *Pharm. Res.*, 12, 96, 1995.

239. Martonen, T. B., Hwang, D., Katz, I., and Yang, Y., Cystic fibrosis: Treatment with a supercomputer drug delivery model, *Adv. Eng. Software*, 28, 359, 1997.

240. Environmental Protection Agency, 40 CFR Part 50, National ambient air quality standards for particulate matter (AD-FRL-5725-2, RIN 2060-AE66), *Federal Register*, 62, 38651, 1997.

241. Ménache, M. G. and Graham, R. C., Conducting airway geometry as a function of age, *Ann. Occup. Hyg.*, 41, 531, 1997.

242. Hofmann, W., Martonen, T. B., and Graham, R. C., Predicted deposition of nonhygroscopic aerosols in the human lung as a function of subject age, *J. Aerosol Med.*, 2, 49, 1989.

243. Martonen, T. B., Graham, R. C., and Hofmann, W., Human subject age and activity level: Factors addressed in biomathematical deposition program for extrapolation modeling, *Health Phys.*, 57, 49, 1989.

244. Warren, D. W., Harifield, W. M., and Dalston, E. T., Effect of age on nasal cross-sectional area and respiratory mode in children, *Laryngoscope*, 100, 89, 1990.

245. Bennett, W. D., Zeman, K. L., Kang, C. W., and Schechter, M. S., Extrathoracic deposition of inhaled, coarse particles in children vs. adults, *Ann. Occup. Hyg.*, 41, 497, 1997.

246. Isaacs, K. K. and Martonen, T. B., Particle deposition in children's lungs: Theory and experiment, *J. Aerosol Med.*, 18, 337, 2005.

247. Harvey, R. P. and Hamby, D. M., Age-specific uncertainty in particulate deposition for 1 micron AMAD particles using ICRP 66 lung model, *Health Phys.*, 82, 807, 2002.

248. Musante C. J. and Martonen, T. B., Computer simulations of particle deposition in the developing human lung, *J. Air Waste Manage. Assoc.*, 50, 1426, 2000.

249. Sweeney, T. D. and Brain, J. D., Pulmonary deposition: Determinants and measurement techniques, *Toxicol. Pathol.*, 19, 384, 1991.

250. Kim, C. S., Methods of calculating lung delivery and deposition of aerosol particles, *Respir. Care*, 45, 695, 2000.

251. Schlesinger, R. B. and Lippmann, M., Particle deposition in casts of the human upper tracheobronchial tree, *Am. Ind. Hyg. Assoc. J.*, 33, 237, 1972.

252. Cohen, B. S., Particle deposition in human and canine tracheobronchial casts: A determinant of radon dose to the critical cells of the respiratory tract, *Health Phys.*, 70, 695, 1996.

253. Phalen, R. F., Oldham, M. J., Beaucage, C. B., Crocker, T. T., and Mortensen, J. D., Postnatal enlargement of human tracheobronchial airways and implications for particle deposition, *Anatom. Rec.*, 212, 368, 1986.

254. Gerde, P., Cheng, Y. S., and Medinsky, M. A., In vivo deposition of ultrafine aerosols in the nasal airway of the rat, *Fund. Appl. Toxicol.*, 16, 330, 1991.

255. Kelly, J. T., Kimbell, J. S., and Asgharian, B., Deposition of fine and coarse aerosols in a rat nasal mold, *Inhal. Toxicol.*, 13, 577, 2001.

256. Martonen, T. B., Measurements of particle dose distribution in a model of a human larynx and tracheo-bronchial tree, *J. Aerosol Sci.*, 14, 11, 1983.

257. Martonen, T. B. and Lowe, J., Assessment of aerosol deposition patterns in human respiratory tract casts, *Aerosols in the Mining and Industrial Work Environments*, Marple, V. A. and Liu, B. Y. H., Eds., Ann Arbor Science, Ann Arbor, MI, 1983.

258. Schlesinger, R. B., Bohning, D. B., Chan, T. L., and Lippmann, M., Particle deposition in a hollow cast of the human tracheobronchial tree, *J. Aerosol Sci.*, 8, 429, 1977.

259. Minocchieri, S., Burren, J. M., Bachmann, M. A., Stern, G., Wildhaber, J., Buob, S., Schindel, R., Kraemer, R., Frey, U. P., and Nelle, M., Development of the premature infant nose throat-model (PrINT-Model)—An upper airway replica of a premature neonate for the study of aerosol delivery, *Pediatr. Res.*, 64, 141, 2008.

260. Janssens, H. M., de Jongste, J. C., Fokkens, W. J., Robben, S. G., Wouters, K., and Tiddens, H. A., The Sophia Anatomical Infant Nose-Throat (Saint) model: A valuable tool to study aerosol deposition in infants, *J. Aerosol Med.*, 14, 433, 2001.

261. Giesel, F. L., Mehndiratta, A., von Tengg-Kobligk, H., Schaeffer, A., Teh, K., Hoffman, E. A., Kauczor, H. U., van Beek, E. J., and Wild, J. M., Rapid prototyping raw models on the basis of high resolution computed tomography lung data for respiratory flow dynamics, *Acad. Radiol.*, 16, 495, 2009.

262. Davies, C. N., Heyder, J., and Subba Ramu, M. C., Breathing of half-micron aerosols I. Experimental, *J. Appl. Physiol.*, 32, 591, 1972.

263. Heyder, J., Gebhardt, J., Heiger, G., Roth, C., and Stahlhofen, W., Experimental studies of the total deposition of aerosol particles in the human respiratory tract, *J. Aerosol Sci.*, 4, 191, 1973.

264. Rosati, J. A., Brown, J. S., Peters, T. M., Leith, D., and Kim, C. S., A polydisperse aerosol inhalation system designed for human studies, *J. Aerosol Sci.*, 33, 1433, 2002.

265. Rosati, J. A., Leith, D., and Kim, C. S., Monodisperse and polydisperse aerosol deposition in a packed bed, *Aerosol Sci. Technol.*, 37, 528, 2000.

266. Laube, B. L., In vivo measurements of aerosol dose and distribution: Clinical relevance, *J. Aerosol Med.*, 9, S77, 1996.

267. Martonen, T. B., Yang, Y., and Dolovich, M., Definition of airway composition within gamma camera images, *J. Thorac. Imaging*, 9, 188, 1994.

268. Martonen, T. B., Yang, Y., Dolovich, M., and Guan, X., Computer simulations of lung morphologies within planar gamma camera images, *Nucl. Med. Commun.*, 18, 861, 1997.

269. Tossici-Bolt, L., Fleming, J. S., Conway, J. H., and Martonen, T. B., An analytical technique to recover the third dimension in planar imaging of inhaled aerosols—2 estimation of the deposition per airway generation, *J. Aerosol Med.*, 20, 127, 2007.

270. Tossici-Bolt, L., Fleming, J. S., Conway, J. H., and Martonen, T. B., Analytical technique to recover the third dimension in planar imaging of inhaled aerosols: (1) impact on spatial quantification, *J. Aerosol Med.*, 19, 565, 2006.

271. Fleming, J. S., Halson, P., Conway, J., Moore, E., Nassim, M. A., Hashish, A. H., Bailey, A. G., and Holgate, S. T., Three-dimensional description of pulmonary deposition of inhaled aerosol using data from multimodality imaging, *J. Nucl. Med.*, 37, 873, 1996.

272. Fleming, J. S., Sauret, V., Conway, J. H., Holgate, S. T., Bailey, A. G., and Martonen, T. B., Evaluation of the accuracy and precision of lung aerosol deposition measurements from single-photon emission computed tomography using simulation, *J. Aerosol Med.*, 13, 187, 2000.

273. Finlay, W. H., Stapleton, K. W., Chan, H. K., Zuberbuhler, P., and Gonda, I., Regional deposition of inhaled hygroscopic aerosols: In vivo SPECT compared with mathematical modeling, *J. Appl. Physiol.*, 81, 374, 1996.

274. Lee, Z., Berridge, M. S., Finlay, W. H., and Heald, D. L., Mapping PET-measured triamcinolone acetonide (TAA) aerosol distribution into deposition by airway generation, *Int. J. Pharm.*, 199, 7, 2000.

275. Dolovich, M. B., Influence of inspiratory flow rate, particle size, and airway caliber on aerosolized drug delivery to the lung, *Respir. Care*, 45, 597, 2000.

276. Dolovich, M. B., Measuring total and regional lung deposition using inhaled radiotracers, *J. Aerosol Med.*, 14, S53, 2001.

277. Fleming, J. S., Nassim, M., Hashish, A. H., Bailey, A. G., Conway, J., Holgate, S., Halson, P., and Moore, E., Description of pulmonary deposition of radiolabeled aerosol by airway generation using a conceptual three dimensional model of lung morphology, *J. Aerosol Med.*, 8, 341, 1995.

278. Fleming, J. S., Hashish, A. H., Conway, J., Hartley-Davies, R., Nassim, M. A., Guy, M. J., Coupe, J., and Holgate, S. T., A technique for simulating radionuclide images from the aerosol deposition pattern in the airway tree, *J. Aerosol Med.*, 10, 199, 1997.

279. Martonen, T. B., Hwang, D., Guan, X., and Fleming, J. S., Supercomputer description of human lung morphology for imaging analysis, *J. Nucl. Med.*, 39, 745, 1998.

280. Schroeter, J. D., Fleming, J. S., Hwang, D., and Martonen, T. B., A computer model of lung morphology to analyze SPECT images, *Comput. Med. Imaging Graph.*, 26, 237, 2002.

281. Martonen, T. B., Deposition patterns of cigarette smoke in human airways, *Am. Ind. Hyg. Assoc. J.*, 53, 6, 1992.

282. Hofmann, W. and Koblinger, L., Monte Carlo modeling of aerosol deposition in human lungs: Part III: Comparison with experimental data, *J. Aerosol Sci.*, 23, 51, 1992.

283. Heyder, J., Gebhardt, J., Rudolf, G., Schiller, C. F., and Stahlhofen, W., Deposition of particles in the human respiratory tract in the size range 0.005–15 µm, *J. Aerosol Sci.*, 17, 811, 1986.

284. Segal, R. A., Martonen, T. B., and Kim, C. S., Comparison of computer simulations of total lung deposition to human subject data in healthy test subjects, *J. Air Waste Manage. Assoc.*, 50, 1262, 2000.

285. Swift, D. L. and Proctor, D. F., Access of air to the respiratory tract, *Respiratory Defense Mechanisms*, Brain, J. D., Proctor, D. F., and Reid, L., Eds., Marcel Dekker, New York, 1977.

286. Heyder, J. and Rudolf, G., Deposition of aerosol particles in the human nose, *Inhaled Part.*, 4, 107, 1975.

287. Foord, N., Black, A., and Walsh, M., Regional deposition of 2.5–7.5 µm diameter inhaled particles in the healthy male non-smoker, *J. Aerosol Sci.*, 9, 343, 1978.

288. Cheng, K. H., Cheng, Y. S., Yeh, H. C., and Swift, D. L., Measurements of airway dimensions and calculation of mass transfer characteristics of the human oral passage, *J. Biomech. Eng.*, 119, 476, 1997.

289. Swift, D. L. and Proctor, D. F., A dosimetric model for particles in the respiratory tract above the trachea, *Ann. Occup. Hyg.*, 32, 1035, 1982.

290. Martonen, T. B. and Zhang, Z., Comments on recent data for particle deposition in human nasal passages, *J. Aerosol Sci.*, 23, 667, 1992.

291. Kelly, J. T., Asgharian, B., Kimbell, J. S., and Wong, B. A., Particle deposition in human nasal airway replicas manufactured by different methods. Part I: Inertial regime particles, *Aerosol Sci. Technol.*, 38, 1063, 2004.

292. Kelly, J. T., Asgharian, B., Kimbell, J. S., and Wong, B. A., Particle deposition in human nasal airway replicas manufactured by different methods. Part II: Ultrafine particles, *Aerosol Sci. Technol.*, 38, 1072, 2004.

293. Cheng, Y. S., Zhou, Y., and Chen, B. T., Particle deposition in a cast of human oral airways, *Aerosol Sci. Technol.*, 31, 286, 1999.

294. Kim, C. S. and Fisher, D. M., Deposition characteristics of aerosol particles in sequentially bifurcating airway models, *Aerosol Sci. Technol.*, 31, 198, 1999.

295. Farkas, A. and Balashazy, I., Quantification of particle deposition in asymmetrical tracheobronchial model geometry, *Comput. Biol. Med.*, 38, 508, 2008.

296. Schlesinger, R. B., Gurman, J. L., and Lippmann, M., Particle deposition within bronchial airways— Comparisons using constant and cyclic inspiratory flows, *Ann. Occup. Hyg.*, 26, 47, 1982.

297. U.S. EPA, Total Risk Integrated Methodology (TRIM) Air Pollutants Exposure Model Documentation (TRIM.Expo/APEX, Version 4.3). Vol. 1: Users Guide. Report no. EPA-452/B-08-001a. Office of Air Quality Planning and Standards, Research Triangle Park, NC. Available at: http://www.epa.gov/ttn/fera/human_apex.html, 2008a.

298. U.S. EPA, Total Risk Integrated Methodology (TRIM) Air Pollutants Exposure Model Documentation (TRIM.Expo/APEX, Version 4.3). Vol. 2: Technical Support Document. Report no. EPA-452/B-08-001b. Office of Air Quality Planning and Standards, Research Triangle Park, NC. Available at: http://www.epa.gov/ttn/fera/human_apex.html, 2008b.

299. Burke, J. M., Zufall, M. J., and Ozkaynak, H. A population exposure model for particulate matter: Case study results for PM2.5 in Philadelphia, PA, *J. Expo. Anal. Environ. Epidemol.*, 11, 470, 2001.

300. U.S. EPA, Exposure model for individuals, 2011, Available at: http://www.epa.gov/heasd/products/emi/emi.html

301. National Institute for Public Health and the Environment (RIVM). Multiple path particle dosimetry model (MPPD v 1.0): A model for human and rat airway particle dosimetry. Bilthoven, the Netherlands. RIVA Report 650010030. Model available online at http://www.ara.com/products/mppd.htm., 2002.

302. Bondesson, E., Bengtsson, T., Borgstrom, L., Nilsson, L. E., Norrgren, K., Olsson, B., Svensson, M., and Wollmer, P., Dose delivery late in the breath can increase dry powder aerosol penetration into the lungs, *J. Aerosol Med.*, 18, 23, 2005.

303. Kleinstreuer, C., Shi, H., and Zhang, Z., Computational analyses of a pressurized metered dose inhaler and a new drug-aerosol targeting methodology, *J. Aerosol. Med.*, 20, 294, 2007.

304. Zhang, Y., Gilbertson, K., and Finlay, W. H., In vivo-in vitro comparison of deposition in three mouth-throat models with Qvar and Turbuhaler inhalers, *J. Aerosol. Med.*, 20, 227, 2007.

305. Coates, M. S., Fletcher, D. F., Chan, H. K., and Raper, J. A., Effect of design on the performance of a dry powder inhaler using computational fluid dynamics. Part 1: Grid structure and mouthpiece length, *J. Pharm. Sci.*, 93, 2863, 2004.

306. Coates, M. S., Chan, H. K., Fletcher, D. F., and Chiou, H., Influence of mouthpiece geometry on the aerosol delivery performance of a dry powder inhaler, *Pharm. Res.*, 24, 1450, 2007.

307. Longest, P. W., Hindle, M., Das Choudhuri, S., and Xi, J. X., Comparison of ambient and spray aerosol deposition in a standard induction port and more realistic mouth-throat geometry, *J. Aerosol Sci.*, 39, 572, 2008.

308. Longest, P. and Hindle, M., Evaluation of the Respimat soft mist inhaler using a concurrent CFD and in vitro approach, *J. Aerosol Med. Pulm. Drug Deliv.*, 22, 99, 2009.

309. Kimbell, J. S., Segal, R. A., Asgharian, B., Wong, B. A., Schroeter, J. D., Southall, J. P., Dickens, C. J., Brace, G., and Miller, F. J., Characterization of deposition from nasal spray devices using a computational fluid dynamics model of the human nasal passages, *J. Aerosol Med.*, 20, 59, 2007.

310. Longest, P. W. and Hindle, M., CFD simulations of enhanced condensational growth (ECG) applied to respiratory drug delivery with comparisons to in vitro data, *J. Aerosol Sci.*, 41, 805, 2010.

311. Hindle, M. and Longest, P. W., Evaluation of enhanced condensational growth (ECG) for controlled respiratory drug delivery in a mouth-throat and upper tracheobronchial model, *Pharm. Res.*, 27, 1800, 2010.

6 Aerosol Chemistry and Physics
An Indoor Perspective

Lara A. Gundel and Hugo Destaillats

CONTENTS

> When Sherlock Holmes mystified his friend Dr. Watson by his amazing deductions, he was utilizing to the fullest degree the data available to him – the muddy boot, the ash of a cigar, the torn ticket. This is a practice which is also basic in the highest forms of scientific research.
>
> **Wilson (1952)**

Research in indoor aerosol physics and chemistry requires the type of careful observations and astute deductions that the fictional detective used so well. Holmes' perspective assists in moving past preconceptions based on outdoor chemistry and presumptions that indoor environments provide sufficient refuge from the assaults of air pollution.

6.1   COMPARISON OF INDOOR AND OUTDOOR AEROSOLS

6.1.1   Importance of Aerosol Exposures

Epidemiologists have established associations between exposures to ambient (outdoor) particles and both human morbidity (disease rate) and mortality (death rate). Health effects of PM include both chronic and acute forms of both respiratory and cardiovascular disease, among others that influence human life spans (Schwartz et al., 2008; Pope et al., 2009). Many of the predicted impacts of climate change can increase exposures to both indoor and outdoor particles (Institute of Medicine, 2011).

6.1.2   Significance of the Indoor Environment to Aerosol Exposure

6.1.2.1   Most Exposure and Inhalation of Outdoor PM Occurs Indoors

People spend 80%–90% of their time indoors, so most human exposure to particles of outdoor origin takes place indoors. Nazaroff et al. (2003) compared mass flow rates in urban and indoor

TABLE 6.1

Mass Flow Rates in Global, Urban, and Indoor Atmospheres

Environment	Mass (kg)	Flow, F (kg day^{-1})	Mass Breathed, Q^a (kg day^{-1})	Intake Fraction^b Ratio Q:F
Global atmosphere	5×10^{18}	—	$\sim 10^{11}$	—
Urban atmosphere^c	$\sim 10^{15}$	$\sim 3 \times 10^{15}$	$\sim 4 \times 10^{10}$	$\sim 10^{-5}$
Indoor atmosphere^d	$\sim 10^{12}$	$\sim 10^{13}$	$\sim 8 \times 10^{10}$	$\sim 10^{-2}$

Source: Adapted from Nazaroff, W.W. et al., *Atmos. Environ.*, 37, 5431, 2003.

[a] Includes air inside and outside of buildings.

[b] The descriptor *intake fraction* was introduced by Bennett et al. (2002).

[c] Sum of all urban environments (globally).

[d] Sum of all indoor environments (globally).

atmospheres to the human breathing rate, as shown in Table 6.1. The last column expresses the ratio of amounts inhaled and amounts emitted as the *intake fraction* (Bennett et al., 2002). This simple calculation shows that a nonreactive compound present indoors is about a thousand times more likely to be inhaled than if the same species is emitted or formed outdoors. Calculations of intake fractions for PM are the first steps in incorporating human exposure to PM into life-cycle assessment of damage to human health. The comprehensive review by Humbert et al. (2011) explains how the modeling approach incorporates types and source strengths of PM, geography, meteorology, time, and activity patterns. Hellweg et al. (2009) provide guidance about generating intake fractions for both gases and PM in the indoor environments. These studies are consistent with one another in concluding that exposures to outdoor PM are thousands of times higher indoors than outdoors for the majority of people in the developed world.

6.1.2.2 In the Developed World Roughly Half of Indoor PM Originates Outdoors

In the early 1990s, the Particle Total Exposure Assessment Methodology (PTEAM) study compared personal exposure to particles smaller than 10 μm in diameter (PM$_{10}$) in Riverside, California, to indoor and outdoor concentrations. By using the data to model PM infiltration and removal processes Ozkaynak et al. (1996) found that over half the indoor particle mass originated outdoors, even in homes with smoking and cooking. As buildings in the developed world have become tighter (less leaky toward outdoor air) the outdoor contributions to indoor PM are decreasing somewhat (Mitchell et al., 2007).

6.1.2.3 Personal Cloud Effect Enhances Exposure to PM

The PTEAM study found that personal exposures were higher than exposures based on indoor or outdoor concentrations (Ozkaynak et al., 1996), and this excess exposure is now called the "personal cloud." When Williams et al. (2003a,b) reported PM$_{2.5}$ (<2.5 μm in diameter) data from a longitudinal study, they confirmed the personal cloud effect and found that mean personal PM$_{2.5}$ exposures were only moderately correlated with ambient PM$_{2.5}$ concentrations. Liu et al. (2003) also found that personal exposure exceeded indoor exposure for susceptible populations in Seattle. These studies raise concerns about the representativeness of central monitoring site data for human exposure assessment.

6.1.2.4 Indoor PM is among the Acute Health Hazards in Homes in the United States

The detailed hazard assessment of contaminants in indoor air by Logue et al. (2010) led to inclusion of PM$_{2.5}$ among the group of acute indoor health hazards, in addition to acrolein, chloroform, formaldehyde, CO, and NO$_2$. Additionally, there is growing recognition that chemical and physical changes in indoor air can contribute to increased exposure to irritants and inflammatory agents (Mitchell et al., 2007). For example, the aerosol-producing reactions of low levels of infiltrated ozone with terpenes (from cleaning products, air fresheners, and fragrances) are associated with increased sensory irritation (Wolkoff et al., 2006). Another example, a review of five studies by Mølhave (2008), concluded that indoor dust can lead to inflammatory and allergic responses in office workers at levels that occur episodically.

6.1.2.5 Introduction to Characterization of Indoor Aerosols

With growing attention to the health effects of inhaled particles, more and more attention has been directed at characterizing indoor aerosols. The growing use of continuous monitors for both fine and ultrafine particles (UFP) has improved the understanding of infiltration of ambient PM and the dynamic behavior of indoor sources of PM (Wallace, 2006; Wallace et al., 2006). Many of the studies cited in the following would not have been possible without the sampling methods and types of instrumentation that have been validated and used indoors by Wallace et al. (2011), among others.

6.1.2.5.1 *PM$_{2.5}$*

The U.S. EPA has set National Ambient Air Quality Standards (NAAQS) for ambient PM$_{2.5}$ at 15 and 35 μg m^{-3} and for annual and daily averages, respectively. In long-term field studies in residences, Wallace, 2006 and Wallace et al. (2006) showed that typical indoor activities like cooking, cleaning, and personal care frequently raised PM$_{2.5}$ concentrations above the NAAQS daily standard. The review of indoor pollutants by Logue et al. (2011) found that the mean indoor PM$_{2.5}$ concentration was 16 μg m^{-3} from 13 studies of homes in the United States and other similar industrialized countries. The 25th, 75th, and 90th percentiles were 9, 20, and 86 μg m^{-3}, respectively. Average PM$_{2.5}$ levels were much lower than outdoors in buildings such as offices and schools that filtered ambient air through mechanical ventilation systems, except when human activities generated both fine and coarse PM (Parker et al., 2008).

6.1.2.5.2 *Ultrafine Particles*

Particles below 100 nm in diameter are typically referred to as UFP, and exposure to them has been increasingly linked to damage to human health (Knol et al., 2009 and included citations). Exposure to outdoor UFP typically occurs when people are near fresh smog or combustion sources like vehicles. After emission, UFP decay faster than PM$_{2.5}$ because they quickly agglomerate with each other, accrete to PM$_{2.5}$ or interact with surfaces. Because of their size they usually do not contribute much to PM$_{2.5}$ mass. With the advent of improved instruments like water-based condensation nucleus counters, researchers have been monitoring number concentrations of UFP down to 6 nm in several types of indoor environments, with emphasis on determining what fractions of indoor UFP originated outdoors and how much indoor UFP is generated by typical human activities, over what timescales. A recent pair of articles by Bhangar et al. (2011) and Mullen et al. (2011) included useful citations to the small but growing literature on the characterization of UFP in both outdoor and indoor environments Their studies focused on UFP in seven California homes and six schools, respectively. The mean exposure concentration to the residents of the seven homes was 17×10^3 cm^{-3} over the time they were at home, with 40% of exposure at home due to infiltrated UFP and the remainder from episodic generation of UFP indoors (Bhangar et al., 2011). Particle number concentrations were about the same in the schools, but most of the indoor UFP came from outdoor sources (Mullen et al., 2011). From wintertime measurements of five homes in northern New York State, McAuley et al. (2010) found that indoor/outdoor ratios averaged 0.3 for UFP when the only indoor aerosol source was infiltrated PM from traffic.

6.1.3 Outdoor Particles

6.1.3.1 Composition PM$_{2.5}$ in the United States

Since roughly one-third to half of PM in homes has infiltrated from outdoors (e.g., Wallace et al., 2003), an overview of outdoor PM composition provides a useful starting point for investigation of indoor aerosol chemistry. Figure 6.1 shows the four-season average PM$_{2.5}$ concentrations for urban

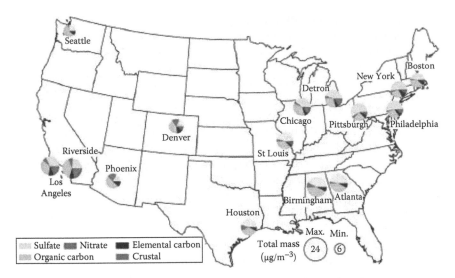

FIGURE 6.1 Average annual $PM_{2.5}$ concentrations for urban areas of the United States in 2008. The diameter of each pie icon scales with its PM concentration. Major PM constituents start with nitrate at the 3 o'clock position and continue clockwise in the order nitrate, elemental carbon, organic carbon, crustal species, and sulfate. (From U.S. Environmental Protection Agency, Our Nation's air, status and trends through 2008, Office of Air Quality Planning and Standards, Research Triangle Park, NC, EPA Publication No. EPA 454/R-09-002, www.epa.gov/airtrends/2010/report/fullreport, 2010.)

areas in the United States for 2008, as reported by the Speciation Trends Network (U.S. EPA, 2010, p. 24). The pie icons show the major constituents of $PM_{2.5}$ that are monitored every third day throughout the United States under the guidance of the EPA: the ions nitrate and sulfate, crustal elements (aluminum, silicon, calcium, iron, and titanium), and carbonaceous components as elemental carbon and organic carbon (Rao et al., 2003).

Monitoring over the last decade shows that urban areas experience higher PM concentrations than rural areas, but the difference is smaller in the eastern United States. East of Denver, Colorado, sulfate accounts for more of the PM mass than the other constituents, followed by organic carbon, nitrate, and elemental carbon. Except in the Los Angeles area of California, organic carbon was the most abundant constituent in the west, followed by sulfate. Figure 6.1 does not show particulate ammonium, but earlier EPA reports indicate that NH_4^+ is ubiquitous and originates from biogenic ammonia that reacts with SO_2 and NO_x (U.S. EPA, 2003). Nationwide differences in ionic composition can be traced to regional differences in concentrations of gas-phase precursors SO_2 (from stationary sources like power plants) and NO_x (primarily from vehicles, western United States).

6.1.3.2 $PM_{2.5}$ Source Apportionment

The first steps in understanding and controlling concentrations of $PM_{2.5}$ require identifying the most important contributors to fine airborne particles and finding out how much each adds to the atmosphere over time and space. For outdoor air, the process depends on knowledge of the composition and behavior of particles produced by each source while they are generated, then dilute with ambient air, and move through the atmosphere. All the while, the source particles are interacting with other pollutants under varying meteorological conditions, with other local pollutants adding to regional air masses.

Vehicle exhaust is ubiquitous and usually among the top three contributors to $PM_{2.5}$. Biomass combustion from residential wood burning, agricultural controlled burns, and forest fires varies by season and region. Emissions from stationary sources like power plants and coke factories, tire wear, plant detritus, spores, halogenated organic compounds from pesticides, PCBs and dioxin-like species also contribute to airborne PM. Secondary organic aerosol (SOA) contributions typically follow seasonal patterns of ozone concentrations. SOA forms when ozone or OH reacts with hydrocarbon gases from combustion and emissions from vegetation, leading to high-molecular-weight, oxygenated compounds that condense on available particles.

Throughout the United States source apportionment is achieved by applying computer models to ambient $PM_{2.5}$ and meteorological data collected at receptor sites in order to quantify the contributions of the dominant sources. The studies published by Srivastava et al. (2007) and Lane et al. (2007) provide useful introductions to source apportionment by showing how the predictions of two very different approaches compare when applied to the same data sets. Sarnat et al. (2008) found that associations of cardiovascular and respiratory morbidity in Atlanta, Georgia with sources of $PM_{2.5}$ agreed for three different approaches to source apportionment. Using 2 years of speciated PM data and data from emergency room visits they found that $PM_{2.5}$ from mobile and biomass sources was associated with both cardiovascular and respiratory visits, but secondary particles were associated only with respiratory visits.

Computer models based on multivariate factor analysis such as positive matrix factorization (PMF) search for patterns in concentrations measured at receptor sites before associating them with known PM sources (often guided by emission inventories such as cited in Lane et al., 2007). An example of this approach applied to Seattle, Washington, showed that wood smoke was the heaviest contributor to $PM_{2.5}$ in 2003–2004 (Wu et al., 2007).

The chemical mass balance (CMB) approach hunts for the vestiges of known molecular markers of individual sources in ambient PM and then reconstructs contributions of each source at receptor sites. CMB models are useful because a large fraction of ambient fine carbonaceous mass has not been traced to individual compounds. For the carbonaceous constituents of $PM_{2.5}$, CMB models in the United States start with detailed emission factors for molecular markers of individual sources based on measurements made in California by Rogge et al. (1991, 1997a,b) and Schauer et al. (1999a,b, 2001, 2002b,c). Soon after these detailed source characterizations became available, Schauer et al. (2002a) constructed and applied a CMB model by Zheng et al. (2002) to a smog episode in southern California to derive the contributions of 11 unique sources from measured concentrations of volatile, semi-volatile, and particulate organic pollutants. They found that vehicle exhaust and SOAs were the most important sources of particles, among others in that episode. CMB models continue to evolve as more molecular markers become available for source apportionment in other regions of the United States with different mixes of local pollutants; for example, new models address agricultural burning in eastern Washington (Jimenez et al., 2007), soil-derived markers of crops grown in the Central Valley, California (Rogge et al., 2007), variability in wood smoke emissions and other emissions from biomass (Robinson et al., 2006a,b), and pollutants from cooking that often contribute to ambient $PM_{2.5}$ (Robinson et al., 2006c). Updating CMB models with new markers of SOAs derived from various gas-phase precursors (Kleindienst et al., 2007) has shown that SOA in southern California differs from SOA in the midwest and eastern regions of the United States (Stone et al., 2009). Researchers have also recognized that CMB models should incorporate oxidation of molecular markers between emission and measurement at receptor sites (Robinson et al., 2006a; Roy et al., 2011).

6.1.4 DIFFERENCES BETWEEN INDOOR AND OUTDOOR ENVIRONMENTS

From a physical chemist's perspective, a building is a leaky reaction vessel with active surfaces. Table 6.2 compares some key physical parameters outdoors and in a typical home in the United States.

Whereas most outdoor atmospheric processes are not constrained by partitions or macroscopic surfaces, the indoor environment is defined by walls and covered with building materials and furnishings. Many of these substances can contribute to the air quality by adsorbing or emitting compounds that are chemically active. Indoor activities also generate gases and particles, leading

TABLE 6.2
Key Physical Parameters for Outdoor and Indoor Air in Homes

Parameter	Urban Atmosphere	Indoor Atmosphere
Residence time	~10 h	~1 h
Light-energy flux	~1000 W m^{-2} (daytime)	~1 W m^{-2}
Surface-volume ratio	~0.01 m^2 m^{-3}	~3 m^2 m^{-3}
Precipitation	~10–150 cm year^{-1}	Absent

Source: From Nazaroff, W.W. et al., *Atmos. Environ.*, 37, 5431, 2003.

to substantially higher indoor concentrations of many volatile organic and semi-volatile (labile) compounds than outdoors. In naturally ventilated buildings, communication between the indoor and outdoor environments takes place via air exchange through doors, windows, and cracks in the building envelope. The residence time of a particle in a room can be shorter or longer than the same particle in an air pocket of the same volume outdoors. The residence times in Table 6.2 are typical for the nighttime urban atmosphere (~10 h) and indoor atmosphere (~1 h, for residential air exchange rates of 0.5–1 h^{-1}). Most large office buildings and many new homes in the United States have mechanical ventilation with recirculation of substantial fractions of the air. Less communication with ambient air than in the past leads to lower infiltration fractions (outdoor/indoor) for PM. Most newer (less leaky) homes in the United States operate at reduced air exchange rates that allow more time for buildup of particles generated indoors since residence times are several hours. Because of much lower levels of ultraviolet light, photochemistry does not play an important role indoors. Thus, dark reactions, especially heterogeneous processes, predominate indoors. Much less seasonal variation in temperature and relative humidity occurs indoors.

6.1.5 EVIDENCE FOR INDOOR AEROSOL PROCESSING AND GENERATION

Figure 6.2 shows important processes involving indoor particles and gases in residences. (Similar processes occur in commercial buildings where a large fraction of the air that the occupants breathe is recirculated Waring and Siegel, 2008.) Airborne particles are shown entering and leaving a building (by infiltration and exfiltration, respectively). Soil particles are tracked indoors as hitchhikers on shoes and clothing. They deposit on floors and carpets, where they can be crushed to smaller size by foot traffic. Compounds can undergo phase change indoors, evaporating from or condensing on particles, depending on vapor pressure and temperature. Many indoor materials act as both sources and sinks for volatile and semi-volatile pollutants, and in typical homes and offices, the building

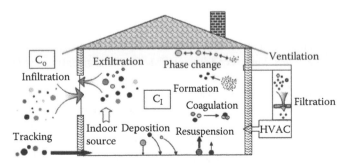

FIGURE 6.2 Processes affecting indoor aerosol concentrations. C_o and C_i represent the outdoor and indoor concentrations, respectively. (Adapted from Thatcher, T.L. et al., *Aerosol Sci. Technol.*, 37, 847, 2003.)

materials and furnishings have much higher exposed surface areas than indoor aerosols. Lots of recent research shows that aerosols can also form indoors when indoor ozone reacts with smog precursor compounds like terpenes from cleaning products and fragrances. Figure 6.2 does not show sorption and desorption of semi-volatile organic compounds (SVOCs) to and from indoor surfaces, airborne particles, and settled dust, but these important processes are discussed in Sections 6.2 through 6.4 and by Weschler and Nazaroff (2010).

UFP from indoor combustion (cooking, heating, smoking, etc.) and indoor chemical reactions inevitably coagulate by collision with each other and infiltrated particles of outdoor origin. Particles are removed by deposition to surfaces, and those larger than $0.8\,\mu m$ can also be resuspended by indoor activities of the occupants (walking, vacuuming).

Because of building characteristics, how people spend time indoors also strongly influences the concentrations and characteristics of $PM_{2.5}$ to which they are exposed. For example, Reff et al. (2005) found infrared signatures of meat cooking (amides) on indoor $PM_{2.5}$ in many homes, and the signatures were even higher in the personal cloud. When there are no indoor sources of PM, such as in unoccupied buildings, indoor PM concentrations are typically less than half of outdoor concentrations. Blondeau et al. (2005) found this to be the case in a study of classrooms in France. During school days the levels of PM increased, and the larger the particles, the greater the increase. The data showed that the school day activities of the occupants caused resuspension of particles that had already settled out on the floors and other surfaces, even when there were no apparent indoor sources of PM (Parker et al., 2008). Similar increases in concentrations of coarse particles were reported for mechanically ventilated schools in Utah by Parker et al. (2008).

Although no U.S.-wide indoor PM speciation trends network operates, more than 20 years of research (Weschler, 2011a,b) has uncovered intriguing differences between indoor and outdoor PM, as illustrated in the following. The first example shows how an ordinary unoccupied home conditions infiltrating aerosol; the second points to the influence of indoor materials on attempts to characterize indoor PM, while the third provides evidence for the counterintuitive notion that fine particles are produced indoors by practices that residents often consider protective and health promoting.

6.1.5.1 Building Envelope

Using real-time concentration and ventilation measurements in an unoccupied house in Fresno, California, Lunden et al. (2003a,b) found that indoor particulate sulfate and soot (black carbon) acted as conservative tracers for infiltration of outdoor $PM_{2.5}$, but indoor $PM_{2.5}$ had much less ammonium nitrate than predicted from the penetration factors for elemental carbon. They also found that the indoor ammonia and nitric acid concentrations were usually lower than the outdoors. The observations are consistent with disruption of the NH_4NO_3 gas/particle equilibrium as the indoor surfaces took up nitric acid and ammonia, as predicted by thermodynamics (Seinfeld and Pandis, 2006). The affinity of indoor surfaces for these species led to evaporative dissociation of infiltrated particulate ammonium nitrate.

6.1.5.2 Indoor Materials

Several groups of investigators have found that conventional measurements of concentrations of indoor carbonaceous particles can exceed indoor $PM_{2.5}$ mass concentrations (Landis et al., 2001; Pang et al., 2002). Pang et al. showed that conservation of mass was not actually violated and that the measured indoor particulate carbon concentrations were reduced substantially when denuders adsorbed semi-volatile organic gases upstream of the particle collection medium (quartz fiber filter). Lunden et al. (2008) confirmed that accounting for the indoor sampling artifacts was essential for understanding gas-to-particle and gas-to-surface partitioning indoors.

Meanwhile, indoor building materials and furnishings have been found to emit organic gases with a wide range of volatilities (e.g., Hodgson et al., 2000, 2002). Rudel et al. (2003) reported high indoor concentrations of phthalate esters (from vinyl flooring). Other indoor air measurements include perfluoroalkyl sulfonamides from surface coatings (Shoeib et al., 2004), fire retardants from polyurethane foam (polybrominated diphenyl ethers [PBDEs]: Sjodin et al., 2001; triethylphosphate: Salthammer et al., 2003), and consumer electronics (triethylphosphate: Carlsson et al., 2000). In many indoor environments, the gas-phase concentrations of SVOC can reach supersaturation, and condensation can occur onto walls, windows (Butt et al., 2004), textiles, airborne particles, air filters, and even quartz fiber filter sampling media (Weschler, 2003). The recent review by Weschler (2011) shows how both indoor use and accumulation of SVOCs have increased from the 1950s to the present. Reservoirs of SVOCs accumulate indoors by sorption to porous materials like wallboard and foam. Unlike more volatile compounds that are removed by ventilation, strongly sorbed SVOC can persist for years (Weschler and Nazaroff, 2008). The ubiquitous presence of SVOC in typical indoor environments can account for the sampling artifacts that often interfere with attempts to assay indoor $PM_{2.5}$ for its content of organics such as described earlier. Infants and children ingest dust while they are close to floors and, thus, receive exposure to many toxic hormone-disrupting SVOCs that accumulate on house dust (Bonvallot et al., 2010). Their toxicity ranking of SVOCs on house dust included phthalates, pesticides, polychlorinated hydrocarbons, and flame retardants, among the most hazardous.

6.1.5.3 Indoor Aerosol Generation

Only within the last decade has the importance of indoor secondary aerosol formation been widely recognized (Weschler, 2011). Whereas the contributions of indoor combustion sources to respirable particles have been investigated for a long time, the groundbreaking study of Weschler and Shields (1999) showed that indoor reactions can form an indoor version of smog. Using a pair of matched offices with the same ambient ozone concentrations and ventilation rates, they tracked increased numbers of fine particles in the room into which D-limonene had been released. Limonene is a biogenic terpene, a cyclic alkene found in lemon and pine oils that are common components of cleaning products. Concerns about the irritancy and other potential health impacts of the gas and aerosol products led to an explosion of work in this area as well as reformulation of some household products and the emergence of green building criteria to limit concentrations of terpenoid precursors to SOA formation.

The stage is now set for deeper exploration of indoor aerosol chemistry and physics. The three aforementioned examples illustrate important areas of current research in indoor aerosol chemistry, and building science. After reviewing common sources of indoor PM, aerosol precursors, and the physical and chemical processes that determine the fate of indoor particles, a discussion of fate and transport begins to show how the three examples are being incorporated into indoor aerosol modeling approaches.

6.2 COMPOSITION OF INDOOR PARTICLES AND AEROSOL PRECURSORS

Overview The PTEAM study identified outdoor air as the greatest contributor to indoor particle mass concentrations in Riverside, California, in the fall of 1990 (Ozkaynak et al., 1996). At that time about 30% of U.S. adults smoked cigarettes, and tobacco smoke and cooking were the next most significant contributors to $PM_{2.5}$ and PM_{10}. Because the prevalence of smoking indoors has decreased, cooking fumes are now more likely to be the major source of indoor-generated PM. Other important indoor combustion sources include wood burning and use of candles, incense, and unvented kerosene heaters (UKHs). In buildings without indoor combustion sources, human activities generate a large fraction of airborne particles by resuspending settled PM and dust. The use of cleaning products and fragrances can lead to the formation of particles by reaction of infiltrated or indoor-generated ozone with reactive sites on terpenes and related compounds. Indoor aerosol formation is discussed more thoroughly in Section 6.4.

Table 6.3 lists significant sources of indoor particles, along with compositional highlights and references. Table 6.4 identifies compounds or chemical classes that are potential chemical tracers for some of these sources. Biogenic aerosols are not discussed in this chapter.

TABLE 6.3

Characteristics of Indoor Particles: Sources, Composition, and Processes

	$PM_{2.5}$	PM_{10}	Processes/Drivers	References
Outdoor particles	OC, NH_4^+, $SO_4^=$, NO_3^-, EC, trace metals	Crustal components: Fe, Ca, Si	Infiltration, deposition	Ozkaynak et al. (1996), Williams et al. (2003a,b)
Combustion	OC, EC		Oxidation, pyrolysis, evaporation, nucleation, condensation	Daisey and Gundel (1991), Manchester-Neesvig et al. (2003)
Cooking	Organic acids >> aldehydes, ketones; UFP			Rogge et al. (1991), Schauer et al. (1999a, 2002b), Andrejs et al. (2002), To et al. (2007), Evans et al. (2008)
Biomass	"Brown carbon"			Dutta et al. (2011), Gao et al. (2009), Padhi and Padhy (2008), Siddiqui et al. (2009), Ward et al. (2008), Colbeck et al. (2010)
Wood fireplaces	Levoglucosan > methoxyphenols > resin acids			Rogge et al. (1998), McDonald et al. (2000), Schauer et al. (2001)
Tobacco smoking	Organic acids>alkanes>N-heterocycles			Rogge et al. (1994), Gorini et al. (2008), Huss et al. (2009), Liu and Zhu (2010), Kavouras et al. (1998), Rumchev et al. (2008)
Candles	High MW alkanes>aldehydes>alkanoic acids>alkenes, >esters; EC; Pb; UFP		Evaporation, condensation; pyrolysis	Fine et al. (1999), van Alphen (1999), Nriagu and Kim (2000), Wasson et al. (2002), Pagels et al. (2009)
Incense	PAH; aldehydes; UFP			Cheng et al. (1995), Jetter et al. (2002), See et al. (2007), Ji et al. (2010)
Kerosene heaters	$PM_{2.5}$, $SO_4^=$, alkanes>PAH			Traynor et al. (1983, 1986, 1990), Leaderer et al. (1990, 1999), Apple et al. (2010)
Biogenic aerosols				
Pets		Allergens in dander, endotoxins	Entrainment	Macher et al. (1999), Heikkines et al. (2004), Song and Liu (2003), Erwin et al. (2003)
Microbes		Endotoxins	Entrainment	Song and Liu (2003), Erwin et al. (2003)
Mites		Allergens	Entrainment	Loan et al. (2003)

Activity	Emission	Process	References
Human activities			
Pesticide use	Flea powder	Entrainment	He et al. (2004), Glytsos et al. (2010), Ferro et al. (2004)
Air fresheners + O_3	Organic aldehydes and acids	Aerosol formation: nucleation, condensation	Nishioka et al. (1996), Becker et al. (2002), Rudel et al. (2003) Weschler and Shields (1999), Nazaroff and Weschler (2004)
Walking (resuspension)	Fe, SVOC, settled dust Dust: crustal elements, mite wastes; SVOC on PM	Abrasion, resuspension	Abt et al. (2000), Lioy et al. (2002), Ferro et al. (2004), Qian and Ferro (2008)
Cleaning activities	Indoor and outdoor PM Dust	Resuspension	Ferro et al. (2002, 2004)
Cleaning products + O_3	Aldehydes and organic acids	Nucleation, condensation	Weschler and Shields (1997), Singer et al. (2006a,b), Corsi and Morrison (2003), Nazaroff and Weschler (2004)
Portable air cleaners	Ozone source		Hubbard et al. (2005), Waring et al. (2008)
Printers			Destaillats et al. (2008), Maddalena et al. (2008), He et al. (2010), He et al. (2007); Morawska et al. (2009), Wensing et al. (2008), Schripp et al. (2008)
Renovation	Abraded building materials; SVOC on dust	Abrasion, resuspension	Kerr and Thi (2001), Rapp et al. (1997), Gohler et al. (2010), Koponen et al. (2011)

TABLE 6.4

Potential Chemical Tracers for Indoor Sources of PM

PM Source	PM$_{2.5}$	PM$_{10}$	References
Indoor source apportionment			Daly et al. (2009), Pekey et al. (2010), Arhami et al. (2010), Clougherty et al. (2011), Lim et al. (2011)
Physical properties of indoor aerosols			Nazaroff et al. (2003)
Size distributions			Harley et al. (2000), Zuraimi (2007)
Particle dynamics			Zuraimi (2007), Nazaroff and Weschler (2004)
Particle motion			
Surface area			
Outdoor particles	SO$_4^=$, EC (if no indoor combustion sources)	Crustal components: Fe, Ca, Si	Ozkaynak et al. (1996), Williams et al. (2003a,b)
Combustion			
Cooking			
Meat	Cholesterol		Rogge et al. (1991)
Seed oils, meat	Triglycerides		Andrejs et al. (2002)
Plant biomass	Levoglucosan		Simoneit (2002)
Softwood (pine)	Retene, guaiacols, diterpenoids		Rogge et al. (1998), McDonald et al. (2000), Simoneit (2002), Schauer et al. (2001)
Hardwood (oak)	Syringols		Rogge et al. (1998), McDonald et al. (2000), Schauer et al. (2001)
Tobacco smoking	N-heterocycles (nicotine), *iso* and *anteiso*-alkanes, solanesol		Rogge et al. (1994), Daisey (1999)
Candles	Wax esters		Fine et al. (1999), Wasson et al. (2002)
Incense			*Tracers not yet proposed*
Kerosene heaters	HONO, excess SO$_4^=$		Leaderer et al. (1990, 1999)
Biogenic aerosols			Macher et al. (1999), Heikkinen et al. (2004)
Pets		Allergens in dander, endotoxins	Song and Liu (2003), Erwin et al. (2003)
Microbes		Endotoxins	Song and Liu (2003), Erwin et al. (2003)
Mites		Allergens	Loan et al. (2003)
Human activities			
Pesticide use	Chlorinated compounds		Rudel et al. (2003)
Air fresheners+O$_3$	Organic aldehydes and acids		*Tracers not yet proposed*
Walking		Crustal elements, mite wastes	Abt et al. (2000), Ferro et al. (2002)
Cleaning activities	Resuspended PM	Resuspended PM, crustal elements, mite wastes, dander	Abt et al. (2000), Ferro et al. (2002)
Cleaning products+O$_3$	Organic aldehydes and acids		Weschler and Shields (1997)
Renovation		Cellulose, lignin, CaSO$_4^=$	

As more evidence has emerged about the health effects of very-small-sized PM, many research groups around the world have been measuring size distributions of UFPs generated by indoor sources, and the tables include some of their work. These efforts have taken advantage of the widespread availability of portable instrumentation for measuring particle number size distributions. For example, He et al. (2004) reported emission rates for many indoor activities in units of both particle mass and particle number, and Glytsos et al. (2010) showed how quickly nanosized particles from many indoor sources agglomerated after they were generated under the same conditions.

6.2.1 COMBUSTION SOURCES

Indoor combustion sources typically operate less efficiently than vehicle engines or industrial boilers, leading to higher emission factors for unburned fuel and partially oxidized products. Semi-volatile combustion-generated gases cool, condense onto existing particles, or nucleate and coagulate with existing particles. The aerosol community is indebted to the late Glenn Cass for leading comprehensive particle characterization efforts at the California Institute of Technology, and many studies from his laboratory are cited in the following. Starting in the early 1990s Rogge et al. (1991, 1998) began reporting detailed chemical characterization of particulate matter from combustion sources that had been operated under carefully controlled conditions, as described by Hildemann et al. (1991). Schauer et al. (1999a,b, 2002a,b,c) expanded these investigations by characterizing both the gas and particulate phases of many of the same sources. In spite of this large body of work and the contributions of many other investigators, identified components typically account for less than half of the fine particulate mass emitted by each combustion source.

The comprehensive source characterization efforts of the 1990s were initially driven by the need to identify and quantify the contribution of each particle source to urban air pollution. The goal was accurate source apportionment so that appropriate control measures could be implemented. With growing recognition and understanding of the health effects of exposure to $PM_{2.5}$, many of the tools developed for the characterization of ambient PM have now been applied to measure human exposure to fine ($PM_{2.5}$) and smaller (ultrafine) particles from indoor combustion sources.

6.2.1.1 Cooking

6.2.1.1.1 Cooking with Oils and Meat

Schauer et al. (2002b) found that frying vegetables in seed oils led primarily to the formation of fine particulate alkenoic and alkanoic acids, whereas the semi-volatile gases contained mostly saturated and unsaturated aldehydes. Emission factors for $PM_{2.5}$ ranged from 13 to 30 mg kg^{-1} of stir-fried vegetables. Andrejs et al. (2002) found that particles larger than about 4 μm were composed primarily of triglyceride components of the original oil. Hildemann et al. (1991) found that charbroiling meat yielded more fine PM (7–40 g kg^{-1} meat) than Schauer et al. (2002b) found from cooking with oils. Rogge et al. (1991) and Schauer et al. (1999a) reported more classes of organic compounds from meat cooking than from the use of seed oils (Schauer et al. 2002b). Of those, saturated and unsaturated fatty acids were the most abundant, as shown in Figure 6.3. However, only about 10% of the particulate organic carbon from meat cooking could be traced to individual compounds.

Cooking with oil produces large numbers (billions to trillions per cooking event) of UFPs (He et al., 2004; Olson and Burke, 2006; Wallace, 2006; Wallace et al., 2006; Evans et al., 2008; Glytsos et al., 2010). In the introduction to their results on particle dose from frying food, Evans et al. (2008) cited studies that point to fumes from cooking hot oil as an important underlying cause of lung cancer in women who have never smoked (e.g., Straif et al., 2006). During the last decade, Yang et al. (2007) found that particles from frying contain mutagenic dienals (aldehydes with two double bonds), among other compounds such as the aromatic amines and polycyclic aromatic hydrocarbons that were found by Schauer et al. (2002b) and To et al. (2007). These classes of compounds cause cytotoxicity and DNA damage. Using measured particle number and mass emission rates for frying a variety of foods in a house Evans et al. (2008) estimated that 20 min of

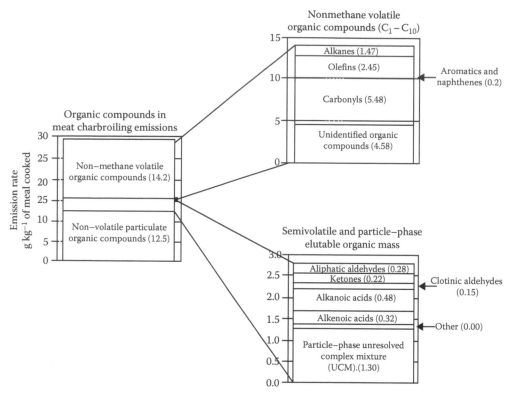

FIGURE 6.3 Chemical composition of emissions from charbroiling meat. (From Schauer, J.J. et al., *Environ. Sci. Technol.*, 33, 1566, 1999a. With permission.)

cooking would lead to at least twice the dose of $PM_{2.5}$ (by mass) as a person would receive during 1 h spent outside in downtown Toronto, Ontario, Canada. A second phase of the study involved monitoring emissions from heating the same amount of cooking oil in five different homes, all of which had exhaust fans to control exposure to PM from cooking before the particles could mix with background air and PM. They found that the average estimated dose was at least an order of magnitude higher than in the first part of their study. The authors attributed the difference to their observation that flow rates for the exhaust fans were up to an order of magnitude lower than the manufacturers' specifications.

6.2.1.1.2 Cooking with Solid Biomass Fuels

As of the end of the first decade of the twenty-first century, solid biomass fuels (SBF) such as wood, charcoal, agricultural waste, and animal dung are the fuel sources for at least a third of the world's population. Widespread exposure to high levels of PM from cooking with SBF has been shown to increase risks of respiratory problems (Padhi and Padhy, 2008; Po et al., 2011), cardiovascular disease (e.g., Dutta et al., 2011), and cancers (e.g., Wu et al., 2004) in Asia and Africa—the developing world (Fullerton et al., 2008).

Besides causing severe health problems and shortening life spans from chronic exposure to smoke from SBF, cooking with SBF in inefficient stoves contributes significantly to global CO_2 and particulate pollution. The U.S. EPA (2010) estimated that replacing one inefficient stove reduces CO_2-equivalent emission by about the same amount as taking one car off the road in the United States. A recent review of worldwide biomass emissions by Akagi et al. (2011) indicates that emissions from open cooking and cooking stoves contribute one teragram per year of black carbon (soot) to the global atmosphere. Table 6.5 shows their compilation of emission factors for major pollutants from combustion of the types of SBF used indoors.

TABLE 6.5

Emission Factors (g kg⁻¹) for Species from Biomass Burning

Compound	Chaparral	Open Cooking	Patsari Stoves	Charcoal Burning	Dung Burning
Carbon dioxide	1700	1500	1600	2400	900
Carbon monoxide	67	77	42	190	110
Methane	2.5	4.9	2.3	5.3	11
Total suspended particulate	15	4.6	3.3	2.4	—
Total particulate carbon	—	—	—	—	23
$PM_{2.5}$	12	6.6	—	—	—
Black carbon	1.3	0.83	0.74	1.0	0.53
Organic carbon	3.7	2.9	1.9	1.3	1.8

Source: Adapted from Akagi, S.K. et al., *Atmos. Chem. Phys.*, 11, 4039, 2011.

During a winter in Tibet, homes that used SBF for heating and cooking averaged roughly four times higher average $PM_{2.5}$ concentrations (200 µg m⁻³) than homes that used methane (Gao et al., 2009). Similarly, Siddiqui et al. (2009) found that indoor cooking with wood led to mean $PM_{2.5}$ concentrations in kitchens of 2.7 mg m⁻³ in semi-rural community in Pakistan. In kitchens with natural gas fuel, $PM_{2.5}$ averaged 0.4 mg m⁻³.

6.2.1.2 Fireplaces

Schauer et al. (2001) found $PM_{2.5}$ emission factors of 5–10 g kg⁻¹ for several types of wood burned in a fireplace, in general agreement with results of other investigators (Hildemann et al., 1991; McDonald et al., 2000). Figure 6.4 illustrates the results of Schauer et al. for the composition of wood smoke from pine and oak burning, at the top and bottom, respectively. When wood burns, its oxygenated organic polymers (cellulose and lignins) decompose to distinctive fragments. Schauer et al. (2001) found that 18%–31% of the fine particulate organic carbon mass from wood burning was levoglucosan, a pyrolysis product of cellulose. Hornig et al. (1984) suggested the use of this sugar anhydride as a wood smoke marker. Methoxyphenols and resin acids are the next most abundant classes of identified components of fine wood smoke particles. Methoxyphenols are pyrolysis products of lignin, a biopolymer that trees synthesize from aromatic vinyl alcohols to strengthen their cellulose frameworks.

Methoxyphenols from different types of wood yield characteristic composition patterns that can be used as tracers (syringols for oak; guaiacols for pine and oak; Rogge et al., 1998; McDonald et al., 2000; Schauer et al., 2001). Methoxyphenols are also found in the gas phase of wood smoke, along with aldehydes, dicarbonyls, alkenes, and other compounds. Many wood smoke components are irritants, some are genotoxic, and others (e.g., terpenes) can react with other indoor pollutants.

6.2.1.3 Tobacco

Although a wide range of tobacco smoke constituents have been identified (Jenkins et al., 2000), the most detailed characterization of secondhand smoke (SHS) (an updated term for environmental tobacco smoke) composition was done by Rogge et al. (1994) who sampled exhaled mainstream smoke and sidestream smoke in a vertical dilution tunnel. $PM_{2.5}$ emission factors are roughly 10–20 mg cig⁻¹ (Hildemann et al., 1991; Gundel et al., 1995, among others). Rogge et al. reported that nitrogen-containing heterocyclic compounds are the most abundant class, when gas- and particulate-phase concentrations of nicotine are included with other particulate N-compounds. Since nicotine is primarily in the gas phase in fresh SHS, and very soon after emission it sorbs to indoor surfaces, its use as a tracer for SHS exposure requires meeting particular conditions (Daisey, 1999). Nicotine, a tobacco-specific compound, is commonly used as a tracer for tobacco smoke exposure indoors (Hammond and

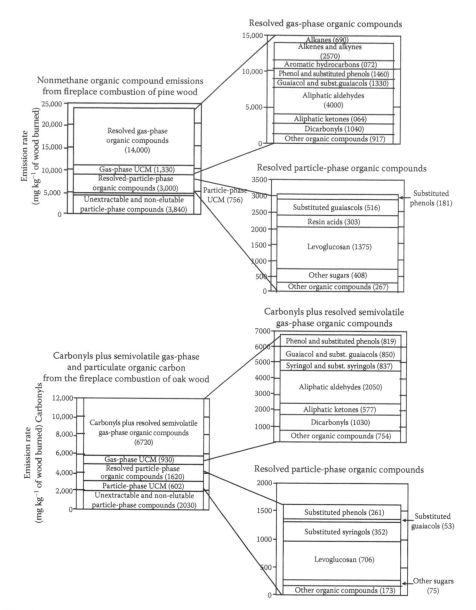

FIGURE 6.4 Chemical composition of smoke from burning pine and oak wood. (From Schauer et al., 2001. With permission.)

Leaderer, 1987; Navas-Acien et al., 2004; Mulcahy et al., 2005; Nebot et al., 2005). When gas-phase nicotine is excluded from the reckoning of Rogge et al. (1994), particulate N-heterocycles are the third most abundant class in ETS particles, after alkanoic acids and alkanes, respectively. Phytosterols (primarily stigmasterol), phenols, alkanols, and branched (*iso*-[1-methyl] and *anteiso*-[2-methyl]) alkanes follow in order of decreasing abundance. Rogge et al. pointed out the potential usefulness of *iso* and *anteiso*-alkanes as SHS tracers, and later Kavouras et al. (1998) quantified them indoors and outdoors in Greece. The carcinogenic components of SHS particles such as *N*-nitrosamines and polycyclic aromatic hydrocarbons are minor constituents. Aldehydes and terpenes are potential reactive gas-phase constituents of SHS and indoor aerosol precursors (Shaughnessy et al., 2001).

Indoor reactions of sorbed nicotine with reactive atmospheric species such as ozone and HONO have been shown to be sources of secondary pollutants of concern (ozone: Destaillats et al.,

2006b; HONO: Sleiman et al., 2010a). The same reactions also produced ultrafine secondary aerosol that was depleted of nicotine and contained oxidation and polymerization by-products (Sleiman et al., 2010b).

6.2.1.4 Candles

Fine et al. (1999) characterized particles from burning candles. They reported fine $PM_{1.8}$ emission factors of 0.5–4 and 2 mg g^{-1}, for paraffin and beeswax, respectively. The highest emission factors and elemental carbon concentrations were found when paraffin candles were "sooting" (emitting visible smoke). For both types of wax, less than a third of the fine particulate organic carbon could be traced to individual compounds. Alkanes, aldehydes, and alkanoic acids were the most abundant classes of compounds in paraffin emissions, reflecting the condensation of unburned alkanes on combustion particles. Emissions from beeswax candles contained the same wax esters, alkanoic acids and alkanes found in the unburned wax, but also decomposition (alkenes) and combustion products (aldehydes). The strong odd carbon number preference found in beeswax was reflected in its fine PM composition. The beeswax candles produced very little elemental carbon.

Lead has been found in emissions from some commercially available candles in the United States and elsewhere (van Alphen 1999; Nriagu and Kim, 2000). Wasson et al. (2002) found lead in 8 pairs among 100 purchased with metal cores or covers of wicks. They reported airborne lead emission rates of 100–1700 µg h^{-1} from individual candles. At an air exchange rate of 0.3 h^{-1} in a 30 m^3 room, it would take only a few minutes for airborne lead concentrations from a single candle at the average emission rate of 550 µg h^{-1} to exceed the EPA's ambient air Pb concentration limit of 1.5 µg m^{-3}. Using this scenario, Wasson et al. estimated that the room concentration of Pb would be around 5 µg m^{-3} for the last 3 h of a 4 h candle burn. Other inorganic constituents have been reported in candle smoke by Pagels et al. (2009). Flame retardants added to the wick were considered to be the source of phosphate and alkali nitrates found in candle PM. Candles burning in sooting conditions were also found to produce high levels of black carbon (between 50% and 90% of the total mass detected on filters and by a scanning mobility particle sizer).

6.2.1.5 Incense

Incense burning can contribute an important source of indoor fine PM (Mannix et al., 1996). See et al. (2007) reported a significant fraction of ultrafine particle emissions from incense burning, between 16% and 55% of the total particle count. The same authors found that up to 6% of these emissions were nanoparticles of diameter <50 nm. Incense smoke has not yet been subjected to the type of comprehensive chemical characterization described earlier for particles from cooking, fireplaces, or cigarette smoke. Because of the variety of ingredients and configurations used to prepare incense around the world, acquiring representative data could be very challenging and expensive. Lung and Hu (2003) reported $PM_{2.5}$ and particulate PAH emission factors for two types of handmade incense, averaging 32 mg g^{-1} $PM_{2.5}$ and 21 µg g^{-1} PAH. Jetter et al. (2002) reviewed the literature and reported particulate emission factors for 23 different types of incense purchased in the United States. They found large numbers of UFP and mass size distributions that typically peaked at around 0.5 µm. $PM_{2.5}$ and PM_{10} emission factors were statistically indistinguishable, with emission factors of 5–56 mg g^{-1} incense and an average emission rate of 42 mg h^{-1}. They also found that the burn time ranged from 14 min for one type of incense cone to 3 h or more, for both a smudge bundle and an incense coil. Average burn time was 43 min. Using an indoor air quality model developed by one of the coauthors (Guo, 2000), they estimated that one unit of incense with the average emission rate and burn time would generate a peak concentration of 0.85 mg m^{-3} if used in a small room (30 m^3) with 0.5 air change per hour. Ji et al. (2010) report the characterization of particulate levels in different rooms within an experimental house: in the proximity of the burning stick, levels were as high as 25,500 particles cm^{-3}; the indoor $PM_{2.5}$ concentration was 197 µg m^{-3}, and the specific surface area concentration was 180 µm^2 cm^{-3}. Significant modifications of PM levels were registered in other rooms within the test house.

6.2.1.6   Unvented Kerosene Heaters and Lamps

Although the sale of portable UKHs has been banned in some U.S. states, at the end of the 20th century, over a million units were sold in the United States each year (Manuel, 1999). Carbon monoxide poisoning and fire are the greatest dangers from such heaters. Increased indoor moisture, kerosene odor, and fine particles are also associated with their use (Traynor et al., 1983, 1986, 1990). Leaderer et al. (1999) found UKHs in about one third of over 200 homes in Virginia and Connecticut whose indoor air was monitored during the winters of 1995–1998. They reported that UKHs added about 40 μg m^{-3} of PM$_{2.5}$ and 15 μg m^{-3} particulate SO$_4^=$ to background indoor levels of 18 and 2 μg m^{-3}, respectively. They also found elevated levels of nitrous acid (HONO) in homes where UKH or gas stoves were operated. In an earlier chamber study, Leaderer et al. (1990) measured emission factors (for K-1 fuel with 0.04% S) of 33–392 and 15–227 μg g^{-1} for PM$_{2.5}$ and SO$_4^=$, respectively, from operation of four types of UKH. The sulfate results suggest that UKHs are a major indoor source of fine acidic aerosol. Leaderer et al. (1999) found evidence that indoor ammonia was neutralized by the acidic aerosol from the UKHs.

Around 1.6 billion people, one quarter of the world's population in 2000, still used kerosene for lighting, together with other low-grade fuels (diesel, propane, biomass) (Mills, 2005). The use of lamps in close proximity to people and indoor environments leads to disproportionately high exposures to PM. Apple et al. (2010) report PM$_{2.5}$ concentrations between 170 and 240 μg m^{-3} and PM$_{10}$ concentrations between 1000 and 2000 μg m^{-3} in kerosene-lighted street vendor kiosks in Kenya. Those levels are ~10 times higher than 24 h guidelines by the WHO.

6.2.2   Human Activities and Consumer Products

6.2.2.1   Pesticides

Organophosphates, chlorinated compounds, and permethrins have been found in indoor PM and dust (Rudel et al., 2003, and references therein; Becker et al., 2002; Berger-Preiss et al., 2002; Rudel et al., 2002). Foot traffic is thought to be a major vector of pesticide transport from outdoor soil particles (Nishioka et al., 1996). Indoor sources of particulate pesticides include flea powder and aerosolized insecticides, among others. Dry flea powder contributes to the coarse particle fraction (2.5–10 μm diameter, authors' comparison to fine powder of known mass size distribution), whereas sprays deliver semi-volatile agents in liquid particles that settle on surfaces or evaporate within minutes (Bukowski and Meyer, 1995). Semi-volatile pesticides like chlorpyrifos, a termiticide, continue to evaporate from indoor surfaces such as carpet for weeks or longer (Stout and Mason, 2003, and references therein).

6.2.2.2   Air Fresheners and Fragrances

Most products sold as air fresheners include fragrances that increase concentrations of VOCs indoors. Frequently, the fragrant compounds include terpenoids such as α-pinene from pine oil or d-limonene from citrus oils. These terpenoids can react with ozone to produce fine PM (Hoffmann et al., 1997; Kamens et al., 1999; Destaillats et al., 2006a; Singer et al., 2006a). Since building envelopes scrub only about half of the infiltrating outdoor ozone, indoor secondary aerosol formation probably occurs to a larger extent than previously recognized (Weschler and Shields, 1997). This is exacerbated by the use of air cleaners that generate ozone (Hubbard et al., 2005).

6.2.2.3   Walking

In their study of real-time particle size distributions in four nonsmoking houses in Boston, Massachusetts, Abt et al. (2000) found that movements of people increased indoor concentrations of large particles (~4 μm median diameter) five times more than they increased concentrations of fine particles (~0.2 μm). Ferro et al. (2002) found that simply walking around indoors increased personal exposure to both fine and coarse particles. Figure 6.5 shows real-time PM$_5$ profiles obtained by Ferro et al. The major contributor is resuspension of settled dust from floors and carpets (Thatcher and Layton, 1995).

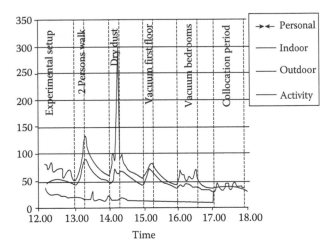

FIGURE 6.5 Concentrations of personal, indoor, and outdoor PM_5 while people walked around a house. (From Ferro, A.R. et al., Exposure to house dust from human activities, *Proceedings of Ninth International Conference on Indoor Air Quality and Climate*, Monterey, CA, June 30–July 2, 2002, Vol. 1, pp. 527–532, 2002. With permission.)

6.2.2.4 Cleaning and Household Activities

Cleaning activities that involve only the movement of equipment and indoor furnishings, rather than the use of cleaning chemicals, have been found to elevate personal exposure to particles. For example, Ferro et al. (2002) reported that 15 min periods of bed making, folding clothes, and vacuuming increased (5 h average) $PM_{2.5}$ and PM_5 personal exposures to 1.4 and 1.6 times the indoor concentrations for the same time periods, as shown in Figure 6.6. Textiles like sheets, towels, and clothes shed fibers, but they can also collect and release airborne particles and indoor dust. Vacuum cleaners without $PM_{2.5}$ filters, as well as dry dusting and sweeping, re-disperse part of the settled dust and indoor PM (Abt et al., 2000).

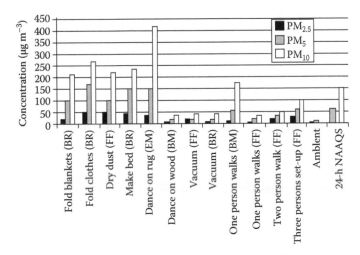

FIGURE 6.6 Personal exposure to three size fractions of PM during indoor activities. BR = first floor bedroom; FF = first floor; BM = basement; walk = continuous vigorous walking and sitting on furniture; vacuum was a canister model with a paper filtration bag. (From Ferro, A.R. et al., Exposure to house dust from human activities, *Proceedings of Ninth International Conference on Indoor Air Quality and Climate*, Monterey, CA, June 30–July 2, 2002, Vol. 1, pp. 527–532, 2002. With permission.)

6.2.2.5 Cleaning Products

Many cleaning products contain in their formulation the same terpenoid ingredients such as α-pinene and D-limonene described in Section 6.2.2.2 that can participate in indoor aerosol chemistry (Weschler and Shields, 1999; Singer et al 2006b). A comparison of personal, indoor, and outdoor exposures to VOCs in three Minnesota communities (Sexton et al., 2004) found that personal exposure to both the aerosol precursors pinene and limonene exceeded their indoor concentrations and that indoor concentrations exceeded outdoor concentrations.

Much less is currently known about indoor concentrations of many other semi-volatile compounds that are used in cleaning products. Rudel et al. (2003) found the semi-volatile disinfectant o-phenylphenol in indoor air and dust, along with several alkylphenols and alkylphenol ethoxylates that are ingredients of detergents. Semi-volatile compounds can adsorb onto and desorb from indoor surfaces and even sampling media, and they were mentioned in the introduction as participants in observed apparent violations of conservation of mass by carbonaceous components of indoor fine mass. SVOC emissions from indoor materials and furnishings will be discussed later in this chapter.

6.2.2.6 Portable Air Cleaners

Portable air cleaners are used widely in residential settings, as well as in the hospitality industry. However, in some cases their pollutant removal benefits may be limited by the generation of secondary pollutants, particularly ozone. In turn, ozone and other concomitant reactive oxygen species can react with terpenoids and other alkenes present in indoor air to generate SOAs. Devices that deliberately generate ozone are considered the most serious problem (Hubbard et al., 2005; Phillips and Jakober, 2006). But high ozone levels and the formation of UFP in the presence of terpenoids have also been reported during the operation of electrostatic precipitators (Alshawa et al., 2007; Jakober and Phillips, 2008; Wensing et al., 2008).

6.2.2.7 Printers, Copiers, and Other Office Equipment

Particulate matter, ozone, and VOC emissions have been reported during idle and operation periods of laser printers, ink-jet printers, and all-in-one machines that include fax, color printer, copier, and scanner (Destaillats et al., 2008). Among those, significant UFP emissions were reported from laser printers and photocopiers, often in association with ozone. Sources of these particulate emissions include heated toner constituents and paper, and likely secondary by-products from ozone reaction with VOCs emitted in the printing process, such as styrene. Little quantitative information is available for size-resolved characterization or chemical composition of particulate matter emitted by office equipment. Schripp et al. (2008) reported particle size distribution of aerosols emitted by nine different printers in a $1 m^3$ glass environmental chamber. UFP (<100 nm) predominated in every case, with number concentrations in the range 10^4–10^6 cm^{-3}. The emission levels varied with printer type but also with the printing cycling and page toner coverage, suggesting that various sources may be present. Recent studies suggest that the high temperature fuser roller plays an important role as a source of UFP (Wensing et al., 2008; He et al., 2010).

6.2.2.8 Renovation

Building renovation includes many activities that can generate large quantities of indoor particles, including removing walls, carpets and flooring materials, sanding wood, stripping paint, installing gypsum board, and spray painting. These activities leave behind micron-sized and visible coarse particles that are responsible for most of the particle mass (Rapp et al., 1997; Kerr and Thi, 2001), and these can be ground to smaller respirable particles by the movement of people. However, operations like sanding and welding generate large numbers of nanoparticles (Göhler et al., 2010; Koponen et al., 2011). Some materials used in renovation (adhesives, sealants) emit semi-volatile species that can adsorb to indoor dust that eventually settles everywhere in the building (Morrison et al., 1998; Hodgson et al., 2000, 2002; Rudel et al., 2003).

6.2.3 INDOOR AEROSOL SOURCE APPORTIONMENT

Indoor source apportionment is critical to understanding the links between PM exposure and health effects, and investigators have used approaches that are similar to those used for PM in outdoor air (Section 6.1.3.2). This review of indoor PM sources began by mentioning that the PTEAM found that smoking and cooking were the largest contributors to indoor particles, after outdoor air. To estimate the contribution of individual sources to indoor and personal PM concentrations found in the PTEAM study, Ozkaynak et al. (1996) used a CMB approach, along with a simple model of indoor pollutant behavior (Koutrakis and Briggs, 1992). They started with mass and elemental concentration data for outdoor, indoor, and personal $PM_{2.5}$ and PM_{10}. Indoor source profiles were derived from reviewing participants' activity reports and multiple linear regression. More recently, Arhami et al. (2010) applied a CMB approach to apportion what they called quasi-ultrafine PM in retirement homes in southern California. Using traffic markers (hopanes, stearanes, and polycyclic aromatic hydrocarbons) they concluded that the main source of indoor particles <250 nm was traffic. They also found that alkanes and alkanoic acids indicated the presence of indoor PM sources.

An advantage of the factor analysis approach is that it does not require *a priori* knowledge of indoor processes. Yakovleva et al. (1999) used PMF to apportion sources of personal exposure to PM, based on PTEAM data. They identified five source profiles that influenced personal PM_{10} exposure: motor vehicles, secondary sulfate, personal activities, resuspended indoor soil, and indoor soil. For example, Figure 6.7 compares the contributions of these factors to the participants' reported activities.

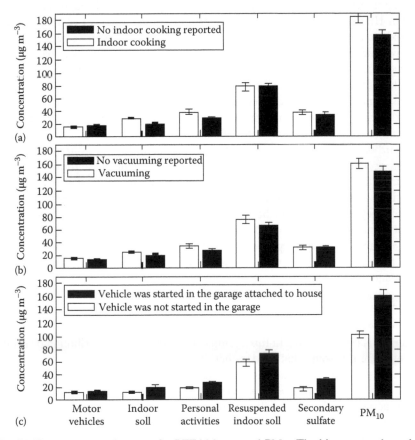

FIGURE 6.7 PMF source apportionment for PTEAM personal PM_{10}: The histograms show the contributions from each computed source factor compared to activity logs: (a) indoor cooking; (b) vacuuming; (c) starting a vehicle in an attached garage. (From Yakovleva, E. et al., *Environ. Sci. Technol.*, 33, 3645, 1999. With permission.)

PMF has been applied worldwide to apportion indoor PM to likely sources (e.g., Kocaeli City, Turkey, Pekey et al., 2010; Seoul, Korea, Lim et al., 2011). Work continues on modeling indoor exposure to PM for epidemiology. For example, using simple measurements of carbonaceous particles and elemental analysis of PM in Boston, Massachusetts, Clougherty et al. (2011) recently showed how factor analysis can be enhanced for indoor source apportionment by incorporating some features of land use regression modeling, along with the effects of ventilation on indoor/outdoor ratios of pollutants. Their work identified three indoor factors, combustion, cleaning, and resuspension, and three outdoor factors, long-range transport, fuel oil/diesel, and road dust/resuspension.

6.3 PHYSICAL PROPERTIES OF INDOOR AEROSOLS

This section presents a brief overview of a few key aspects of aerosol physics that influence the indoor behavior and fate of airborne particles. Hinds presents a clear and thorough discussion of aerosol physical properties in his classic text (Hinds, 1999). Seinfeld and Pandis (2006) and Friedlander (2000) also introduce aerosol physics with many examples from ambient air.

6.3.1 Size Distributions

Particle size and mass are key physical parameters that influence aerosol behavior, for example, the penetration of aerosols through building envelopes (Thatcher et al., 2003), deposition of particles in the human respiratory system (Chapter 1 of Hinds, 1999), and flow through sampling instruments (Chapter 10 of Hinds, 1999). Many studies indicate that fine particles (<2.5 μm diameter) appear to be more toxic to cells per unit mass than coarse particles with diameters between 2.5 and 10 μm. Because many sources contribute to airborne particles in an urban area, and particle sizes vary in time and space, *size distribution functions* are very useful tools to describe the size dependence of particle number, surface area, and volume (or mass) concentrations.

Aerosol instrumentation design depends on understanding the size dependence of aerosol behavior, particularly aerosol flow dynamics. Number size distributions are measured with particle counters that depend on aerosol electrical mobility and optical properties. The size dependence of light scattering by airborne particles is used in optical particle counters, for example.

Nucleation, accumulation, and coarse size modes: Particles that arise from nucleation are small (~5–50 nm), and large numbers of UFP (<0.1 μm; nucleation mode: 2–70 nm) are typically found near combustion sources. Atmospheric lifetimes of nucleation mode particles are minutes or less. After formation they coagulate when they collide with each other and with larger particles, and reach the accumulation mode between about 0.1 and 1 μm. Airborne particles between about 2 and 10 μm are part of the coarse mode. Resuspended soil particles, sea spray, plant debris, pollen, and spores contribute to the coarse mode. Outdoor lifetimes of accumulation and coarse mode particles are days to weeks and minutes to days, respectively (Seinfeld and Pandis, 2006).

6.3.1.1 Number

The top part of Figure 6.8 shows a typical urban aerosol *number size distribution*, as illustrated by Seinfeld and Pandis (1998, p. 431). In this example, the number size distribution peaks in the nucleation mode, even for this somewhat aged, rather than fresh, aerosol.

Normal or bell-shaped distributions of particle size are symmetrical about the average particle diameter. The standard deviation describes the spread of particle diameter around the mean. Since most airborne aerosols are not normally distributed in size, but tail off with increasing size, *log normal* distributions are more appropriate descriptors. The logarithms of particle diameters, rather than the particle diameters, are normally distributed, and the geometric standard deviation describes the spread of the log of particle diameters about the median. Many ambient aerosol number size distributions can be described as the sum of modes that are each log-normally distributed (Hinds, 1999; Seinfeld and Pandis, 2006).

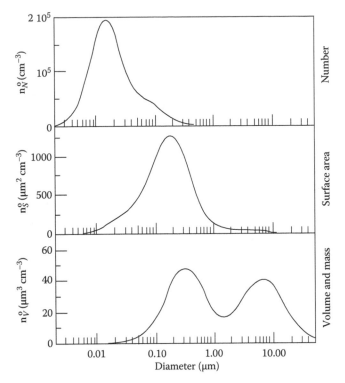

FIGURE 6.8 Three descriptions of the size distribution of "typical" urban airborne particles. The top panel shows *number* size distribution; the middle shows *surface area* size distribution, and the lower panel shows *volume* size distribution. The *mass* size distribution has the same shape as the volume distribution. (From Seinfeld, J.H. and Pandis, S.N., *Atmospheric Chemistry and Physics*, 2nd edn., Wiley-Interscience, New York, 1232 p., 2006. With permission.)

Sampling applications: Particle number size distributions are measured with electrical mobility analyzers, condensation nuclei counters, differential mobility analyzers, electron microscopes, and optical counters.

6.3.1.2 Surface Area

The middle section of Figure 6.8 shows the *surface area size distribution* for the typical urban aerosol example. The peak in the surface area distribution occurs at larger particle size, within the accumulation mode. Interactions between particles and fluids take place at particle surfaces. Adsorption, absorption, condensation, and reactions all occur at surfaces or in liquid films on particle surfaces. The size dependence of concentrations of aerosol constituents can be used to infer mechanisms of aerosol processing. For example, if the size dependence of concentrations of semi-volatile compounds such as polycyclic aromatic hydrocarbons or photochemical reaction products matches the size dependence of surface area, rather than particle number or mass, sorption or condensation is responsible.

Sampling surface area: The epiphaniometer is an example of an aerosol instrument that measures particle surface area size distributions.

6.3.1.3 Volume and Mass

The lower section of Figure 6.8 shows the *mass size distribution* of the same urban aerosol example. There are two peaks, one in the accumulation mode and one in the coarse mode. This is a bimodal distribution. The minimum in mass concentration around 2 μm is usually quite pronounced in urban and rural air. Sometimes trimodal distributions are seen very near sources of fresh aerosol, when enough numbers of new tiny particles are present to contribute to the mass size distribution (Seinfeld and Pandis, 2006).

Particle mass and size are key physical parameters that influence the movement of particles in air. Diffusion rate depends on particle mass, as do the effects of inertia and fluid viscosity on particle behavior in moving air. During air movement through tubes (e.g., the respiratory system, sampling tubing) the heavier the particle, the more likely the particle is to be caught in a bend because it cannot keep up with the airflow. This property leads to particle deposition in bronchial tubes. Size selective inlets, impactors, and virtual impactors take advantage of this property to separate particles based on their mass.

Instruments based on particle mass: Particle mass size distributions are measured directly by aerosol impactors. Mass size distributions can be integrated to yield total mass concentrations that can be compared directly to mass concentrations determined gravimetrically from filters. If particle number and mass size distributions are known, the size dependence of particle density can be calculated. Frequently, the particle density is estimated for spherical particles of the same aerodynamic diameter as the measured distribution, and the calculated density is used to convert from mass to number size distributions and vice versa. Geometric surface area estimates depend only on the aerodynamic diameter and can be derived from either the mass or number size distributions. However, actual surface areas may be larger due to particle shape and porosity.

6.3.2 Particle Dynamics

The behavior of indoor particles depends on their size, the forces they experience, and their interactions with surfaces in the indoor environment. These factors are embedded in the mass balance equations used in Section 6.5 to describe their fate and transport. The objective of this section is to illustrate how basic aerosol concepts apply in the indoor environment, starting with a room with dimensions $3.65\,m \times 3.65\,m \times 3\,m$. This room has a total volume of $40\,m^3$ and an air exchange rate of $1\,h^{-1}$.

6.3.2.1 Particle Motion

The flow rate at one air exchange per hour through the example room is $11.1\,cm^3\,s^{-1}$. A cube of air with this volume has a linear velocity of $2.23\,cm\,s^{-1}$. A quick way to get a sense of the behavior of particles in air at this velocity is to look at their Reynolds numbers (ratios of inertial to frictional forces), relaxation times (for adjustment of particle velocity to applied force), and terminal settling velocities in still air (Table 6.6). When particle Reynolds numbers are <1 the particles experience laminar flow. That is, they keep up with the streamlines, and gravity has little effect on them. In this example, the particles move with the airflow unless they are large enough to be visible to the human eye. Respirable particles (diameter <2.5 μm) would remain entrained in room air movement even at much higher air exchange rates. The relaxation times show that they quickly adjust to the air exchange rate, and the settling velocities show that respirable particles do not deposit onto the floor quickly under the influence of gravity.

Deposition: Airborne particles deposit to building materials and other indoor surfaces after collision and adhesion. As outdoor air infiltrates through cracks in the building envelope, particles below 0.1 μm in diameter are lost because of diffusion much more efficiently than larger particles. However, larger particles deposit by impaction when they cannot keep up with the airflow around turns and obstacles. Once inside the building, both diffusion and gravitational settling contribute to particle loss. Table 6.6 shows that diffusion is responsible for more deposition to a horizontal surface than gravitational settling, for particles smaller than 0.2 μm, over the 100 s period examined by Hinds (1999, p. 162).

TABLE 6.6

Properties of Particles with Unit Density in a room with Volume 40 m³ and 1 ACH at Atmospheric Pressure and 20°C

Diameter (μm)	Re[a] at 2.23 cm s⁻¹	Relaxation Time[b] (s)	Terminal Settling Velocity[c] (cm s⁻¹)	Diffusion Coefficient[d] (cm² s⁻¹)	Ratio, Diffusion to Settling[e]
0.01	0.00015	6.8×10^{-9}	7.0×10^{-6}	5.2×10^{-4}	390
0.10	0.0015	8.8×10^{-8}	8.8×10^{-5}	6.7×10^{-6}	3.4
1.0	0.015	3.6×10^{-6}	3.5×10^{-3}	2.7×10^{-7}	1.7×10^{-2}
10	0.15	3.1×10^{-4}	0.29	2.4×10^{-8}	5.5×10^{-5}
100	1.5	3.2×10^{-2}	17	2.4×10^{-9}	2.2×10^{-7}

[a] Reynolds number, Re = 6.6 Vd, where V is the particle velocity and d its diameter, in cgs units. Hinds (1999, p. 28). Re is shown for particle velocity and is equated to the linear flow velocity in the example room, 2.23 cm s⁻¹.
[b] Hinds (1999, p. 112, Table 5.1).
[c] In still air, Hinds (1999, p. 51, Table 3.1).
[d] Hinds (1999, p. 153, Table 7.1).
[e] Cumulative deposition of particles over 100 s, Hinds (1999, p. 162, Table 7.5).

6.3.2.2 Particle Formation and Phase Partitioning

6.3.2.2.1 Nucleation

Nucleation describes the process of particle formation from gas-phase precursor compounds. *Homogeneous nucleation* starts when the air becomes supersaturated with precursor gas molecules that collide with each other. No preexisting particles are necessary. Supersaturation means that the actual vapor pressure of a compound is higher than its equilibrium vapor pressure, as combustion exhaust expands and cools, for example. After multiple collisions, molecular clusters or "embryos" are formed. They grow by further collisions with gas molecules, and when they reach a critical diameter that depends on the extent of supersaturation and gas molecular properties, some of them become stable enough to form nanoparticles. Homogeneous nucleation requires supersaturation ratios of 2–10 (ratios of actual vapor pressure to equilibrium vapor pressure (Hinds, 1999, p. 279). Particles continue to grow by condensation as long as supersaturation continues.

6.3.2.2.2 Condensation

Heterogeneous nucleation or *nucleated condensation* occurs at much lower supersaturation when preexisting nuclei are present to provide some initial surface area for adsorption of gas-phase molecules. Above the oceans water vapor condenses on soluble nuclei such as sodium chloride crystals at low supersaturation. When mixtures of gases are present, UFP may appear at supersaturation ratios somewhat lower than 1. SOA forms when concentrations of oxidized products of the gas-phase reactions of unsaturated hydrocarbons with atmospheric oxidants reach supersaturation and condense onto existing particles.

6.3.2.2.3 Coagulation

After particles are released into the indoor environment, they will bump into each other at a rate that is proportional to the square of their number concentration. The number concentration will decrease rapidly as the particles agglomerate, and their mass concentration will remain the same until they become too large to remain suspended in the air. Hinds (1999) suggests that coagulation can usually be neglected in industrial hygiene applications when the particle number concentration is below 10^6 cm⁻³.

6.3.2.2.4   Sorption/Desorption

While supersaturation is necessary for particle formation entirely from gas molecules, the extent of gas *ad*sorption onto existing dry particles or *ab*sorption into liquid films on particles depends not only on the extent of saturation, but also on the amount of available surface area. Sorption and desorption influence the size and mass of particles in both the nucleation and accumulation size modes.

Once gas-phase concentrations of low-volatility compounds (from direct primary emission or secondary reactions) exceed their saturation vapor pressures, they will distribute between the gas phase and available surfaces that can be described using thermodynamic concepts. The partition coefficient K_p is the most commonly used parameter for describing gas/particle partitioning at equilibrium, primarily because of its log-linear relationship to compound vapor pressure, p_L^o. Although a compound's vapor pressure at the temperature of interest has the greatest influence on partitioning, the interaction between compound structure and the sorptive medium plays an important role (e.g., compound size and polarity versus *ad*sorptive affinity or *ab*sorptive capacity). Plots of log K_p versus log p_L^o can provide information on the nature of the partitioning.

Using Pankow's nomenclature (Pankow, 1999 and Seinfeld and Pankow, 2003), the equilibrium partitioning of an SVOC to an environmental surface S can be represented most simply by

$$G + S = P \tag{6.1}$$

where G and P represent the gas and particulate phases of the sorbing molecule. Since the gas and particulate phases of the molecule are usually collected on an adsorbent and filter, respectively, their concentrations have been conveniently represented by A and F in the literature (ibid.). If the sorbing surface is total suspended particulate matter, its concentration can be represented by TSP. (G/P theory takes the same form for size-segregated particles, but TSP is used here to be consistent with Pankow's development.) At equilibrium, the gas/particle partitioning constant K_p for *ad*sorption of the SVOC compound i onto the solid surface of a particle can be expressed as in Equation 6.2:

*Ad*sorption to a solid surface

$$K_p = \frac{F_i}{A_i \cdot TSP} \tag{6.2}$$

Pankow (1987) showed that Langmuir adsorption theory predicts that K_p (at constant temperature) is inversely proportional to the vapor pressure of i. If i is a solid, the subcooled liquid vapor pressure p_L^o is used.

$$K_p = \frac{N_s a_{tsp} T e^{(Q_l - Q_v)/RT}}{1600\, p_L^o} \quad \text{adsorption} \tag{6.3}$$

N_s and a_{tsp} are terms for the number of adsorption sites per unit area and the surface area of the particles, respectively. For compounds of the same class, with similar enthalpies of desorption and vaporization Q among the members, plots of log K_p versus log p_L^o will be linear slope of −1, as shown in Equation 6.4.

$$\log K_p = -\log p_L^o + \log \frac{N_s a_{tsp} T e^{(Q_l - Q_v)/RT}}{1600} \quad \text{adsorption} \tag{6.4}$$

Semi-volatile compounds can also *ab*sorb into liquid particles such as secondhand tobacco smoke or liquid (organic and/or water) films on particles with solid cores, as Pankow (1994) has shown. Gas/particle partitioning of SVOCs in urban areas is better explained as *ab*sorption than *ad*sorption. The *ab*sorptive partitioning of SVOC i into a liquid organic layer on a particle is like a gas dissolving in a liquid (Finlayson-Pitts and Pitts, 1999, p. 417), and the measured partitioning coefficient for *ab*sorption takes the same form as Equation 6.2.

$$K_p = \frac{F_{i,om}}{A_i \cdot TSP} \quad \text{absorption into a liquid film or droplet} \tag{6.5}$$

$F_{i,om}$ represents the particle-associated concentration of i in air as measured from a filter, with explicit recognition that i has dissolved in liquid organic material, om, on the particle. For absorption into liquid films on particles, Pankow (1994) showed that K_p is proportional to the weight fraction of om to TSP.

$$K_p = \frac{f_{om} 760 RT}{MW_{om} p_L^0 \gamma 10^6} \quad \text{adsorption} \tag{6.6}$$

K_p is inversely proportional to the product of p_L^0 and the activity coefficient γ of i in the liquid phase. Vapor pressure is the most important factor influencing K_p, followed by activity coefficient and molecular weight. If the activity coefficient does not vary much across members of a class of SVOCs, plots of K_p versus $\log p_L^0$ will have a slope of -1 for both adsorption and absorption, as in Equations 6.4 and 6.7.

$$\log K_p = -\log p_L^0 + \log \frac{f_{om} 760 RT}{MW_{om} \gamma 10^6} \tag{6.7}$$

Recent contributions to gas/surface partitioning theory address observed deviations from the predictions of Equations 6.4 and 6.7. Jang et al. (1997) applied a comprehensive thermodynamic approach to calculate group contributions to activity coefficients for adsorption of SVOC into nonideal organic films. This allows calculation of activity-normalized partitioning coefficients, $K_{p,g}$. Goss and Schwarzenbach (1998) argued that slope deviations from -1 in $\log K_p$ versus $\log p_L^0$ plots do not necessarily indicate nonequilibrium conditions, and they outlined how these deviations can be used to identify types of sorbate/sorbent interactions and thus characterize sorption processes. For example, they showed how acid/base interactions can influence gas/surface partitioning polar SVOC. Harner and Bidleman (1998) demonstrated that using laboratory-derived octanol/air partitioning coefficients circumvents the need to estimate activity coefficients for compounds absorbed in the organic films that coat urban particles. Mader and Pankow (2000, 2001a,b) expanded partitioning theory to quartz and Teflon filter materials that are used to collect particles, thus tackling the sticky problem of SVOC adsorption artifacts in PM sampling on filters.

6.3.3   SURFACE AREA IN THE INDOOR ENVIRONMENT

From the perspective of indoor air quality and personal exposure, understanding the influence of surface area on the behavior of airborne particles includes recognizing the importance of indoor surfaces on aerosol processing. The surface-to-volume ratio is 2 orders of magnitude greater indoors than outdoors. Weschler's instructive discussion (Weaschler, 2003) of indoor surface area will be recounted briefly here. The example is based on a room with volume $40\,m^3$ and a carpet of $10\,m^2$.

6.3.3.1   Particles

In a now classic study of gas/particle partitioning, Liang et al. (1997) inferred the surface area of their airborne particles as $2\,m^2\,g^{-1}$. Using their data, Weschler (2003) calculated that surface loading of a semi-volatile compound with subcooled liquid vapor pressure of 10^{-6} atm (hexadecane or phenanthrene) onto airborne particles (at their total suspended particle concentration $20\,\mu g\,m^{-3}$) was $2.1 \times 10^3\,\mu g\,m^{-2}$. From the data of Naumova et al. (2003), Weschler calculated the mass loading of the same compounds per unit of particulate surface area as $7.7 \times 10^3\,\mu g\,m^{-2}$ in a room of volume $40\,m^3$.

6.3.3.2 Indoor Materials

Morrison and Nazaroff (2000) found that the actual surface area of nylon carpet (with pad) was about 100 times its nominal surface area, and Weschler used their data to calculate the surface loading capacity for hexadecane or phenanthrene in the example room. He found that the surface loading ($7.1 \times 10^3\,\mu g\,m^{-2}$) for the carpet was similar to that found by Naumova et al. (2003) for fine particles. Therefore, indoor surfaces, with their much higher total surface area than indoor particles, are very likely to be important for any indoor aerosol process that depends on gas/surface interactions.

Weschler estimated that the surface area of the particles would be $1.6 \times 10^{-3}\,m^2$, compared to the room's nominal surface area of $36\,m^2$, not counting surface porosity or furnishings. The carpet and pad contributed actual surface area of $1000\,m^2$. Although the surface area of the airborne particles is miniscule compared to the surface area of the room and its furnishings, it is the surface of the particles that makes contact with the respiratory systems of the occupants. Particle motion provides the vector for moving low-volatility pollutants from one indoor surface to another: Weschler showed that particle deposition is the main route for transport of the plasticizer diethylhexylphthalate (DEHP; vapor pressure 1.9×10^{-10} atm at 25°C) to the indoor surfaces.

6.3.3.3 Adsorption Indoors

Indoor aerosol behavior is strongly influenced by the presence of indoor surfaces because building materials and furnishings act as reservoirs of sorbed compounds. At equilibrium, sorption to indoor surfaces follows Equation 6.1. Won et al. (2000, 2001) determined that partitioning coefficients of organic gases to indoor materials depend on p_L^o, the subcooled liquid vapor pressure. Weschler (2003) showed that the molecular weight and the negative log of vapor pressure are linearly related for a large number of organic compounds found indoors, as shown in Figure 6.9. Thus, for compounds whose vapor pressures are unknown, partitioning behavior can be predicted from knowledge of only the molecular weight.

Table 6.7, taken from Weschler (2003), shows how a range of common organic pollutants partition in the example room. The lower the vapor pressure, the more significant is the particulate-bound fraction on the transport and fate of the compound, even if that fraction is very small. Inhalation

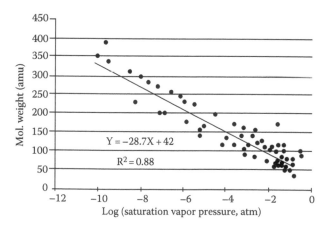

FIGURE 6.9 Molecular weight versus the log of the vapor pressure for nonpolar compounds. (From Won, D. et al., *Environ. Sci. Technol.*, 34, 4193–4198, 2000; Weschler, C.J., *Atmos. Environ.*, 37, 5455, 2003. With permission.)

TABLE 6.7

Distribution of Selected Organic Compounds between the Gas Phase and the Surfaces of Airborne Particles, a Carpet, and Walls within a Typical Room

Compound	Mol. Weight (amu)	Vapor Pressure at 25°C (atm)	Assumed Gas-Phase Concentration ($\mu g\ m^{-3}$)	Mass in Gas Phase (μg)	Mass on Particles (μg)	Mass on Carpet (μg)	Mass on Walls (μg)
MTBE	88	3.2E − 01	10	400	2.3E − 5	17	19
Toluene	92	3.7E − 02	10	400	1.4E − 4	100	70
Ethylbenzene	106	1.3E − 02	10	400	3.6E − 4	260	140
Propylbenzene	120	4.5E − 03	10	400	8.9E − 4	610	260
Naphthalene	128	1.0E − 04	5	200	1.2E − 2	7400	1390
Acenaphthene	154	5.9E − 06	5	200	0.13	8.0E + 4	8000
Hexadecane	226	9.1E − 07	5	200	0.66	3.8E + 5	2.6E + 4
Phenanthrene	178	1.4E − 06	1	40	0.093	5.4E + 4	4000
Octadecane	254	2.5E − 07	1	40	0.40	2.3E + 5	1.1E + 4
Pyrene	202	7.6E − 08	1	40	1.1	6.2E + 5	2.4E + 4
Heneicosane	296	8.7E − 09	0.5	20	3.6	1.9E + 6	4.6E + 4
Chrysene	228	5.0E − 09	0.5	20	5.8	3.0E + 6	6.4E + 4
Tetracosane	338	2.8E − 10	0.01	0.4	1.4	6.9E + 6	7800
DEHP	390	1.9E − 10	0.07	3.0	14	6.7E + 6	6.9E + 4
Pentacosane	352	8.7E − 11	0.01	0.4	3.8	1.8E + 6	1.6E + 4

Source: From Weschler, C.J., *Atmos. Environ.*, 37, 5455, 2003.

Values derived for a $3 \times 3.65 \times 3.65\ m^3$ room containing $20\ \mu g\ m^{-3}$ of airborne particles (TSP), a $10\ m^2$ carpet with pad, and painted gypsum board walls. See text for further details.

effects then depend on aerosol dynamics. Accurate assessment of human exposure to indoor particulate organics is hindered by the use of sampling methods that cannot exclude adsorption of gas-phase semi-volatile compounds by the sampling medium (Pang et al., 2002).

6.4 CHEMICAL AND PHYSICAL PROCESSES RELEVANT TO INDOOR AEROSOL CHEMISTRY

This section intends to show that thorough investigation of indoor aerosol physics and chemistry, as well as health effects, must include the role of the building envelope in conditioning infiltrating ambient particles. Indoor materials, furnishings, and consumer products also act as emission sources and sorption inks for semi-volatile species. Because of the likelihood of finding high indoor concentrations of semi-volatile organic species, some measurement methods that were developed for outdoor air need to be modified for use indoors. The unexpected and possibly counterintuitive phenomena introduced in Sections 6.1.5.1 through 6.1.5.3, related to the building envelope, indoor materials, and aerosol production, respectively, provide useful examples to illustrate indoor aerosol chemistry.

6.4.1 REEQUILIBRATION OF INFILTRATED PARTICLES

Section 6.1.5 mentioned a recent comparison of the composition of outdoor and indoor particles and gases that showed the influence of the building envelope. Using real-time concentration and ventilation measurements in an unoccupied house in Fresno, California, Lunden et al. (2003a,b) found that indoor particulate sulfate and soot (black or elemental carbon) acted as conservative tracers for infiltration of outdoor $PM_{2.5}$, but indoor $PM_{2.5}$ had much less ammonium nitrate than predicted from the penetration factors for elemental carbon. The indoor gas-phase concentration of nitric acid was usually lower than outdoors, but the indoor ammonia levels were frequently higher than outdoors.

The observations of Lunden et al. (2003a) are consistent with the disruption of the NH_4NO_3 gas/particle equilibrium as gas-phase nitric acid reacted with the indoor surfaces, causing evaporative dissociation of aerosol NH_4NO_3 that originated outdoors. The surfaces took up nitric acid more readily than ammonia. Lunden et al. showed that the observed evaporation rates for NH_4NO_3 (0.3 to $18\,h^{-1}$) were in the same range as ventilation rates and higher than particle deposition rates. They found that the evaporation rate depended strongly on temperature, as well as gas-phase concentrations of ammonia and nitric acid. More detail about the modeling approach for this study is given in Section 6.5 that follows.

6.4.2 Neutralization of Acidic Particles

Indoor ammonia provides an example of the occupants' influence on indoor aerosol chemistry that was not mentioned in the introduction to this chapter. Humans and their pets generate ammonia that can neutralize aerosol acidity. Some cleaning products also contribute ammonia. By comparing field measurements (Leaderer et al., 1999) with results of environmental chamber studies (Leaderer et al., 1990) Leaderer et al. found evidence that indoor ammonia neutralized much of the increased aerosol acidity that originated as sulfur in the fuel of UKHs.

6.4.3 Emission and Partitioning of Semi-Volatile Organic Compounds

As mentioned in Section 6.1, significant concentrations of semi-volatile organic constituents of building materials, furnishings, and consumer products have been reported in indoor air and dust. Ventilation rate influences the indoor concentrations of VOCs more than SVOC, whereas concentrations of SVOC are controlled to a greater extent by sorption and (re)emission (Singer et al., 2004). Increasing attention is being paid to associations between indoor SVOC and respiratory symptoms (e.g., phthalates: Bornehag et al., 2004). Possible disruptors of human endocrine systems, besides phthalates, include constituents of textile surface treatments such as perfluoroalkyl sulfonamides (Shoeib et al., 2004) and fire retardants such as PBDEs (Sjödin et al., 2001; Kemmlein et al., 2003) and triethylphosphate (Carlsson et al., 2000; Salthammer et al., 2003).

In many indoor environments, the gas-phase concentrations of SVOCs can reach supersaturation, and condensation can occur onto walls, windows (Butt et al., 2004), textiles, airborne particles, air filters, and even quartz fiber filter sampling media (Mader and Pankow, 2000, 2001a,b; Weschler, 2003; Xu and Little, 2006). Whereas Figure 6.2 illustrates processes that influence indoor aerosol concentrations, Figure 6.10 shows interactions of SVOCs with indoor surfaces, including emission from furnishings (indoor sources), sorption, desorption, and surface reactions.

As shown in Table 6.7 carpets and walls can function as large reservoirs of semi-volatile compounds such as the plasticizer DEHP that will partition to airborne particles and settled dust. Partitioning of indoor SVOC to sampling media such as fiberglass or quartz filters (Mader and Pankow, 2000, 2001a,b) can account for the large positive sampling artifacts that have been observed for organic carbon concentrations (Landis et al., 2001). Pang et al. (2002) minimized the indoor positive artifact by removing the gas-phase SVOC before the particles reached the collection filter.

Weschler and Nazaroff (2008) examined the factors affecting equilibrium partitioning of SVOCs to all indoor surfaces, and concluded that equilibrium was reached faster for smaller particles than for extended surfaces. The authors developed a framework to identify exposure pathways to indoor SVOCs, in which the rate of equilibration and potential for transport from and to surfaces was

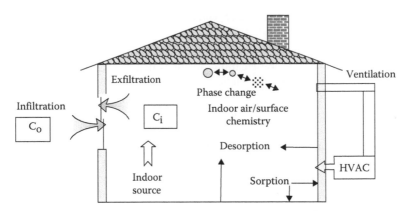

FIGURE 6.10 Indoor processes involving the interaction of volatile and semi-volatile species with indoor surfaces (building materials, furnishings, and particles).

assessed using the octanol/air partitioning coefficient. In a more recent article, the same authors reviewed data from 19 studies reporting measurements of dustborne and airborne SVOCs in more than 1000 buildings, which included 66 different SVOCs whose octanol/air partition coefficients span more than 5 orders of magnitude (Weschler and Nazaroff, 2010). The authors verified that the octanol/air partition coefficient is a strong predictor of the abundance of a particular compound in settled dust relative to its gas-phase concentration, and predicted dustborne mass fractions that correlated well with measured mass fractions. Also, for SVOCs with high octanol/air partitioning coefficients, they observed that settled dust likely did not have sufficient time to equilibrate with the gas-phase levels. This approach was also verified in the case of the partitioning of phthalate esters (plastic additives) measured in settled dust, by using those values to estimate their concentration in airborne particles (Weschler et al., 2008).

6.4.4 Indoor Aerosol Formation

In addition to combustion (Section 6.2.1), at least two other pathways for indoor aerosol generation have been identified: nucleation from reemitted SVOC and secondary aerosol formation from reaction of unsaturated hydrocarbons with ozone or hydroxyl radicals.

6.4.4.1 Semi-Volatile Organic Compounds Reemission

Johansson et al. (1993) observed the formation of small (10–20 nm) particles in a closed room (100 m³) that had been sealed after the ventilation phase of experiments with high concentrations of secondhand tobacco smoke (SHS). They also showed how the particle size distribution evolved as the particles coagulated. Apte et al. (2004) confirmed the phenomenon of nanoparticle ingrowth in an environmental chamber whose walls had been conditioned with environmental tobacco smoke. Like Johansson et al., they could only observe particle formation when the chamber had been sealed after ventilation with particle-free air. As soon as unfiltered ambient air was admitted to the chamber, the size distribution changed as the small nucleation-mode particles were scavenged by collision with ambient accumulation mode particles. These observations can now be explained as instances of SVOC reemission from the walls of the conditioned smoking rooms, followed by particle nucleation, condensation, and coagulation.

6.4.4.2 Reactive Organics and Ozone

Indoor chemistry is a relatively young discipline that developed over the past two decades, principally describing the role of ozone and other reactive atmospheric species in gas-phase and heterogeneous reactions that lead to the formation of secondary pollutants and aerosol particles

(Weschler, 2011). Weschler and Shields (1997) showed that indoor aerosol generation was possible when unsaturated hydrocarbons react with ozone under typical building airflows and residence times. In 1999, they provided the first experimental evidence for the counterintuitive notion that fine particles are produced indoors by cleaning practices that residents often consider protective and health promoting. Using a pair of matched offices with the same ambient ozone concentrations and ventilation rates, they tracked increased numbers of fine particles in a room into which d-limonene had been released (Weschler and Shields, 1999; Weschler 2000). Limonene is a biogenic terpenoid, a cyclic alkene found in lemon and pine oils that are common components of cleaning products. Shaughnessy et al. (2001) compared the reactivity of hydrocarbons found in SHS with ozone and showed that both limonene and dimethylfuran in SHS would react with 50 ppb ozone indoors even at higher than typical residential air exchange rates. Several recent studies have investigated the influence of air exchange rates on indoor aerosol generation from reactions of unsaturated hydrocarbons. Sorenson and Weschler (2002) applied computational fluid dynamics; Weschler and Shields (2003) showed how the new particle number and size distributions depended on air exchange rates. Sarwar et al. (2003) extended the work to predict how indoor secondary aerosol formation depends not only on ventilation rates, but also building temperature, indoor terpene emission rates, as well as outdoor ozone and particle concentrations. A review by Nazaroff and Weschler (2004) summarized the available data on reaction rates of terpenes and related compounds with ozone, OH, and nitrate radical. Mixing ozone with the vapor emissions of common household products and with air freshener emissions in a laboratory chamber experiment led to the formation of an initially large number of UFP that subsequently grew by condensation and agglomeration (Destaillats et al., 2006a). Experimental factors such as the air exchange rate, the RH, and the presence of seed particles had a strong influence in the observed particle dynamics (Coleman et al., 2008).

In parallel with the explosion of work on indoor secondary aerosol formation, rapid progress is occurring in the identification of reaction mechanisms for the oxidation of terpenoids by ozone. Figure 6.11, kindly provided by P. Ziemann, shows a condensed version of the likely reaction pathway for generation of the products that are shown in Figure 6.12, which is based on the work of Yu et al. (1999). While ozone/terpenoid chemistry is considered the most significant chemical source of indoor aerosols, more recently other systems have been studied. For example, Sleiman et al. (2010a) described the formation of UFP in the reaction of ozone with nicotine vapor, as well as with constituents of secondhand tobacco smoke (that included nicotine).

6.4.5 Multimedia Modeling for the Indoor Environment

Bennett and Furtaw (2004) introduced the use of fugacity, the tendency of pollutants to escape from an environmental compartment to another, as a promising approach for exploring indoor chemical and physical process. The upper part of Figure 6.13 shows a home floor plan from their study of the behavior of pesticides. The lower part shows a corresponding diagram that highlights multimedia processes indoors. This type of multimedia model has been successfully used to understand the fate and transport of pollutants as they move among outdoor environmental "compartments" (soil, sediments, water, and air) (Cowan et al., 1995). Such a framework shows promise for clarifying the role of indoor aerosol chemistry in human exposure to toxic and hazardous pollutants, and it builds on the mass balance approach described in Section 6.5.

6.5   FATE AND TRANSPORT

The concentration of aerosols indoors is the result of several dynamic processes where the production of aerosols (source terms) is balanced by various removal or transformation mechanisms. These processes were shown schematically in Figure 6.2.

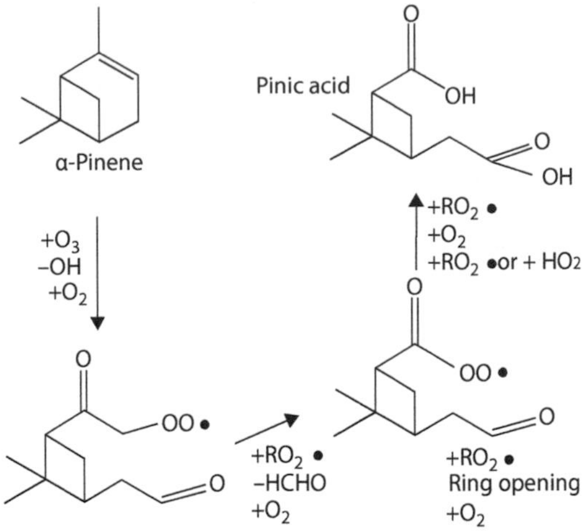

Aerosol Chemistry and Physics 161

FIGURE 6.11 Mechanism of the reaction of α-pinene with ozone. (Courtesy of Paul Ziemann, American Association for Aerosol Research, Mt. Laurel, NJ.)

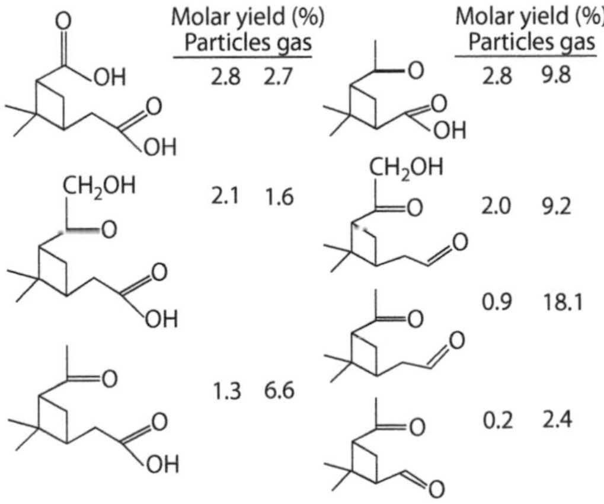

FIGURE 6.12 Products of the reaction of α-pinene with ozone, identified by GC-MS with double derivatization. (From Yu, J. et al., *J. Atmos. Chem.*, 34, 2007, 1999; Courtesy of Paul Ziemann American Association for Aerosol Research, Mt. Laurel, NJ.)

6.5.1 MODEL OF INDOOR AEROSOL BEHAVIOR

Definitions:

Outdoor aerosol concentration, Co (dp) ($\mu g\ m^{-3}$).

Interior volume of building, V (m^3).

Infiltration (as flow), Q(in) ($m^3\ h^{-1}$); $Q/V = \lambda_{in}$ (h^{-1}) (as rate)

Aerosol penetration, P(d_p).

Ventilation (forced outside airflow via HVAC mechanical systems), Q(oa) ($m^3\ h^{-1}$).

Total mechanical system flow, Q(t) = Q(oa) + Q(r) (recirculation); in most residences with forced air systems, there is no explicit ventilation (outside air) flow so Q(t) = Q(r).

Aerosol filtration efficiency, $\varepsilon(d_p)$; transport through the filter is [1 − $\varepsilon(d_p)$].

Exfiltration/Mechanical Exhaust, Q(ex) = Q(in) + Q(oa) ($m^3\ h^{-1}$).

Indoor deposition rate, kj(d_p) (h^{-1}); the value of k depends upon the orientation of surface j; kj = $v_d(d_p)$ A_j/V, where v_d is the particle-size-dependent deposition velocity and A_j is the area of surface j.

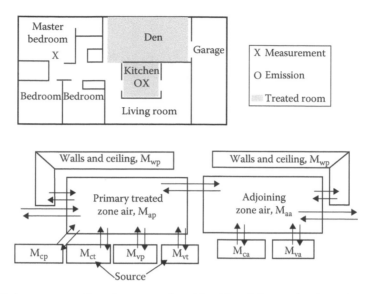

FIGURE 6.13 Indoor processes from the perspective of "multimedia" modeling, including fugacity analysis. (From Bennett, D.H. and Furtaw, E.J., *Environ. Sci. Technol.*, 38, 2142, 2004. With permission.)

Indoor resuspension, $R(d_p)$ ($\mu g \ h^{-1}$).

Indoor aerosol generation, $S_i (d_p)$ ($\mu g \ h^{-1}$); examples include cooking, smoking, etc.

Transformation processes T ($\mu g \ h^{-1}$); these include particle formation via gas-particle conversion; particle size change due to coagulation; particle size change from accumulation (hygroscopic growth) or loss (desiccation) of water; particle formation or loss due to phase change.

Tracking processes; these are not a direct source of airborne particles, rather they contribute to the source term for indoor resuspension. Material may be transported from outdoors to indoors or from room to room—typically via the soles of shoes or attached to clothing or other objects. The uptake and release processes that affect transport and mass transfer are poorly defined or quantified at present.

Using these concepts and definitions, the generalized mass-balance equation for aerosol concentrations in indoor air is given by

$$V\frac{dC_i(d_p)}{dt} = Q(in)P(d_p)C_o(d_p) + Q(oa)[1-\varepsilon(d_p)]C_o(d_p) + R(d_p) + S_i(d_p)$$

$$- Q(ex)C_i(d_p) - C_i(d_p)\sum_j v_j(d_p)A_j - Q(r)[1-\varepsilon(d_p)]C_i(d_p) + T \qquad (6.8)$$

Note that in this description we have explicitly incorporated particle size (d_p is the particle diameter) as an important variable for most of the parameters in the equation. As will be discussed later, some of these parameters are known to be highly dependent upon particle size. The first four terms in Equation 6.8 describe the indoor particle sources—the first term accounts for infiltrating outdoor air as an aerosol source, the second term provides for air brought into the building via a mechanical (HVAC) system, the third term represents resuspension of materials collected on the floor, and the fourth term describes the generation of particles indoors (e.g., combustion, cooking). Resuspension can, in principle, occur from materials collected on other surfaces, including other horizontal surfaces like a table top, but the dominant source of resuspension is that due to occupant activity (such as walking, vacuuming, etc.) on the floor surface. Note that all of the source terms are particle size dependent.

Three sink terms are shown here: the first term is removal of particles by the airflow out of the building (either due to exfiltration or exhaust), the second term is deposition to surface j, and the third term accounts for filtration removal due to air recirculation in a heating/cooling system with a filter. Only the first of these terms is independent of particle size. The second term—particle deposition to various indoor surfaces—is dependent upon both particle size and surface orientation while the third term incorporates filtration efficiency, which is inherently particle size dependent.

Other terms, represented here by T, can be added to the equation to account for particle transformation processes (Nazaroff and Cass, 1989) that can either add or remove particles or shift the particle size spectrum. Coagulation, for example, removes small particles and creates (fewer) larger ones, although for this process to be significant, the number concentrations of small particles need to be elevated. Coagulation has been observed in studies of environmental tobacco smoke where small particle concentrations exceed $\sim 10^5$ cm^{-3} (Klepeis et al., 2003). Studies of hygroscopic growth of various indoor aerosols have shown mixed effects—in some cases very little change in particle size was observed, while in other situations, for example, wood smoke, the mass median diameter increased (Dua et al., 1995; Dua and Hopke, 1996).

As a final example of transformation processes, phase change has been observed to alter particle concentrations indoors. In particular, ammonium nitrate aerosol, a major component of ambient aerosols in the western United States, is volatile and exists in equilibrium with its gas-phase constituents, ammonia and nitric acid. When these aerosols enter buildings via infiltration or ventilation, the temperature and relative humidity conditions may change, driving the equilibrium toward dissociation into the gas-phase species. These species in turn interact with indoor surfaces—especially nitric acid—leading to additional equilibrium dissociation. Under these conditions, indoor concentrations of ammonium nitrate particles are significantly reduced (Lunden et al., 2003a,b).

In order to illustrate the main features of the mass balance equation and to keep the number of variables tractable, we drop the filtration and transformation terms from Equation 6.8 and simplify some of the variables to yield

$$\frac{dC_i(d_p)}{dt} = \frac{\lambda_{in}P(d_p)C_o(d_p) + [R(d_p) + S_i(d_p)]}{V - \lambda_{in}C_i(d_p) - C_i(d_p)\sum_j k_j(d_p)} \qquad (6.9)$$

If we assume that the indoor space is well mixed and that the various terms vary slowly with time, the average indoor concentration can be approximated by the steady-state solution

$$C_i(d_p) = \frac{\lambda_{in}P(d_p)C_o(d_p) + [R(d_p) + S_i(d_p)]/V}{\lambda_{in} + \sum_j k_j(d_p)} \qquad (6.10)$$

Equation 6.10 is useful in "well-controlled" situations, such as laboratory-based experiments or where indoor sources, for example, are operated to produce high concentrations of aerosols so that variable contributions from outdoors, etc. can be neglected. This equation does help illustrate the balance between typical indoor aerosol sinks and sources.

However, in "real-world" situations, as would be the case in examining or estimating the transport and fate of aerosols in actual buildings, two important parameters in Equation 6.9 are often time-varying, outdoor aerosol concentrations and infiltration rates—especially in houses where mechanical systems are not used to supply ventilation air. If there is sufficient time-series information on the variability of these two parameters, then a "forward-marching" approach with a small time step, Δt, can be used, as has been recently demonstrated in the analysis of particle penetration data (Thatcher et al., 2003). The form of this equation is

$$C_i(d_p, t_2) = C_i(d_p, t_1) + P(d_p)C_o(d_p, t_1)\lambda_v(t_1)\Delta t + [R(d_p) + S(d_p)]/V$$

$$-C_i(d_p, t_1)\left[\lambda_v(t_1) + \sum_j k_j(d_p)\right]\Delta t \tag{6.11}$$

This example treats R and S as constant in time, but these could be restated to incorporate their time variability by using the form $R(d_p, t_1)\Delta t$. The use of this forward-marching approach requires that information/data are available for each parameter at each time step.

The key parameters determining aerosol transport and fate in buildings are discussed in the following.

6.5.2 Airflow and Aerosol Transport through Penetrations in Building Envelopes

Airflow through building envelopes occurs as a result of temperature and wind-driven pressure differences between the inside and outside of the building. Houses are often kept "closed" during wintertime heating periods and in some regions, during summertime cooling periods, thus limiting airflows to gaps, holes, or other inadvertent penetrations in the building shell, created as a result of (poor) construction practices, settling, and/or aging of building components, etc. Since these heating and cooling seasons often coincide with periods of high ambient aerosol concentrations, transport of particles into the building during these periods will occur through building leaks.

Few measurements have been made to determine the nature and significance of most building leaks. On average, the largest air leakage values are for leaks in the walls and floor (35%), in the ceiling (15%), and around windows and doors (15%) (Diamond and Grimsrud, 1984). Not all of these leaks will conduct aerosol into the building interior under most operating conditions. High leaks, such as those in the ceiling or openings such as fireplace or furnace chimneys, are usually locations for exfiltration. The physical dimensions of many such penetrations are poorly character-ized and the flows across the building envelope may proceed through tortuous pathways, making *a priori* prediction of aerosol penetration efficiency difficult.

Many studies have used indoor and outdoor mass measurements to provide broad categorical pen-etration factors. In a summary paper (Wallace, 1996), penetration factors of ~1 are derived for both fine ($PM_{2.5}$) and coarse (PM_{10}) mode particles, based on a statistical analysis of data from the PTEAM study. However, these estimates do not take into account the details in the size dependence of the deposition rates nor can they account for any potential size-dependent effect of the penetration process itself. Because these values are based on mass, the results will heavily depend upon the underlying size distribution of the aerosols and, in particular, the populations of the largest particle-size fractions.

Similar results have been reported by Thatcher and Layton (1995), based on experimental mea-surements of penetration; however, this study is limited to summertime measurements in one house and to particles larger than $1\,\mu m$. In contrast, recently published results from a series of controlled chamber experiments show a significant decline in the penetration factor as the test aerosol becomes larger than $2\,\mu m$ in diameter (Lewis, 1995). For particles larger than ~$6\,\mu m$, the penetration fraction was measured to be essentially zero.

A set of experiments were performed in two residential buildings to examine both aerosol pene-tration and deposition indoors (Thatcher et al., 2003). Continuous, size-resolved data were collected during a three-step experimental procedure:

1. Artificially enhancing indoor particle concentrations indoors and following the decay in concentrations.
2. Rapidly reducing particle concentrations through induced exfiltration by pressurizing the dwellings with HEPA-filtered air.
3. Following the rebound in infiltrating particle concentrations when overpressurization was stopped. Penetration factors as a function of particle size are shown in Figure 6.14.

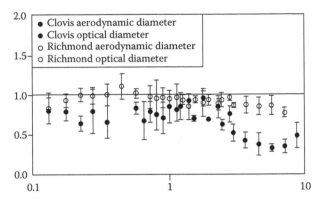

FIGURE 6.14 Particle penetration in two dwellings as measured by both optical and aerodynamic particle size instruments. (From Thatcher, T.L. et al., *Aerosol Sci. Technol.*, 37, 847, 2003.)

The dwelling labeled "Richmond" is an older structure with a much larger overall leakage area than the "Clovis" dwelling. As can be seen in Figure 6.2, the tighter structure had lower overall penetration factors and particles larger than ~3 μm had penetration factors as low as ~0.3. In contrast, penetration factors for the Richmond structure were close to 1 as a function particle size, with the lowest penetration factor equal to ~0.8.

6.5.3 Indoor Aerosol Deposition Rates

Indoor aerosol deposition rates as a function of particle size have been developed both empirically and the results summarized by Thatcher et al. (2002) as shown in Figure 6.15. An important feature of these curves is the strong dependence of the deposition decay rates on particle size. There is a minimum deposition rate for particle diameters between 0.1 and 0.2 μm ($k \sim 0.03\,h^{-1}$ under quiescent conditions), but the rates increase rapidly for particles both larger ($k \sim 0.3\,h^{-1}$ and $k \sim 18\,h^{-1}$ for 1.5 and 5 μm diameter particles, respectively) and smaller ($k \sim 0.15\,h^{-1}$ for 0.06 μm diameter particles).

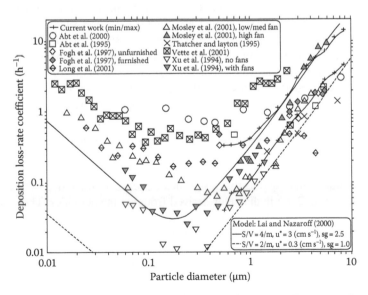

FIGURE 6.15 Particle deposition loss rates as reported in several studies. Current work refers to Thatcher et al. (2002), from which the figure originated. The model results cover a range of surface-to-volume ratios, internal air motion (friction velocities), and particle densities.

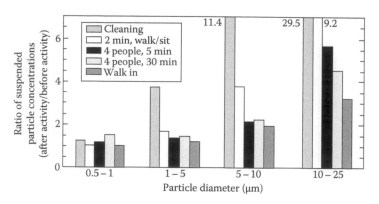

FIGURE 6.16 Ratios of suspended particle concentrations after human activities to concentrations before activity in a single-family residence. (From Thatcher, T.L. and Layton, D.W., *Atmos. Environ.*, 29, 1487, 1995.)

These deposition rates are a combination of deposition due to diffusion (more important for small particles) and that due to gravitational settling (more important for large particles). For a horizontal, upward facing surface, the two deposition mechanisms are about equal for $\sim$0.2 μm diameter, unit-density spheres. In contrast, diffusion is about 400 times more important for 0.01 μm particles, while settling is about 60 times more important for 1 μm particles (Hinds, 1999, p. 162). There are also effects due to the turbulence of the air within the enclosure, as reported by Lewis (1995), Xu et al. (1994), and Thatcher et al. (2002), and as shown in Figure 6.16.

Thus, the indoor aerosol dynamics depends strongly on both particle size and surface orientation. These differences may also have important implications for depositional losses in building leaks, depending upon the flow rates and particle velocities.

6.5.4 Resuspension Rate of Particles on Carpets/Floors

Only limited data are available on the rates with which particles on floors are suspended into air (expressed as a fraction of particulate loading on floor surfaces suspended per unit time) by human activities. Early work on resuspension indoors focused on the movement of radionuclides from floors to air. Healy (1971) developed a time-weighted-average resuspension rate of 5×10^{-4} h^{-1} for a house. This is comparable to the value of Murphy and Yocom (1986), who selected a value of 10^{-4} h^{-1}. Measurements of particles in indoor air using optical particle counters demonstrate that resuspension is a function of human activities as well as particle size. Kamens et al. (1991), for example, showed that the increase in suspended particles over the course of a day in a house corresponded to the activities of the residents. The apparent resuspension threshold of particles from floor surfaces is about 2 μm (Thatcher and Layton, 1995). The study of Clayton et al. (1993) supports this relationship. They found that the concentrations of fine particulate matter (i.e., particles under 2.5 μm in diameter) in the main living area of a sample of houses were highly correlated with the outdoor levels recorded at fixed monitors; however, the PM$_{10}$ concentrations only had a correlation coefficient of only 0.37 with fixed-site monitors. This suggests that a significant portion of the PM$_{10}$ particles collected were suspendable particles over 2 μm in diameter derived from human activities. The resuspension rates increase significantly with aerosols larger than $\sim$2 μm, as illustrated in Figure 6.16.

The mass loadings of soil/dust on floors in the literature vary from 0.136 to 0.870 g m^{-2}. The geometric mean of the mass loadings is 0.42 g m^{-2} with a geometric standard deviation of 1.88. Most of the data on dust loadings are based on studies dealing with lead contamination of the indoor environment (see, for example, Gulson et al., 1995). The mass loading of particulate matter on carpets/floors represents the mass available for resuspension of particles into indoor air. Floor dust also serves as a contact medium for infants/toddlers who crawl on floors and have hand-to-mouth behaviors that result in the ingestion of dusts. An unresolved issue is the relative suspendability of particles on carpets versus bare floors.

REFERENCES

Abt, E., Helen H., Suh, H.H., and Catalano, P. Relative contribution of outdoor and indoor particle sources to indoor concentrations, *Environ. Sci. Technol.*, *34*, 3579–3587 (2000).

Akagi, S.K., Yokelson, R.J., Wiedinmyer, C., Alvarado, M.J., Reid, J.S., Karl, T., Crounse, J.D., and Wennberg, P.O. Emission factors for open and domestic biomass burning for use in atmospheric models, *Atmos. Chem. Phys.*, *11*, 4039–4072 (2011).

Alshawa, A., Russell, A.R., and Nizkorodov, S.A. Kinetic analysis of competition between aerosol particle removal and generation by ionization air purifiers, *Environ. Sci. Technol.*, *41*, 2498–2504 (2007).

Andrejs, B., Fauss, J., Weigl, M., and Rietschel, P. Ventilation in kitchen—Aerosol concentration and key components in the vapor, *Proceedings of the Ninth International Conference on Indoor Air Quality and Climate*, Monterey, CA, June 30–July 5, 2002, Vol. 3, pp. 292–297 (2002).

Apple, J., Vicente, R., Yarberry, A., Lohse, N., Mills, E., Jacobson, A., and Poppendieck, D. Characterization of particulate matter size distributions and indoor concentrations from kerosene and diesel lamps, *Indoor Air*, *20*, 399–411 (2010).

Apte, M.G., Gundel, L.A., Dod, R.L., Russell, M.I., Singer, B.C., Sohn, M.D., Sullivan, D.P., Chang, G.-M., and Sextro, R.G. Indoor measurements of environmental tobacco smoke, Final Report to the Tobacco Related Disease Research Program, Lawrence Berkeley National Laboratory Report LBNL-49148 (2004).

Arhami, M., Minguill, C., Polidori, A., Schauer, J.J., Delfino, R.J., and Sioutas, C. Organic compound characterization and source apportionment of indoor and outdoor quasi-ultrafine particulate matter in retirement homes of the Los Angeles Basin, *Indoor Air*, *20*, 17–30 (2010).

Becker, K., Seiwert, M., Schulz, C., Kaus, S, Krause, C., and Seifert, B. German environmental survey 1998 (GerESIII); Pesticides and other pollutants in house dust, *Proceedings of the Ninth International Conference on Indoor Air Quality and Climate*, Monterey, CA, June 30–July 5, 2002, Vol. 4, pp. 883–887 (2002).

Bennett, D.H. and Furtaw, E.J. Fugacity–based indoor residential pesticide fate model, *Environ. Sci. Technol.*, *38*, 2142–2152 (2004).

Bennett, D.H., McKone, T.E., Evans, J.S., Nazaroff, W.W., Margni, M.D., Jolliet, O., and Smith, K.R., *Environ. Sci. Technol.*, *36*, 206A–211A (2002).

Berger-Preiss, E., Levsen, K., Leng, G., Idel, H., and Ranft, U. Indoor monitoring of homes with wool carpets, treated with permethrin, *Proceedings of the Ninth International Conference on Indoor Air Quality and Climate*, Monterey, CA, June 30–July 5, 2002, Vol. 1, pp. 1021–1025 (2002).

Bhangar, S., Mullen, N.A., Hering, S.V., Kreisberg, N.M., and Nazaroff, W.W., Ultrafine particle concentrations and exposures in seven residences in northern California, *Indoor Air*, *21*, 132–144 (2011).

Blondeau, P., Iordache, V., Poupard, O., Genin, D., and Allard, F., Relationship between outdoor and indoor air quality in eight French schools, *Indoor Air*, *15*, 2–12 (2005).

Bonvallot, N., Mandin, C, Mercier, F., Le Bot, B., and Glorennec, P. Health ranking of ingested semi-volatile organic compounds in house dust: An application to France. *Indoor Air*, *20*, 458–472 (2010).

Bornehag, G.G., Sundell, J., Weschler, C.J., Sigsgaard, T., Lundgren, B., Hasselgren, M., and Hägerhed-Engman, L. The association between asthma and allergic symptoms in children and phthalates in house dust: A nested case-control study, *Environ. Health Perspect.*, *112*, 1393–1397 (2004).

Bukowski, J.A. and Meyer, L.W. Simulated air levels of volatile organic compounds following different methods of indoor insecticide application, *Environ. Sci. Technol.*, *29*, 673–676 (1995).

Butt, C.M., Diamond, M.L., and Truong, J., Spatial distribution of polybrominated diphenyl ethers in Southern Ontario as measured in indoor and outdoor window organic films, *Environ. Sci. Technol.*, *38*, 724–731 (2004).

Byrne, M.A., Goddard, A.J.H., Lange, C., and Roed, J. Stable tracer aerosol deposition measurements in a test chamber, *J. Aerosol Sci.*, *26*, 645–653 (1995).

Carlsson, H., Nilsson. U., and Ostman, C. Video display units: An emission source of the contact allergenic flame retardant triphenyl phosphate in the indoor environment, *Environ. Sci. Technol.*, *34*, 3885–3889 (2000).

Cheng, Y.S., Bechtold, W.E., Yu, C.C., and Hung I.F. Incense smoke—Characterization and dynamics in indoor environments, *Aerosol Sci. Technol.*, *23*, 271–281 (1995).

Clayton, C.A., Perritt, R.L., Pellizzari, E.D., Wallace, L.A., Ozkaynak, H., and Spengler, J.D. Particle Total Exposure Assessment Methodology (PTEAM) Study: Distributions of aerosols and elemental concentrations in personal, indoor, and outdoor samples in a southern California community, *J. Expo. Anal. Environ. Epidemiol.*, *3*, 227–250 (1993).

Clougherty, J.E., Houseman, E.A., and Levy, J.I. Source apportionment of indoor residential fine particulate matter using land use regression and constrained factor analysis, *Indoor Air*, *21*, 53–66 (2011).

Colbeck, I., Nasir, Z.A., and Ali, Z. Characteristics of indoor/outdoor particulate pollution in urban and rural residential environment of Pakistan, *Indoor Air*, *20*, 40–51 (2010).

Coleman, B.K., Lunden, M.M., Destaillats, H., and Nazaroff, W.W. Secondary organic aerosol from ozone-initiated reactions with terpene-rich household products, *Atmos. Environ.*, *42*, 8234–8245 (2008).

Corsi, R.L. and Morrison, G. Smog and lemons: Discovering indoor air chemistry, *EM: Air and Waste Management Association's Magazine for Environmental Managers*, 14–20 (May 2003).

Cowan, C.E., Mackay, D., Feijtel, T.C.J., van de Meent, D., Di Guardo, A., Davies, J., and Mackay, N. (Eds.) *The Multimedia Fate Model: A Vital Tool for Predicting the Fate of Chemicals*, SETAC Press, Penascola, FL, 78 p. (1995).

Daisey, J.M. Tracers for assessing exposure to environmental tobacco smoke: What are they tracing? *Environ. Health Perspect.*, *107(Suppl. 2)*, 319–327 (1999).

Daisey, J.M. and Gundel, L.A. Tracing the sources of indoor aerosols using evolved gas analysis, *Aerosol Sci. Technol.*, *14*, 25–32 (1991).

Daly, B.J., Schmid, K., and Riediker, M. Contribution of fine particulate matter sources to indoor exposure in bars, restaurants, and cafes, *Indoor Air*, *20*, 204–212 (2009).

Destaillats, H., Lunden, M.M., Singer, B.C., Coleman, B.K., Hodgson, A.T., Weschler, C.J., and Nazaroff, W.W. Indoor secondary pollutants from the use of household products in the presence of ozone. A bench scale study, *Environ. Sci. Technol.*, *40*, 4421–4428 (2006a).

Destaillats, H., Maddalena, R.L., Singer, B.C., Hodgson A.T., and McKone, T.E. Indoor pollutants emitted by office equipment. A review of reported data and information needs, *Atmos. Environ.*, *42*, 1371–1388 (2008).

Destaillats, H., Singer, B.C., Lee, S.K., and Gundel, L.A. Effect of ozone on nicotine desorption from model surfaces: Evidence for heterogeneous chemistry, *Environ. Sci. Technol.*, *40*, 1799–1805 (2006b).

Diamond, R.C. and Grimsrud, D.T. Manual on indoor air quality, EPRI Report EM-3469, Electric Power Research Institute, Palo Alto, CA (1984).

Dua, S.K. and Hopke, P.K. Hygroscopic growth of assorted indoor aerosols, *Aerosol Sci. Technol.*, *24*, 151–160 (1996).

Dua, S.K., Hopke, P.K., and Raunemaa, T. Hygroscopic growth of consumer spray products, *Aerosol Sci. Technol.*, *23*, 331–340 (1995).

Dutta, A., Mukherjee, B., Das, D., Banerjee, A., and Ray, M.R. Hypertension with elevated levels of oxidized low-density lipoprotein and anticardiolipin antibody in the circulation of premenopausal Indian women chronically exposed to biomass smoke during cooking, *Indoor Air*, *21*, 165–176 (2011).

Erwin, E.A., Woodfolk, J.A., Custis, N., and Platts-Mills, T.A.E. Animal danders, *Immunol. Allergy Clin. North Am.*, *23*, 469–481 (2003).

Evans, G.J., Peers, A., and Sabaliauskas, K. Particle dose estimation from frying in residential settings, *Indoor Air*, *18*, 499–510 (2008).

Ferro, A.R., Kopperud, R.J., and Hildemann, L.M. Exposure to house dust from human activities, *Proceedings of Ninth International Conference on Indoor Air Quality and Climate*, Monterey, CA, June 30–July 2, 2002, Vol. 1, pp. 527–532 (2002).

Ferro, A.R., Kopperud, R.J., and Hildemann, L.M. Source strengths for indoor human activities that resuspend particulate matter, *Environ. Sci. Technol.*, *38*, 1759–1764 (2004).

Fine, P.M., Cass, G.R., and Simoneit B.R.T. Characterization of fine particle emissions from burning church candles, *Environ. Sci. Technol.*, *33*, 2352–2362 (1999).

Finlayson-Pitts, B.J. and Pitts J.N. *Chemistry of the Upper and Lower Atmosphere: Theory, Experiments, and Applications*, Academic Press, San Diego, CA, 969 p. (1999).

Fogh, C.L., Byrne, M.A., Roed, J., and Goddard, A.J.H. Size-specific indoor aerosol deposition measurements and derived I/O concentration ratios, *Atmos. Environ.*, *31*, 2193–2203 (1997).

Friedlander, S.K. *Smoke, Dust and Haze: Fundamentals of Aerosol Behavior*, 2nd edn., Oxford University Press, New York, 407 p. (2000).

Fujii, M., Shinohara, N., Lim, A., Otake, T., Kumagai, K., and Yanagisawa, Y. A study on emission of phthalate esters from plastic materials using a passive flux sampler, *Atmos. Environ.*, *37*, 5495–5504 (2003).

Fullerton, D.G., Bruce, N., and Gordon, S.B. Indoor air pollution from biomass fuel smoke is a major health concern in the developing world, *Trans. R. Soc. Trop. Med. Hyg.*, *102*, 843–851 (2008).

Gao, X., Yu, Q., Gu, Q., Chen, Y., Ding, K., Zhu, J., and Chen L. Indoor air pollution from solid biomass fuels combustion in rural agricultural area of Tibet, China, *Indoor Air*, *19*, 198–205 (2009).

Glytsos, T., Ondráček, J., Džumbová, L., Kopanakis, I., and Lazaridis, M. Characterization of particulate matter concentrations, during controlled indoor activities, *Atmos. Environ.*, *44*, 1539–1549 (2010).

Göhler, D., Stintz, M., Hillemann, L., and Vorbau, M. Characterization of nanoparticle release from surface coatings by the simulation of a sanding process, *Ann. Occup. Hyg.*, *54*, 615–624, (2010).

Gorini, G., Moshammer, H., Sbrogiò, L., Gasparrini, A., Nebot, M., Neuberger, M., Tamang, E., Lopez, M.J., Galeone, D., and Serrahima, E. Italy and Austria before and after study: Second-hand smoke exposure in hospitality premises before and after 2 years from the introduction of the Italian smoking ban, *Indoor Air*, *18*, 328–334 (2008).

Goss, K.-U. and Schwarzenbach, R.P. Gas/solid and gas/liquid partitioning of organic compounds: Critical evaluation of the interpretation of equilibrium constants, *Environ. Sci. Technol.*, *32*, 2025–2032 (1998).

Gulson, B.L., Davis, J.J., Mizon, K.J., Korsch, M., and Bawden-Smith, J. Source of lead in soil and dust and the use of dust fallout as a sampling medium, *Sci. Total Environ.*, *166*, 245–262 (1995).

Gundel, L.A., Mahanama, K.R.R., and Daisey, J.M. Semi-volatile and particulate polycyclic aromatic hydrocarbons in environmental tobacco smoke: Cleanup, speciation and emission factors, *Environ. Sci. Technol.*, *29*, 1607–1614 (1995).

Guo, Z. Simulation tool kit for indoor air quality and inhalation exposure (IAQX), version 1.0, user's guide, U.S. EPA Report No. EPA-600/R-00-094, National Risk Management Research Laboratory, Research Triangle Park, NC (2000).

Hammond, S.K. and Leaderer, B.P. A diffusion monitor to measure exposure to passive smoking, *Environ. Sci. Technol.*, *21*, 494–497 (1987).

Harley, N.H., Chittaporn, P., Fisenne, I.M., and Pamela Perry, P. ^{222}Rn decay products as tracers of indoor and outdoor aerosol particle size, *J. Environ. Radioactivity*, *51*, 27–35, (2000).

Harner, T. and Bidleman, T.F. Octanol-air partition coefficient for describing particle/gas partitioning of aormatic compounds in urban air, *Environ. Sci. Technol.*, *32*, 1494–1502 (1998).

He, C., Morawska, L., Hitchins, J., and Gilbert, D., Contribution from indoor sources to particle number and mass concentrations in residential houses, *Atmos. Environ.*, *38*, 3405–3415 (2004).

He, C., Morawska, L., and Taplin, L. Particle emission characteristics of office printers, *Environ. Sci. Technol.*, *41*, 6039–6045 (2007).

He, C., Morawska, L., Wang, H., Jayaratne, R., McGarry, P., Johnson, G.R., Bostrom, T., Gonthier, J., Authemayou, S., and Ayoko, G. Quantification of the relationship between fuser roller temperature and laser printer emissions, *J. Aerosol Sci.*, *41*, 523–530 (2010).

Healy, J.W. and Dennis, A.J. (Eds.) *Surface Contamination: Decision Levels*, Los Alamos Scientific Laboratory, Los Alamos, NM, 1971.

Heikkinen, M.S.A., Hjelmroos-Koski, M.K., Häggblom, M.M., and Macher, J.M. *Aerosols Handbook: Measurement, Dosimetry, and Health Effects*, (Ruzer, L.S. and Harley, N.H., eds.), CRC Press, Boca Raton, FL (2004).

Hellweg, S., Demou, E., Bruzzi, R., Meijer, A., Rosenbaum, R.K., Huijbregts, M.A.J., and McKone, T.E. Integrating human indoor air pollutant exposure within life cycle impact assessment, *Environ. Sci. Technol.*, *43*, 1670–1679 (2009).

Hildemann, L.M., Markowski, G.R., and Cass, G.R. Chemical composition of emissions from urban sources of fine organic aerosol, *Environ. Sci. Technol.*, *25*, 744–759 (1991).

Hinds, W.C. *Aerosol Technology*, 2nd edn., Wiley-Interscience, New York, 483 p. (1999).

Hodgson, A.T., Beal, D., and McIlvaine, J.E.R. Sources of formaldehyde, other aldehydes and terpenes in a new manufactured house, *Indoor Air*, *12*, 2135–2242 (2002).

Hodgson, A.T., Rudd, A.F., Beal, D., and Chandra, S. Volatile organic compound concentrations and emission rates in new manufactured and site-built houses, *Indoor Air*, *10*, 178–192 (2000).

Hoffmann, T., Odum, J.R., Bowman, F., Collins, D., Klockow, D., Flagan, R.C., and Seinfeld, J.H. Formation of organic aerosols from the oxidation of biogenic hydrocarbons, *J. Atmos. Chem.*, *26*, 189–222 (1997).

Hornig, J.F., Soderberg, R.H., Barefoot III, A.C., and Galasyn, J.F. Woodsmoke analysis: Vaporization losses of PAH from filters and levoglucosan as a distinctive marker for woodsmoke, Cooke, M. (Ed.) *Polynuclear Aromatic Hydrocarbons*, Battelle Press, Columbus, OH 561–568 pp. (1985).

Hubbard, H.F., Coleman, B.K., Sarwar, G., and Corsi, R.L. Effects of an ozone-generating air purifier on indoor secondary particles in three residential dwellings, *Indoor Air*, *15*, 432–444 (2005).

Humbert, S., Marshall, J.D., Shaked, S., Spadaro, J.V., Nishioka, Y., Preiss, P., McKone, T.E., Horvath, A., and Jolliet, O. Intake fraction for particulate matter: Recommendations for life cycle impact assessment, *Environ. Sci. Technol.*, *45*, 4808–4816 (2011).

Huss, A., Kooijman, C., Breuer, M., Böhler, P., Zünd, T., Wenk, S., and Röösli, M. Fine particulate matter measurements in Swiss restaurants, cafés and bars: What is the effect of spatial separation between smoking and non-smoking areas? *Indoor Air*, *20*, 52–60 (2009).

Institute of Medicine (IOM), *Climate Change, the Indoor Environment, and Health*, The National Academies Press, Washington, DC, 261 p. (2011).

Jakober, C. and Phillips, T. Evaluation of ozone emissions from portable indoor air cleaners: Electrostatic precipitators and ionizers; Staff Technical Report to the California ARB, www.arb.ca.gov/research/indoor/esp_report.pdf (2008).

Jang, M., Kamens, R.M., Leach, K.B., and Strommen, M.R. A thermodynamic approach to group contribution methods to model the partitioning of semi-volatile organic compounds on atmospheric particulate matter, *Environ. Sci. Technol.*, *31*, 2805–2811 (1997).

Jenkins, R.A., Guerin, M.R., and Tompkins, B.A. *The Chemistry of Environmental Tobacco Smoke: Composition and Measurement*, 2nd edn., Lewis, Boca Raton, FL, 467 p. (2000).

Jetter, J.J., Guo, Z.S., McBrian, J.A., and Flynn, M.R. Characterization of emissions from burning incense, *Sci. Total Environ.*, *295*, 51–67 (2002).

Ji, X., Le Bihan, O., Ramalho, O., Mandin, C., D'Anna, B., Martinon, L., Nicolas, M., Bard, D., and Pairon, J.C. Characterization of particles emitted by incense burning in an experimental house, *Indoor Air*, *20*, 47–58 (2010).

Jimenez, J.R., Claiborn, C.S., Dhammapala, R.S., and Simpson, C.D. Methoxyphenols and levoglucosan ratios in $PM_{2.5}$ from wheat and Kentucky bluegrass stubble burning in eastern Washington and northern Idaho, *Environ. Sci. Technol.*, *41*, 7824–7829 (2007).

Johansson, J., Olander, L., and Johansson, R. Long-term test of the effect of room air cleaners on tobacco smoke, *Proceedings of the sixth International Conference on Indoor Air Quality and Climate*, Vol. 6, Helsinki, Finland, pp. 387–392 (1993).

Kamens, R.M., Jang, M., Chien, C.J., and Leach, K. Aerosol formation from the reaction of α-pinene and ozone using a gas-phase kinetics and gas-particle partitioning theory, *Environ. Sci. Technol.*, *35*, 1394–1405 (1999).

Kamens, R., Lee, C.-T., Wiener, R., and Leith, D. A study to characterize indoor particles in three non-smoking homes, *Atmos. Environ.*, *25A*, 939–948 (1991).

Kavouras, I.G., Stratigakis, N., and Stephanou, E.G. *Iso-* and *anteiso*-alkanes: Specific tracers of environmental tobacco smoke in indoor and outdoor particle-size distributed urban aerosols, *Environ. Sci. Technol.*, *32*, 1369–1377 (1998).

Kemmlein, S., Hahn, O., and Jann, O. Emissions of organophosphate and brominated flame retardants from selected consumer products and building materials, *Atmos. Environ.*, *37*, 5485–5493 (2003).

Kerr, G. and Thi, L.C.N., Identification of contaminants, exposures and control options for construction/renovation activities, Final Report for ASHRAE Research Project 961-TRP, 220 p. (2001).

Kleindienst, T.E., Jaoui, M., Lewandowski, M., Offenberg, J.H., Lewis, C.W., Bhave, P.V., and Edney, E.O. Estimates of the contributions of biogenic and anthropogenic hydrocarbons to secondary organic aerosol at a Southeastern US location. *Atmos. Environ.*, *41*, 8288–8300 (2007).

Klepeis, N.E., Apte, M.G., Gundel, L.A., Sextro, R.G., and Nazaroff, W.W. Determining size-specific emission factors for environmental tobacco smoke particles, *Aerosol Sci. Technol.*, *37*, 780–790 (2003).

Knol, A.B., de Hartog, J.J., Boogaard, H., Slottje, P., van der Sluijs, J.P., Lebret, E., Cassee, F.R. et al. Expert elicitation on ultrafine particles: Likelihood of health effects and causal pathways, *Part. Fibre Toxicol.*, *6*, 19 (2009), doi: 10.1186/1743-8977-6-19.

Koponen, I.K., Jensen, K.J., and Schneider, T. Comparison of dust released from sanding conventional and nanoparticle-doped wall and wood coatings, *J. Expo. Sci. Environ. Epidemiol.*, *21*, 408–418 (2011).

Koutrakis, P. and Briggs, S.L.K. Source apportionment of indoor aerosols in Suffolk and Onondaga Counties, New York, *Environ. Sci. Technol.*, *26*, 521–527 (1992).

Lai, A.C.K. and Nazaroff, W.W. Modeling indoor particle deposition from turbulent flow onto smooth surfaces, *J. Aerosol. Sci.*, *31*, 463–476 (2000).

Landis, M.S., Norris, G.A., Williams, R.W., and Weinstein, J.P. Personal exposures to $PM_{2.5}$ mass and trace elements in Baltimore, MD, USA, *Atmos. Environ.*, *35*, 6511–6524 (2001).

Lane, D.A. (Ed.) *Gas and Particle Phase Measurements of Atmospheric Organic Compounds*, Vol. 2 of Advances in Environmental, Industrial and Process Control Technologies, Gordon and Breach, Amsterdam, the Netherlands, 402 p. (1999).

Lane, T.E., Pinder, R.W., Shrivastava, M., Robinson, A.L., and Pandis, S.N. Source contributions to primary organic aerosol: Comparison of the results of a source-resolved model and the chemical mass balance approach, *Atmos. Environ.*, *41*, 3758–3776 (2007).

Leaderer, B.P., Boone, P.M., and Hammond, S.K. Total particle, sulfate and acidic aerosol emissions from kerosene space heaters. *Environ. Sci. Technol.*, *24*, 908–912 (1990).

Leaderer, B.P, Naeher, L., Jankun, T., Balenger, K., Holford, T.R., Toth, C., Sullivan, J., Wolfson, J.M., and Koutrakis, P. Indoor, outdoor, and regional summer and winter concentrations of PM_{10}, $PM_{2.5}$, SO_4^{2-}, H^+, NH_4^+, NO_3^-, NH_3, and nitrous acid in homes with and without kerosene space heaters, *Environ. Health Perspect.*, *107*, 223–231 (1999).

Lewis, S. Solid particle penetration into enclosures, *J. Hazard. Mater.*, *43*, 195–216 (1995).

Liang, C., Pankow, J.F., Odum, J.R., and Seinfeld, J.H. Gas/particle partitioning of semivolatile organic compounds to model inorganic, organic and ambient smog aerosols, *Environ. Sci. Technol.*, *31*, 3086–3092 (1997).

Lim, J.-M., Jeong, J.-H., Lee, J.-H., Moon, J.-H., Chung, Y.-S., and Kim, K.-H. The analysis of $PM_{2.5}$ and associated elements and their indoor/outdoor pollution status in an urban area, *Indoor Air*, *21*, 145–155 (2011).

Lioy, P.J., Freeman, N.C.G., and Millette, J.R. Dust: A metric for use in residential and building exposure assessment and source characterization, *Environ. Health Perspect.*, *110*, 969–983 (2002).

Liu, L.J.S., Box, M., Kalman, D., Kaufman, J., Koenig, J., Larson, T., Lumley, T., Sheppard, L., and Wallace, L. Exposure assessment of particulate matter for susceptible populations in Seattle, *Environ. Health Perspect.*, *111*, 909–918 (2003).

Liu, S. and Zhu, Y. A case study of exposure to ultrafine particles from secondhand tobacco smoke in an automobile, *Indoor Air*, *20*, 412–423 (2010).

Loan, R., Siebers, R., Fitzharris, P., and Crane, J. House dust-mite allergen and cat allergen variability within carpeted living room floors in domestic dwellings, *Indoor Air*, *13*, 232–236 (2003).

Logue, J.M., McKone, T.E., Sherman, M.H., and Singer, B.C. Hazard assessment of chemical air contaminants measured in residences, *Indoor Air*, *21*, 92–109 (2011).

Long, C.M., Suh, H.H., Catalano, P.J., and Koutrakis, P. Using time- and size-resolved particulate data to quantify indoor penetration and deposition behavior, *Environ. Sci. Technol.*, *35*, 2089–2099 (2001).

Lunden, M.M., Kirchstetter, T.W., Thatcher, T.T., Hering, S.V., and Brown, N.J. Factors affecting the indoor concentrations of carbonaceous aerosols of outdoor origin, *Atmos. Environ.*, *42*, 5660–5671 (2008).

Lunden, M.M., Revzan, K.L., Fischer, M.L., Thatcher, T.L., Littlejohn, D., Hering, S.V., and Brown, N.J. The transformation of outdoor ammonium nitrate aerosols in the indoor environment, *Atmos. Environ.*, *37*, 5633–5644 (2003a).

Lunden, M.M., Thatcher, T.L., Hering, S.V., and Brown, N.J. Use of time and chemically resolved particulate data to characterize the infiltration of outdoor $PM_{2.5}$ into a residence in the San Joaquin Valley, *Environ. Sci. Technol.*, *37*, 4724–4732 (2003b).

Lung, S.-C.C. and Hu, S.-C. Generation rates and emission factors of particulate matter and particle-bound polycyclic aromatic hydrocarbons of incense sticks, *Chemosphere*, *50*, 673–679 (2003).

Macher, J.M., Ammann, H., Burge, H.A., Milton, D.K., and Morey, P.R. (Eds.). *Bioaerosols: Assessment and Control*, American Conference of Governmental Industrial Hygienists (ACGIH), Cincinnati, OH (1999).

Maddalena, R., Destaillats, H., Russell, M.L., Hodgson, A.T., Hammond, S.K., and McKone, T.E. Emissions measurements to characterize residential exposures to indoor pollutants from printers, *Epidemiology*, *19*, S265–S266 (2008).

Mader, B.T., Flagan, R.C., and Seinfeld, J.H. Sampling atmospheric carbonaceous aerosols using a particle trap impactor/denuder sampler, *Environ. Sci. Technol.*, *35*, 4857–4867 (2001).

Mader, B.T. and Pankow, J.F. Gas/solid partitioning of semivolatile organic compounds (SOCs) to air filters. 1. Partitioning of polychlorinated dibenzodioxins, polychlorinated dibenzofurans and polycyclic aromatic hydrocarbons to quartz fiber filters, *Atmos. Environ.*, *34*, 4879–4887 (2000).

Mader, B.T. and Pankow, J.F. Gas/solid partitioning of semivolatile organic compounds (SOCs) to air filters. 2. Partitioning of polychlorinated dibenzodioxins, polychlorinated dibenzofurans and polycyclic aromatic hydrocarbons to Teflon membrane filters, *Atmos. Environ.*, *35*, 1217–1223 (2001a).

Mader, B.T. and Pankow, J.F. Gas/solid partitioning of semivolatile organic compounds (SOCs) to air filters. 3. An analysis of gas adsorption artifacts in measurements of atmospheric SOCs and organic carbon (OC) when using Teflon membrane filters and quartz fiber filters, *Environ. Sci. Technol.*, *35*, 3422–3432 (2001b).

Manchester-Neesvig, J.B., Schauer, J.J., and Cass, G.R. The distribution of particle-phase organic compounds in the atmosphere and their use for source apportionment during the southern California children's health study, *J. Air Waste Manage. Assoc.*, *53*, 1065–1079 (2003).

Mannix, R.C., Nguyen, K.P., Tan, E.W., Ho, E.E., and Phalen R.F. Physical characterization of incense aerosols, *Sci. Total Environ.*, *193*, 149–158 (1996).

Manuel, J., A healthy home environment? *Environ. Health Perspect.*, *107*, A352–A357 (1999).

McAuley, T.R., Fisher, R., Zhou, X., Jaques, P.A., and Ferro, A.R. Relationships of outdoor and indoor ultrafine particles at residences downwind of a major international border crossing in Buffalo, NY, *Indoor Air*, *20*, 298–308 (2010).

McDonald, J.D., Zielinska, B., Fujita, E.M., Sagebiel, J.C., Chow, J.C., and Watson, J.G. Fine particle and gaseous emission rates from residential wood combustion, *Environ. Sci. Technol.*, *34*, 2080–2091 (2000).

Mills, E. The specter of fuel-based lighting, *Science*, *308*, 1263–1264 (2005).

Mitchell, C.S., Zhang, J., Sigsgaard, T., Jantunen, M., Lioy, P.J., Samson, R., and Karol, M.H. Current state of the science: Health effects and indoor environmental quality, *Environ. Health Perspect.*, *115*, 958–964 (2007).

Mølhave, L. Inflammatory and allergic responses to airborne office dust in five human provocation experiments, *Indoor Air*, *18*, 261–270 (2008).

Morawska, L., He, C., Johnson, G., Jayaratne, R., Tunga Salthammer, T., Wang, H., Uhde, E. et al. An investigation into the characteristics and formation mechanisms of particles originating from the operation of laser printers, *Environ. Sci. Technol.*, *43*, 1015–1022 (2009).

Morawska, L. and Salthammer, T. (Eds.) *Indoor Environment: Airborne Particles and Settled Dust*, Wiley, New York, 350 p. (2003).

Morrison, G.C., Nazaroff, W.W., Cano-Ruiz, J.A., Hodgson, A.T., and Modera, M.PT. Indoor air quality impacts of ventilation ducts: Ozone removal and emissions of volatile organic compounds, *J. Air Waste Manage. Assoc.*, *48*, 941–952 (1998).

Morrison, G.C. and Nazaroff, W.W. The rate of ozone uptake on carpets: Experimental studies, *Environ. Sci. Technol.*, *36*, 4963–4968 (2000).

Mosley, R.B., Greenwell, D.J., Sparks, L.E., Guo, Z., Tucker, W.G., Fortmann, R., and Whitfield, C. Penetration of ambient fine particles into the indoor environment, *Aerosol Sci. Technol.*, *34*, 127–136 (2001).

Mulcahy, M., Evans, D.S., Hammond, S.K., Repace, J.L., and Byrne, M. Secondhand smoke exposure and risk following the Irish smoking ban: An assessment of salivary cotinine concentrations in hotel workers and air nicotine levels in bars, *Tob. Control*, *14*, 384–388 (2005).

Mullen, N.A., Liu, C., Zhang, Y., Wang, S., and Nazaroff, W.W. Ultrafine particle concentrations and exposures in four high-rise Beijing apartments, *Atmos. Environ.*, *45*, 7574–7582, doi: org/10.1016/j.atmosenv.2010.07.060 (2011).

Murphy, B.L. and Yocom, J.E. Migration factors for particulates entering the indoor environment, *Proceedings of the 79th Annual Meeting of the Air Pollution Control Association*, June 22–27, Paper 86–7.2, APCA, Pittsburgh, PA, (1986).

Naumova, Y.Y., Offenberg, J.H., Eisenreich, S.J., Meng, Q., Polidori, A., Turpin, B.J., Weisel, C.P. et al. Gas/particle distribution of polycyclic aromatic hydrocarbons in coupled outdoor/indoor atmospheres, *Atmos. Environ.*, *37*, 703–719 (2003).

Navas-Acien, A., Peruga, A., Breysse, P., Zavaleta, A., Blanco-Marquizo, A., Pitarque, R., Acuña, M. et al. Secondhand tobacco smoke in public places in Latin America, 2002–2003. *J. Am. Med. Assoc.*, *291*, 2741–2745 (2004).

Nazaroff, W.W. and Alvarez-Cohen, L. *Environmental Engineering Science*, Wiley, New York, 690 p. (2001).

Nazaroff, W.W. and Cass, G.R. Mathematical modeling of indoor aerosol dynamics, *Environ. Sci. Technol.*, *23*, 157–166 (1989).

Nazaroff, W.W. and Weschler, C.J. Cleaning products and air fresheners: Exposure to primary and secondary air pollutants, *Atmos. Environ.*, *38*, 28411–2865 (2004).

Nazaroff, W.W., Weschler, C.J., and Corsi, R.L. Indoor air chemistry and physics, *Atmos. Environ.*, *37*, 5431–5453 (2003).

Nebot, M., Lopez, M.J., Gorini, G., Neuberger, M., Axelsson, S., Pilali, M., Fonseca, C. et al. Environmental tobacco smoke exposure in public places of European cities. *Tob. Control*, *14*, 60–63 (2005).

Nishioka, M.G., Burkholder, H.M., Brinkman, M.C., Gordon S.M., and Lewis, R.G. Measuring transport of lawn-applied herbicide acids from turf to home: Correlation of dislodgeable 2,4-D turf residues with carpet dust and carpet surface residues, *Environ. Sci. Technol.*, *30*, 3313–3320 (1996).

Nriagu, J.O. and Kim, M.J. Emissions of lead and zinc from candles with metal-core wicks, *Sci. Total Environ.*, *250*, 37–41 (2000).

Olson, D. and Burke, J. PM$_{2.5}$ source strengths cooking RTP PM panel study, *Environ. Sci. Technol.*, *40*, 163–169 (2006).

Ozkaynak, H., Hue, J., Spengler, J., Wallace, L., Pellizari, E., and Jenkins, P. Personal exposure to airborne particles and metals; results from the particle TEAM study in Riverside CA, *J. Expo. Anal. Environ. Epidemiol.*, *6*, 57–78 (1996).

Padhi, B.K. and Padhy, P.K. Domestic fuels, indoor air pollution, and children's health, *Ann. NY Acad. Sci.*, *1140*, 209–217 (2008).

Pagels, J., Wierzbicka, A., Nilsson, E., Isaxon, C., Dahl, A., Gudmundsson, A., Swietlicki, E., and Bohgard, M., Chemical composition and mass emission factors of candle smoke particles, *Aerosol Sci.*, *40*, 193–208 (2009).

Pang, Y., Gundel, L.A., Larson, T., Finn, D., Liu, S. (L.-J.), and Claiborn, C.S. Development and evaluation of a personal particulate organic and mass sampler (PPOMS), *Environ. Sci. Technol.*, *36*, 5205–5210 (2002).

Pankow, J.F. Review and comparative analysis of the theories on partitioning between the gas and aerosol particulate phases in the atmosphere, *Atmos. Environ.*, *21*, 2275–2283 (1987).

Pankow, J.F. An absorption model of gas/particle partitioning of organic compounds in the atmosphere, *Atmos. Environ.*, *28*, 185–188 (1994).

Pankow, J.F. Fundamentals and mechanisms of gas/particle partitioning in the atmosphere, Chapter 3, Lane, D.A. (Ed.) *Gas and Particle Phase Measurements of Atmospheric Organic Compounds*, Vol. 2 of *Advances in Environmental, Industrial and Process Control Technologies*, Gordon and Breach, Amsterdam, 402 p. (1999).

Parker, J.L., Larson, R.R., Eskelson, E., Wood, E.M., and Veranth, J.M. Particle size distribution and composition in a mechanically ventilated school building during air pollution episodes. *Indoor Air*, *18*, 386–393 (2008).

Pekey, B., Bozkurt, Z.B., Pekey, H., Doğan, G., Zararsiz, A., Efe, N., and Tuncel, G. Indoor/outdoor concentrations and elemental composition of $PM_{10}/PM_{2.5}$ in urban/industrial areas of Kocaeli City, Turkey, *Indoor Air*, *20*, 112–125 (2010).

Phillips, T. and Jakober, C. Evaluation of ozone emissions from portable indoor "air cleaners" that intentionally generate ozone, Staff Technical Report to the California ARB, www.arb.ca.gov/research/indoor/o3g-rpt.pdf (2006).

Po, J.Y.T., FitzGerald, J.M., and Carlsten, C. Respiratory disease associated with solid biomass fuel exposure in rural women and children: Systematic review and meta-analysis, *Thorax*, *66*, 232–239 (2011).

Pope, C.A., Ezzati, M., and Dockery, D.W. Fine-particulate air pollution and life expectancy in the United States, *N. Engl. J. Med.*, *360*, 376–386 (2009).

Qian, J. and Ferro, A.R. Resuspension of dust particles in a chamber and associated environmental factors, *Aerosol Sci. Technol.*, *42*, 566–578 (2008).

Rao, V., Frank, N., Rush, A., and Dimmick, F. Chemical speciation of $PM_{2.5}$ in urban and rural areas, National Air Quality and Emissions Trends Report, 2003 Special Studies, S13, US EPA, www.epa.gov/airtrends/aqtrnd03/pdfs/2_chemspecofpm25.pdf (2003).

Rapp, A.O., Brandt, K., Peek, R.-D., and Schmitt, U. Quantitative measurement and analysis of wood dust collected in German woodworking companies, *Eur. J. Wood Wood Prod.*, *55*, 141–147 (1997).

Reff, A., Turpin, B.J., Porcja, R.J., Giovennetti, R., Cui, W., Weisel, C.P., Zhang, J. et al. Functional group characterization of indoor, outdoor, and personal $PM_{2.5}$: Results from RIOPA, *Indoor Air*, *15*, 53–61 (2005).

Robinson, A.L., Donahue, N.M., and Rogge, W.F. Photochemical oxidation and changes in molecular composition of organic aerosol in the regional context, *J. Geophys. Res. Atmos.*, *111*, D03302, doi: 10.1029/2005JD006265 (2006a).

Robinson, A.L., Subramanian, R., Donahue N.M., Bernardo-Bricker, A., and Rogge, W.F. Source apportionment of molecular markers and organic aerosol. 2. Biomass smoke, *Environ Sci Technol.*, *40*, 7811–7819 (2006b).

Robinson, A.L., Subramanian, R., Donahue, N.M., Bernardo-Bricker, A., and Rogge, W.F. Source apportionment of molecular markers and organic aerosol. 3. Food cooking emissions, *Environ. Sci. Technol.*, *40*, 7820–7827 (2006c).

Rogge, W.F., Hildemann, L.M., Mazurek, M.A., and Cass, G.R. Sources of fine organic aerosol. 6. Cigarette smoke in the urban atmosphere, *Environ. Sci. Technol.*, *28*, 1375–1388 (1994).

Rogge, W.F., Hildemann, L.M., Mazurek, M.A., Cass, G.R., and Simoneit, B.R.T. Sources of fine organic aerosol. 1. Charbroilers and meat cooking operations, *Environ. Sci. Technol.*, *25*, 1112–1125 (1991).

Rogge, W.F., Hildemann, L.M., Mazurek, M.A., Cass, G.R., and Simoneit, B.R.T. Sources of fine organic aerosol. 2. Non-catalyst and catalyst-equipped automobiles and heavy-duty diesel trucks, *Environ. Sci. Technol.*, *27*, 636–651 (1993a).

Rogge, W.F., Hildemann, L.M., Mazurek, M.A., Cass, G.R., and Simoneit, B.R.T. Sources of fine organic aerosol. 3. Road dust, tire debris and organometallic brake lining dust: Reads as sources and sinks, *Environ. Sci. Technol.*, *27*, 1892–1904 (1993b).

Rogge, W.F., Hildemann, L.M., Mazurek, M.A., Cass, G.R., and Simoneit, B.R.T. Sources of fine organic aerosol. 4. Particulate abrasion products from leaf surfaces of urban plants, *Environ. Sci. Technol.*, *27*, 2700–2711 (1993c).

Rogge, W.F., Hildemann, L.M., Mazurek, M.A., Cass, G.R., and Simoneit, B.R.T. Sources of, fine organic aerosol. 5. Natural gas home appliances, *Environ. Sci. Technol.*, *27*, 2736–2744 (1993d).

Rogge, W.F., Hildemann, L.M., Mazurek, M.A., Cass, G.R., and Simoneit, B.R.T. Sources of fine organic aerosol. 7. Hot asphalt roofing tar pot fumes, *Environ. Sci. Technol.*, *31*, 2726–2730 (1997a).

Rogge, W.F., Hildemann, L.M., Mazurek, M.A., Cass, G.R., and Simoneit, B.R.T. Sources of fine organic aerosol. 8. Boilers burning No. 2 distillate fuel oil, *Environ. Sci. Technol.*, *31*, 2731–2737 (1997b).

Rogge W.F., Hildemann, L.M., Mazurek M.A., Cass G.R., and Simoneit B.R.T. Sources of fine organic aero-sol. 9. Pine, oak and synthetic log combustion in residential fireplaces, *Environ. Sci. Technol.*, *32*, 13–22 (1998).

Rogge, W.F., Medeiros, P.M., Bernd R.T., and Simoneit, B.R.T. Organic marker compounds in surface soils of crop fields from the San Joaquin Valley fugitive dust characterization study, *Atmos. Environ.*, *41*, 8183–8204 (2007).

Roy, A.A., Wagstrom, K.M., Adams, P.J., Pandis, S.N., and Robinson, A.L. Quantification of the effects of molecular marker oxidation on source apportionment estimates for motor vehicles, *Atmos. Environ.*, *45*, 3132–3140 (2011).

Rudel, R.A., Brody, J.G., Spengler, J.D., Vallarino, J., Geno, P.W., Sun, G., and Yau, A. Identification of selected hormonally active agents and animal mammary carcinogens in commercial and residential air and dust samples, *J. Air Waste Manage. Assoc.*, *51*, 499–513 (2002).

Rudel, R.A., Camann, D.E., Spengler, J.D., Korn, L.R., and Brody J.G. Alkylphenols, pesticides, phthalates, polybrominated diphenyl ethers, and other endocrine-disrupting compounds in indoor air and dust, *Environ. Sci. Technol.*, *37*, 4543–4553 (2003).

Rumchev, K., Jamrozik, K., Stick, S., and Spickett, J. How free of tobacco smoke are 'smoke-free' homes? *Indoor Air*, *18*, 202–208 (2008).

Salthammer, T. (Ed.) *Organic Indoor Air Pollutants: Occurrence, Measurement, Evaluation*, Wiley, New York, 344 p. (1999).

Salthammer, T., Fuhrmann, F., and Uhde, E. Flame retardants in the indoor environment—Part II: Release of VOCs (triethylphosphate and halogenated degradation products from polyurethane, *Indoor Air*, *13*, 49–52 (2003).

Sarnat, J.A., Marmur, A., Klein, M., Kim, E., Russell, A.G., Sarnat, E.T., Mulholland, J.A., Hopke, P.K., and Tolbert, P.E. Fine particle sources and cardiorespiratory morbidity: An application of chemical mass balance and factor analytical source-apportionment methods, *Environ. Health Perspect.*, *116*, 459–466 (2008).

Sarwar, G., Corsi, R., Allen, D., and Wechsler, C. The significance of secondary organic aerosol formation and growth in buildings: Experimental and computational evidence, *Atmos. Environ.*, *37*, 1365–1381 (2003).

Schauer, J.J., Fraser, M.P., Cass, G.R., and Simoneit, B.R.T. Source reconciliation of atmospheric gas-phase and particle-phase pollutants during a severe photochemical episode, *Environ. Sci. Technol.*, *36*, 3806–3814 (2002a).

Schauer, J.J., Kleeman, M.J., Cass, G.R., and Simoneit, B.R.T. Measurement of emissions from air pollution sources. 1. C_1 through C_{29} organic compounds from meat charbroiling, *Environ. Sci. Technol.*, *33*, 1566–1577 (1999a).

Schauer, J.J., Kleeman, M.J., Cass, G.R., and Simoneit, B.R.T. Measurement of emissions from air pollution sources. 2. C_1 through C_{30} organic compounds from medium duty diesel trucks, *Environ. Sci. Technol.*, *33*, 1578–1587 (1999b).

Schauer, J.J., Kleeman, M.J., Cass, G.R., and Simoneit, B.R.T. Measurement of emissions from air pollution sources. 3. C_1–C_{29} organic compounds from fireplace combustion of wood, *Environ. Sci. Technol.*, *35*, 1716–1728 (2001).

Schauer, J.J., Kleeman, M.J., Cass, G.R., and Simoneit, B.R.T. Measurement of emissions from air pollution sources. 4. C_1–C_{27} organic compounds from cooking with seed oils, *Environ. Sci. Technol.*, *36*, 567–575 (2002b).

Schauer, J.J., Kleeman, M.J., Cass, G.R., and Simoneit, B.R.T. Measurement of emissions from air pollution sources. 5. C_1–C_{32} organic compounds from gasoline-powered motor vehicles, *Environ. Sci. Technol.*, *36*, 1169–1180 (2002c).

Schauer, J.J., Rogge, W.F., Hildemann, L.M., Mazurek, M.A., and Cass, G.R. Source apportionment of air-borne particulate matter using organic compounds as tracers, *Atmos. Environ.*, *30*, 3837–3855 (1996).

Schripp, T., Wensing, M., Uhde, E., Salthammer, T., He, C., and Morawska, L. Evaluation of ultrafine particle emissions from laser printers using emission test chambers, *Environ. Sci. Technol.*, *42*, 4338–4343 (2008).

Schwartz, J., Coull, B., Laden, F., and Ryan, L. Effect of dose and timing of dose on the association between airborne particles and survival, *Environ. Health Perspect.*, *116*, 64–69, doi: 10.1289/ehp.9955 (2008).

See, S.W., Balasubramanian, R., Man, U., and Joshi, U.M. Physical characteristics of nanoparticles emitted from incense smoke, *Sci. Technol. Adv. Mater.*, *8*, 25–32 (2007).

Seinfeld, J.H. and Pandis, S.N. *Atmospheric Chemistry and Physics*, 2nd edn., Wiley-Interscience, New York, 1232 p. (2006).

Seinfeld, J.H. and Pankow, J.F. Organic atmospheric particulate material, *Ann. Rev. Phys. Chem.*, *54*, 121–140 (2003).

Sexton, K., Adgate, J.L., Ramachandran, G., Pratt, G.C., Mongin, S.J., Stock, T.H., and Morandi, M.T., Comparison of personal, indoor and outdoor exposures to hazardous air pollutants in three urban communities, *Environ. Sci. Technol.*, *38*, 423–430 (2004).

Shaughnessy, R.J., McDaniels, T.J., and Weschler, C.J. Indoor chemistry: Ozone and volatile organic compounds found in tobacco smoke, *Environ. Sci. Technol.*, *35*, 2758–2764 (2001).

Shoeib, M., Harner, T., Ikonomou, M., and Kannan, K. Indoor and outdoor air concentrations and phase partitioning of perfluoroalkyl sulfonamides and polybrominated diphenyl ethers, *Environ. Sci. Technol.*, *38*, 1313–1320 (2004).

Siddiqui, A.R., Lee, K., Bennett, D., Yang, X., Brown, K.H., Bhutta, Z.A., and Gold, E.B. Indoor carbon monoxide and $PM_{2.5}$ concentrations by cooking fuels in Pakistan, *Indoor Air*, *19*, 75–82 (2009).

Simoneit, B.R.T. Biomass burning—A review of organic tracers for smoke from incomplete combustion, *Appl. Geochem.*, *17*, 129–162 (2002).

Singer, B.C., Coleman, B.K., Destaillats, H., Hodgson, A.T., Weschler, C.J., and Nazaroff, W.W. Indoor secondary pollutants from cleaning product and air freshener use in the presence of ozone, *Atmos. Environ.*, *40*, 6696–6710 (2006a).

Singer, B.C., Destaillats, H., Hodgson, A.T., and Nazaroff, W.W. Cleaning products and air fresheners: Emissions of terpenoids and ethylene-based glycol ethers, *Indoor Air*, *16*, 179–191 (2006b).

Singer, B.C., Revzan, K.L., Hotchi, T., Hodgson, A.T., and Brown, N.J. Sorption of organic gases in a furnished room, *Atmos. Environ.*, *38*, 2483–2494 (2004).

Sjödin, A., Carlsson, H., Thuresson, K., Sjolin, S., Bergman, A., and Ostman, C., Flame retardants in indoor air at an electronics recycling plant and at other work environments, *Environ. Sci. Technol.*, *35*, 448–454 (2001).

Sleiman, M., Destaillats, H., Smith, J.D., Liu, C., Ahmed, M., Wilson, K.R., and Gundel, L. Secondary organic acrosol formation from ozone-initiated reactions with nicotine and secondhand tobacco smoke, *Atmos. Environ.*, *44*, 4191–4198 (2010b).

Sleiman, M., Gundel, L.A., Pankow, J.F., Jacob III, P., Singer, B.C., and Destaillats, H. Formation of carcinogens indoors by surface-mediated reactions of nicotine with nitrous acid, leading to potential thirdhand smoke hazards, *Proc. Natl Acad. Sci. U.S.A.*, *107*, 6576–6581 (2010a).

Song, B.J. and Liu, A.H. Metropolitan endotoxin exposure, allergy and asthma, *Curr. Opin. Allergy Clin. Immunol.*, *3*, 331–335 (2003).

Sorenson, D.N. and Weschler, C.J. Modeling gas phase reactions in indoor environments using computational fluid dynamics, *Atmos. Environ.*, *36*, 9–18 (2002).

Srivastava, M.K., Subramanian, R., Roggge, W.F., and Robinson, A.L. Sources of organic aerosol: Positive matrix factorization of molecular marker data and comparison of results from different source apportionment models, *Atmos. Environ.*, *41*, 9353–9369 (2007).

Stone, E.A., Zhou, J., Snyder D.C., Rutter, A.P., Mieritz, M., and Schauer, J.J. A comparison of summertime secondary organic aerosol source contributions at contrasting urban locations, *Environ. Sci. Technol.*, *43*, 3448–3454, 2009.

Stout II, D.M. and Mason, M.A. The distribution of chlorpyrifos following a crack and crevice type application in the US EPA Indoor Air Quality Research House, *Atmos. Environ.*, *37*, 5539–5549 (2003).

Straif, K., Baan, R., Grosse, Y., Secretan, B., El Ghissassi, F., and Cogliano, V. Carcinogenicity of household solid fuel combustion and of high-temperature frying, *Lancet Oncol.*, *7*, 977–978 (2006).

Thatcher, T.L., Lai, A.C.K., Moreno-Jackson, R., Sextro, R.G., and Nazaroff, W.W. Effects of room furnishings and air speed on particle deposition rates indoors, *Atmos. Environ.*, *36*, 1811–1819 (2002).

Thatcher, T.L. and Layton, D.W. Deposition, resuspension and penetration of particles within a residence, *Atmos. Environ.*, *29*, 1487–1497 (1995).

Thatcher, T.L., Lunden, M.M., Revzan, K.L., Sextro, R.G., and Brown, N.J. A concentration rebound method for measuring particle penetration and deposition in the indoor environment, *Aerosol Sci. Technol.*, *37*, 847–864 (2003).

To, W.M., Lau, Y.K., and Yeungg, L.L. Emission of carcinogenic components from commercial kitchens in Hong Kong, *Indoor Built Environ.*, *16*, 29–38 (2007).

Traynor, G.W., Allen, J.R., Apte, M.G., Girman, J.R., and Hollowell, C.D. Pollutant emissions from portable kerosene-fired space heaters, *Environ. Sci. Technol.*, *17*, 369–371 (1983).

Traynor, G.W., Apte, M.G., Carruthers, A.R., Dillworth, J.F., Grimsrud, D.T., and Thompson, W.T. Indoor air pollution and inter-room transport due to unvented kerosene-fired space heaters. *Environ. Int.*, *13*, 159–166 (1986).

Traynor, G.W., Apte, M.G., Sokol, H.A., Chuang, J.C., Tucker, W.G., and Mumford, J.L. Selected organic pollutant emissions from unvented kerosene space heaters, *Environ. Sci Technol.*, *24*, 1265–1270 (1990).

U.S. Environmental Protection Agency, Latest findings on national air quality: 2002 status and trends, Office of Air Quality and Standards, Air Quality Strategies and Standards Division, Research Triangle Park, NC, EPA Publication No. EPA 454/K-03-001, 31 pp., http://www.epa.gov/airtrends/2002_airtrends_final.pdf (2003).

U.S. Environmental Protection Agency, Our Nation's air, status and trends through 2008, Office of Air Quality Planning and Standards, Research Triangle Park, NC, EPA Publication No. EPA 454/R-09-002, www.epa.gov/airtrends/2010/report/fullreport (2010).

Van Alphen, M. Emission testing and inhalation exposure-based risk assessment for candles having Pb metal wick cores, Sci. Total Environ., 244, 53–65 (1999).

Vette, A.F., Rea, A.W., Lawless, P.A., Rodes, C.E., Evans, G., Highsmith, V.R., and Sheldon, L. Characterization of indoor-outdoor aerosol concentration relationships during the Fresno PM exposure studies, Aerosol Sci. Technol., 34, 118–126 (2001).

Wallace, L. Indoor particles: A review, J. Air Waste Manage. Assoc., 46, 98–126 (1996).

Wallace, L. Indoor sources of ultrafine and accumulation mode particles: Size distributions, size-resolved concentrations, and source strengths, Aerosol Sci. Technol., 40, 348–360 (2006).

Wallace, L., Mitchell, H., O'Connor, G.T., Lucas Neas, L., Lippmann, M., Kattan, M., Koenig, J. et al. Particle concentrations in inner-city homes of children with asthma: The effect of smoking, cooking and outdoor pollution, Environ. Health Perspect., 111, 1265–1272 (2003).

Wallace, L.A., Wheeler, A.J., Kearney, J., Van Ryswyk, K., You, H., Kulka, R.H., Rasmussen, P.T.E., Brook, J.R., and Xu, X. Validation of continuous particle monitors for personal, indoor, and outdoor exposures, J. Expo. Sci. Environ. Epidemiol., 21, 49–64 (2011).

Wallace, L., Williams, R., Rea, A., Croghan, C. Continuous weeklong measurements of personal exposures and indoor concentrations of fine particles for 37 health-impaired North Carolina residents for up to four seasons, Atmos. Environ., 40, 399–414 (2006).

Ward, T. and Noonan, C. Results of a residential indoor PM2.5 sampling program before and after a woodstove changeout, Ind. Air, 18, 408–415 (2008).

Waring, M.S. and Siegel, J.A. Particle loading rates for HVAC filters, heat exchangers, and ducts, Indoor Air, 18, 209–224 (2008).

Waring, M., Siegel, J.A., and Corsi, R.L. Ultrafine particle removal and generation by portable air cleaners, Atmos. Environ., 42, 5003–5014 (2008).

Wasson, S.J., Guo, Z.S., McBrian, J.A., and Beach, L.O. Lead in candle emissions, Sci. Total Environ., 296, 159–174 (2002).

Wensing, M., Schripp, T., Uhde, E., and Salthammer, T. Ultra-fine particles release from hardcopy devices: Sources, real-room measurements and efficiency of filter accessories, Sci. Total Environ., 407, 418–427 (2008).

Weschler, C.J. Ozone in indoor environments: Concentration and chemistry, Indoor Air, 10, 269–288 (2000).

Weschler, C.J. Indoor/outdoor connections exemplified by processes that depend on an organic compound's saturation vapor pressure, Atmos. Environ., 37, 5455–5465 (2003).

Weschler, C.J. Changes in indoor pollutants since the 1950s, Indoor Air, 43, 153–169 (2011a).

Weschler, C.J. Commemorating 20 years of Indoor Air. Chemistry in indoor environments: 20 years of research, Indoor Air, 43, 1–14 (2011b).

Weschler, C.J. and Nazaroff, W.W. Semivolatile organic compounds in indoor environments, Atmos. Environ., 42, 9018–9040 (2008).

Weschler, C.J. and Nazaroff, W.W. SVOC partitioning between the gas phase and settled dust indoors, Atmos. Environ., 44, 3609–3620 (2010).

Weschler, C.J., Salthammer, T., and Fromme, H. Partitioning of phthalates among the gas phase, airborne particles and settled dust in indoor environments, Atmos. Environ., 48, 1449–1460 (2008).

Weschler, C.J. and Shields, H.C. Potential reactions among indoor pollutants, Atmos. Environ., 31, 3487–3495 (1997).

Weschler, C.J. and Shields, H.C. Indoor ozone reactions as a source of indoor particles, Atmos. Environ., 33, 2301–2312 (1999).

Weschler, C.J. and Shields, H.C. Experiments probing the influence of air exchange rates on secondary organic aerosols derived from indoor chemistry, Atmos. Environ., 37, 5621–5631 (2003).

Williams, R., Suggs, J., Rea, A., Leovic, K., Vette, A., Croghan, C., Sheldon, L. et al. The Research Triangle Park particulate matter panel study: PM mass concentration relationships, Atmos. Environ., 37, 5349–5363 (2003a).

Williams, R., Suggs, J., Rea, A., Sheldon, L., Rodes, C., and Thornburg, J. The Research Triangle Park particulate matter panel study: Modeling ambient source contribution to personal residential PM mass concentrations, Atmos. Environ., 37, 5365–5378 (2003b).

Wilson, E.B. Introduction to Scientific Research, McGraw-Hill, New York, p. 148 (1952).

Wolkoff, P., Clausen, P.A., Wilkins, C.K., and Nielsen, G.D. Formation of strong airway irritants in terpene/ozone mixtures, *Indoor Air*, *10*, 82–91 (2000).

Wolkoff, P., Wilkins, C.K., Clausen, P.A., and Nielsen, G.D. Organic compounds in office environments—Sensory irritation, odor, measurements and the role of reactive chemistry, *Indoor Air*, *16*, 7–19 (2006).

Won, D., Corsi, R.L., and Rynes, M. New indoor carpet as adsorptive reservoir for volatile organic compounds, *Environ. Sci. Technol.*, *34*, 4193–4198 (2000).

Won, D., Corsi, R.L., and Rynes, M. Sorptive interactions between VOCs and indoor materials, *Indoor Air*, *11*, 246–256 (2001).

Wu, C.F., Larson, T.V., Wu, S.Y., Williamson, J., Westberg, H.H., and Liu, L.J. Source apportionment of PM(2.5) and selected hazardous air pollutants in Seattle, *Sci. Total Environ.*, *386*, 42–52 (2007).

Wu, M.-T., Lee, L.-H., Ho, C.-K., Wu, S.-C., Lin, L.-Y., Cheng, B.-H., Liu, C.L., Yang, C.-Y, Tsai, H.-T., and Wu, T.-N. Environmental exposure to cooking oil fumes and cervical intraepithelial neoplasm, *Environ. Res.*, *94*, 25–32 (2004).

Xu, Y. and Little, J.C. Predicting emissions of SVOCs from polymeric materials and their interaction with airborne particles, *Environ. Sci. Technol.*, *40*, 456–461 (2006).

Xu, M., Nematollahi, M., Sextro, R.G., and Gadgil, A.J. Deposition of tobacco smoke particles in a low ventilation room, *Aerosol Sci. Technol.*, *20*, 194–206 (1994).

Yakovleva, E., Hopke, P.K., and Wallace, L. Receptor modeling assessment of particle total exposure assessment methodology data, *Environ. Sci. Technol.*, *33*, 3645–3652 (1999).

Yang, H., Chien, S., Lee, H., Chao, M., Luo, H., Hsieh, D., and Lee, W. Emission of trans, trans-2,4-decadienal from restaurant exhaust to the atmosphere, *Atmos. Environ.*, *41*, 5327–5333 (2007).

Yu, J., Cocker, D.R. III, Griffin, R.J., Flagan, R.C., and Seinfeld, J.H. Gas-phase ozone oxidation of monoterpenes: Gaseous and particulate products, *J. Atmos. Chem.*, *34*, 2007–2058 (1999).

Zheng, M., Cass, G.R., Schauer, J.J., and Edgerton, E.S. Source apportionment of $PM_{2.5}$ in the southeastern United States using solvent-extractable organic compounds as tracers, *Environ. Sci. Technol.*, *36*, 2361–2371 (2002).

Zuraimi, M.S., Weschler, C.J., Tham, K.W., and Fadey, M.O. The impact of building recirculation rates on secondary organic aerosols generated by indoor chemistry, *Atmos. Environ.*, *41*, 5213–5223 (2007).

7 Chemical Analyses of Particle Filter Deposits*

Judith C. Chow and John. G. Watson

CONTENTS

7.1 INTRODUCTION

Although continuous measurement methods for ambient particulate matter (PM) have advanced (Burtscher, 2002; McMurry, 2000; Solomon and Sioutas, 2008; Wang et al., 2009; Watson et al., 1998; Wexler and Johnston, 2008), sampling through size-selective inlets onto filter media with subsequent laboratory analysis is still the most widely applied method for PM characterization

---

* The identification of commercial suppliers as examples should not be considered an endorsement of these vendors or their products, although the authors have found these products to be reliable and cost-effective.

(Watson and Chow, 2011). PM filter analyses are used to identify and quantify source contributions (Watson and Chow, 2005, 2007; Watson et al., 2002, 2008), estimate visibility impairment (Watson, 2002), evaluate adverse effects on human health (Mauderly and Chow, 2008; Pope and Dockery, 2006), and determine causes of material damage (Hu et al., 2009a,b). PM concentrations for mass, elements, water-soluble ions, and organic and elemental carbon (OC and EC, respectively) are measured on filters acquired at ~400 sites in urban and nonurban U.S. networks (Flanagan et al., 2006; Hansen et al., 2006; Watson, 2002). More useful information could be obtained from these and other filter samples at a small incremental cost. Source samples acquired by dilution sampling systems (ASTM, 2008; Deuerling et al., 2010; England et al., 2007a,b; Hildemann et al., 1989) are also amenable to the same measurements described here and are becoming more in demand for multi-pollutant air quality management strategies (Chow and Watson, 2011; Chow et al., 2010a; Hidy and Pennell, 2010).

This chapter summarizes and updates relevant information from previously published reviews (Chow, 1995; Chow et al., 2002a, 2007c, 2008; Fehsenfeld et al., 2004; Landsberger and Creatchman, 1999; Lodge, 1989; Solomon et al., 2001; Vincent, 1994; Watson and Chow, 1994; Wilson et al., 2002) and draws from recent experience of the authors. Chow et al. (2008) provide more detailed lists of chemical species, detection limits, and comparison studies that are not repeated here.

7.2   FILTER SAMPLING

Sampling systems, filter media, and analysis methods must be considered as a unit and complemented with comprehensive sample handling and analysis protocols. A multi-channel sampling configuration is shown in Figure 7.1. Sampling system and inlet details are specified by Watson and Chow (2011). $PM_{2.5}$ and PM_{10} (particle fractions less than 2.5 and 10 μm aerodynamic diameter, respectively) inlets are commonly used for 16.7 L/min flow rates (1 m³/h), which provide ample material for analyses when 24 h mass concentrations exceed ~5 μg/m³. Lower flow rates (e.g., 5–7 L/min) may be needed for higher PM concentrations (>50 μg/m³) to minimize filter clogging (Davies, 1970). An even lower flow rate (e.g., 2–5 L/min) is used for microscopic analysis as the particles should not overlap for accurate automated detection.

The Figure 7.1 system consists of four parallel channels using flow rates of 5–16.7 L/min to acquire PM deposits on front filters along with gaseous precursors for ammonia (NH_3) (Appel et al., 1988; Cao et al., 2009; Chow et al., 1993a; Ferm, 1979) and sulfur dioxide (SO_2) (Axelrod and Hansen, 1975; Huygen, 1963) on impregnated backup filters. Impregnated filters also can be used to collect nitrogen dioxide (NO_2) (Gotoh, 1980; Hedley et al., 1994), hydrogen sulfide (H_2S) (de Santis et al., 2006; Duckworth, 1971; Nash and Leith, 2010), ozone (O_3) (Grosjean and Hisham, 1992; Miwa et al., 2009; Monn and Hangartner, 1990), and some volatile organic compounds (VOCs) (Kume et al., 2008; Mason et al., 2011; Miller et al., 2010; Shields and Weschler, 1987; Stock et al., 2008). Figure 7.1 includes a quartz-fiber backup filter to evaluate organic vapors adsorbed onto the front quartz-fiber filter (Cheng et al., 2009, 2010; Chow et al., 2010b; Vecchi et al., 2009; Watson et al., 2009) that may positively bias the OC concentration. A backup nylon-membrane or sodium chloride (NaCl)-impregnated cellulose-fiber filter can be used to determine ammonium nitrate evaporation during sampling (Chow et al., 2002b, 2005b). Denuders (Kitto and Colbeck, 1999) may be placed downstream of the inlet to remove gases from the airstream.

No single filter medium is appropriate for all analyses, so it is necessary to sample on multiple substrates for chemical speciation. Ringed Teflon®-membrane filters (Pall Corporation, Ann Arbor, MI, part number R2PJ047; Whatman Inc., Clifton, NJ, part number 7592-104) consist of a thin, porous polytetrafluoroethylene (PTFE) Teflon-membrane sheet stretched across a polymethylpentane ring. The thin membrane collapses without the ring, and the filter cannot be accurately sectioned into smaller pieces. The white membrane is nearly transparent (thickness of 30–46 μm) and has been used to estimate light transmission/absorption (b_{abs}) (Campbell et al., 1995; Chow et al., 2009, 2010c; Moosmüller et al., 2009). PTFE Teflon membrane absorbs negligible amounts of

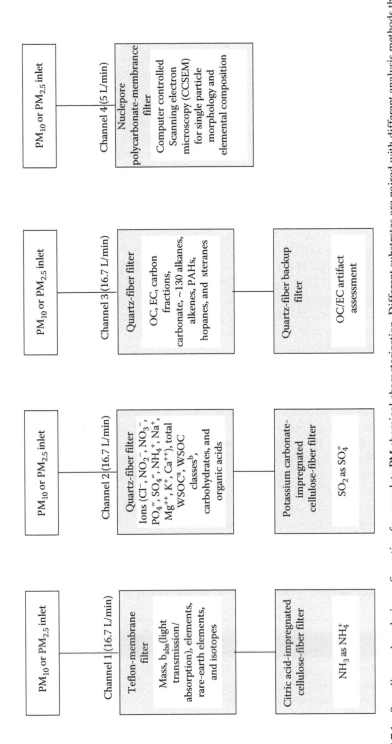

FIGURE 7.1 Sampling and analysis configuration for complete PM chemical characterization. Different substrates are paired with different analysis methods that are optimized for the material collected on each filter. Measurement methods include source marker species as well as hazardous air pollutants. Subsets of this overall configuration can be used at different locations and for different time periods. Different flow rates can be used depending on the inlet selected to obtain filter mass deposits of 1000–1500 μg that are optimal for the chemical analyses. [a]Water-soluble organic carbon (WSOC); [b]Including neutral compounds (NC), mono-and dicarboxylic acids (MDA), and polycarboxylic acids (PA, mostly humic-like substances [HULIS]).

water or gases and has inherently low contamination levels, but chemicals have been found in some batches by acceptance testing (Chow, 1995). Carbon cannot be measured on Teflon-membrane filters because of their high carbon content, although aerosol carbon has been inferred from hydrogen measurements (Kusko et al., 1989). A variation on this filter consists of a PTFE Teflon-membrane filter mounted on a woven PTFE mat instead of a support ring (Pall Corporation, part number P5PJ001). The membrane and mat sides look the same, and care must be taken to prevent mounting it upside down on a filter holder resulting in particles drawn through the mat rather than onto the surface of the membrane.

Quartz-fiber filters (Pall Corporation, part number 7202; Whatman Inc., part number 1851-047) consist of a tightly woven mat of quartz (SiO_2). These filters can withstand high temperatures without decomposing, making them suitable for different thermal analyses. Quartz-fiber filters adsorb organic vapors passively before and after sampling, and actively during sampling (Chow et al., 2010b; Sutter et al., 2010; Watson et al., 2009), so it is important to prefire them for at least 3 h at ~900°C and keep them sealed prior to and after sampling. The Whatman QM/A quartz-fiber filter contains a 5% borosilicate glass binder that reduces its friability, but the binder also includes trace elements and catalysts that may bias thermal carbon analyses (Lin and Friedlander, 1988). The Pallflex® Tissuquartz™ 2500 QAT-UP filter (Pall Corporation, part number 7202) is pure quartz and undergoes a distilled water washing (thus the "UP," or "ultra-pure," designation).

Cellulose-fiber filters (Whatman, Inc., Clifton, NJ, part number 31 ET and 41) consist of a tightly woven paper mat. These filters are hygroscopic and require precise RH control in the filter processing environment to obtain accurate mass measurements (Demuynck, 1975). Cellulose-fiber filters are most useful for impregnation with gas-absorbing compounds. The Whatman 31ET filters are thicker than the Whatman 41 filters (0.5 versus 0.22 mm), so they absorb more of the impregnation solution and have a higher collection capacity.

Etched polycarbonate-membrane filters (Nuclepore™, Whatman, Inc., Clifton, NJ, part number 111107) are constructed from a thin polycarbonate sheet through which pores of uniform diameter (e.g., 0.4 µm pore size) have been produced by radioactive particle penetration and chemical etching (Spurny et al., 1969). These are the best filter media for single particle analyses by electron microscopy because particles are easily distinguished from the flat filter surface.

Glass-fiber filters should never be used for aerosol measurements. They consist of a mat of borosilicate glass filaments with a high alkalinity that reacts with nitrogen oxide (NO_x), nitric acid (HNO_3), and SO_2 (Coutant, 1977; Spicer and Schumacher, 1979). Blank levels for many elements are high and variable (Witz et al., 1983).

When filters are received from the vendor, one or two out of each 100 filters should be analyzed for all species to verify that preestablished specifications have been met (Fehsenfeld et al., 2004). Filters also are visually examined on a light table prior to labeling for discoloration, pinholes, creases, separation of ring, loose material, or other defects.

7.3   SAMPLE PROCESSING

Figure 7.2 illustrates sample handling procedures and analysis methods for different observables. Since filter mass deposits are on the order of ~1 mg and individual species concentrations are in the nanogram to microgram range, potential contamination must be minimized. Filter handling in a laboratory laminar flow hood is always preferable to loading and unloading in the field. Watson and Chow (2011) and Lippmann (2001) describe different types of filter holders in common use. Chain-of-custody includes filter preparation and acceptance testing, followed by pre-sampling gravimetric analyses for mass and light transmission/absorption (b_{abs}) analyses as a surrogate for black carbon (BC) (Chow et al., 2010c), if desired. Prior to sampling, filter packs are assembled and transported to the field. After sampling, filter packs are disassembled, followed by post-sampling gravimetric and b_{abs} analyses.

Filters can be stored in individual Petri slides (PD1504700; Millipore, Billerica, MA) that accommodate 47, 37, and 25 mm diameter filters, the sizes most commonly used for ambient and source sampling. Bar-coded labels can be affixed to the Petri slide with duplicates placed on the filter holder and field data sheet. It is best to use a filter identifier (one for each type in Figure 7.1) followed by a sample identifier that is the same for each of the seven filters. The sample identifier can contain information about the network, sampling site, and sampling date if desired. Not more than one filter should be removed from its holder or Petri slide at a time to minimize the risk of swapping of filters among different samples. The same filter ID should be transferred to analysis vials and Laboratory Information Management Systems (LIMS) so that the different analyses can be assembled into a unified database.

For unknown reasons, some batches of ringed Teflon-membrane filters have yielded variable (by up to 100 μg/filter over a few days) blank masses (e.g., Tombach et al., 1987). This may be due to evaporation of the adhesive between the ring and Teflon membrane soon after manufacture. A 1 month storage period in a controlled environment, followed by 1 week of equilibration, is recommended before gravimetric analysis. A sample of each batch of 100 prefired, quartz-fiber filters should be tested for carbon blank levels prior to sampling, and filter batches with carbon levels exceeding $1.5 \mu g/cm^2$ for OC, or $0.5 \mu g/cm^2$ for EC, should be refired or rejected. All prefired filters should be sealed and stored in a refrigerator ($<4°C$) prior to and after field sampling.

Cellulose-fiber filters are immersed in the appropriate impregnating solution for approximately 30 min. These disks are then removed and placed in large Petri dishes for drying in a vacuum oven for 5–10 min. One hundred of the dried impregnated filters are immediately sealed in polyethylene bags and placed under refrigeration for later loading into filter holders. One sample from each lot of filters should be analyzed prior to field sampling to assure that filter batches have not been contaminated and that the impregnating solution concentration level has adequate capacity for specific gas sampling. Impregnated filters also are used for passive gas sampling (Cox, 2003), and their blank levels will increase with exposure to ambient air.

Some analyses require the filter to be cut into sections for different extraction and analysis procedures. This can be done with a standard paper cutter to which a half-circle or quarter-circle polycarbonate jig has been attached. The jig should have the diameter of the filter to allow the exact fraction desired to be cut. The blade is cleaned with methanol (MeOH) and a laboratory wipe between filter cuttings. The unanalyzed filter portion is archived under refrigeration in its original Petri slide. The remaining filter section is placed in a polystyrene extraction vial with a screw cap (e.g., 17×120 mm vial, Greiner #188271). Each vial is labeled with a bar code sticker containing the filter ID code. The extraction tubes are placed in tube racks and extraction solutions are added. The extraction vials are capped and sonicated for 60 min, shaken for 60 min, and then aged overnight to assure complete extraction of the deposited material in the solvent. The ultrasonic bath water is monitored to prevent temperature increases from dissipation of ultrasonic energy. After extraction, these solutions are stored under refrigeration prior to analysis.

After completion and validation of nondestructive x-ray fluorescence (XRF) analysis, the entire Teflon-membrane filter is submitted to a strong acid extraction (Anzano and Ruiz-Gil, 2005; Feng and Barratt, 1994; Kaasalainen and Yli-Halla, 2003; Link et al., 1998; Melaku et al., 2005; Rao et al., 2008; Silveira et al., 2006; Wang et al., 1995). U.S. EPA Method 200.8 (U.S.EPA, 1994) is modified with the addition of hydrofluoric acid (HF) to separate the metals from their mineral oxides. The filter is cut into pieces using ceramic scissors to allow the filter to fit inside a 68 mL digestion vessel. Ethanol (0.2 mL) is added to wet the filter, along with 2 mL of a concentrated (68%) HNO_3 mixture (1:1 v/v) with distilled-deionized water (DDW), 5 mL of concentrated (34%) hydrochloric acid (HCl) mixture with DDW (1:4 v/v), and 0.1 mL of concentrated (49%) HF with DDW. The digestion vessels with a reflux cap are then placed in a hot block (Environmental Express, Mt. Pleasant, SC) located in a clean-air enclosure and gently heated from room temperature to 113°C, which maintains the sample at 95°C. After 90 min, the samples are removed, cooled

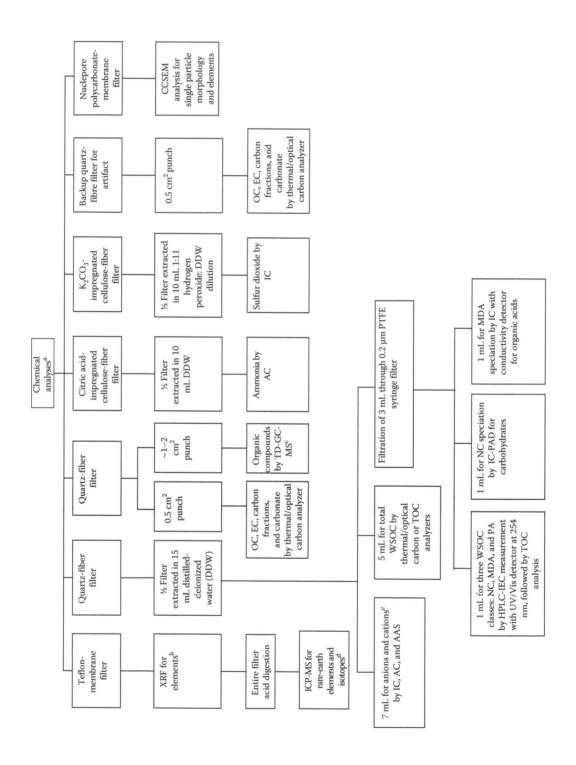

FIGURE 7.2 Flow diagram for filter processing and chemical analyses. [a]*Analytical instruments*: AAS, atomic absorption spectroscopy; AC, automated colorimetry; CCSEM, computer-controlled scanning electron microscopy; ELSD, evaporative light scattering detector; HPLC-IEC, high-performance liquid chromatography with an ion exchange column; IC, ion chromatography; IC-PAD, IC with pulsed amperometric detector; ICP-MS, inductively coupled plasma—mass spectrometry; PTFE, polytetrafluoroethylene; SEC, size-exclusion chromatography; TD-GC-MS, thermal desorption-gas chromatography-mass spectrometry; TOC, total organic carbon; UV/VIS, ultraviolet/visible light detector; XRF, x-ray fluorescence. *Observables*: OC, organic carbon; EC, elemental carbon; HULIS, humic-like substances; MDA, mono- and dicarboxylic acids; NC, neutral/basic compounds; PA, polycarboxylic acids; WSOC, water soluble organic carbon. [b]51 elements, Table 7.1; [c]124 organic compounds, see Table 7.2; [d]Cs, Ba, La, Ce, Pr, Nd, Sm, Eu, Gd, Tb, Dy, Ho, Er, Tm, Yb, Lu, Pb[204, 205, 206, 207, 208]. anions include: [e]Cl$^-$, NO$_2$, NO$_3^-$, PO$_4^=$, SO$_4^=$ (by IC); cations include NH$_4^+$ (by AC); Na$^+$, Mg^{++}, K$^+$, and Ca^{++} (by AAS), K$_2$CO$_3$; potassium carbonate.

to room temperature, and diluted to 50 mL with high-purity DDW. The samples are then capped, shaken for 10 s, and aged for 12 h prior to metal analyses.

7.4 LABORATORY ANALYSES

Laboratory analyses designated in Figure 7.2 are briefly described in the following subsections. The methods described here were selected because they can be practically and cost-effectively implemented for a large number of samples using commonly available sampling systems. Each one is implemented following a standard operating procedure (SOP) that includes (1) summary of measurement method, principles, expected accuracy and precision, and the assumptions for validity; (2) materials, equipment, reagents, and suppliers; (3) individuals responsible for performing each part of the procedure; (4) traceability path, primary standards or reference materials, tolerances for transfer standards, and schedule for transfer standard verification; (5) start-up, routine, and shutdown operating procedures and an abbreviated checklist; (6) data forms; (7) routine maintenance schedules, maintenance procedures, and troubleshooting tips; (8) internal calibration and performance testing procedures and schedules; (9) external performance auditing schedules; and (10) references to relevant literature and related SOPs. Although not mentioned in every subsection, each method requires periodic analyses of calibration standards, performance standards, audit standards, solution blanks, filter blanks, and sample replicates. These are used to identify issues that arise during analysis and to quantify the accuracy, precision, and validity of the individual sample concentrations.

7.4.1 FILTER WEIGHING GRAVIMETRY FOR MASS

The mass of the deposit is determined as the weight difference between the exposed and unexposed filter. Filter weighing should be performed on an electro-microbalance (e.g., Mettler XP6; Toledo, OH) with ±0.001 mg sensitivity (Allen et al., 2001; Feeney et al., 1984; Lawless and Rodes, 1999; Swanson and Kittelson, 2008; Yanosky and MacIntosh, 2001). Unexposed and exposed Teflon-membrane filters are equilibrated at 21.5°C ± 1.5°C and 35% ± 5% RH for a minimum of 24 h prior to weighing (Brown et al., 2006; Kajino et al., 2006; McInnes et al., 1996), as specified by U.S. EPA (1997) for $PM_{2.5}$ compliance monitoring. These equilibration conditions appear to reduce variability to acceptable levels (Chow et al., 2006), although there is evidence that liquid water is still retained when water-soluble compounds dominate the PM deposit. The charge on each filter is neutralized by exposure to a polonium source (Po_{210}) for 60 s prior to the filter being placed on the balance pan (Engelbrecht et al., 1980; Swanson and Kittelson, 2008; Tsai et al., 2002). The balance is zeroed and spanned with class 1 weights (50, 100, and 200 mg) prior to weighing each batch of filters. After every 10 filters are weighed, the calibration and tare are rechecked. The filters should be reweighed by a separate technician. Since the initial weight can never be recovered, 100% reweighing of the unexposed filters is a good insurance policy. An unexposed filter should be reweighed when the deviation exceeds the tolerance of ±10 μg/filter for a 47 or 37 mm diameter filter. For exposed filters, 30% reweighing is recommended, and an entire batch should be reweighed when deviations exceed ±15 μg/filter. If pre- and postexposure weighing are performed at locations with different elevations, there must be accounting for the change in filter buoyancy, which is ~20 μg for a 1524 m (5000 ft) elevation change (Rasmussen et al., 2010).

7.4.2 LIGHT TRANSMISSION/ABSORPTION (b_{ABS}) ANALYSIS AS A BLACK CARBON SURROGATE

Due to the potential effects of BC on climate (Chow et al., 2010d; Jacobson, 2002), the b_{abs} measurement is being used more frequently as a surrogate for BC measurements. This is most conveniently done as an accompaniment to the pre- and postexposure Teflon-membrane filter weighing. A photography densitometer (e.g., Tobias TBX-10, Ivyland, PA) can be adapted to this purpose by affixing a circular jig over the broad bandwidth incandescent light source that accommodates the filter diameter. The jig includes a stop for the light detector arm so that the photodetector is positioned at the same distance above the filter deposit, and the detector is calibrated with neutral density filters (Kodak, Rochester, NY). A new transmissometer (Sootscan OT21, McGee Scientific Instruments, Berkeley, CA) measures 47 mm filters at 370 and 880 nm wavelengths simultaneously. Emissions from the smoldering phase of biomass burning have higher absorption efficiencies at ~350 nm than those from higher temperature combustion (e.g., diesel engine exhaust have higher absorption efficiencies at ~880 nm), and the ratios of transmittance through the filter at these wavelengths have been used to distinguish contributions from pollution sources (Sandradewi et al., 2008). The optical density ($OD = \ln(I_b/I_l)$, where I_b is the light intensity transmitted through a blank filter and I_l is the light intensity transmitted through the loaded filter) is converted to b_{abs} (in inverse megameters, Mm^{-1}). This is not a true measure of light absorption in the atmosphere due to multiple scattering within the filter and changes in particle shape and orientation after filter collection (Horvath, 1993), but it is highly correlated (correlation $R > 0.8$) with BC (Chow et al., 2010c).

7.4.3 X-Ray Fluorescence Analysis for Elements

Elements are important source markers, and some of them are believed to be toxic. XRF (Dzubay and Nelson, 1974; Giauque, 1973; Grennfelt et al., 1971) irradiates the thin aerosol deposit on the Teflon membrane with photons of 1,000–30,000 eV generated by an x-ray tube. This removes inner shell electrons, and a characteristic x-ray photon is emitted as an outer-shell electron drops to the vacant lower-energy level. The XRF analyzer is calibrated with thin-film standards (Baum et al., 1977; Billiet et al., 1980; Dzubay et al., 1977; Greenberg, 1979; Heagney and Heagney, 1979). Characteristic peaks for each element must be separated from background and overlapping peaks; and corrections must be made for x-ray absorption with the particles, deposit matrix, and filter (Watson et al., 1999). Teflon membranes are preferred for these analyses because the deposit does not penetrate as deeply into the filter as with fibrous filters.

Older XRF analyzers used a broad range of excitation energies (Watson et al., 1999), but newer units (e.g., PANalytical Epsilon 5, Almelo, The Netherlands) use secondary targets to generate excitation energies close to the absorption edges of four or five elements. For the Epsilon 5, a side window x-ray tube with dual scandium (Sc)/tungsten (W) anodes excites secondary x-rays from up to 11 secondary targets (i.e., Al, Ca, Ti, Fe, Ge, Zr, Mo, Ag, Cs, Ba, and Ce), or an aluminum oxide (Al_2O_3) Barkla target (Heckel, 1995), which in turn emits polarized x-rays used to excite elements in the sample. The fluoresced photons are detected by a solid-state silicon (Si) or germanium (Ge) detector. Each photon that enters the detector generates an electrical charge with magnitude proportional to the photon's energy. Electrical signals from the detector are sorted into energy channels, counted, and displayed. Table 7.1 shows an example of excitation conditions optimized for different sets of elements. Analysis times, primary x-ray voltage and currents, and secondary targets can be selected to minimize background and peak overlaps.

7.4.4 Inductively Coupled Plasma-Mass Spectrometry
Analysis for Elements and Isotopes

Inductively coupled plasma-mass spectrometry (ICP-MS) (e.g., Thermo Elemental X-7, Waltham, MA) (Herner et al., 2006; Komarek et al., 2008; Ohno and Hirata, 2004; Polak et al., 2006) complements, rather than replaces XRF, by obtaining lower detection limits for certain elements and isotopes identified in the caption of Figure 7.2. A portion of the acid-digested sample is nebulized into a plasma torch, which ionizes the dissolved elements. These ionized components are directed to a quadrupole MS where they are sorted by their mass-to-charge ratios. Ions are detected by an electron multiplier and tallied for the elements sought. The electron multiplier response is calibrated against known solution standards.

TABLE 7.1

Example of XRF Excitation Conditions That Optimize Detection Limits and Minimize Spectral Interferences for Different Elements

Elements Quantified	X-Ray Line Detected	Secondary Target	Analysis Time (s)	Primary X-Ray Tube Voltage (kV)	Primary X-Ray Tube Current (mA)
Na, Mg, Al, Si, P, S, Cl, K, Ca, Sc	Kα	Ca	400	40	15
Cs, Ba, La, Ce	Kα	Al_2O_3	200	100	6
Ca, Sc, Ti, V, Cr	Kα	Fe	400	40	15
Mn, Fe, Co, Ni, Cu, Zn	Kα	Ge	400	75	8
Ga, As, Se, Br, Rb,	Kα	Zr	200	100	6
Hf, Ta, W, Ir, Au, Hg, Tl, Pb	Lα	Zr	200	100	6
Zr, Nb, Mo	Kα	Ag	200	100	6
Sr, Y	Kα	Mo	200	100	6
U	Lα	Mo	200	100	6
Pd, Ag, Cd, In, Sn, Sb	Kα	BaF_2	200	100	6
Sm, Eu, Tb	Kα	Al_2O_3	200	100	6

Laser ablation ICP-MS (Chin et al., 1999; Coedo et al., 2005; Durrant, 1999; Gligorovski et al., 2008; Owega et al., 2002; Panne et al., 2001; Tan et al., 2002; Tanaka et al., 1998) is being evaluated as an alternative to hard acid digestion. A high-powered pulsed laser vaporizes a portion of the filter deposit for direct injection into the ICP-MS. This could be applied to only a portion of the Teflon-membrane filter, leaving the remainder for additional characterization. The method requires calibration standards, equivalence testing, and optimization with respect to reproducibility, standardization, and detection limits.

7.4.5 Ion Chromatography Analysis for Anions and Cations

Water-soluble inorganic ions are directly emitted by sources and form from directly emitted gases—with nitrate (NO_3^-), sulfate ($SO_4^=$), and ammonium (NH_4^+) contributing large quantities to $PM_{2.5}$ and PM_{10} concentrations (Malm et al., 2002; Sisler and Malm, 1994). Ion chromatography (IC) (e.g., Dionex ICS-3000, Sunnyvale, CA) sends a water-extracted sample through an ion exchange column to separate the ions retention time for individual quantification in a conductivity detector. Prior to detection, the column effluent enters a suppressor column where chemical composition of the component is altered, resulting in a matrix of low conductivity (Chow and Watson, 1999; Mulik and Sawicki, 1978, 1979). The ions are identified by their elution/retention times and quantified by the conductivity peak area. Peak areas are related to concentrations from standard solutions (e.g., ERA, Arvada, CO) for quantification.

Chloride (Cl^-), nitrate (NO_2^-), NO_3^-, phosphate ($PO_4^=$), and $SO_4^=$ are the commonly measured anions using a guard column (AG14 column, Cat. No. 046134) plus an anion separator column (AS14 column, 250×4 mm ID, Cat. No. 046129) with a strong basic anion exchange resin, and an anion micro-membrane suppressor. The anion eluent consists of 0.0035 M sodium carbonate (Na_2CO_3) and 0.001 M sodium bicarbonate ($NaHCO_3$) prepared in DDW. Lithium (Li^+), sodium (Na^+), ammonium (NH_4^+), potassium (K^+), magnesium (Mg^{++}), and calcium (Ca^{++}) are commonly measured cations using a guard column (CG16 column, 5×50 mm, Cat. No. 057574) and a separator column (CS16, 5×250 mm, Cat. No. 057573). The cation columns use a hydrophilic, high-capacity, and carboxylate functionalized cation exchange resin. For cations, a micro-membrane suppressor, is used to reduce background conductivity. The eluent is 30 Nm methanesulfonic acid (MSA).

7.4.6 AUTOMATED COLORIMETRY ANALYSIS FOR AMMONIUM

Automated colorimetry (AC) (Fung et al., 1979; Sandell, 1959) has long been used to quantify water-soluble anions and cations by reacting them with reagents that create a colored solution. Light at a characteristic wavelength is directed through the solution and detected by a photomultiplier. Beer's Law relates the liquid's absorbance to concentration of the ion in the sample as determined by reactions with known solution standards. Modern ACs (e.g., Astoria Analyzer Automated Colorimetric System, Astoria Pacific, Clackamas, OR) are automated to the extent that solution vials can be placed in an autosampler from which the instrument retrieves the desired volume, mixes it with appropriate reagents, quantifies the absorbance, and applies the appropriate calibration curve. The sample can be split among different analysis modules to obtain a variety of anions and cations, including Cl^-, NO_3^-, $SO_4^=$, and NH_4^+. The method is especially applicable to large numbers of samples (Mueller et al., 1983).

As a complement to anion IC, AC applies the indophenol method (Bolleter et al., 1961; Rommers and Visser, 1969) in which water-soluble NH_4^+ is reacted with phenol and alkaline sodium hypochlorite to produce indophenol, a blue dye. The reaction is catalyzed by the addition of sodium nitroprusside, and absorbance is measured at 630 nm. The system determines carryover by analysis of a low concentration standard following a high concentration. Formaldehyde has been found to interfere with measurements when it is present in an amount that exceeds 20% of the NH_4^+ content. H_2S interferes with measurements when it is present in concentrations that exceed 1 mg/mL. NO_3^- and $SO_4^=$ also potentially interfere when present at levels that exceed 100 times the NH_4^+ concentration. These levels are rarely exceeded in ambient samples. Precipitation of the hydroxides of heavy metals such as magnesium (Mg) and calcium (Ca) is prevented by addition of a sodium citrate/sodium potassium tartrate buffer.

7.4.7 ATOMIC ABSORPTION SPECTROPHOTOMETRY ANALYSIS FOR MONOATOMIC CATIONS

Atomic absorption spectrophotometry (AAS) (Butler et al., 2009; Fernandez, 1989) has long been used for individual elemental analysis, but the capability of ICP-MS makes it more attractive for multi-element applications. Modern AAS units (e.g., SpectrAA 880 Double Beam AAS, Atomic Absorption Spectrometer, Varian, Palo Alto, CA), however, are sensitive and cost-effective for a small number of elements, such as water-soluble Na^+, Mg^{++}, K^+, and Ca^{++}. Water extracts of 1–2 mL are aspirated into an air/acetylene flame at approximately 0.5 mL/min. A hollow-cathode lamp emits wavelengths appropriate for each species: 589 nm for Na^+, 285.2 nm for Mg^{++}, 766.5 nm for K^+, and 422.7 nm for Ca^{++}. Transmitted light is detected by a photomultiplier with and without the presence of the sample, and the reduction of transmitted light absorbed at each characteristic wavelength is related to the concentration via Beer's Law and quantified by similar absorption for known solution standards.

7.4.8 THERMAL/OPTICAL ANALYSIS BY REFLECTANCE AND TRANSMITTANCE (TOR AND TOT) FOR ORGANIC AND ELEMENTAL CARBON

OC and EC are important components of most combustion sources (Chow et al., 2004b, 2011a; Sahu et al., 2011; Watson et al., 1994, 2001a). More than 20 methods (Watson et al., 2005) separate OC from EC fractions by heating the sample to various temperatures, oxidizing the evolved carbon to carbon dioxide (CO_2), and quantifying evolved CO_2 directly with an infrared absorption detector, or reducing it to methane (CH_4) for more sensitive carbon detection by a flame ionization detector (FID). Although these methods produce comparable results for total carbon (TC = OC + EC) (Watson and Chow, 2002), they provide different values for the division between OC and EC (Schmid et al., 2001). Since EC is typically the smaller of the two, it is the least comparable among the methods.

U.S. ambient networks (Flanagan et al., 2006; Hansen et al., 2006; Watson, 2002) and most of the source profiles in the U.S. EPA's SPECIATE library (U.S. EPA, 2008) have applied the IMPROVE_TOR or IMPROVE_A_TOR (Chow et al., 1993b, 2004a, 2005a, 2007a, 2011b) method (DRI Model 2001 Thermal/Optical Carbon Analyzer, Atmoslytic, Inc., Calabasas, CA). This protocol heats the sample in steps, first in an inert 100% helium (He) atmosphere, to obtain OC evolving between ambient temperature ($\sim25°C$) and $140°C$ (OC1), $140°C–280°C$ (OC2), $280°C–480°C$ (OC3), and $480°C–580°C$ (OC4). The analysis atmosphere is then changed to a 98% $He/2\%$ O_2 composition, producing three EC fractions at $580°C$ (EC1), $580°C–740°C$ (EC2), and $740°C –840°C$ (EC3). The evolved carbon peaks, detected by an FID as CH_4, are defined by their return to a stable baseline; and analysis time depends on the composition of each peak rather than on preset windows. These carbon fractions have been found useful as source markers (Cao et al., 2005b; Kim and Hopke, 2004; Lee et al., 2003; Maykut et al., 2003). Reflectance from and transmittance through the sample of a 633 nm laser beam is monitored before, during, and after analysis to detect darkening of the filter deposit as some of the OC chars to EC in the inert He atmosphere.

The optical pyrolysis fraction (OP) is defined as carbon evolved between the time O_2 is added and the reflected or transmitted signal returns to its original value. This OP is added to the sum of the OC fractions and subtracted from the sum of the EC fractions to account for OC that changed to EC during analysis. Chow et al. (2004a) discovered that transmitted light is dominated by charring of the adsorbed organic vapors within the filter, while reflected light is dominated by charring of the surface deposit. EC by TOR is typically higher and less sensitive to the temperature program than EC by TOT, so the default EC is that by TOR. By summing the different temperature fractions and selecting OP by TOR or TOT, however, an EC value can be achieved from a single IMPROVE measurement that approximates many of the EC values obtained by other methods (Watson et al., 2005).

During analysis, a $0.5\,cm^2$ circle is removed from the quartz-fiber filter with a precision punch and placed in a sample boat that is inserted into the analysis oven and immersed in the appropriate carrier gas. If carbonate carbon ($CO_3^=$) is suspected to be in the sample (Cao et al., 2005a; Chow and Watson, 2002; Li et al., 2008), it can be removed and quantified by injecting $20\,\mu L$ of $0.4\,M$ HCl through the septum port onto the sample punch. The CO_2 evolved is measured as $CO_3^=$ and after $600\,s$, the carbon analysis method for OC and EC specified earlier is initiated. The FID response is calibrated by analyzing samples of known amounts of CH_4, CO_2, sucrose, and potassium hydrogen phthalate (KHP).

7.4.9 THERMAL ANALYSIS FOR WATER-SOLUBLE ORGANIC COMPOUNDS

Water-soluble organic compounds (WSOC) are related to aerosol radiative forcing in clouds (Novakov and Corrigan, 1996; Novakov and Penner, 1993), secondary organic aerosol (SOA), and hygroscopic aerosol (Saxena and Hildemann, 1996). WSOC can be measured on the water extract with a total organic carbon (TOC) analyzer (e.g., Shimadzu TOC-VCSH with ASI-V autosampler, Columbia, MD) or with the OC/EC carbon analyzer by replacing the quartz boat used for the filter punch with a platinum boat that can retain $20–100\,\mu L$ of the water extract. The extract is evaporated at $50°C$, followed by ramping the temperature to $900°C$ in a 98% $He/2\%$ O_2 atmosphere. Peak integration and calibration standards are the same as they are for the OC/EC analyses.

For the TOC analyzer, $\sim5\,mL$ of the quartz-fiber filter sample extract is placed in 9 mL prebaked glass tubes. Once injected, the sample is mixed with 1.5% of $2\,M$ HCl and the mixture is sparged in zero air for 1.5 min. Acidification and agitation of the sample eliminate interference from inorganic carbon (i.e., carbonates). The sample is then conveyed to the combustion chamber, where the extract is catalytically oxidized at $680°C$ to CO_2. Combustion products in zero-air carrier gas move into a dehumidifier for moisture removal and cooling and then into a halogen scrubber before progressing to the sample cell for nondispersive infrared (NDIR) detection. Moisture and chlorine gas cause positive interference in CO_2 measurements. NDIR peaks are integrated for comparison with KHP and glucose standards that span the range of expected concentrations.

7.4.10 HIGH-PERFORMANCE LIQUID CHROMATOGRAPHY-UV/VIS FOR THREE MAJOR WSOC CLASSES

Since WSOC consists mostly of polar compounds, organic solvent extraction, derivatization, and identification using GC-MS is needed for a large range of compounds (Rogge et al., 1993; Samy et al., 2010; Yang et al., 2005). A simpler, but useful, analysis divides WSOC from a portion of the water extract into three classes (i.e., neutral/basic compounds [NC], mono- and di-carboxylic acids [MDA], and poly-carboxylic acid [PA]) using high-performance liquid chromatography (HPLC) or IC with UV-Vis detection (Chang et al., 2005; e.g., Decesari et al., 2000).

A 1 mL portion of the water extract is injected into an HPLC with vacuum degasser, quaternary pump, anion exchange column (5 × 5 mL, GE HiTrap DEAE FF), UV/Vis diode array detector (at 254 nm) (Engling et al., 2006; Falkovich et al., 2005), and fraction collector (Agilent 1200 series, Santa Clara, CA). The initial mobile phase is pure water, isocratically operated to elute the NC fraction. After 16 min, the mobile phase is switched to a 0.04 M sodium hydroxide (NaOH) solution to elute the trapped MDA fraction. The mobile phase is then switched to a 1 M NaCl; the strong ionic interactions elute PA fractions within ~16 min. Measuring the ratio of absorbance in UV/Vis chromatograms at 250 nm (E_2) over 340 nm (E_3) for the PA fraction provides structural information of HULIS (Duarte et al., 2005; Krivacsy et al., 2008). Sampling sites with aromatic precursors resulting from biomass burning have lower ratios (e.g., $E_2/E_3 = 2.88$; Chamonix, France, in December 2007), consistent with an inverse correlation between E_2/E_3 and HULIS aromaticity. In contrast, summer urban aerosol has high ratios (e.g., $E_2/E_3 = 5.90$; Marseilles, France, in July 2008), suggesting abundant aliphatic HULIS. SOA of biogenic and anthropogenic origin is known to contribute in the aliphatic HULIS (Decesari et al., 2007).

7.4.11 IC WITH PULSED AMPEROMETRIC DETECTOR ANALYSIS FOR CARBOHYDRATES AND IC WITH CONDUCTIVITY DETECTOR ANALYSIS FOR ORGANIC ACIDS

Seventeen carbohydrates from C_3 to C_{12} can be detected and quantified by IC with a pulsed amperometric detector (PAD): glycerol, erythritol, arabinose, xylose, arabitol, xylitol, levoglucosan, mannosan, galactose, fructose, glucose, inositol, sorbitol, mannitol, trehalose, sucrose, and maltitol. Levoglucosan, mannosan, and galactose derive from biomass burning (Mazzoleni et al., 2007; Rinehart et al., 2006; Ward et al., 2006). Quantification of these compounds (e.g., Dionex ICS 3000 series, IC, Sunnyvale, CA) uses a CarboPac MA1 analytical column (4 × 250 mm, Cat. No.044066), a CarboPac MA1 guard column (4 × 50 mm, Cat. No.044067), and a GM-4 (2 mm, Cat. No.049135) gradient mixer, with a 600 mM NaOH eluent at a flow rate of 0.400 mL/min. Detection is carried out with an electrochemical detector (ECD; e.g., Dionex ED50) in PAD mode equipped with an electrochemical cell (e.g., Dionex ED50/ED50A), utilizing a gold working electrode and PAD. The analytes are identified by their retention times and quantified by amperometric peak areas.

In the MDA class, four monocarboxylic acids from C_1 to C_4 (i.e., formic, acetic, methanesulfonic, and lactic) and five dicarboxylic acids from C_2 to C_5 (i.e., oxalic, malonic, maleic, succinic, and glutaric) are analyzed by IC with a conductivity detector. Oxalic acid is the most abundant organic acid in aerosol samples and has been found to correlate positively with the concentration of HULIS (Samburova et al., 2005). For the aforementioned nine organic acids, the IC is equipped with an IonPac AS11-HC (4 × 250 mm, Cat. No. 052960) analytical column, an IonPac AG11-HC (4 × 50 mm, Cat. No. 052962) guard column, and an ATC-3 anion trap column (4 mm, Cat. No. 059660) with a 100 mM NaOH eluent at a flow rate of 1.5 mL/min. An ECD (e.g., Dionex ED50) in conductivity mode is used along with a detection stabilizer conductivity cell (e.g., Dionex DS3, Cat. No. 044130). A 4 mm suppressor (e.g., Dionex ASRS-300), operated in the anion self-regenerating suppression mode, is also included in this configuration to suppress the eluent signal. The analytes are identified by their retention times and quantified by conductivity peak areas.

The HPLC-UV/Vis, IC-PAD, and IC-conductivity detection methods each require 1ml of DDW which may require a larger sample deposit or more parallel filters than illustrated in Figure 7.1.

7.4.12 THERMAL DESORPTION-GAS CHROMATOGRAPHY-MASS SPECTROMETRY ANALYSIS FOR NONPOLAR ORGANIC COMPOUNDS

Organic compounds are important source markers (Chow et al., 2007b; Daisey et al., 1986; Labban et al., 2006; Zielinska et al., 2008) and have potentially adverse health effects (Mauderly and Chow, 2008; McDonald et al., 2004). OC consists of thousands of organic compounds, more than is practical or desirable to measure. Table 7.2 lists functional groups and some of the specific compounds that can be practically measured with current technology.

Thermal desorption-gas chromatography-mass spectrometry (TD-GC-MS) (Chow et al., 2007c; Falkovich and Rudich, 2001; Greaves et al., 1985; Hays et al., 2003; Ho and Yu, 2004; Schnelle-Kreis et al., 2005) has returned results that are comparable to those from solvent extraction methods (Ho et al., 2008). TD-GC-MS is a cost-effective alternative approach for qualitative and quantitative analysis of nonpolar organic compounds on aerosol-loaded filters (Ho and Yu, 2004).

Filter punches are removed following the TOR procedure and spiked with 1 μL of two internal standard solutions (e.g., $nC_{16}D_{34}$ and $nC_{24}D_{50}$ for alkanes and phenanthrene-d_{10}, and chrysene-d_{12} for polycyclic aromatic hydrocarbons [PAHs]) to normalize the MS response. The punches are divided with a clean, sharp blade to facilitate loading of the filter pieces into a Pyrex glass tube that is matched to the size of the GC-MS (e.g., Agilent 6890 GC with Model 5973 or 5975 model MS, Santa Clara, CA) injection port. A small amount of prebaked glass wool holds the filter parts in position after placing the glass tube loaded with the filter. For in-injection port TD-GC-MS, the injector port temperature is hold at 50°C prior to analysis. After placing the glass tube loaded with the filter, the septum cap is closed, and the injector port temperature is increased to 275°C for 11 min to desorb the organic materials in splitless mode. The GC oven is maintained at 30°C during sample heating to focus the released organic analytes on the head of the GC column. After the sample temperature achieves 275°C, the GC oven temperature is retained at 30°C for 2 min, increased at a rate of 10°C/min to 120°C, followed by an increase of 8°C/min to 310°C, and then held at 310°C for 20 min. An HP-5 ms capillary column (5% diphenyl/95% dimethylsiloxane; 30 m long × 0.25 mm I.D. × 0.25 μm film thickness; Agilent Technologies, Santa Clara, CA) separates the peaks in an ultrahigh purity (99.9999%) He carrier gas. The MS detector scans from 50 to 650 amu at 230°C and 70 eV electron ionization. Species are identified by comparing mass spectra and retention times with standards that also relate the peak areas to concentration levels, after normalization to internal standards. Periodic cleaning is needed for the injection port and gold seal, the column head where the eluted sample is focused, and the ion source after every 50–100 samples, depending on sample loading (Ho et al., 2011). Although normalization to the internal standard accounts for changes in the instrument response, the detection limits degrade after 50–100 sample runs especially for samples containing high amounts of polar organic compounds.

7.4.13 COMPUTER-CONTROLLED SCANNING ELECTRON MICROSCOPY FOR SINGLE-PARTICLE MORPHOLOGY AND COMPOSITION

Shapes and compositions of individual particles are often useful for evaluating source contributions (Casuccio et al., 1989) and the effects of nonspherical particles on radiative transfer (Chakrabarty et al., 2009). Scanning electron microscopy (SEM) (Burleson et al., 2004; Jambers et al., 1995; Maynard, 2000) uses electrons (rather than light as in an optical microscope) to form magnified images. Electrons provide for better feature resolution, wider range of magnification, and a greater depth-of-field than is available in the conventional optical microscope for particles with diameters of 0.2–1 μm. Computer-controlled scanning electron microscopy (CCSEM) couples the SEM with software that locates a particle, obtains an image, conducts an elemental analysis, records the

TABLE 7.2

Organic Compounds by Functional Group using Thermal Desorption-Gas Chromatography-Mass Spectrometry Analysis

Functional Group and Analysis Method	Quantified Compounds (Carbon Numbers in Parentheses)
PAHs	Acenaphthylene (C12), acenaphthene (C12), fluorene (C13), phenanthrene (C14), anthracene (C14), fluoranthene (C16), pyrene (C16), benzo[a]anthracene (C18), chrysene (C18), benzo[b]fluoranthene (C20), benzo[j+k]fluoranthene (C20), benzo[a]fluoranthene (C20), benzo[e]pyrene (C20), benzo[a]pyrene (C20), perylene (C20), indeno[1,2,3-cd]pyrene (C22), dibenzo[a,h]anthracene (C22), benzo[ghi]perylene (C22), coronene (C24), dibenzo[a,e]pyrene (C24), 1-methyl fluorene (C14), dibenzothiophene (C12), 9-fluorenone (C13), 1-methylphenanthrene (C15), 2-methylphenanthrene (C15), 9-methylanthracene (C15), 3,6-dimethyl phenanthrene (C16), methylfluoranthene (C17), retene (C18), benzo[ghi]fluoranthene (C18), benzo(c)phenanthrene (C18), benzo[b]naphtho[1,2-d]thiophene (C16), cyclopenta[cd]pyrene (C18), benz[a]anthracene-7,12-dione (C18), methylchrysene (C19), benzo(b)chrysene (C22), picene (C22), anthanthrene (C22)
n-Alkanes	Pentadecane (n-C15), hexadecane (n-C16), heptadecane (n-C17), octadecane (n-C18), nonadecane (n-C19), eicosane (n-C20), heneicosane (n-C21), docosane (n-C22), tricosane (n-C23), tetracosane (n-C24), pentacosane (n-C25), hexacosane (n-C26), heptacosane (n-C27), octacosane (n-C28), nonacosane (n-C29), triacontane (n-C30), hentriacontane (n-C31), dotriacontane (n-C32), tritriacontane (n-C33), tetratriacontane (n-C34), pentatriacontane (n-C35), hexatriacontane (n-C36), heptatriacontane (n-C37), octatriacontane (n-C38), nonatriacontane (n-C39), tetracontane (n-C40)
Iso/anteiso-alkanes	Iso-nonacosane (iso-C29), anteiso-nonacosane (anteiso-C29), iso-triacontane (iso-C30), anteiso-triacontane (anteiso-C30), iso-hentriacontane (iso-C31), anteiso-hentriacontane (anteiso-C31), iso-dotriacontane (iso-C32), anteiso-dotriacontane (anteiso-C32), iso-tritriacontane (iso-C33), anteiso-tritriacontane (anteiso-C33)
Methyl-alkanes	2-Methylnonadecane (C20), 3-methylnonadecane (C20)
Branched-alkanes	Pristane (C19), phytane (C20), squalane (C30)
Cycloalkanes	Octylcyclohexane (C14), decylcyclohexane (C16), tridecylcyclohexane (C19), n-heptadecylcyclohexane (C23), nonadecylcyclohexane (C25)
Alkenes	1-Octadecene (C18)
Hopanes	22,29,30-Trisnorneophopane (Ts) (C27), 22,29,30-trisnorphopane (C27), αβ-norhopane (C29αβ-hopane), 22,29,30-norhopane (29Ts), αα- + βα-norhopane (C29αα- + βα-hopane), αβ-hopane (C30αβ -hopane), αα-hopane (30αα-hopane), βα-hopane (C30βα -hopane), αβS-homohopane (C31αβS-hopane), αβR-homohopane (C31αβR-hopane), αβS-bishomohopane (C32αβS-hopane), αβR-bishomohopane (C32αβR-hopane), 22S-trishomohopane (C33), 22R-trishomohopane (C33), 22S-tetrahomohopane (C34), 22R-tetrahomohopane (C34), 22S-pentashomohopane(C35), 22R-pentashomohopane(C35),
Steranes	ααα 20S-Cholestane (C27), αββ 20R-cholestane (C27), αββ 20s-cholestane (C27), ααα 20R-cholestane (C27), ααα 20S 24S-methylcholestane (C28), αββ 20R 24S-methylcholestane (C28), αββ 20S 24S-methylcholestane (C28), ααα 20R 24R-methylcholestane (C28), ααα 20S 24R/S-ethylcholestane (C29), αββ 20R 24R-ethylcholestane (C29), αββ 20S 24R-ethylcholestane (C29), ααα 20R 24R-ethylcholestane (C29)

results, and moves on to the next particle. Modern CCSEMs have the capability to locate particles deposited on a consistent background and acquire sizes, elemental spectra, and images (Casuccio et al., 1983; Chen et al., 2004; Coz et al., 2008; Cprek et al., 2007; Langmi and Watt, 2003; Poelt et al., 2002). These particles then can be automatically classified by size and spectra, but the images must be examined manually to confirm the classification. These CCSEM analyses can be done off-line, however, as the images and spectra are recorded during analysis.

Signals from electron bombardment of the particle include secondary and backscattered electrons along with characteristic x-rays. These are detected in synchronization with the position of the beam to provide size, morphological, and chemical information. The secondary electron signal yields an image with a three-dimensional perspective, high depth-of-field, and the appearance of overhead illumination. The backscattered electron signal provides compositional information because the signal intensity depends on the atomic number elements in the feature being examined. The x-ray spectrum is the same as that obtained by XRF, which uses a photon to ionize the atom in place of the SEM electron. Due to different shapes of particles and different mixtures, SEM x-ray spectra provide semiquantitative elemental concentrations; but the ratios of peak areas are similar for similar particles types, thereby providing a "fingerprint" for that type of particle. In other cases, particle shapes may be similar, but they are distinguishable by their elemental profiles. This is typically the case for spherical particles that form during high-temperature combustion and occur naturally as pollens.

It is not practical to characterize every one of the million or more particles on a single air filter. The number of particles within a category follows a Poisson distribution, for which the counting error is proportional to the square root of the number of counts. At least 100 particles must be counted within a category to obtain a counting error (1 standard deviation) of less than ±10%. If 10 categories are desired, at least 1000 particles must be characterized, and this is not practical by manual methods.

As shown in Figure 7.1, polycarbonate-membrane filters are among the best choices for CCSEM because they have smooth and uniform surfaces with uniformly sized holes that can be recognized and ignored by the CCSEM. The ringed Teflon-membrane filter has a nonuniform, fibrous appearance under the SEM, which interferes with automated particle detection. Inert particles can be removed from the Teflon-membrane filter and redeposited on a smooth surface for CCSEM by sonicating the filter in a solvent, such as acetone, and filtering the suspension through another filter such as etched polycarbonate. This results in loss of soluble species, but it is appropriate for nonsoluble minerals. Particles also must be coated with a conducting substance—usually carbon, gold, or platinum.

7.4.14 Data Management and Validation

Modern analysis software allows data to be output into comma delimited, Excel, or Access files. Microsoft Access (Redmond, WA) is a useful and commonly available relational database that allows data from field sampling and various laboratory analyses to be unified. Each record should contain at least the analysis ID, date and time of analysis, analyst ID, value obtained by the analysis, an analysis flag (e.g., first measurement, replicate measurement, standard, blank, or audit standard), and a validation flag (indicating an unusual situation). Simple programs can be written in the database language to convert instrument output (e.g., $\mu g/mL$, $\mu g/cm^2$) to μg/sample. Replicate analyses performed during the time period of the original analyses are used to estimate precision of the measurement for different concentration levels (Watson et al., 2001b). Field blanks or backup filters (see Figure 7.1) are used to estimate biases for passive deposition or adsorption, with averages subtracted and standard deviations propagated (Chow et al., 2010b). The remaining concentration is then divided by sampling flow rate times the sample duration to obtain ambient concentration, with uncertainties propagated to the final concentration (Watson et al., 2001b).

Once the concentrations have been calculated, data can be further evaluated for physical consistency (Chow et al., 1994): (1) the sum of measured species must be less than the mass concentrations; (2) water-soluble K^+, Cl^-, and $SO_4^=$—measured by AAS, IC, or AC on quartz-fiber filters—should be less than total K, Cl, and three times S concentrations measured by XRF on Teflon-membrane filters, respectively; (3) cations should balance anions when all water-soluble ions are measured. Additional internal consistency tests can be applied depending on the overlap of species measured by different multicomponent methods. Filter and extract remnants should be archived under refrigeration so that reanalyzes can be performed to resolve discrepancies.

7.5 SUMMARY AND CONCLUSIONS

A brief overview has been given of practical and cost-effective methods to analyze PM filter samples, pointing out some of the available methods and common pitfalls with reference to more complete and alternative treatments. Different filter media must be coupled with appropriate analysis techniques, but more than one method can be applied to each sample or extract to obtain a maximum amount of information. The methods described here are appropriate for both source and receptor samples, and they quantify most of the marker species that allow source contributions to be estimated by receptor modeling. Several of these components have been or are being implicated in effects on human health and climate, and it is to be expected that there will be greater demand for their measurement in the future. There are many opportunities to add additional methods for a larger range of compounds, and this chapter should serve as a starting point, not an end, for researchers desiring a more complete chemical characterization of ambient aerosols.

ACKNOWLEDGMENTS

Thanks are due to Steven Kohl, Dana Trimble, Ed Hackett, Brenda Cristani, Stephanie Salke, and Jerome Robles of DRI's Environmental Analysis Facility (EAF) for their assistance in gathering information for this chapter. Jo Gerrard assisted with assembling and Roger Kreidberg for editing the chapter. Partial support for chapter preparation was provided by the Nazir and Mary Ansari Foundation.

REFERENCES

Allen, R.; Box, M.; Liu, L.J.S.; Larson, T.V. (2001). A cost-effective weighing chamber for particulate matter filters. *J. Air Waste Manage. Assoc.*, **51**(12):1650–1653.

Anzano, J.M.; Ruiz-Gil, M. (2005). Comparison of microwave acid digestion with the wet digestion and ashing methods for the determination of Fe, Mn, and Zn in food samples by flame AAS. *At. Spectrosc.*, **26**(1):28–33.

Appel, B.R.; Tokiwa, Y.; Kothny, E.L.; Wu, R.; Povard, V. (1988). Evaluation of procedures for measuring atmospheric nitric acid and ammonia. *Atmos. Environ.*, **22**(8):1565–1573.

ASTM (2008). WK752 Test method for determination of $PM_{2.5}$ mass and species emissions from stationary combustion sources by dilution sampling, Prepared by American Society for Testing Materials International, Conshohocken, PA.

Axelrod, H.D.; Hansen, S.G. (1975). Filter sampling method for atmospheric sulfur dioxide at background concentrations. *Anal. Chem.*, **47**:2460–2461.

Baum, R.M.; Willis, R.D.; Walter, R.L.; Gutknecht, W.F.; Stiles, A.R. (1977). Solution-deposited standards using a capillary matrix and lyophilization. In *X-Ray Fluorescence Analysis of Environmental Samples*, Dzubay, T. G. (Ed.), Ann Arbor Science: Ann Arbor, MI, pp. 165–173.

Billiet, J.; Dams, R.; Hoste, J. (1980). Multielement thin film standards for x-ray fluorescence analysis. *X-Ray Spectrom.*, **9**(4):206–211.

Bolleter, W.T.; Bushman, C.J.; Tidwell, P.W. (1961). Spectrophotometric determination of ammonia as indophenol. *Anal. Chem.*, **33**(4):592–594.

Brown, A.S.; Yardley, R.E.; Quincey, P.G.; Butterfield, D.M. (2006). Studies of the effect of humidity and other factors on some different filter materials used for gravimetric measurements of ambient particulate matter. *Atmos. Environ.*, **40**:4670–4678.

Burleson, D.J.; Driessen, M.D.; Penn, R.L. (2004). On the characterization of environmental nanoparticles. *J. Environ. Sci. Health A Tox. Hazard Subst. Environ. Eng.*, **39**(10):2707–2753.

Burtscher, H. (2002). Novel instrumentation for the characterization of ultrafine particles. *J. Aerosol Med.*, **15**(2):149–160.

Butler, O.T.; Cook, J.M.; Davidson, C.M.; Harrington, C.F.; Miles, D.L. (2009). Atomic spectrometry update. Environmental analysis. *J. Anal. At. Spectrom.*, **24**(2):131–177.

Campbell, D.E.; Copeland, S.; Cahill, T.A. (1995). Measurement of aerosol absorption coefficient from Teflon filters using integrating plate and integrating sphere techniques. *Aerosol Sci. Technol.*, **22**(3):287–292.

Cao, J.J.; Lee, S.C.; Zhang, X.Y.; Chow, J.C.; An, Z.S.; Ho, K.F.; Watson, J.G.; Fung, K.K.; Wang, Y.Q.; Shen, Z.X. (2005a). Characterization of airborne carbonate over a site near Asian dust source regions during spring 2002 and its climatic and environmental significance. *J. Geophys. Res. Atmos.*, **110**(D03203):1–8. doi:10.1029/2004JD005244.

Cao, J.J.; Wu, F.; Chow, J.C.; Lee, S.C.; Li, Y.; Chen, S.W.; An, Z.S. et al. (2005b). Characterization and source apportionment of atmospheric organic and elemental carbon during fall and winter of 2003 in Xi'an, China. *Atmos. Chem. Phys.*, **5**:3127–3137. 1680-7324/acp/2005-5-3127.

Cao, J.J.; Zhang, T.; Chow, J.C.; Watson, J.G.; Wu, F.; Li, H. (2009). Characterization of atmospheric ammonia over Xi'an, China. *AAQR*, **9**(2):277–289.

Casuccio, G.S.; Janocko, P.B.; Lee, R.J.; Kelly, J.F.; Dattner, S.L.; Mgebroff, J.S. (1983). The use of computer controlled scanning electron microscopy in environmental studies. *J. Air Pollut. Control Assoc.*, **33**(10):937–943.

Casuccio, G.S.; Schwoeble, A.J.; Henderson, B.C.; Lee, R.J.; Hopke, P.K.; Sverdrup, G.M. (1989). The use of CCSEM and microimaging to study source/receptor relationships. In *Transactions, Receptor Models in Air Resources Management*, Watson, J. G. (Ed.), Air & Waste Management Association: Pittsburgh, PA, pp. 39–58.

Chakrabarty, R.K.; Garro, M.A.; Chancellor, S.; Herald, C.; Moosmüller, H. (2009). FracMAP: A user-interactive package for performing simulation and orientation-specific morphology analysis of fractal-like solid nano-agglomerates. *Comp. Phys. Commun.*, **180**(8):1376–1381.

Chang, H.; Herckes, P.; Collett, J.L., Jr. (2005). On the use of anion exchange chromatography for the characterization of water soluble organic carbon. *Geophys. Res. Lett.*, **32**(1):L01810. doi:10.1029/2004 GL021322.

Chen, Y.; Shah, N.; Huggins, F.E.; Huffman, G.P.; Linak, W.P.; Miller, C.A. (2004). Investigation of primary fine particulate matter from coal combustion by computer-controlled scanning electron microscopy. *Fuel Proc. Technol.*, **85**:743–761.

Cheng, Y.; He, K.B.; Duan, F.K.; Zheng, M.; Ma, Y.L.; Tan, J.H. (2009). Measurement of semivolatile carbonaceous aerosols and its implications: A review. *Environ. Int.*, **35**(3):674–681.

Cheng, Y.; Lee, S.C.; Ho, K.F.; Fung, K. (2010). Positive sampling artifacts in particulate organic carbon measurements in roadside environment. *Environ. Mon. Assess.*, **168**(1–4):645–656.

Chin, C.-J.; Wang, C.-F.; Jeng, S.-L. (1999). Multi-element analysis of airborne particulate matter collected on PTFE-membrane filters by laser ablation inductively coupled plasma mass spectrometry. *J. Anal. At. Spectrom.*, **14**:663–668.

Chow, J.C. (1995). Critical review: Measurement methods to determine compliance with ambient air quality standards for suspended particles. *J. Air Waste Manage. Assoc.*, **45**(5):320–382.

Chow, J.C.; Bachmann, J.D.; Kinsman, J.D.; Legge, A.H.; Watson, J.G.; Hidy, G.M.; Pennell, W.R. (2010a). Multipollutant air quality management: Critical review discussion. *J. Air Waste Manage. Assoc.*, **60**(10):1154–1164.

Chow, J.C.; Doraiswamy, P.; Watson, J.G.; Chen, L.-W.A.; Ho, S.S.H.; Sodeman, D.A. (2008). Advances in integrated and continuous measurements for particle mass and chemical composition. *J. Air Waste Manage. Assoc.*, **58**(2):141–163.

Chow, J.C.; Engelbrecht, J.P.; Freeman, N.C.G.; Hashim, J.H.; Jantunen, M.; Michaud, J.P.; de Tejada, S.S. et al. (2002a). Chapter one: Exposure measurements. *Chemosphere*, **49**(9):873–901.

Chow, J.C.; Fujita, E.M.; Watson, J.G.; Lu, Z.; Lawson, D.R.; Ashbaugh, L.L. (1994). Evaluation of filter-based aerosol measurements during the 1987 Southern California Air Quality Study. *Environ. Mon. Assess.*, **30**(1):49–80.

Chow, J.C.; Watson, J.G. (1999). Ion chromatography in elemental analysis of airborne particles. In *Elemental Analysis of Airborne Particles, Vol. 1*, Landsberger, S., Creatchman, M. (Eds.), Gordon and Breach Science: Amsterdam, the Netherlands, pp. 97–137.

Chow, J.C.; Watson, J.G. (2002). $PM_{2.5}$ carbonate concentrations at regionally representative interagency monitoring of protected visual environment sites. *J. Geophys. Res. Atmos.*, **107**(D21):ICC 6-1-ICC 6-9. doi: 10.1029/2001JD000574.

Chow, J.C.; Watson, J.G. (2011). Air quality management of multiple pollutants and multiple effects. *Air Qual. Clim. Change J.*, **43**(3):26–32.

Chow, J.C.; Watson, J.G.; Bowen, J.L.; Frazier, C.A.; Gertler, A.W.; Fung, K.K.; Landis, D.; Ashbaugh, L.L. (1993a). A sampling system for reactive species in the western United States. In *Sampling and Analysis of Airborne Pollutants*, Winegar, E. D., Keith, L. H. (Eds.), Lewis Publishers: Ann Arbor, MI, pp. 209–228.

Chow, J.C.; Watson, J.G.; Chen, L.-W.A.; Arnott, W.P.; Moosmüller, H.; Fung, K.K. (2004a). Equivalence of elemental carbon by thermal/optical reflectance and transmittance with different temperature protocols. *Environ. Sci. Technol.*, **38**(16):4414–4422.

Chow, J.C.; Watson, J.G.; Chen, L.-W.A.; Chang, M.C.O.; Robinson, N.F.; Trimble, D.L.; Kohl, S.D. (2007a). The IMPROVE_A temperature protocol for thermal/optical carbon analysis: Maintaining consistency with a long-term database. *J. Air Waste Manage. Assoc.*, **57**(9):1014–1023.

Chow, J.C.; Watson, J.G.; Chen, L.-W.A.; Lowenthal, D.H.; Motallebi, N. (2011a). Source profiles for black and organic carbon emission inventories. *Atmos. Environ.*, **45**(31):5407–5414.

Chow, J.C.; Watson, J.G.; Chen, L.-W.A.; Paredes-Miranda, G.; Chang, M.-C.O.; Trimble, D.L.; Fung, K.K.; Zhang, H.; Yu, J.Z. (2005a). Refining temperature measures in thermal/optical carbon analysis. *Atmos. Chem. Phys.*, **5**(4):2961–2972. 1680-7324/acp/2005-5-2961.

Chow, J.C.; Watson, J.G.; Chen, L.W.A.; Rice, J.; Frank, N.H. (2010b). Quantification of $PM_{2.5}$ organic carbon sampling artifacts in US networks. *Atmos. Chem. Phys.*, **10**(12):5223–5239.

Chow, J.C.; Watson, J.G.; Doraiswamy, P.; Chen, L.W.A.; Sodeman, D.A.; Lowenthal, D.H.; Park, K.; Arnott, W.P.; Motallebi, N. (2009). Aerosol light absorption, black carbon, and elemental carbon at the Fresno Supersite, California. *Atmos. Res.*, **93**(4):874–887.

Chow, J.C.; Watson, J.G.; Edgerton, S.A.; Vega, E. (2002b). Chemical composition of PM_{10} and $PM_{2.5}$ in Mexico City during winter 1997. *Sci. Total Environ.*, **287**(3):177–201.

Chow, J.C.; Watson, J.G.; Green, M.C.; Frank, N.H. (2010c). Filter light attenuation as a surrogate for elemental carbon. *J. Air Waste Manage. Assoc.*, **60**(11):1365–1375.

Chow, J.C.; Watson, J.G.; Kuhns, H.D.; Etyemezian, V.; Lowenthal, D.H.; Crow, D.J.; Kohl, S.D.; Engelbrecht, J.P.; Green, M.C. (2004b). Source profiles for industrial, mobile, and area sources in the Big Bend Regional Aerosol Visibility and Observational (BRAVO) Study. *Chemosphere*, **54**(2):185–208.

Chow, J.C.; Watson, J.G.; Lowenthal, D.H.; Chen, L.-W.A.; Tropp, R.J.; Park, K.; Magliano, K.L. (2006). $PM_{2.5}$ and PM_{10} mass measurements in California's San Joaquin Valley. *Aerosol Sci. Technol.*, **40**(10):796–810.

Chow, J.C.; Watson, J.G.; Lowenthal, D.H.; Chen, L.-W.A.; Zielinska, B.; Mazzoleni, L.R.; Magliano, K.L. (2007b). Evaluation of organic markers for chemical mass balance source apportionment at the Fresno supersite. *Atmos. Chem. Phys.*, **7**(7):1741–2754. http://www.atmos-chem-phys.net/7/1741/2007/acp-7-1741-2007.pdf

Chow, J.C.; Watson, J.G.; Lowenthal, D.H.; Chen, L.-W.A.; Motallebi, N. (2010d). Black and organic carbon emission inventories: Review and application to California. *J. Air Waste Manage. Assoc.*, **60**(4):497–507.

Chow, J.C.; Watson, J.G.; Lowenthal, D.H.; Magliano, K.L. (2005b). Loss of $PM_{2.5}$ nitrate from filter samples in central California. *J. Air Waste Manage. Assoc.*, **55**(8):1158–1168.

Chow, J.C.; Watson, J.G.; Pritchett, L.C.; Pierson, W.R.; Frazier, C.A.; Purcell, R.G. (1993b). The DRI Thermal/Optical Reflectance carbon analysis system: Description, evaluation and applications in U.S. air quality studies. *Atmos. Environ.*, **27A**(8):1185–1201.

Chow, J.C.; Watson, J.G.; Robles, J.; Wang, X.L.; Chen, L.-W.A.; Trimble, D.L.; Kohl, S.D.; Tropp, R.J.; Fung, K.K. (2011b). Quality assurance and quality control for thermal/optical analysis of aerosol samples for organic and elemental carbon. *Anal. Bioanal. Chem.*, **401**(10):3141–3152.

Chow, J.C.; Yu, J.Z.; Watson, J.G.; Ho, S.S.H.; Bohannan, T.L.; Hays, M.D.; Fung, K.K. (2007c). The application of thermal methods for determining chemical composition of carbonaceous aerosols: A review. *J. Environ. Sci. Health A*, **42**(11):1521–1541.

Coedo, A.G.; Padilla, I.; Dorado, M.T. (2005). Determination of minor elements in steelmaking flue dusts using laser ablation inductively coupled plasma mass spectrometry. *Talanta*, **67**(1):136–143.

Coutant, R.W. (1977). Effect of environmental variables on collection of atmospheric sulfate. *Environ. Sci. Technol.*, **11**(9):873–878.

Cox, R.M. (2003). The use of passive sampling to monitor forest exposure to O_3, NO_2 and SO_2: A review and some case studies. *Environ. Pollut.*, **126**(3):301–311.

Coz, E.; Artinano, B.; Robinson, A.L.; Casuccio, G.S.; Lersch, T.L.; Pandis, S.N. (2008). Individual particle morphology and acidity. *Aerosol Sci. Technol.*, **42**(3):224–232.

Cprek, N.; Shah, N.; Huggins, F.E.; Huffman, G.P. (2007). Distinguishing respirable quartz in coal fly ash using computer-controlled scanning electron microscopy. *Environ. Sci. Technol.*, **41**(10):3475–3480.

Daisey, J.M.; Cheney, J.L.; Lioy, P.J. (1986). Profiles of organic particulate emissions from air pollution sources: Status and needs for receptor source apportionment modeling. *J. Air Pollut. Control Assoc.*, **36**(1):17–33.

Davies, C.N. (1970). The clogging of fibrous aerosol filters. *J. Aerosol Sci.*, **1**:35–39.

Decesari, S.; Facchini, M.C.; Fuzzi, S.; Tagliavini, E. (2000). Characterization of water-soluble organic compounds in atmospheric aerosol: A new approach. *J. Geophys. Res.*, **105**(D1):1481–1489.

Decesari, S.; Mircea, M.; Cavalli, F.; Fuzzi, S.; Moretti, F.; Tagliavini, E.; Facchini, M.C. (2007). Source attribution of water-soluble organic aerosol by nuclear magnetic resonance spectroscopy. *Environ. Sci. Technol.*, **41**(7):2479–2484.

Demuynck, M. (1975). Determination of irreversible absorption of water by cellulose filters. *Atmos. Environ.*, **9**:523–528.

Deuerling, C.F.; Maguhn, J.; Nordsieck, H.O.; Warnecke, R.; Zimmermann, R. (2010). Measurement system for characterization of gas and particle phase of high temperature combustion aerosols. *Aerosol Sci. Technol.*, **44**(1):1–9.

Duarte, R.M.B.O.; Pio, C.A.; Duarte, A.C. (2005). Spectroscopic study of the water-soluble organic matter isolated from atmospheric aerosols collected under different atmospheric conditions. *Anal. Chim. Acta.*, **530**(1):7–14.

Duckworth, S. (1971). A simple technique for measuring mean H_2S concentration. *J. Air Pollut. Control Assoc.*, **21**(5):280–283.

Durrant, S.F. (1999). Laser ablation inductively coupled plasma mass spectrometry: Achievements, problems, prospects. *J. Anal. At. Spectrom.*, **14**:1385–1403.

Dzubay, T.G.; Lamothe, P.J.; Yoshuda, H. (1977). Polymer films as calibration standards for x-ray fluorescence analysis. *Adv. X-Ray Anal.*, **20**:411.

Dzubay, T.G.; Nelson, R.O. (1974). Advances in x-ray analysis. *Advances in X-ray Analysis Vol.18*, Pickles, W. L., Barrett, C. S., Newkirk, J. B., Ruud, C. O. (Eds.), Plenum Press: New York, pp. 619–629.

Engelbrecht, D.R.; Cahill, T.A.; Feeney, P.J. (1980). Electrostatic effects on gravimetric analysis of membrane filters. *J. Air Pollut. Control Assoc.*, **30**(4):391–392.

England, G.C.; Watson, J.G.; Chow, J.C.; Zielinska, B.; Chang, M.-C.O.; Loos, K.R.; Hidy, G.M. (2007a). Dilution-based emissions sampling from stationary sources: Part 1. Compact sampler, methodology and performance. *J. Air Waste Manage. Assoc.*, **57**(1):65–78.

England, G.C.; Watson, J.G.; Chow, J.C.; Zielinska, B.; Chang, M.-C.O.; Loos, K.R.; Hidy, G.M. (2007b). Dilution-based emissions sampling from stationary sources: Part 2. Gas-fired combustors compared with other fuel-fired systems. *J. Air Waste Manage. Assoc.*, **57**(1):79–93.

Engling, G.; Carrico, C.M.; Kreldenweis, S.M.; Collett, J.L.; Day, D.E.; Malm, W.C.; Lincoln, E.; Hao, W.M.; Iinuma, Y.; Herrmann, H. (2006). Determination of levoglucosan in biomass combustion aerosol by high-performance anion-exchange chromatography with pulsed amperometric detection. *Atmos. Environ.*, **40**(Suppl.2):S299–S311.

Falkovich, A.H.; Graber, E.R.; Schkolnik, G.; Rudich, Y.; Maenhaut, W.; Artaxo, P. (2005). Low molecular weight organic acids in aerosol particles from Rondonia, Brazil, during the biomass-burning, transition and wet periods. *Atmos. Chem. Phys.*, **5**:781–797.

Falkovich, A.H.; Rudich, Y. (2001). Analysis of semivolatile organic compounds in atmospheric aerosols by direct sample introduction thermal desorption GC/MS. *Environ. Sci. Technol.*, **35**(11):2326–2333.

Feeney, P.; Cahill, T.; Olivera, J.; Guidara, R. (1984). Gravimetric determination of mass on lightly-loaded membrane filters. *J. Air Pollut. Control Assoc.*, **34**(4):376–377.

Fehsenfeld, F.C.; Hastie, D.; Chow, J.C.; Solomon, P.A. (2004). Particle and gas measurements. In *Particulate Matter Science for Policy Makers, A NARSTO Assessment*, McMurry, P. H., Shepherd, M. F., Vickery, J. S. (Eds.), Cambridge University Press: Cambridge, U.K., pp. 159–189.

Feng, Y.; Barratt, R.S. (1994). Digestion of dust samples in a microwave oven. *Sci. Total Environ.*, **143**(2,3):157–161.

Ferm, M. (1979). A method for determination of atmospheric ammonia. *Atmos. Environ.*, **13**:1385–1393.

Fernandez, F.J. (1989). Atomic absorption spectroscopy. In *Methods of Air Sampling and Analysis*, 3rd edn., Lodge, J. P., Jr. (Ed.), Lewis Publishers: Chelsea, MI, pp. 83–89.

Flanagan, J.B.; Jayanty, R.K.M.; Rickman, E.E., Jr.; Peterson, M.R. (2006). $PM_{2.5}$ Speciation Trends Network: Evaluation of whole-system uncertainties using data from sites with collocated samplers. *J. Air Waste Manage. Assoc.*, **56**(4):492–499.

Fung, K.K.; Heisler, S.L.; Price, A.; Nuesca, B.V.; Mueller, P.K. (1979). Comparison of ion chromatography and automated wet chemical methods for analysis of sulfate and nitrate in ambient particulate filter samples. In *Ion Chromatographic Analysis of Environmental Pollutants*, Sawicki, E., Mulik, J. D. (Eds.), Ann Arbor Science Publishers, Inc.: Ann Arbor, MI, pp. 203–209.

Giauque, R.D. (1973). Trace element analysis using x-ray tubes and semiconductor. *Trans. Am. Nucl. Soc.*, **17**:104.

Gligorovski, S.; Elteren, J.T.; Grgic, I. (2008). A multi-element mapping approach for size-segregated atmospheric particles using laser ablation ICP-MS combined with image analysis. *Sci. Total Environ.*, **407**(1):594–602.

Gotoh, T. (1980). Physical examination of a method for determination of nitrogen dioxide in the atmosphere by using triethanolamine filter paper. *Taiki Osen Gakkaishi*, **15**(8):334–341.

Greaves, R.C.; Barkley, R.M.; Sievers, R.E. (1985). Rapid sampling and analysis of volatile constituents of airborne particulate matter. *Anal. Chem.*, **57**(14):2807–2815.

Greenberg, R.R. (1979). Trace element characterization of the NBS urban particulate matter standard reference material by instrumental neutron acitration analysis. *Anal. Chem.*, **51**:2004.

Grennfelt, P.; Akerstrom, A.; Brosset, C. (1971). Determination of filter-collected airborne matter by x-ray fluorescence. *Atmos. Environ.*, **5**(1):1–6.

Grosjean, D.; Hisham, M.W.M. (1992). A passive sampler for atmospheric ozone. *J. Air Waste Manage. Assoc.*, **42**(2):169–173.

Hansen, D.A.; Edgerton, E.; Hartsell, B.; Jansen, J.; Burge, H.; Koutrakis, P.; Rogers, C. et al. (2006). Air quality measurements for the aerosol research and inhalation epidemiology study. *J. Air Waste Manage. Assoc.*, **56**(10):1445–1458.

Hays, M.D.; Smith, N.D.; Kinsey, J.; Dong, Y.; Kariher, P. (2003). Polycyclic aromatic hydrocarbon size distributions in aerosols from appliances of residential wood combustion as determined by direct thermal desorption—GC/MS. *J. Aerosol Sci.*, **34**(8):1061–1084.

Heagney, J.M.; Heagney, J.S. (1979). Thin film x-ray fluorescence calibration standards. *Nucl. Instrum. Methods*, **167**(1):137–138.

Heckel, J. (1995). Using Barkla polarized x-ray radiation in energy dispersive x-ray fluorescence analysis (EDXRF). *J. Trace Microprobe Tech.*, **13**(2):97–108.

Hedley, K.J.; Shepson, P.B.; Barrie, L.A.; Bottenheim, J.W.; MacTavish, D.C.; Anlauf, K.G.; Mackay, G.I. (1994). An evaluation of integrating techniques for measuring atmospheric nitrogen dioxide. *Int. J. Environ. Anal. Chem.*, **54**(3):167–181.

Herner, J.D.; Green, P.G.; Kleeman, M.J. (2006). Measuring the trace elemental composition of size-resolved airborne particles. *Environ. Sci. Technol.*, **40**(6):1925–1933.

Hidy, G.M.; Pennell, W.R. (2010). Multipollutant air quality management: A critical review. *J. Air Waste Manage. Assoc.*, **60**(6):645–674.

Hildemann, L.M.; Cass, G.R.; Markowski, G.R. (1989). A dilution stack sampler for collection of organic aerosol emissions: Design, characterization and field tests. *Aerosol Sci. Technol.*, **10**(10–11):193–204.

Ho, S.S.H.; Chow, J.C.; Watson, J.G.; Ng, L.P.T.; Kwok, Y.; Ho, K.F.; Cao, J.J. (2011). Precautions for in-injection port thermal desorption-gas chromatography/mass spectrometry (TD-GC/MS) applied to aerosol filter samples. *Atmos. Environ.*, **45**(7):1491–1496.

Ho, S.S.H.; Yu, J.Z. (2004). In-injection port thermal desorption and subsequent gas chromatography-mass spectrometric analysis of polycyclic aromatic hydrocarbons and *n*-alkanes in atmospheric aerosol samples. *J. Chromatogr. A*, **1059**(1–2):121–129.

Ho, S.S.H.; Yu, J.Z.; Chow, J.C.; Zielinska, B.; Watson, J.G.; Sit, E.H.L.; Schauer, J.J. (2008). Evaluation of an in-injection port thermal desorption-gas chromatography/mass spectrometry method for analysis of non-polar organic compounds in ambient aerosol samples. *J. Chromatogr. A*, **1200**(2):217–227. doi:10.1016/j.chroma.2008.05.056.

Horvath, H. (1993). Atmospheric light absorption—A review. *Atmos. Environ.*, **27A**(3):293–317.

Hu, T.F.; Lee, S.C.; Cao, J.J.; Chow, J.C.; Watson, J.G.; Ho K.F.; Ho, W.K.; Rong, B.; An, Z.S. (2009a). Characterization of winter airborne particles at Emperor Qin's terra-cotta museum, China. *Sci. Total Environ.*, **407**:5319–5327.

Hu, T.F.; Lee, S.C.; Cao, J.J.; Ho, W.K.; Ho, K.F.; Chow, J.C.; Watson, J.G.; Rong, B.; An, Z.S. (2009b). Atmospheric deterioration of Qin brick in an environmental chamber at Emperor Qin's Terracotta Museum, China. *J. Archaeol. Sci*, **36**(11):2578–2583.

Huygen, C. (1963). The sampling of sulfur dioxide in air with impregnated filter paper. *Anal. Chim. Acta*, **28**:349.

Jacobson, M.Z. (2002). Control of fossil-fuel particulate black carbon plus organic matter, possibly the most effective method of slowing global warming. *J. Geophys. Res.*, **107**(D19):ACH 16-1–ACH 16-22. doi:10.1029/2001JD001376.

Jambers, W.; DeBock, L.; Vangrieken, R. (1995). Recent advances in the analysis of individual environmental particles—A review. *Analyst*, **120**:681–692.

Kaasalainen, M.; Yli-Halla, M. (2003). Use of sequential extraction to assess metal partitioning in soils. *Environ. Pollut.*, **126**(2):225–233.

Kajino, M.; Winwarter, W.; Ueda, H. (2006). Modeling retained water content in measured aerosol mass. *Atmos. Environ.*, **40**(27):5202–5213.

Kim, E.; Hopke, P.K. (2004). Improving source identification of fine particles in a rural northeastern US area utilizing temperature-resolved carbon fractions. *J. Geophys. Res. Atmos.*, **109**(D09204):1–13. doi:2003JD004199.

Kitto, A.M.; Colbeck, I. (1999). Filtration and denuder sampling techniques. In *Analytical Chemistry of Aerosols*, Spurny, K. R. (Ed.), CRC Press LLC: Boca Raton, FL, pp. 103–132.

Komarek, M.; Ettler, V.; Chrastny, V.; Mihaljevic, M. (2008). Lead isotopes in environmental sciences: A review. *Environ. Int.*, **34**(4):562–577.

Krivacsy, Z.; Kiss, G.; Ceburnis, D.; Jennings, G.; Maenhaut, W.; Salma, I.; Shooter, D. (2008). Study of water-soluble atmospheric humic matter in urban and marine environments. *Atmos. Res.*, **87**(1):1–12.

Kume, K.; Ohura, T.; Amagai, T.; Fusaya, M. (2008). Field monitoring of volatile organic compounds using passive air samplers in an industrial city in Japan. *Environ. Pollut.*, **153**(3):649–657.

Kusko, B.H.; Cahill, T.A.; Eldred, R.A.; Matsuda, Y.; Miyake, H. (1989). Nondestructive analysis of total nonvolatile carbon by Forward Alpha Scattering Technique (FAST). *Aerosol Sci. Technol.*, **10**:390–396.

Labban, R.; Veranth, J.M.; Watson, J.G.; Chow, J.C. (2006). Feasibility of soil dust source apportionment by the pyrolysis-gas chromatography/mass spectrometry method. *J. Air Waste Manage. Assoc.*, **56**(9):1230–1242.

Landsberger, S.; Creatchman, M. (1999). *Elemental Analysis of Airborne Particles*. Gordon and Breach: Newark, NJ.

Langmi, H.W.; Watt, J. (2003). Evaluation of computer-controlled SEM in the study of metal-contaminated soils. *Mineral. Mag.*, **67**(2):219–231.

Lawless, P.A.; Rodes, C.E. (1999). Maximizing data quality in the gravimetric analysis of personal exposure sample filters. *J. Air Waste Manage. Assoc.*, **49**(9):1039–1049.

Lee, P.K.H.; Brook, J.R.; Dabek-Zlotorzynska, E.; Mabury, S.A. (2003). Identification of the major sources contributing to $PM_{2.5}$ observed in Toronto. *Environ. Sci. Technol.*, **37**(21):4831–4840.

Li, X.X.; Cao, J.J.; Chow, J.C.; Han, Y.M.; Lee, S.C.; Watson, J.G. (2008). Chemical characteristics of carbonaceous aerosols during dust storms over Xi'an in China. *Adv. Atmos. Sci.*, **25**(5):847–855.

Lin, C.I.; Friedlander, S.K. (1988). A note on the use of glass fiber filters in the thermal analysis of carbon containing aerosols. *Atmos. Environ.*, **22**(3):605–607.

Link, D.D.; Walter, P.J.; Kingston, H.M. (1998). Development and validation of the new EPA microwave-assisted leach method 3051A. *Environ. Sci. Technol.*, **32**(22):3628–3632.

Lippmann, M. (2001). Filters and filter holders. In *Air Sampling Instruments for Evaluation of Atmospheric Contaminants*, 9th edn., Cohen, B. S., McCammon, C. S. J. (Eds.), ACGIH: Cincinnati, OH, pp. 281–314.

Lodge, J.P. (1989). *Methods of Air Sampling and Analysis*. Lewis Publishers, Inc.: Chelsea, MI.

Malm, W.C.; Schichtel, B.A.; Ames, R.B.; Gebhart, K.A. (2002). A ten-year spatial and temporal trend of sulfate across the United States. *J. Geophys. Res. Atmos.*, **107**(D22):ACH 11-1–ACH 11-20. doi:10.1029/2002JD002107.

Mason, J.B.; Fujita, E.M.; Campbell, D.E.; Zielinska, B. (2011). Evaluation of passive samplers for assessment of community exposure to toxic air contaminants and related pollutants. *Environ. Sci. Technol.*, **45**(6):2243–2249.

Mauderly, J.L.; Chow, J.C. (2008). Health effects of organic aerosols. *Inhal. Toxicol.*, **20**(3):257–288. doi: 10.1080/08958370701866008.

Maykut, N.N.; Lewtas, J.; Kim, E.; Larson, T.V. (2003). Source apportionment of $PM_{2.5}$ at an urban IMPROVE site in Seattle, Washington. *Environ. Sci. Technol.*, **37**(22):5135–5142.

Maynard, A.D. (2000). Overview of methods for analysing single ultrafine particles. *Philos. Trans. R. Soc. Lond. Ser. A*, **358**(1775):2593–2610.

Mazzoleni, L.R.; Zielinska, B.; Moosmüller, H. (2007). Emissions of levoglucosan, methoxy phenols, and organic acids from prescribed burns, laboratory combustion of wildland fuels, and residential wood combustion. *Environ. Sci. Technol.*, **41**(7):2115–2122.

McDonald, J.D.; Eide, I.; Seagrave, J.; Zielinska, B.; Whitney, K.; Lawson, D.R.; Mauderly, J.L. (2004). Relationship between composition and toxicity of motor vehicle emission samples. *Environ. Health Perspect.*, **112**(15):1527–1538.

McInnes, I.M.; Quinn, P.K.; Covert, D.S.; Anderson, T.I. (1996). Gravimetric analysis, ionic composition, and associated water mass of the marine aerosol. *Atmos. Environ.*, **30**(6):869–884.

McMurry, P.H. (2000). A review of atmospheric aerosol measurements. *Atmos. Environ.*, **34**(12–14):1959–1999.

Melaku, S.; Dams, R.; Moens, L. (2005). Determination of trace elements in agricultural soil samples by inductively coupled plasma-mass spectrometry: Microwave acid digestion versus aqua regia extraction. *Anal. Chim. Acta*, **543**(1–2):117–123.

Miller, L.; Lemke, L.D.; Xu, X.H.; Molaroni, S.M.; You, H.Y.; Wheeler, A.J.; Booza, J. et al. (2010). Intra-urban correlation and spatial variability of air toxics across an international airshed in Detroit, Michigan (USA) and Windsor, Ontario (Canada). *Atmos. Environ.*, **44**(9):1162–1174.

Miwa, T.; Maruo, Y.Y.; Akaoka, K.; Kunioka, T.; Nakamura, J. (2009). Development of colorimetric ozone detection papers with high ultraviolet resistance using ultraviolet absorbers. *J. Air Waste Manage. Assoc.*, **59**(7):801–808.

Monn, C.; Hangartner, M. (1990). Passive sampling of ozone. *J. Air Waste Manage. Assoc.*, **40**(3):357–358.

Moosmüller, H.; Chakrabarty, R.K.; Arnott, W.P. (2009). Aerosol light absorption and its measurement: A review. *J. Quant. Spectrosc. Radiat. Transfer*, **110**:844–878.

Mueller, P.K.; Hidy, G.M.; Watson, J.G.; Baskett, R.L.; Fung, K.K.; Henry, R.C.; Lavery, T.F.; Warren, K.K. (1983). The Sulfate Regional Experiment (SURE): Report of findings (Vols. 1, 2, and 3). Report Number EA-1901, Prepared by Electric Power Research Institute, Palo Alto, CA.

Mulik, J.; Sawicki, E. (1978). *Ion Chromatographic Analysis of Environmental Pollutants: Volume I*. Ann Arbor Science Publishers: Ann Arbor, MI.

Mulik, J.; Sawicki, E. (1979). *Ion Chromatographic Analysis of Environmental Pollutants: Volume II*. Ann Arbor Science Publishers: Ann Arbor, MI.

Nash, D.G.; Leith, D. (2010). Use of passive diffusion tubes to monitor air pollutants. *J. Air Waste Manage. Assoc.*, **60**(2):204–209.

Novakov, T.; Corrigan, C.E. (1996). Cloud condensation nucleus activity of the organic component of biomass smoke. *Geophys. Res. Lett.*, **23**:2141–2144.

Novakov, T.; Penner, J.E. (1993). Large contribution of organic aerosols to cloud-condensation-nuclei concentrations. *Nature*, **365**:823–826.

Ohno, T.; Hirata, T. (2004). Advances in elemental and isotopic analyses by ICP-mass spectrometry and their applications to geochemistry. *Bunseki Kagaku*, **53**(7):631–644.

Owega, S.; Evans, G.J.; Jervis, R.E.; Tsai, J.; Fila, M.; Tan, P.V. (2002). Comparison between urban Toronto PM and selected materials: Aerosol characterization using laser ablation/ionization mass spectrometry (LAMS). *Environ. Pollut.*, **120**(1):125–135.

Panne, U.; Neuhauser, R.E.; Theisen, M.; Fink, H.; Niessner, R. (2001). Analysis of heavy metal aerosols on filters by laser-induced plasma spectroscopy. *Spectrochim. Acta Part B*, **56**(6):839–850.

Poelt, P.; Schmied, M.; Obernberger, I.; Brunner, T.; Dahl, J. (2002). Automated analysis of submicron particles by computer-controlled scanning electron microscopy. *Scanning*, **24**(2):92–100.

Polak, J.; Mestek, O.; Suchanek, M. (2006). Uncertainty of determination of palladium in road dust sample by inductively coupled plasma mass spectrometry. *Accredit. Qual. Assur.*, **10**(11):627–632.

Pope, C.A., III; Dockery, D.W. (2006). Critical Review: Health effects of fine particulate air pollution: Lines that connect. *J. Air Waste Manage. Assoc.*, **56**(6):709–742.

Rao, C.R.M.; Sahuquillo, A.; Sanchez, J.F.L. (2008). A review of the different methods applied in environmental geochemistry for single and sequential extraction of trace elements in soils and related materials. *Water Air Soil Pollut.*, **189**(1–4):291–333.

Rasmussen, P.E.; Gardner, H.D.; Niu, J.J. (2010). Buoyancy-corrected gravimetric analysis of lightly loaded filters. *J. Air Waste Manage. Assoc.*, **60**(9):1065–1077.

Rinehart, L.R.; Fujita, E.M.; Chow, J.C.; Magliano, K.L.; Zielinska, B. (2006). Spatial distribution of $PM_{2.5}$ associated organic compounds in central California. *Atmos. Environ.*, **40**(2):290–303.

Rogge, W.F.; Mazurek, M.A.; Hildemann, L.M.; Cass, G.R.; Simoneit, B.R.T. (1993). Quantification of urban organic aerosols at a molecular level: Identification, abundance and seasonal variation. *Atmos. Environ.*, **27A**(8):1309–1330.

Rommers, P.J.; Visser, J. (1969). Spectrophotometric determination of micro amounts of nitrogen as indophenol. *Analyst*, **94**:653–658.

Sahu, M.; Hu, S.; Ryan, P.H.; LeMasters, G.; Grinshpun, S.A.; Chow, J.C.; Biswas, P. (2011). Chemical compositions and source identification of $PM_{2.5}$ aerosols for estimation of a diesel source surrogate. *Sci. Total Environ.*, **409**(13):2642–2651.

Samburova, V.; Szidat, S.; Hueglin, C.; Fisseha, R.; Baltensperger, U.; Zenobi, R.; Kalberer, M. (2005). Seasonal variation of high-molecular-weight compounds in the water-soluble fraction of organic urban aerosols. *J. Geophys. Res. Atmos.*, **110**(D23):doi:10.1029/2005JD005910.

Samy, S.; Mazzoleni, L.R.; Mishra, S.; Zielinska, B.; Hallar, A.G. (2010). Water-soluble organic compounds at a mountain-top site in Colorado, USA. *Atmos. Environ.*, **44**(13):1663–1671.

Sandell, E.B. (1959). *Colorimetric Determination of Traces of Metals*. John Wiley & Sons: New York.

Sandradewi, J.; Prevot, A.S.H.; Weingartner, E.; Schmidhauser, R.; Gysel, M.; Baltensperger, U. (2008). A study of wood burning and traffic aerosols in an Alpine valley using a multi-wavelength Aethalometer. *Atmos. Environ.*, **42**(1):101–112.

de Santis, F.; Allegrini, I.; Bellagotti, R.; Vichi, F.; Zona, D. (2006). Development and field evaluation of a new diffusive sampler for hydrogen sulphide in the ambient air. *Anal. Bioanal. Chem.*, **384**(4):897–901.

Saxena, P.; Hildemann, L.M. (1996). Water-soluble organics in atmospheric particles: A critical review of the literature and application of thermodynamics to identify candidate compounds. *Atmos. Chem.*, **24**(1):57–109.

Schmid, H.P.; Laskus, L.; Abraham, H.J.; Baltensperger, U.; Lavanchy, V.M.H.; Bizjak, M.; Burba, P. et al. (2001). Results of the "Carbon Conference" international aerosol carbon round robin test: Stage 1. *Atmos. Environ.*, **35**(12):2111–2121.

Schnelle-Kreis, J.; Sklorz, M.; Peters, A.; Cyrys, J.; Zimmermann, R. (2005). Analysis of particle-associated semi-volatile aromatic and aliphatic hydrocarbons in urban particulate matter on a daily basis. *Atmos. Environ.*, **39**(40):7702–7714.

Shields, H.C.; Weschler, C.J. (1987). Analysis of ambient concentrations of organic vapors with a passive sampler. *J. Air Pollut. Control Assoc.*, **37**(9):1039–1045.

Silveira, M.L.; Alleoni, L.R.F.; O'Connor, G.A.; Chang, A.C. (2006). Heavy metal sequential extraction methods—A modification for tropical soils. *Chemosphere*, **64**(11):1929–1938.

Sisler, J.F.; Malm, W.C. (1994). The relative importance of soluble aerosols to spatial and seasonal trends of impaired visibility in the United States. *Atmos. Environ.*, **28**(5):851–862.

Solomon, P.A.; Norris, G.; Landis, M.S.; Tolocka, M.P. (2001). Chemical analysis methods for atmospheric aerosol components. In *Aerosol Measurement: Principles, Techniques, and Applications*, 2nd edn., Baron, P., Willeke, K. (Eds.), John Wiley & Sons: New York, pp. 261–293.

Solomon, P.A.; Sioutas, C. (2008). Continuous and semicontinuous monitoring techniques for particulate matter mass and chemical components: A synthesis of findings from EPA's particulate matter supersites program and related studies. *J. Air Waste Manage. Assoc.*, **58**(2):164–195.

Spicer, C.W.; Schumacher, P.M. (1979). Particulate nitrate: Laboratory and field studies of major sampling interferences. *Atmos. Environ.*, **13**:543–552.

Spurny, K.R.; Lodge, J.P., Jr.; Frank, E.R.; Sheesley, D.C. (1969). Aerosol filtration by means of Nuclepore filters: Structural and filtration properties. *Environ. Sci. Technol.*, **3**:453–464.

Stock, T.H.; Morandi, M.T.; Afshar, M.; Chung, K.C. (2008). Evaluation of the use of diffusive air samplers for determining temporal and spatial variation of volatile organic compounds in the ambient air of urban communities. *J. Air Waste Manage. Assoc.*, **58**(10):1303–1310.

Sutter, B.; Bemer, D.; Appert-Collin, J.C.; Thomas, D.; Midoux, N. (2010). Evaporation of liquid semi-volatile aerosols collected on fibrous filters. *Aerosol Sci. Technol.*, **44**(5):395–404.

Swanson, J.; Kittelson, D. (2008). A method to measure static charge on a filter used for gravimetric analysis. *Aerosol Sci. Technol.*, **42**(9):714–721.

Tan, P.V.; Fila, M.S.; Evans, G.J.; Jervis, R.E. (2002). Aerosol laser ablation mass spectrometry of suspended powders from PM sources and its implications to receptor modeling. *J. Air Waste Manage. Assoc.*, **52**(1):27–40.

Tanaka, S.; Yasushi, N.; Sato, N.; Fukasawa, T.; Santosa, S.J.; Yamanaka, K.; Ootoshi, T. (1998). Rapid and simultaneous multi-element analysis of atmospheric particulate matter using inductively coupled plasma mass spectrometry with laser ablation sample introduction. *J. Anal. At. Spectrom.*, **13**:135–140.

Tombach, I.H.; Malm, W.C.; Pitchford, M.L. (1987). Performance criteria for monitoring visibility-related variables. In *Transactions, Visibility Protection: Research and Policy Aspects*, Bhardwaja, P. S. (Ed.), Air Pollution Control Association: Pittsburgh, PA, pp. 529–540.

Tsai, C.J.; Chang, C.T.; Shih, B.H.; Aggarwal, S.G.; Li, S.N.; Chein, H.M.; Shih, T.S. (2002). The effect of environmental conditions and electrical charge on the weighing accuracy of different filter materials. *Sci. Total Environ.*, **293**(1–3):201–206.

U.S. EPA (1994). Method 200.8: Determination of trace elements in waters and wastes by inductively coupled plasma—Mass spectrometry. Report Number Revision 5.4, Prepared by U.S. EPA, Cincinnati, OH.

U.S. EPA (1997). National ambient air quality standards for particulate matter: Final rule. *Fed. Reg.*, **62**(138):38651–38760. http://www.epa.gov/ttn/amtic/files/cfr/recent/pmnaaqs.pdf

U.S. EPA (2008). SPECIATE Version 4.2, Prepared by U.S. Environmental Protection Agency, Research Triangle Park, NC, http://www.epa.gov/ttn/chief/software/speciate/index.html

Vecchi, R.; Valli, G.; Fermo, P.; D'Alessandro, A.; Piazzalunga, A.; Bernardoni, V. (2009). Organic and inorganic sampling artefacts assessment. *Atmos. Environ.*, **43**(10):1713–1720.

Vincent, J.H. (1994). Measurement of coarse aerosols in workplaces—A review. *Analyst*, **119**(1):13–18.

Wang, X.L.; Chancellor, G.; Evenstad, J.; Farnsworth, J.E.; Hase, A.; Olson, G.M.; Sreenath, A.; Agarwal, J.K. (2009). A novel optical instrument for estimating size segregated aerosol mass concentration in real time. *Aerosol Sci. Technol.*, **43**:939–950.

Wang, C.F.; Chen, W.H.; Yang, M.H.; Chiang, P.C. (1995). Microwave decomposition for airborne particulate matter for the determination of trace elements by inductively coupled plasma mass spectrometry. *Analyst*, **120**(6):1681–1686.

Ward, T.J.; Rinehart, L.R.; Lange, T. (2006). The 2003/2004 Libby, Montana $PM_{2.5}$ source apportionment research study. *Aerosol Sci. Technol.*, **40**(3):166–177.

Watson, J.G. (2002). Visibility: Science and regulation—2002 Critical review. *J. Air Waste Manage. Assoc.*, **52**(6):628–713.

Watson, J.G.; Chen, L.-W.A.; Chow, J.C.; Lowenthal, D.H.; Doraiswamy, P. (2008). Source apportionment: Findings from the U.S. Supersite Program. *J. Air Waste Manage. Assoc.*, **58**(2):265–288.

Watson, J.G.; Chow, J.C. (1994). Particle and gas measurements on filters. In *Environmental Sampling for Trace Analysis*, Markert, B. (Ed.), VCH: New York, pp. 125–161.

Watson, J.G.; Chow, J.C. (2002). Comparison and evaluation of in-situ and filter carbon measurements at the Fresno Supersite. *J. Geophys. Res.*, **107**(D21):ICC 3-1–ICC 3-15. doi: 10.1029/2001JD000573.

Watson, J.G.; Chow, J.C. (2005). Receptor models. In *Air Quality Modeling—Theories, Methodologies, Computational Techniques, and Available Databases and Software. Vol. II—Advanced Topics*, Zannetti, P. (Ed.), Air and Waste Management Association and the EnviroComp Institute: Pittsburgh, PA, pp. 455–501.

Watson, J.G.; Chow, J.C. (2007). Receptor models for source apportionment of suspended particles. In *Introduction to Environmental Forensics*, 2nd edn., Murphy, B., Morrison, R. (Eds.), Academic Press: New York, pp. 279–316.

Watson, J.G.; Chow, J.C. (2011). Ambient aerosol sampling. In *Aerosol Measurement: Principles, Techniques and Applications*, 3rd edn., Kulkarni, P., Willeke, K., Baron, P. A. (Eds.), John Wiley & Sons, Inc., Hoboken, NJ, pp. 591–614.

Watson, J.G.; Chow, J.C.; Chen, L.-W.A. (2005). Summary of organic and elemental carbon/black carbon analysis methods and intercomparisons. *AAQR*, **5**(1):65–102. http://aaqr.org/

Watson, J.G.; Chow, J.C.; Chen, L.-W.A.; Frank, N.H. (2009). Methods to assess carbonaceous aerosol sampling artifacts for IMPROVE and other long-term networks. *J. Air Waste Manage. Assoc.*, **59**(8):898–911.

Watson, J.G.; Chow, J.C.; Frazier, C.A. (1999). X-ray fluorescence analysis of ambient air samples. In *Elemental Analysis of Airborne Particles, Vol. 1*, Landsberger, S., Creatchman, M. (Eds.), Gordon and Breach Science: Amsterdam, the Netherlands, pp. 67–96.

Watson, J.G.; Chow, J.C.; Houck, J.E. (2001a). $PM_{2.5}$ chemical source profiles for vehicle exhaust, vegetative burning, geological material, and coal burning in northwestern Colorado during 1995. *Chemosphere*, **43**(8):1141–1151.

Watson, J.G.; Chow, J.C.; Lowenthal, D.H.; Pritchett, L.C.; Frazier, C.A.; Neuroth, G.R.; Robbins, R. (1994). Differences in the carbon composition of source profiles for diesel- and gasoline-powered vehicles. *Atmos. Environ.*, **28**(15):2493–2505.

Watson, J.G.; Chow, J.C.; Moosmüller, H.; Green, M.C.; Frank, N.H.; Pitchford, M.L. (1998). Guidance for using continuous monitors in $PM_{2.5}$ monitoring networks. Report Number EPA-454/R-98-012, Prepared by U.S. Environmental Protection Agency, Research Triangle Park, NC, http://www.epa.gov/ttn/amtic/pmpolgud.html

Watson, J.G.; Turpin, B.J.; Chow, J.C. (2001b). The measurement process: Precision, accuracy, and validity. In *Air Sampling Instruments for Evaluation of Atmospheric Contaminants*, 9th edn., Cohen, B. S., McCammon, C. S. J. (Eds.), American Conference of Governmental Industrial Hygienists: Cincinnati, OH, pp. 201–216.

Watson, J.G.; Zhu, T.; Chow, J.C.; Engelbrecht, J.P.; Fujita, E.M.; Wilson, W.E. (2002). Receptor modeling application framework for particle source apportionment. *Chemosphere*, **49**(9):1093–1136.

Wexler, A.S.; Johnston, M.V. (2008). What have we learned from highly time-resolved measurements during EPA's Supersites program, and related studies? *J. Air Waste Manage. Assoc.*, **58**(2):303–319.

Wilson, W.E.; Chow, J.C.; Claiborn, C.S.; Fusheng, W.; Engelbrecht, J.P.; Watson, J.G. (2002). Monitoring of particulate matter outdoors. *Chemosphere*, **49**(9):1009–1043.

Witz, S.; Smith, M.M.; Moore, A.B., Jr. (1983). Comparative performance of glass fiber hi-vol filters. *J. Air Pollut. Control Assoc.*, **33**(10):988–991.

Yang, H.; Yu, J.Z.; Ho, S.S.H.; Xu, J.H.; Wu, W.S.; Wan, C.H.; Wang, X.D.; Wang, X.R.; Wang, L.S. (2005). The chemical composition of inorganic and carbonaceous materials in $PM_{2.5}$ in Nanjing, China. *Atmos. Environ.*, **39**(20):3735–3749.

Yanosky, J.D.; MacIntosh, D.L. (2001). A comparison of four gravimetric fine particle sampling methods. *J. Air Waste Manage. Assoc.*, **51**(6):878–884.

Zielinska, B.; Campbell, D.E.; Lawson, D.R.; Ireson, R.G.; Weaver, C.S.; Hesterberg, T.W.; Larson, T.; Davey, M.; Liu, L.J.S. (2008). Detailed characterization and profiles of crankcase and diesel particulate matter exhaust emissions using speciated organics. *Environ. Sci. Technol.*, **42**(15):5661–5666.

8 Health Effects of Ambient Ultrafine Particles

Beverly S. Cohen

CONTENTS

8.1 INTRODUCTION

More than 90% of all airborne particles are generally found in nuclei less than 150 nm in diameter [1–3]. Number concentrations in an urban environment vary, depending on local sources of particles and gaseous precursors, season, and weather. In a boreal forest in Finland, the measured concentration of ultrafine particles (UFPs) was about 1000 cm^{-3} [4]. When there is substantial vehicular traffic, the number concentration is frequently on the order of 10^4–10^5 particles cm^{-3}. In a quiet rural environment, or indoors in an undisturbed clean room, the concentration more typically ranges from a few hundred to a few thousand particles cm^{-3}. These particles, on average, represent only about one half of 1% of the total airborne particulate volume and mass. Thus, they have not generally been regarded as important contributors to the toxic effects of inhaled ambient air. However, recent evidence suggests that they may have an important role in health decrements associated with ambient particulate matter (PM). Other evidence suggests that particle number may also be a better metric than mass on which to base risk estimates for certain occupational diseases [2,5].

Various boundaries are in current use for both the upper and lower limits for the diameter of UFPs. UFPs include those in the atmospheric nuclei mode, plus some of the smaller particles classified as "accumulation" mode particles. The lower limit is generally regarded as about 1 nm, but some reports refer to those up to 10 nm as "nanometer" particles, reserving ultrafine for particles with diameters larger than 10 nm. A convenient upper limit of 100 nm is frequently used. This is because a 100 nm diameter particle is an acceptable lower size limit to the accumulation mode, and an upper limit to the nucleation mode (modes are defined in Chapter 2). Also, the term "manufactured nanoparticles" generally refers to particles smaller than 100 nm. In this size range, particle aerodynamic behavior is dominated by Brownian diffusion, and particle size is adequately described by a thermodynamic diameter. The thermodynamic diameter is the diameter of a sphere of unit density that would have the same diffusion coefficient in air as the particle of interest. An upper limit value of 150 nm is chosen in this chapter, because it represents the particle size at which gravitational and inertial effects are of little importance when particles are inhaled. Thus, UFP behavior in human airways is dominated by diffusion.

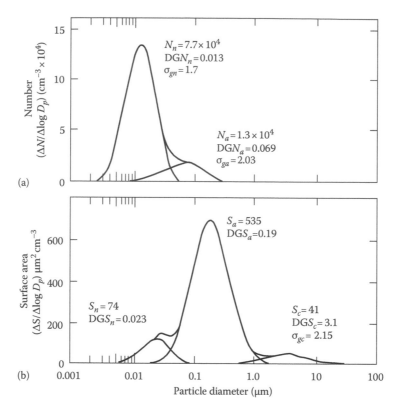

FIGURE 8.1 Frequency distribution of particles averaged over 1000 measured ambient particle size distribution in the United States: (a) by number and (b) by surface area. DGN and DGS, geometric mean diameter by number and surface, respectively; D_p, geometric diameter. (From U.S. EPA, Air quality for particulate matter, Vols. I, II, III, EPA/600/P-95/001aF, EPA600/P-95/001bF, EPA/600/P-95/001cF, 1996.)

The grand average concentration of over 1000 particle size distributions of ambient particles measured in the United States [6] is shown in Figure 8.1. The size distributions are shown as a number or surface concentration. It is clear that UFPs dominate the atmosphere when particles are counted and provide a dominant fraction of particle surface. The number in the nuclei mode is clearly an order of magnitude greater than the number in the larger size ranges. Taking an upper boundary diameter at 150 nm, almost all observed particles are in this ultrafine size range.

In the 1997 revision of the U.S. National Ambient Air Quality Standards, the U.S. EPA set 15 μg m³ as an average annual mass concentration that should not be exceeded for particles with diameters less than 2.5 μm in aerodynamic diameter.

The annual average standard was retained in their 2006 revision, although the 24 h average limit was reduced from 65 to 35 μg m³. The number of airborne unit density particles per cm³ of a specific diameter that would result in this mass concentration is shown in Table 8.1. The number of 0.1 μm particles would be 3 orders of magnitude greater than the number of 1 μm diameter particles.

Very detailed data have become available on the ambient number concentration segregated by particle size. This resulted from the development of the scanning mobility particle size analyzer (SMPS). The data are acquired into narrow size classes by automatically selecting particles in a specific size range that penetrate a differential mobility analyzer (DMA) and counting them with a condensation nucleus counter (CNC). Automation of this combination of instruments has permitted semi-continuous acquisition of particle size spectra.

A few examples of such data are shown in Figures 8.1 through 8.3. These size distribution measurements were made with an SMPS system (TSI, St. Paul, MN) averaging three scans every 30 min

TABLE 8.1

Number Concentration of Unit Density Monodisperse Particles at a Mass Concentration of 15 µg m⁻³

Particle Diameter (µm)	Number Concentration (Number cm⁻³)
0.01	28,600,000
0.05	229,000
0.1	28,600
0.15	8,490
0.2	3,580
0.5	229
1	28.6
1.5	8.5
2.5	1.8

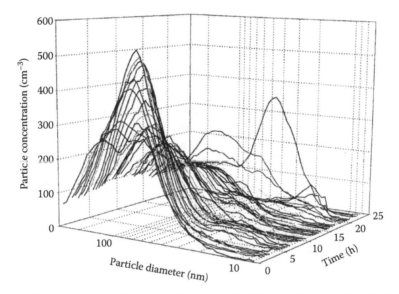

FIGURE 8.2 Particle size spectra collected over the course of a 24 h period at a rural location in Tuxedo, NY.

for the duration of the measurements. The particle size range was from 7 to 300 nm in 32 size bins. Each clearly demonstrates both the temporal variability and the dominance of UFPs when ambient particles are counted.

8.2 FORMATION

UFPs are formed by condensation reactions of precursor atmospheric gases and by nucleation of gas-phase species. The boundaries of the ultrafine region, as previously noted, are not exact, but the upper boundary of the nuclei mode is between 50 and 100 nm, whereas the lower boundary of the accumulation mode is roughly 100–200 nm. Nuclei mode particles are formed by nucleation, condensation, and coagulation. The nuclei mode is sometimes further split into the nucleation (diameter 10 nm) and Aitkin modes to indicate that only the smallest particles are formed directly by nucleation. The particles are created when gas-phase species form condensed-phase species with very low equilibrium vapor pressure. Those UFPs that are in the lower size range of the accumulation mode are primarily formed by condensation and coagulation, and also from evaporation of fog and cloud droplets in which dissolved gases have reacted [2,4].

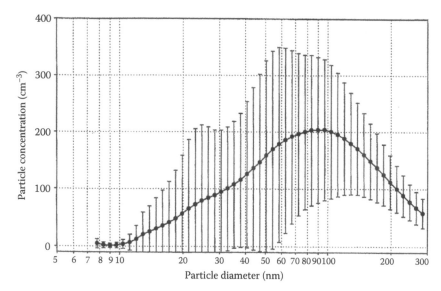

FIGURE 8.3 Average particle size distribution (mean and standard deviation) collected for 7 days (July 14–21, 1999) in Tuxedo, NY. Standard deviation represents the variability of the number detected by sampling every 30 min (NIEM outdoor, SMPS: sampling every 30 min, three scans per sample).

Nucleation events in the atmosphere in which large concentrations of nanometer-sized particles form and grow over a period of hours occur about one quarter of the days per year. Concentrations of particles smaller than 50 nm can be increased by a factor of 10 or more when nucleation occurs. A nucleation event is shown in Figure 8.4, which displays the development of the particle number concentration over 24 h at a rural location in New York.

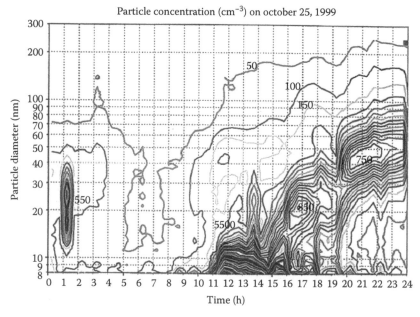

FIGURE 8.4 Nucleation event showing the development of a very high concentration of particles less than 10 nm. The contours are lines of equal particle concentration in particles cm^{-3}. The interval between contours is 50 particles cm^{-3}.

8.3 COMPOSITION

There are few data on the composition of ambient UFPs. Because they are a very small fraction of the total ambient particle mass, specialized methods are required to isolate and analyze them. Reported studies indicate that they are composed, in large part, of organic and elemental carbon together with trace metals, ammonium, and sulfates [7]. Hughes et al. [8] reported the chemical composition of particles between 56 and 97 nm in diameter measured at Pasadena, CA. They reported that the largest fraction is organic compounds, followed by elemental carbon, then trace metals and sulfates. This is not unexpected since combustion sources are major contributors to ambient UFPs. Detailed size distributions of ambient sulfates, ammonium, trace metals, and sulfuric acid in ambient particles smaller than 100 nm were measured by Hazi [9]. Although most of the acid and sulfate mass was measured in the 0.38 μm midpoint diameter fraction, the ultrafine fraction (0.1 μm) was found to have lower pH. Iron, zinc, and sulfur were the dominant trace elements of the nine measured in the ultrafine fraction. Iron is the most abundant metal measured in UFPs [7,9,10].

Measurements of the number of strong-acid UFPs in New York City and Tuxedo, NY, have been reported by Cohen et al. [11]. The number was measured by the deposition of ambient particles less than 100 nm in diameter onto detectors that are coated with a 20 nm thick layer of iron applied by vapor deposition. The interaction of acid droplets and iron is detected by scanning the detector surface topography with an atomic force microscope. They report that the fraction of UFPs that were acidic varied from 10% to 88% for the different seasons and sites. The average concentrations of acidic particles ranged from about 100–1500 particles cm^{-3} over the sampling periods.

Both vehicular and stationary combustion sources contribute UFPs to the atmosphere. Gasoline and diesel vehicles are major sources. Emission testing indicates that organic carbon comprises roughly 70%–90% of the carbon emitted from gasoline vehicles and 40%–50% from diesel vehicles [12]. Sulfur has also been reported as a major component of emissions from combustion vehicles [13]. A detailed understanding of emissions from mobile sources is documented in [14].

High concentrations of ultrafines are found on roadways and people are exposed to them in vehicles and alongside the roadways. However, because of particle growth, they have very short residence times, and number concentrations diminish very rapidly with distance from the road. In some cases, particle number concentrations at 50 m from a busy roadway were five to eight times those measured at 100 and 500 m [15].

Recent technological advances provide excellent single-particle data via aerosol time-of-flight mass spectrometry (ATOFMS) instruments [16]. However, data reduction is arduous because a vast amount of data is collected during a sampling session. Methods for analysis of the data continue to be developed. These spectrometers are used for continuous single-particle measurements of size and composition simultaneously [17]. Particles are size selected as they enter the inlet of the spectrometer. They are then disintegrated by a laser beam, and the resulting ions are driven down separate positive and negative ion channels for identification of the charged fragment masses. In some systems, particle concentrators are being used to increase the concentration at the inlet. The concentrators have been developed for use in testing the effects of inhalation exposure to ambient particles. Measurements with these time-of-flight instruments have been undertaken in cities located in different geological areas of the United States to characterize fine and ultrafine urban PM.

8.4 LUNG DEPOSITION

UFPs deposit in the respiratory tract primarily by diffusion, with deposition increasing as particle size decreases. The total lung deposition of UFPs measured in human volunteers confirms diffusion as the mechanism [18,19]. When the smallest UFPs are inhaled, they deposit very efficiently in the nasal and oral passages. The high deposition in the extrathoracic region was determined in human nasal/oral casts and in human volunteers [20,21].

Those particles that penetrate to the thorax are also very efficiently deposited. The deposition of UFPs in the human tracheobronchial airways was measured in a hollow airway cast for particle sizes between 40 and 200 nm by Cohen et al. [22]. The data indicated higher deposition for the smaller diameter particles, again confirming diffusion as the primary deposition mechanism. Additionally, this work demonstrated that deposition in the tracheobronchial airways was higher than predicted by diffusional deposition in a tube, assuming a fully developed or parabolic flow profile. Subsequent studies showed that the deposition of iodine vapors with a diffusion coefficient of 0.08 cm^2 s^{-1}, used as a surrogate for an approximately 1.8 nm particle, agreed with the theoretical prediction of diffusional deposition [23]. The same research group also showed that the charge on UFPs enhances deposition in the airway replica [24,25], that is, charged particles and particles in charge equilibrium have higher deposition efficiency in the tracheal region as compared to neutral particles of the same size. The ratio of deposition efficiency for charge-equilibrium and neutral particles was 1.6 and 2.7 for 20 and 125 nm particles, respectively [26]. This is important because most ultrafine ambient particles carry one, or a few charges.

The deposition in airway casts for 1.75, 10, and 40 nm particles was measured at flow rates corresponding to respiratory minute volumes at rest and during moderate exercise [27]. Replicate casts of the upper tracheobronchial airways of 3, 16, and 23 year old humans were used, including the larynx, trachea, and bronchial airways down to generations 5–8. The deposition of the 1.75 nm particle was substantially higher than that of the 10 and 40 nm particles. The dependence of particle deposition on the flow rate was relatively weak, and deposition efficiencies were only slightly higher at the lower flow rates. The deposition models for diffusion from parabolic flow underestimated aerosol deposition, whereas the diffusion deposition predicted for plug flow overestimated the tracheobronchial deposition. This is in agreement with the earlier studies [22].

8.5 TOXICOLOGY

Submicrometer particles with demonstrated health effects include diesel exhaust, radon progeny, cigarette smoke, metal fumes, acidic aerosols, and trace metals. When generated, these primary aerosol particles are ultrafine. Additionally, biofragments such as endotoxin extend into this size range. Table 8.2 shows the toxic effects demonstrated when UFPs are inhaled.

A substantial body of experimental data on the toxicity of ambient particles has developed in recent years as an indirect result of The Clean Air Act enacted by the United States in 1970. This act established the Environmental Protection Agency and mandated the setting of Primary Ambient Air Quality Standards that would protect the public against adverse health effects of ubiquitous pollutants (such as ambient PM) with an adequate margin of safety. The Clean Air Act Amendments of 1977 then required that the air quality standards be reviewed at 5 year intervals and revised as necessary.

Evidence has accumulated that implicates UFPs as a cause of the adverse effects of exposure to ambient PM, but the mechanisms are as yet unclear. Mechanisms that have been proposed for the induction of lung injury include irritant signaling [28], acid effects [29], and inflammation [30]. It is clear that UFPs generally exhibit greater toxic potency than larger particles of the same material [31]. Table 8.3 provides a summary of experiments that compared responses to UFPs with response to larger diameter particles of the same material.

Further research into the effects of UFPs has been stimulated by the introduction of engineered manufactured nanoparticles into the economy, inevitably spreading them into the environment and biosphere as a whole [32]. A working definition of engineered UFPs is given by Warheit (2010), as those with diameters of roughly 100 nm that exhibit a property that is uniquely different from that of the bulk counterpart. Given the relatively similar size of cellular components, the unique properties can have substantial implications for biological response.

Other suggestions for the increased potency are that (1) they are biologically more reactive, (2) there is a much higher number and surface area for the same total particle mass, and (3) they deposit

TABLE 8.2
Toxic Effects of Ultrafine Particles

Inhalation Exposure	Result
High concentration metal or polymer fume (occupational exposures)	Fever, diffusion impairment, respiratory symptoms
Aggregated ultrafines (TiO$_2$, carbon black, diesel soot)	Epithelial cell proliferation, occlusion of inter-alveolar pores, impairment of alveolar macrophages, chronic pulmonary inflammation, pulmonary fibrosis, induction of lung tumors
11 nm CuO at 10^9/cm^3 60 min (hamsters) [54]	Fourfold increase in pulmonary resistance. Particles dispersed throughout lung (interstitium, alveolar capillaries, pulmonary lymphatics)
Teflon (PTFE) fume (26 nm) 10^6/cm^3, 10–30 min [33]	Highly inflammatory/mortality
TiO$_2$ 1000 μg/m^3, 7 h [55]	Oxidative stress in lung
PTFE fumes, whole body inhalation, 1, 2.5, or 5×10^5 particles/cm^3, 18 nm, rat, 15 min, analysis 4 h postexposure [56]	Increased PMN, mRNA of MnSOD, and MT, IL-1α, IL-1β, IL-6, MIP-2, TNF-α mRNA of MT and IL-6 expressed around all airways and interstitial regions; PMN expressed IL-6, MT, TNF-α; AM and epithelial cells were actively involved
PTFE fumes, whole body inhalation, 1, 2.5, or 5.3×10^5 particles/cm^3, 18 nm, mice, C57BL/6J, 8 weeks and 8 months old, mice 30 min exposure analysis 6 h following exposure [57]	Increased PMN, lymphocytes, and protein levels in old mice over young mice; increased TNF-α mRNA in old mice over young mice; no difference in LDH and β-glucuronidase
CdO fumes, 8 nm, rats and mice, 1–3 h [58]	Mice created more metallothionein than rats, which may be protective of tumor formation

Source: Adapted from U.S. EPA, Air quality for particulate matter, Vols. I, II, III, EPA/600/P-95/001aF, EPA/600/P-95/001bF, EPA/600/P-95/001cF, 1996.

with very high efficiency in lungs. Additionally, it has been demonstrated that UFPs are more rapidly transferred to the interstitium than are fine particles of the same composition, and exhibit a greater accumulation in the regional lymph nodes and a greater retention in the lung [33,34]. Thus, UFPs may penetrate the epithelium better and they may be less effectively cleared by macrophages, and the nonphagocytosed particles may penetrate the interstitium in a few hours. Those with low solubility appear to be significantly more inflammatory in the lung than are larger-sized particles of the same composition [35].

In addition to lung injury, evidence has accumulated that adverse cardiac effects result from inhalation of PM 2.5, the fine particle fraction of ambient air. In the laboratory, single exposures of mice to concentrated PM 2.5 were shown to increase the frequency of cardiac arrhythmias in aged male rats [36]. Long-term exposure demonstrated adverse effects on cardiac function such as changes in heart rate variability and exacerbation of atherosclerosis in susceptible mice [37]. Most studies have focused on PM 2.5, but cardiac effects associated with the ultrafine fraction have been demonstrated [38].

The potential for inhaled nanoparticles to become widely distributed in the body was demonstrated by [39] who reported dose-dependent pulmonary inflammatory responses to inhaled nickel hydroxide nanoparticles. The particles were approximately 40 nm in diameter and composed of agglomerates of 5 nm primary particles. Using both short- and long-term inhalation exposures, they found a clearance half-time of approximately 1 day for the nickel. This rapid clearance contrasts with that measured for larger nickel-containing particles and may demonstrate the increased solubility of nanoparticles due to their large surface area. Short-term exposure in mice has also been shown to induce acute endothelial disruption and alter vasoconstriction and

TABLE 8.3
Studies Comparing Ultrafine and Fine Particles Composed of the Same Material

Particle Type	Size	Exposure	Endpoint	Reference
TiO_2 anatase	Ultrafine 20 nm Fine 250 nm	Inhalation 23 mg/m³ 7 h/day 5 days/12 weeks	Inflammation in BAL Lung lymph node burdens Slowed clearance	[59]
TiO_2 rutile	Ultrafine 12 nm Fine 230 nm	Instillation of 500 μg	Interstitialization measured as unlavageable fraction	[59]
Al_2O_3	Ultrafine 20 nm Fine 500 nm	Instillation 500 μg	Lung lymph node burden	[60]
Carbon black	Ultrafine 20 nm Fine 200—250 nm PM_{10}	Instillation of 50—125 μg in 0.2 mL	Increased PMN, protein, and LDH following PM_{10}, greater response with ultrafine CB but not CB; decreased GSH level in BAL; free radical activity (deplete supercoil DNA); leukocytes from treated animals produced greater NO and TNF	[61–63]
TiO_2	Ultrafine 20 nm Fine 300 nm	Instillation of 10,000 μg particles/kg BW	Increased inflammatory indicators in BAL Pathology changes in alveolar ducts Increased pulmonary retention of ultrafine	[63]
TiO_2	Ultrafine 20 nm Fine 250 nm	Instillation 500 μg	Increased acute inflammation indicators in BAL at 24 h with ultrafines	[33]
MnO_2	Surface area of 0.16, 0.5; 17, 62 mg²/g	Intratracheal instillation; in vitro 0.037, 0.12, 0.75, 2.5 mg/animal	LDH, protein, and cellular recruitment increased with increasing surface area; Freshly ground particles had enhanced cytotoxicity	[65]
TiO_2	Ultrafine 21 nm Fine 250 nm	Intratracheal inhalation and intratracheal instillation	Inflammation produced by intratracheal inhalation (both severity and persistence) was less than that produced by instillation; ultrafine particles produced greater inflammatory response than fine particles for both dosing methods	[33]
TiO_2	Ultrafine 21 nm Fine 250 nm	Intratracheal inhalation and intratracheal instillation (rat) Inhalation at 125 μg/m³ Instillation at 500 μg for fine 750 μg for ultrafine Inhalation exposure 2 h; sacrificed at 0, 1, 3, and 7 days postexposure for both techniques	MIP-2 increased in lavage cells but not in supernatant in those groups with increased PMN (more in instillation than in inhalation; more in ultrafine than in fine); TNF-α levels had no correlation with either particle size or dosing methods	[67]
H_2SO_4 and O_3	Ultrafine 60 nm Fine 300 nm	Inhalation, nose-only 500 μg/m³ H_2SO_4 aerosol (two different particle sizes) with or without 0.6 ppm O_3 Rats, Sprague—Dawley, male, 250—300 g 4 h/day for 2 days	The volume percentage of injured alveolar septae was increased only in the combined ultrafine acid/O_3, animals BrdU labeling in the periacinar region was increased in a synergistic manner in the combined fine acid/O_3 animals	[68]

Source: Expanded from Donaldson, K. et al., *J. Aerosol Sci.*, 29, 553, 1998.

vasorelaxation [40]. In long-term studies of nano-nickel hydroxide exposure, the research group demonstrated the upregulation of genes associated with oxidative stress in heart tissue. Their work suggests that at levels that are at least occupationally, if not environmentally, relevant, chronic exposure to inhaled nano-NH significantly exacerbates atherosclerosis [41].

It is unlikely that all components of PM are equally toxic. Cohen et al. [11] reported that candidates for especially active components of ambient PM are H^+, ultrafine number, and soluble transition metals. While all three have some supporting toxicological evidence consistent with known mechanisms of toxicity, only H^+ and UFPs have produced effects at exposure levels that could occur in ambient air [2]. Perhaps the most likely candidate is a hybrid of H^+ and UFPs, that is, acid-coated UFPs.

Chen et al. [42] exposed guinea pigs to varying amounts of sulfuric acid layered onto 10^8 ultrafine (90 nm) carbon core particles cm^{-3}, and to a constant (300 μg m^{-3}) concentration of acid layered onto 10^6, 10^7, or 10^8 particles cm^{-3}. Indicators of irritant potency on macrophages harvested from the lungs of exposed animals clearly showed an increased response to a constant dose of acid when it was divided into an increased number of particles, as well as a response to an increased dose of acid at a constant number concentration. Oberdörster et al. [43] reported that nonreactive UFPs do not appear to cause inflammation in young healthy rats.

Evidence, some of which was noted earlier, supports the hypothesis that the number of UFPs that deposit per unit surface of the epithelial lining of the human respiratory system is an important determining factor affecting lung injury [28,31,42,44–49], and that the resulting alveolar inflammation is able to provoke attacks of acute respiratory illness in susceptible individuals [30]. In particular, the work of Peters et al. [47] and Wichmann et al. [50] suggest that at ambient levels, the number concentration of inhaled particles may be a significantly more important determinant of the risk than inhaled mass measures. Very recent evidence has also shown that UFPs are associated with human mortality [50].

8.6 EPIDEMIOLOGICAL STUDIES

Epidemiological research has determined that ambient fine particles (PM < 2.5 μm) are responsible for much of the cardiac mortality and also for the lung cancer excess seen in the U.S. population [51] but less evidence is available related to the risk from ambient UFPs (PM < 0.1 μm).

Although long-term animal exposures with concentrated ambient particles in this size range demonstrate adverse effects on pulmonary and cardiac function, and toxicology has documented potential mechanisms for cardiac and vascular damage, the risk to people from ambient UFPs is still unclear.

A recent epidemiological study of the effects of ultrafine ambient particles on cardiac function in cyclists has demonstrated changes in heart rate variability [52].

A European "expert elicitation," convened to quantitatively examine available data relating concentration of ambient UFPs to human health, reported the "estimated percentage decrease in all-cause mortality with a permanent 1000 particles/cm^3 decrease in UFP concentration ranged between 0.1% and 1.2%, with a median of 0.30 However, there were substantial uncertainties [53]. The report suggests that although it is very unlikely that there is no effect, there is as yet insufficient evidence to exclude the possibility.

8.7 SUMMARY

UFPs constitute the largest share of particles in the atmosphere by both number and surface area. These particles, which are formed by condensation reactions of precursor atmospheric gases and by nucleation of gas-phase species, range in size from approximately 1 to 150 nm. Studies indicate that they are composed, in large part, of organic and elemental carbon, together with trace metals, ammonium, and sulfates, but there are few data on the temporal or spatial variations in the composition.

High concentrations of ultrafines are emitted from vehicles and other combustion sources and people are exposed to them in vehicles and alongside roadways. These small particles deposit very efficiently in the respiratory system, and when inhaled, the evidence derived from both experimental and occupational exposures indicates that they are more toxic than larger particles of the same composition. Some epidemiological studies associate ambient UFPs with morbidity and mortality [12]. No mechanism has as yet been demonstrated by which these very small particles exert their toxicity on people or in animals at exposure levels that could occur in ambient air.

REFERENCES

1. National Research Council (NRC), *Airborne Particles*, National Academy of Sciences, Washington, DC, 1977.
2. U.S. EPA, Air quality for particulate matter, Vols. I, II, III, EPA/600/P-95/001aF, EPA/600/P-95/001bF, EPA/600/P-95/001cF, 1996.
3. Whitby, K.T., Husar, R.B., and Liu, B.Y.H., The aerosol size distribution of Los Angeles smog, *J. Colloid Interface Sci.*, 39, 177, 1972.
4. Makela, J.M. et al., Observations of ultrafine aerosol particle formation and growth in boreal forest, *Geophys. Res. Lett.*, 24, 1219, 1997.
5. McCawley, M.A., Kent, M.S., and Berakis, M.T., Ultrafine beryllium number concentration as a possible metric for chronic beryllium disease risk, *Appl. Occup. Environ. Hyg.*, 16, 631, 2001.
6. Whitby, K.T., The physical characteristics of sulfur aerosols, *Atmos. Environ.*, 12, 135, 1978.
7. Cass, G.R. et al., The chemical composition of atmospheric ultrafine particles, *Philos. Trans. R. Soc. Lond. Ser. A*, 358, 2581, 2000.
8. Hughes, L.S. et al., Physical and chemical characterization of atmospheric ultrafine particles in the Los Angeles area, *Environ. Sci. Technol.*, 32, 1153, 1998.
9. Hazi, Y., Measurements of acidic sulfates and trace metals in fine and ultrafine ambient particulate matter: Size distribution, number concentration and source region, PhD thesis, New York University School of Medicine, New York, 2001.
10. Gone, J.K., Olmez, I., and Ames, M.R., Size distribution and probable sources of trace elements in submicron atmospheric particulate material, *J. Radioanal. Nucl. Chem.*, 244, 133, 2000.
11. Cohen, B.S. et al., Field validation of nanofilm acid detectors for assessment of H^1 in indoor and outdoor air and measured ambient concentrations of ultrafine acid particles, HEI Report, Health Effects Institute, Boston, MA, 2004.
12. U.S. EPA, Particulate matter air quality criteria document, Vols. 1 and 2, June 2003.
13. Gertler, A.W. et al., Preliminary results of a tunnel study to characterize mobile source particulate emissions, Presented at *PM 2000; Particulate Matter and Health—The Scientific Basis for Regulatory Decision-Making, Specialty Conference and Exhibition*, January, Charleston, SC, Air and Waste Management Association, Pittsburgh, PA, 2000.
14. Seigneur, C., Current understanding of ultrafine particulate matter emitted from mobile sources, *J. Air Waste Manage. Assoc.*, 59(1), 3–17, 2009.
15. Kleinman, M.T. et al., Exposure to concentrated fine and ultrafine ambient particles near heavily trafficked roads induces allergic reactions in mice, in *AAAR PM Meeting*, March 31–April 4, 2003, Pittsburgh, PA, 2003.
16. Prather, K.A., Chemical characteristics of ambient PM2.5, in *AAAR'97, Abstracts of the Sixteenth Annual Conference*, October 13–17, 1997, Denver, CO, American Association for Aerosol Research, Cincinnati, OH, Abstract No. 9SE3, 1997, p. 302.
17. Zhao, Y. et al., Using ultrafine concentrators to increase the hit rate of single particle mass spectrometers, in *AAAR PM Meeting*, March 31–April 4, 2003, Pittsburgh, PA, 2003.
18. Schiller, C. et al., Deposition of monodisperse insoluble aerosol particles in the 0.005 to 0.2 μm size range within the human respiratory tract, *Ann. Occup. Hyg.*, 32, 41, 1988.
19. Jaques, P.A. and Kim, C.S., Measurement of total lung deposition of inhaled ultrafine particles in healthy men and women, *Inhal. Toxicol.*, 12, 715, 2000.
20. Cheng, Y.S. et al., Deposition of thoron progeny in human head airways, *Aerosol Sci. Technol.*, 18, 359, 1993.

21. Cheng, Y.S. et al., Nasal deposition of ultrafine particles in human volunteers and its relationship to airway geometry, *Aerosol Sci. Technol.*, 25, 274, 1996.

22. Cohen, B.S., Sussman, R.G., and Lippmann, M., Ultrafine particle deposition in a human tracheobronchial cast, *Aerosol Sci. Technol.*, 12, 1082, 1990.

23. Li, W., Xiong, J.Q., and Cohen, B.S., The deposition of unattached radon progeny in a tracheobronchial cast as measured with iodine vapor, *Aerosol Sci. Technol.*, 28, 502, 1998.

24. Cohen, B.S. et al., Deposition of inhaled charged ultrafine particles in a simple tracheal model, *J. Aerosol Sci.*, 26, 1149, 1995.

25. Cohen, B.S., Xiong, J.Q., and Li, W., The influence of charge on the deposition behavior of aerosol particles with emphasis on singly charged nanometer sized particles, in *Aerosol Inhalation, Lung Transport and Deposition and the Relation to the Environment*, Marijnissen, J. and Gradon, L. (Eds.), Recent Research and Frontiers, Kluwer Scientific Press, Dordrecht, the Netherlands, 1996, p. 153.

26. Cohen, B.S. et al., Deposition of charged particles on lung airways, *Health Phys.*, 74, 554, 1998.

27. Smith, S., Cheng, Y.-S., and Yeh, H.C., Deposition of ultrafine particles in human tracheobronchial airways of adults and children, *Aerosol Sci. Technol.*, 35, 697, 2001.

28. Hattis, D. et al., Acid particles and the tracheobronchial region of the respiratory system—An "irritation-signaling" model for possible health effects, *JACPA*, 37, 1060, 1987.

29. Lippmann, M., Background on health effects of acid aerosols, *Environ. Health Perspect.*, 79, 3, 1989.

30. Seaton, A.W. et al., Particulate air pollution acute health effects, *Lancet*, 345, 176, 1995.

31. Donaldson, K., Li, X.Y., and MacNee, W., Ultrafine (nanometer) particle mediated lung injury, *J. Aerosol Sci.*, 29, 553, 1998.

32. Oberdorster, G., Stone, V., and Donaldson, K., Toxicology of nanoparticles: A historical perspective, *Nanotoxicology*, 1(1), 2–25, 2007.

33. Oberdörster, G. et al., Role of the alveolar macrophage in lung injury: Studies with ultrafine particles, *Environ. Health Perspect.*, 97, 193, 1992.

34. Oberdörster, G. et al., Association of particulate air pollution and acute mortality: Involvement of ultrafine particles, *Inhal. Toxicol.*, 7, 111, 1995.

35. Driscoll, K.E., Macrophage inflammatory proteins: Biology and role in pulmonary inflammation, *Exp. Lung Res.*, 20, 473, 1994.

36. Nadziejko, C., Fang, K., Narciso, S., Zhong, M., Su, W.C., Gordon, T., Nadas, A., and Chen, L.C., Effect of particulate and gaseous pollutants on spontaneous arrhythmias in aged rats, *Inhal. Toxicol.*, 16, 373–380, 2004.

37. Chen, L.C. and Nadziejko, C., Effects of subchronic exposures to concentrated ambient particles (CAPs) in mice: V. CAPs exacerbate aortic plaque development in hyperlipidemic mice, *Inhal. Toxicol.*, 17, 217–224, 2005.

38. Tong, H., Cheng, W.Y., Samet, J.M., Gilmour, M.I., and Devlin, R.B., Differential cardiopulmonary effects of size-fractionated ambient particulate matter in mice. *Cardiovasc. Toxicol.*, 10(4), 259–267, December 2010.

39. Gillespie, P.A., Kang, G.S., Elder, A., Gelein, R., Chen, L., Moreira, A.L., Koberstein, J., Tchou-Wong, K.M., Gordon, T., and Chen, L.C., Pulmonary response after exposure to inhaled nickel hydroxide nanoparticles: Short and long-term studies in mice, *Nanotoxicology*, 4(1), 106–119, March 2010.

40. Cuevas, A.K., Liberda, E.N., Gillespie, P.A., Allina, J., and Chen, L.C. Inhaled nickel nanoparticles alter vascular reactivity in C57BL/6 mice. *Inhal. Toxicol.*, 22(Suppl. 2):100–106, 2010.

41. Kang, G.S., Gillespie, P.A., Gunnison, A., Moreira, A.L., Tchou-Wong, K.M., and Chen, L.C., Long-term inhalation exposure to nickel nanoparticles exacerbated atherosclerosis in a susceptible mouse model. *Environ. Health Perspect.*, 119(2), 176–181, February 2011. Epub September 22, 2010.

42. Chen, L.C. et al., Number concentration and mass concentration as determinants of biological response to inhaled irritant particles, *Inhal. Toxicol.*, 7, 577, 1995.

43. Oberdörster, G. et al., Acute pulmonary effects of ultrafine particles in rats and mice, HEI Report 96, Health Effects Institute, Boston, MA, 2000.

44. Amdur, M.O. and Chen, L.C., Furnace-generated acid aerosols: Speciation and pulmonary effects, in *Symposium on the Health Effects of Acid Aerosols*, Research Triangle Park, NC, October 1987; *Environ. Health Perspect.*, 79, 147, 1989.

45. Oberdörster, G. et al., Particulate air pollution and acute health effects, *Lancet*, 345, 176, 1995.

46. Hattis, D.S., Abdollahzadeh, S., and Franklin, C.A., Strategies for testing the "irritation-signaling" model for chronic lung effects of fine acid particles, *J. Air Waste Manage. Assoc.*, 40, 322, 1990.

47. Peters, A. et al., Respiratory effects are associated with the number of ultrafine particles, *Am. J. Respir. Crit. Care Med.*, 155, 1376, 1997.

48. Wichmann, H.-E. and Peters, A., Epidemiological evidence of the effects of ultrafine particle exposure, *Philos. Trans. R. Soc. Ser. A*, 358, 2751, 2000.

49. Penttinen, P. et al., Number concentration and size of particles in urban air: effects on spirometric lung function in adult asthmatic subjects, *Environ. Health Perspect.*, 109, 319, 2001.

50. Wichmann, H.-E. et al., Daily mortality and fine and ultrafine particles in Erfurt, Germany, Part I. role of particle number and particle mass, HEI Report 98, Health Effects Institute, Boston, MA, 2000.

51. Pope, C.A., Burnett, R.T., Thurston, G.D., Thun, M.J., Calle, E.E., Krewski, D., and Godleski, J.J., Cardiovascular mortality and long-term exposure to particulate air pollution: Epidemiological evidence of general pathophysiological pathways of disease, *Circulation*, 109, 71–77, 2004.

52. Weichenthal, S., Kulka, R., Dubeau, A., Martin, C., Wang, D., and Dales, R., Traffic-related air pollution and acute changes in heart rate variability and respiratory function in urban cyclists, *Environ. Health Perspect.* June 14, 2011 [Epub ahead of print].

53. Hoek, G., Boogaard, H., Knol, A. et al., Concentration response functions for ultrafine particles and all-cause mortality and hospital admissions: Results of a European expert panel elicitation, *Environ. Sci. Technol.*, 44(1), 476–482, January 1, 2010.

54. Stearns, R.C. et al., Detection of ultrafine copper oxide particles in the lungs of hamsters by electron spectroscopic imaging, in *Proceedings of the International Conference of Electron Microscopy*, ICEM 13, Paris, France, July 1994, p. 763.

55. MacNee, W. et al., Pro-inflammatory effect of particulate air pollution (PM10) *in vivo* and *in vitro*, *Ann. Occup. Hyg.*, 41(Suppl. 1), 7, 1997.

56. Johnston, C.J. et al., Characterization of the early pulmonary inflammatory response associated with PTFE fume exposure, *Toxicol. Appl. Pharmacol.*, 140, 154, 1996.

57. Johnston, C.J. et al., Pulmonary inflammatory responses and cytokine and antioxidant mRNA levels in the lungs of young and old C57BL/6 mice after exposure to Teflon fumes, *Inhal. Toxicol.*, 10, 931, 1998.

58. McKenna, I.M. et al., Expression of metallothionein protein in the lungs of Wistar rats and C57 and DBA mice exposed to cadmium oxide fumes, *Toxicol. Appl. Pharmacol.*, 153, 169, 1998.

59. Ferin, J., Oberdorster, G., and Penney, D.P., Pulmonary retention of ultra-fine and fine particles in rats, *Am. J. Respir. Cell Mol. Biol.*, 6, 535, 1992.

60. Ferin, J. et al., Increased pulmonary toxicity of ultra-fine particles? I. Particle clearance, translocation, morphology, *J. Aerosol Sci.*, 21, 381, 1990.

61. Li, X.Y. et al., Free radical activity and pro-inflammatory effects of particulate air pollution (PM_{10}) *in vivo* and *in vitro*, *Thorax*, 51, 1216, 1996.

62. Li, X.Y. et al., *In vivo* and *in vitro* proinflammatory effects of particulate air pollution (PM_{10}), in *Proceedings of the Sixth International Meeting on the Toxicology of Natural and Man-Made Fibrous and Non-Fibrous Particles*, Driscoll, K. E. and Oberdorster, G. (Eds.), Lake Placid, NY, September 1996, p. 1279; *Environ. Health Perspect.* Suppl., 105(5), 1279–1283.

63. Li, X.Y. et al., Short term inflammatory responses following intratracheal instillation of fine and ultrafine carbon black in rats, *Inhal. Toxicol.*, 11, 709, 1999.

64. Driscoll, K. and Maurer, Cytokine and growth factor release by alveolar macrophages: Potential biomarkers of pulmonary toxicity, *Toxicol. Pathol.*, 19(4 Part 1), 398, 1991.

65. Lison, D. et al., Influence of particle surface area on the toxicity of insoluble manganese dioxide dusts, *Arch. Toxicol.*, 71, 725, 1997.

66. Osier, M. and Oberdorster, G., Intratracheal inhalation vs. intratracheal instillation: Differences in particle effects, *Fundam. Appl. Toxicol.*, 40, 220, 1997.

67. Osier, M., Baggs, R.B., and Oberdorster, G., Intratracheal instillation versus intratracheal inhalation: influence of cytokines on inflammatory response, in *Proceeding of the Sixth International Meeting on the Toxicology of Natural and Man-Made Fibrous and Non-Fibrous Particles*, Driscoll, K. E. and Oberdorster, G. (Eds.), Lake Placid, NY, September 1996, p. 1265; *Environ. Health Perspect.* Suppl., 105(5), 1265–1271.

68. Kimmel, T.A., Chen, L.C., Bosland, M.C., and Nadziejko, C. Influence of droplet size on structural and functional changes to the rat lung caused by acute exposure to sulfuric acid and ozone. *Toxicol. Appl. Pharmacol.*, 144, 348–355, 1997.

9 Nanoparticle Cell Penetration

Steven M. Hankin and Craig A. Poland

CONTENTS

9.1 INTRODUCTION

Following inhalation, nanoparticles must cross cellular barriers to enter the body further. In order for a substance to enter a cell's interior, it must pass through the diffuse layer surrounding the cell and the plasma membrane which segregates the internal and external environments of a cell and regulates the entry and exit of substances into and out of the cell. The uptake of substances is accomplished via a variety of processes that can be described as active (energy dependent) or passive (energy independent). There are a number of clearly defined mechanisms for crossing the plasma membrane that include diffusion, facilitated diffusion, active transport, and endocytosis. Understanding the specific processes and physicochemical factors controlling the ability of nanoparticles to cross barriers, in particular epithelial cells, is key to understanding the intracellular fate as well as the potential for distribution of nanoparticles around the body.

The respiratory system can be divided into upper (nasal cavity, pharynx, and larynx) and lower (trachea, primary bronchi, and alveoli) sections. The respiratory system is lined with a barrier of epithelial cells, whose structure and function differ between parts of the pulmonary system. Epithelial cells form a confluent barrier in healthy tissues and control the movement of substances in both directions across the epithelium. Damage to the epithelium by toxicants or disease can lead to an increase in permeability and, therefore, toxicant absorption into the body. Toxicants may also be taken up directly into the cell and then pass through the epithelial cell into the interstitium or cardiovascular system. This is known as the transcellular route of absorption (Figure 9.1). The upper section of the respiratory system presents a potential route of entry to the central nervous system (CNS) via transport along nerves. In addition, if inhaled nanoparticles cross the lung epithelium and become bloodborne, they may have the potential to gain access to the blood–brain barrier (BBM).

9.2 PARTICLE INTERACTIONS WITH CELLS OF THE RESPIRATORY SYSTEM

The size of many nanoparticles is similar to biological macromolecules such as proteins and DNA, as well as biological structures such as bacteria and viruses, all of which are readily taken up by cells, including lung cells. Vesicular transport exists in lung epithelial cells, since the existence of

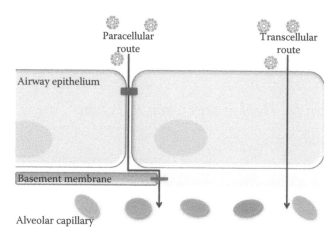

FIGURE 9.1 Diagram of paracellular and transcellular route of transfer through the respiratory epithelium.

large numbers of membrane vesicles within the type I alveolar epithelial cells has been recognized for many years (reviewed in Gumbleton, 2001). However, the function of these vesicles remains unclear. It has been hypothesized that vesicular transport may provide a route of transport for solutes from the alveolar to interstitial surfaces of the epithelial cells. Recovery of the lung from edema requires the removal of protein across the alveolar epithelium, and this is believed to occur via endocytic transcytosis routes as well as via paracellular routes (Hastings et al., 2004). Endocytosis is the more important mechanism at low protein concentrations, but the paracellular route becomes increasingly important on exposure to high protein concentrations. During edema or acute lung injury, it is suggested that transcellular routes become more active in order to clear protein from the lung surface (Hastings et al., 2004). Therefore, it is feasible that nanoparticles could cross this barrier by both transcellular and paracellular routes, especially in a diseased lung.

Phagocytic uptake is thought to be restricted to specialized cells such as macrophages and neutrophils (Conner and Schmid, 2003) and is generally responsible for the uptake of large materials (greater than 0.5 μm) such as bacteria or cell debris (Khalil et al., 2006). In the lung this is especially important as inhaled air provides a route of delivery of foreign particles and microorganisms into the body; therefore, macrophages are essential for maintaining a clear, sterile, and functioning respiratory surface. During phagocytosis, the cell recognizes ligands via cell surface receptors. Receptor binding then triggers the polymerization and rearrangement of the actin cytoskeleton to form membrane extensions so that the plasma membrane surrounds the material to be internalized (Liu and Shapiro, 2003; Perret et al., 2005; Khalil et al., 2006). The phagosome that is formed then fuses with lysosomes so that the cargo can be degraded (Perret et al., 2005).

When the alveolar macrophages' ability to phagocytose becomes impaired, the integrity of the epithelium can also become compromised. Non-phagocytosed particles interact with epithelial cells, which can lead to necrosis or apoptosis (Iyer et al., 1996) as well as activation leading to the release of pro-inflammatory cytokines (Driscoll et al., 1995, 1996; Finkelstein et al., 1997). On exposure to high particle concentrations, used in many toxicology studies, lung overload is induced in susceptible species (i.e., rats) (Bermudez et al., 2002), resulting in compromised epithelial cells, Type I cell hyperplasia and inflammation. The continuing presence of non-phagocytosed particles on the injured, activated and partially denuded epithelial surface will enhance their likelihood of being transferred to the interstitium. As these particles become interstitialized, they are likely to interact with interstitial macrophages that reside in close contact with fibroblast, epithelial, and endothelial cells (Adamson and Hedgecock, 1995). The proximity of the macrophages to these cells means that any mediators released by the interstitial macrophage can have a detrimental impact on the interstitial basement membrane and other interstitial cells leading to interstitial fibrosis. These interstitialized particles, either free or phagocytosed, may eventually be transported to the lymph nodes (Lehnert, 1993). Current evidence suggests that the transportation of inhaled particles to the lymph nodes is also size dependent (Oberdörster et al., 1988).

It has been demonstrated that a number of different entry mechanisms have been proposed to explain the uptake of nanoparticles by a variety of cell types. However, it is also worth acknowledging that different types of endocytosis can operate simultaneously (Rejman et al., 2004) so that more than one type of internalization pathway could contribute to their uptake and that if one uptake pathway does not function, another can take over.

9.3 TRANSLOCATION ACROSS THE PULMONARY INTERSTITIUM: THE INFLUENCE OF SIZE, SURFACE AREA, AND ASPECT RATIO

Although the lung consists of many cell types, the epithelial and macrophage cells play a primary role in controlling the entry of substances into the body via the lung. A number of studies have demonstrated ability for nanoparticles to cross the lung epithelium. Probably the first study to do so was published by Oberdörster's group and demonstrated that following instillation or inhalation of TiO_2, nanoparticles (25 nm) to a greater extent than fine particles (250 nm), could be found within the lung interstitium (Ferin et al., 1990a,b, 1992). Ferin et al. (1992) identified that "particles not phagocytosed by alveolar macrophages in the alveoli were taken up by alveolar type I epithelial cells, which was probably the first step for interstitial access of particles." This interstitialization and reduced clearance was associated with an elevated lung inflammatory response, but it is worth noting that the exposure concentrations used in this study were very high (500 µg instilled and up to 5 mg/rat lung burden following inhalation). A number of *in vivo* studies conducted by different groups using different particles have also identified similar results. For example, using electron microscopy (EM), it has been demonstrated that carbon black particles instilled into the mouse lung were accumulated in gaps between the cytoplasmic processes of alveolar epithelial cells, allowing transfer across the epithelial barrier into blood (Shimada et al., 2006). In a study exposing rats to iridium-192 nanoparticles (Semmler-Behnke et al., 2007), it was suggested that nanoparticles are much less actively phagocytosed by alveolar macrophages than larger particles, but that they are effectively removed from the lung epithelium into the interstitium. This is at odds with *in vitro* studies that suggest that nanoparticles are taken up by macrophages (Clift et al., 2008), although it has been suggested that nanoparticles decrease the subsequent phagocytosis of micron-sized particles (Renwick et al., 2001). Further evidence is required before it is confirmed that diffusion is a viable uptake route for nanoparticles into or across cells.

It is anticipated that modifying the specific surface area and reactivity of the particles, such as their ability to cause inflammation, would have effects on their ability to become interstitialized. That is, the more inflammogenic the particles, the more they are likely to interstitialize. An inhalation study using differently sized poorly soluble particles, TiO_2 and $BaSO_4$ (Tran et al., 1999), demonstrated a linear relationship between the particle burden expressed in units of surface area and the number of neutrophil cells (indicative of inflammation) found in the bronchoalveolar lavage (BAL) fluid, suggesting that high specific surface area of particles enhances inflammation (Stoeger et al., 2006; Duffin et al., 2007), leading to interstitialization. Since surface area is inversely related to size it could be hypothesized that increased surface area results in increased uptake into cells and penetration of cell barriers, but it is not possible at this time to assess whether such an observation is size or surface-area driven.

When considering the propensity of particles to enter the interstitium and potentially translocate, a consideration not only of the primary particle size but also of the aggregation/agglomeration state should be made. Indeed, it has been noted that in experimental systems of lung administration with particles prone to agglomeration such as carbon nanotubes (CNTs), the degree of dispersion can have marked effects on the type of response seen. In the 2008 study by Mercer et al. (2008), they tested the hypothesis of whether exposure to more dispersed single-walled carbon nanotube (SWCNT) structures would alter pulmonary distribution and response. Using pharyngeal aspiration as a route of lung exposure, they noted that a highly dispersed solution of SWCNTs led to a low observation of macrophage phagocytosis (potentially indicating low macrophage uptake) and no formation of granulomatous lesions that are common after lung administration of CNTs (Lam et al., 2004; Warheit et al., 2004; Ma-Hock et al., 2009). The authors concluded that dispersed SWCNTs are rapidly incorporated into the alveolar interstitium and that they produce an increase in collagen deposition.

It is obvious to see that the aggregation/agglomeration state of a particle will have effects on the zone of deposition within the lung based on alteration of the aerodynamic diameter of aggregate/agglomerate in relation to the primary particle and this in turn will affect the clearance rate and structures with which the particle can interact in the lung. In addition, the size of the agglomerate and propensity for disagglomeration/aggregation may also affect the potential for trans- or paracellular translocation and associated pathology (e.g., localized airway granuloma formation versus interstitial fibrosis or perhaps). Inhaled particles such as air pollution (Mitchev et al., 2002) and high aspect ratio particles (e.g., asbestos fibers) (Boutin et al., 1996) are known to translocate from the lung into the pleural cavity and even the peritoneal cavity. Here, material such as long fibers may be retained at points of egress from the pleural cavity where they can be associated with the formation of pleural pathologies. This translocation and retention of high aspect ratio materials has raised questions about high aspect ratio nanoparticles such as CNTs (Donaldson et al., 2010). Indeed such concern has been further exacerbated by the observation of subpleural deposition of CNT in the lungs (Ryman-Rasmussen et al., 2009) and even pleural transfer of CNT from the lung into the pleural space (Mercer et al., 2010). This has now suggested a shift in the question away from simply "can high aspect ratio nanoparticle translocate into the pleural space," to "what do they do when they reach there?". Indeed studies with asbestos and mineral fibers have shown that longer fiber length is associated with greater inflammogenicity (Davis et al., 1986; Moalli et al., 1987; Donaldson et al., 1989; Ye et al., 1999) and evidence is emerging that certain multiwalled CNTs exhibit length-dependent inflammogenic/pathogenic effects (Poland et al., 2008). Recent studies have tried to address the issue of pleural retention and resultant effects of high aspect ratio nanoparticles, which has suggested that long fibrous particles are retained in the pleural space with resultant stimulation of inflammation and fibrosis. In contrast, short fibers or particulates are able to escape the pleural space and localize in the mediastinal lymph nodes (Murphy et al., 2011).

9.4 TRANSLOCATION TO THE CIRCULATORY SYSTEM

Although there is a lack of any substantial epidemiology on the effects of exposure to engineered nanoparticles, the best guide as to likely effects on the cardiovascular system comes from studies such as the American Cancer Society studies (Pope et al., 2003) and the implication of combustion-derived nanoparticles (CDNP) in such adverse cardiovascular effects of $PM_{10}/PM_{2.5}$. As regards mechanisms, the evidence is accumulating for an impact on the endothelium and on atherothrombosis (Brook et al., 2003; Mills et al., 2007b). As a result, there is considerable interest in investigating how or if engineered nanoparticles might similarly impact cardiovascular disease.

Very little is known of the role of structure on translocation to the blood insofar as only low-toxicity, low-solubility nanoparticles have been studied. However, analogously to the situation with interstitialization, the increased permeability caused by inflammation could aid penetration to the blood. Changing the size and surface reactivity, factors that would influence inflammogenicity, may be considered likely modifiers of the potential of any nanoparticle to translocate to the blood and subsequent cardiovascular effects.

From the few studies of the toxicokinetics of a range of different nanoparticles in rats following inhalation (Kreyling et al., 2002, 2004, 2007; Oberdörster et al., 2002), there is some evidence that inhaled nanoparticles of various types of materials (e.g., iridium) enter the blood at a rate of about 1% of the deposited dose (Kreyling et al., 2007). Similar values (1%–2% of the deposited dose entering the blood) have been reported for TiO_2 nanoparticles following instillation (Chen et al., 2006). Lower concentrations (0.05%) of inhaled gold nanoparticles were reported to enter the blood following instillation (Takenaka et al., 2006). Results from electron microscope morphometry (Mühlfeld et al., 2007) have suggested that there is rapid transfer of 22 nm TiO_2 nanoparticles (count median diameter) into the blood via the interstitium, within a very short space of time following deposition. Although the mass fraction entering the blood may be small, the particle number that reaches these other targets can be very great (Kreyling et al., 2007) and these organs are not necessarily equipped, as the lungs are, to deal with large numbers of particles and so the effects may be more severe.

The endothelium is the key cell that particles must cross in order to gain access to the blood. Many studies have made the assumption that particles definitely cross the endothelium and have reported effects on the endothelium and cellular elements of the blood. Carbon black nanoparticles and PM_{10}/$PM_{2.5}$ affect endothelial cells in ways that enhance coagulation (Gilmour et al., 2005) and similarly pro-thrombotic effects in liver endothelium following treatment *in vivo* have been reported (Khandoga et al., 2004). The endothelium, atherosclerotic lesions, and the clotting system are key targets for nanoparticles in circulatory system. The potential cardiovascular impact of engineered nanoparticles is a major concern given the data showing adverse cardiovascular as lead effects of PM_{10}/$PM_{2.5}$ (Schwartz and Morris, 1995) and recent work on diesel inhalation showing effects on the endothelium that are very likely occurring via a mechanism involving oxidative stress (Mills et al., 2005, 2009). There is experimental evidence that deposition of PM_{10}/$PM_{2.5}$ in the lungs accelerates and worsens atherosclerotic plaque development in animal models (Suwa et al., 2002; Lippmann et al., 2005; Sun et al., 2005). The only data that indicate that engineered nanoparticles might have similar effects are the finding that intratracheally administered CNT causes oxidative stress and mitochondrial dysfunction in the aortae of mice (Li et al., 2007). This does suggest that effects on the vascular wall might be a generic effect of nanoparticles. These effects could be driven by the oxidative stress from the particle or the inflammation resulting from the particles being deposited in the lungs. However, an alternative explanation is that the nanoparticles enter the blood and cause a direct effect on the plaques by affecting the overlying endothelium of entering the atherosclerotic lesion and affecting the stability of the plaque. Paradoxically, given this concern, nanoparticulate iron has been used to image plaques where the particles have been found to enter the macrophages in the plaque with, perhaps surprisingly, no adverse effects reported (Trivedi et al., 2006).

9.5 TRANSLOCATION TO THE BRAIN

The present interest among the toxicology community in engineered nanoparticles translocating to non-pulmonary organs, in particular the CNS, was stimulated in part by a 2002 editorial (Oberdörster and Utell, 2002) and the subsequent acknowledgment of the need for research in a number of reviews (see, e.g., Borm and Kreyling, 2004; Hoet et al., 2004; Oberdörster et al., 2005; Gwinn and Vallyathan, 2006; Peters et al., 2006).

The translocation of inhaled nanoparticles to the CNS has been postulated to occur via three pathways: (1) across the BBB after their translocation into the blood circulation from deposits anywhere in the respiratory tract, (2) via the olfactory nerve from deposits on the olfactory mucosa and uptake into the olfactory bulb, and (3) via paracellular or perineural pathways across the olfactory mucosa and ethmoid bone into cerebrospinal fluid (Oberdörster et al., 2004). Translocation to the brain via the olfactory nerve has been a focus of recent research, given the estimated 80% probability of inhaled NPs of ~1 nm in size depositing in the nasopharyngeal region (ICRP, 1994).

Research on the translocation of nanoparticles to the CNS is widely cited to originate from studies in the 1940s using 30 nm polio virus particles (Bodian and Howe, 1941) and the 1970s using 50 nm colloidal gold particles (De Lorenzo, 1970) instilled intranasally. These early studies revealed the olfactory nerve and olfactory bulbs as portals of entry to the CNS. A historical perspective and a summary of findings from more recent translocation studies are presented in a 2005 review of nanotoxicology (Oberdörster et al., 2005). Within the area of neuronal uptake and translocation, the review highlights the size-dependent findings from inhalation studies with manganese oxide particles in rats that showed a predominance for uptake in the olfactory bulb, compared to the lung.

In a study of the effect of nanoparticle size, shape, concentration, and incubation time on cellular uptake kinetics (Chithrani et al., 2006), it is speculated that nonspecific adsorption of serum proteins mediates the uptake of the nanoparticles and that the presence of these proteins on the surface of the nanoparticles dictates uptake half-life, rates, and amount.

The role of the surface groups and the charge they carry is paramount to the behavior of nanomaterials *in vivo*. Surface groups can make the material hydrophilic, hydrophobic, lipophilic, lipophobic, or catalytically active or passive. In a study to evaluate the effect of neutral, anionic, and cationic charged nanoparticles on BBB integrity and permeability using nanoparticles composed of emulsified wax (Lockman et al., 2004), neutral and low concentrations of anionic nanoparticles were found to have no effect on BBB integrity, whereas, high concentrations of anionic and cationic nanoparticles disrupted the BBB.

Nanoparticles have been used for pharmaceutical and medical applications for over 30 years and those developed for pharmacological uses are currently made from a wide array of materials such as poly(alkylcyanoacrylates); poly(methylidene malonate); polyesters such as poly (lactic acid), poly(glycolic acid), poly(ε-caprolactone), and their copolymers; polysaccharides; and proteins. The choice of nanoparticle materials is based on biodegradability, intrinsic immunogenicity, and toxicity.

A recent review (Teixido and Giralt, 2008) of the role played by peptides in BBB interactions highlighted that the binding of a peptide-coated nanoparticle to a given receptor can result in the nanoparticle being transported across a barrier, often with a mechanism other than that expected of the coating. Hydrophilic surfactants have been shown to reduce nanoparticle absorption by reticuloendothelial system organs that alters biodistribution of the nanoparticles. Coating of colloidal nanoparticles with block copolymers such as poloxamers and poloxamines induces a steric repulsion effect, minimizing the adhesion of particles to the surface of macrophages, which in turn results in the decrease of phagocytic uptake and in significantly higher levels in the blood and organs including the brain, intestine, and kidneys among others. Surface PEGylation increases the blood half-life of nanoparticles and polysorbate-80 improves BBB transport of nanoparticles.

In a comparable review of nanocarrier-based CNS delivery systems (Tiwari and Amiji, 2006), a number of mechanisms proposed for the BBB transport of polymeric solid and lipid nanoparticles were summarized. The authors state that it is possible that combination of some or all of the mechanisms may act to facilitate transport. The various mechanisms proposed include the following:

1. Adhesion of nanoparticles to the inner endothelial cells of brain capillaries and the subsequent transport by passive diffusion, possibly by a larger concentration gradient. Nanoparticle degradation products may also act as adsorption enhancers, thus contributing to increased passive diffusion.
2. Surfactants used in coating of nanoparticles may solubilize the endothelial cell membrane lipids, thus enhancing the transport across the BBB.
3. Surfactant-coated nanoparticles, particularly polysorbate coated, administered intravenously, become further coated with absorbed plasma proteins, especially, apolipoprotein E (Apo-E), leading to this final product being mistaken for low-density lipoprotein (LDL) particles by the cerebral endothelium and internalized by the LDL uptake system. Solid lipid nanoparticles may also transport drugs across the BBB by this mechanism.
4. Components of nanoparticle structures might open the tight junctions of the brain capillary endothelial cells, and allow the penetration of surfactant-coated nanoparticles into the CNS.

5. Excipients used in the manufacture of nanoparticles (e.g., polysorbate 80) may inhibit the drug efflux system and improve the drug absorption across brain capillary endothelial cells.
6. Nanoparticles might be endocytosed or transcytosed through the brain capillary endothelial cells.

The chemical identity of the surface of a nanoparticle is key to many of these possible mechanisms and is the basis of many pharmacological approaches using the activation of natural transport routes to penetrate the BBB.

9.6 CONCLUSION

One of the major concerns regarding the possible toxic effects of nanoparticles is the capacity of these materials to penetrate cells and potentially translocate to other cells, tissues, and organs remote from the portal of entry to the body. This is considered to be a necessary step in the movement of particles deposited in the lung, entering the blood, acting upon cells in other tissues, and manifesting ultimately in a physiological response.

The ability to enter the interstitium from the airspaces seems to be a fundamental property of nanoparticles. However, the exact role that interstitialization has in human toxicity is not yet well understood. The blood and the cardiovascular system are significant potential targets for adverse effects of engineered nanoparticles. CDNP appear to be able to enhance atherosclerosis, and the possibility that this could be a general property of nanoparticles is being examined. Size and chemical structure are likely to be important and might act through the effect that they have on lung inflammation and oxidative stress. In the case of translocation to the blood and direct interaction with plaques, size is likely to be important.

The unique biokinetic behavior exhibited by nanoparticles, including cellular endocytosis, transcytosis, and neuronal and circulatory translocation and distribution, makes them desirable for therapeutic and diagnostic medical applications, but may also convey potential toxicity. The routes of entry of nanoparticles to the CNS are becoming increasingly recognized, although the influence of physicochemical properties on the mechanisms remains to be fully elucidated. There is evidence of uptake in the CNS both of particles translocating from the lungs and via the olfactory bulb. It is expected that transport of nanoparticles across the BBB is possible by either passive diffusion or by carrier-mediated endocytosis (Hoet et al., 2004). Moreover, as the BBB is defective in a number of locations (e.g., pineal gland, pituitary gland, area postrema, choroid plexus), nanoparticle entry in these areas may be possible. Surface-modified particles can interact with the receptors, leading to uptake by endothelial cells. Also, other processes such as tight junction modulation or P-glycoprotein (Pgp) inhibition also may play a role (Kreuter, 2001).

Understanding the role of nanoparticle structural features, effects of modifications to the structure and the ability to produce inflammation is key to elucidating the factors that render a particle able to enter the interstitium.

Investigation of the mechanism(s) of nanoparticle translocation is an active research area. Further detail and discussion of the aspects highlighted in this chapter and newly emerging findings are available in the peer-reviewed literature, including publications on the biokinetics of inhaled particles (Geiser and Kreyling, 2010), macrophage clearance of inhaled micro- and nanoparticles (Geiser, 2010), and translocation of multiwall CNTs (Reddy et al., 2010), quantum dots (Geys et al., 2009), and polystyrene nanoparticles (Yacobi et al., 2010), to suggest but a few.

ACKNOWLEDGMENT

The contribution is acknowledged of coauthors (Vicki Stone, Ken Donaldson, Lang Tran, Bryony Ross, and Qasim Chaudhry) of the report entitled "Cell Penetration: A study to identify the physicochemical factors controlling the capacity of nanoparticles to penetrate cells" upon which this chapter has drawn.

REFERENCES

Adamson, I.Y.R. and Hedgecock, C. 1995. Patterns of particle deposition and retention after instillation to mouse lung during acute injury and fibrotic repair, *Exp. Lung Res.*, 21, 695–709.

Bermudez, E., Mangum, J.B. et al. 2002. Long-term pulmonary responses of three laboratory rodent species to subchronic inhalation of pigmentary titanium dioxide particles, *Toxicol. Sci.*, 70(1), 86–97.

Bodian, D. and Howe, H.A. 1941. The rate of progression of poliomyelitis virus in nerves, *Bull. Johns Hopkins Hosp.*, 69(2), 79–85.

Borm, P.J. and Kreyling, W. 2004. Toxicological hazards of inhaled nanoparticles—Potential implications for drug delivery, *J. Nanosci. Nanotechnol.*, 4(5), 521–531.

Boutin, C., Dumortier, P. et al. 1996. Black spots concentrate oncogenic asbestos fibers in the parietal pleura. Thoracoscopic and mineralogic study, *Am. J. Respir. Crit. Care Med.*, 153(1), 444–449.

Brook, R.D., Brook, J.R., and Rajagopalan, S. 2003. Air pollution: The 'Heart' of the problem, *Curr. Hypertens. Rep.*, 5, 32–39.

Chen, J., Tan, M., Nemmar, A., Song, W., Dong, M., Zhang, G., and Li, Y. 2006. Quantification of extra-pulmonary translocation of intratracheal-instilled particles in vivo in rats: Effect of lipopolysaccharide, *Toxicology*, 222, 195–201.

Chithrani, B.D., Ghazani, A.A., and Chan, W.C.W. 2006. Determining the size and shape dependence of gold nanoparticle uptake into mammalian cells, *Nano Lett.*, 6, 662–668.

Clift, M.J.D., Rothen-Rutishauser, B., Brown, D.M., Duffin, R., Donaldson, K., Proudfoot, L., Guy, K., and Stone, V. 2008. The impact of different nanoparticle surface chemistry and size on uptake and toxicity in a murine macrophage cell line, *Toxicol. Appl. Pharmacol.*, 232, 418–427.

Conner, S.D. and Schmid, S.L. 2003. Regulated portals of entry into the cell, *Nature*, 422, 37–44.

Davis, J.G., Addison, J., Bolton, R.E., Donaldson, K., Jones, A.D., and Smith, T. 1986. The pathogenicity of long versus short fiber samples of amosite asbestos administered to rats by inhalation and intraperitoneal injection, *Br. J. Exp. Pathol.*, 67, 415–430.

De Lorenzo, A.J.D. 1970. The olfactory neuron and the blood-brain barrier, in *Taste and Smell in Invertebrates*, G.E.W. Wolstenholme and J. Knight (Eds.), Churchill, London, U.K., p. 151.

Donaldson, K., Brown, G.M., Brown, D.M., Bolton, R.E., and Davis, J.G. 1989. Inflammation generating potential of long and short fiber amosite asbestos samples, *Br. J. Ind. Med.*, 46, 271–276.

Donaldson, K., Murphy, F.A. et al. 2010. Asbestos, carbon nanotubes and the pleural mesothelium: A review and the hypothesis regarding the role of long fibre retention in the parietal pleura, inflammation and mesothelioma, *Part. Fibre Toxicol.*, 7(1), 5.

Driscoll, K.E., Carter, J.M., Howard, B.W., Hassenbein, D.G., Pepelko, W., Baggs, R.B., and Oberdörster, G. 1996. Pulmonary inflammatory chemokine and mutagenic responses in rats after subchronic inhalation of carbon black, *Toxicol. Appl. Pharmacol.*, 136/2, 372–380.

Driscoll, K.E., Hassenbein, D.G., Carter, J.M., Kunkel, S.L., Quinlan, T.R., and Mossman, B.T. 1995. TNFα and increased chemokine expression in rat lung after exposure, *Toxicol. Lett.*, 82/83, 483–489.

Duffin, R., Tran, L., Brown, D., Stone, V., and Donaldson, K. 2007. Proinflammogenic effects of low-toxicity and metal nanoparticles in vivo and in vitro: Highlighting the role of particle surface area and surface reactivity, *Inhal. Toxicol.*, 19, 849–856.

Ferin, J., Oberdörster, G., and Penney, D.P. 1992. Pulmonary retention of ultrafine and fine particles in rats, *Am. J. Respir. Cell. Mol. Biol.*, 6, 535–542.

Ferin, J., Oberdöster, G., Penney, D.P., Soderholm, S.C., Gelein, R., and Piper, H.C. 1990a. Increased pulmonary toxicity of ultrafine particles? I. Particle clearance, translocation, morphology, *J. Aerosol. Sci.*, 21, 381–384.

Ferin, J., Penney, D.P., Soderholm, S., Gelein, R., and Piper, H.C. 1990b. Increased pulmonary toxicity of ultra-fine particles? 1 Particle clearance, translocation, morphology, *J. Aerosol Sci.*, 21, 381–384.

Finkelstein, J.N., Johnston, C., Barrett, T., and Oberdörster, G. 1997. Particulate-cell interactions and pulmonary cytokine expression, *Environ. Health Perspect.*, 105(Suppl. 5), 1179–1182.

Geiser, M. 2010. Update on macrophage clearance of inhaled micro- and nanoparticles, *J. Aerosol Med. Pulm. Drug Deliv.*, 23(4), 207–217.

Geiser, M. and Kreyling, W.G. 2010. Deposition and biokinetics of inhaled nanoparticles, *Part. Fibre Toxicol.*, 7, 2.

Geys, J., De Vos, R., Nemery, B., and Hoet, P.H. 2009. In vitro translocation of quantum dots and influence of oxidative stress, *Am. J. Physiol. Lung Cell Mol. Physiol.*, 297, L903–L911.

Gilmour, P.S., Morrison, E.R., Vickers, M.A., Ford, I., Ludlam, C.A., Greaves M., Donaldson, K., and MacNee, W. 2005. The procoagulant potential of environmental particles (PM10), *Occup. Environ. Med.*, 62, 164–171.

Gumbleton, M. 2001. Caveolae as potential macromolecule trafficking compartments within alveolar epithelium, *Adv. Drug Deliv. Rev.*, 49, 281–300.

Gwinn, M.R. and Vallyathan, V. 2006. Nanoparticles: health effects—Pros and cons, *Environ. Health Perspect.*, 114(12), 1818–1825.

Hastings, R.H., Folkesson, H.G., and Matthay, M.A. 2004. Mechanisms of alveolar protein clearance in the intact lung, *Am. J. Physiol. Lung Cell Mol. Physiol.*, 286, L679–L689.

Hoet, P.H., Bruske-Hohlfeld, I., and Salata, O.V. 2004. Nanoparticles—Known and unknown health risks, *J. Nanobiotechnol.* [electronic resource], 2(1), 12.

ICRP. 1994. Human respiratory tract model for radiological protection, International Commission on Radiological Protection, Publication No. 66.

Iyer, R., Hamilton, R.F., Li, L., and Holian, A. 1996. Silica induced apoptosis mediated via scavenger receptor in human alveolar macrophages, *Toxicol. Appl. Pharmacol.*, 141, 84–92.

Khalil, I.A., Kogure, K., Akita, H., and Harashima, H. 2006. Uptake pathways and subsequent intracellular trafficking in nonviral gene delivery, *Pharmacol. Rev.*, 58, 32–45.

Khandoga, A., Stampfl, A., Takenaka, S., Schulz, H., Radykewicz, R., Kreyling, W., and Krombach, F. 2004. Ultrafine particles exert prothrombotic but not inflammatory effects on the hepatic microcirculation in healthy mice in vivo, *Circulation*, 109, 1320–1325.

Kreuter, J. 2001. Nanoparticulate systems for brain delivery of drugs, *Adv. Drug Deliv. Rev.*, 47(1), 65–81.

Kreyling, W.G., Moller, W., Semmler-Behnke, M., and Oberdörster, G. 2007. Particle dosimetry: Deposition and clearance from the respiratory tract and translocation to extra-pulmonary sites, Chapter 3, in *Particle Toxicology*, Donaldson, K. and Borm, P. (Eds.), CRC Press, Boca Raton, FL, pp. 47–74.

Kreyling, W., Semmler, M., Erbe, F., Mayer, P., Takenaka, S., Oberdörster, G., and Ziesenis, A. 2002. Minute translocation of inhaled ultrafine insoluble iridium particles from lung epithelium to extrapulmonary tissues, *Ann. Occup. Hyg.*, 46(Suppl. 1), 223–226.

Kreyling, W., Semmler, M., and Moller, W. 2004. Dosimetry and toxicology of ultrafine particles, *J. Aerosol Med.*, 17, 140–152.

Lam, C.W., James, J.T. et al. 2004. Pulmonary toxicity of single-wall carbon nanotubes in mice 7 and 90 days after intratracheal instillation. *Toxicol. Sci.*, 77(1), 126–134.

Lehnert, B.E. 1993. Defense mechanisms against inhaled particles and associated particle-cell interactions, in *Health Effects of Mineral Dusts*, Vol. 28, G.D. Guthrie, Jr. and B.T. Mossman (Eds.), Mineralogical Society of America, Washington, DC, pp. 427–469.

Li, Z., Hulderman, T., Salmen, R., Chapman, R., Leonard, S.S., Young, S.H. Shvedova, A., Luster, M.I., and Simeonova, P.P. 2007. Cardiovascular effects of pulmonary exposure to single-wall carbon nanotubes, *Environ. Health Perspect.*, 115, 377–382.

Lippmann, M., Hwang, J.S., Maciejczyk, P., and Chen, L.C. 2005. PM source apportionment for short-term cardiac function changes in ApoE–/– mice, *Environ. Health Perspect.*, 113, 1575–1579.

Liu, J. and Shapiro, J.I. 2003 Endocytosis and signal transduction: Basic science update, *Biol. Res. Nurs.*, 5(2), 117.

Lockman, P.R., Koziara, J.M., Mumper, R.J., and Allen, D.D. 2004. Nanoparticle surface charges alter blood-brain barrier integrity and permeability, *J. Drug Target.*, 12(9–10), 635–641.

Ma-Hock, L., Treumann, S. et al. 2009. Inhalation toxicity of multiwall carbon nanotubes in rats exposed for 3 months, *Toxicol Sci.*, 112(2), 468–481.

Mercer, R.R., Hubbs, A.F. et al. 2010. Distribution and persistence of pleural penetrations by multi-walled carbon nanotubes. *Part. Fibre Toxicol.*, 7, 28.

Mercer, R.R., Scabilloni, J. et al. 2008. Alteration of deposition pattern and pulmonary response as a result of improved dispersion of aspirated single-walled carbon nanotubes in a mouse model, *Am. J. Physiol. Lung Cell. Mol. Physiol.*, 294(1), L87–L97.

Mills, N.L., Donaldson, K., Hadoke, P.W., Boon, N.A., MacNee, W., Cassee, F.R., Sandström, T., Blomberg, A., Newby, D.E. 2009. Adverse cardiovascular effects of air pollution, *Nature Clinical Practice Cardiovascular Medicine*, 6, 36–44.

Mills, N.L., Tornqvist, H., Robinson, S.D., Gonzalez, M., Darnley, K., MacNee, W., Boon, N.A. et al. 2005. Diesel exhaust inhalation causes vascular dysfunction and impaired endogenous fibrinolysis, *Circulation*, 112, 3930–3936.

Mills, N.L., Tornqvist, H., Robinson, S.D., Gonzalez, M.C., Soderberg, S., Sandstrom, T., Blomberg, A., Newby, D.E., and Donaldson, K. 2007b. Air pollution and atherothrombosis, *Inhal. Toxicol.*, 19(1), 81–89.

Mitchev, K., Dumortier, P. et al. 2002. 'Black Spots' and hyaline pleural plaques on the parietal pleura of 150 urban necropsy cases, *Am. J. Surg. Pathol.*, 26(9), 1198–1206.

Moalli, P.A., Macdonald, J.L., Goodglick, L.A., and Kane, A.B. 1987. Acute injury and regeneration of the mesothelium in response to asbestos fibers, *Am. J. Pathol.*, 128, 426–445.

Mühlfeld, C., Geiser, M., Kapp, N., Gehr, P., and Rothen-Rutishauser, B. 2007. Re-evaluation of pulmonary titanium dioxide nanoparticle distribution using the "relative deposition index": Evidence for clearance through microvasculature, *Part. Fibre Toxicol.*, 4, 7.

Murphy, F.A., Poland, C.A. et al. 2011. Length-dependent retention of carbon nanotubes in the pleural space of mice initiates sustained inflammation and progressive fibrosis on the parietal pleura, *Am. J. Pathol.* 178(6), 2587–2600.

Oberdörster, G., Morrow, P.E., and Spurny, K. 1988. Size dependent lymphatic clearance of amosite fibres in the lung, *Ann. Occup. Hyg.*, 32, 149–156.

Oberdörster, G., Oberdörster, E., and Oberdörster, J. 2005. Nanotoxicology: An emerging discipline evolving from studies of ultrafine particles, *Environ. Health Perspect.*, 113(7), 823–839.

Oberdörster, G., Sharp, Z., Atudorei, V., Elder, A., Gelein, R., Kreyling, W., and Cox, C. 2004. Translocation of inhaled ultrafine particles to the brain, *Inhal. Toxicol.*, 16(6–7), 437–445.

Oberdörster, G., Sharp, Z., Atudorei, V., Elder, A.P., Gelein, R., Lunts, A.K., Kreyling, W., and Cox, C. 2002. Extrapulmonary translocation of ultrafine carbon particles following whole-body inhalation exposure of rats, *J. Toxicol. Environ. Health A*, 65, 1531–1543.

Oberdörster, G. and Utell, M.J. 2002. Ultrafine particles in the urban air: To the respiratory tract—and beyond? *Environ. Health Perspect.*, 110(8), A440–A441.

Perret, E., Lakkaraju, A., Deborde, S., Schreiner, R., and Rodriguez-Boulan, E. 2005. Evolving endosomes: How many varieties and why? *Curr. Opin. Cell Biol.*, 17(4), 423–434.

Peters, A., Veronesi, B., Calderon-Garciduenas, L., Gehr, P., Chen, L.C., Geiser, M., Reed, W., Rothen-Rutishauser, B., Schurch, S., and Schulz, H. 2006. Translocation and pontential neurological effects of fine and ultrafine particles a critical update, *Particle and Fibre Toxicology*, 3, 13.

Poland, C., Duffin, R., Kinloch, I., Maynard, A.D., Wallace, W., Seaton, A. et al. 2008. Carbon nanotubes show asbestos-like pathogenic effects that are length-dependent, *Nat. Nanotechnol.*, 3, 423–428.

Pope, C.A., Burnett, R.T., Thurston, G.D., Thun, M.J., Calle, E.E., Krewski, D., and Godleski, J. 2003. Cardiovascular mortality and long-term exposure to particulate air pollution: epidemiological evidence of general pathophysiological pathways of disease, *Circulation*, 109, 71–77.

Reddy, A.R., Krishna, D.R., Reddy, Y.N., and Himabindu, V. 2010. Translocation and extra pulmonary toxicities of multi wall carbon nanotubes in rats, *Toxicol. Mech. Methods*, 20, 267–272.

Rejman, J., Oberle, V., Zuhorn, I.S., and Hoekstra, D. 2004. Size-dependent internalization of particles via the pathways of clathrin- and caveolae-mediated endocytosis, *Biochem. J.*, 377(1), 159–169.

Renwick, L.C., Donaldson, K., and Clouter, A. 2001. Impairment of alveolar macrophage phagocytosis by ultrafine particles, *Toxicol. Appl. Pharmacol.*, 172, 119–127.

Ryman-Rasmussen, J.P., Cesta, M.F. et al. 2009. Inhaled carbon nanotubes reach the subpleural tissue in mice. *Nat. Nanotechnol.*, 4(11), 747–751.

Schwartz, J. and Morris, R. 1995. Air pollution and hospital admissions for cardiovascular disease in Detroit, Michigan, *Am. J. Epidemiol.*, 142, 23–35.

Semmler-Behnke, M., Takenaka, S., Fertsch, S., Wenk, A., Seitz, J., Mayer, P., Oberdörster, G., and Kreyling, W.G. 2007. Efficient elimination of inhaled nanoparticles from the alveolar region: Evidence for interstitial uptake and subsequent reentrainment onto airways epithelium, *Environ. Health Perspect.*, 115, 728–733.

Shimada, A., Kawamura, N., Okajima, M., Kaewamatawong, T., Inoue, H., and Morita, T. 2006. Translocation pathway of the intratracheally instilled ultrafine particles from the lung into the blood circulation in the mouse, *Toxicol. Pathol.*, 34, 949–957.

Stoeger, T., Reinhard, C., Takenaka, S., Schroeppel, A., Karg, E., Ritter, B., Heyder, J., and Schulz, H. 2006. Instillation of six different ultrafine carbon particles indicates a surface area threshold dose for acute lung inflammation in mice, *Environ. Health Perspect.*, 114, 328–333.

Sun, Q., Wang, A., Jin, X., Natanzon, A., Duquaine, D., Brook, R.D., Aguinaldo, J.-G.S. et al. 2005. Long-term air pollution exposure and acceleration of atherosclerosis and vascular inflammation in an animal model, *JAMA*, 294, 3003–3010.

Suwa, T., Hogg, J.C., Quinlan, K.B., Ohgami, A., Vincent, R., and van Eeden, S.F. 2002. Particulate air pollution induces progression of atherosclerosis, *J. Am. Coll. Cardiol.*, 39, 935–942.

Takenaka, S., Karg, E., Kreyling, W.G., Lentner, B., Moller, W., Behnke-Semmler, M., Jennen, L. et al. 2006. Distribution pattern of inhaled ultrafine gold particles in the rat lung, *Inhal. Toxicol.*, 18, 733–740.

Teixido, M. and Giralt, E. 2008. The role of peptides in blood-brain barrier nanotechnology, *J. Pept. Sci.*, 14(2), 163–173.

Tiwari, S.B. and Amiji, M.M. 2006. A review of nanocarrier-based CNS delivery systems, *Curr. Drug Deliv.*, 3(2), 219–232.

Tran, C.L., Cullen, R.T., Buchanan, D., Jones, A.D., Miller, B.G., Searl, A., Davis, J.M.G., and Donaldson, K. 1999. Investigations into the pulmonary effects of low toxicity dusts. Part II—Investigation and prediction of pulmonary response to dust, Contract Research Report 216/1999, Health and Safety Executive, Suffolk, U.K.

Trivedi, R.A., Mallawarachi, C., King-Im, J.M., Graves, M.J., Horsley, J., and Goddard, M.J. 2006. Identifying inflamed carotid plaques using in vivo USPIO-enhanced MR imaging to label plaque macrophages, *Arterioscler. Thromb. Vasc. Biol.*, 26, 1601–1606.

Warheit, D.B., Laurence, B.R. et al. 2004. Comparative pulmonary toxicity assessment of single-wall carbon nanotubes in rats, *Toxicol. Sci.*, 77(1), 117–125.

Yacobi, N.R., Malmstadt, N., Fazlollahi, F., DeMaio, L., Marchelletta, R., Hamm-Alvarez, S.F., Borok, Z., Kim, K.J., and Crandall, E.D. 2010. Mechanisms of alveolar epithelial translocation of a defined population of nanoparticles, *Am. J. Respir. Cell. Mol. Biol.*, 42, 604–614.

Ye, J., Shi, X., Jones W., Rojanasakul, Y., Cheng, N., Schwegler-Berry, D. Baron, P., Deye, G.J., Li, C., and Castranova, V. 1999. Critical role of glass fiber length in TNF-alpha production and transcription factor activation in macrophages, *Am. J. Physiol.*, 276, L426–L434.

10 High Aspect Ratio Nanomaterials

Characterization and Toxicology

Steven M. Hankin and Craig A. Poland

CONTENTS

10.1 INTRODUCTION

High aspect ratio nanomaterials (HARN) are a class of new materials whose novel properties are being investigated and applied in a variety of applications as diverse as structural composites, electronics, and medicine. HARN can be classed as "one-dimensional" nanoscale building blocks; other one-dimensional particles include nanobelts and nanoribbons—these are nanostructures with high surface area and high aspect ratio (of greater than 10:1, length-to-diameter).

However, concern exists about the potential health effects of high aspect ratio nanoparticles, derived primarily from toxicology studies of industrial fibers, including asbestos. This chapter provides an overview of the characterization techniques and the paradigm underpinning the toxicity of HARN.

10.2 CHARACTERIZATION

10.2.1 CHALLENGES TO THE CHARACTERIZATION OF HARN

In general, challenges that exist with the characterization of nanoparticles also exist with HARN, which will ultimately have a big impact on the success of the analysis and the end results obtained. It is reasonable to assume that such challenges can be magnified when it comes to HARN because of the higher aspect ratio property compared to other nanoparticles. The high aspect ratio property implies a much increased surface area, which will mean increase in surface interaction and activity, compared to other nanoparticles.

There are boundaries as to what we classify as fibers that are to be considered an inhalation hazard. The World Health Organization (WHO) defines a fiber as a particle longer than 5 μm, less than 3 μm in width and having an aspect ratio of >3:1 (WHO/EURO Technical Committee for Monitoring and Evaluating MMMF, 1985; WHO, 1997). This is not a health-based criterion but the one based on a practical definition of a respirable fiber.

Sample preparation is one of the most critical steps toward successful characterization of nanoparticles, in which there are many variables to consider when designing a method for preparation. The first step to consider in sample preparation is the need to have "reliable" sampling, such that "sample collected from bulk represents the physical and chemical characteristics of the entire sample" (NIST 960-1, 2001). Most nanoparticles are received in the powder form or in liquid form, that is, the form of a stable suspension. Powder sampling is more difficult, as there is a natural tendency for nanoparticles to aggregate and unlike in solution phase, it is more difficult to control surface charges on particles; some general guidelines on powder sampling can be found in Allen (2004a,b). The next steps in the sample preparation will be governed by the requirements of individual methods, which may require specialized conditions for measurement; each will represent their own unique challenges. Ideally, samples for analysis should be free from (1) the inherent aggregation problems associated with nanoparticles and (2) other contaminants not associated with the nanoparticles being characterized. However, to achieve such goals is not trivial. For example, to successfully disperse nanoparticles in a liquid media, sonication methods often need to be employed. However, this has the potential to change the size distribution of the HARN and introduce defects (e.g., Islam et al., 2001). Furthermore, the stability of dispersions over time (Vaisman et al., 2006) also influences the outcome of the analysis as does the "state" of the sample required for analysis, that is, whether the nanoparticles should be fixed onto a solid substrate, suspended in liquid media, or aerosolized (solid or liquid aerosols). Last, some surface techniques (such as conventional electron microscopy, Brunauer, Emmert, and Teller [BET], and x-ray photoelectron spectroscopy [XPS]) are not applicable to the analysis of samples dispersed in liquids.

Characterization of nanomaterials can be complex, as there are many different material attributes that need to be considered. The need to characterize nanoparticles for toxicological evaluation is made even more complicated by the need to analyze as close as possible to the "as-dosed" form, that is, to what is required under the toxicological investigation. This is not easy as (1) quantities used in the analysis are normally smaller and (2) the state of particles is likely to change under the conditions of the analysis. Overall, it is the experimental conditions used in toxicological studies, which will eventually determine the choice of techniques used for characterization. For example, an inhalation study may require the need to characterize the nanoparticles in dry powder aerosol form.

The main challenge is to identify those techniques deemed to be most "suitable" for characterization. This overview focuses on those techniques that are well established or commercially available analytical techniques, and those techniques that yield chemical/physical property information that have been linked/hypothesized in some way to toxicological activities. The techniques highlighted fall into two categories: imaging (high resolution, with the capability of probing individual HARN) and nonimaging techniques (often involves the measurement of a collection/ensemble of HARN).

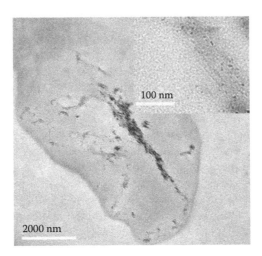

FIGURE 10.1   TEM image shows CNTs (dark areas) within a cell nucleus.

10.2.2   IMAGING TECHNIQUES

This group of techniques is extremely advantageous in that they have potential to yield highly resolved images, which allows direct visualization of nanoparticles and thus will yield information on size and shape. It is a general consensus that properties related to the three-dimensional structure of nanoparticles have a profound influence on their toxicity and the small size has the potential to enter cells and ultimately cause cell death (Porter et al., 2007), as shown in Figure 10.1.

The ability to provide images on the nanoscale, with sensitivity that reaches the individual nanoparticle level, is a powerful tool used in the study of cellular interactions of nanomaterials.

Imaging techniques have been widely used for investigating structure and morphology of nanoparticles. The amount of detail present in the image will ultimately be dependent on the instrument's spatial resolution; a high-resolution microscope, for example, will result in images that can reveal very small defects or anomalies. Although these tools are extremely useful, they all share the same disadvantages, in that they do not have a wide field of view, are relatively expensive techniques (to purchase and maintain), require the need of specially trained analysts, and have no potential for automation.

10.2.2.1   Electron-Based Microscopies

These tools are able to produce highly magnified, resolved images of objects with a much greater depth of field in comparison to conventional optical microscopes (Egerton, 2005). Scanning electron microscopy (SEM) and transmission electron microscopy (TEM) are the two most common techniques for nanoparticle characterization. SEM creates an image by scanning a tightly focused electron beam over the sample and detecting the secondary electrons from the sample onto the screen; each point on the screen will then correspond to a pixel (picture element). TEM, on the other hand, forms an image using a system of lenses. Unlike SEM, the electron beam passes entirely through the sample and is subsequently collected to appear on a screen, generating a "transmission" electron image (Reimer, 1993). There are advantages and disadvantages associated with both techniques. TEM, for example, has a far greater spatial resolution than SEM, but suffers from lengthy, time-consuming sample preparation. One difficulty is getting the specimen sample thin enough for analysis using TEM. Another disadvantage with TEM is the need for a very intense electron beam (with energy in the range ~200 to 300 keV) (Kiang et al., 1996) compared to SEM (~20 keV); this may pose some challenges to the structural and thermal stability of nanoparticles during analysis when under the influence of high-energy electron irradiation.

10.2.2.2 Scanning Probe Microscopies

These are techniques that acquire an image by raster scanning an atomically sharp, microscopic probe capable of sensing height changes as small as 0.1 Å (10 pm) (Sakurai, 2000). The two most widely used high-resolution scanning probe microscopy (SPM) techniques are atomic force microscopy (AFM) and scanning tunneling microscopy (STM). In an AFM, it is the cantilever deflection signal versus probe base position that results in the image, whereas in an STM, it is the variation in current as the probe passes over the surface that is translated into an image. Unlike electron microscopes, which have the potential to destroy or modify sample structure in the process due to the use of electron beam, SPM techniques are noninvasive as they are able to create highly resolved three-dimensional images (providing both in-plane as well as height features) without the need for an electron source. Again, sample preparation is critical; it is particularly important for the nanoparticles under analysis to have a greater affinity to the flat substrate surface than the sensor probe tip. If there is weak adhesion between the nanoparticle and substrate, then the image acquired either shows a reduced resolution or contains "artifacts" (e.g., streaking) (Bonnell, 2001).

10.2.3 NONIMAGING TECHNIQUES

This is a group of techniques that are "nonimaging" in nature, and most have much lower detection sensitivity than the aforementioned "imaging"-based techniques.

An essential property of interest is the size of the nanoparticles under analysis, as the association between particle size and toxicity is well founded (Oberdorster et al., 2005). The "state of aggregation" is a difficult parameter to quantify but is potentially significant for toxicological evaluation. This parameter can be used to describe the degree to which particles are agglomerated (loosely held together in groups or clusters by attractive interparticle forces, the most fundamental being Van der Waals forces); particle agglomerate size may play a crucial role in the uptake of such particles inside the body by macrophages (Rudt and Muller, 1992).

In addition to size, surface area is becoming of increasing importance in relation to particle toxicity. In industry, surface area characterization of nanoparticles is needed as they can be correlated to surface-related phenomena such as catalyst activity, and electrostatic properties that can influence the processing and behavior of nanoparticles. Other surface properties, apart from simply surface area, are emerging as important from a toxicological point of view, as they may provide mechanistic details in the uptake, persistence, and biological activity of HARN inside living cells. Such properties include surface charge (zeta-potential measurements) and surface chemistry (Gill et al., 2007). The techniques now overviewed can be divided into two groups, based on whether they characterize chemical or physical properties.

10.2.3.1 Chemical Property Information

The molecular composition and structure of the surface of nanoparticles will ultimately define its chemistry. Techniques used for chemical characterization are mainly spectroscopic in nature. Spectroscopic techniques measure the interaction between a probe and matter, yielding a "spectrum," that is, a response plot as a function of wavelength. Such techniques can be very useful in identifying the "chemical class" of various components of an analyte under study.

10.2.3.1.1 Vibrational Spectroscopy

This is a tool used to probe the "vibrational states" of a molecule and, hence, for the determination of molecular structure. This can be achieved in several ways. In infrared (IR) spectroscopy, the molecule will be exposed to a frequency range of IR light and ultimately the measure of wavelength-intensity of IR absorption by the molecule; for a vibrational motion to be IR active, the dipole moment of the molecule must change. Raman spectroscopy is a measurement of wavelength-intensity of inelastically scattered light from molecules; for a molecule to be Raman active, there must be a change in the polarizability of the molecule. A Raman spectrum of a molecule gives complementary information to its corresponding IR spectrum and both techniques are powerful tools for nondestructive characterization of nanoparticles. Both are suited for routine analysis; they can be operated by technicians, have a relatively quick analysis time, and have well-established frequency standards (Laserna, 1996).

10.2.3.1.2 X-Ray Photoelectron Spectroscopy

This is a surface analysis technique, requiring ultrahigh vacuum conditions, that bombards the sample with monoenergetic soft x-rays (~1 to 2 keV), causing electrons to be ejected. The number and kinetic energy of the ejected "photoelectrons" (from a depth of 1 to 10 nm of the material being analyzed) are then simultaneously measured. Spectra are capable of showing elemental composition, empirical formula, chemical state, and electronic state of the elements. XPS has the ability to analyze nonconducting materials with minimum charging effects (unlike techniques such as SEM), with excellent inter-element resolution. Some of disadvantages include poor lateral resolution, a relatively weak signal, and possible nonuniqueness of its chemical shift information (Watts and Wolstenholme, 2003). XPS and secondary ion mass spectroscopy (Lee et al., 2007), in particular, have been extensively used for characterizing nanoparticles. These are very powerful surface analysis techniques and have been used to give information-rich spectra detailing the "chemical state" at the surface; XPS, in particular, has the advantage of being quite oxidation specific but this is dependent on the particular element which is analyzed. XPS has been used to probe information relating to structural modification due to chemical interaction with organic compounds or gases adsorption and sidewall functionalization of HARN. A recent review by Powers and coworkers indicated that the technique is applicable to correlating biomaterial surface properties to physiological endpoints (Powers et al., 2007).

10.2.3.2 Physical Property Information

10.2.3.2.1 Surface Area

The BET technique is a traditional method for the measurement of surface area and other characteristics such as pore size and pore distribution of nanoparticles. The technique is based on the addition of a known volume of gas (the adsorbate) and the subsequent gas adsorption onto the solid material (at cryogenic temperatures), resulting in a direct relationship between the pressure and the volume of gas in the sample vessel. By measuring the reduced pressure due to adsorption, the ideal gas law can then be used to determine the volume of gas adsorbed by the sample and, subsequently, the surface area of the sample, which is reported as the specific surface area (i.e., surface area per unit mass, usually m^2/g) (Lowell et al., 2004).

10.2.3.2.2 Zeta-Potential Measurements

This is a method to probe "surface charge" information of nanoparticles in a liquid suspension. Theoretically, zeta potential is the electric potential in the interfacial double layer, that is, layer at the location of the "slipping plane" versus a point in the bulk fluid away from the interface. It has been recognized that the zeta potential is a very good index of the magnitude of the interaction between nanoparticles, and the measurements of zeta potential are commonly used to assess the stability of colloidal systems. One way to determine zeta-potential measurement is to obtain the electrophoretic mobility of the particle and this can be done through the combination of laser Doppler velocimetry and phase analysis light scattering of the sample, under the influence of an applied electric field (Hunter, 1981).

10.2.3.2.3 Photon Correlation Spectroscopy

A common method used to probe size distribution of nanoparticles in the submicron range is based on photon correlation spectroscopy (PCS) (more commonly called dynamic light scattering). This nondestructive method determines particle size by measuring the rate of fluctuations in laser light intensity scattered by particles as they diffuse through a fluid. However, the size information obtained from PCS will be the average diffusion coefficient of the particles and the size is correlated to the equivalent sphere diameter. Hence, PCS is more likely to be successful in measuring size if the nanoparticle under investigation is close to being spherical. In the case of HARN, size information will be limited and the information obtained will be neither the length nor the width of the particle. Nonetheless, it is possible to probe the state of dispersion, that is, evidence of agglomeration, taking place in the suspension (relative to the nanoparticles if in the dispersed state, that is, as close to the "primary" particle size distribution as possible). For example, it has been used by Lee et al. (2005) for measuring polydispersity and the stability of single-walled CNTs.

10.2.3.2.4   Scanning Mobility Particle Sizer

If the aerosolized form is to be analyzed, then a different technique is needed. For example, the scanning or fast mobility particle sizers are able to present a particle size distribution of the aerosolized material, typically for particles between 5 and 500 nm. In the case of the scanning mobility particle sizer, particles are classified with an electrostatic classifier, whose voltage is "scanned" and used to deliver singly charged, monodisperse aerosols of known size and composition to a condensation particle counter or electrometer, yielding particle size distribution information (Berne and Pecora, 2000). Again, the interpretation of the results for nonspherical/agglomerated nanoparticles is not straightforward (Van Gulijk et al., 2004). Limited research has been done in relation to the characterization of HARN using this technique and again its suitability is questionable as the nanoparticles under investigation move away from the idealized spherical model.

10.3   TOXICOLOGY OF HARN

There has been considerable concern about the potential health effects that could be caused by occupational or environmental exposure to HARN. This concern and specifically the link between carbon nanotubes (CNTs) and asbestos has been raised as early as 1998 in the journal *Science* under the title "Nanotubes: The next asbestos?" (Service, 1998). This article was a mere 7 years after the seminal papers by Iijima (1991; Iijima and Ichihashi, 1993), which reignited the interest in CNT and paved the way for other HARN. The basis for this comparison of CNT and asbestos is the morphological similarity that exists between the two forms of material and the general suspicion of all new industrially used fibers. This is due to the pandemic of disease caused by asbestos exposure and the ensuing considerable legal and financial fallout from the use of asbestos that has left both the public and governments suspicious of new industrial fibers. It is worth considering that this suspicion has not just been leveled at HARN, but has also been considered for other industrially relevant fibers. For example, the organic fiber para-aramid, which is more commonly known by its trade name, Kevlar (Donaldson, 2009), has also been under considerable scrutiny for toxicological effects due to its morphological similarity to asbestos.

10.3.1   PRINCIPLES

In drawing a comparison between HARN and asbestos (or indeed any other fiber such as refractory ceramic fibers [RCF] or man-made vitreous fiber [MMVF]), the focus is on *fiber* toxicity, but a potential *particle* effect for fibrous materials is still possible as many types of HARN, such as CNTs, can be found in fibrous (e.g., as singlet fibers or agglomerated "nanoropes") and non-fibrous forms (e.g., very short rodlike particles, tightly curled spherical bundles, or agglomerates) as shown in Figure 10.2.

Currently, information on the toxicity of HARN exists mainly for CNTs and reviews of the paradigm which forms the basis of fiber toxicity as it relates to CNT has been conducted recently (Donaldson et al., 2006, 2010, 2011).

Using the WHO guidelines for counting asbestos fibers (WHO, 1997), a particle should be considered a fiber if it has a length of >5 μm, a diameter of <3 μm, and an aspect ratio greater than 3:1.

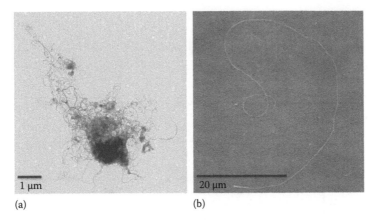

(a)   (b)

FIGURE 10.2 Morphological structure of MWCNT as shown by TEM (a) and SEM (b).

Aspect ratio is defined as the relationship between length and diameter of a particle or specifically the ratio of the longest dimension to the shortest dimension. An aspect ratio of 1:1 would equate to a spherical particle. The increased disparity between length and diameter results in an elongated or fibrous shape. Aspect ratio, therefore, defines a fiber by describing this relationship. As mentioned, the WHO definition of a fiber requires it to have an aspect ratio of greater than 3:1 and a length greater than 5 μm. However, nanoparticles with a high aspect ratio are not necessarily *fibers*. For example, a nanoparticle with a diameter of 40 nm and length of 3 μm has a very large aspect ratio (75:1) but would not be classified as a *fiber* according to the WHO definition. To identify a nanoparticle as having a high aspect ratio is, therefore, not solely sufficient to classify it as a fiber and if it does not meet *all* the WHO criteria (minimum diameter, length, and aspect ratio) then it would not be counted as a respirable fiber under the WHO method.

Another point of comparison that should be borne in mind is the level of contaminating metals found both with asbestos and HARN. The mineralogical nature of asbestos and the nature of the environment in which they form mean that asbestos fibers are contaminated with various metals; the specific elemental constituents of asbestos fibers can act as a fingerprint of their source. Amphiboles such as crocidolite are known to be contaminated by iron in the form of ferric oxide (Fe_2O_3) (Virta and Geological Survey (U.S.), 2002), which has been suggested as a contributing factor to its toxicity and carcinogenicity (Broaddus et al., 1997; Unfried et al., 2002). Depending on the synthesis method, substrates, and catalysts used, CNTs can also be contaminated with metals such as iron (Fe), nickel (Ni), or cobalt (Co). These impurities are often removed during processing, but amounts can still remain in the finished product and may contribute to the observed toxicity, as has been shown *in vitro* with single-walled CNTs (Kagan et al., 2006). Other forms of HARN may also be contaminated with reactive metals or may indeed be composed of such metals (e.g., nickel nanowires), which can also present "conventional" elemental/molecular toxicity.

10.3.2 FIBER PATHOGENICITY PARADIGM

The basis of fiber toxicity has been the subject of scrutiny over the past 30 years and while asbestos exposure leading to disease has provided the driving force toward understanding the pathogenicity of fibers, it is this difference in pathogenicity of different types of fibrous materials that has facilitated the understanding and development of structure–activity relationships. Some fibers, such as asbestos, have been shown to be carcinogenic (IARC group 1 [IARC, 1987]); others such as RCFs are considered only possible carcinogens (IARC group 2b [WHO IARC, 2002]), while many others such as soluble glass fibers are considered nonpathogenic.

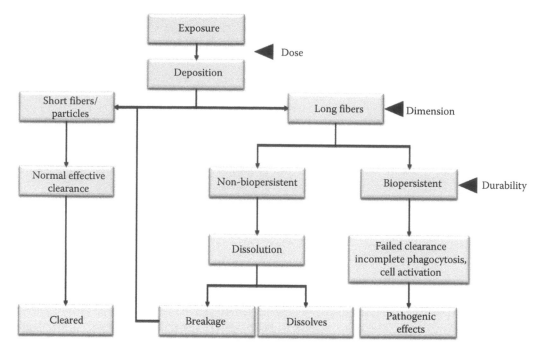

FIGURE 10.3 Role of the 3Ds in fiber pathogenicity. (Adapted from Donaldson, K. et al., *Toxicol. Sci.*, 92(1), 5, 2006.)

The fiber pathogenicity paradigm (FPP) depends on three critical features that are required for a fibrous particle to present a fiber-type health hazard. These attributes consist of, most crucially, exposure to the particle in question (*dose*), the role of aspect ratio and length (*dimension*), and the persistence of a particle in the biological environment and its resistance to breakage and dissolution (*durability*) (Donaldson and Tran, 2004). These components of dose, dimension, and durability, or the 3 Ds provide the cornerstone of the FPP as summarized in Figure 10.3.

10.3.2.1 Dose

Exposure to an agent—particle, chemical, or biological—is required to elicit a response. As the absence of exposure leads to an absence of response, exposure or dose is arguably the most important of the 3Ds when defining the health risk of fibers. It is necessary to first consider how exposure may lead to dose. External exposure is simply the presence of particles or some other form of exogenous material found in the environment external to the body (i.e., a *source*), however the presence of a substance in this external environment does not necessarily mean that it shall gain entry into the body. For a substance present as a source to gain entry to the body, a *pathway* is required, which can lead to the presentation of an internal dose that may go on to cause any effect. If a particle or HARN is respirable, it may gain access to the lungs leading to an internal dose. This does not necessarily mean that there will be a subsequent biological effect, as the dose may still be too low or the substance may not be toxic and/or cleared rapidly (removal of dose), or deposit in a region not leading to toxicity.

In the case of studies relating the incidence of disease, the establishment of dose is usually based on quantitative or semiquantitative estimates of external exposure from occupational/industrial hygiene assessments using air monitoring. This only allows an accurate measurement of the *external* exposure and generally only gives a proxy measure of the *internal* dose, which is what would lead to a pathological effect. While actual internal measurements of dose, zones of deposition, and clearance rates are difficult in humans, mathematical models based on lung structure and airflow have aided significantly in bridging the gap between exposure and internal dose (Tran et al., 1999). This is increasingly important when considering the different aerodynamic characteristics and deposition patterns dictating the dosimetry of nano-sized materials in the lung.

10.3.2.1.1 Lung Deposition

Based on the understanding of the role of particle size in dictating the respirability of a particle, it seems unlikely that a particle of >5 μm could be respirable, let alone a fiber reaching upward of 50 μm. However, fibers that are in excess of the 3 μm diameter cutoff for respirability *do* reach the periphery of the lung due to a peculiarity of fiber aerodynamics. Aerodynamic diameter (D_{ae}) is proportional to the fiber diameter and not the length. In a laminar flow of air, a fiber aligns itself axially due to airflow across the fiber surface, allowing it to travel along, aligned with the airways (Morgan, 1995), presenting a small D_{ae}. Because of this, small D_{ae} fibers many times longer than the cutoff diameter for a spherical particle can deposit in various zones of the respiratory system, including the alveolar region. Using the theory of Cox (Cox, 1970) to establish the aerodynamic diameter of a cylindrical fiber, Jones calculated that the ratio of aerodynamic diameter to geometric diameter is approximately 2.5–3 over a wide range of aspect ratios based on a fiber settling perpendicular to its axis (Jones, 1993). More simply put, the D_{ae} of a fiber is 2.5–3 times the actual diameter largely irrespective of length.

As the aerodynamic diameter of a fiber is approximately three times the actual diameter, it is difficult to know how this may relate to the deposition of fibrous nanoparticles in the respiratory tract. For particles <0.5 μm in diameter, the appropriate metric relating to deposition is the diffusion equivalent diameter rather than aerodynamic diameter. However, nanofibers can aggregate to form nanoropes (Donaldson et al., 2006), so it remains to be elucidated whether aerodynamic diameter or diffusion equivalent diameter dictates the particle behavior in the airflow of the lung.

10.3.2.1.2 Measuring Dose

The metric by which dose is measured is also an important concept. Ideally, we would measure the biologically effective dose (BED), which is the quantity that drives the adverse (biological) effect. The BED refers to the fraction of a total dose that may cause an adverse effect. In particle toxicology, the difference between the *total* dose and the *effective* dose can be stark. In the case of asbestos fibers (and a hypothesis for HARN), this would be the proportion of the total dose which is both biopersistent and sufficiently long to cause problems with clearance mechanisms. While a measure of the BED would be the preferred metric for measuring dose, most often mass is used as a surrogate due to the relative ease of its measurement. Other metrics have been put forward as more closely approaching the BED. In the case of nano-sized particles that may have a low mass, surface area is often seen as a driving factor and is suggested to be the BED and so the preferred metric (Brown et al., 2001; Duffin et al., 2002; Maynard and Maynard, 2002). This requires further study and method development but, needless to say, the low-density, high surface area, and high fiber number of nanofibers means that as a dose metric, mass is actually likely to be the least useful.

10.3.2.2 Dimension

The issue of dimension relates not only to the ability of a fiber to penetrate the distal lung and deposit in the alveolar region, it also dictates the ease with which alveolar macrophages can clear material from this region of the lung. The origins of the importance of length again lie in the comparison of pathogenic and nonpathogenic fibers.

In 1981, Stanton published a seminal paper in the world of fiber toxicology dealing with the relationship between particle dimension and the development of mesothelioma, a rare tumor of the chest wall, which is almost always associated with exposure to asbestos (Stanton et al., 1981). A range of 72 experiments were conducted using a wide variety of respirable and durable minerals ranging in size and chemical and structural attributes, which were introduced to the chest (pleural) cavity of rats. Following this, they observed that the formation of tumors correlated most closely with fiber attributes based on length and diameter. Specifically, they found that fibers that measured <0.25 μm in diameter and >8 μm in length correlated well with the formation of malignant mesenchymal neoplasms while short fibers did not cause mesothelioma. It is worth noting that ~8 μm is the maximal size of pore-like structures called stomata on the parietal pleura, the main route of removal from the plural space. The use of several forms of mineral fibers of differing physicochemical properties also enabled Stanton to conclude that carcinogenicity of the fibers was dependent on dimension and durability rather than on physicochemical properties. The importance of this study and its relevance to future fiber toxicology and the setting of fiber counting standards such as that used by the WHO and Occupational Safety & Health Administration (OSHA), cannot be overstated. Stanton, however, has not been the only one to demonstrate the importance of fiber length, shown in subsequent years in various experimental systems (Davis et al., 1986; Adamson and Bowden, 1987a,b; Davis and Jones, 1988; Donaldson et al., 1989; Goodglick and Kane, 1990).

The basis of the Stanton hypothesis most likely relates to the role fiber length plays in clearance from its site of deposition. The airways are constructed in such a way that it prevents larger particles from reaching the lung altogether or reaching beyond the larger, conducting airways. Long fibers or particles depositing in these larger airways are removed rapidly from the lung via the action of constantly beating hair-like cilia lining these airways (the mucociliary escalator), which propel particles and mucus out of the lung. Small particles that can negotiate the airways and deposit in the proximal alveolar region are phagocytosed with ease by alveolar macrophages, either persisting within the macrophage if durable or moved up onto the mucociliary escalator and ejected from the lung. Fibers depositing in these non-ciliated regions, due to the aforementioned unusual aerodynamic behavior, must be dealt with by alveolar macrophages. The issue of impaired clearance occurs when a fiber is longer than the maximal length that a macrophage can comfortably enclose (~15 to 20 μm). In this situation, the macrophage will try unsuccessfully to phagocytose the fiber and become "frustrated" as shown in Figure 10.4.

The action of *frustrated phagocytosis* leads to the release of various toxic components, which are crucial to the normal microbiocidal activity of the cells as part of the innate immune defense, which can damage surrounding cells and cause inflammation. Inflammation is an important process and the body's inflammatory cells are critical in maintaining a healthy biological environment. However, these same protective mechanisms can become damaging, possibly leading to further damage and disease, when inappropriately activated, uncontrolled, or occurring repeatedly.

If the deposited fibers cannot be cleared via the normal macrophage-mediated means, they may persist in the alveoli or may translocate out of the lung into the cavity that surrounds the lung (the pleural cavity), where they may be retained and drive a pathogenic response.

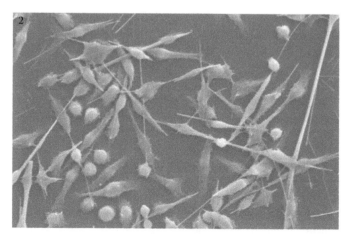

FIGURE 10.4 Frustrated phagocytosis. SEM image of mouse alveolar macrophages trying to phagocytose fibrous material and forming the typical elongated morphology of a frustrated macrophage. (Image copyright SAFENANO, Edinburgh, U.K.)

10.3.2.3 Durability

The durability of a particle is its ability to resist damage or modification that would otherwise alter its structure. A particle's biodurability is defined as its ability to resist damage or modification that would substantially alter its structure in a biological environment such as in regions of the lungs. A material's biopersistence relates to the ability of a particle to endure the biological environment once it has deposited there and persist.

The importance of biopersistence in the FPP arises from the observation that different natural and man-made fibers had very different lung retention times, and those which were most durable had the highest fibrogenic and carcinogenic potential (Bignon et al., 1994). Within the lung, less durable fibers may split longitudinally as seen with fibrils of chrysotile asbestos, or break transversely as in the case of glass fibers forming short fibers, and be more easily cleared by the lung's natural defenses. They may also dissolve in the milieu of the lung due to leaching of certain structural components or due to the acidic environment found within cellular compartments (lysosomes) inside of specialist immune cells such as macrophages.

The influence of characteristics such as susceptibility to dissolution and breakage was shown by Searl and colleagues by comparing the length fraction of biopersistent amosite asbestos against non-biopersistent MMVF-10 between 3 days and 12 months post-inhalation in rats. In the case of the MMVF-10 sample, the number of short fibers increased after 12 months, indicating breakage of the long fibers adding to the population of short fibers (Searl et al., 1999). Even within the asbestos family of minerals there are differences in biopersistence. Chrysotile asbestos has been shown to be less biopersistent than the amphibole forms of asbestos such as crocidolite and tremolite. This has been attributed to the layer of a magnesium hydroxide or brucite between the silicate sheets of chrysotile, which is more prone to dissolution, causing the layers to unravel and break. As a concept, exposure to a fiber with a dimension that allows penetration of the lung but does not allow clearance by macrophages leaves only one route of removal (resulting in a reduction of dose), namely, dissolution or breakage. Therefore, exposure to a biopersistent fiber that will not dissolve or break means that it shall persist in the lung environment where it may trigger pathological effects.

In summary, the FPP, could perhaps most simply be viewed in terms of a "hazard triangle" with each of the criteria of dose, dimension, and durability making up a wall of the triangle (Figure 10.5). Together, these criteria form a potentially pathogenic fiber, but removal of any one results in a removal of a fiber toxicity hazard (although not necessarily the particle toxicity hazard). For example, if we remove the "dimension" criteria by making a fiber very short, it would be more easily cleared via normal clearance mechanisms and does not cause frustrated phagocytosis. Removal of "durability" means that the long fibers do not persist and either break or dissolve, leading to removal of fibers and reduction of dose.

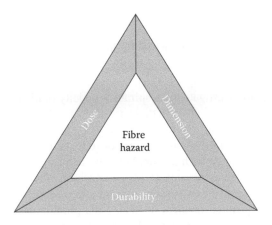

FIGURE 10.5 The fibre pathogenicity triangle.

10.3.3 Evidence Base for Harn

When considering the evidence for adverse effects of exposure to a material such as nanomaterials, various resources are available within the peer-reviewed literature. Ideally, the most robust data would be derived from large numbers of human subjects exposed within well-controlled relevant exposure scenarios. However, this is very rarely ethically acceptable or feasible, and alternatives or surrogates for this type of data have to be used, with all the caveats and limitations associated with information from animals or models. These surrogates may include epidemiological information if available but due to the early stage of much of the nanotechnology industry, there is currently no epidemiological information available for HARN. Moreover, as epidemiological evidence is most often retrospective (except, e.g., in the case of a prospective cohort study), its strength is in identifying causal links and trends but is less than ideal from the point of prevention.

When trying to understand the potential health effects of a new substance, a great deal of importance is placed upon animal data as they still provide insight into the complex interactions which culminate in toxic responses. A great deal of *in vivo* toxicology focuses on rodent species with both rats and mice being the most commonly used, and typically the basis for many international accepted test guidelines such as the OECD guidelines for the testing of chemicals. These species obviously have substantial differences to humans both in the mechanisms of particle deposition based primarily on differences in breathing patterns and airway morphometry (Hofmann et al., 2000), but also in their relative sensitivities to particle-induced effects (Bermudez et al., 2004). These differences need to be taken into account when extrapolating from the surrogate back to a human equivalent dose (HED) and numerous modeling efforts and approaches have been published to facilitate this (Anjilvel and Asgharian, 1995; Hofmann et al., 2000). When considering lung exposure, the gold standard is inhalation exposure to an aerosol that most faithfully replicates real exposure, although there are still likely to be considerable differences in the aerosol parameters (e.g., the fluctuating nature of particle concentration and size distribution) between a workplace and experimental exposure system. In addition, there is much discussion among researchers regarding the ability to generate some forms of airborne HARN, such as CNTs, in an industrial or consumer setting. This scepticism is driven by the observation that CNTs exhibit a high propensity to aggregate. In order to address this question, Maynard et al. (2004) investigated unrefined single-walled CNTs agitated in a controlled laboratory setting, and found that such agitation resulted in fine particle release into air. A subsequent study of four production facilities found airborne concentrations in the vicinity of single-walled CNT production equipment to be relatively low ($<53\,\mu g/m^3$). While this is a low concentration when compared to the occupational exposure limits for other nuisance particles such as TiO_2, it is relatively high compared to ambient air pollution (associated with morbidity and mortality). Furthermore, in comparison to fibers, exposure limits are set according to the number of fibers per cubic centimeter of air. The relationship between ug/m^3 and number of CNTs per cubic centimeter is currently unknown, but is likely to vary between different CNT types and sources.

It is important to note that it is challenging to produce inhalable or respirable aerosols of CNTs because of the tendency of these materials to form aggregates, but efforts are being made to develop more efficient methods for aerosolizing CNTs (Baron et al., 2008). In relation to the use of gold standard inhalation studies to investigate the respiratory toxicity of high aspect ratio nanoparticles, there have now been several studies, but again these are all focused on CNTs and the results of these studies are summarized in Table 10.1.

From the results gained from inhalation exposure of rodents to CNTs, it appears that low-level exposure ($<0.5\,mg/m^3$) is associated with minimal effects such as transient lung inflammation and the formation of small granulomas at sites of fiber deposition. The formation of granulomas is typical of a foreign-body response and is seen in many circumstances where a foreign material cannot be removed (e.g., implanted medical device). Such granulomas are typically formed of a collection of inflammatory cells such as macrophages and may even contain multinucleated giant

TABLE 10.1

Summary of Respiratory Effects if Inhalation Exposure to Carbon Nanotubes

		Sample Information			Test Conditions					
HARN	Morphology	Diameter[a] (nm)	Length[a] (µm)	Mass Median Aerodynamic Diameter (µm)	Species	Exposure Period	Exposure Conc. (mg/m³)	Observation Points	Result	Reference
MWCNT	Agglomerates and individual tubes	—	0.5–40	183	M	6h	30	1 day, 2 week, 6 week, 14 week	Subpleural fibrosis	Ryman-Rasmussen et al. (2009a)
	Agglomerates and individual tubes	30–50	0.3–50	0.71	M	6h	100	1 day, 14 day	Acute inflammation (resolving). Airway fibrosis in allergically sensitized mice only	Ryman-Rasmussen et al. (2009b)
	Spherical agglomerates	10–16	—	1.9, 2.0	R	6h	11, 241	1 week,4 week, 13 week	Acute inflammatory response with some association to the presence of contaminating catalyst (cobalt), which regressed with time	Ellinger-Ziegelbauer et al. (2009)
	Spherical agglomerates	~10	0.2–0.3	NA, -,3.05, 2.74,3.42	R	90 days (6h/day, 5 day/week, 13 week)	0.1, 0.4, 1.5, 6.0	0 week, 4 week, 13 week, 26 week	No detectable systemic toxicity. Evidence of particle overload phenomena at 1.5 mg/m³+ with resultant inflammation and fibrosis, borderline changes at 0.4 mg/m³	Pauluhn (2010a)

(continued)

TABLE 10.1 (continued)
Summary of Respiratory Effects if Inhalation Exposure to Carbon Nanotubes

		Sample Information				Test Conditions				
	Morphology	Diameter[a] (nm)	Length[a] (μm)	Mass Median Aerodynamic Diameter (μm)	Species	Exposure Period	Exposure Conc. (mg/m³)	Observation Points	Result	Reference
HARN	Spherical agglomerates	5–15	0.1–10	2.0, 1.5, 0.7	R	90 days (6h/day, 5 day/week, 13 week)	0.1,0.5,2.5	0	Increased lung weights, inflammation, granuloma formation at 0.5 and 2.5 mg/m³ and granuloma formation at 0.1 mg/m³	Ma-Hock et al. (2009)
	Agglomerates and individual tubes	10–20	5–15	0.7–1 (for 0.3 and 1.0 mg/m³), 1.8	M	7 day, 14 day (6h/day)	0.3, 1.0, 5	0	Evidence of particle-laden alveolar macrophages yet no inflammation or tissue damage observed. Evidence of systemic immunosuppression after 14 days	Mitchell et al. (2007)
	Agglomerates and individual tubes (>70% individual fibers)	63	1.1	—	R	4 week (6h/day, 5 day/week)	0.37	3 day, 4 week, 12 week	Minimal transient inflammation and no granulomatous lesions in the lung	Morimoto et al. (2011)
SWCNT	Small fiber-like agglomerates	0.8–1.2	0.1–1	4.2 (count mode aerodynamic diameter ~240 nm)	M	4 days (5h/day)	5.53	1 day, 1 week, 4 week	Inflammation and fibrosis	Shvedova et al. (2008)

[a] Primary particle only.
NA, Not applicable

cells (fused macrophages), which again are typical of a foreign-body response (Poland et al., 2008). These are held together with extracellular matrix proteins such as collagen and together form a persistent structure, walling off the foreign object.

Higher levels of exposure lead to a greater acute inflammatory response, fibrosis, and further granuloma formation at sites of particle deposition. These dose-response-related effects have led some authors to suggest a no or lowest observable effect level of $0.1\,mg/m^3$ for some CNTs (Ma-Hock et al., 2009; Pauluhn, 2010a), which has formed the basis of a proposed OEL of $0.05\,mg/m^3$ (time-weighted average) for a specific Baytubes® CNT sample (Pauluhn, 2010b). Despite several inhalation studies being undertaken, no study has yet addressed the role of length, in the sense that none has compared long (greater than $\sim20\,\mu m$) to short ($<10\,\mu m$) nanotubes.

There is a wide variability in the physicochemical characteristics between samples of CNTs, and the understanding of how such properties influence the toxicity observed is still emerging. Hence, it is premature to draw conclusions on the toxicity of all CNTs from a single or even multiple studies. This creates a problem as there are numerous different CNTs, each with different morphologies, sizes, lengths, and surface characteristics, and inhalation studies are costly and require specialist facilities in order to be carried out. Therefore, the more common exposure method of lung (intratracheal) instillation or (pharyngeal) aspiration is used to deliver an immediate dose of particles to the lung in a liquid bolus (Warheit et al., 2004; Lam et al., 2004). However, as the delivery of dose in this way does not reflect true exposure, the results gained are subject to further caveats and uncertainty. In addition, as the particles are in solution, this in itself may further modify the dose by causing agglomeration of particles resulting in the instillation of non-respirable (and hence irrelevant) particles into the lung. This extra uncertainty is reflected in how such data may be used. Currently, data obtained by instillation/aspiration methods are not considered suitable for extrapolation to a HED for the purposes of deriving a human exposure limit (ECHA, 2008). Instead, these data are useful for understanding the nature of the hazard and the resultant effects of particles in the lung.

There have now been two studies comparing exposure of the lung by CNTs using these two methods. In the first study, Li et al. found that exposure by instillation could lead to the presence of large, non-respirable aggregates, resulting in typical foreign-body responses, while inhalation led to a lower dose of smaller material in the alveolar region due to the elutriating effect of the lung. Thus, the responses gained by each method were different possibly due to aggregation state and lung distribution (Li et al., 2007). It was noted that exposure by inhalation generated an enhanced cellular inflammation, protein, and LDH release over instillation (two- to fourfold increase) with large increases in fibrosis also seen (Shvedova et al., 2008).

The next source of information on toxicity comes from *in vitro* assays which typically use immortalized cells grown in an artificial medium supplemented with various proteins and growth factors to stimulate their continued growth and division. The benefits of such a system are that it is very controllable with various aspects of the experiment manipulated with relative ease both quickly and cheaply with minimal ethical issues. When considering the caveats associated with *in vitro* toxicology and its relevance to human exposure, one can see many issues reflected in its use primarily as a highly useful screening tool and a source of mechanistic information, but it is currently not suitable for direct extrapolation to deriving a human exposure limit and is of limited use in risk assessment.

When considering HARN toxicity, the vast majority of research is directed toward CNT toxicity, which reflects the current higher usage of CNT as an industrial product as compared to other forms of HARN. There are now numerous exposure studies to CNTs using a variety of routes, including dermal (Murray et al., 2009), intravenous (Singh et al., 2006), and oral (Folkmann et al., 2009), which reflect the potential exposure routes and medical applications of CNT.

In relation to HARN toxicity, there may also be variability in the durability of different materials, which may result in differing toxicity. It is conceivable that a soluble nanowire, such as a silver nanowire, will dissolve rapidly in the complex biological environment of the lung leading to a reduction in dose. In contrast, CNTs are hypothesized to be very durable due to their structure. There have only been a few studies which have directly addressed the issue of CNT durability, although several studies have reported CNTs persisting within tissues for extended periods (Elgrabli et al., 2008; Pauluhn, 2010a). In the study by Osmond-McLeod et al. (2011), the durability of several CNT samples consisting of both MWCNT and SWCNT was evaluated using artificial system based on the use of a ringer's solution (Gambles) developed to mimic the low pH environment of being digested by a cell. The authors found that to the most part, the CNTs were indeed durable over a period of 6 months in solution with little loss of pathogenic activity when tested. They did note that one sample did lose a significant portion of its mass, which was reflected in a significant reduction in its pathogenicity, which aids the contention that the FPP is relevant and applicable to HARN where they meet its criteria.

10.4 CONCLUSIONS

In summary, the responsible technological development and application of HARN needs to be accompanied with adequate and relevant characterization and safety assessment.

The complete chemical characterization of nanoparticles often requires the concurrent application of several techniques. In addition to those highlighted, others, including inductively coupled plasma (ICP) mass spectrometry, nuclear magnetic resonance (NMR), UV–vis, and fluorescence spectroscopy, are widely used to characterize the composition of nanoparticles (Powers et al., 2007).

Techniques such as Raman and IR spectroscopy have been popular choices for the characterization of HARN to determine impurities, surface functionalization/chemistry, and catalytic properties. Raman spectroscopy, in particular, has been widely used for the characterization of CNTs to obtain a variety of information to include structural, chirality, surface modification/functionalization, and in the determination of their vibration and electronic energies.

When considering the potential toxicity of HARN, it is crucially important to consider what form the dose is taking. It should be stressed that the adequate physicochemical characterization of a sample is of paramount importance. Failure to do this can result in inappropriate comparison and potentially an under- or overestimation of the relative risks. Likewise, it is nonsensical to say that all forms of CNT exhibit fiber pathogenicity akin to asbestos, as not all CNTs are fibers and even those which are would be subject to the same criteria which dictate asbestos toxicity. Other important considerations are the similarities and differences in the behavior of particles in the real environment compared to those in a test system during hazard assessment (Oller and Oberdörster, 2010).

HARN are available in many different forms based on a wide range of attributes, including chemical composition, diameter, length, number of walls (for CNT), surface area, level of contaminants, without even approaching the various forms of surface functionalizations that are possible. As such, it is impossible currently to fully assess each form of HARN (or even that of a subset such as CNT) for adverse health effects using *in vivo* models. To identify the forms that may pose a hazard and focus attention on these, it is important to identify valid structure–activity relationships and *in vitro* techniques, which may help predict toxicity. The FPP is one such structure–activity relationship, which, if used correctly to access truly fibrous material with appropriate controls, could help identify those HARN that may pose a health risk.

REFERENCES

Adamson, I.Y. and Bowden, D.H. 1987a. Response of mouse lung to crocidolite asbestos. 1. Minimal fibrotic reaction to short fibres. *J. Pathol.*, 152(2), 99–107.
Adamson, I.Y. and Bowden, D.H. 1987b. Response of mouse lung to crocidolite asbestos. 2. Pulmonary fibrosis after long fibres. *J. Pathol.*, 152(2), 109–117.
Allen, T. 2004a. *Particle Size Measurement, Vol 1: Powder Sampling and Particle Size Measurement*, 5th edn. London, U.K.: Chapman and Hall.
Allen, T. 2004b. *Particle Size Measurement, Vol 2: Surface Area and Pore Size Determination*, 5th edn. London, U.K.: Chapman and Hall.

Anjilvel, S. and Asgharian, B. 1995. A multiple-path model of particle deposition in the rat lung. *Fundam. Appl. Toxicol.*, 28(1):41–50.

Baron, P.A., Deye, G.J., Chen, B.T., Schwegler-Berry, D.E., Shvedova, A.A. and Castranova, V. 2008. Aerosolization of single-walled carbon nanotubes for an inhalation study. *Inhal. Toxicol.* 20(8):751–760.

Berne, B.J. and Pecora, R. 2000. *Dynamic Light Scattering: With Applications to Chemistry, Biology, and Physics*, Unabridged edition. Mineola, New York: Dover Publications.

Bermudez, E., Mangum, J.B., Wong, B.A., Asgharian, B., Hext, P.M., Warheit, D.B., and Everitt, J.I. 2004, Pulmonary responses of mice, rats, and hamsters to subchronic inhalation of ultrafine titanium dioxide particles. *Toxicol. Sci.*, 77(2), 347–357.

Bignon, J., Saracci, R., and Touray, J.C. 1994. Introduction: INSERM-IARC-CNRS workshop on biopersistence of respirable synthetic fibers and minerals. *Environ. Health Perspect.*, 102(Suppl 5), 3–5.

Bonnell, D.E. 2001. *Scanning Probe Microscopy and Spectroscopy: Theory, Techniques, and Applications*, 2nd edn. New York: Wiley VCH.

Broaddus, V.C., Yang, L., Scavo, L.M., Ernst, J.D., and Boylan, A.M. 1997. Crocidolite asbestos induces apoptosis of pleural mesothelial cells: Role of reactive oxygen species and poly(ADP-ribosyl) polymerase. *Environ. Health Perspect.*, 105(Suppl 5), 1147–1152.

Brown, D.M., Wilson, M.R., MacNee, W., Stone, V., and Donaldson, K. 2001, Size-dependent proinflammatory effects of ultrafine polystyrene particles: A role for surface area and oxidative stress in the enhanced activity of ultrafines. *Toxicol. Appl. Pharmacol.*, 175(3), 191–199.

Cox, R.G. 1970. The motion of long slender bodies in a viscous fluid Part 1. General theory. *J. Fluid Mech. Digital Arch.*, 44(04), 791–810.

Davis, J.M., Addison, J., Bolton, R.E., Donaldson, K., Jones, A.D., and Smith, T. 1986. The pathogenicity of long versus short fibre samples of amosite asbestos administered to rats by inhalation and intraperitoneal injection. *Br. J. Exp. Pathol.*, 67(3), 415–430.

Davis, J.M. and Jones, A.D. 1988. Comparisons of the pathogenicity of long and short fibres of chrysotile asbestos in rats. *Br. J. Exp. Pathol.*, 69(5), 717–737.

Donaldson, K. 2009. The inhalation toxicology of p-aramid fibrils. *Crit. Rev. Toxicol.*, 39(6), 487–500.

Donaldson, K., Aitken, R., Tran, L., Stone, V., Duffin, R., Forrest, G., and Alexander, A. 2006. Carbon nanotubes: A review of their properties in relation to pulmonary toxicology and workplace safety. *Toxicol. Sci.*, 92(1), 5–22.

Donaldson, K., Brown, G.M., Brown, D.M., Bolton, R.E., and Davis, J.M. 1989. Inflammation generating potential of long and short fibre amosite asbestos samples. *Br. J. Ind. Med.*, 46(4), 271–276.

Donaldson, K., Murphy, F.A., Duffin, R., and Poland, C.A. 2010. Asbestos, carbon nanotubes and the pleural mesothelium: A review and the hypothesis regarding the role of long fibre retention in the parietal pleura, inflammation and mesothelioma. *Part. Fibre Toxicol.*, 7(1), 5.

Donaldson, K., Murphy, F., Schinwald, A., Duffin, R., and Poland, C.A. 2011. Identifying the pulmonary hazard of high aspect ratio nanoparticles to enable their safety-by-design. *Nanomed. (Lond.)*, 6(1), 143–156.

Donaldson, K. and Tran, C.L. 2004. An introduction to the short-term toxicology of respirable industrial fibres. *Mutat. Res.*, 553(1–2), 5–9.

Duffin, R., Clouter, A., Brown, D.M., Tran, C.L., MacNee, W., Stone, V., and Donaldson, K. 2002. The importance of surface area and specific reactivity in the acute pulmonary inflammatory response to particles. *Ann. Occup. Hyg.*, 46(Suppl 1), 242–245.

ECHA. 2008. *Guidance on information requirements and chemical safety assessment. Chapter R.8: Characterisation of Dose [Concentration]—Response for Human Health*. Helsinki, Finland: European Chemicals Agency.

Egerton, R.F. 2005. *Physical Principles of Electron Microscopy: An Introduction to TEM, SEM, and AEM*. New York: Springer.

Elgrabli, D., Floriani, M., Abella-Gallart, S., Meunier, L., Gamez, C., Delalain, P., Rogerieux, F., Boczkowski, J., and Lacroix, G. 2008 Biodistribution and clearance of instilled carbon nanotubes in rat lung. *Part. Fibre Toxicol.*, 9(5), 20.

Ellinger-Ziegelbauer, H. and Pauluhn, J. 2009. Pulmonary toxicity of multi-walled carbon nanotubes (Baytubes) relative to alpha-quartz following a single 6h inhalation exposure of rats and a 3 months post-exposure period. *Toxicology*, 266(1–3), 16–29.

Folkmann, J.K., Risom, L., Jacobsen, N.R., Wallin, H., Loft, S., and Moller, P. 2009. Oxidatively damaged DNA in rats exposed by oral gavage to C60 fullerenes and single-walled carbon nanotubes. *Environ. Health Perspect.*, 117(5), 703–708.

Gill, S. et al. 2007. Nanoparticles: Characteristics, mechanisms of action, and toxicity in pulmonary drug delivery—A review. *J. Biomed. Nanotechnol.*, 3(2), 107–119.

Goodglick, L.A. and Kane, A.B. 1990. Cytotoxicity of long and short crocidolite asbestos fibers in vitro and in vivo. *Cancer Res.*, 50(16), 5153–5163.

Hofmann, W., Asgharian, B., Bergmann, R., Anjilvel, S., and Miller, F.J. 2000. The effect of heterogeneity of lung structure on particle deposition in the rat lung. *Toxicol. Sci.*, 53(2), 430–437.

Hunter, R.J. 1981. *Zeta Potential in Colloid Science*. New York: Academic Press.

IARC. 1987. *Asbestos: IARC Monographs. Supplement 7.* France: IARC.

Iijima, S. 1991. Helical microtubules of graphitic carbon. *Nature*, 354(6348), 56–58.

Iijima, S. and Ichihashi, T. 1993. Single-shell carbon nanotubes of 1-nm diameter. *Nature*, 363(6430), 603–605.

Islam, M.F., Zhang, J., Mei, B., Johnson, A.T., and Yodh, A.G. 2001. Dispersion and characterization of carbon nanotubes. *American Physical Society, Annual March Meeting*, March 12–16, 2001, Washington State Convention Center, Seattle, WA, Meeting ID: MAR01, abstract #C20.006.

Jones, A.D. 1993. Respirable industrial fibres: Deposition, clearance and dissolution in animal models. *Ann. Occup. Hyg.*, 37(2), 211–226.

Kagan, V.E., Tyurina, Y.Y., Tyurin, V.A., Konduru, N.V., Potapovich, A.I., Osipov, A.N., Kisin, E.R. et al., 2006. Direct and indirect effects of single walled carbon nanotubes on RAW 264.7 macrophages: Role of iron. *Toxicol. Lett.*, 165(1), 88–100.

Kiang, C.H. et al., Structural modification of single-layer carbon nanotubes with an electron beam. *J. Phys. Chem.*, 1996, 100(9), 3749–3752.

Lam, C.W., James, J.T., McCluskey, R., and Hunter, R.L. 2004. Pulmonary toxicity of single-wall carbon nanotubes in mice 7 and 90 days after intratracheal instillation. *Toxicol. Sci.*, 77(1), 126–134.

Laserna, J.J. 1996. *Modern Techniques in Raman Spectroscopy*. Amsterdam, the Netherlands: Elsevier Science.

Lee, H.C., Alegaonkar, P.S. et al. 2007. Growth of carbon nanotubes: Effect of Fe diffusion and oxidation. *Philos. Mag. Lett.* 87(10): 767–780.

Lee, J.Y., and Kim, J.S. et al. 2005. Electrophoretic and dynamic light scattering in evaluating dispersion and size distribution of single-walled carbon nanotubes. *J. Nanosci. Nanotechnol.*, 5(7), 1045–1049.

Li, J.G., Li, W.X., Xu, J.Y., Cai, X.Q., Liu, R.L., Li, Y.J., Zhao, Q.F., and Li, Q.N. 2007. Comparative study of pathological lesions induced by multiwalled carbon nanotubes in lungs of mice by intratracheal instillation and inhalation. *Environ. Toxicol.*, 22(4), 415–421.

Lowell, S., Shields, J.E., Thomas, M.A., and Thommes, M. 2004. *Characterization of Porous Solids and Powders: Surface Area, Pore Size and Density*. Particle Technology Series. New York: Springer, 1st edn., Corr. 2nd printing edition, 2006.

Ma-Hock, L., Treumann, S., Strauss, V., Brill, S., Luizi, F., Mertler, M., Wiench, K., Gamer, A.O., van Ravenzwaay, B., and Landsiedel, R. 2009. Inhalation toxicity of multi-walled carbon nanotubes in rats exposed for 3 months. *Toxicol. Sci.*, 112, 273–275.

Maynard, A.D., Baron, P.A., Foley, M., Shvedova, A.A., Kisin, E.R., and Castranova, V. 2004. Exposure to carbon nanotube material: Aerosol release during the handling of unrefined single-walled carbon nanotube material. *J. Toxicol. Environ. Health A*, 67(1), 87–107.

Maynard, R.L. and Maynard, R.L. 2002. Ambient aerosol exposure-response as a function of particulate surface area: Re-interpretation of historic data using numerical modelling. *Ann. Occup. Hyg.*, 46(1), 444–449.

Mitchell, L.A., Gao, J., Wal, R.V., Gigliotti, A., Burchiel, S.W., and McDonald, J.D. 2007. Pulmonary and systemic immune response to inhaled multiwalled carbon nanotubes. *Toxicol. Sci.*, 100(1), 203–214.

Morgan, A. 1995. Deposition of inhaled asbestos and man-made mineral fibres in the respiratory tract. *Ann. Occup. Hyg.*, 39(5), 747–758.

Morimoto, Y., Hirohashi, M., Ogami, A., Oyabu, T., Myojo, T., Todoroki, M., Yamamoto, M. et al. 2011. Pulmonary toxicity of well-dispersed multi-wall carbon nanotubes following inhalation and intratracheal instillation. *Nanotoxicology*, Epub ahead of print.

Murray, A.R., Kisin, E., Leonard, S.S., Young, S.H., Kommineni, C., Kagan, V.E., Castranova, V., and Shvedova, A.A. 2009. Oxidative stress and inflammatory response in dermal toxicity of single-walled carbon nanotubes. *Toxicology*, 257(3), 161–171.

NIST. 2001. Recommended practice guide, particle size characterization, National Institute of Standards & Technology, Document 960-1. Downloadable from http://www.nist.gov/public_affairs/practiceguides/SP960-1.pdf

Oberdorster, G., Oberdorster, E., and Oberdorster, J. 2005. Nanotoxicology: An emerging discipline evolving from studies of ultrafine particles. *Environ. Health Perspect.*, 113(7), 823–839.

Oller, A.R. and Oberdörster, G. 2010. Incorporation of particle size differences between animal studies and human workplace aerosols for deriving exposure limit values. *Regul. Toxicol. Pharmacol.*, 57(2–3), 181–194.

Osmond-McLeod, M.J., Poland, C.A., Murphy, F., Waddington, L., Morris, H., Hawkins, S.C., Clark, S., Aitken, R., McCall, M.J., and Donaldson, K. 2011. Durability and inflammogenic impact of carbon nanotubes compared with asbestos fibres. *Part. Fibre Toxicol.* 8(1), 15.

Pauluhn, J. 2010a. Subchronic 13-week inhalation exposure of rats to multiwalled carbon nanotubes: toxic effects are determined by density of agglomerate structures, not fibrillar structures. *Toxicol. Sci.*, 113(1), 226–242.

Pauluhn, J. 2010b. Multi-walled carbon nanotubes (Baytubes): Approach for derivation of occupational exposure limit. *Regul. Toxicol. Pharmacol.*, 57, 78–89.

Poland, C., Duffin, R., Kinloch, I., Maynard, A.D., Wallace, W., Seaton, A., Stone, V., Brown, S., MacNee, W., and Donaldson, K. 2008. Carbon nanotubes introduced into the abdominal cavity of mice show asbestos-like pathogenicity in a pilot study. *Nat. Nanotechnol.*, 3, 423–428.

Porter, A., Gass, M. Muller, K., Skepper, J.N., Midgley, P.A., and Welland, M. 2007. Direct imaging of single-walled carbon nanotubes in cells. *Nat. Nanotechnol.*, 2, 713–717.

Powers, K.W., Brown, S.C. et al. 2006. Research strategies for safety evaluation of nanomaterials. Part VI. Characterization of nanoscale particles for toxicological evaluation. *Toxicol. Sci.*, 90(2), 296–303.

Powers, K.V., Palazuelos, M. et al. 2007. Characterization of the size, shape, and state of dispersion of nanoparticles for toxicological studies, Nanotoxicology 1(1), 42–51.

Reimer, L. 1993. *Transmission Electron Microscopy: Physics of Image Formation and Microanalysis*, 3rd edn.. New York: Springer-Verlag.

Rudt, S. and Muller, R.H. 1992. In vitro phagocytosis assay of nanoparticles and microparticles by chemiluminescence .1. Effect of analytical parameters, particle-size and particle concentration. *J. Controll. Release*, 22(3): 263–271.

Ryman-Rasmussen, J.P., Cesta, M.F., Brody, A.R., Shipley-Phillips, J.K., Everitt, J.I., Tewksbury, E.W., Moss, O.R. et al. 2009a. Inhaled carbon nanotubes reach the subpleural tissue in mice. *Nat. Nanotechnol.*, 4(11), 747–751.

Ryman-Rasmussen, J.P., Tewksbury, E.W., Moss, O.R., Cesta, M.F., Wong, B.A., and Bonner, J.C. 2009b. Inhaled multiwalled carbon nanotubes potentiate airway fibrosis in murine allergic asthma. *Am. J. Respir. Cell Mol. Biol.*, 40(3), 349–358.

Sakurai, T.E. 2000. *Advances in Scanning Probe Microscopy (Advances in Materials Research, 2)*. New York: Springer Verlag.

Searl, A., Buchanan, D., Cullen, R.T., Jones, A.D., Miller, B.G., and Soutar, C.A. 1999. Biopersistence and durability of nine mineral fibre types in rat lungs over 12 months. *Ann. Occup. Hyg.*, 43(3), 143–153.

Service, R.F. 1998. Chemistry: nanotubes: The next asbestos? *Science*, 281(5379), 941.

Shvedova, A.A., Kisin, E., Murray, A.R., Johnson, V.J., Gorelik, O., Arepalli, S., Hubbs, A.F. et al., 2008. Inhalation vs. aspiration of single-walled carbon nanotubes in C57BL/6 mice: inflammation, fibrosis, oxidative stress, and mutagenesis. *Am. J. Physiol. Lung Cell. Mol. Physiol.*, 295(4), L552–L565.

Singh, R., Pantarotto, D., Lacerda, L., Pastorin, G., Klumpp, C., Prato, M., Bianco, A., and Kostarelos, K. 2006. Tissue biodistribution and blood clearance rates of intravenously administered carbon nanotube radiotracers. *Proc. Natl Acad. Sci. USA*, 103(9), 3357–3362.

Stanton, M.F., Layard, M., Tegeris, A., Miller, E., May, M., Morgan, E., and Smith, A. 1981. Relation of particle dimension to carcinogenicity in amphibole asbestoses and other fibrous minerals. *J. Natl Cancer Inst.*, 67(5), 965–975.

Stuart, B.H. 2004. *Infrared Spectroscopy: Fundamentals and Applications*. New York: Wiley.

Tran, C.L., Jones, A.D., Cullen, R.T., and Donaldson, K. 1999. Mathematical modelling of the retention and clearance of low-toxicity particles in the lung. *Inhal. Toxicol.*, 11(12), 1059–1076.

Unfried, K., Schurkes, C. and Abel, J. 2002. Distinct spectrum of mutations induced by crocidolite asbestos: Clue for 8-hydroxydeoxyguanosine-dependent mutagenesis in vivo. *Cancer Res.*, 62(1), 99–104.

Vaisman, L., Wagner, H.D., and Marom, G. 2006. The role of surfactants in dispersion of carbon nanotubes. *Adv. Colloid Interface Sci.*, 128, 37–46.

Van Gulijk, C., Marijnissen, J.C.M. et al. 2004. Measuring diesel soot with a scanning mobility particle sizer and an electrical low-pressure impactor: Performance assessment with a model for fractal-like agglomerates. *J. Aerosol Sci.* 35(5): 633–655.

Virta, R.L. and Geological Survey (U.S.). 2002. *Asbestos Geology, Mineralogy, Mining, and Uses*. Version 1.0 edn., Reston, VA: U.S. Department of the Interior, U.S. Geological Survey.

Warheit, D.B., Laurence, B.R., Reed, K.L., Roach, D.H., Reynolds, G.A., and Webb, T.R. 2004. Comparative pulmonary toxicity assessment of single-wall carbon nanotubes in rats. *Toxicol. Sci.*, 77(1), 117–125.

Watts, J.F. and Wolstenholme, J. 2003. *An Introduction to Surface Analysis by XPS and AES*, 2nd rev. edn. New York: Wiley.

WHO. 1997. *Determination of Airborne Fibre Number Concentrations: A Recommended Method by Phase Contrast Optical Microscopy*. Geneva, Switzerland: World Health Organisation.

WHO/EURO Technical Committee for Monitoring and Evaluating MMMF. 1985. Reference methods for measuring airborne man-made mineral fibres (MMMF). Copenhagen: Work Health Organisation, http://whqlibdoc.who.int/euro/ehs/Euro_EHS_4.pdf

WHO IARC, I.LF. 2002. IARC monographs on the evaluation of carcinogenic risks to humans. Volume 81 Man-made vitreous fibres. Lyon, France: IARC Press, http://monographs.iarc.fr/ENG/Monographs/vol81/index.php

11 Research and Development of a New Safe Form of Drugs

T.G. Tolstikova, A.A. Onischuk, I.V. Sorokina,
A.M. Baklanov, V.V. Karasev, V.V. Boldyrev, V.M. Fomin,
M.V. Khvostov, A.O. Bryzgalov, and G.A. Tolstikov

CONTENTS

11.1 PART I. PHYSICAL-CHEMICAL AND TOXIC-PHARMACOLOGICAL CHARACTERISTICS OF AEROSOLS FORM OF DRUGS

T.G. Tolstikova, A.A. Onischuk, I.V. Sorokina, A.M. Baklanov, V.V. Karasev, V.V. Boldyrev, and V.M. Fomin

11.1.1 INTRODUCTION

It is expected that the use of nanoparticulate materials will be of high importance in future science and technology. There is increasing optimism that nanotechnology, as applied to medicine, will bring significant advances in the treatment of disease. The administration of drugs directly into the respiratory tract has been used in a number of therapeutic areas. The field of aerosolized drug application includes treatment of lung diseases, like asthma, chronic obstructive pulmonary disease, cystic fibrosis, and lung cancer (Hickey, 2004). Aerosol delivery has also expanded into the field of systemic drug delivery (Laube, 2005). The optimal target in the lungs for delivery of drugs to the systemic circulation is the alveolar region. For rapid delivery, alveolar drug administration has a number of advantages including large absorptive surface area, easy permeability of the alveolar walls resulting in the fast passage from the alveolar airspace to the pulmonary capillary bed, and a direct connection between the pulmonary circulation and the systemic circulation. Aerosol delivery is much more efficient than the peroral administration because of low oral bioavailability for some drugs such as zanamivir (Fenton et al., 1977) or amantadine hydrochloride (Skyler, 2007). In the case of systemic targeting, the advantages of aerosol delivery with respect to peroral treatment include the possibility of avoiding losses in the gastrointestinal tract as well as metabolic destruction in the liver. In contrast to injection therapy, inhalation therapy is noninvasive, and so it is a more convenient and safe route leading to an improved treatment outcome. On the other hand, the aerosol treatment has no limitations for the use of water-insoluble drugs giving evident advantage with respect to injection therapy. Approximately one third of the modern drugs are water-insoluble or poorly water-soluble. Many currently available injectable formulations of such drugs can cause side effects that originate from the detergents and other agents used for their solubilization. Besides, water-solubility problems delay or completely block the development of many new drugs and other biologically useful compounds. Thus, the lung deposition route can be a good alternative for the administration of poorly soluble substances.

The modern devices which are available on the market for therapeutic aerosol delivery systems can be subdivided into three groups: nebulizers, dose-metering inhaler systems, and dry powder inhalers (Baron and Willeke, 2001; Hickey, 2004). All these inhalers have evident shortcomings (Hickey, 2004). One of the main problems of jet nebulizers is their limited portability due to the need for compressed gas supply. Besides, it is important that the nebulizer solutions not be stored in nebulizer reservoirs due to the possible growth of bacteria. The disadvantages of pressurized dose-metering inhaler systems include high aerosol velocity (which results in substantial oropharyngeal deposition), limited single dose etc. Different facilities including extension tubes and spacers were applied to overcome some of these problems; however, new problems have appeared such as large spacer volume and particle electric charge (Rubin and Fink, 2005). One of the problems of dry powder inhalers is that the dry powder formulation should be appropriately prepared. The powder properties are shelf life dependent. The crucial question is hygroscopicity. Thus, aerosol properties and the lung deposition efficiency can be functions of the powder storage time.

A more serious problem related to traditional inhalers is that all these devices are able to generate the particles as small as a few microns in diameter. However, the alveolar deposition efficiency is a strong function of the particle size. Particles 10–20 nm in size deposit in the alveolar region about four times more efficiently than those several microns in diameter (Hinds, 1999; Hickey, 2004; Oberdörster et al., 2005; Wong, 2007). Moreover, the nanoparticles are easily transported across the membranes (protein-mediated transport). However, the translocation from the alveolar region to the blood circulation depends essentially on size. Thus, the 80 nm particles translocate about one order of magnitude less effectively than the 20 nm ones. The micron-sized particles are excluded from this transportation because of their large size. Also, since microparticles are rapidly phagocytosed by lung surface macrophages, they are only shortly available for protein-mediated transport (Geiser and Kreyling, 2010). Thus, alternative ways of particle synthesis are to be developed for generation of nanosized medical particles with high concentration. The promising route seems to be the homogeneous nucleation from supersaturated vapor.

Approximately one third of the modern drugs are water-insoluble or poorly water-soluble. Many currently available injectable formulations of such drugs can cause side effects that originate from detergents and other agents used for their solubilization. Besides, water-solubility problems delay or completely block the development of many new drugs and other biologically useful compounds. Thus, the lung deposition route can be a good alternative for the administration of poorly soluble substances. Nonsteroid anti-inflammatory drugs (NAD) like indomethacin and ibuprofen which have low water solubility are considered for the lung delivery (Rabinowitz and Zaffaroni, 2004). Indomethacin is a well-known drug for use against a wide range of diseases such as rheumatoid arthritis, spondylosis, and chondrosis. Ibuprofen is a nonsteroidal, chiral, anti-inflammatory drug that inhibits the enzyme cyclooxygenase and thus acts as an analgesic (Meade et al., 1993). It is most often prescribed to treat rheumatoid arthritis and pain. However, the side effects from these medicines can cause serious disorders such as bleeding and perforation of gastrointestinal tract, depression, drowsiness, mental disorder, increased blood pressure, congestive heart failure etc. One can hope that the aerosol lung administration of NAD may be an alternative route which would diminish side effects and decrease the therapeutic dose. However, new side effects like pulmonary emphysema are possible. Therefore, it is necessary to estimate both the therapeutic benefits and possible risks of aerosol administration.

In this chapter we study the evaporation–condensation formation of indomethacin and ibuprofen nanoparticles and its anti-inflammatory and analgesic effects, as well as side effects on outbred male mice.

11.1.2 Synthesis of Nanoaerosol

The inhalation scheme includes a flow aerosol generator, inhalation chambers for mice, filters, diluters, flow control equipment and aerosol spectrometer (Figure 11.1). The horizontal

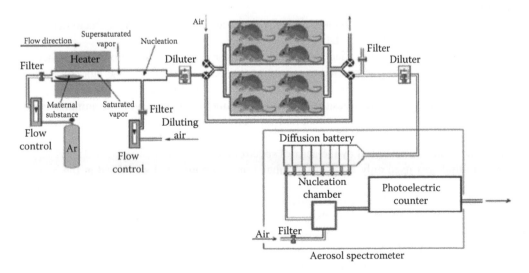

FIGURE 11.1 Scheme of the experimental set-up for inhalation experiments.

evaporation–condensation aerosol generator is made of a molybdenum glass tube (with the inner diameter to be in the range from 0.8 to 1.5 cm) with an outer heater. Argon flow is supplied to the inlet of the generator through the Petrianov's high efficiency aerosol filter (Kirsh et al., 1975) at the rate of 8 cm³/s (at standard temperature and pressure). The original substance (indomethacin, ICN-190217 or racemic ibuprofen from Ratiopharm [Germany]) is put into the hot zone inside the tube. The saturated vapor is formed inside the generator. The temperature drops down at the outlet of the heated zone resulting in vapor supersaturation. The flow of argon + supersaturated vapor is mixed with the air flow (with the ratio 1:13, respectively) downstream resulting in the homogeneous nucleation. Then the aerosol is diluted up to the necessary concentration using an aerosol diluter (based on flow splitting, filtering one of the subflows and mixing the subflows again). The final aerosol is admitted into the plastic whole-body (WB) exposure chambers (each of 1500 cm³ volume). Two chambers were used in parallel; thus, the aerosol flow rate was about 55 cm³/s in each chamber. The temperature in the chambers was 295 K which corresponded to the room temperature, and the relative humidity was in the range of 50%–70%. Each chamber contained four mice during the experiment. Outbred laboratory male mice were used. Their age and weight were 12 weeks and 25–30 g, respectively. The inhalation time in all the experiments was 20 min. The mice were free to move along the chamber during the aerosol exposure. The experiments with mice were conducted in the Pharmacology Research Laboratory (Institute of Organic Chemistry, Novosibirsk) which was accredited as satisfying to the international standards ISO/IEC 17025-2000, approval code ROSS RU.0001.514430; No000269. All studies were carried out in accordance with the Guideline for the Care and Use of Laboratory Animals (Geneva Convention for the Protection of Animals, 1986).

The aerosol concentration and size distribution were measured with the aerosol spectrometer designed and built at the Institute of Chemical Kinetics and Combustion, Novosibirsk, Russia (Ankilov et al., 2002). This aerosol spectrometer consists of an automatic diffusion battery, condensation chamber, and photoelectric counter. The spectrometer measured the aerosol number concentration and particle size distribution at the chamber inlet and outlet during the exposure.

To analyze the nanoparticles, the aerosol was passed through the glass fiber filters (Shleicher & Schuell, GF 6). Then the deposit was dissolved in ethanol.

The chemical composition of ibuprofen nanoparticles was analyzed by means of high-performance liquid chromatography with Milikhrom-1 coupled to a computer through a 14 bit analog-to-digital converter. For this purpose, the aerosol particles were sampled by passing the aerosol flux through the Petrianov high-efficiency aerosol filter. Then the deposit was dissolved in acetonitrile.

The crystal phase analysis of nanoparticles was carried out using x-ray diffractometer system Bruker-AXS D8 Discover with GADDS Area Detector.

Special experiments were carried out to determine the lung-deposited dose. Nose-only exposure (NOE) glass chambers were used to minimize the skin or fur effect. The laboratory animals were confined so that only the nose was exposed to the aerosol (Figure 11.2). The aerosol flow was switching between two parallel lines (Figure 11.3). One line was loaded with the animals; the other one had empty chambers. Each line contained six chambers in tandem. The aerosol depletion due to mice breathing was determined by comparing the particle concentration at the outlets of loaded and unloaded lines as measured by the aerosol spectrometer.

Each animal was used only once in the inhalation procedure and sacrificed at the end of the experiment. The mice were separated into three groups. The animals of the first group (untreated) were not treated by medicine (neither orally nor by aerosol inhalation); the animals of the second group (oral) were treated orally with the water-Tween suspension of indometicin or ibuprofen with the dose varied from 8×10^{-3} to 170 mg per kg bodyweight (bw); the animals from the third group (aerosol group) were subjected to the aerosol inhalation. The mice from all the three groups were put to the WB chambers for 20 min. The mice from the first and the second groups were exposed to pure air; the mice of the third group were exposed to the aerosol (the average lung-deposited mass was determined from the NOE chamber experimental data) (Nada et al., 2005).

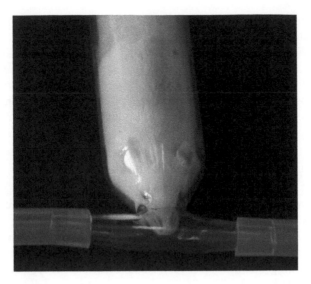

FIGURE 11.2 NOE chamber.

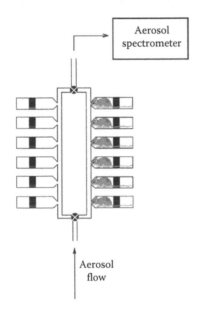

FIGURE 11.3 Scheme of the experimental set-up for the lung-deposited dose measurements.

The analgesic effect of ibuprofen was estimated in an "acetic acid writhing" test (Koster et al., 1959). One hour after the chamber exposure, 0.1 mL of 0.75% acetic acid solution in water was injected intraperitoneally to all animals. Five minutes after the injection, the number of writhes (i.e., abdominal constriction followed by dorsiflexion and stretching of hind limbs) occurring during a 3 min period was measured. Each animal was observed by one observer. The observations were performed blindly with respect to the treatment regime.

The anti-inflammatory effect of indomethacin nanoparticles was also studied. One hour after the chamber exposure, 0.05 mL of 0.1% histamine solution in water was injected into the subplanar surface of the mouse hind paw. Six hours later the mice were killed by cervical dislocation. Then the mouse's paws were cut off at the ankle joint and weighed. The ratio of the difference in weight between the treated and untreated hind paws to the weight of the untreated hind paw was used as an index of paw edema.

A histologic analysis was performed to observe the aerosol effect on the mice lungs morphology. The mice were killed 6.3 h after exposure. Lungs were fixed in 4% paraformaldehyde in phosphate buffer (pH 7.2–7.4). The fixed tissues were treated in a standard way using histological equipment "MICROM" (Carl Zeiss) and then embedded to paraffin. Sections 3–4 μm thick were stained with hematoxylin and eosin. Slides were examined under the light microscope Axioskop 40 (Carl Zeiss).

11.1.2.1 Aerosol Size, Concentration, and Composition

A typical example of the mean particle size and number concentration (as measured at the WB chamber inlet) is shown in Figure 11.4 as a function of generator temperature. As seen in the plot, the range of the particle diameter is 20–100 nm under the standard operating conditions. Changing the vapor to air mixing conditions, it was possible to decrease the mean particle diameter to 10 nm. The particle size distribution was well described by the lognormal function. For the whole range of the particle diameters, the geometric standard deviation was 1.4. The typical particle size distribution is shown in Figure 11.5.

Chromatographic analysis of the aerosol particles was performed to make sure that there was no thermal decomposition of the maternal substance during evaporation–condensation. The chromatographic analysis showed that the chromatograms from nanoparticles of both indometacin and

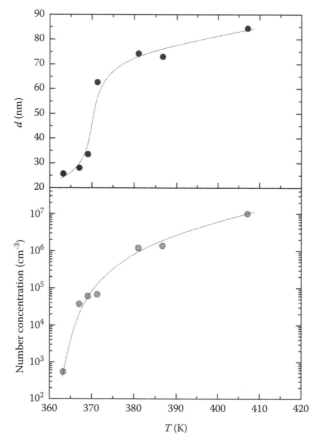

FIGURE 11.4 Mean diameter d of ibuprofen particles and number concentration versus temperature in the saturated vapor zone (as measured by the aerosol spectrometer).

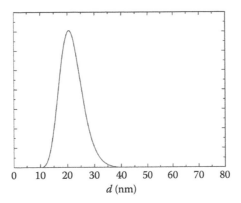

FIGURE 11.5 Typical diameter distribution for ibuprofen nanoparticles (as measured by the aerosol spectrometer).

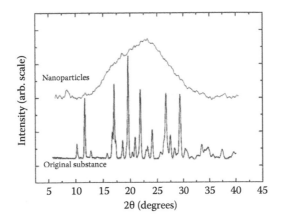

FIGURE 11.6 Comparison between the chromatograms of original ibuprofen powder and nanoparticles (mean particle diameter $d = 85\,nm$) formed by evaporation-nucleation.

ibuprofen were identical to that from the original powder. Thus, one may suppose that the nanoparticles are chemically identical to the original substance.

The crystal phase analysis of nanoparticles was carried out using x-ray diffractometer system Bruker-AXS D8 Discover with GADDS Area Detector. The powder x-ray diffraction patterns of indomethacin nanoparticles and γ indomethacin original powder are compared in Figure 11.6. The XRD pattern of nanoparticulate material contains a broad halo uncorrelated with the crystalline peaks, which is typical for the amorphous indomethacin (Hermsdorf et al., 2007). Thus, the evaporation–condensation route results in the formation of amorphous nanoparticles.

The powder x-ray diffraction patterns of ibuprofen nanoparticles and original powder are compared in Figure 11.7. Both XRD patterns from nanoparticles and original substance correspond to racemic ibuprofen (Stahly et al., 1997; Hermsdorf et al., 2007; Lee et al., 2008), that is, the nanoparticles form the same crystal phase as the original substance. The small difference in the peak relative intensities between the curves a and b in the range of $17 < 2\theta < 21$ is probably related to the difference in distribution of crystallographic orientations in the micro-sized original powder and nanoparticles.

11.1.2.2 Lung-Deposited Dose

To determine the lung-deposited dose, we measured the fraction α of particles consumed per chamber due to mouse breathing:

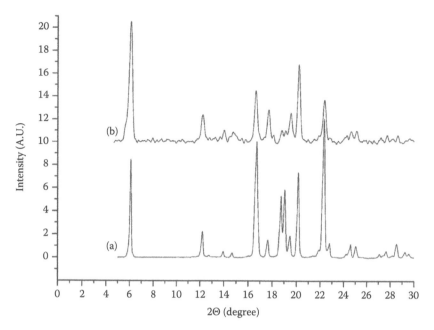

FIGURE 11.7 X-ray diffraction patterns of (a) original ibuprofen powder and (b) ibuprofen nanoparticles formed via evaporation-nucleation route (mean particle diameter d = 85 nm).

$$\alpha = 1 - \left(\frac{n_{out}}{n_{out}^{(0)}} \right)^{1/N} \tag{11.1}$$

where
   n_{out} and $n_{out}^{(0)}$ are aerosol number concentrations at the outlets of the loaded and unloaded NOE lines, respectively
   N is the number of chambers in the tandem (see Figure 11.3)

The rate D [s^{-1}] of particle lung deposition per mouse can be written as

$$D \approx F \alpha n \tag{11.2}$$

where
   F (cm^3/s) is the flow rate through the NOE chambers tandem
   n is the arithmetic mean between the loaded NOE line inlet and outlet particle concentrations

One can also use the relative deposition rate:

$$D_0 = \frac{D}{n} = F \alpha \tag{11.3}$$

Figure 11.8 shows the relative deposition rate D_0 as a function of the mean particle diameter. The logistic regression analysis applied to the relative deposition rate data showed that there was a statistically significant correlation between the particle diameter and the relative deposition rate (R^2 = 0.96). The fitted curve is shown as a solid line. Using the function $D_0(d)$ we determined the lung-deposited dose (weight of the deposited particles) for the WB chambers inhalation experiments:

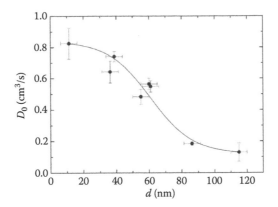

FIGURE 11.8 Relative deposition rate per mouse versus mean particle diameter (ibuprofen). Bars indicate standard error. Line is logistic regression analysis result.

$$\text{Dose} = D_0 n_{av}^{WB} mt \qquad (11.4)$$

where
  n_{av}^{WB} is the arithmetic mean between the WB chamber inlet and outlet particle concentrations
  m is the mean particle mass
  t is the inhalation time

Note that the lung-deposited dose is a function of the product fV_T (where f and V_T are average mouse breathing frequency and tidal volume, respectively) which for the WB inhalation experiments can differ from that of the NOE chamber by 20% (Schaper and Brost, 1991; Hamelmann et al., 1997; Vijayaraghavan, 1997; Currie et al., 1998). Therefore, we assume that the accuracy of Equation 11.4 is also about 20%.

To demonstrate the validity of our measurements of the lung-deposited dose, we evaluated the particle deposition efficiency which follows from Equation 11.5:

$$D_0 \approx fV_T \varepsilon \qquad (11.5)$$

where ε is the lung deposition efficiency, that is, the ratio of the difference between the numbers of inhaled and exhaled particles to the number of inhaled particles. From Equations 11.3 and 11.5 we get

$$\varepsilon \approx \frac{F}{fV_T}\left[1-\left(\frac{n_{out}}{n_{out}^{(0)}}\right)^{1/N}\right] \qquad (11.6)$$

We used the mouse breathing frequency and the mice tidal volume as equal to $f = 5.0\,\text{s}^{-1}$ and $V_T = 0.16\,\text{cm}$ (Fairchild, 1972; Meade et al., 1993; Onischuk et al., 2008). The particle deposition efficiency as a function of the mean particle diameter is shown in Figure 11.9. One can see that ε tends to unity at small diameter values, which is in good agreement with numerical simulations for the particle lung deposition (Wong, 2007). The aerosol depletion in both the empty and mice-occupied WB chambers was measured in special experiments and found to be independent of the mean particle diameter within the range of $35 < d < 120\,\text{nm}$. To this end aerosol concentration outlet $(n_{out}^{(0)})$ to inlet $(n_{in}^{(0)})$ ratio was measured with the aerosol spectrometer for the empty chamber to be $(n_{out}^{(0)}/n_{in}^{(0)} = 0.94 \pm 0.01)$. To measure the aerosol depletion in the occupied chambers due to the

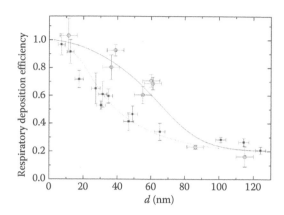

FIGURE 11.9 Mouse respiratory efficiency (ε) versus mean particle diameter; circles, the data for ibuprofen; squares, the data from the indomethacin nanoparticle inhalation experiments. Bars, indicate standard error; lines are given as eye guides. There is some discrepancy between the deposition efficiencies measured for indomethacin and ibuprofen particles. One of the possible reasons for this discrepancy may arise from different hygroscopic properties of indomethacin and ibuprofen nanoparticles.

fur deposition, two chambers were put in parallel, the same way as it was done in the inhalation experiments. One chamber was empty and the other one was occupied by four mice. The ratio of the particle concentration at the outlet of the occupied chamber (n_{out}) to that of the empty one $n_{out}^{(0)}$ was measured to be ($n_{out}/n_{out}^{(0)} = 0.91 \pm 0.02$). Thus, the total aerosol depletion in the occupied chamber was about 15%. When evaluating the lung-deposited dose the average between inlet and outlet concentrations was used.

11.1.2.3 Analgesic Effect of Ibuprofen Nanoaerosol

It is convenient to use the relative analgesic index (RAI), that is, the ratio between the mean number of writhes for the aerosolized group (or oral group) and that for the untreated group. During 1 day, one to three WB exposure runs were provided for aerosolized or oral groups and one run for the animals from group 1 (eight animals per one run). Thus, during 1 day for all the runs of group 2 or 3, one and the same run of group 1 was considered as a match. Figure 11.10 shows the RAI for the aerosolized animals as well as RAI for the orally treated animals versus the lung-deposited dose. The RAI data for both the aerosolized and oral groups were analyzed for the dose–response relationship. The fitted curves are shown as the solid lines. Both aerosol and oral treatments give no analgesic effect (RAI is about unity) at small lung-deposited or orally deposited doses (less than 10^{-6} and 1 mg per kg bw for the aerosol and oral treatments, respectively). The regression analysis applied to the RAI results showed that there was a statistically significant dose response ($R^2 = 0.91$ and 0.96 for aerosolized and oral groups, respectively). Note that the lung-deposited dose was varied by changing the heating temperature in the generator of nanoparticles; an increase in temperature results in an increase of both the mean particle diameter and number concentration. Therefore, the RAI points for different dose magnitudes correspond to different mean particle diameters. One can also see from Figure 11.10 that the aerosol treatment is more effective (gives the same RAI for a less dose) than the oral administration. The body-delivered doses for aerosol and oral treatments differ by 3 and 5 orders of magnitude at the lung-deposited dose 10^{-1} and 10^{-4} mg per kg bw, respectively. One of the reasons for this difference is probably the high level of metabolism of ibuprofen being administered orally (Mutschler and Derendorf, 1995).

11.1.2.4 Anti-Inflammatory Effect of Indomethacin Nanoaerosol

Table 11.1 gives an example of the edema index for the groups 1 through 3 in the indometacin nanoaerosol inhalation. The particle number concentration and mean diameter were 7×10^5 cm^{-3}

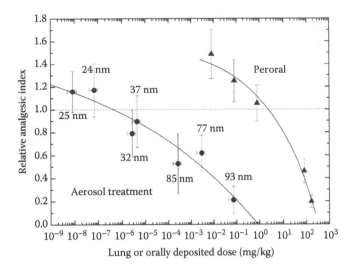

FIGURE 11.10 RAI (the ratio between the mean number of writhes for the aerosolized group and that for untreated group) versus the lung-deposited dose (circles). Triangles are RAI values for peroral treatment. Bars indicate standard error. The fitted dose–response curves are shown as solid lines. Mean particle diameter is indicated for each inhalation point.

and 37 nm, respectively, in these experiments which corresponded to the average lung-deposited dose of 1.4×10^{-5} mg per mouse (5.1×10^{-4} mg per kg bw). One can see a considerable anti-inflammatory effect from both aerosol and oral forms of indomethacin—the *mean edema index* (MEI) for the groups 2 and 3 is considerably less than that for the control group 1. The indomethacin aerosol form is more effective than the indomethacin peroral treatment (MEI = 7.1% and 12.2%, respectively), while the lung-deposited dose is a few orders of magnitude less than the peroral dose. It is more convenient to use the *relative edema index* (REI), that is, the ratio between the mean edema indexes for the aerosolized group (or peroral group) and untreated group. Figure 11.11 shows the mean REI for the aerosolized animals versus the lung-deposited dose as well as REI for the orally treated animals. These data were obtained during six experimental trials. Each trial involved one untreated group, one oral group and two aerosolized groups treated with particles of different size. Thus, there are 12 points for the inhalation treatments and one point for the peroral treatment (averaged over six treatments). In total, the data presented in Figure 11.11 involve about 210 animals. The REI data were analyzed for the dose–response relationship. The fitted curve is shown as the solid line. The regression analysis applied to the REI results showed that there was a statistically significant dose response ($R^2 = 0.91$). The EC50 was 2.7×10^{-6} mg per kg bw. The triangle symbol gives the mean REI for peroral animals for comparison. One can see that the aerosol administration is more effective (gives less REI) than the peroral treatment even at the lung-deposited dose 6 orders of magnitude less than the peroral dose. The mean particle diameter is indicated for each point in the graph. It is hardly possible to find any dependence on the particle size. Thus, the points for the nearby diameters (e.g., 21, 22, 24 and 32, 35, 37 nm) seem to be following the basic tendency within the experimental accuracy. On the other hand, the points differing essentially by size (180 and 37 nm; 35 and 4–5 nm) also follow the same tendency.

11.1.2.5 Morphology of the Lungs

A histologic analysis was performed to observe possible hemodynamic abnormalities and pulmonary edema after the aerosol treatment. In the case of indometacin nanoparticle treatment the animals were exposed to two kinds of aerosol of $d = 200$ nm, with the lung-deposited dose of 2.3×10^{-5}

TABLE 11.1
Edema Index for Treated and Untreated Animals

Group Number	Animal Number	Weight of Paw (mg) Treated	Weight of Paw (mg) Untreated	Edema Index (%)	Mean Edema Index (%) Relative Edema Index
1 (untreated)	1	193	160	20.6	20.3 ± 1.9
	2	186	150	24.0	
	3	187	145	29.0	
	4	150	130	15.4	
	5	203	177	14.7	
	6	170	149	14.1	
	7	204	168	21.4	
	8	179	145	23.4	
2 (peroral)	1	187	167	12.0	12.2 ± 1.6
	2	183	173	11.6	
	3	173	160	8.1	
	4	178	156	14.1	
	5	174	162	7.0	
	6	170	154	10.4	
	7	204	176	15.9	
	8	174	163	6.7	
	9	179	144	24.3	
	10	152	170	11.8	
3 (aerosol group)	1	200	177	13.0	7.3 ± 1.2
	2	171	157	8.9	
	3	159	155	2.3	
	4	160	147	8.4	
	5	178	170	4.7	
	6	157	151	4.0	
	7	179	165	8.5	
	8	158	171	8.2	0.36 ± 0.06

Standard errors are shown in the last column.

mg per kg bw (group 3.1) and $d = 9$ nm, with the dose $= 1.2 \times 10^{-5}$ mg per kg bw (group 3.2). The aerosol number concentrations in the WB chambers were 7.1×10^2 and 7.4×10^5 cm^{-3}, respectively. The group 1 included again untreated mice being exposed to the pure air in the chambers. Each group consisted of 8–10 animals. The lungs of animals from both group 3.1 and group 3.2 as well as from the group 1 have a normal structure without any destructive and hemodynamic pathologic changes (Figure 11.12). However, a moderate venous and arterial hyperemia was observed for all the animals in group 3.1. Two animals of this group have revealed the focal emphysematous dilatation of the respiratory bronchioles without any visible vascular bed reduction (Figure 11.13). All the animals from group 3.2 have demonstrated the vascular and capillary hyperemia (which was considerably stronger than in the case of group 3.1). A homogeneous venous deposition (presumably fibrin) was observed. Besides, typical emphysematous signs, stronger than those in group 3.1, were observed, that is, the dilatation of bronchioli and alveolar channels, alveolar wall thinning and partial capillary bed reduction (Figure 11.14).

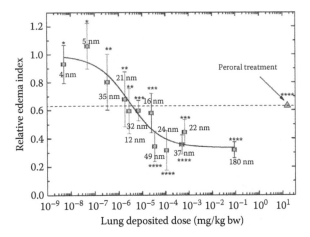

FIGURE 11.11 Mean REI versus the lung-deposited dose (circles). Bars indicate standard error. The particle mean diameter is shown at every point. The levels of statistical significance of differences of the edema indexes between the aerosolized and untreated groups were calculated using Students t-test and are indicated as *$1 < P < 0.5$, **$0.5 < P < 0.05$, ***$0.05 < P < 0.001$, ****$P < 0.001$. The fitted dose–response curve is shown as solid line. Triangle is REI for peroral treatment; this point is a mean value throughout six experimental trials (each trial included groups of 8–10 animals).

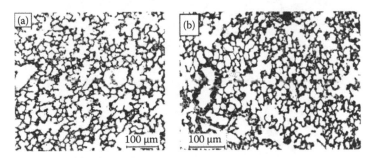

FIGURE 11.12 Representative sections from the lungs of untreated animal (group 1) (a) and the animal treated by the indometacin aerosol of $d = 200$ nm (group 3.1) (b).

In the case of ibuprofen treatment, the animals were exposed to nanoparticles of $d = 100$ nm, with the lung-deposited dose of 5.5×10^{-3} mg per kg bw (group 3.1, 16 animals), and $d = 75$ nm, with the lung-deposited dose of 2.9×10^{-3} mg per kg bw (group 3.2, 16 animals). A moderate venous hyperemia was observed for seven animals from group 3.1 and for all the animals from group 3.2 (Figure 11.15). The other nine animals of group 3.1 have demonstrated more pronounced venous and arterial hyperemia (Figure 11.16). A homogeneous venous deposition (presumably fibrin) was observed. Typical emphysematous signs occurred in the lungs of those animals, that is, the dilatation of bronchioli and alveolar channels, alveolar wall thinning, and partial capillary bed reduction.

11.1.3 Conclusions

The anti-inflammatory action analgesic effect and side pulmonary effects caused by the inhalation of indomethacin and ibuprofen nanoparticles, respectively, were investigated. To this end, an evaporation–condensation system was developed which was able to generate aerosol nanoparticles within the size range of $3 < d < 200$ nm. The chromatographic analysis showed that the aerosol particles were chemically identical to the maternal substance (i.e., there was no thermal decomposition

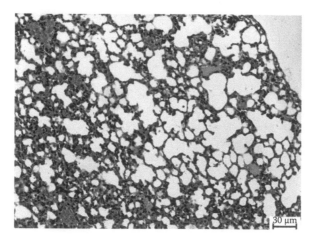

FIGURE 11.13 Cross section of the lung from a mouse treated with 200 nm indometacin nanoparticles showing enlarged airspaces.

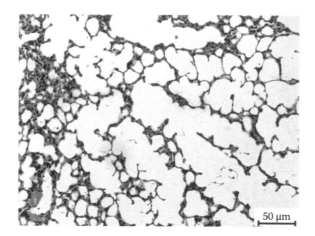

FIGURE 11.14 Cross section of the lungs of mice treated with 9 nm indometacin nanoparticles.

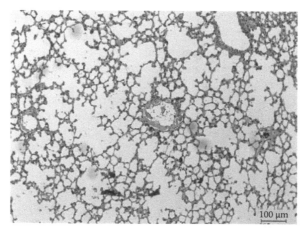

FIGURE 11.15 Cross section of the lung from a mouse treated with 100 nm ibuprofen nanoparticles showing enlarged airspaces.

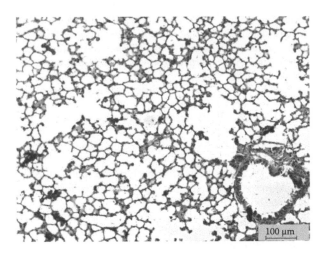

FIGURE 11.16 Cross section of the lung from a mouse treated with 100 nm ibuprofen nanoparticles showing a pronounced venous and arterial hyperemia as well as dilatation of bronchioli and alveolar channels, alveolar wall thinning, and partial capillary bed reduction.

during the substance evaporation). The x-ray diffraction analysis showed that the indomethacin nanoparticles were amorphous and the ibuprofen nanoparticles have the crystalline structure identical to that of the maternal substance.

The lung-deposited dose versus nanoparticle diameter was measured using the NOE chambers. The mice lung deposition efficiency was evaluated as a function of the particle diameter changing from about unity at $d = 10$ nm to about 0.2 at $d = 100$ nm.

The dose-dependent effect of aerosolized indometacin and ibuprofen was studied in comparison with the oral treatment. It was found that aerosol administration is much more effective than the oral delivery; thus, the aerosol treatment needs the dose a 3–6 orders of magnitude less than the oral one at the same analgesic or anti-inflammatory effect.

However, the lung histology analysis for the mice treated with the small particles ($d = 9$ nm) at dose as small as 10^{-5} mg per kg bw has revealed emphysematous signs like the dilatation of bronchioles and alveolar channels, alveolar wall thinning and partial capillary bed reduction.

REFERENCES

Ankilov, A., Baklanov, A., Colhoun, M., Enderle, K.-H., Gras, J., Junlanov, Yu., Kaller, D. et al. 2002. Intercomparison of number concentration measurements by various aerosol particle counters, *Atmospheric Research*, 62:177–207.

Baron, P.A. and Willeke K.A. (Eds.) 2001. *Aerosol Measurement. Principles, Techniques, and Applications*, Wiley–Interscience, New York.

Currie, W.D., van Schaik, S., Vargas, I., and Enhorning, G. 1998. Breathing and pulmonary surfactant function in mice 24 h after ozone exposure. *European Respiratory Journal*, 12:288–293.

Fairchild. G.A. 1972. Measurement of respiratory volume for virus retention studies in mice. *Applied Microbiology*, 24:812–818.

Fenton, R.J., Bessell, C., Spilling, C.R., and Potter, C.W. 1977. The effects of peroral or local aerosol administration of 1-aminoadamantane hydrochloride (amantadine hydrochloride) on influenza infections of the ferret. *Journal of Antimicrobial Chemotherapy*, 3:463–472.

Geiser, M. and Kreyling, W.G. 2010. Deposition and biokinetics of inhaled nanoparticles. *Particle and Fibre Toxicology*, 7:2–18.

Hamelmann, E., Schwarze, J., Takeda, K., Oshiba, A., Larsen, G.L., Irvin, C.G., and Gelfand, E.W. 1997. Noninvasive measurement of airway responsiveness in allergic mice using barometric plethysmography. *American Journal of Respiratory and Critical Care Medicine*, 156:766–775.

Hermsdorf, D., Jauer, S., and Signorell, R. 2007. Formation and stabilization of ibuprofen nanoparticles by pulsed rapid expansion of supercritical solutions. *Molecular Physics*, 105(8):951–959.

Hickey, A.J. (Ed.) 2004. *Pharmaceutical Inhalation Aerosol Technology*, Informa Health Care, London, U.K.

Hinds, W.C. 1999. *Aerosol Technology. Properties, Behavior, and Measurement of Airborne Particles*, 2nd edn., John Wiley & Sons, Inc., New York.

Kirsh, A.A., Stechkina, I.B., and Fuchs, N.A. 1975. Efficiency of aerosol filters made of ultrafine polydisperse fibres. *Journal of Aerosol Science*, 5:119–124.

Koster, R., Anderson, M., and Deber, E.I. 1959. Acetic acid for analgesic screening. *Federation Proceedings*, 18:412–414.

Laube, B.L. 2005. The expanding role of aerosols in systemic drug delivery, gene therapy, and vaccination. *Respiratory Care*, 50:1161–1176.

Lee, T., Chen, Y.H., and Wang, Y.W. 2008. Effects of homochiral molecules of (S)-(+)-ibuprofen and (S)-(–)-sodium ibuprofen dihydrate on the crystallization kinetics of racemic (R,S)-(±)-sodium ibuprofen dihydrate. *Crystal Growth and Design*, 8(2):415–426.

Meade, E.A., Smith, W.L., and DeWittm D.L. 1993. Differential inhibition of prostaglandin endoperoxide synthase (cyclooxygenase) isozymes by aspirin and other non-steroidal antiinflammatory drugs. *Journal of Biological Chemistry*, 268(9):6610–6614.

Mutschler, E. and Derendorf, H. 1995. *Drug Actions. Basic Principles and Therapeutic Aspects*, Medpharm, Stuttgart, Germany, pp. 170–171.

Nada, A.H., Al-Saidan, S.M., and Mueller, B.W. 2005. Crystal modification for improving physical and chemical properties of ibuprofen. *Pharmaceutical Technology*, 29:90–101.

Oberdörster, G., Oberdörster, E., and Oberdörster, J. 2005. Nanotoxicology: An emerging discipline evolving from studies of ultrafine particles. *Environmental Health Perspectives*, 113:823–839.

Onischuk, A.A., Tolstikova, T.G., Sorokina, I.V., Zhukova, N.A., Baklanov, A.M., Karasev, V.V., Dultseva, G.G., Boldyrev, V.V., and Fomin, V.M. 2008. Anti-inflammatory effect from indomethacin nanoparticles inhaled by male mice. *Journal of Aerosol Medicine and Pulmonary Drug Delivery*, 21(3):231–244.

Rabinowitz, J.D. and Zaffaroni, A.C. 2004. Delivery of nonsteroidal antiinflammatory drugs through an inhalation route. United States Patent 7087217.

Rubin, B.K. and Fink, J.B. 2005. Optimizing aerosol delivery by pressurized metered-dose inhalers. *Respiratory Care*, 50:1191–1200.

Schaper, M. and Brost, M.A. 1991. Respiratory effects of trimellitic anhydride aerosols in mice. *Archives of Toxicology*, 65:671–677.

Skyler, J.S. 2007. Pulmonary insulin delivery—State of the art 2007. *Diabetes Technology and Therapeutics*, 9:S1–S3.

Stahly, G.P., Mckenzie, A.T., Andres, M.C., Russell, C.A., Byrn, S.R., and Johnson, P. 1997. Determination of the optical purity of ibuprofen using x-ray powder diffraction. *Journal of Pharmaceutical Sciences*, 86(8), 970–971.

Vijayaraghavan, R. 1997. Modifications of breathing pattern induced by inhaled sulphur mustard in mice. *Archives of Toxicology*, 71:157–164.

Wong, B.A. 2007. Inhalation exposure systems: Design, methods and operation. *Toxicologic Pathology*, 35:3–14.

11.2 PART II. STUDY OF THE TOXIC-PHARMACOLOGICAL PROPERTIES OF NANOSTRUCTURED DRUG FORMS

T.G. Tolstikova, M. Khvostov, A. Bryzgalov, and G.A. Tolstikov

11.2.1 INTRODUCTION

Modern medicine operates a broad spectrum of drugs to treat patients with different diseases. However, a wider application of most of the drugs is impeded by their low bioavailability because of insufficient water-solubility and different toxic effects.

Thereby, the development of water-soluble, low-toxic, safe and high-effective drugs is very significant for modern pharmacology and medicine.

Nowadays, in medicinal chemistry rapidly develops the area that is dedicated to the improvement of well-known drugs properties by complexation with different synthetic and plant compounds. Such complexation results in the drug's water-solubility enhancement, bioavailability improvement, and stability increase. There are many examples of such complexing compounds, for instance, cyclodextrins, PEG, pectin, and chitosan.

One of the promising approaches in this area is the usage of carbohydrate-containing plant metabolites as a complexing agent. Striking examples of such agents are glycyrrhizic acid derived from licorice root and arabinogalactan that contains up to 12%–15% in wood of larches *Larix sibirica* and *Larix gmelinii*. Other interesting agents are the metabolites of South American plant *Stevia rebaudiana* Bertoni stevioside and rebaudiside A. The important advantages of the aforementioned compounds are their low toxicity ($LD_{50} > 5000$ mg/kg), ability to form complexes, and huge amount of raw materials suitable for their industrial isolation.

Aforesaid compounds have been used to synthesize complexes with different well-known drugs. This chapter describes the results of pharmacological investigation of these compounds.

11.2.1.1 Glycyrrhizic Acid

18βH-glycyrrhizic acid (GA) (Figure 11.17), a triterpene glycoside produced by widespread plants, Fabaceae *Glycyrrhiza glabra* L., *Glycyrrhiza uralensis* Fisch, and *Glycyrrhiza korshinskyi* Grig (licorice), is one of the few plant metabolites combining availability with a unique variety of pharmacological activities (Baltina, 2003; Tolstikov et al., 1997a, 2007a). Glycyrrhizic acid, the molecule of which consists of hydrophobic (aglycon) and hydrophilic (carbohydrate chain) parts, manifests the properties typical for micelle-forming substances (James et al., 1962; Gilbert and James, 1964; Azaz and Segal, 1980; Kondo et al., 1986; Maskan, 1999; Romanenko and Murinov, 2001). According to Azaz and Segal (1980), the micellar critical concentration (MCC) of GA water solution is 10^{-3} M. The study of rheological properties of GA solutions by the viscometric method resulted in a conclusion that their properties were similar to those of Newton liquids (Romanenko and Murinov, 2001). The solution viscosity at the concentration lower than 3×10^{-5} M does not change. With increase of the concentration above that for MCC ($>3 \times 10^{-5}$ M) the solution viscosity and turbidity increase. Thus, GA associates existing in water solution start to group into micelles. When ethanol is added to water solution, at 20 volume percent of ethanol MMC increases to 10^{-4} M. Apparently, the presence of ethanol prevents micelle formation at GA concentration ranging from 3×10^{-5} to 1×10^{-4} M.

GA association in solutions was studied by NMR methods (Kornievskaya et al., 2008). In NMR spectra of H^1 for water–methanol solutions (20% MeOH) as GA concentration increases, the lines of methyl groups corresponding to aglycone broaden. The authors suggest that this is due to substantial change of the relaxation time of GA methyl protons, which is caused by the formation of larger structures as the concentration increases. Using pulse NMR-spectroscopy, the spin–spin relaxation of GA methyl protons depending on GA concentration in the solution was investigated. The experimental data indicate that small micelles with aggregation number $M \leq 10$ may form. This applies to

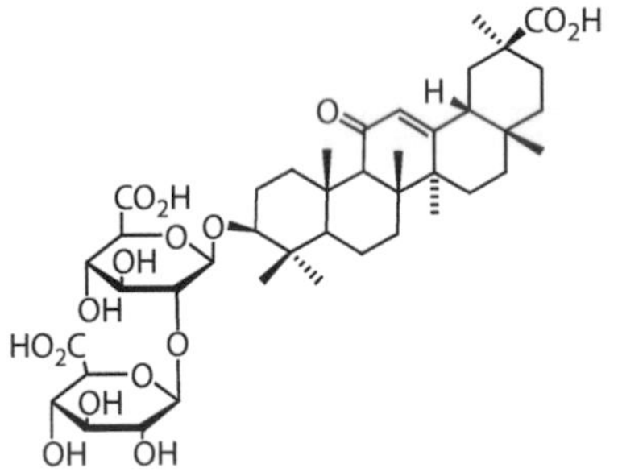

FIGURE 11.17 18βH-glycyrrhizic acid.

water–methanol (20% MeOH), neutral or acid solutions with GA concentrations ≤0.5 mM. In 50% water solution, micellation does not occur.

The conclusion about small-size associates formation is based on the fact that the dependence of GA mole fraction in micellar state on the concentration does not drastically change in the MCC range.

The study of water solutions by small-angle x-ray scattering method showed that at the concentration of about 1.0 mM glycoside forms rod-like associates with the radius ~1.4 nm and length up to 60 nm. The associates are stabilized by intermolecular hydrogen bonds.

11.2.1.2 Complexes of Pharmacons with Glycyrrhizic Acid

The pioneering works on complexation are Soltesz et al. (1963) and Krasova et al. (1978), where the solubilizing effect of glycyrrhizic acid and its monoammonium salt, glycyrrham, was investigated for a series of water-insoluble pharmacons. In the earlier-stated works as well as in Sasaki et al. (1988), it was demonstrated that it is possible to obtain water solutions of practically water insoluble antibiotics such as oxytetracycline, nystatin, actinomycin C, corticosteroids such as hydrocortisone and prednisolone, and sulfazine, a sulfanilamide drug. GA demonstrates solubilizing properties at concentrations higher than MMC (Krasova et al., 1978). It should be noted that none of the earlier-mentioned works discusses the ability of glycyrrhizic acid to form complexes with pharmacons. For the first time the issue was discussed in Tolstikov et al. (1990), where GA complexes were called new carriers for drugs. In earlier studies, the solution method of complex synthesis prevailed. Nowadays, the method of mechanochemical solid-phase synthesis is employed to obtain the complexes with water-insoluble medicines; it allows achieving an even higher solubility compared to liquid-phase synthesis (Dushkin et al., 2001).

The mechanochemical processing is performed in an AGO-2 planetary mill. Processing routine: acceleration of grinding bodies, 60 g; mass of processed mixture, 3 g; cylinder size, 40 mL; grinding bodies—steel balls, 6 mm in diameter; load size, 75 g. Time of processing ranges from 3 to 10 min. A longer processing may lead to partial chemical decomposition of samples, while a shorter processing period may result in insufficient sample homogenizing.

After the mechanochemical processing, the water solutions of GA were characterized by the method of gel-chromatography (Figure 11.18). In all ranges of studied concentrations (0.0001–0.5 wt.%), peaks of high molecular formations, with molecular mass of ~46–67 kDa, are observed, while the molecular mass of GA amounts to 836.96 Da. It has been established that glycyrrhizic acid is almost fully self-associated into micellae, the most stable ones being the micellae with MM = ~66 kDa, consisting of almost 80 GA molecules (Dushkin et al., 2010).

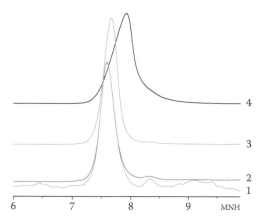

FIGURE 11.18   Gel-chromatograms of GA solution. 1, Concentration in the tested solution—0.001% weight; 2, 0.01%; 3, 0.1%; 4, 0.5%.

During the mechanochemical synthesis of solid dispersions of GA, its mass excess of 10:1 is used, which corresponds to mole ratio with pharmacons (2.5:1–4:1). During the solution of the obtained dispersions, a significant increase in pharmacon water solubility is observed, which proves the high effectiveness of GA as a solubilizer and the effectiveness of the mechanochemical method in the obtaining of water-soluble solid dispersions.

11.2.1.3   Complexes of Nonsteroid Anti-Inflammatory Drugs with Glycyrrhizic Acid

Complexes of GA with acetylsalicylic acid (ASA), ortophenum (OF), butadionum (BD), and indo-metacin (IM) with the composition (GA:nonsteroid anti-inflammatory drugs [NSAID]) 1:1 and 2:1 were synthesized by both the solution (Tolstikov et al., 1991a) and solid-phase (Dushkin et al., 2001) methods. Complex of GA with ibuprofen (IP) with the composition 4:1 was synthesized by solid-phase method. Complexation was confirmed by spectral analysis. In the IR-spectra of the complexes the spectral lines of hydroxylic and carbonylic groups of glycoside are shifted to the short waves.

All the earlier-stated complexes demonstrate anti-inflammatory (AI) activity at the doses lower than the source pharmacon. The dose range of the complexes (LD_{50}/ED_{50}) is 3–11 times higher than that of the initial NSAID (Tolstikov et al., 1991a) (Table 11.2).

Complexes with ASA and OF (GA: ASA, GA: OF) produce an expressed AI effect in six models of acute inflammation induced by carragenin, formalin, histamine, serotonin, agar (Difko), tripsin, as well as at chronic inflammation ("cotton pellet" and "pocket" granulomas) of intact and adrenal-ectomized animals (Bondarev et al., 1991). A complex of GA with IM 1:1 has a more pronounced AI effect as compared with initial drug if administered in equal doses (10 mg/kg).

At complexation of NSAID with GA, the potentiation of other kinds of biological activity (anal-gesic, antipyretic) was observed (Baltina et al., 1988a,b,c; Bondarev et al., 1991). At electrical and thermal irritation, the analgesic effect of GA complex with OF was more pronounced than that of OF (57.5 ± 2.0 and 43.2 ± 2.6), and at thermal pain irritation, it exceeded the effect of amidopyrine (23.4 ± 1.1 and 18.5 ± 1.4). The analgesic effect of GA complex with analgin (AN) was 11.4 stronger than that of AN (Baltina et al., 1992). The analgesic effect of complex GA:ASA exceeds the effect of aspirin at thermal pain irritation. At acetylcholine convulsions the analgesic dose range of GA complexes with ASA and OF is 3 and 2.3 times broader than that of NSAID, respectively. In the acetum convulsion model the therapeutic ratio of complex GA:ASA was four times higher than that of aspirin. Complexes GA:ASA and GA:OF have pronounced antipyretic action. Their antipyretic dose range was two times broader than that of initial pharmacons (Baltina et al., 1991b, 1992; Bondarev et al., 1991).

Thus, water-soluble complexes GA:ASA and GA:OF have pronounced AI and analgesic effects with the pharmacological spectrum and therapeutic dose range exceeding those of initial NSAIDs. The complexes have pronounced membrane stabilizing action that is expressed in a decreased accu-mulation of primary and secondary lipid peroxidation products in animals with chronic inflammation.

TABLE 11.2
Anti-Inflammatory Dose Range of Glycyrrhizic Acid Complexes and NSAID

Compounds	Dose Range of the Complex[a]	Dose Range of Initial NSAID
GA:ASA (1:1)	4500/82 = 54.8	1900/98 = 19.4
GA:OF (1:1)	1750/12.5 = 140	310/8 = 33.7
GA:BD (1:1)	3150/62 = 50.8	880/56 = 15.7
GA:AN (1:1)	8000/68 = 117.6	570/55 = 10.3

[a] LD_{50}/ED_{50}; ED_{50}, effective dose.

TABLE 11.3
Acute Toxicity of Complexes GA with NSAID in Mice (Per os)

Drugs	LD_{50} (mg/kg)	LD_{50} (mg/kg) Pharmacon
GA:ASA (1:1)	4500	1900
GA:OF (1:1)	1750	310
GA:BD (1:1)	3150	880
GA:AN (1:1)	8000	570

Complexes with GA have less irritant effect on the mucous coat of stomach as compared with original NSAID. Thus, complex GA:ASA stimulates ulcerous lesions reparation and complex GA:OF shows low ulcerogenic activity. Both complexes decrease the level of prostaglandins E_1 and E_2 in the blood of animals with chronic inflammation. The complexes can be recommended as anti-inflammatory drugs for clinical trials, including the trial in patients with stomach and duodenal ulcers.

Acute toxicity of GA complexes with NSAID is 2–14 times lower as compared with original drugs (see Table 11.3) (Baltina et al., 1991b, 1992; Bondarev et al., 1991; Tolstikov et al., 1991a).

Complexation of IP with glycyrrhizic acid by the mechanochemical method allows lowering the dose by 10 times while preserving the high analgetic activity, but only in the model of chemical irritation of peritoneum. This also demonstrates the preservation of its anti-inflammatory activity for this pharmacon in the complex.

At the complexation of NSAID with GA synergetic effect can be observed, manifested by an increase in water-solubility and biological activity and simultaneous decrease of toxicity and ulcerogenic action on gastrointestinal tract. Comparative study of pharmacological activity of molecular complexes GA:ASA and GA:OF synthesized by the solution and mechanochemical methods showed that solid-phase synthesis opens an economical and technologically more suitable way of making highly effective NSAI drugs (Dushkin et al., 2001).

11.2.1.4 Complexes of Prostaglandins with Glycyrrhizic Acid as a New Group of Uterine Tonics

Prostaglandins have been widely used in medicine and veterinary practice due to their ability to stimulate uterine muscles in small doses.

The use of prostaglandin in veterinary practice is one of the most important advantages of prostaglandin studies. Prostaglandin derivatives are widely used in pigs try to synchronize farrowing. Prostaglandins are also highly effective in preparing cows and mares for artificial insemination. Introduction of prostaglandin drugs helped to resolve the problems of postpartum complications in cows and mares.

Cloprostenol is one of the main drugs used in veterinary practice. Along with other prostaglandins, it is obtained by multistage synthesis, which results in high prices of the drug. The reduction of therapeutic active dose is an important task. Besides, it is necessary to increase the stability of labile prostaglandins in the finished drugs.

Both problems were successfully resolved by the complexation of prostaglandins with GA. GA complexes with prostaglandins of the E and F series (PGE_1, PGE_2, $PGF_{2\alpha}$, sulprostone [SP] and cloprostenol) were obtained and investigated.

Unlike the initial prostanoids, the complexes synthesized by the solution method in the form of water-soluble amorphous substances are highly stable both in solid state and in solutions (Tolstikov et al., 1991c). Complexation has been proved by spectral analysis methods and chromatographic homogeneity.

TABLE 11.4

Uterine Tonic Activity of Prostaglandins and Their Complexes with GA (1:1) in Rats *In Vitro*, Phosphate Buffer $C = 10^{-8}$ g/mL (Polyakov et al., 2006): PGE_1, PGE_2, SP, $PGF_{2\alpha}$

Substance	Change of Uterine Contraction Amplitude (%)	P	Changes in Uterus Tonus (%)	P
GA:PGE1	53.4 ± 5.0	<0.002	49.4 ± 1.2	<0.002
PGE1	24.3 ± 1.5	<0.05	30.7 ± 2.2	<0.05
GA:SP	150.0 ± 11.0	<0.001	135.0 ± 10.0	<0.001
SP	50.0 ± 5.0	<0.001	115.0 ± 9.5	<0.001
GA:PGE2	63.5 ± 6.0	<0.001	40.7 ± 4.0	<0.002
PGE2	20.0 ± 2.8	<0.05	33.5 ± 2.4	<0.05
GA:PGF2α	55.6 ± 5.0	<0.001	61.0 ± 5.6	<0.001
PGF2α	27.8 ± 1.5	<0.05	39.4 ± 5.3	<0.02

In experiments on the uterus of rats and guinea-pigs, complexes GA:PGE_1 (1:1) and GA:$PGF_{2\alpha}$ (1:1) double uterine contraction amplitude as compared with sodium PGE_1 at the same concentrations (10^{-8} g/mL). SP and PGE_2 in the form of complexes with GA (1:1) strengthen uterine contraction amplitude by three times while increasing uterus tonus (Tolstikov et al., 1991b) (Table 11.4).

On the basis of GA complexes with cloprostenol, a known synthetic luteolytic prostaglandin, a highly effective veterinary drug "Chlatraprostin" was developed, the active agent dosage of which is five times lower than that accepted in the world practice. In comparison with imported analogues the drug is cheaper and its action is more physiological.

"Chlatiram" containing amino acid tyrosine along with GA and cloprostenol is even more efficacious (Tolstikov et al., 1997b). Its effect is stronger than that of Estrofan, the well-known veterinarian drug, at a 100 times reduced dose of prostaglandin.

11.2.1.5 Complexes of Glycyrrhizic Acid with Cardiovascular Drugs

The structure and pharmacological properties of GA complexes with antiarrhythmic drugs allapinin (hydrobromide of lappaconitine, diterpenoid alkaloid) (LA) and antihypertensive drug nifedipine were investigated.

The process of complexation of lappaconitine with glycyrrhizic acid in solutions was investigated by photo-CIDNP method (Polyakov et al., 2005; Kornievskaya et al., 2007a,b,c). Photolytic destruction of lappaconitine, which slows down sharply in the presence of GA, was chosen as a reaction simulating the complexation effect. The reaction leads to protection of the complex-bonded alkaloid. The experiment demonstrating that complexation results in the photodestruction slowdown is a sufficiently correct model of pharmacon metabolism *in vivo* for the complex of glycyrrhizic acid with lappaconitine base in water–methanol solution (20% MeOH), stoichiometry 1:1, stability constant $K_s = 2 \times 10^5 \pm 0.13 \times 10^5$ M^{-1}. In pure methanol the constant is lower by an order, $K_s = 1.3 \times 10^4$ M^{-1}. Complex of lappaconitine hydrobromide with GA has the stability constant $K_s = 2.6 \times 10^3$ M^{-1}.

It should be noted that the average values of the stability constants of pharmacon complexes with cyclodextrins are about 10^3 M^{-1}.

Allapinin has been included in the list of antiarrhythmic drugs. It is recommended at different forms of cardiac rhythm disturbance, especially in ventricular arrhythmias, paroxysmal atrial fibrillation, and monofocus atrial tachycardia (Sadritdinov and Kurmukov, 1980; Mashkovsky, 2005). The disadvantage of allapinin is its high toxicity.

In a special series of experiments, it was shown that complexes GA:LA produce an effect on antiarrhythmic activity and that complex GA:LA of composition 4:1, patented as "alaglysine," has the most potent antiarrhythmic effect (Bryzgalov et al., 2005). In the calcium chloride and aconite arrhythmia models, alaglysine demonstrated high antiarrhythmic activity and the antiarrhythmic index (LD_{50}/ED_{50}) higher than that of all known antiarrhythmic drugs. Alaglysine is 10 times less toxic than allapinin. In an extended study of alaglysine antiarrhythmic activity using the calcium chloride and aconite arrhythmia models, the complex at the doses of 0.125 and 0.250 mg/kg as such did not affect the ECG parameters. Intravenous injection of the agent at the dose of 0.125 mg/kg before the injection of calcium chloride lethal dose blocked the development of arrhythmia in 80% of the rats. When administrated after the arrhythmogen ($CaCl_2$), alaglysine at the dose of 0.250 mg/ kg reverses the arrhythmia development in 50% of the animals. In the adrenaline arrhythmia model, a single administration of alaglysine at the doses of 0.125 and 0.250 mg/kg prevents the development of profound arrhythmia. The ECG parameters reverse to norm in 50% (dose 0.125 mg/kg) and in 100% (dose 0.250 mg/kg) of the animals. In the model, ED_{50} is 0.125 mg/kg for alaglysine and 0.290 for allapinine, as follows from the effective dose of alaglysine, it contains 15 times less lappaconitine than the effective dose of allapinine.

Nifedipine (NF) (2.6-dimethyl-3.5-dicarbomethoxy-4-(2′-nitrophenyl)-1.4-dihydropyridine) belongs to calcium antagonists. It dilates coronary and peripheral (primarily arterial) vessels and diminishes myocardial oxygen need. Nifedipine produces an insignificant negative inotropic effect and a very weak antiarrhythmic effect. The lukewarm attitude of physicians to this drug has recently changed into a revived interest. This is substantially due to the appearance of novel dosage forms of Nifedipine and to additional data characterizing the drug as an affordable and active agent for arresting hypertensive crises. The mere fact of the continuing development of novel dosage forms of Nifedipine indicates that as the simplest derivative of 1.4-dihydropyridine with proven production technology Nifedipine will be used in the future as a reliable alternative of more expensive drugs of this type (Pogosova, 2006).

The study of complexation of GA with Nifedipine using NMR method showed the existence of complex with the molecular composition GA:NF = 2:1 in water–alcohol solutions at the concentrations of GA ≤ 0.5 mM. At higher GA concentrations associates are formed, which include equal number of molecules of both substances. The complex is highly stable as indicated by its stability constant ($K_s = 1.2 \times 10^5$ M^{-1}) (Polyakov et al., 2006).

Nevertheless, the preliminary study of hypertensive activity of Nifedipine convincingly demonstrated that GA:NF = 4:1 is an optimal composition. The complex synthesized by the mechanochemical activation method is an amorphous substance consisting of 20–50 μm vitreous particles. Complexation was also confirmed by radiographic and thermogravimetric data, according to which traces of crystalline phase of both reagents disappear. Water-solubility of Nifedipine in the complex increases 8.5 times.

The study of antihypertensive action showed that at intravenous administration of the complex in water solution to rats the desired effect is manifested at the dose of Nifedipine 10 times lower than usual dose (Tolstikova et al., 2006).

As stated earlier, an important consequence of complexation could be the enhancement of pleiotropic properties of a pharmacon. For Nifedipine this is its antiarrhythmic action. However, the use of Nifedipine as an antiarrhythmic drug at an effective dose to achieve the desired positive effect on cardiac rhythm is impossible, for at this dose Nifedipine provokes a decrease in arterial pressure and may cause hypotensive crisis. The study of antiarrhythmic action of GA: NF complex showed that a protective antiarrhythmic effect can be reached at a dose 29 times lower than that providing antihypertensive effect.

Thus, GA: NF 4:1 complex is promising for the development of the first universal drug capable of arresting hypertension and arrhythmia and safe at parenteral administration.

In the comparative study of the mechanisms of action of Nifedipine and its complex with GA, the complexation effect was confirmed *in vitro* experiments with the use of isolated neurons of peripharyngeal ganglia of mollusk *Lymnaea stagnalis*. The neuron technique allows tracing the effect of the agents introduced in the intercellular space on the transport of Ca^{+2} ions using generation of calcium channels action potentials (Zapara et al., 1988, 1999). Nifedipine was found to fully block the responses at the dose of 3.0 mM. For GA:NF 4:1 complex this effect was found to be manifested at pharmacon concentration 30 times lower (0.1 mM) than that of the original drug. The blockade of responses caused by the original drug and their restoration after neurons washing out proceed faster than in the experiments with the complex. This indicates stronger bonding of the complex with the receptors and, therefore, the prolonged action of the complex. Comparative study of the effect of Nifedipine and its complex with GA on the dynamics of calcium current amplitude confirmed the conclusions on the enhancement of complex–receptor affinity (Tolstikova et al., 2007b).

11.2.1.6 Complex of Glycyrrhizic Acid with Indirect Oral Anticoagulant Warfarin

Warfarin (WF) is widely used in modern medicine as an indirect oral anticoagulant for long-term administration. The number of applications for the indirect anticoagulants in the world has grown by 10 times in the last 10 years, which is explained by the large number of indications for their usage. The main disadvantage of such therapy is the development of hemorrhages, often large ones, particularly during the induction period (or dose selection), as well as the difficulty of dose control (Levine et al., 2001; Epstein et al., 2010).

The complex WF:GA with the mass ratio 1:10 was synthesized. This component ratio had been previously selected in our other works and determined to be optimal for complexes with GA (Tolstikov et al., 2007b; Tolstikova et al. 2009).

The definition of solubility is considered to be an indicator of complex stability. In the complex of glycyrrhizic acid with warfarin, we managed to increase the solubility of the latter by 14 times (Table 11.5), which indicated the formation of a stable complex.

During the trial experiments with animals we determined the optimal complex dose to be 20 mg/kg (warfarin dose being 2 mg/kg), which clearly stimulated the increase in prothrombin time (PT). The first measurement of PT parameters was performed in 6 h after a single intragastric administration. This time interval was chosen on the basis of warfarin's pharmacologic properties; namely, the authors Zhu and Shin (2005) had previously proven that the maximum concentration of warfarin metabolites in the blood plasma of rats is reached in 6–12 h.

According to our research, PT definitely alters at this temporal point only in groups of positive control—WF and WF (Nycomed, tablets) inserted in 2 mg/kg doses, which corresponds with the content of warfarin in complex with GA. However, this length of PT is not clinically significant, as it does not secure the necessary increase in hemopexis time.

A more significant increase in PT was observed in 24 h after a single intragastric administration of the comparator agents; in case of the examined complex, the PT increased (24.88 ± 1.77 s) only in 30 h (two administrations) after the start of the experiment and reached the values corresponding to positive control WF (PT 42.0 ± 1.60 s) only in 54 h (three administrations, PT 38.8 ± 1.31 s) (Figure 11.19).

As a result, the complexation of warfarin with GA did not allow reducing its effective dose, which may be explained through the complex structure. Using the method of gel-penetrating chromatography, we discovered that one of the mechanisms of complexation of pharmacons with GA, particularly of warfarin, is the involvement of their molecules in micellae/self-associates of GA. This kind of structure serves as protection from fast metabolism and slows down the binding of warfarin molecule's active centers with the receptors, and therefore the basic activity manifests

TABLE 11.5
Solubility in Water of the WF:GA (1:10)

Agents	WF Solubility in Water (g/L)	Increase in Solubility (Times)
WF	0.021	—
WF:GA (1:10)	0.295	14

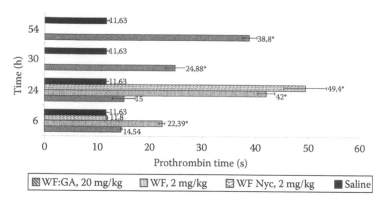

FIGURE 11.19 Prothrombin time of the WF:GA (1:10). *Significantly different ($P<0.05$) from control (saline).

itself only after a longer period of time and cumulation of warfarin in the organism. However, the modern anticoagulant medicine should act faster (Aken et al., 2001) and because of this, to obtain the desired effect we should change a complexing agent (see more in Section 11.2.1.14).

11.2.1.7 Complexes of Glycyrrhizic Acid with Psychotropic Drugs

Complexes with the following pharmacons were obtained for the examination: fluoxetin (an antidepressant), phenibut, buspiron, diazepam, oxazepam, medazepam (anxiolytics), and azaleptin (neuroleptic). Of these, the GA complexes with fluoxetin, phenibut, and buspiron are the most well-researched ones.

Fluoxetin (FL) (N-methyl-3-(4′-trifluoromethylphenoxy)-3-phenylpropylamin hydrochloride) is known to cause an antidepressant action connected with inhibition of serotonin capture by neurons in CNS. Antidepressants are known to have such disadvantages as rather high doses, a narrow range of therapeutic index, high toxicity and long elimination period, which provide a negative effect on kidney functions and damage hepatocytes.

The study of GA complexes with fluoxetin was conducted with the main aim to determine whether the side effects of this pharmacon could be decreased. Preliminary tests showed that the complex with the composition GA:FL 4:1 was the most active. This complex, one weight party of which contains 0.072 parts of pharmacon, is patented under the name Fluoglyzine (FG) (Tolstikov et al., 2004). The complexes with the compositions 4:1 and 2:1 are characterized by LD_{50} higher than 5000 mg/kg as compared with 248 mg/kg for Fluoxetin.

Fluoglyzine in the Porsolt test at a single and prolonged administration was found to have a more pronounced antidepressant action as compared with Fluoxetine. In comparison with Fluoxetine, Fluoglyzine produces a stronger inhibiting effect on serotonin structures. For example, in the test with 5-oxytriptophan, fluoglyzine inhibits chloral hydrate action to a greater extent than Fluoxetine. Both drugs produce no anxiolytic effect.

In the social depression model in mice, Fluoglyzine was shown to have an antidepressant effect similar to that of Fluoxetine, which was manifested in a twofold increase of communicability of individuals (the number and duration of contacts) in response to a known and strange partner. The dose of Fluoxetine in complex GA:FL 4:1 is 1.08 mg/kg as compared with 15 mg/kg for the sample drug. Fluoglyzine, similarly to fluoxetine, prevents a decrease of glucose content in blood and lowers the intensity of peroxidation, thereby normalizing antioxidant status of depressed individuals (Avgustinovich et al., 2004; Kovalenko et al., 2004; Tolstikova et al., 2004).

To elucidate the action mechanism of complexes GA: FL 4:1 and 1:1 as compared with fluoxetine, their effect on the content of catecholamines and their precursors in different parts of cerebrum was studied at a single and therapeutic administration at the dose of 25 mg/kg. A decrease of fluoxetine dose by 4 times (complex 1:1) and 17 times (complex 4:1) in complexes with GA was found to lower the effect on serotonin metabolism and to activate dopamine metabolism in cerebrum (Shishkina et al., 2005, 2006).

In Shishkina et al. (2006), the nootropic activity of fluoxetine was established for the first time. Both complexes GA: FL 1:1 and 4:1 demonstrate the same effect but at a slightly lower level than reference drugs.

Fluoxetine (30 µM) is known to suppress epileptiform activity, which is manifested by a 50% decrease in the amplitude of the electric potential oscillation, which reflects the activity of nerve cells induced by electric stimulation of hippocampus under the action of picrotoxin.

Fluoglyzine (complex 4:1), similar to fluoxetine, suppresses bicuculline-induced epileptiform activity in the sections of rat hippocampus, demonstrating antiepileptic activity (Shishkina et al., 2006).

Phenibut (PhB) (γ-amino-β-phenylbutyric acid hydrochloride) is a nootropic sedative medication, relieving tension and anxiety and improving sleep. In clinical practice it is used for asthenic syndrome, anxious neurotic conditions, sleep disorders and as an anti-naupathia preparation in pre-surgery procedures. As a nootropic drug Phenibut has serious flaws, for example, it provokes sleepiness and allergic reactions.

Complexes GA: PhB 2:1 and 4:1, the toxicity of which is twice as lower than that of Phenibut, produce a cognitive effect similar to that of the pharmacon and GABA. Unlike Phenibut and GABA, the complexes increase mnestic capabilities in animals by 20% and decrease sedative effect (Shishkina et al., 2006).

Buspirone (BSP) is an anxiolytic drug. It is characterized by high affinity to pre- (agonist) and postsynaptic (partial agonist) serotonin receptors of the 5-HT_{1A} subtype. It impedes the synthesis and release of serotonin, the activity of serotonergic neurons, particularly in the dorsal raphe nucleus. It selectively blocks (antagonist) the pre- and postsynaptic D2-dopamine receptors (has a moderate affinity) and increases the speed of activations of dopamine neurons in the mesencephalon. The pharmacological properties of the PSP:GA complex (1:10) were studied on the model of social depression in sexually mature male mice of the C57BL/6J line. It was shown that neither buspirone nor the complex influenced the level of glucose and the protein content in blood (Table 11.6).

Earlier it had been shown that the stress of social interaction in mice causes the increase in glucose content in the blood plasma of the victims (Avgustinovich et al., 2003). It is known that a prolonged emotionally negative social stress increases the need in glucose, as it serves as the main source of energy for nerve cells. Considering this, it may be said that neither buspirone nor the complex of buspirone with glycyrrhizic acid influences the mobilization of the energetic and structural resources of the body caused by chronic social stress.

TABLE 11.6
Effects of Chronic Agents Dosing on the Biochemical Blood Indices in Anxious-Depressed Mice

Blood Indices	Solvent	Buspiron	Buspiron + GA
Glucose (mmol/L)	5.53 ± 0.035	5.52 ± 0.049	5.29 ± 0.154
Protein (g/L)	18.32 ± 2.093	21.46 ± 2.843	19.14 ± 2.755
Triglyceride (µmol/L)	1.27 ± 0.394	1.24 ± 0.314	0.90 ± 0.259
Cholesterol (mmol/L)	1.69 ± 0.554	2.97 ± 1.054	2.09 ± 0.512
MDA (µmol/L)	61.05 ± 1.567	53.25 ± 4.219[a]	53.76 ± 3.877
Catalase (U/L)	32.21 ± 2.842	38.39 ± 1.975[a]	38.81 ± 2.703[a]

[a] Significantly different ($P < 0.05$) from solvent.

According to our observations, buspirone and BSP: GA did not influence the content of choles-terol and triglyceride in the blood of anxious-depressive mice either (Table 11.6). Therefore, the chronic introduction of buspirone and BSP: GA does not affect the markers of lipid exchange.

One of the nonspecific factors determining the functional changes during social stress is the increase in lipid peroxidation (LP) and the responsive activation of the body's antioxidant system, an important component of which is the catalase ferment (Pshennikova, 2000). Avgustinovich et al. (2003) had previously observed the decrease in catalase activity in anxious-depressive mice. The dis-covered and confirmed increase in catalase activity during the chronic introduction of buspirone and BSP: GA complex, as well as the decrease in malondialdehyde (MDA) content in the group of animals with buspirone introduction (Table 11.6), proves the activation of the antioxidant system in "sick" animals. Considering that buspirone is primarily used as an anxiolytic drug, one may suspect the involvement of the LP mechanisms in the process of the anxious status formation in mammals. It is especially important that an equally strong effect is observed in cases of both buspirone and its com-plex with GA, in which the buspirone content is essentially 10 times lower. It shows that the usage of BSP: GA complex may prove more effective than that of buspirone, and that it may serve as a method of reduction of the side effects observed during the prolonged introduction of the anxiolytic to patients.

11.2.1.8 Complexes of Glycyrrhizic Acid with Antitumor Drugs

Synthesis by the solution method of complexes of glycyrrhizic acid with 5-fluorouracil, fluorofur, and rubomicine hydrochloride with composition 1:1 is described in (Baltina et al., 2003). Composition of the complexes, which are chromatographically homogeneous amorphous substances, was proved by spectral methods. Complexation resulted in water solubility and lower toxicity. The complex of fluorofur has an antitumor action on Pliss lymphosarcoma, melanoma B-16, and Heren's carcinoma. The efficiency indices for these diseases are equal to 3.05, 2.11, and 1.7, respectively. Inhibition of growth for these tumors is 67.2%, 53.4%, and 87.1%, respectively.

11.2.1.9 Complexes of Glycyrrhizic Acid with Antimicrobial Drugs

The 1:1 complexes of GA with antibiotics such as levomycetin, sulfapyridazin, sulfadimethoxine, sulfamonomethoxine, sulfadimesine, sulgin, sulfanilamides; and drugs isoniaside and furacillin were synthesized by the solution method. All the complexes are chromatographically homogeneous amorphous substances; complexation is proved by IR spectral data indicating that the spectral lines of hydroxylic and carbonylic groups are shifted to the short-wave band.

Comparative data are available on the antimicrobial action of the complexes. In staphylococcosis, on the 10th day after contagion the survival rate was the highest (90%) in the group of animals receiv-ing the complex of GA with levomycetin at the dose of 50 mg/kg. In the group receiving only levo-mycetin the survival rate was 30%. The survival rate among the animals infected with *Pseudomonas aeruginosa*, *Proteus vulgaris*, and *Escherichia coli* was about 80% in the group receiving the com-plex. In the group receiving levomycetin the survival rate ranged between 20% and 50%. The com-plex was shown to be able to stimulate humoral and cellular immunity (Kondratenko et al., 2003).

11.2.1.10 Complexation with Hypocholesterolemic Agent Symvastatin

The inhibitors of 3-hydroxi-3-methylglutaril-CoA reductase (3HMG-CoA reductase), so-called statins, are known as efficacious drugs lowering low-density lipoprotein secretion, which explains their wide use in anti-atherosclerosis therapy. At the same time, most statins are known to have side effects. That is why the development of safer drugs with a prolonged action is a present-day necessity. An NMR study of symvastatin (SMS) behavior in solutions in the presence of GA led to a conclusion on the formation of stable complexes. Complex GA: SMS = 4:1 was synthesized, which demonstrated stability in water solutions at GA concentration more than 0.2 mM (Tolstikov et al., 2007b). The complex patented under the name "symvaglysine" (SMG) (Vavilin et al., 2008) showed 3HMG-CoA reductase noncompetitive inhibiting activity. Effective at the doses containing one third of statin's normal dose, symvaglysin is a more efficacious and safer agent as compared with symvastatin.

Thus, symvaglysine is a noncompetitive inhibitor/proinhibitor of 3HMG-CoA reductase, arresting the synthesis of cholesterol in the microsomal fraction of rat liver *in vitro* on par with symvastatin. At the inhibition constant ranging between 100 and 300 nM, symvaglysine inhibits the formation of mevalonat by 37.7%–42.0%. After a 14 day course of treatment of hypercholesterolemic rats with symvaglysine at the doses containing one third of symvastatin normal dose, symvaglysine decreases total cholesterol level by 31%–33%, which is comparable to the effect of symvastatin in the therapeutic dose.

Higher safety of symvaglysine follows from the fact that after 14 days of intake at the doses containing symvaglysine the content of which was two to five times lower the increase of creatine phosphokinase level in blood of hypercholesterolemic rats was two to five times lower (Ragino et al., 2008; Vavilin et al., 2008).

11.2.1.11 Antidote and Antiradical Activity of Complexes of Glycyrrhizic Acid with Uracil Derivatives

GA forms stable complexes with uracil derivatives with the molecular composition 1:1. The complex with 2-tiouracil is a dehydrate, complexes with salts are crystallized with one molecule of water. Complexes with 5-oxi-6-methyl-uracil and aminouracil are waterless (Tolstikov et al., 1996). All complexes are 1.8–1.9 times less toxic than uracil and belong to moderately or low toxic substances. By the range of antidote action determined using the model of male mice poisoning with sodium nitrite, the complexes exceeded initial uracils on average twice and cystamin—2.6–17.4 times. Complexes with uracils have a lower value of effective dose (ED_{50}) than cystamin, that is, they are more efficacious than the reference drug. Complexes with 2-tiouracil and 5-oxi-6-methyluracil, although having a higher value of ED_{50} as compared with cystamin, exceed it in the range of pharmacological action.

The experiment results were evaluated using the survival and mortality rates in the control and test groups 48 h after poisoning. The control animals received physiological solution and test animals were injected complexes of GA and initial uracils at the doses of 6.75, 12.5, 25.0, 50.0, 100.0, and 200.0 mg/kg.

Antiradical activity of GA complexes *in vitro* was studied by chemiluminescence method by determining the reaction rate constant (K_7) for the interaction of ethylbenzol peroxide radicals with the compounds studied. It was found out that the antiradical activity of GA complexes with aminopyrimidines and 2-tiouracil was on average 10 times higher than that of ionol (Table 11.7).

High antidote activity of GA complexes with uracil derivatives in combination with their antiradical action makes these compounds promising for the development of drugs protecting living beings from toxic exposures.

TABLE 11.7
Antiradical Activity of Uracils and Their Complexes

Uracils	Constant K_7 M/l	
	Complex with GA	Pyrimidine
2-Thiouracil	$1.3 \pm 0.6 \times 10^5$	$1.3 \pm 0.6 \times 10^5$
6-Amino-2-thiouracil	$4.3 \pm 1.3 \times 10^5$	$1.3 \pm 0.4 \times 10^5$
6-Aminouracil	$4.3 \pm 1.3 \times 10^5$	$1.3 \pm 0.4 \times 10^5$
6-Methyl-2-thioutacil	$9.1 \pm 2.7 \times 10^4$	$9.1 \pm 2.7 \times 10^4$
5-Oxi-6-methyluracil	$2.6 \pm 0.8 \times 10^4$	$2.6 \pm 0.8 \times 10^4$
5-Aminouracil	$2.0 \pm 0.6 \times 10^5$	$2.0 \pm 0.6 \times 10^5$
GA	$1.3 \pm 0.4 \times 10^2$	
Ionol	$2.3 \pm 0.6 \times 10^4$	

11.2.1.12   *Stevia rebaudiana* (Bertoni)

S. rebaudiana (Bertoni) Bertoni is a perennial shrub of the Asteraceae (Compositae) family native to certain regions of South America (Paraguay and Brazil). It is often referred to as "the sweet herb of Paraguay." Stevioside, the main sweet component in the leaves of *S. rebaudiana* (Bertoni) Bertoni tastes about 300 times sweeter than sucrose (0.4% solution).

Structures of the sweet components of *Stevia* occurring mainly in the leaves are given in Figures 11.20 and 11.21. Their content varies between 4% and 20% of the dry weight of the leaves depending on the cultivar and growing conditions. Stevioside is the main sweet component. Other compounds present but in lower concentration are as follows: steviolbioside, rebaudioside A, B, C, D, E, F, and dulcoside A (Kennelly, 2002; Starrat et al., 2002). As the objects of pharmacological research, stevioside and rebaudioside A have been thoroughly investigated. In particular, they were shown to be absolutely safe ($LD_{50} > 10,000$ mg/kg for stevioside and 8,000 mg/kg for rebaudioside), a hypotensive activity has been shown for stevioside and the lack of such for rebaudioside (Koyama et al., 2003), metabolism in human gut organisms has been investigated (Tolstikova et al., 2005).

Recently, an antiarrhythmic effect has been demonstrated for both glycosides (Starrat et al., 2002). For example, preventive intravenous administration of these glycosides at the dose of 0.120 mg/kg secures survival of up to 50% of the animals given a lethal dose of calcium chloride. In the adrenaline arrhythmia model with similar administration regimen, only stevioside was shown to be effective. The survival rate was 50%. In the same arrhythmia models, both glycosides administrated at the dose of 0.120 mg/kg after the administration of the arrythmogen failed to block arrhythmia that had already developed (Tolstikova et al., 2009).

11.2.1.13   Complexes with Stevioside and Rebaudioside

Complexes of lappaconitine with stevioside were made with the molecular ratios ST:LA = 1:1, 4:1, 8:1, and 16:1. Their antiarrhythmic activity was studied using the earlier-stated arrhythmia models.

As can be seen from Table 11.8, complexes with the composition 1:1 and 4:1 when administered preventively ensure full protection from the development of calcium chloride and adrenaline

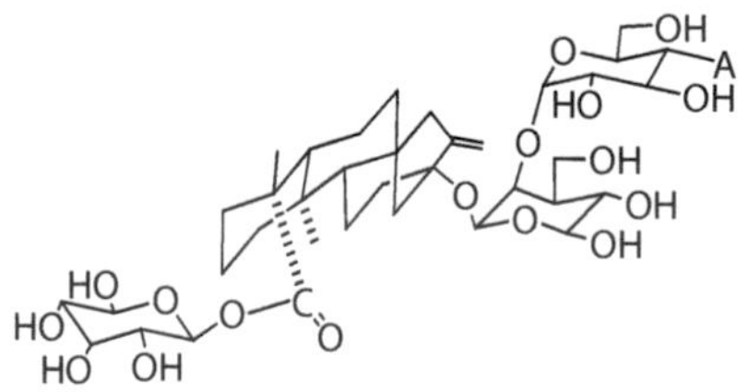

FIGURE 11.20   Structure of stevioside.

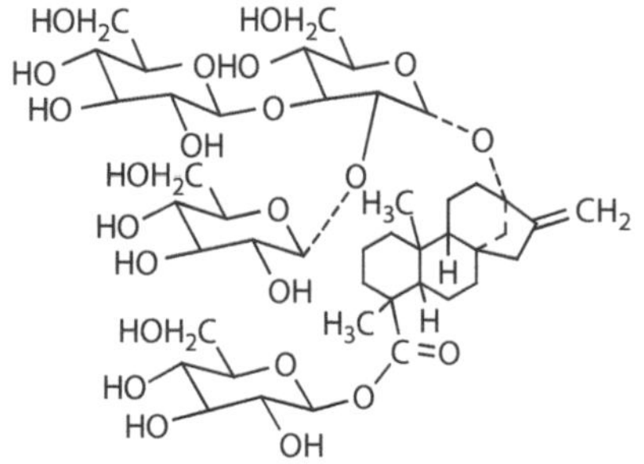

FIGURE 11.21   Structure of rebaudioside.

TABLE 11.8

Antiarrhythmic Activity of Stevioside and Its Complexes with Lappaconitine

Agent	Dose (mg/kg)	Rat Survival Rate (%)			
		Regimen of Administration			
		Complex + Arrhythmogen		Arrhythmogen + Complex	
		$CaCl_2$	Adrenaline	$CaCl_2$	Adrenaline
Stevioside	0.120	40	50	0	0
Complex 1:1	0.150	100	100	20	0
Complex 4:1	0.150	100	100	20	0
Complex 8:1	0.135	80	100	100	0
Complex 16:1	0.128	80	40	20	20
Lappaconitine	0.290	50	0	0	0

Arrhythmogens: $CaCl_2$ (250 mg/kg), adrenaline (0.3 mg/kg) administered intravenously.

arrhythmias in animals. Both complexes showed low activity in the developed calcium arrhythmia model and no activity at all in the adrenaline arrhythmia model.

Noteworthy results were obtained for the complex with composition 8:1, which ensured full protection of animals when administered preventively and after the development of calcium chloride arrhythmia. In the calcium chloride arrhythmia model the complex with composition 16:1 retains high activity (80%) only when administered preventively.

The data of Table 11.9 on the quantitative characteristics of the complexes indicate that complexation of lappaconitine with stevioside, like in the case of glycyrrhizic acid, allows decreasing the therapeutic dose of the pharmacon significantly.

It should be noted that the molecule of stevioside, unlike GA, has no free carboxyl groups. Association of glycoside is likely to occur only due to the bonding of carbohydrate fragments. However, it is not impossible that *in vivo* ester bonds hydrolysis takes place with the formation of acid diglucoside capable of forming associates acting as pharmacon complexing agents.

Complexation of stevioside with fluoxetine allows decreasing the pharmacon toxicity by 32 times. For example, at intragastrical administration LD_{50} of fluoxetine is 248 mg/kg, while LD_{50} of complex 4:1 is higher than 8000 mg/kg.

The study of pharmacological properties of complexes of fluoxetine with stevioside ST:FL with composition 1:1 and 4:1 demonstrated a reduction of depressive effect of L-DOPA on emotional and motor activity in mice, similar to that of fluoxetine. The effect of complexes was shown to depend on the dose. Both complexes act on the dopaminergic system in the same way as fluoxetine.

TABLE 11.9

Quantitative Characteristics of Stevioside–Lappaconitine Complexes

Molecular Composition	Dose (mg/kg)	Lappacontine Content in the Dose (mg)	Decrease in Therapeutic Dose (Times)
Complex 1:1	0.150	0.063	4.6
Complex 4:1	0.150	0.023	12.0
Complex 8:1	0.135	0.011	16.0
Complex 16:1	0.128	0.006	48.0

The effect of fluoxetine complexation with stevioside is also manifested in a reduced sedative action of the pharmacon on the background of chloral hydrate, a somnific drug.

Similar to GABA, complex ST:FL = 4:1 containing a 10 times reduced dose of the pharmacon produces a positive effect on the mnestic abilities in animals (Shishkina et al., 2006).

Complexes of rebaudioside (RB) with nifedipine with the RB:NF ratio 2:1 and 4:1, containing 9 and 18 wt.% of the pharmacon, respectively, were synthesized using the solution method. Doses ensuring high antiarrhythmic and standard antihypertensive activity were found for the complexes. The overall (100%) antiarrhythmic effect was demonstrated by both complexes containing a 10- and 5-times reduced dose of nifedipine, respectively. Antiarrhythmic activity similar to that of nifedipine was shown for complex RB:NF 2:1 at a five-times reduced dose of the pharmacon.

11.2.1.14 Arabinogalactan

Arabinogalactans (Figure 11.22) belong to polysaccharides, which are widespread in plants (Clarke et al., 1979; Ovodov, 1998; Arifkhodzhaev, 2000). Larch arabinogalactans, the content of which in the wood reaches 15%, stand apart from these compounds. Arabinogalactan macromolecule is widely branched. The backbone chain consists of D-galactose segments connected by β-(1–3)-glycoside bonds. Side branches connected with the backbone chain by β-(1–6) linkages are D-galactose blocks connected by β-(1–6)-bonds and D-arabinose blocks connected by β-(1–3)-bonds. In arabinogalactan commercially derived from Western larch *Larix occidentalis*, the galactose to arabinose fragments ratio increases to 7:1 from 2.33:1 with growth of molecular mass (Prescott and Graman, 1997).

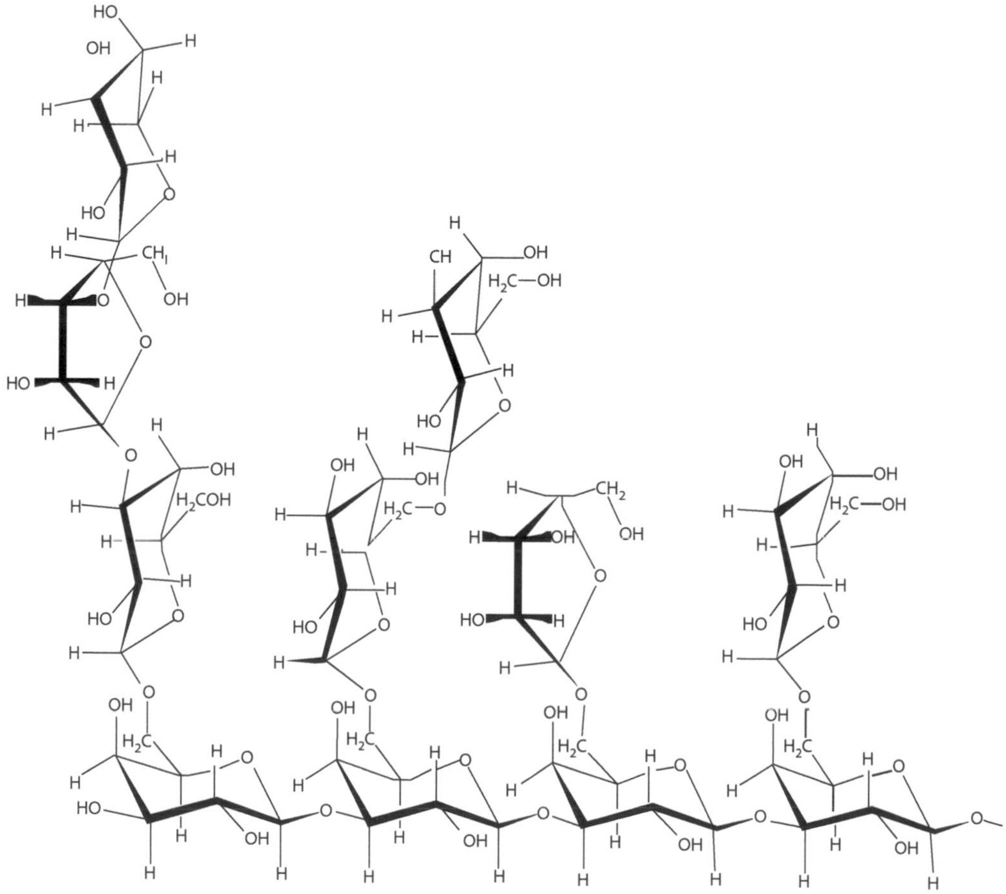

FIGURE 11.22 Arabinogalactan.

Arabinogalactan (AG) of Russian larches *L. sibirica* and *L. gmelinii* is characterized by a mono-modal curve of molecular mass distribution with a maximum within 13–18 kDa. Practicable methods have been proposed to extract high purity arabinogalactan (Tsvetaeva and Paskerova, 1962; Antonova and Tyukavkina, 1976; Medvedeva et al., 2003, 2004). Oxidative destruction of arabinogalactan derived from Siberian larch *L. sibirica* in water–peroxide solution under the action of molecular oxygen made it possible to synthesize the polymeric and oligomeric products enriched with carbonyl and carboxyl groups. Destruction proceeds with the breaking of glycoside bonds, cycle opening and oxidation of the anomer carbon atom to carboxyl group, and detachment of formic acid molecules in a stepwise manner (Borisov et al., 2002, 2004; Monakov et al., 2004; Mudarisova et al., 2005).

The usage of AG as a complexing agent is relatively new. It was chosen due to its chemical composition, the physical and chemical properties of which resemble those of the previously studied glycosides—GA, ST, RB.

AG complexes with various pharmacons have been obtained; in all cases, a significant increase in solubility of poorly soluble substances (by 2–21 times) is observed, which demonstrates the high effectiveness of AG as a complexing agent. All synthesized complexes were characterized by physical and chemical properties previously used for the aforementioned GA complexes.

11.2.1.15   Complexes with Arabinogalactan

Arabinogalactan and its oxidates were used for complexation of 5-aminosalicylic acid, an antituberculous compound. The complexes were synthesized in the form of water-soluble amorphous substances. The authors believe that bonding into complexes occurs due to the acid-base interaction between the pharmacon aminogroup and carboxyl groups of oxidates. As for arabinogalactan itself, complexation occurs through pharmacon retention in the side chains by intermolecular hydrogen bonds and interaction of aminopharmacon with a small number of carboxyl groups, which are present in the natural polysaccharide. As a result, complexes have been synthesized with the content of 5-aminosalicylic acid as follows: arabinogalactan, 1.4%; polymer oxidates, 8.5%; and oligomer oxidades, 16.0%. Pharmacon complexes with arabinogalactan and oligomer oxidades show high *in vitro* activity against mycobacteria (Badykova et al., 2005, 2006).

In Dushkin et al. (2008a,b) it was shown for the first time that the complexation effect discovered during investigations of complexes of terpenoid glycosides with pharmacons is also manifested by complexes of arabinogalactan as a reduction of the therapeutic dose of pharmacon and appearance of additional positive pharmacological properties.

Complexation of AG with IM, a NSAID, tranquilizers sibazon (7-chloro-2,3-dihydro-1-methyl-5-phenyl-1H-1,4-benzodiazepin-2-on) and mezapam (7-chloro-2,3-dihydro-1-methyl-5-phenyl-1H-1,4-benzodiazepin), azaleptin (8-chloro-11(4-methyl-1-piperazinyl)5H-dibenzo-[b,e]-1,4-diazepin), a neuroleptic, was studied in the three modes of complex synthesis.

The first mode is intensive stirring of components in water solution. The second is the mechano-chemical activation of a mixture of solid components by the method described in Shakhtshneider and Boldyrev (1999). The third mode is the mechanochemical pretreatment of arabinogalactan followed by the mechanochemical activation of the mixture of pharmacon with treated AG. The samples of initial AG have unimodal molecular mass distribution with average molecular mass of 13.5 kDa. AG treated mechanochemically is characterized by an intricate chromatogram with several maximums in the interval of 1.2–30 kDa. The average MM value is 5 kDa. Although the chromatographic characteristics of AG treated separately or together with pharmacon do not differ significantly, the water solubility of the pharmacon is the highest in the composition of complexes synthesized by in second mode.

It was shown that water solubility for the complexes with the AG:pharmacon having the weight ratio of 10:1 increases for IM, sibazon, mezapam, azaleptin by 9.9, 2.4, 19.1, and 20.5 times, respectively. For complexes of AG: pharmacon 20:1 these values are as follows: IM, 16.8; sibazon, 3.0; mezapam, 46.8; azaleptin, 38.8. All modes of complex synthesis have to be compared in terms of solubility ratio of one of the pharmacons. For example, in complex AG: mezapam 20:1 the solubility ratios are 10.8, 46.8, and 17.5 for the first, second, and third modes, respectively. Therefore, mechanochemical activation of the solid component of the mixture produces optimal conditions for the synthesis. The complexes are powders consisting of vitreous particles sized 2–50 μm.

No traces typical of the crystalline phase of pharmacons can be found in the x-ray photographs and thermograms of complexes, indicating that pharmacon molecules have undergone molecular dispersion into the polysaccharide matrix, that is, complexation has taken place. The stability of intermolecular complexes of AG with pharmacons increases in the series sibazon < IM < mezapam < azaleptin. Complexation proceeds mainly due to the intermolecular hydrogen bonds and Coulomb interaction at the ionization of pharmacon molecules. The investigation by the NMR-relaxation method showed that mobility of the pharmacon molecules dramatically decreases in solutions, which is evidence of complex existence (Fielding, 2000).

The study of pharmacological properties of the complexes produced the following main results. The basic activity of indomethacin within the complexes remains high at the doses of pharmacon reduced by 10 (complex 10:1) and 20 (complex 20:1) times. Destructive mucosal involvement of the stomach halves. The basic activity of sibazone within the complexes holds at the dose reduced by 10 times. Besides, the anxiolythic effect of pharmacon increases.

For complexes of AG with mezapam the standard antianxious activity was observed at the pharmacon doses reduced by 20 times. The complexation effect for the complex of AG with azaleptin is manifested in the halved dose and amplification of the sedative component.

As was discussed earlier in the part about glycyrrhizic acid complexes, pharmacological properties of anticoagulant drug warfarin in the WF:GA complex were not as desired. Therefore, the obtained results of examination of WF: GA complex prompted us to replace the complexing agent and we selected AG. We assumed that a complexing agent with such a structure should interact differently with warfarin.

The WF:AG complex was synthesized in a similar way and with the same component ratio, 1:10, as the complex with GA. This component ratio had previously been determined as effective for other pharmacons with AG (Tolstikova et al., 2009).

The same scheme was used in order to secure the accuracy of the experiment. No increase in PT was observed in 6 h after a single intragastric administration, just as in the previous experiment. However, in 24 h after the experiment start (one administration), some significant differences were observed. PT increased by more than two times, compared to the control (phys. solution), and amounted to 30.03 s. Although this indicator is lower than that of the comparator agent (42 s) in a 2 mg/kg dose, it still may be considered sufficient for the manifestation of the effect of increase in hemopexis time (Figure 11.23).

We can suppose that this 28.5% difference between WF and WF:AG may be beneficial in the reduction of hemorrhages during the dose selection period, that is, we can reduce undesirable side effect of warfarin therapy. Moreover, the efficacy of WF and WF:AG is the same in 48 h after a single administration (Figure 11.24).

Thus, it may be said that while the dissolubility of the WF:AG clathrate is lower than that of the WF:GA clathrate, the structure of clathrates with AG allows preserving the anticoagulative potential of warfarin and increase its solubility in water.

Thus, the study of pharmacological properties shows that the results for AG complexes agree well with those for complexes with terpenoids (glycyrrhizic acid, stevioside, rebaudioside). In both cases complexation results in enhanced basic activity, reduced dose, appearance of new properties of pharmacons.

Therefore, the discovered effect of complexation of drugs by plant carbohydrate-containing metabolites allows reducing the toxicity and the effective dose (by 5–100 times), and to improve the pharmacological properties, which may lead to the increase in safety of the therapy of many socially significant diseases. Besides, this method may lower the pharmacoeconomical production expenses and make the course of treatment cheaper for the patients.

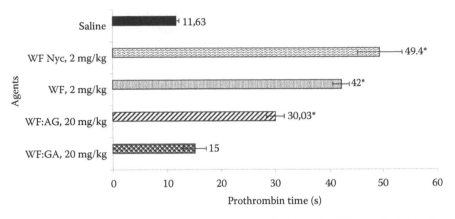

FIGURE 11.23 Prothrombin time of the WF:AG (1:10). * Significantly different (P<0.05) from control (saline).

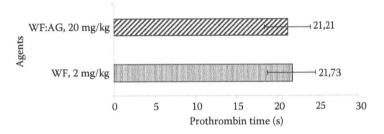

FIGURE 11.24 Prothrombin time after 48 h from single dosing of the WF:AG (1:10) and WF.

REFERENCES

Aken H.V., Bode, Ch., Darius, H. et al. 2001. State-of-the-art review: Anticoagulation: The present and future. *Clinical and Applied Thrombosis/Hemostasis*, 7:195–204.

Antonova, G.F. and Tyukavkina, N.A. 1976. Synthesis of high-clean arabinogalactan from larch wood. *Wood Chemistry*, 4:60.

Arifkhodzhaev, A.O. 2000. Galactans and galactan containing polysaccharides of higher plants. *Chemistry of Natural Compounds*, 3:89–96.

Avgustinovich, D.F., Kovalenko, I.L., Sorokina, I.V. et al. 2003. Changes in the rats blood biochemical indices during the experimental depression development. *Bulletin of Experimental Biology and Medicine*, 135(4):406–409.

Avgustinovich, D.F., Kovalenko, I.L., Sorokina, I.V. et al. 2004. Ethological study of antidepressant effect of fluaglisine in conditions of chronic social stress in mice. *Bulletin of Experimental Biology and Medicine*, 137(1):99–103.

Azaz, E. and Segal, R. 1980. Glycyrrizin as gelling agent. *Pharmaceutica Acta Helvetiae*, 55(7):183–186.

Badykova, L.A., Mudarisova, R.H., Khamidullina, G.S. et al. 2006. Modification of arabinogalactan and its oxidized forms by n-aminosalicilic acid. *Journal of Applied Chemistry*, 79(10):1647–1649.

Badykova, L.A., Mudarisova, R.H., Tolstikova, T.G. et al. 2005. Modification of poly- and oligosaccharides based on arabinogalactan by 5-aminosalicilic acid. *Chemistry of Natural Compounds*, 3:219–222.

Baltina, L.A. 2003. Chemical modification of glycyrrhizic acid as a route to new bioactive compounds for medicine. *Current Medicinal Chemistry*, 10:155–171.

Baltina, L.A., Davydova, V.A., Boldyrev, A.V. et al. 1988. The complex compound of sodium salt of 2-[(2,6-dichlorophenyl)-amino]phenylacetic acid with glycyrrhizic acid that possesses anti-inflammatory activity. Inventors certificate of USSR No 1566699.

Baltina, L.A., Davydova, V.A., Boldyrev, A.V. et al. 1988. The complex compound of acetylsalicylic acid with glycyrrhizic acid that possesses anti-inflammatory and antipyretic activity. Inventors certificate of USSR No 1566700.

Baltina, L.A., Murinov, Yu.I., Ismagilova, A.F. et al. 2003. Production and antineoplastic action of complex compounds of β-glycerrhizic acid with some antineoplastic drugs. *Pharmaceutical Chemistry Journal*, 37(11):3–4.

Baltina, L.A., Davydova, V.A., Sharipova, F.V. et al. 1988. The complex compound of diphenyl-4-n-butylpirazolidindion-3,5 with glycyrrhizic acid that that possesses anti-inflammatory activity. Inventors certificate of USSR No 1566698.

Bondarev, A.I., Bashkatov, S.A., Davydova, V.A. et al. 1991. Antiinflammatory and analgesic action of complexes of antiflogistics with glycyrrhizic acid. *Pharmacology and Toxicology*, 5:47–50.

Borisov, I.M., Shirokova, E.N., Babkin, V.A. et al. 2002. About mechanism of arabinogalactan peroxidation in aquatic environment. *Izvestiya of RAS*, 383:774–776.

Borisov, N.M., Shirokova, E.N., Mudarisova, R.H. et al. 2004. Peroxide oxidation of arabinogalactan: Kinetics and products. *Izvestiya of RAS, Series: Chemistry*, 2:305–311.

Bryzgalov, A.O., Tolstikova, T.G., Sorokina, I.V. et al. 2005. Antiarrhythmic activity of alaglizin. *Experimental and Clinical Pharmacology*, 68:24–28.

Clarke, A.E., Anderson, R.I., and Stone, B.A. 1979. Form and function of arabinogalactans and rabinogalactan-proteins. *Phytochemistry*, 18:521–540.

Dushkin, A.V., Karnatovskaya, L.M., Chabueva, E.N. et al. 2001. Production and analysis of ulcerogenic action of quick-dissolving solid dispersed systems based on acetylsalicylic acid and biologically active licorice derivatives. *Pharmaceutical Chemistry Journal*, 11:18–21.

Dushkin, A.V., Metelyova, E.S., Tolstikova, T.G. et al. 2008a. Mechanochemical production and pharmacological action of water soluble complexes of arabinogalactan and drugs. *Izvestiya of RAS, Series: Chemistry*, 6:1274–1282.

Dushkin A.V., Meteleva E.S., Tolstikova T.G. et al. 2010. Complexation of pharmacons by glycyrrhizic acid as a route to the development of the Preparation with enhanced efficiency. *Journal of Chemistry for Sustainable Development*, 4:517–525.

Dushkin, A.V., Tolstikova, T.G., Tolstikov, G.A. et al. 2008b. Water-soluble compositions and method of its production. RF Patent RU 2006143081A.

European convention for the protection of vertebrate animals used for experimental and other scientific purposes, strsbourg, 1986.

Epstein, R.S., Moyer, Th.P., Aubert, R.E. et al. 2010. Warfarin genotyping reduces hospitalization rates. *Journal of the American College of Cardiology*, 55:2804–2812.

Fielding, L. 2000. Determination of association constants (Ka) from solution NMR data. *Tetrahedron*, 56:6151–6166.

Gilbert, R.J. and James, K.C. 1964. Preparation and surface-active properties of glycyrrhizic acid and its salts. *Journal of Pharmacy and Pharmacology*, 16:394–399.

James, K.C. and Stanford, J.B. 1962. The solubilising properties of liquorice. *Journal of Pharmacy and Pharmacology*, 5:445–450.

Kennelly, E.J. 2002. Sweet and non-sweet constituents of *Stevia rebaudiana* (Bertoni) Bertoni. In *Stevia, the Genus Stevia. Medicinal and Aromatic Plants—Industrial Profiles*, A.D. Kinghorn (Ed.), pp. 68–85., Taylor and Francis, London, U.K.

Kondo, M., Minamino, H., and Okiyama, G. 1986. Physicochemical properties and applications of α- and β-glycyrrhizins, natural surface active agents in licorice root extract. *Society of Cosmetic Chemists*, 37(3):177–189.

Kondratenko, R.M., Baltina, L.A., Mustafina, S.R. et al. 2003. Complex compounds of glycerrhizic acid with antimicrobial drugs. *Pharmaceutical Chemistry Journal*, 37(9):32–35.

Kornievskaya, V.S., Kruppa, A.I., Polyakov, N.E. et al. 2007a. Effect of glycyrrhizic acid on lappaconitine phototransformation. *Journal of Physical Chemistry*, 111(39):11447–11452.

Kornievskaya, V.S., Kruppa, A.I., Polyakov, N.E. et al. 2007b. Research of influence near areas on the reaction possibilities of radicals of model organic compound of methyl ether N-acetyl-antranil acid by methods H NMR and CPN. *Vestnik NGU. Series: Physics*, 2(2):21–29.

Kornievskaya, V.S., Kruppa, A.I., and Leshina, T.V. 2008. NMR and photo-CIDNP investigations of the glycyrrhizinic acid micelles influence on solubilized molecules. *Journal of Inclusion Phenomena and Macrocyclic Chemistry*, 60:123–130.

Kovalenko, I.L., Avgustinovich, D.F., Sorokina, I.V. et al. 2004. Influence of chronic introduction of fluoglisine on a biochemical blood values in mice in chronic social stress conditions. *Issues of Medical, Biological and Pharmaceutical Chemistry*, 11:44–47.

Koyama, E., Kitazawa, K., Izawa, O. et al. 2003. In vitro metabolism of the glycosidic sweeteners, stevia mixture and enzymatically modified stevia in human intestinal microflora. *Food Chemistry and Technology*, 41:359–374.

Krasova, T.G., Bashura, G.S., Murav'ev, I.A. 1978. Hydrocortisone and prednisolone solubilization in aqueous solutions of glyciram. *Pharmacy*, 27(5):32–35.

Levine, M.N., Raskob, G., Landefeld, S. et al. 2001. Hemorrhagic complications of anticoagulant treatment. *Chest* 119:108S–121S.

Mashkovsky, M.D. 2005. *Medicinal Agents*, Novaya Volna, Moscow, Russia.

Maskan, M. 1999. Rheological behavior of liqorice (*Glycyrrhiza glabra*) extract. *Journal of Food Engineering*, 39:389–393.

Medvedeva, E.N., Babkin, V.A., Makarenko, O.A. et al. 2004. The production of high-clean arabinogalactan from larch and examination of its immunomodulatory property. *Chemistry of Plant Stock*, 4:17–24.

Medvedeva, E.N., Babkin, V.A., and Ostroukhova, L.A. 2003. Arabinogalactan from larch—Properties and usage perspectives (review). *Chemistry of Plant Stock*, 1:27–37.

Monakov, Yu.B., Borisov, I.M., Zimin, Yu.S. et al. 2004. *Peroxides at the Beginning of the Third Millennium*, Nova Science Publishers, New York.

Mudarisova, R.H., Badykova, L.A., Tolstikova, T.G. et al. 2005. Modification of arabinogalactan and its oxidized forms by 5-aminosalicilic acid. *Journal of Applied Chemistry*, 78(10), 1721–1724.

Ovodov, Yu.S. 1998. Polysaccharides of flower plants: Structure and physiological. *Bioorganic Chemistry*, 24(7):483–501.

Prescott, J.H. and Graman, E.V. 1997. New molecular weight forms of arabinogalactan from *Larix occidentalis*. *Carbohydrate Research*, 301:89–93.

Pogosova, G.V. 2006. Nifedipine in treatment of cardiovascular system diseases: New about well known. *Clinical Pharmacology and Therapeutics*, 3:2–6.

Polyakov, N.E., Khan, V.K., Taraban, M.B. et al. 2005. Complexation of lappaconitine with glycyrrhizic acid: Stability and reactivity studies. *Journal of Physical Chemistry B*, 109(51):24526–24530.

Polyakov, N.E., Leshina, T.V., Salakhutdinov, N.F. et al. 2006. Host-guest complexes of carotenoids with beta-glycyrrhizic acid. *Journal of Physical Chemistry B*, 110(13):6991–6998.

Pshennikova, M.G. 2000. Stress phenomenon: Emotional stress and it's role in pathology. *Pathol Physiol Exp Ther.*, 4:21–31.

Ragino, Yu.I., Vavilin, V.A., Salakhutdinov, N.F. et al. 2008. Examination of the hypocholesteremic effect and safety of simvaglisin on the model of hypercholesterolemia in rabbits. *Bulletin of Experimental Biology and Medicine*, 3:285–287.

Romanenko, T.V. and Murinov, Yu.I. 2001. Some features of diluted glycyrrhizic acid solutions flux. *Journal of Physical Chemistry*, 75(9):1601–1604.

Sadritdinov, F.S. and Kurmukov, A.G. 1980. *Pharmacology of Plant Alkaloids and Their Application in Medicine*, Medicine of UzSSR, Tashkent, Uzbekistan.

Sasaki, Y., Mizutani, K., Kasai, R. et al. 1988. Solubilizing properties of glycyrrhizin and its derivatives solubilization of saikosaponin-a, the saponin of bupleuri radix. *Chemical and Pharmaceutical Bulletin*, 36(9):3491.

Shakhtshneider, T.P. and Boldyrev, V.V. 1999. *Reactivity of Molecular Solids*, John Wiley & Sons, England, U.K.

Shishkina, G.T., Dygalo, N.N., Yudina, A.M. et al. 2005. Effects of fluoxetine and its complexes with glycyrrizhinic acid on behavior and brain monoamine levels in rats. *I. P. Pavlov Journal of Higher Nervous Activity*, 55(2):207–212.

Shishkina, G.T., Dygalo, N.N., Yudina, A.M. et al. 2006. The effects of fluoxetine and its complexes with glycerrhizic acid on behavior in rats and brain monoamine levels. *Neuroscience and Behavioral Physiology*, 36(4):329–333.

Soltesz, G. and Uri, G. 1963. Solubilisierende Wirkung des Monoammonium glycyrrhizinats auf Antibiotika und einige andere Stoffe. *Naturwissenschaften*, 50(13–24):691.

Starrat, A.N., Kirby, C.W., Pocs, R. et al. 2002. Rebaudioside F, a diterpene glycoside from *Stevia rebaudiana*. *Phytochemistry*, 59:367–370.

Tolstikov, G.A., Baltina, L.A., Grankina, V.P. et al. 2007a. *Licorice: Biodiversity, Chemistry and Application in Medicine*, Academic Publishing House "GEO," Novosibirsk, Russia.

Tolstikov, G.A., Baltina, L.A., Murinov, Y.I. et al. 1991a. Complexes of with non-steroid anti-inflammatory drugs as new transport forms. *Pharmaceutical Chemistry Journal*, 2:29–32.

Tolstikov, G.A., Baltina, L.A., Schults, E.E. et al. 1997a. Glycyrrhizin acid. *Bioorganic Chemistry*, 23(9):691–709.

Tolstikov, G.A., Murinov, Y.I., and Baltina, L.A. 1990. Complexes of β-glycyrrhizic acid with drugs as a novel transport forms. *Pharmaceutical Chemistry Journal*, 24(8):26–27.

Tolstikov, G.A., Murinov, Y.I., Baltina, L.A. et al. 1991b. *Collected Works "The Study and Use of Licorice in the USSR*. Alma-Ata, Kazakhstan: Galym.

Tolstikov, G.A., Myshkin, V.A., Baltina, L.A. et al. 1996. Antidotal and antiradical activity of complexes of β-glycyrrhizic acid with pyrimidine derivatives. *Pharmaceutical Chemistry Journal*, 30(5):320–322.

Tolstikov, G.A., Nikitin, Yu.P., Lyakhovich, V.V. et al. 2007b. Drug with hypocholesterolemic effect Symvaglysin. RF Patent RU2308947 C1.

Tolstikov, G.A., Tolstikov, G.A., Murinov, Yu.I. et al. 1991c. Complexes of β-glycyrrhizic acid with prostaglandins—New class of uterotonic agents. *Pharmaceutical Chemistry Journal*, 3:42–44.

Tolstikov, A.G., Tolstikov, G.A., Murinov, Yu.I. et al. 1997b. RF Patent No 2123336.

Tolstikov, A.G., Tolstikova, T.G., Sorokina, I.V. et al. 2004. "Fluoglizin" is the drug for treatment different forms of depression. RF Patent 2232574.

Tolstikova, T.G., Bryzgalov, A.O., Sorokina, I.V. et al. 2007a. To the nature of the effect of glycosidic clathration of pharmacons. *Doklady Akademii Nauk*, 416(1):336–337.

Tolstikova, T.G., Bryzgalov, A.O., Sorokina, I.V. et al. 2005. Stevioide is a novel stimulator of pharmacon clathration with glycosides. *Doklady Akademii Nauk*, 403(2):274–276.

Tolstikova, T.G., Bryzgalov, A.O., Sorokina, I.V. et al. 2007b. Increase in pharmacological activity of drugs in their clathrates with plant glycosides. *Letters in Drug Design and Discovery*, 4:168–170.

Tolstikova, T.G., Khvostov, M.V., and Bryzgalov A.O. 2009. The complexes of drugs with carbohydrate-containing plant metabolites as pharmacologically promising agents. *Mini Reviews in Medicinal Chemistry*, 9:1317–1328.

Tolstikova, T.G., Sorokina, I.V., Bryzgalov, A.O. et al. 2006. The usage of new approach of known drugs complexation with plant glycosides in prophylaxis and relief of acute arterial hypertension (experimental study). *Rational Pharmacotherapy in Cardiology*, 1:55–58.

Tolstikova, T.G., Sorokina, I.V., Kovalenko, I.L. et al. 2004. Influence of clathrate formation to the pharmacon activity in complexes with glycyrrhizic acid. *Doklady Akademii Nauk*, 394(2):707–709.

Tolstikova, T.G., Tolstikov, A.G., and Tolstikov, G.A. 2007c. On the way to low-dose drugs. *Bulletin of RAS*, 77(10):867–874.

Tsvetaeva, I.P. and Paskerova, E.D. 1962. Change in structure of arabinogalactan from *Larix sibirica* subject to its extraction regime from wood. *Journal of Applied Chemistry*, 35(5):1128–1131.

Vavilin, V.A., Salakhutdinov, N.F., Ragino, Yu.I. et al. 2008. Hypocholesteremic properties of complex compound of simvastatin with glycerrhizic acid (simvaglizin) in experimental models. *Biomedical Chemistry*, 54(3):301–313.

Zapara, T.A., Ratushnyak, A.S., and Shtark, M.B. 1988. Local changes in the transmembrane ion currents in plastic reorganization of electrogenesis in isolated neurons of the pond snail. *Zhurnal Vysshei Nervnoi Deyatelnosti Imeni I P Pavlova*, 38(1):140–145.

Zapara, T.A., Simonova, O.G., Ratushnyak, A.S. et al. 1999. Influence of morphine and biologically active compound (BAV-S) on electric parameters of isolated neurons. *Bulletin of Experimental Biology and Medicine*, 10:392.

Zhu, X. and Shin, W.G. 2005. Gender differences in pharmacokinetics of oral warfarin in rats. *Biopharmaceutics and Drug Disposition*, 26:147–150.

12 Bioaerosols

Janet M. Macher, Jeroen Douwes,
Brad Prezant, and Tiina Reponen

CONTENTS

12.1 INTRODUCTION

Bioaerosols are those airborne particles that originated from living organisms (e.g., bacteria, protists, plants, fungi, and animals) or that depend on living organisms (e.g., viruses). Bioaerosols may consist of entire microscopic structures, for example, viruses, intact bacterial cells and spores, protozoa and their cysts, fungal cells and spores, and plant pollen grains and spores. Cell fragments may be present in indoor and outdoor air and are also considered bioaerosols, for example, airborne particles of decayed microbial, plant, and animal matter; wood and grain dusts; the droppings and dried body parts of arthropods; and particles of larger animal skin, saliva, feces, and urine. The term biological agent refers to any substance of biological origin that is capable of producing an effect on humans, for example, infectious agents, bacterial DNA, peptidoglycans, endotoxin, exotoxins, mycotoxins, and $(1 \rightarrow 3)$-β-D-glucans and allergens from bacteria, pollen, fungi, dust mites, cockroaches, and so forth. Bioaerosols may elicit responses similar to those caused by nonbiogenic particles (e.g., a hypersensitivity, irritant, or inflammatory response) as well as unique reactions (e.g., infectious diseases and toxicoses). The respiratory tract responds to injury, including that caused by biological agents, in a limited number of ways, for example, rhinosinusitis, pharyngitis, laryngitis, upper airway obstruction, alveolitis, pulmonary edema, asthma, chronic obstructive pulmonary disease (COPD), bronchitis, and pulmonary infection.[1] Bioaerosols occur as airborne particles in a size range of ~0.02–100 μm; thus, different cells, spores, pollen grains, and biological fragments may deposit in all regions of the human airways. Smaller bioaerosols remain airborne for longer periods and travel further distances, which makes it possible for one person in a shared space to expose a large number of individuals to an infectious agent. Microorganisms, plants, and animals are important components of surface waters and soil. Although not particulate in form, plants and animals release gases and vapors, for example, oxygen, carbon dioxide, methane, and water. Emissions from microorganisms (microbial volatile organic compounds [MVOCs]) account for the earthy smell that follows a rain shower or comes from freshly turned garden soil. The distinctive flavors of certain foods and beverages as well as the less pleasant aromas of decay, body odor, and moldy buildings are also volatile microbial metabolites. The complex interactions of MVOCs and bioaerosols with other airborne particles are seldom studied but are likely important for a comprehensive understanding of the effects of biological agents on human health and comfort. While it is impossible to cover this broad topic in great depth in a single chapter, we discuss health effects, dosimetry, and bioaerosol measurement with illustrations and examples from the vast literature on the subject.

12.2 HEALTH EFFECTS OF BIOAEROSOLS IN INDOOR AND OUTDOOR ENVIRONMENTS

Microorganisms and other bioaerosols are ubiquitous in the ambient environment and their concentrations may be elevated in damp and poorly ventilated buildings as well as in industrial settings where microbial, plant, or animal products are handled. Respiratory diseases and symptoms are the most common health effects associated with infectious aerosols (Section 12.2.1). Noninfectious bioaerosols may cause asthma, hay fever (rhinitis), organic dust toxic syndrome (ODTS), hypersensitivity pneumonitis (HP), chronic bronchitis, and airflow obstruction (Section 12.2.2). Bioaerosol exposure also may contribute to building-related symptoms (BRS, sick building syndrome, SBS), and cancer (Section 12.2.3.2).

Interest in bioaerosol exposures has grown, in particular due to a global increase in allergies and asthma as well as concerns about damp indoor environments. In addition to changes in building design and operation to increase energy efficiency, several new industrial activities have emerged that elevate bioaerosol exposures. Examples include waste recycling, waste sorting, and organic waste collection and composting as well as the production of purified biological substances such as microbial enzymes for food processing (e.g., α-amylase in commercial bakeries) and detergents. Many of these enzymes are potent allergens that can cause asthma and rhinitis. Furthermore, the widespread nontherapeutic use of antibiotics as growth promotants in livestock has accelerated the development of antibiotic resistant pathogens that may increase the risk of severe infectious diseases in workers who handle and process these animals.[2]

There is also heightened awareness of the potential use of bioaerosols for bioterrorism, in particular viruses, bacteria, rickettsia, and fungi.[3] For instance outside well-defined geographical areas and some occupations, anthrax was a relatively unknown disease until 2001 when spores were spread by mail causing five deaths in the United States. Also, the global transmission of influenza is a re-emerging concern and many experts wonder not if, but when, the next serious pandemic will occur. The influenza pandemic of 1918–1919 that killed nearly 50 million people worldwide and the 2009 pandemic influenza A (H1N1) outbreak underscore this concern.

Despite the recognition of the importance of bioaerosols, and with the exception of some infectious diseases, the precise role of biological agents in the induction, aggravation, and progression of symptoms and diseases is only poorly understood. Paradoxically, indoor exposure to moderate bioaerosol concentrations has been suggested to reduce the risk of developing hypersensitivity and allergic asthma in early life (Section 12.2.4). This protective effect is consistent with the "hygiene hypothesis," which postulates that growing up in a less microbiologically hygienic environment may decrease the risk of developing respiratory allergies (Section 12.2.4).[4–12] However, the evidence for a protective effect from microbial exposure has not been consistent (Section 12.2.4).[13–19] In the next sections we present an overview of the health effects associated with bioaerosol exposures.

12.2.1 Infectious Diseases

Most of the agents responsible for respiratory infections are spread through the air, primarily from person to person (anthroponoses) but also from living animals (zoonoses) and the abiotic environment (e.g., soil and water) or decaying plant or animal matter (sapronoses).[20] Viruses are responsible for most childhood respiratory tract infections, although bacteria (e.g., *Mycoplasma* and *Chlamydia* species) also infect children at particular ages (Table 12.1). A World Health Organization (WHO) study estimated that lower respiratory tract infections accounted for over 6% of the total, worldwide burden of disease and that tuberculosis alone represented 2% of the burden.[21] Tuberculosis exceeds all other chronic infections in incidence and prevalence, and *Mycobacterium tuberculosis* ranks seventh among all causes of death.[22] Globally, tuberculosis incidence peaked in 2004, with the most cases in Asia (55%) and Africa (30%) and small proportions in the Eastern Mediterranean Region (7%), the European Region (5%), and the Region of the Americas (3%).[22]

TABLE 12.1

Particle Sizes and Health Effects of Airborne Biological Agents

Biological Agent	Type Size (μm)	Location	Reservoir[a]	Health Effect	Susceptible Population[b]
Infectious agents					
Viruses					
Adenovirus	Droplets of respiratory discharges: 0.07–0.09	Oropharynx	H	Acute febrile respiratory disease Acute viral rhinitis (common cold) Oropharyngitis (sore throat)	H, H1, H3
Coronavirus	Droplets of nasal and throat secretions: 0.10–0.15	Nasopharynx	H	Acute febrile respiratory disease Acute viral rhinitis (common cold)	H
Enteroviruses	Droplets of nasopharyngeal secretions: 0.03	Oropharynx	H	Oropharyngitis (sore throat)	H
Hanta virus	Droplet nuclei of rodent excreta: 0.08–0.12	Lung	A1	Hantavirus pulmonary syndrome	H, H4, H5
Influenza viruses type A, B, C	Droplet nuclei of respiratory secretions: 0.08–0.12	Nasopharynx Larynx Bronchi	H, A2, A3	Influenza (A: pandemic; A,B: epidemic) Pneumonia	H, H6, H7, H8
Measles virus	Droplet nuclei of respiratory secretions: 0.12–0.25	Bronchi	H	Measles (rubeola)	H, H7, H6, H9
Mumps virus	Droplets of saliva: 0.15–0.20	Salivary glands	H	Mumps (infectious parotitis)	H
Parainfluenza virus	Droplets of respiratory secretions: 0.15–0.20	Larynx-trachea Lung	H	Croup: obstruction of the upper airway in young children; hoarseness in adults Acute febrile respiratory disease Pneumonia	H1, H2
Respiratory syncytial virus (RSV)	Droplets of respiratory secretions: 0.10–0.35	Bronchioles Lung	H	Acute febrile respiratory disease Pneumonia	H1, H2
Rhinoviruses	Droplets, droplet nuclei of sputum, saliva, or nasal or throat secretions: 0.02–0.27	Nasopharynx	H	Acute febrile respiratory disease Acute viral rhinitis (common cold)	A
Rubella virus	Droplets of nasopharyngeal secretions: 0.07	Pharynx	H	Rubella (German measles)	H, H7, H9

Organism	Source / droplet size (µm)	Site of infection	Code	Disease	Survival code
Varicella-zoster virus	Droplet nuclei of respiratory secretions, vesicular fluid: 0.15–0.20	Mucosa of the upper respiratory tract or eyes	H	Varicella (chicken pox)	H, H7, H9
Other respiratory viruses	Droplet nuclei of nasal and throat secretions	Nasopharynx; Lung	H	Pneumonia	H1
Bacteria					
Bacillus anthracis	Aerosol of spores; Spores: 0.8×1.2–1.7	Lung	A4	Anthrax (woolsorter's disease); Pneumonia	H10
Bordetella pertussis	Droplets of respiratory secretions: 0.2–1	Ciliated epithelial cells in the upper respiratory tract	H	Pertussis (whooping cough)	H, H7
Burkholderia pseudomallei	Droplet nuclei: 0.5–1.0×1–5	Lung	E1, E2	Melioidosis; Pneumonia	H10
Chlamydia pneumoniae	Uncertain, possibly droplet nuclei: ~0.3	Lung	H	Pneumonia	H
Chlamydia psittaci	Droplet nuclei of dried feces, secretions, feathers, tissues: ~0.3	Lung	A2a	Psittacosis; Pneumonia	H, H4
Chlamydia trachomatis	Uncertain, possibly droplet nuclei: ~0.3	Lung	H	Pneumonia	H1
Corynebacterium diphtheriae	Droplets of throat secretions: 0.3–0.8×1.5–8.0	Throat (toxin passes into bloodstream)	H	Diphtheria	H
Coxiella burnetii	Droplet nuclei: 0.2×0.7	Lung	A	Q fever; Pneumonia	H, H4
Francisella tularensis	Droplet nuclei: 1–2	Lung	A	Tularemia; Pneumonia	H
Haemophilus influenzae	Droplets of nasopharyngeal secretions: 0.3×1	Middle ear and paranasal sinuses, Epiglottis, Bronchi	H	Meningitis; Pneumonia	H1
Legionella spp.	Droplet nuclei: 0.3–0.9×2	Lung	E1	Legionnaires' pneumonia (Legionnaires' disease)	H, H2, H11, H12
Atypical mycobacteria	Droplet nuclei of contaminated water or soil or respiratory secretions: 0.2–0.6×1–10	Lung	E1, E2, H, A	Mycobacteriosis; Pneumonia	H12
Mycobacterium tuberculosis	Droplet nuclei of respiratory secretions: 1–5	Lung	H	Tuberculosis; Subacute or chronic pneumonia	H, H1, H2, H7, H8

(continued)

TABLE 12.1 (continued)
Particle Sizes and Health Effects of Airborne Biological Agents

Biological Agent	Type Size (μm)	Location	Reservoir[a]	Health Effect	Susceptible Population[b]
Mycoplasma pneumoniae	Droplets of nasopharyngeal secretions: 0.1–0.2×1–2	Bronchi	H	Pneumonia	H, H1, H2
Neisseria meningitidis	Droplets of nasopharyngeal secretions: 0.6–1.0	Lung	H	Pneumonia	H
Nocardia	Airborne spores: 0.5–1.0	Lung	E2	Nocardiosis Pneumonia	H12
Streptococcus pneumoniae (pneumococcus)	Droplets of nasopharyngeal secretions: <2	Middle ear and paranasal sinuses, Bronchi	H	Pneumonia Meningitis	H, H1, H2
Group A streptococcus (*Streptococcus pyogenes*)	Droplets of nasopharyngeal secretions: <2	Oropharynx, Middle ear and paranasal sinuses	H	Streptococcal sore throat Pneumonia	H
Yersinia pestis	Droplet nuclei of respiratory secretions: 0.5–0.8×1–3	Lung	A, H	Pneumonic plague Pneumonia	H
Fungi					
Aspergillus spp.	Airborne conidia: 2–5	Lung	E2, E3	Pneumonia	H12
Blastomyces dermatitidis	Airborne conidia: 2–10	Lung	E2	Blastomycosis Subacute pneumonia	H13
Coccidioides immitis	Airborne arthroconidia: 2–4×3–6	Lung	E2	Coccidioidomycosis Subacute pneumonia	H13
Cryptococcus neoformans	Yeast cells: 4–8	Lung	E2, A2	Cryptococcosis Subacute pneumonia	H4
Histoplasma capsulatum	Airborne microconidia: 2–5	Lung	A2, A2	Histoplasmosis Subacute pneumonia	H4
Zygomycetes (phycomycetes)	Spores	Lung	E2, E3	Zygomycosis (phycomycosis) Pneumonia	H12
Protists					
Pneumocystis carinii	Droplet nuclei of respiratory secretions Trophozoite: 1–5	Lung	H (uncertain)	Pneumonia	H1, H12

Organism	Size (μm)	Location	Allergen	Disease	Ref.
Amoebae					
Acanthamoeba spp.	Precyst: 5–8; Cysts: 8; Trophozoite: 15–45; Cysts: 10–25	Nasal passages	E1, E2	Granulomatous amebic encephalitis (GAE)	H10
Balamuthia spp.	Trophozoite: 12–60; Cysts: 6–30	Nasal passages	E1, E2	Granulomatous amebic encephalitis (GAE)	H10
Naegleria fowleri	Trophozoite: 10–35; Cysts: 7–15	Nasal passages	E1, E2	Primary amebic meningoencephalitis (PAM)	H10
Aeroallergens					
Bacterial antigens	Cells and cell fragments	Lung, mucous membranes	E1, E3, E4	BRS; HP	H14
Bacillus subtilis	Spores: 0.8×1.1; Cells: 0.8×2.2–4.5	Lung	E5	HP: detergent worker's lung	H14
Mycobacterium immunogenum	Bacterial cells or cell fragments: 0.2–0.6×1–10	Lung	E6	HP	H15
Saccharopolyspora (Faenia) rectivirgula	Spores	Lung	P1	HP: farmer's lung	H14
Thermoactinomyces candidus	Spores	Lung	E1	HP: humidifier lung	H14
Thermoactinomyces sacchari	Spores	Lung	E7	HP: bagassosis	H14
Thermoactinomyces vulgaris	Spores	Lung	E8	HP: grain worker's lung, mushroom worker's lung	H14
Fungal allergens	Spores, cell fragments	Lung, mucosal membranes	E1, E3, E4	Allergic conjunctivitis, rhinitis, sinusitis, asthma, BRS	H14
Aspergillus clavatus	3.0–4.5×2.5–4.5	Lung	E9	HP: malt worker's lung	H14
Aureobasidium pullulans	8–12×2–5	Lung	E10	HP: sequoiosis	H14
Graphium spp.	4–7				
Cryptostroma corticale	4–6.5×3.5–4	Lung	E11	HP: maple bark stripper's lung	H14

(continued)

TABLE 12.1 (continued)
Particle Sizes and Health Effects of Airborne Biological Agents

Biological Agent	Type Size (µm)	Location	Reservoir[a]	Health Effect	Susceptible Population[b]
Penicillium commune (P. casei, P. camembertii)	3.0–4.5×4.0–5.0	Lung	E12	HP: cheese worker's lung	H14
Penicillium glabrum (P. frequentans)	3.0–3.5	Lung	E13	HP: suberosis	H14
Aspergillus flavus	3–8	Lung	E	Allergic bronchopulmonary aspergillosis	H16
Aspergillus fumigatus	2–3.5				
Aspergillus terreus	1.5–2.5				
Penicillium spp.	2–3.5	Lung	E	Allergic bronchopulmonary mycosis	H16
Amebic allergens					
Acanthamoeba castellanii	Trophozoite: 15–45 long cysts: 10–25	Lung	E1, E2	HP: humidifier lung	H
Naegleria gruberi	Trophozoite: 10–35 long Cysts: 7–15	Lung	E1, E2	HP: humidifier lung	H
Plant allergens	Pollen grains, plant fragments	Lung, Uncertain	P, E4	Allergic conjunctivitis, rhinitis, sinusitis, asthma	H14
Ambrosia, ragweed	17–19×16–20				
Artemisia, mugwort	20–23×19–21				
Betulaceae, birch family	20–29×18–29				
Cryptomeria japonica, Japanese cedar, Sugi	36–35				
Cupressaceae, juniper family	23–31×21–29				
Fagaceae, beech family (Quercus, oaks)	32×29				
Juglandaceae, walnut family	35–41×32–38				
Mimosoidae, mimosa subfamily (Acacia; Prosopis, mesquites)	Acacia: 40×60 Prosopis: 30–27				

Morus, mulberry	18–17				
Oleaceae, olive family	*Olea:* 21–19				
Poaceae, grass family	Wild grass: 28–36×26–33; Cultivated grass: >42				
Parietaria, pellitory	12–15				
Orbicules, plant fragments	1.5–2				
Arthropod allergens	Cockroach or house dust mite excreta and body fragments	E4	Lung, Uncertain	Allergic conjunctivitis, rhinitis, sinusitis, asthma, BRS	H14
Blatella germanica, Bla ε1, cockroach allergen	Roach allergens: >5				
Dermatophagoides farince, Der f1, *Dermatofagoides pteronyssinus, Der* p1, dust mite allergens	Mite allergens: 10–35				
Avian proteins	Particles of bird droppings, feathers	A2	Lung	HP: Bird breeder's lung, asthma	H4, H14
Mammalian allergens	Particles of cat, dog, or rodent skin, saliva, or urine	A, E4	Lung, Uncertain	Allergic conjunctivitis, rhinitis, sinusitis, asthma; BRS; HP: animal handler's lung	H4, H14
Felis domesticus, Fel d1, cat allergen	Cat and dog allergens: <5				
Canis familiaris, Can f1 allergen					
Rattus norvegicus Rat n1, rat allergen					
Muscus musculus, Mus m1,2, house mouse allergen					
Cavia porcellus, Cav p1, guinea pig allergen					

(continued)

TABLE 12.1 (continued)
Particle Sizes and Health Effects of Airborne Biological Agents

Biological Agent	Type Size (µm)	Location	Reservoir[a]	Health Effect	Susceptible Population[b]
Biological toxins and inflammatory agents					
Bacterial endotoxin	Gram-negative bacterial cells and cell fragments	Lung	E1, E3	Respiratory inflammation	H15
				Humidifier fever	
				Cottonworker's lung (byssinosis)	
				Mill fever	
				Grain fever	
				BRS	
Legionella spp.	Bacterial cells or cell fragments	Lung	E1	Pontiac fever	H15
Fungal glucan	Spores, cell fragments	Lung	E3	Airway inflammation	H15
				Mucous membrane irritation	
				Fatigue	
				Anti-tumor activity	
Fungal toxins	Spores, cell fragments	Lung, uncertain	E3	Toxic and irritant effects	H15
Organic dust	Fungal spores, plant, animal, and microbial cell fragments	Airways and alveoli	E14, E15	ODTS	H15
Cotton, hemp, jute, and flax dust	Microorganisms, cell fragments	Lung	P	Mill fever	H11, H15
				Acute airway obstruction (byssinosis)	
				Chronic bronchitis	
Grain (oat, wheat, barley) dust	Microorganisms, cell fragments	Lung	P	Irritation,	H11, H15
Flour dust				Grain fever	
				Asthma,	
				Chronic bronchitis	
				Obstructive lung disease	
Wood (hard woods: beech and oak; soft wood) dust	Dust particles	Lung	P	Irritation	H15
		Nose		Lung function	
				Asthma	
				Cancer (nasal)	

Vegetable oil mists (castor bean, sesame, acacia, cashew nut)	Sprays and mists	Lung	P	Nuisance particles Respiratory irritation Asthma	H15

Sources: APHA, *Control of Communicable Diseases Manual*, 19th edn., Heymann, D.L. (Ed.), American Public Health Association, Washington, DC, 2008; Salyers, A.A. and Whitt, D.D., Introduction to infectious diseases, and the lung, a vital but vulnerable organ, in *Microbiology: Diversity, Disease, and the Environment*, Fitzgerald Science Press, Bethesda, MD, pp. 289, 315, 2001; Hjelmroos, M. et al., *Airborne Allergens. Interactive Identification of Allergenic Pollen Grains and Fungal Spores*, Institute of Respiratory Medicine, University of Sydney, New South Wales, Australia, 1999. Samson, R.A. et al., *Introduction to Food- and Airborne Fungi*, 6th edn., Centraalbureau voor Schimmelcultures, Utrecht, the Netherlands, 2000, Larone, D.H., *Medically Important Fungi: A Guide to Identification*, 4th edn., ASM Press, Washington, DC, 2002; Watanabe, T., *Pictorial Atlas of Soil and Seed Fungi, Morphologies of Cultured Fungi and Key to Species*, 3rd edn., CRC Press, Boca Raton, FL, 2010; ASM, *Manual of Clinical Microbiology*, 9th edn., Murray, P.R. (Ed.), ASM Press, Washington, DC, 2007.

[a] H, humans; A, animals: A1, infected deer mice (*Peromyscus maniculatus*); A2, birds or bird droppings; A2a, psittacine birds (parakeets, parrots); A2b, pigeon droppings; A3, swine; A4, infected animals, pelts, and hides; E, environment; E1, water; E2, soil; E3, organic matter; E4, settled dust; E5, detergent enzyme; E6, contaminated water-based MWF; E7, moldy sugar cane fiber; E8, moldy grain, compost; E9, moldy malt; E10, moldy redwood dust; E11, moldy maple bark; E12, moldy cheese; E13, moldy cork dust; E14, moldy silage, compost, wood chips; E15, sewage sludge; P, plants; P1, moldy hay.

[b] H, humans; H1, infants and young children; H2, elderly persons; H3, nonimmune military recruits; H4, animal handlers, animal care; H5, farmers; H6, teachers; H7, healthcare workers; H8, public safety workers; H9, childcare workers; H10, unknown or uncertain; H11, smokers; H12, immunocompromised persons; H13, workers with soil contact; H14, hypersensitive persons (repeated exposures required for HP); H15, highly exposed workers and other persons; H16, asthmatics.

Acute respiratory infections (ARIs) are the most common of human illnesses, which we take for granted because they occur so often and are usually mild. However, among children under 5 years of age, approximately 17% of all deaths (1.7 million in 2004) are due to ARIs, primarily pneumonia, which is the largest single cause of death in this age group.[21] Respiratory tract infections in childhood may have long-term sequelae, including loss of lung function after severe episodes of lower respiratory tract infection, the development of asthma or bronchiectasis (Section 12.2.2), and an increased risk of developing COPD in adulthood.[23]

Three-quarters of ARI deaths in children under 5 years of age are in Africa and Southeast Asia. Respiratory syncytial viral (RSV) is believed to be the most important viral cause of ARI, responsible for an estimated 66,000–199,000 deaths in children under 5 years of age (99% in developing countries).[24] The higher childhood mortality in developing countries probably reflects poorer nutrition and immunization practices and more frequent low birth weight, crowding, and exposure to indoor and outdoor air pollutants.[25] Disadvantaged population groups bear the greatest burden, but communicable diseases occurring anywhere are of concern to everyone, given the connectedness of all parts of the world via voluntary and involuntary human migrations and rapid international air travel.[26]

Outbreaks of previously unrecognized airborne infectious diseases seriously affect the global economy, for example, Legionnaires' disease in the 1970s, severe acute respiratory syndrome (SARS) in 2002–2003, and the 2009 pandemic influenza A (H1N1) more recently. Communicable disease outbreaks are designated epidemics when the occurrence in a community or region clearly exceeds normal expectancy, and epidemics are designated pandemics when they occur worldwide or over a wide geographic area.[27] The origins of emerging infectious diseases also are significantly correlated with socio-economic, environmental, and ecological factors. Unfortunately, the majority of scientific and surveillance efforts focus on countries from which new diseases are least likely to originate.[28]

The economic cost of lost productivity from the common cold has been estimated at over $25 billion in the United States, and mean productivity losses in Sweden from allergic rhinitis and the common cold have been estimated at 5.1 days €653/worker/year.[29,30] Of the common ARIs, influenza produces the most severe illness and accounts for the greatest number of days of restricted activity in the United States.[31,32] The total economic burden in the United States of annual influenza epidemics has been estimated at $87 billion.[33]

Influenza strains with high transmissibility and pathogenicity, such as H5N1 and H1N1, continue to evolve and may be responsible for the next global pandemic.[34] As of August 6, 2010, a total of 214 countries and overseas territories or communities reported laboratory confirmed cases of pandemic influenza H1N1 2009, including over 18,449 deaths.[35] This pandemic virus disproportionately affected children, who in turn posed a risk for secondary household transmission, especially to caregivers and siblings.[36,37] The household transmissibility of this influenza virus was lower than past pandemic strains, with most transmission occurring shortly before or after symptom onset. The efficacy of hand hygiene, respiratory protection, cough etiquette, isolation, social distancing, and travel restrictions to prevent transmission has been difficult to evaluate.[37,38] There is some evidence that the wearing of face masks or respirators during illness can protect contacts and help reduce transmission of respiratory tract infections; however, there is insufficient evidence that the use of respiratory protection protects wearers from infection.[39,40]

12.2.1.1 Transmission of Infectious Agents

Inhalation is the most important and efficient route by which infectious agents enter the human body, and infections contracted by this route are the most difficult to control.[41] The great pandemics in human history (e.g., plague and influenza) as well as many common childhood infections (e.g., measles, rubella, varicella, and mumps) are acquired from the respiratory secretions of others (Table 12.1). Transmission by air allows an infectious agent to reach a larger number of potential hosts than would be possible if infected individuals had to come into direct contact to transfer microorganisms from person to person. Humans release particles from the larynx, mouth, and throat during normal breathing, speaking, singing, and expectorating, but sneezing and coughing expel droplets with greater efficiency. Most of the droplets from a sneeze originate in the mouth and do not carry infectious agents. However, an increase of nasal secretions due to an upper respiratory tract infection can trigger sneezing and explosive expulsion of infectious particles from that region. Coughing is an efficient means of transmitting viruses and bacteria if an infection causes an increase of mucus secretions that, in turn, induce the cough reflex. A sneeze has been estimated to generate millions of aerosol particles and a cough many thousands.[42] The fate of these particles depends on ambient environmental conditions (e.g., humidity and air movement) and particle size. Influenza viral RNA has been detected in exhaled human breath (generation rate: <3.2–20 particles/min; 87% of particles <1 μm) in some but not all studies.[43,44]

Droplet contact involves large particles (>5 μm) that humans release from the nasopharyngeal region or that contaminated bodies of water generate through sprays or splashes.[45] Laboratory, surgical, and dental procedures can also produce aerosols of blood, saliva, enamel, dentin, bone, and tissue, some of which contain infectious agents.[46] Particles ranging from 0.3 to 20 μm have been measured from surgical procedures, oral operations, laser smoke, and dermabrasions, their concentration and size being dependent on the aerosol generation method.[47] After water evaporates, the dried residue of a droplet is called a droplet nucleus. Infectious aerosols are readily transmitted where people congregate, for example, in transportation vehicles and day care centers, schools, health clinics, correctional facilities, dormitories, military barracks, and shelters for the homeless.

12.2.1.2 Infectious Dose

Respiratory tract infections are not equally communicable.[48] The outcome of a human exposure to an infectious agent depends on the number of microorganisms encountered, the virulence of the agent, the strength of the body's defenses, and the ability of the pathogen to avoid or interfere with these defenses.[49] Infectivity and virulence can be expressed in quantitative terms. The 50% infectious dose (ID_{50}) is estimated from the number of microorganisms needed to initiate infection in half of the exposed subjects. The virulence of an infectious agent can be measured as the 50% lethal dose (LD_{50}) (the number of microorganisms or amount of toxin required to kill half of the infected animals). Endpoints other than half also may be considered. In an outbreak investigation, the infectivity of a particular strain of an agent can be determined from the observed attack rate, that is, the number of cases of clinically apparent disease divided by the number of susceptible persons. Host defenses that determine the severity of an encounter with an infectious agent include mucosal immune factors, which may block adherence and local proliferation, and humoral or cellular immune responses, which may contain proliferation and cell invasion, resulting in an asymptomatic infection rather than a more serious illness.[48]

An early study demonstrated the relationship between infectious dose, particle size, and pulmonary deposition site.[50] This group showed that the number of 2–3 μm droplet nuclei of *M. tuberculosis* to which rabbits were exposed approximately equaled the number of infectious foci (tubercles) that developed in their lungs. In contrast, only 6% of 13-μm particles caused tubercles because few particles of this size reach the deep lung and the bacterium is innocuous in the upper respiratory tract.

It is difficult to measure the number of infectious particles to which people are exposed and determine the number of viable organisms in each infectious particle; therefore, Wells introduced the idea of quantal infection.[51] He noted that the number of persons who become infected bears a Poisson relationship to the number of infective particles that they breathe. Approximately 63% of persons homogeneously exposed to an airborne agent will be infected when, on average, each of them has breathed one infective particle or "quantum." The infectivity of the organism in combination with the strength of the host's defenses determines the number of organisms required to induce a quantal response.

Some of the information on infectious doses for inhalation exposure has been modeled from outbreak situations or calculated from experimental exposures and surveillance data.[52] For example, the dynamics of measles and tuberculosis transmission have been analyzed using the concept of infectious quanta in mass-balance equations similar to those applied to the study of other environmental contaminants.[53–57] Investigations of outbreaks in which the number of infected persons, their duration of exposure, and the indoor ventilation rate were known have allowed estimation of the average air concentrations of infectious agents and their generation rates. Infectious doses may be as low as ten adenoviruses or *Coxiella burnetii*, *Franciscella tularensis*, or *M. tuberculosis* cells, 10^2 for the SARS coronavirus, and as high as 10^3–10^4 for *Bacillus anthracis* spores.[41,58,59] In animals, the LD_{50} for *B. anthracis* was found to increase with increasing aerosolized particle diameter.[60] Increasingly sophisticated mathematical models and data from animal experiments provide a promising approach for the assessment of human risk.[61]

Modeling allows examination of the effects of environmental factors such as ventilation rate, air cleaning, room occupancy, and respiratory protection as well as public health interventions such as vaccination, isolation, and contact tracing.[62,63] A review of publications on the association between building ventilation and the transmission of airborne infections concluded that there is strong and sufficient evidence to demonstrate an association between ventilation, air movement, and transmission but that there is insufficient data to specify minimum ventilation requirements to prevent aerosol spread.[64]

Model-based cost-effectiveness analyses can provide mathematical projections to better understand the key factors that affect outcomes, for example, dependence on the interplay among exposure rate, transmissibility (secondary attack rate), case-fatality rate, and risk of transmission from atypical cases.[65] A predictive model integrating influenza viral kinetics (target cell-limited model), indoor aerosol transmission potential (Wells-Riley mathematical equation), and a population dynamic model (susceptible–exposed–infected–recovery model) was applied to a proposed susceptible population to estimate influenza infection risk.[42] Another group applied a generalizable environmental infection transmission system (EITS) framework to provide a theoretical basis for understanding and modeling intervention efficacy in realistically detailed situations involving diverse venues where transmission may take place.[66] EITS provides a way to conceptualize the extent to which transmission is a density- or frequency-dependent contact process, for example, transmission through thoroughly mixed air or infrequently touched fomites generates density-dependent patterns while transmission through frequently touched fomites, such as doorknobs, generates frequency-dependent patterns.

12.2.2 Respiratory Diseases other than Aerosol Transmissible Infections

Respiratory symptoms are the most common health effects associated with noninfectious bioaerosol exposure and generally result from airway inflammation caused by toxins, pro-inflammatory agents, or allergens (Section 12.2.3). Many of the symptoms or conditions described in detail in the following are not exclusively associated with bioaerosols but also can be caused by exposure to a number of nonbiological agents, for example, cigarette smoke, ambient air pollutants, and chemicals. The current section describes the most important of these respiratory diseases and symptoms and briefly discusses causal factors. The following sections discuss the health effects of specific biological agents in more detail.

12.2.2.1 Asthma

Asthma is a heterogeneous chronic inflammatory disorder of the airways involving airflow limitation that is at least partly reversible and results in recurrent episodes of symptoms such as wheezing, breathlessness, chest tightness, and cough. There is increasing evidence that the allergic phenotype of asthma may account for only half of all asthma; therefore, exposures that induce a nonallergic inflammatory response may play equally important roles in asthma development and exacerbation.[67,68] In the last few decades, Western countries have experienced an epidemic of asthma prevalence and incidence that appears to have commenced after the Second World War and has only recently peaked and begun to decline for reasons not fully understood.[69] WHO estimates that some 300 million persons currently suffer from asthma, which is the most common chronic disease among children.[70] In 2010, the annual economic cost of asthma in the United States was estimated to be >$15 billion in direct costs and >$5 billion in indirect costs such as lost productivity.[71]

Viral infections are a common cause of asthma exacerbation.[72] In fact, viral respiratory infections are detected in the majority of asthma exacerbations (80%–85% in children and 75%–80% in adults); of these, ~60% are from rhinoviruses.[73] There also is a strong association between viral infections and hospital admission for asthma in both children and adults. In addition, viral infections may be involved in asthma development, but the evidence is less clear.[74–76] The mechanisms of viral-induced asthma are poorly understood but it has been speculated that impairment of innate immune responses may play a crucial role.[73]

The environmental exposures most commonly associated with asthma in the general population include a wide range of bioaerosols such as nonallergenic microbial agents including bacterial endotoxin;[68,70] house dust mite, pet, and cockroach allergens;[77] and allergenic and nonallergenic fungal agents (Section 12.3.2).[78,79] Many of these exposures are elevated in damp indoor environments. Several structured expert panels have reviewed the literature on dampness and asthma and have concluded that there is sufficient evidence of an association between exposure to damp environments and asthma symptoms and exacerbation in sensitized persons but that for asthma development there is only limited or suggestive evidence.[70,80] Another expert panel, investigating housing interventions concluded that there is sufficient evidence that mold and moisture remediation reduces exposure to asthma triggers.[81]

It is commonly assumed that fungal exposure plays a major role in explaining the association between damp homes and asthma, but the evidence is based primarily either on self-reported or investigator-assessed visible mold and water damage. Therefore, it is unclear whether fungi and other microbial agents are causally related to asthma or simply are markers of dampness and that nonmicrobial dampness-related agents are responsible. Of those studies that have included quantitative exposure measurements, few have found significant associations,[15,82,83] which may be due to problems related to exposure assessment (Section 12.6.1).[70,78] However, a recent publication was the first study showing an association between quantitatively measured mold and asthma development.[84] Early exposure to molds at one but not 7 years of age significantly increased the risk for asthma at 7 years of age. If these associations are causal and fungi play a major role, then mold exposure has been estimated to be associated with an approximately 30%–50% increase in a variety of respiratory and asthma-related health outcomes.[80] The same authors estimated that, if reported associations were causal, 21% of current asthma cases in the United States would be attributable to dampness and mold in housing, for an annual cost of $3.5 billion.[85] Thus, if the underlying associations identified in epidemiological studies truly are causal, dampness and mold have enormous health and social costs worldwide.

Bioaerosol exposure is also a cause of occupational asthma, for example, agricultural and related industries, waste recycling, wood processing and furniture making, biotechnology and primary enzyme production, and the bakery industry (Section 12.3.3).[86] More than 250 agents have been identified as causes of occupational asthma,[87] divided broadly into those that may cause allergic asthma (e.g., enzymes, latex, and other allergens from microorganisms, animals, and plants) and inflammatory agents that may cause nonallergic asthma (e.g., bacterial endotoxin, bacterial DNA, and possibly $(1 \rightarrow 3)$-β-D-glucans).

In the occupational literature the nonallergic phenotype is sometimes referred to as "irritant-induced asthma" or "asthma-like syndrome." Occupational asthma is the most common work-related respiratory disease in developed countries, with approximately 10%–15% of cases of adult asthma attributable to occupational factors.[88] Estimates of the total proportion of adult asthma thought to be occupational in origin range from 2% to 15% in the United States, 15% in Japan,[89] 5% in Spain,[90]

2%–3% in New Zealand,[91] 2%–6% in the United Kingdom,[92] although one study estimated 28%,[93] and 17% and 29% in Finland for men and women, respectively.[94]

12.2.2.2 Rhinitis

Rhinitis is caused by allergic or nonallergic inflammation in the mucosa of the upper respiratory tract. Typical symptoms include sneezing, rhinorrhea, and nasal congestion (runny or stuffy nose) often accompanied by tears, itchy eyes, and red and swollen eyelids (conjunctivitis). One of the consequences of the nasal obstruction associated with rhinitis is greater oral breathing, which bypasses some of the body's defenses and may lead to greater susceptibility to airborne infectious agents (Section 12.5.1).[95]

As with asthma, rhinitis has increased in the last few decades, and approximately half of all rhinitis appears to be allergic.[96] Episodic symptoms are the hallmark of "hay fever" or allergic rhinitis caused by pollen or fungi. The list of pollen types that may cause hay fever is long and includes ragweed; Russian thistle; English plantain; timothy, rye, and other grasses; and birch, maple, oak, and willow trees.[77] The outdoor environment is the primary source of pollen exposure, although indoor occupational allergy to flower pollen also has been documented.[97] While rhinitis is associated with asthma they are separate conditions.[98]

Nonallergic rhinitis in the general population is common and is most likely caused by air pollutants and other chemical exposures. Bioaerosols also may play a role, for example, endotoxins and fungi, but most studies linking bioaerosol exposure and nonallergic rhinitis have been conducted in the occupational environment (Section 12.3.3). Occupational, nonallergic rhinitis is sometimes referred to as "mucous membrane irritation" and is characterized by dry cough and irritation of the eyes, nose, and throat. It is among the most commonly observed occupational conditions associated with occupations with high bioaerosol exposure, for example, farming.

12.2.2.3 Organic Dust Toxic Syndrome

ODTS is a nonallergic, acute, febrile illness, characterized by fever, shivering, dry cough, chest tightness, dyspnea, headache, muscle and joint pains, fatigue, nausea, and general malaise.[86] The symptoms resemble those of influenza but usually disappear in 1 day. ODTS is also known as "toxic alveolitis" or "inhalation fever." Historically, this syndrome of symptoms was named after the particular environment in which it was observed, for example, "mill," "grain," or "humidifier" fever or "silo unloaders syndrome." ODTS is based on an acute, intense, airway inflammation characterized by neutrophilia, typically requiring 4–8 h to manifest fully following exposure. No prior sensitization is required and, with sufficiently high exposures, all exposed persons may develop the disease. Repeated exposures can induce adaptation or tolerance, and ODTS may not reappear until there is an unusually high re-exposure or after a period of absence, for example, a holiday.

ODTS is relatively common where workers are highly exposed to bioaerosols, but may also occur in the home or office environment where heavily microbially contaminated humidifiers or air-conditioning systems are in use. The disease is particularly common among farmers and workers in the textile fiber processing industry. Microorganisms and bacterial endotoxin are believed to be the main causes.[86]

12.2.2.4 Hypersensitivity Pneumonitis

HP is used synonymously with extrinsic allergic alveolitis, "farmer's lung," or "pigeon breeder's lung." HP is a generic term used to describe a restrictive, obstructive or mixed restrictive/obstructive, pulmonary condition with delayed, febrile, systemic symptoms, manifested by an influx of inflammatory cells to the lung parenchyma and the formation of granulomas there. Prior sensitization is required, which is believed to involve Type-III or -IV hypersensitivity, but the immunological mechanisms are still only partly understood.[99]

Typical symptoms of acute HP include systemic ones, such as fever and chills, as well as chest tightness, dyspnea, and cough, and closely resemble those of ODTS.[100] HP symptoms develop 4–8 h after inhalation of the offending substance and usually last for several days. In the subacute form, dyspnea and cough are manifested gradually over several weeks or months, and the condition can progress to severe respiratory impairment. The chronic form of HP is often afebrile and associated with dyspnea, malaise, weakness, weight loss, and cough. Pulmonary function abnormalities range from diffusion defects with restrictive dysfunction to varying degrees of obstructive dysfunction. Very severe disease can result in fibrosis and even death.[99]

HP is a rare condition, generally only described in case reports, particularly in relation to exposure to thermophilic microorganisms. The most well known is farmer's lung disease, which occurs most often in farmers exposed to one or more of four thermophilic, mycelial bacteria growing in hay or straw: *Saccharopolyspora rectivirgula*, *Saccharomonospora viridis*, *Thermoactinomyces sacchari*, or *Thermoactinomyces vulgaris*. Evidence from Finland and France suggests that fungi may play a role in farmer's lung disease in these regions.[101,102] HP also has been described for grain workers, mushroom growers, bird breeders, sawmill workers, and compost workers. Other examples include buildings with heavily contaminated humidifiers, but this is very rare.

HP case reports were noticed in the early 1990s among machinists exposed to water-based metalworking fluids (MWFs) treated with biocides (Section 12.3.3). These outbreaks often were associated with species of non-tuberculous *Mycobacterium*. *Mycobacterium immunogenum* is believed to be infectious and a cause of "hot tub lung" in immunocompromised persons.[103] "Hot tub lung" has been described as an HP-like, granulomatous, lung disease. Characteristics of this condition suggest an immunologic mechanism similar to HP, but other evidence suggests an infectious response.[103]

Reports of HP-like symptoms in lifeguards and pool workers have suggested HP or an HP-like condition, but it is unclear if these symptoms are the result of irritant effects from pool chemicals (trichloramines), bacterial growth and aerosolization due to poor facility design and maintenance, elevated endotoxin levels, or a combination of these or other factors.[104,105]

12.2.2.5 Chronic Bronchitis

Chronic bronchitis is most often defined as a nonallergic condition characterized by cough with phlegm most days for at least 3 months of the year for a consecutive period of at least 2 years. Chronic bronchitis may present with or without airway obstruction and symptoms such as dyspnea, chest tightness, and wheezing. The underlying mechanism is an increase in mucus-secreting glands and alterations in the characteristics of the mucus itself. Smoking is one of the most important causes of chronic bronchitis. However, bioaerosol exposure may play a role as well, as has been demonstrated in several occupational environments such as farming, the cotton and compost industry, and other occupations where workers are exposed to high dust levels. Whether nonoccupational indoor exposures play a role is not clear.[86]

12.2.2.6 Airflow Obstruction

Several studies have shown associations between chronic airflow obstruction—measured by spirometry—and bioaerosol exposures in industrial and agricultural workers. Associations were particularly strong for endotoxin exposure.[106] A decline in forced expiratory volume in 1 s (FEV$_1$), which is indicative of airway obstruction, also has been shown in occupants of damp and moldy homes,[107] although findings have not been consistent.[78] Indoor $(1 \rightarrow 3)$-β-D-glucan and endotoxin levels as well as floor dampness also have been shown to be associated with greater variability in peak expiratory flow (PEF; reduced PEF is another marker of airway obstruction). Airflow obstruction is associated with both allergic and nonallergic disease.

12.2.2.7 Pulmonary Hemorrhage

Pulmonary hemorrhage is a condition characterized by elevated levels of hemosiderin, an iron-containing pigment, in lung tissues. The condition can be fatal and is caused by diffuse bleeding or hemorrhage in the alveoli. Idiopathic pulmonary hemorrhage (IPH) describes this condition when the cause is unknown. A U.S. Centers for Disease Control and Prevention (CDC) report suggested that indoor exposure to *Stachybotrys chartarum* was a factor in a cluster of acute IPH in ten infants presenting in 1993–1994 at a children's hospital in Cleveland, OH.[108–110] One infant died of severe respiratory failure, and the episode received wide press coverage. The findings were criticized and the CDC reassessed the data because of shortcomings in differential diagnosis and data collection, analysis, and reporting.[111]

A hemolytic protein (stachylysin) has since been identified in *S. chartarum*, and the role of this fungus in disrupting pulmonary surfactant synthesis has been explored in animal models.[112] The inflammatory, allergic, and cytotoxic potential of the *S. chartarum*-associated mycotoxins and protein factors (hemolysins, proteases, glucans, and spirocyclic drimanes) and interaction with other co-factors (glucans and endotoxins) have also been discussed.[113]

Reports from the U.S. Institute of Medicine (IOM) and WHO concluded that available case-report information, taken together, constituted inadequate or insufficient information to determine whether an association exists between acute IPH and exposure to *S. chartarum*.[78,79] To date the role of *S. chartarum* in adverse health problems is unresolved and controversial.[114] Despite the enormous media attention and the seriousness of pulmonary hemorrhage, it is important to note that in terms of public health relevance the Cleveland cluster and other incidents of infant IPH are extremely rare. This is in sharp contrast with other well-established and much more common respiratory health effects associated with damp buildings.

12.2.3   Other Adverse Health Effects

12.2.3.1   Building-Related Illnesses and Symptoms

Building-related illnesses (BRIs) are diagnosable diseases with known etiologies that are frequently accompanied by documented physical signs and laboratory findings, for example, acute viral infections, legionellosis, or tuberculosis; carbon monoxide poisoning; irritant-induced or exacerbated asthma or rhinitis; or HP. BRSs or SBS are a combination of nonspecific symptoms, in the absence of diagnosed disease, related to the building in which people live or work.[78] BRSs commonly include (but are not limited to) irritation of the skin and eyes, nasal itching and dryness, headache, fatigue, prolonged sore throat, hoarseness, dry cough, and chest discomfort. Less common are nausea, vomiting, difficulty with concentration, joint pain, and low-grade fever. A high frequency of respiratory tract infections and enhanced or abnormal odor perception also can be a component of BRS. Although BRSs are believed to be relatively common, no accurate estimates of the incidence or prevalence are available due primarily to the lack of objective criteria to diagnose the syndrome.

BRSs may be related to the inflammatory effects of inhaled indoor particles, including bacteria, fungi, their structural components, and other organic substances (i.e., the organic fraction of indoor dust) causing localized and systemic nonspecific inflammation, but the evidence for this is weak.[115] Reactive chemistry involving indoor-generated oxidants such as ozone has been suggested as a contributing cause.[116] Environmental factors (chemical contaminants, inadequate ventilation, odor, and thermal discomfort) and personal factors (gender as well as stress and other psychological factors) are likely contributing causes.[117]

12.2.3.2   Cancer

Workers frequently exposed to livestock and grain (farmers) as well as to animal carcasses being processed for meat (abattoir workers and butchers) are at increased risk for several cancers including lung, hematologic, lip, stomach, prostate, connective tissue, and brain cancer. Hypothesized explanations involve exposure to pesticides, zoonotic viral or bacterial agents that are able to cross species barriers (e.g., herpes, avian leucosis, and papilloma viruses and *Brucella* and *Leptospira* species), mycotoxins, and other yet unidentified biological exposures.[86]

12.2.4 POTENTIAL PROTECTIVE EFFECTS OF MICROBIAL AGENTS ON ALLERGIES AND ASTHMA

The "hygiene hypothesis" postulates that growing up in a more hygienic environment may increase the risk of developing allergies.[11] This hypothesis was developed to explain epidemiological findings that overcrowding and unhygienic conditions were associated with a lower prevalence of allergies, eczema, hay fever, and asthma.

Confirming findings from cross-sectional studies of significant inverse associations between indoor endotoxin concentrations and atopic sensitization, hay fever, and atopic asthma, a prospective birth cohort study found an inverse association between endotoxin and fungi in dust at 3 months of age and doctor diagnosed asthma and persistent wheeze at age 4.[10] For infants born to atopic parents in another study, indoor exposure to high β-$(1 \rightarrow 3)$-D-glucan concentrations (>60 µg/g) was associated with a decreased risk for recurrent wheeze.[12,118]

However, several large studies have not found a protective effect or even have seen a positive association between less hygienic environments and respiratory health. For instance, the U.S. National Survey of Allergens and Endotoxin in Housing found an endotoxin exposure-dependent increase in diagnosed asthma, wheeze, and asthma medication use in adults.[13] Animal studies also have shown mixed evidence regarding the possible protective effect of endotoxin for atopy and airways hyperreactivity.[68] Part of these inconsistencies may be related to the timing of exposure, with early life exposures protective and exposures later in life a risk factor for these conditions. Alternatively, it may be that endotoxin exposure prevents allergic asthma but at higher exposures causes nonallergic asthma.[119]

There are other limitations of the current hygiene hypothesis. In particular, although housing conditions in the United States have not become more hygienic for inner city populations, their asthma prevalence has increased significantly. The rise is greatest among African Americans living in poverty, which is in contrast to previous findings showing a positive association between affluence and asthma prevalence. Also in the past few years, asthma prevalence appears to have dropped in many high income countries with no indication that hygiene has deteriorated.[71] Further, epidemiological studies have assessed exposure to a limited suite of bioaerosols and the causal agents responsible for immunomodulation may remain unknown.

12.3 SOURCES AND EXPOSURES

12.3.1 BIOAEROSOLS FROM OUTDOOR SOURCES

A variety of environments, including soil, surface water, and plants, are natural reservoirs of biological agents. In number, bacteria are the most abundant organisms in surface soils, where their concentration can exceed 10^{10} cells/g, of which typically 10^7–10^8 cells are culturable on standard media.[120,121] Total emissions of bacteria to the atmosphere have been estimated to be 1400 Gg/year with an upper bound of 4600 Gg/year, originating mainly from grasslands, shrubs, and crops.[122] Estimates of 590 and 100,000 bacterial species/g of soil have been obtained through empirical models of species frequency distribution.[123] Protozoa and algae are also abundant in soil (10^3–10^6 cells/g) along with fungi (typically 10^5–10^6 cells/g), but the overall biomass of fungi is greatest because of their larger cell sizes.

Soil microorganisms can be aerosolized during mechanical disturbances, such as the tilling of agricultural fields, soil excavation, the mechanical turning of compost, landfill site operation, and dust storms.[124,125] Many plant pathogens are transmitted by air, for example, during crop harvesting, but these agents do not appear to pose a hazard for people. However, some human infections result from exposure to soil dust contaminated with infectious bacteria or fungi, for example, *Burkholderia pseudomallei*: melioidosis; *Aspergillus* species: aspergillosis, *Coccidioides immitis*: coccidioidomycosis; *Cryptococcus neoformans*: cryptococcosis; and *Histoplasma capsulatum*: histoplasmosis (Table 12.1).

The activated sludge systems in wastewater treatment plants can generate large amounts of bio-aerosols.[126–128] Irrigation with reclaimed water and the soil application of solid waste materials from water treatment processes also can aerosolize microorganisms and particles of organic matter.[128,129] The Florida red tides are an example of a growing environmental issue, that is, that of harmful algal blooms that may adversely affect asthmatic persons exposed to organisms in onshore winds.[130] Of related interest are studies of aerosolized brevitoxin from *Karenia brevis*, which can cause respiratory distress, coughing, and eye irritation in humans.[131] The severity of pulmonary responses to brevitoxin depends on the combined interaction of the total air toxin load, the specific toxins aerosolized, and the size of airborne toxin particles.[132] Airborne cyanobacteria (blue-green algae) and other algae also deserve attention because of potential health risks for occupationally exposed persons.[133,134]

Ambient air is comprised of suspended solid and liquid particles with relatively stable size distributions, and particles of biological origin are part of this mixture. The fraction of the particle number concentration of outdoor PM originating from living sources has been estimated to be 37% on a rooftop in Mainz, Germany,[135] and 4%–20% over the North Sea.[136–138] Carbon from fungi accounted for up to 10% of the coarse (2.5–10 μm) fraction of organic carbon (OC) in samples from the Austrian Alps.[139] Average concentrations of bacteria and fungi were 1.2×10^4 cells/m^3 and 7.3×10^2 spores/m^3, corresponding to 0.03% and 0.9% of OC, respectively. Measurements in Vienna, Austria, were higher, where fungal OC concentrations ranged from 22 to 677 ng/m^3, 2%–14% of the aerosol OC mass concentration, 1%–7% of the PM$_{10}$ concentration, and predominant contributors to the coarse aerosol OC concentration (mean: 60%).[140]

Air samples and budget calculations for actively wet-spore-discharging Ascomycota and Basidiomycota suggested that they may account for a large proportion of coarse PM in tropical rain-forest regions during the wet season (0.7–2.3 μg/m^3).[141] Using the average abundance of mannitol as a tracer for basidiospores, these authors estimated their global average emission rate to be ~17 Tg/year. Global average atmospheric abundance and emission rate for total fungi were estimated to be a factor of three higher, that is, ~1 μg/m^3 and ~50 Tg/year, respectively. This rate can be compared to estimates of ~47 Tg/year for anthropogenic primary organic aerosols and 12–70 Tg/year for secondary organic aerosols. Another model used mannitol as a fungal spore tracer and estimated that they contribute 23% of total primary emissions of organic aerosol, or 7% of the fine-mode source.[142] These authors estimated annual, mean, simulated, surface concentrations of primary biological aerosol particles (PBAP) over vegetated regions to be in the range of 0.1–0.7 μg m^{-3} (PM$_{2.5}$) and 0.4–3.0 μg m^{-3} (PM$_{10}$), with the highest concentrations in the tropics, where primary PBAP may be the dominant source of organic aerosol. The source strength of biogenic material (from plants and microorganisms) emitted as PM$_{2.5}$ has been estimated at 65 Tg/year.[143–145]

Analyses at urban, rural, and high-alpine locations showed that the mass fraction of DNA in PM$_{2.5}$ was ~0.05% (average concentration: ~7 ng/m^3).[146] Bacterial sequences were from Proteobacteria along with Actinobacteria and Firmicutes. Fungal sequences were characteristic of Ascomycota and Basidiomycota, which actively discharge their spores. Plant sequences could be attributed to green plants and moss spores, while animal DNA for only one unicellular eukaryote was detected (a protist). Another study found that >99% of small-subunit rRNA sequences could be identified as belonging to bacteria, fungi, plants, or metazoa (multicellular animals).[147] Airborne bacteria were diverse (367 unique taxonomic units). Bacteria assigned to the CFB (Cytophaga-Flavobacterium-Bacteroides or Bacteroidestes group) and proteobacterial groups were the most abundant. Airborne fungi were much less diverse, with ~97% of sequences classified as Ascomycota (>90% Hypocreales order) and 3% as Basidiomycota (genera *Paecilomyces, Fusarium, Acremonium, Trichoderma,* and *Cordyceps*). Plants accounted for 13%–38% of rRNA sequences with 95% of them assigned to the genus *Pinus* (pine trees) and 5% to the genera *Poa, Abies, Alnus,* and *Equisetum* (bluegrass, fir, alder, and horsetail, respectively). Metazoan sequences were identified as dipteran flies (genus *Ornithoica* or *Ceratitis*) or mites (genus *Chortoglyphus*).

12.3.1.1 Seasonal, Diurnal, and Geographic Variability in Ambient Bioaerosol Concentrations

ARIs follow regular seasonal patterns. For example, prior to the World Trade Center attack, deaths in the United States from pneumonia and influenza peaked predictably on February 17 ± 2 days. However, following September 11, 2001 the season was delayed due to travel restrictions and decreased air travel.[148] Domestic air volume in November (related to the Thanksgiving holiday in the United States) predicted the rate of influenza spread, and international travel influenced the timing of flu mortality.

Fluctuations in bioaerosol concentrations are influenced by climate and weather (resulting in diurnal and seasonal cycles) and by local sources (resulting in regional variation). For example, hot, dry conditions, length of drought, and mean annual rainfall have been associated with the incidence of coccidioidomycosis caused by a soil fungus.[149,150] Weather also affects respiratory allergic symptoms directly, by acting on the airways, and indirectly, through episodic fluctuations in aeroallergen and air pollutant concentrations.[151]

12.3.1.1.1 Time of Day

Diurnal patterns in bioaerosol concentrations are caused by changes in RH and air and surface temperatures as well as fluctuations in wind speed and turbulence, all of which affect the emission, suspension, and removal of pollen grains, fungal spores, and other bioaerosols from the atmosphere.[145,152,153] Many plants (e.g., mugwort and sorrel) have their peak hour of pollen release just before noon whereas other plant species (e.g., grasses) peak in the early afternoon.[153] Diurnal cycles also have been observed for bacterial and fungal air concentrations (maxima in the early morning and before nightfall), but patterns are species and microclimate specific.[154] In a tropical rainforest during the wet season, fungi as a percentage of coarse PM have been estimated at ~25% during the daytime and ~45% at night.[141] Modeling estimations indicate that the highest fungal deposition dose would occur between 11 p.m. and 5 a.m. and that the greatest number of spores would deposit in the alveolar-interstitial region.[155]

A northern California study compared the size characteristics of PM, protein, endotoxin, and $(1 \rightarrow 3)$-β-D-glucan indoors and outdoors simultaneously.[156] The largest mass fraction of protein was associated with particles <2.5 µm, while the largest mass fractions of endotoxin and glucan were associated with the coarse fraction. $PM_{10-2.5}$ concentrations of protein, endotoxin, and glucan tended to be higher indoors than outdoors. Indoor pets were associated with elevated indoor PM and bioaerosols. However, there were no statistically significant associations between bioaerosols and dust levels in homes, suggesting that dust loading may not serve as a satisfactory surrogate for human inhalation exposure. Outdoors protein concentration, as a marker of primary biological particles, was observed to follow the same temporal patterns as $PM_{2.5}$ concentration.[157]

12.3.1.1.2 Time of Year

Seasonal patterns in outdoor bioaerosol concentrations are caused by temperature, moisture availability, and hours of daylight. Studies have found that culturable bacteria are more prevalent in summer than winter in regions where dry, dusty summer conditions and associated agricultural or human activities contrasted with wet winter conditions and snow cover.[158–162] A study in Colorado observed higher concentrations of total airborne bacteria in summer than in winter both indoors and outdoors, higher endotoxin concentrations outdoors in summer, but no seasonal endotoxin difference indoors.[163] Indoor bacterial concentrations typically are higher than outdoor levels throughout the year and are associated primarily with human shedding, whereas fungal concentrations are slightly lower indoors in buildings without interior sources.[164]

Outdoor fungi are dominated by local sources and are often found year-round with maxima in spring, summer, or fall.[165–167] Particularly in temperate northern regions, outdoor concentrations of fungi decrease in winter due to subfreezing temperatures and snow cover, for example, a New York study found 26-fold higher concentrations of indoor culturable fungi in summer than in winter.[166] Water vapor can condense on cold indoor surfaces and this, along with releases of liquid water, may create conditions that allow microbial growth with both qualitative and quantitative changes in the bioaerosol composition inside buildings.[168–170]

The fungal fraction of OC mass concentration has been shown to be lower in spring than summer (2%–5% vs. 8%–14%).[140] The relative frequency of occurrence of individual fungal species, that is, the proportion of samples in which the species were detected, has allowed comparisons of species richness and seasonal variability for Ascomycota and Basidiomycota.[171] For example, in central Europe the prominent allergenic ascospore groups, *Cladosporium* and *Alternaria* species, occurred most frequently in summer and fall, whereas *Penicillium* species and the plant pathogen *Blumeria graminis* were found most often in winter and spring. The fall maximum for basidiospores primarily was due to species that were found only once during the year and can be attributed to mushroom fruiting and enhanced plant decomposition during that season. In the extratropics, where vegetation cover varies significantly throughout the year, simulated fungal emissions and air concentrations exhibited a peak in late local summer and early fall.[142]

Airborne pollen are found predominantly during their respective pollination periods from early spring to fall when plants are producing and releasing whole grains. Spring typically is the main pollen season for those trees and shrubs that release airborne pollen grains. Pollen dehiscence (opening of mature anthers) occurs before foliage is fully grown, which otherwise could impair the release and distribution of pollen grains. Normally, the primary pollen season for trees is no longer than 2–3 weeks. Herbs and grasses can flower several times during each growing season; thus, airborne pollen grains from these plants can be found throughout the year in snow-free parts of the world.

Over a 27 year period in western Liguria, Italy, a progressive increase in pollen-season duration has been noted for *Parietaria* species (pellitory) (+85 days), olive (+18 days), and cypress (+18 days).[172] The authors attributed this increase to an overall advance of the seasons' start dates. In addition to a longer season, the total pollen load throughout the season has progressively increased as well (~25% on average) for the groups mentioned earlier but not for grasses. Percentages of patients sensitized to these outdoor allergens increased over this 27 year period while the rate for house dust mite sensitivity remained stable. These behaviors paralleled the constant increase in direct radiation, temperature, and number of days >30°C. Climate changes may affect the quality and amount of airborne allergenic pollen and the prevalence and severity of hypersensitivity and inflammatory responses.[151,173] However, the direct assessment of such an effect requires long observation periods for restricted geographic areas.[172] Dispersion of microorganisms in dust may play a significant role in the biogeographical distribution of pathogenic and nonpathogenic species as long-range atmospheric transport routes and concentrations shift through time due to climatic and geologic change.[174]

12.3.1.1.3 Geographic Region

Bioaerosol concentrations, in particular bacterial concentrations, are generally higher in urban than rural atmospheres in the absence of local sources such as animal houses, agricultural operations, waste treatment plants, or composting facilities. In contrast, fungal contributions to total OC, PM_{10}, and the coarse fraction were approximately half at a traffic-dominated urban versus a suburban sampling site.[140] A study in Boulder, Colorado, observed more phylogenetic similarity between bacteria collected from geographically distant sites (Colorado, Texas, Antarctica, and France) than between bacteria collected in Boulder over 8 days.[147] From this finding the authors concluded that outdoor air may harbor similar types of bacteria regardless of location and that short-term temporal variability can be large and must be considered when designing a sampling plan. Another study found that bacterial diversity was higher in a rural than in urban and alpine locations.[146] The distribution of pollen-producing plants is a result of natural floristic patterns, but landscaping has significantly changed the air biota in many parts of the world.

12.3.2 Bioaerosols in the Indoor Environment

Exposure to inhalable aerosols containing infectious viruses or bacteria is more common in winter when people spend more time indoors, ventilation rates are reduced, and ill persons cough and sneeze in proximity to susceptible individuals.

12.3.2.1 Viruses, Bacteria, and Bacterial Endotoxin

Humans are the primary source of indoor viral pathogens, but very little is known about the aerodynamic size of infectious virus particles. See Section 12.2.1.1 for a discussion of indoor transmission of infectious viruses and bacteria.

Only a few studies have focused on bacteria in damp indoor environments. Some of these identified *Streptomycetes* species on damp or wet indoor surfaces or released from moisture-damaged building materials.[175–177] Streptomycetes are Gram-positive, spore-forming actinobacteria that can produce a wide range of metabolites including some toxins such as valinomycin.[178–182]

Environmental mycobacteria also have been found to be common in moisture-damaged buildings, and their occurrence was shown to increase with an increasing degree of fungal damage.[183] Concentrations of total viable bacteria in indoor environments may range between 10^1 and 10^3 colony-forming units (CFU)/m^3.[184] These levels most likely represent the degree of building occupancy and ventilation efficiency. From the literature it is not clear what "typical" bacterial air concentrations are in damp versus non-damp indoor environments.

Endotoxins are composed of lipopolysaccharides (LPS) and lipooligosaccharides (LOS), cell wall components of Gram-negative bacteria. Endotoxins are nonallergenic, with strong pro-inflammatory properties. They are present in many occupational environments, ambient air,[185] and house dust.[13,68]

Endotoxin concentrations in house dust range from a few to several thousand Endotoxin Units (EU)/mg.[68] Concentrations expressed per square meter vary even more. Studies have found no evidence for a positive relationship between endotoxin in house dust and observed dampness or mold,[186–188] but a positive association was found between moldy odor and the concentration of airborne endotoxin.[189] Only a few studies have focused on indoor air concentrations, for example, a mean airborne endotoxin concentration of 0.64 EU/m^3 was reported in 15 homes in Boston, MA (mean dust concentration: 44–105 EU/mg).[190] The mean inhalable endotoxin concentration in nine Belgian homes was similar, that is, 0.24 EU/m^3.[191] Higher concentrations of airborne endotoxin were measured in the Midwestern United States and New Orleans (geometric means, 4.2 and 3.0 EU/m^3, respectively).[189,192]

Animal and human toxicity tests with endotoxin and LPS have shown that inhaled endotoxin causes inflammation with the release of different cytokines and increased production of oxygen metabolites. Alveolar macrophages and bronchial epithelial cells are the primary targets; endothelial cell damage also has been observed. The concentrations of cytokines peak a few hours after inhalation.

Subjects exposed to endotoxin in inhalation experiments experience clinical effects such as fever, shivering, arthralgia, influenza-like symptoms (malaise), blood leukocytosis, neutrophilic airway inflammation, asthma symptoms (such as dry cough, dyspnea, and chest tightness), and bronchial obstruction as well as dose-dependent lung function impairment (measured as FVC, FEV_1, and flow-volume variables) and decreased lung diffusion capacity.[68] Subjects with increased bronchial hyperresponsiveness or asthma may be more likely to develop symptoms when challenged with endotoxin, but large differences in airway responsiveness to inhaled endotoxin also exist in healthy (nonallergic) subjects.[193]

Endotoxin has been suggested to play a role in nonoccupational respiratory disease, most notably asthma (Section 12.2.2.1). Several studies showed a positive association between endotoxin in house dust and symptoms, lung function, and medication use in children with pre-existing asthma.[14] A positive association between indoor endotoxin and respiratory symptoms also has been reported in infants[194] and adults.[13] In contrast, several other studies have suggested that domestic endotoxin exposure in early life may protect against allergy development and new onset asthma (Section 12.2.4), and it has even been suggested that endotoxin may reverse pre-existing atopic sensitization and related diseases.[9] Thus, the role of nonoccupational endotoxin exposure in allergy and asthma development is currently unclear.

12.3.2.2 Fungal Spores and Hyphal Fragments

Human challenge studies of school personnel with BRI have shown no observed effect levels (NOEL) of 4×10^3 and 8×10^3 spores/m^3 for *Trichoderma harzianum* and *Penicillium chrysogenum*, respectively.[195] For *Alternaria alternata* and *Penicillium* species, lowest observed effect levels (LOEL) for airway obstruction in patients with mild allergic asthma were found to be 1×10^4 and 2×10^4 spores/m^3, respectively (the applied doses were calculated to equal an 8 h exposure).[196] In various worker populations (e.g., wood workers, waste handlers, and farmers), a variety of respiratory effects such as symptoms, lung function changes, and increased inflammatory airway markers have been observed at concentrations >10^5 spores/m^3. Experimental studies have further shown that the inflammatory effects do not depend on fungal (actinomycete) viability,[197] although there may be differences in fungal allergenicity.[198]

For indoor air quality assessments, concentrations measured in test environments are typically compared to baseline data from reference areas or to data reported in the literature.[199] The extremes (~95th percentiles) of Finnish data distributions of indoor, culturable bacteria and fungi (5000 CFU/m^3 and 500 CFU/m^3, respectively) have been recommended as indicators of the presence of abnormal indoor sources or insufficient ventilation (but not health risk) in urban and suburban residences in a subarctic climate.[164] The Finnish National Guidelines of Indoor Air Quality incorporated these values for the interpretation of bioaerosol sampling results.[200] However, the general consensus is that it is not possible to set health-based, numeric concentration limits for bacteria or fungi in indoor and occupational environments.[79,201–207]

In addition to spores, fungi also release even smaller fragments.[208] These are derived from broken or fractured spores and hyphae and can be categorized into submicrometer particles (<1 μm) or larger fungal fragments (>1 μm). Like spores, hyphal fragments are known to contain allergens[209] and mycotoxins.[210] Therefore, both spores and fungal fragments may play a role in mold-related adverse health effects, and indoor exposures to fungal fragments may be at least as important as fungal spores (Section 12.4.1.3).

12.3.2.3 Fungal (1 → 3)-β-D-Glucans

(1 → 3)-β-D-glucans are nonallergenic water-insoluble structural cell wall components of most fungi, some bacteria, most higher plants, and many lower plants. They consist of glucose polymers with variable molecular weight and degree of branching. Methods to analyze (1 → 3)-β-D-glucans in environmental samples have not been standardized; therefore, measurements are not comparable across studies making dose estimations difficult.

Elevated levels of (1 → 3)-β-D-glucans have been demonstrated in buildings with mold problems and several occupational settings including sawmills, the paper industry, waste handling and recycling, and farming.[211] In Sweden and Switzerland, glucan concentrations in airborne dust vigorously generated from settled dust from buildings with fungal problems ranged from ~10 to >100 ng/m^3 using a *Limulus* Amebocyte Lysate (LAL) assay (Section 12.6.4.3).[212] Air concentrations in buildings with no obvious fungal problems were close to 1 ng/m^3. In the Netherlands and Germany, mean (1 → 3)-β-D-glucans concentrations in house dust determined with a specific enzyme immunoassay (Section 12.6.4.4) were highly comparable, with concentrations of ~1–2 mg/g dust and 0.5–1 mg/m^2, respectively.[213–217]

One study suggested that glucan exposure was associated with an increased risk of atopy, similar to that observed in some animal studies, but this finding was not confirmed in a smaller study,[211] and some studies even have shown a protective effect on respiratory health outcomes.[12,118] The evidence with regard to the potential effects of glucan on airway inflammation is mixed. *In vitro* studies have demonstrated the potential of glucans to induce inflammatory responses, and several studies in laboratory animals have shown that sufficiently high glucan exposure may cause airway inflammation. However, in most population and human challenge studies no such association was found.[211] Thus, the currently available data do not permit conclusions to be drawn regarding an association between environmental glucan exposure and specific adverse health effects, nor is it clear from the available evidence which specific inflammatory mechanisms might underlie the presumed health effects.[211]

12.3.2.4 Mycotoxins

Mycotoxins, or fungal toxins, are low molecular weight biomolecules produced by fungi, some of which are potent inhibitors of protein synthesis in both animals and humans. Mycotoxins have been studied extensively for their ingestion toxicity,[218] and mycotoxin residues in human food crops and animal feed are extensively regulated throughout the world.[219] Mycotoxins have been detected in fungal spores, mycelia, and dust from water-damaged buildings as well as in the serum of building occupants, although the health effects associated with mycotoxin inhalation is controversial.[220,221] It has been suggested that pulmonary inflammation and alveolar hemorrhage in infant rats, used as a model for human infant lungs, is related to the mycotoxin content of *S. chartarum* spores.[114]

Several mycotoxins are potent carcinogens (e.g., aflatoxin from *Aspergillus flavus* and *Aspergillus parasiticus*) (Section 12.2.3.2). Other important mycotoxins are the trichothecenes produced by *Fusarium* and *Stachybotrys* species, fumonisins and zearalenone produced by *Fusarium* species, and ochratoxin A produced by *Aspergillus ochraceus* and *Penicillium verrucosum*. Many mycotoxins are immunotoxic. However, the trichothecene mycotoxins are immunostimulating at low doses. Numerous other mycotoxins have been classified possessing distinct chemical structures and reactive functional groups, including primary and secondary amines, hydroxyl or phenolic groups, lactams, carboxylic acids, and amides.

Only a few studies have assessed indoor mycotoxin concentrations. Several mycotoxins produced by *S. chartarum* and *A. versicolor* (i.e., macrocyclic trichothecenes, trichodermin, sterigmatocystin, and satratoxin G) could be detected in dust and building materials from dwellings with current or historic dampness or water damage.[222] Other studies found increased levels of macrocyclic trichothecenes in flooded dwellings known to be contaminated with *S. chartarum*.[223,224] These studies demonstrate that mycotoxins are detectable in the indoor environment and may be higher in buildings affected by mold or dampness. However, to date it is not clear whether airborne mycotoxin concentrations in damp buildings are sufficiently high to cause adverse health effects (Section 12.2.2).

While early efforts at understanding the toxic health effects of inhaled fungal particles focused on mycotoxins, the contribution of mycotoxins versus other fungal proteins in causing inflammation is unclear. Current thinking is that a wide range of inflammatory substances are present in fungal particles, and that health effects likely result both from mycotoxin-producing species as well as other anamorphic Trichocomaceae found in damp indoor environments.[225]

12.3.2.5 Fungal Volatile Organic Components

Several fungi produce volatile metabolites comprising a mixture of compounds that can be common to many species, although some fungi also produce compounds that are genus or species specific. To date, more than 200 VOCs have been identified from different bacteria and fungi (Section 12.6.4.5).[226–230] However, no large field studies have been conducted comparing MVOCs in contaminated and control buildings, and MVOCs have only rarely been measured in health surveys. Therefore, data on air concentrations and dose–response relationships are scarce, and the utility of measuring VOCs to detect hidden mold currently is doubtful.[231]

12.3.2.6 Indoor Allergens

Many fungal species have been described as producers of Type I allergens. IgE sensitization to the most common outdoor and indoor fungi (e.g., *Alternaria*, *Penicillium*, *Aspergillus*, and *Cladosporium* species) is strongly associated with allergic respiratory disease, especially asthma (Section 12.2.2).

Cladosporium herbarum, *A. alternata*, and *Aspergillus fumigatus* have been shown to produce a variety of allergens including several major allergens, that is, *Cla h*I, *Alt a*I and II, and *Asp f*I and III. Allergens can be found in spores, hyphae, and fungal fragments. At high concentrations fungi also may be involved in Type III and IV allergic reactions including HP (Section 12.2.2.4). Due to difficulties with the manufacture and standardization of fungal allergen extracts, commercial assays are available for only a limited number of indoor fungal allergens (including *Alternaria* species allergens) (Section 12.6.4.4). Therefore, exposure data on indoor fungal allergens (as opposed to fungal spores) are scarce (Section 12.6.2). Other species have been suggested to be related to asthma in the general population and in workers, but these fungi are not commonly included in allergy tests and their contribution to fungal allergy is not known. Some studies have shown a higher prevalence of fungal sensitization in subjects living in damp buildings and in severe asthmatics. Also, in a large European multi-center study, an association between mold sensitization (*A. alternata* and *C. herbarum*) and asthma severity was demonstrated,[232] and allergic responses to molds have been shown in relation to outdoor exposures to *Alternaria* species in desert environments. Thus, exposure to fungal allergens may be related to asthma; however, the evidence that allergic responses play a major role in asthma development is limited.[78] This may be because mold allergens and IgE directed against these allergens are very difficult to measure because the natural production of mold allergen is highly variable. Another possible explanation is that resting spores are less allergenic than spores that have germinated.[198] Few studies have examined fungal allergy in highly exposed occupational populations, but specific IgE to the prevalent species is remarkably absent in some of the studies.[233]

The major allergens produced by the house dust mite, *Dermatophagoides pteronyssinus* (called *Der p*1 and *Der p*2) are proteases present in high amounts in fecal pellets, which contain digestive enzymes and partially digested food (primarily human skin flakes).[77,234] Another house dust mite species, *Dermatophagoides farinae* produces *Der f*1 as its major allergen. Elevated levels of these allergens have been detected in house dust, mattress dust, and bedding collected in damp homes. The major cat (*Felis domesticus*) and dog (*Canis familiaris*) allergens are proteins called *Fel d*1 and *Can f*1, respectively. These allergens are present in cat and dog dander, saliva, and urine, and can be found in house dust, dust in other buildings and transportation vehicles, and on clothes of people that keep pets or are regularly exposed to them. Two major cockroach (*Blatella germanica*) allergens have been identified called *Bla g*1 and *Bla g*2, suggested to come from feces or body parts.[77]

Indoor allergens such as pet and house dust mite allergens can exacerbate pre-existing asthma in sensitized subjects,[77] and epidemiological studies have shown an association between indoor allergen exposure and specific atopic sensitization (Sections 12.2.2 and 12.2.3).[235] For example, many studies have shown an association between dust mite allergen levels in house dust and dust mite specific sensitization. There also is a strong association between house dust mite sensitization and asthma. Moreover, several challenge studies have demonstrated that dust mite allergen inhalation in allergic asthmatics could produce a strong asthmatic response, and asthma in mite allergic subjects has been shown to significantly improve when they move from high to low exposure areas.

However, few studies have shown a clear association between allergen exposure and new-onset asthma.[235] From cross-sectional studies, weighted averages of the population attributable risks for new onset asthma in children has been estimated to be 11% for *Fel d*1, 6% for *Can f*1, 4% for *Der p*1, and −4% for *Bla g*2.[67] Nor have intervention studies minimizing exposure to indoor allergens in early life proven effective in preventing asthma,[67] and international patterns of asthma prevalence do not "fit" the international patterns of allergen exposure (Sections 12.2.2.1).[98]

These observations suggest that different factors may be important in allergy exacerbation than are involved in new-onset asthma development. Aside from the possibility that indoor allergen exposure *per se* is not a major cause of asthma,[67] studies might have suffered from unmeasured or unknown confounding from other indoor inhalation exposures such as fine particles, ozone, or pro-inflammatory compounds such as endotoxin or glucan.

12.3.3 Bioaerosols in Occupational Environments

Occupations in the public and private service sectors can place workers in close proximity to clients in shared spaces, resulting in a greater risk for communicable disease transmission.[236] This is especially true for diseases with annual outbreaks, such as influenza, acute viral gastroenteritis, and colds. As a consequence of treating patients with these diseases, health care workers are at increased risk of exposure to viral and bacterial bioaerosols. Zoonotic diseases with airborne transmission such as Q-fever, psittacosis, histoplasmosis, and multidrug-resistant *Staphylococcus aureus* (MRSA) may affect farmers, animal handlers, pet store clerks, veterinarians, zookeepers, and abattoir workers.[236] The influenza virus can be transmitted to humans from birds or swine or from avian species via swine. Some infectious diseases are associated with episodic exposures arising from engineering errors, maintenance failures, or unusual exposures. For example, outbreaks and individual cases of pneumonia caused by *Legionella pneumophila* have occurred in connection with contaminated cooling towers and evaporative condensers, untreated whirlpool spas, dental office apparatus, and produce misters in grocery stores, but often the source is not identified.[237,238]

Anthrax, although of current concern as a biological weapon, may be one of the first infectious, respiratory diseases to be associated with certain occupations. Inhalation anthrax has been associated with exposure to *B. anthracis* spores in contaminated animal products, such as during the industrial processing of wool (woolsorters' disease).[239] The intentional distribution of spores through mail established a new and unusual route of occupational exposure from contaminated letter-sorting machines.[240] In October 2001, four cases of inhalation anthrax occurred in employees at a Washington, DC, postal processing and distribution center. These cases were part of a multistate outbreak of inhalation and cutaneous anthrax associated with intentional distribution of envelopes containing *B. anthracis* spores to media and federal government offices. Together, these represent the first reported cases of inhalation anthrax in postal workers and the first outbreak of inhalation anthrax caused by occupational exposure in the United States since 1957.

Opening letters containing a *B. anthracis* simulant has been observed to spread the spores throughout a building in <4.5 min, and potential mitigation techniques, such as closing the office door or shutting off the ventilation system, were not effective.[241] One concern is possible "weaponization" of infectious particles through the addition of surface-active materials that would cause them to separate from each other and to maintain a uniform and inhalable particle size. Threats of bioterrorism and biowarfare have heightened the need for accurate and sensitive methods for the measurement of airborne infectious agents and have encouraged the development of rapid detection methods (Section 12.6.3.4).

Microbial contamination is a persistent problem in a variety of manufacturing processes.[242–248] One setting that has been investigated extensively is the metalworking industry, for example, facilities that manufacture automobile components, where rapidly moving parts can generate droplets of contaminated cutting fluid. Water-based MWFs provide an excellent medium for the growth of a variety of bacteria and fungi to concentrations >10^6/mL despite the expanded use of biocides and the implementation of other control measures.[248–252] *Mycobacterium* species and other Gram-positive bacteria may be pervasive MWF contaminants and cause HP in exposed workers.[248,253,254] Using an established animal model for HP, *M. immunogenum* in MWFs has been shown to induce HP, and co-exposure with endotoxin can augment the severity of this response (Section 12.2.2).[255]

Exposure to organic dusts that contain microbes or microbial cell fragments is of concern in industrial and agricultural environments.[47,256,257] Gram-negative bacteria and associated endotoxin are among the primary biological contaminants of concern, with inhalation of endotoxin resulting in both reversible (asthmatic) and chronic airway obstruction, respiratory symptoms (symptoms of asthma, bronchitis, and byssinosis), and increased airway responsiveness.[258] Consensus recommendations have been published for the measurement of endotoxin concentration,[259–261] and health-based occupational exposure limits have been proposed.[262,263] Workers are also exposed to airborne bacteria and endotoxins as well as fungi in large animal confinement buildings and sheds and in industrial food-processing facilities.[264] Exposure levels and bioaerosol composition can vary widely in occupational environments. Very little is known about exposures to airborne mycotoxins and respiratory health effects in occupational settings. Mycotoxins of *Fusarium*, *Aspergillus*, and *Penicillium* species are known to be present in the inhalable fraction of airborne corn, cotton, and grain dust. However, it is not clear whether these components contribute to respiratory symptoms in these industries. Serum albumin adducts of aflatoxin B_1 have been detected in animal feed workers who had handled aflatoxin B_1-containing raw material.[265] These findings may explain the elevated risk of liver and biliary cancer observed in a study of animal feed workers in Denmark.[266] Also, warnings of climatic conditions favorable for fungal crop infections have been found to correlate with hormone dependent cancers (as well as reproductive effects) in Norwegian farmers, and similar associations have been found for lip cancer.[267,268] Other studies have shown that fungal warnings were correlated with the trichothecene mycotoxin concentration in fine grain dust and with DNA of toxigenic *Fusarium* species in personal air samples.[269,270]

Potent IgE-binding allergens are found in biotechnology facilities that produce enzymes derived from fungi and bacteria for use in washing powders and both human and animal foods. Populations at risk are workers in the enzyme-producing industries and in detergent-manufacture and food-processing establishments where enzyme preparations are used or incorporated into consumer products. Another well-known IgE-binding allergen is plant pollen, which may cause allergies in green house workers. Latex allergens have received extensive attention during the last decade with high numbers of healthcare and hospital workers becoming sensitized to latex gloves produced from the sap of the rubber tree *Hevea brasiliensis*.

Several animal proteins (e.g., dust mite, cat, mouse, and rat allergens) also are known to have strong allergenic properties (Section 12.3.2). In particular, it is well established that laboratory animal workers are at risk of developing occupational IgE-mediated allergy to mouse and rat allergens. In addition to IgE-binding allergens, workers may be exposed to IgG-binding allergens. These fungal and actinomycetes allergens are assumed to be involved in the pathogenesis of HP or farmer's lung (Section 12.2.2.4). Airborne animal proteins from crustacean processing and exposure to aerosolized fish proteins can result in "crab asthma" and other allergic asthma in the commercial fishing and processing industries.[271]

12.4 BIOAEROSOL SIZE DISTRIBUTIONS

Bioaerosols vary greatly in diameter from nanometer-sized viruses and cell fragments to single cells, cell agglomerates, and aeroallergen particles in the micrometer range. Viruses are generally 20–300 nm in physical diameter, intact bacteria 0.3–10 μm, fungal spores 1–100 μm, and pollen grains 5–200 μm (Table 12.1). However, information is incomplete on the aerodynamic size of airborne viruses, bacteria, pollen, fungi, and allergens, and very little information is available on the size distributions of cell fragments (Section 12.4.1.3). The concentration and distribution of bioaerosols vary considerably with atmospheric and indoor conditions and between and within sites. The dynamic properties of all airborne particles depend on their physical characteristics such as physical size, shape, surface features, density, hygroscopicity, and electrostatic charge (Section 12.5). Electrostatic charges on particles may enhance their deposition on oppositely charged surfaces or through image charging (induction of an opposite charge on a surface by a charged particle). The amount of charge that particles carry depends on the aerosol generation mechanism, environmental conditions, and the time that has elapsed between aerosolization and measurement. For example, a net negative charge has been observed on laboratory-generated bacterial aerosols.[272,273] However, few studies have measured naturally occurring charges on biological particles.

Most airborne particles have a thin layer of water molecules on their surface and can absorb or lose water. Substantial changes in the particle size of hygroscopic cells can alter their deposition behavior. Hygroscopic particle growth that is rapid enough to occur during passage through the respiratory tract will increase deposition in the regions where settling and impaction are important and decrease deposition in regions where diffusion dominates, for example, the alveolar region. Although fungal spores are assumed to be hydrophobic, an almost 30% growth in the size of some spores has been seen at relative humidities (RHs) >90%, and particle sizes from 0.5 to 2 μm have been estimated to be most affected by hygroscopic growth.[274] An immediate increase in particle diameter and separation of fungal and actinomycete spores was observed as they passed briefly through a test chamber of warm, humid air (38°C, 95% RH) compared with exposure to cooler, dryer air (20°C, 40% RH).[275]

12.4.1 Aerodynamic Diameters of Airborne Biological Agents

An optical particle counter and filter sampling have been used to characterize exhaled particles and indirectly showed that most exhaled influenza virus particles were in droplets <1 μm.[43] Air sampling also has shown that >50% of influenza A virus-containing particles are <4 μm.[276,277] These findings confirm the existence of airborne virus in the fine particle size range. A fraction (~10%) of the pollen from angiosperms (flowering plants) and all of that from gymnosperms (conifers and other plants that produce seeds within cones) are dispersed by wind,[278,279] but not all of these pollen are allergenic. Pollen grains and plant spores that rely on wind transport (anemophilous species) are generally smaller (~50 μm) than those transported by other means, for example, by insects (entomophilous species). However, there is considerable overlap, and some strictly wind-pollinated plants (e.g., the pine family) produce larger pollen grains.[152] The texture or ornamentation of the outer pollen wall differs between wind- and insect-borne grains. Anemophilous pollen are often nearly spherical and have relatively smooth surfaces. Conversely, entomophilous pollen tend to have more elaborate surface features, which help them adhere[152] but also affect their dynamic shape factor and thus their aerodynamic diameter (d_a) when airborne.

The density of common airborne fungal spores has been estimated to vary from 0.56 to 1.44 g/cm^3 and that for pollen grains from 0.39 to 1.1 g/cm^3.[280] Size comparisons for fungal and actinomycete spores by four size measurement methods have found disagreement.[281] The largest discrepancy was seen for wet spores of *Cladosporium cladosporioides* for which the volume-equivalent physical diameter by optical microscope was 4.0 μm, that is, $d_a = 4.0$ μm if density = 1 g/cm^3, whereas the d_a measured with an aerosol particle sizer (APS) was 1.8 μm. Thus, d_a cannot be estimated accurately without information on cell density, which may vary with hydration status.

For a given genus or species of microorganism or plant, the variability (geometric standard deviation [GSD]) in spore or pollen grain diameter can be small and close to what is considered monodisperse (GSD < 1.2).[282] For example, using a six-stage impactor, the following GSDs were measured for culturable fungi: *Penicillium* species 1.1–1.4; yeasts 1.1–1.5; *Cladosporium* species 1.1–1.9; and *Aspergillus* species 1.4–1.9.[283] Even smaller size variations have been observed in the laboratory with direct-reading particle counters for fungal spores aerosolized from pure cultures.[274,275,284] However, measurement of the dimensions of individual spores by microscope has shown much higher variability.[285] Part of the variability in d_a measurements may be due to the occurrence of cell clusters or chains or the attachment of biological matter to other particles. Comparisons of measurements made with a multi-stage viable impactor and an aerodynamic particle sizer have revealed that fungal particles in chains behave aerodynamically like single fungal particles.[275,286] The effect of several forms of particle aggregation on the average d_a of a cluster of *P. chrysogenum* spores (physical size: 2.8 × 3 μm) has been considered for two particle densities (0.5 and 1 g/cm^3).[281] The average d_a of a unit-density, single spore was 45% larger than one of half that density (2.9 vs. 2.0 μm). For unit-density eight-cell aggregates, a chain of spores had an 18% smaller d_a than a compact aggregate of spores (4.4 vs. 5.2 μm).

12.4.1.1 Allergens

Air sampling has provided some information on the size of indoor allergens of health significance, for example, cat and dog: ~5 μm; cockroach: ~5 μm; and house dust mite: 10–35 μm.[234,287–291] Higher resuspension rates were calculated for two dust mite allergens ($Der f1$: $1 \times 10^{-5}/m^3$; $Der p1$: $2 \times 10^{-6}/m^3$) than for cat dander ($8 \times 10^{-7}/m^3$), dog dander ($3 \times 10^{-7}/m^3$), or endotoxin ($4 \times 10^{-7}/m^3$).[292] When sieved house dust was reaerosolized by vacuuming the carpeted floor of a test chamber, 9% of airborne cat allergen was detected on particles <5 μm, 30% on particles 5–10 μm, and >60% on particles 10–20 μm.[293] The allergenic content of the PM was calculated to be 0.03%–0.10% on a mass basis, with a suggestion of somewhat higher cat allergen concentrations in submicrometer particles. Only a small fraction of allergen particles may be <5 μm and able to enter the lungs. However, this variable particle fraction is assumed to produce inflammatory responses in hypersensitive persons.[78]

12.4.1.2 Culturable Bacteria and Fungi

Bacterial endo- and exospores resist environmental stress much better than vegetative cells and are considerably smaller in size.[294] For example, in Polish homes the highest concentrations of typically single-spore, thermophilic, actinomycete spores were found in the 1.1–2.1 μm size range.[295] In contrast, the maximum concentrations of mesophilic Gram-positive and Gram-negative bacteria were in the 3.3–7.0 μm size range, larger than the individual cell sizes, suggesting aggregation. Cell aggregation has been shown to increase viability of airborne bacteria and may be a survival strategy.[296]

The size distributions of culturable airborne fungi have been studied in many environments. The highest fungal air concentrations typically are found on the stages of size-segregating samplers that correspond to the diameters of intact cells or cell agglomerates.[124,297,298] With a six-stage impactor and malt extract agar (MEA), the highest concentrations of the major fungal genera were observed in the 2.1–3.3 μm size range in residences with mold problems and control homes.[283] The highest concentrations were found in the same size range for samples on dichloran glycerol 18 agar (DG-18) from moisture-damaged and reference schools, but on MEA, the highest concentrations were in the smaller size range.[298,299] These differences likely are caused by size differences in the fungal species that grow on the two culture media as well as disparities in the fungal flora in schools and homes.

12.4.1.3 Fragments

All particle sizes in Table 12.1 are for intact bioaerosol forms, but fragments of biological materials may be found as particles much smaller than their original structures. For example, airborne endotoxin has been found in particles smaller than intact Gram-negative cells.[300–303] The modes of fungal-fragment size distributions have been observed to vary from 0.5 to 1.5 μm, depending on the species and aerosolization method.[304,305] With an electrostatic, low pressure impactor, fungal fragments were detected in particles as small as $d_a = 30$ nm.[306] These findings have important implications for bioaerosol exposure measurement and dosimetry because pollen and microbial fragments behave differently in air samplers, control devices, and the human respiratory tract than would be expected given the size of their original pollen grains, cells, spores, or hyphae.

Antigen content has been measured in whole fungal spores and fragments entrained in air that had passed over contaminated ceiling tiles or agar surfaces placed in a test chamber.[307] Common antigens were identified in both fragments and spores suggesting their shared origin and similar potential importance to human health. In later studies, fungal fragments have been shown to contain allergens,[209] mycotoxins,[210,223] and fungal glucans.[308–310] Therefore, both fungal spores and fragments may play a role in mold-related adverse health effects. Currently it is not known if fungal fragments originate primarily from hyphae or spores. However, a recent study suggested that broken or fractured spores may be the main source.[304] The smaller size of fungal spores and fragments allows them to reach the lower airways and trigger asthma, but a different exposure mechanism must be involved for larger pollen grains.[232]

High concentrations of birch pollen and antigen occasionally are found before and after the pollination period. It has been suggested that allergy symptoms that occur outside the pollination period may be explained by exposure to allergenic particles smaller than intact pollen grains.[311–314]

In Japan, plant antigens were found in two particle size ranges (0.5–1.4 and 29–40 µm, the latter the size of intact pollen grains)[315] as have been ragweed and grass pollen.[316–318] The antigenic activity of the smaller particle size fraction was eight times greater than that of the larger fraction,[315] suggesting a greater allergenicity for the smaller particles (Section 12.3.1).

There is evidence that during thunderstorms, pollen is concentrated by changes in airflow, the grains are ruptured by osmotic shock, and each grain releases hundreds of allergen-containing starch granules that are small enough to be respired.[173,319–322] This hypothesis has been confirmed in laboratory studies.[323] A significant interaction has been seen between the effects of grass pollen and weather conditions on emergency room visits for asthma in England, with the increase most marked on days of light rainfall.[324] A relationship also has been noted between thunderstorms and asthma admissions in Canada.[325] However, while fungal air concentrations increased twofold, changes in pollen and air pollutants on thunderstorm days were relatively small.

12.5 RESPIRATORY DOSIMETRY

Exposure refers to the amount of a biological, chemical, or physical agent to which persons come into contact in such a way that they develop the relevant outcome, which may be adverse (e.g., a respiratory disease) or beneficial (e.g., immunity).[27] For bioaerosols, exposure usually is measured as the air concentration of an agent and the time the person spends in the contaminated environment; and respiratory dose refers to the amount of the biological agent that deposits in the airways. In some situations measurements of external exposure and internal dose may be strongly correlated, for example, in some occupational environments. In other situations, air concentration can vary widely due to an individual's lifestyle and activities, and dose may be correlated only weakly with measured air concentration, for example, house dust mite allergen.

Bioaerosol exposures may be chronic if they occur repeatedly over extended time periods, for example, as experienced by the occupants of microbially contaminated buildings. Acute exposures, on the other hand, take place over relatively short time intervals, often during periods of elevated air concentrations, for example, in agricultural and manufacturing workplaces and outdoors during peak pollen seasons (Section 12.3.1.1). The risk of becoming sensitized to an allergen may be much greater if the exposure is prolonged or intense, therefore, measurement of total cumulative exposure may be important to understand a dose–response relationship. On the other hand, once a person has become sensitized, the intensity of an allergen exposure may be crucial in provoking a response even if the encounter is brief.

More accurate estimations of bioaerosol exposures can be made if both particle size and breathing-zone air concentration are known so that the dose of an agent that deposits in a particular area of the respiratory tract can be determined (Section 12.5.1). Indoor air quality data have been coupled with a compartmental lung model to estimate indoor/outdoor/personal exposure relationships for airborne fungi.[155]

12.5.1 RESPIRATORY DEPOSITION AND CLEARANCE

Generally the same physical principles govern the respiratory deposition of biological and nonbiological particles. A semiempirical model has been applied to size distribution data from homes to predict that for nasal breathing 30%–50% of fungal spores would deposit in the nose and 30%–40% in the alveoli, whereas for oral breathing 70% would deposit in the alveoli.[283] However, some aspects of bioaerosols may cause inaccuracies in modeling respiratory deposition. For example, the size of a hygroscopic particle can increase after it has been inhaled (Section 12.4.1). A 30% increase in spore size due to hygroscopic growth has been calculated to cause a 20% increase in respiratory deposition, primarily in the bronchi.[274] Likewise, uncertainty in the measurement of biological particle size can introduce inaccuracy in dose estimation. For example, errors in size measurement were found to result in overestimation of respiratory deposition of *C. cladosporioides* spores by a factor of 1.2–1.8 (Section 12.4.1).[281] For certain microbial species, exposure may occur mainly through fragments rather than intact cells or spores (Section 12.4.1.3). The ratio of fungal spores depositing in the respiratory tract to the indoor concentration has been estimated to be higher for the extrathoracic region of the lung (0.70–0.80) than for the bronchial (0.41–0.60), bronchiolar (0.12–0.40), or alveolar–interstitial (0.01–0.24) regions.[155]

As for deposition, the mechanisms that clear other particles also operate for those of biological origin (see Chapters 5 and 24). However, differences have been noted among microbial species that have not yet fully been explained. A review of studies on lung clearance concluded that the spores of most culturable fungi are eliminated within a few days but that for others clearance was not complete even after a week.[233] Furthermore, alveolar macrophages were able to either kill fungi or prevent germination with the exception of *A. fumigatus*.

12.5.2 Dose–Response Relationships

Animal studies can provide exposure–response data under well-controlled and defined conditions when a biological agent is delivered at an environmentally relevant concentration and route of exposure. Experimental inhalation studies have been used to estimate infectivity in humans, define the pathogenesis of airborne diseases, study the mechanisms by which biological components exert their effects, and understand the action of anti-inflammatory agents and antibiotics.[326,327] The number of inhalation tests on humans or animals as well as field exposure studies is increasing, especially for endotoxin and glucans.[86] Based on what was known, toxin concentrations in spores and fragments of fungi that grow on damp building materials have been estimated to be on the order of $\sim 10^{-5}$ moles.[225] They observed time- and toxin-dependent transcription and expression of inflammation-associated genes and inflammatory responses in mouse lungs intratracheally exposed to eight fungal toxins (4×10^{-5} moles toxin/kg lung weight). These observations could provide a biological basis for some of the inflammatory health effects that occupants of damp buildings experience.

However, extrapolation from animal models to humans is not straightforward due to interspecies differences.[327] There is relative certainty about potential risks for some bioaerosols, such as infectious agents, whereas the uncertainty is much greater for other bioaerosols, such as endotoxin and other bacterial and fungal toxins. Furthermore, some effects that have been inferred from cross-sectional population studies require further confirmation preferably from longitudinal studies. The contribution of biological agents to the adverse health effects observed for PM has yet to be evaluated.

12.5.3 Interactions

Complicating studies of human responses to bioaerosol exposures are simultaneous exposures to gaseous and particulate air pollutants that may irritate or damage the respiratory tract, altering people's sensitivity to bioaerosols. Any practice or condition that impairs the natural defenses of the lung increases a person's susceptibility to infection.[328] For example, influenza is often a precursor to secondary bacterial pneumonia because the virus temporarily destroys ciliated cells.

Smoking also depresses the effectiveness of the ciliated cell defense. This damage might explain why persons with a long history of smoking have a higher risk of bacterial and fungal infections as well as other adverse health effects as compared with nonsmokers. The presence of (1) culturable bacteria and fungi in tobacco; (2) 3-hydroxy fatty acids (3-OHFA), muramic acid, and ergosterol in tobacco and tobacco smoke; and (3) bacterial metagenomes from environmental bacteria and commensals as well as the potential presence of human pathogens in tobacco may contribute to these observed effects.[329,330]

The contribution of biological agents to the adverse health effects that are seen with exposure to airborne PM is only beginning to be evaluated. Airborne particles can provide adsorptive and absorptive surfaces for inorganic and organic gases and vapors. Particles of vehicle exhaust emissions are generally 50–200 nm in diameter and are the most abundant ambient particles by number. The association between traffic-related particle exposure and persistent wheezing at 36 months of age can be modified by exposure to endotoxin, a finding that supports prior toxicological studies demonstrating a synergistic production of reactive oxygen species after coexposure to diesel exhaust particles and endotoxin.[331]

12.6 ENVIRONMENTAL MEASUREMENT AND ANALYTICAL TECHNIQUES

12.6.1 EXPOSURE ASSESSMENT

Assessment of exposure to environmental agents in indoor and outdoor air plays a central role in epidemiological studies seeking to characterize population risks, screening studies aimed at identifying individuals at risk, and interventions designed to reduce risk.[77] However, the assessment of bioaerosol exposures presents challenges distinct from those for inorganic aerosols and chemical agents. Pathogenic microorganisms may be hazardous at extremely low concentrations while other organisms may become important health hazards only at orders of magnitude higher concentrations.

Even when air samples are collected to determine bioaerosol concentrations, exposure typically is estimated from a readily measured surrogate of an active agent, for example, fungal spores, ergosterol, glucans, or mannitol to represent fungal allergens or toxins, whole pollen grains for plant allergens, guanine for house dust mite allergens, or CFUs of Gram-negative bacteria for endotoxin. Certain bacteria and fungi that have been observed in moisture-damaged buildings have been suggested as "indicator species."[332] Other researchers have used the indoor presence of hydrophilic (water-loving) fungi as an indicator of water damage and as possible markers of increased risk of building-related respiratory diseases.[333]

12.6.2 BIOAEROSOL MEASUREMENT

Bioaerosols are collected for a wide variety of purposes, for example, to measure inhalation exposure, characterize indoor and outdoor environments, identify emissions from work activities, and evaluate the effectiveness of control methods. Table 12.2 lists some commercially available bioaerosol instruments. For reviews of bioaerosol samplers and study design, see Cox and Wathes,[334] AIHA,[205,335] ACGIH,[201] Hurst,[336] and Muilenberg.[337]

12.6.3 BIOAEROSOL SAMPLERS

Most bioaerosol samplers are based on active collection using a pre-determined air flow. Because active air sampling may not be feasible in population-based studies, some investigators have substituted breathing-zone, dust sampling platforms for passive, long-term collection of settled dust.[338,339] Another alternative to active air sampling is a specially designed aerosolization chamber that collects particles from contaminated surfaces released by air currents and vibration, that is, the Fungal-Spore Source Strength Tester (FSSST)[340] and the Particle-Field and Laboratory Emission Cell (P-FLEC).[341]

12.6.3.1 Inertial Sampling

Impactors operate on the principle of drawing air through a nozzle and forcing the jet to turn sharply after exiting the inlet. Particles larger than the d_{50} cutoff diameter are collected with at least 50% efficiency. Many single- and multiple-jet impactors have been developed to collect bioaerosols (single-stage and cascade impactors, Table 12.2). A positive-hole correction may be applied to the colony counts obtained with multiple-hole agar impactors to account for coincidental impaction at high particle concentrations.[342,343] The impaction stage typically is removable to allow easy

TABLE 12.2

Bioaerosol Samplers

Sampler	Principle of Operation	Sampling Rate (L/min)	Manufacturer/Supplier	Commercial Name	Application
Slit agar impactors					
Rotating slit or slit-to-agar impactors	Impaction onto agar on rotating surfaces	28	Barramundi Corp.	Mattson-Garvin Air Sampler	C
		175–700	Casella Ltd.	Casella Airborne Bacteria Sampler	C
Impactors					
Single-stage impactors	Impaction onto agar	10, 20	Burkard Manufacturing Co., Ltd.	Burkard Portable Air Sampler for Agar Plates	C
		28	Thermo Scientific	Single Stage N6 (Viable Andersen Cascade Impactor)	C
		28	Aerotech Laboratories, Inc.	Aerotech 6 Bioaerosol Sampler	C
		28	Zefon International	Zefon A-6	C
		28	EMS	ems E-6 Sampler	C
		28	EMS	Biocassette[a]	C
		28	SKC Inc.	Biostage Bioaerosol Impactor	C
		30–120	Zefon International	Bioculture Microbial Air Sampler	C
		28, 71, 142	Veltek Associates, Inc.	Sterilizable Microbiological Atrium (SMA): Micro Sampler and MicroPortable Viable Air Sampler	
		100 or 180	International PBI; Scientific Products Corp.; Bioscience International	Surface-Air-Sampler (SAS): Super 100 Sampler, HiVac Impact, HiVAC Petri, Super 180 Sampler, SAS Duo, SAS Isolator, SAS PCR	C
		1000	Bioscience International	SAS Pinocchio Super II Air Sampler	C
		100	F.W. Parrett Ltd	MicroBio Air Samplers: MB1, MB2	C
		100, 200	Microbiology International	Sampl' Air Air Sampler	C
		100	Merck KGaA/VWR Scientific Products Corp.	Merck Air Sampler MAS 100	C
		140, 180	Millipore Corp.	M Air T, Millipore Air Tester	C

Category	Collection method	Manufacturer/Supplier	Device	Flow rate	Type
	Impaction onto rotating drum with tape strip or glass slide	Burkard Manufacturing Co., Ltd	Burkard Recording Volumetric Spore Trap	10	M
		Lanzoni, S.R.L.	Lanzoni Volumetric Pollen and Particle Sampler	10	M
	Impaction onto moving glass slides	Allergen LLC; McCrone Microscopes and Accessories; EMS	Allergenco Air Sampler (MK-3)	15	M
		Burkard Manufacturing Co., Ltd	Burkard Continuous Recording Air Sampler	10	M
		Lanzoni, S.R.L.	Lanzoni Volumetric Pollen and Particle Sampler	10	M
	Impaction onto stationary glass slide	Burkard Manufacturing Co., Ltd	Burkard Personal Volumetric Air Sampler	10	M
		Zefon International; Aerotech Laboratories, Inc.; McCrone Microscopes and Accessories; SKC Inc.	Air-O-Cell Sampling Cassette (disposable)	15	M
		EMS	Cyclex-D[a]	20	M
		EMS	Allergenco-D; Allergenco-D Positrack[a]	15	M
		EMS	Biosis Slit Impaction Air Sampler	15	M
		EMS	Micro-5; Micro-5 Positrack[a]	5	M
		Zefon International	Moldsnap[a]	5	M
		Zefon International	Via-Cell[a]	15	H, M, O
	Impaction onto collection disk	ICX Technologies	BioBadge	40	H, M, O
	Impaction onto rotating rods	Sampling Technologies, Inc., SDI Innovations	Rotorod	48	M
Personal aeroallergen sampler	Impaction on a protein-binding membrane held in a nasal insert	Available for research only	Inhalix	Dependent on breathing rate	O
Cascade impactors	Impaction onto agar or other surface	Thermo Scientific	Andersen Two- and Six-Stage Viable Sampler/Cascade Impactor	28.3	C
		Thermo Scientific	Andersen Personal Cascade Impactor, Series 290 Marple Personal Cascade Impactor	2	C

(continued)

TABLE 12.2 (continued)
Bioaerosol Samplers

Sampler	Principle of Operation	Sampling Rate (L/min)	Manufacturer/Supplier	Commercial Name	Application
Liquid impingers					
All-glass impingers	Impingement into liquid	12.5	Ace Glass Inc.; Hampshire Glassware; Millipore Corp.	All-Glass Impingers (AGI): AGI-4, AGI-30	C, M, O
Three-stage impingers	Impingement into liquid	10, 20, 50	Burkard Manufacturing Co., Ltd	Burkard Multiple-Stage Liquid Impinger	C, M, O
		10, 20, 50	Hampshire Glassware	Hampshire Glass Three-Stage Impinger	C, M, O
Centrifugal samplers					
Centrifugal agar impactors	Impaction onto agar in plastic strips	40, 50, 100	Biotest Diagnostics Corp.	Reuter Centrifugal Samplers (RCS): Standard RCS, RCS Plus, RCS Isolator, and RCS High Flow Microbial Air Sampler	C
Wetted cyclone samplers	Tangential impingement into liquid	50–55, 167, 500	Hampshire Glassware	AEA Technology PLC Aerojet Cyclones	C, M, O
		167	F.W. Parrett Ltd.	MicroBio MB3 Portable Cyclone	C, M, O
		100–800	MidWest Research Institute	SpinCon High-Volume Cyclonic Liquid Sampler	C, M, O
		450 or 1000	InnovaTek Inc.	BioGuardian Air Sampler	C, M, O
		300	Life Safety Systems	OMNI 3000	C, M, O
		1200	Bioscience International	SAS Cyclone	C, M, O
		150	ICX Technologies	BioXC	C, M, O
		325	Research International	SASS 2300	C, M, O
		40	Research International	SASS 2400	C, M, O
		150	ICX Technologies	BioXC	C, M, O
Dry cyclone sampler	Reverse flow cyclone	20	Burkard Manufacturing Co., Ltd.	Burkard Cyclone Sampler	H, M, O
Three-jet, tangential sampler	Tangential dry impaction or impingement into liquid	12.5	SKC Inc.	BioSampler	C, M, O
Wet electrostatic precipitation sampler	Electrostatic deposition on collector tube followed by washdown	300	Zaromb Research Corporation	Portable High-Throughput Liquid Assisted Air Sampler (PHTLAAS-APASS2)	C, H, M, O

Type	Description		Manufacturer	Product	C, H, M, O
	Electrostatic deposition on collector tube followed by washdown	500	Zaromb Research Corporation	*WEP-2*	C, H, M, O
Rotating arm impactor	Impaction onto rotating arm with liquid rinse	125	MesoSystems Technology. Inc.	BioCapture Air Sampler	C, H, M, O
Filter samplers					
Filter holder	Inhalable sampler for filter collection	4	SKC Inc.	Button Sampler	H, M, O
		2	SKC Inc.	IOM Sampler	H, M, O
	Collection on electrostatically charged polymer disc	50–300	Research International	SASS 3100	H, M, O
	Collection on gelatin filter	30–50	Microbiology International	Sartorius AirPort MD8	H, M, O
	Collection on gelatin filter	33–133	Microbiology International	Sartorius MD AirScan	H, M, O
Real- or near-real-time sampler					
Time-of-flight spectrometer	Aerodynamic size: 0.5–15µm, fluorescence characteristics of individual particles at one UV emission wavelength	1	TSI Inc.	UV-APS (FLAPS 1)	O
	Aerodynamic size: 0.8–10µm, fluorescence characteristics of individual particles at two UV emission wavelengths	1	TSI Inc.	(FLAPS)™III	O
	Fluorescence detection of particles at one UV emission wavelength; trigger for secondary aerosol sampler (e.g., BioXC) and subsequent identification	3	ICX Technologies	IBAC Biological Aerosol Threat Monitor	O
Optical particle counter	Optical size and shape: 0.5–15µm, fluorescence characteristics of individual particles at two UV emission wavelengths	33	Biral	Vero Tect Bio-detector	O
Impaction of particles on regenerated substrate	Fluorescence detection of particles at one UV emission wavelength	40	ICX Technologies	AirSentinel	O

C, culture of sensitive and hardy microorganisms, for example, vegetative bacterial and fungal cells and spores; H, culture of hardy microorganisms only, for example, spore-forming bacteria and fungi; M, microscopic examination of collected particles; O, other assay, for example, immunoassays, bioassays, chemical assays, or molecular detection methods.

a Disposable sampler.

replacement of the collection medium, which may be an agar plate, coated microscope slide, filter, or tape. Several models of disposable impactors have recently become commercially available (Table 12.2).

High impact velocity can injure microorganisms, and bacterial and fungal survival has been found to decrease with increased impaction velocity or sampling time.[344,345] Bacterial culturability also has been shown to decrease with increased jet-to-plate distance, possibly because the increased jet dissipation desiccates a larger fraction of the agar surface.[346]

Liquid impingers collect particles by impaction, diffusion, and interception followed by suspension in a collection fluid. One impinger uses three curved jets that induce a swirling motion in the capture liquid, which increases collection efficiency and decreases particle re-entrainment (three-jet, tangential sampler, Table 12.2).[347] Liquid volumes from 5 to 50 mL may be used with fluids as diverse as sterile distilled water and nonevaporating mineral oil.

Centrifugal samplers and cyclones collect particles by impaction onto a solid medium or into a liquid. In cyclones, air tangentially enters a cylindrical or inverted conical chamber, spins down along the chamber walls, flows up through the center, and exits at the top (wetted and dry cyclone samplers, Table 12.2). Large particles deposit on the cyclone walls and very large particles fall to the bottom. Liquid is often pumped into a cyclone's inlet to wash the particles into a collection container.

12.6.3.2 Filtration

Filter samplers collect particles of all sizes, the upper limit depending on the filter holder's inlet characteristics and addition of precollectors. Filter cassettes developed for sampling of inhalable particles, for example, the Button sampler,[348] have commonly been used for both area and personal bioaerosol sampling (filter samplers, Table 12.2).

Most filters have their minimum collection efficiency at the particle size of 0.01–0.03 μm (see Chapter 24).[343,349,350] Many types of filter materials can be used for bioaerosol collection, for example, polycarbonate, polytetrafluoroethylene (PTFE, Teflon®), mixed cellulose ester (MCE), and gelatin filters. The choice of filter material depends mainly on the type of analysis that follows. Porous membrane filters are often chosen for culturing and immunostaining of bioaerosols. Polycarbonate and PTFE filters typically are used when the collected material must be washed from a filter.[350,351] Dehydration of vegetative bacterial cells during filter sampling significantly reduces their viability, but hardy bacterial and fungal spores are not as susceptible to damage.[352–354] Gelatin filters can reduce the desiccation effect caused by filter sampling.[350,352] Capillary pore filters have smooth, flat surfaces suitable for the examination of particles with an optical or scanning electron microscope (SEM, see Section 12.6.4.1). MCE filters can be made transparent for analysis by light microscope.[355] Investigators are increasingly using membrane filters to collect bioaerosols in conjunction with analyses not based on culture because of their ease of use and high collection efficiency (Sections 12.6.4.3 through 12.6.4.6). However, some of these filters have shown poor recovery for specific agents such as endotoxin for which glass fiber filters have been recommended.[356]

12.6.3.3 Electrostatic Precipitation

The electrostatic precipitator is an example of an instrument long used for particle collection that has been adapted for laboratory sampling of bioaerosols. Collection efficiency for bacteria onto rectangular agar dishes was observed to vary from 50% to 90% depending on air flow rate and applied voltage.[272,273] Recently, a Wet Electrostatic Precipitator (WEP) was developed for bioaerosol collection.[357] An electrostatic field charges particles in an air stream and forces them to the wall of the WEP collection tube from which they are washed and concentrated by a recirculating liquid. An Electrostatic Precipitator with Superhydrophobic Surface (EPSS) achieved a concentration rate of 10^6 for latex particles.[358] The EPSS subsequently was found to be compatible with polymerase chain reaction (PCR)-based sample analysis and to reach a collection efficiency of 72% for *P. fluorescens* and *B. subtilis*.[359]

12.6.3.4 Future Directions in Bioaerosol Sampling

Several real-time instruments for bioaerosol measurement are under development, primarily for biodefense applications. Four models of direct-reading instruments currently are available commercially. The first commercial unit was the ultraviolet aerodynamic particle size spectrometer (UV-APS, time-of-flight spectrometer, Table 12.2). The biological origin of the particle is confirmed by detection of ultraviolet irradiation-induced fluorescence. UV-APS relies on optical detection of particles; thus, its use is not feasible for particles <0.1 μm. Furthermore, fluorophores decay after microbial death, thus, UV-APS is best suited for measurement of viable microorganisms.[360,361]

Three other real-time instruments are modifications of the UV-APS (FLAPS-III, IBAC Biological Aerosol Threat Monitor, and Vero Tect Bio-detector; Table 12.2). FLAPS-III measures particle size using light scattering and fluorescence emission at two distinct wavelengths, which increases specificity for biological particles. The IBAC has been made rugged and packaged for military applications. It can be connected to a concentrator and used as a trigger for a secondary aerosol sampler with subsequent identification of biological particles. The Vero Tect Bio-detector measures particle shape in addition to measuring the particle size and fluorescence emission at two wavelengths.

Many near real-time instruments take advantage of the progress in analytical methods that can provide results within minutes. One instrument collects aerosol particles for a predefined period onto a low-fluorescence substrate (AirSentinel, Table 12.2). The spot is exposed to UV excitation and the emitted fluorescence is detected. After each collect-spot-interrogate cycle is completed, the collection substrate is regenerated to its original condition and a new cycle can begin. Wetted wall cyclones with continuous liquid outflow can be connected to a bioassay unit that is more specific than intrinsic fluorescence, for example, immunoassay or PCR.[362] However, a concentrated aerosol is required for these samplers to achieve sufficient sensitivity. The EPSS can transfer particles into water droplets as small as 5 μL and achieve concentration rates of up to 10^6.[358]

Many conventional samplers take advantage of non-culture methods for bioaerosol analysis, and personal sampling is becoming more convenient for individual exposure measurement. For example, personal samplers have been used to quantify viruses with real-time PCR.[277,363] An unconventional personal sampler fits into a test subject's nostrils (Intra-nasal air sampler, INAS) and has been used to measure allergen and fungal exposures.[209] Several particle sensors under development eventually may be adapted to biological particle detection, for example, an array biosensor that can interrogate multiple samples simultaneously for multiple targets has been miniaturized and automated for portability and on-site use by untrained persons.[364]

12.6.4 Sample Analysis

12.6.4.1 Microscopy

Light microscopy still is one of the most powerful methods available for the qualitative and quantitative analysis of those pollen grains and fungal spores that have unique morphological features. The lower limit of resolution for this method is 1 μm. Quality control evaluations using an intercalibration test for airborne pollen monitoring demonstrated the important role of operator training and the need for standards in pollen sample analysis by microscope.[365] The British Aerobiology Federation, the Pan-American Aerobiology Association, the Italian Organization for Standardization, and the American Society for Testing and Materials have published standard methods for counting pollen grains and fungal spores.[366–369] An evaluation of the Italian method found that for the same sample size, confidence intervals vary in relation to pollen abundance in terms of number of grains or species.[370] The sample size suggested by the standard (20% of the target surface) may result in errors in pollen counts ranging from 7% to 55% of the mean value, and in missing 22%–54% of the taxa present on a slide.

12.6.4.2 Cultivation-Based Methods for Viruses, Bacteria, and Fungi

In cultivation-based (culture- or growth-based) methods, airborne microorganisms are collected in a liquid, on a filter, or directly on a semisolid growth medium. After an appropriate incubation period, the resulting plaques (viruses) or visible colonies (bacteria and fungi) can be counted and the isolates identified. One of the primary limitations of cultivation-based methods is that they provide a measure of only those organisms that are able to grow in the laboratory. Furthermore, aggregation of bacterial or fungal cells may lead to errors in the enumeration of the total number of culturable organisms. This is especially true with direct agar impactors as opposed to samplers that collect particles into liquid, that is, a CFU may result from the growth of one or a cluster of several cells.

12.6.4.3 Biological Assays

The effects of bioaerosols on other biological systems can be measured by different methods, among them whole animal exposure, infectivity assays for viruses, bioassays utilizing prokaryotic or eukaryotic cell lines, and nonspecific toxicity, cytotoxicity, and genotoxicity assays. One of the most commonly used biological assays, the LAL assay, measures bacterial endotoxin and fungal glucans. The LAL assay is an *in vitro* biological test that uses a lysate of blood cells from the horseshoe crab (*Limulus polyphemus*). The lysate contains a serine protease that triggers an enzyme cascade when activated by endotoxin or glucan. Kits for kinetic chromogenic and turbidimetric assays are available commercially. The enzyme activation with endotoxin and glucans occurs through different pathways (Factor C and D, respectively). Therefore, in commercially available kits, the competing pathway is depleted to make the assay sensitive only for endotoxin or glucan.

Variations have been found between LAL preparations from different manufacturers and within reagent lots from single producers. Variation between laboratories is even larger than within laboratories, but can be decreased with standardized extraction protocols (Section 12.6.4.4).[371] LAL response varies for LPS molecules from different microorganisms, and the amount of glucan detected per fungal spore varies widely between species.[372,373] The assumption that responses to different endotoxins correspond with toxic potencies has not been proven. A recombinant factor C (rFC) assay that uses a reagent produced from the cDNA of the Mangrove horseshoe crab (*Carcinoscorpius rotundicauda*) is an alternate, although less utilized assay, for endotoxin.[374] Recent studies have shown that the rFC and LAL assays give similar estimates of endotoxin concentration.[375,376] Furthermore, the results were found to correlate with 3-OHFA measurements (a surrogate for total endotoxin) from gas chromatography–mass spectrometry (GC–MS) analysis (Section 12.6.4.5).[377]

A method for detection and quantification of fungal biomass based on fluorogenic detection of β-*N*-acetylhexosaminidase enzyme activity has been developed (the MycoMeter test, Mycometer Inc., Copenhagen, Denmark)[378] and used to measure mold contamination.[379] Another rapid method for the quantification of microbial biomass is based on the detection of adenosine triphosphate (ATP), a basic energy molecule present in all living organisms. This method has been applied in laboratory-based tests where other living cells do not interfere.[380]

A variety of other assays (many of which use continuous cell lines) have been developed to detect overall toxicity, cytotoxicity, or mutagenicity and to screen for the presence of particular toxins, for example, aflatoxin from *A. flavus* in air samples from food-processing plants.[381] A boar sperm cell, motility inhibition assay also has been used to detect bacterial depsipeptide and other toxins in foods and indoor environments.[382,383] The assay is sensitive to mitochondrial toxins that inhibit sperm motility (e.g., valinomycin) but is relatively insensitive to toxins that affect protein or nucleic acid synthesis (e.g., many mycotoxins). A protein synthesis inhibition assay was used to study the trichothecene toxicity of airborne particles in a home heavily contaminated with *Stachybotrys* species.[384] The highest toxin activity was found during renovation, but detectable air concentrations also were detected before and after these activities.

12.6.4.4 Immunoassays

Antibody-based immunoassays, particularly enzyme-linked immunosorbent assays (ELISA), are widely used for the measurement of aeroallergens and allergens in settled dust, using enzymatic reactions or radio-immunoassays (RIA) for detection. Important advantages of immunoassays include (1) the stability of most of the measured components, allowing longer air sampling times, and freezing of samples for storage prior to analysis; (2) the incorporation of standards; (3) the possibility of testing reproducibility; and (4) specificity and sensitivity.

To date, the house dust mite allergens, *Der p*1, *Der f*1, and *Der p/f*2 have been most widely investigated and the methods have been well described.[385–387] Methods for assessment of exposure to rodent,[388–390] cockroach,[391] and storage mite allergens[392] also have been published. Recently, a fluorescent multiplex assay was developed to measure several indoor allergens simultaneously in one assay.[393] It is based on monoclonal antibodies that are covalently coupled to fluorescent microspheres. A commercially available assay kit simultaneously measures eight indoor allergens: *Der p*1, *Der f*1, Mite Group 2 (*Der p*2, *Der f*2, and *Eur m*2, *Euroglyphus maynei*), *Fel d*1, *Can f*1, *Rat n*1 (*Rattus norvegicus*, rat), *Mus m*1 (*Mus musculus*, mouse), and *Bla g*2 (Maria™).

Methods for measurement of fungal allergens are not widely available due mainly to the fact that fungal allergen production is highly variable and dependent on factors such as substrate and temperature (Section 12.3.2.6). However, specific immunoassays have been developed to measure fungal glucans[213,394,395] and extracellular polysaccharides,[396] with the latter assay allowing partial identification of fungal genus. These assays are experimental and as yet have not routinely been applied nor become commercially available.

12.6.4.5 Chemical Assays

Chemical analyses for bioaerosols have gained expanded utility with the development of increasingly sensitive instruments that allow analysis of smaller amounts of material. Because preservation of cell viability is not an issue in chemical assays, a wider variety of methods and longer sampling times can be used to collect bioaerosols.

The cell walls and membranes of fungi and bacteria contain unique chemical components by which they can be identified and quantified.[397,398] These compounds include lipids and various proteins and peptides, peptidoglycan in bacterial cell walls, teichoic acid in the thick peptidoglycan layer of Gram-positive bacteria, LPS in the outer membranes of Gram-negative bacteria, and ergosterol in the cytoplasmic membranes of fungi. Aerosolized toxins and volatile metabolites also can be analyzed by standard chemical methods.

Cellular fatty acids are commonly used to identify bacteria after they have been grown in culture. For example, tuberculostearic acid, a component of lipoarabinomannan in the cell walls of coryneform bacteria (e.g., *Mycobacterium*, *Rhodococcus*, and *Corynebacterium* species), readily can be analyzed by GC.[397] Fatty acid analysis also has been use for house dust and air samples.[174,399] The benefits of these assays are that they address biomass; the drawbacks are the costs of the analyses, overlapping of agent profiles, questions in conversion of biomass to cell counts, and possible compositional shift from variance in growth conditions.[174] Chemical LPS analysis is generally based on detection of 3-OHFA by high-performance liquid chromatography (HPLC) or GC–MS.[400] Chemical analysis of endotoxin has the advantage of being insensitive to variations in biological activity, but these techniques are two to three orders of magnitudes less sensitive than the LAL assay (Section 12.6.4).[263]

Ergosterol is a principal sterol in the membranes of fungal hyphae and spores and has been used to measure fungal biomass. One specific drawback to ergosterol analysis is that it is not species specific. Ergosterol is stable in air-dried conditions and can be extracted in basic aqueous methanol followed by microwave heating and analysis by HPLC, GC, GC–MS, or MS/MS.[399,401,402]

VOC profiles also show a potential for identification of environmental microorganisms (Section 12.3.2.5). Air samples collected on sorbent material and analyzed by thermal desorption-GC and

HPLC have been used to detect metabolic activity of fungi and actinomycetes on building materials in a test chamber.[227] Activity occurred at 90%–99% RH, and the main measurable VOCs were 3-methyl-1-butanol, 1-pentanol, 1-hexanol, and 1-octen-3-ol. Ventilation system filters colonized with fungi have been found to release acetone, hexane, and organic compounds that uncolonized filters did not emit.[226] Selected VOCs have been monitored in cleaned, flooded homes and an unflooded residence as possible indicators of fungal metabolism.[403] Three alcohols and one ketone were detected in significant concentrations in some flooded homes (70–2700 ng/m^3), the most common being 3-methyl-1-butanol along with 2-octen-1-ol, 2-heptanone, and 1-octen-3-ol. However, MVOC levels did not correlate with bacterial or fungal concentrations and could not be linked to specific sources or used to quantify microbial air concentration.

The sugar alcohol mannitol has been suggested as a suitable molecular tracer for basidiospores and PM potassium for ascospores.[141] Mannitol concentration also has been used as a biotracer in a global model simulation of PBAP from fungi (GEOS-Chem).[142] Proteins are characteristic of particles of biological origin and have been measured in indoor and outdoor air with a bicinchoninic acid assay.[156,157] Algal toxins (e.g., brevitoxins from *K. brevis*) have been measured with LC–MS.[134]

12.6.4.6 Molecular Genetic Assays

Molecular genetic techniques have received increased attention as diagnostic tools in the study of airborne microorganisms and can provide genus-, species-, or strain-specific identification. These procedures allow precise detection and quantification of specific organisms or genes. PCR assays amplify target nucleic acid sequences. Genus- and species-specific nucleic acid probes have been designed that can be used for microscopic visualization.[404] PCR-based methods have been used to detect bacteria on contaminated surfaces,[405] *L. pneumophila* in filter and impinger samples,[406] *M. tuberculosis* in air samples from patient isolation rooms,[407] the aerodynamic size range of airborne mycobacteria associated with whirlpools,[408] and *Penicillium roquefortii* in a miniature cyclone sampler.[409] Solid-phase PCR from filtered impinger samples was found to be more sensitive than a cultivation-based method for aerosolized *Escherichia coli*.[410] Bacterial CpG DNA has been analyzed using quantitative PCR specific for bacterial ribosomal RNA.[411]

Mold-specific PCR (MQPCR) has been applied in various environments, including homes, schools, and hospitals.[85,299,412–414] Assays have been developed for >130 fungal species, and the method is based on the detection of pre-selected species groups. A panel of 36 indicator fungi has been proposed for quantitative evaluation of mold burden in U.S. homes.[415] Other molecular genetic methods such as cloning and sequencing or denaturing gradient gel electrophoresis (DGGE) have been used as research tools to investigate dynamic changes in microbial community diversity. Such methods have been employed to characterize microbial contaminants in MWFs,[416] track microorganisms aerosolized from floodwater,[417] examine the diversity of indoor bacterial communities,[418] and compare microbial composition in indoor and outdoor environments.[419]

12.7 CONCLUDING REMARKS

Particles of microbial, plant, and animal origin cover a wide range of sizes and deposit in all regions of the human airways. Some of the respiratory effects of bioaerosol exposure are similar to reactions to other inflammatory particles, but certain responses are unique, for example, infection as a result of microbial multiplication after inhalation. The economic burden of bioaerosol-related diseases in the community and workplaces is very great, but public health management of new and reemergent infectious diseases requires national and international cooperation. Increased support of research on the aerosol transmission of infectious agents, exposure measurement, engineering controls, personal protection, and clean-up procedures is expected to benefit other areas of public health preparedness as well.

Occupational risk assessment studies of bioaerosols have been limited by the lack of accepted reference values for specific agents other than those for which occupational exposure limits have been determined, for example, cellulose, wood dust, cotton dust, grain dust, nicotine, pyrethrum, starch, subtilisins, sucrose, turpentine, and vegetable oil mist.[201] Better human exposure–response data and information from animal models eventually may lead to the establishment of numeric exposure limits for more biological agents. Nevertheless, avoidance of indoor dampness and prompt removal of contaminated materials provide better protection from indoor mold than would adherence to air concentration limits, and selection of less allergenic plants for landscaping projects would help protect persons hypersensitive to those pollen.

The fields of aerobiology and bioaerosol research would advance more rapidly if it were convenient to measure personal exposure and inhaled dose over extended time periods rather than to estimate exposures from the concentration of biological agents in settled dust or small-volume, grab air samples collected with stationary samplers. The continued development and wider availability of rapid methods not based on microbial cell multiplication or visual particle recognition will provide better exposure information. It is not clear to what level microorganisms and pollen grains need to be identified (e.g., class, family, genus, or species) to understand dose–response relationships nor to what extent measurements of marker compounds can be used for risk assessment (e.g., glucan or ergosterol for fungi and muramic acid or fatty acids for bacteria). Therefore, we look forward to wider availability of methods that measure exposure to biologically active agents rather than surrogates.

Continued research is needed to gain a clearer understanding of the positive and negative effects of exposure to biological agents at different ages and stages of lung development, and more effort should be made to understand the mechanisms of bioaerosol action and fate in the respiratory tract. Particles of biological origin account for substantial fractions of airborne PM, and the interactions of bioaerosol and other air pollutants deserve greater attention.

REFERENCES

1. Balmes, J.R. and Scannell, D.H., Occupational lung diseases, in *Occupational and Environmental Medicine*, 2nd edn., LaDou, J. (Ed.), Appleton & Lange, Stamford, CT, 1997, Chap. 20.
2. Graveland, H., Wagenaar, J.A., Heesterbeek, H. et al., Methicillin resistant *Staphylococcus aureus* ST398 in veal calf farming: Human MRSA carriage related with animal antimicrobial usage and farm hygiene. *PLoS One*, 5, e10990, doi:10.1371/journal.pone.0010990, 2010.
3. Whitehouse, C.A., Schmaljohn, A.L., and Dembek, Z.F., Emerging infectious diseases and future threats, in *Medical Aspects of Biological Warfare*, Dembek, A.F. (Ed.), Office of the Surgeon General United States Army, Falls Church, VA, and Borden Institute, Walter Reed Army Medical Center, Washington, DC, On U.S. Army Medical Department, Borden Institute Online Information Server, 2007, Chap. 25 (www.bordeninstitute.army.mil/published_volumes/biological_warfare/BW-ch25.pdf, website accessed December 15, 2010).
4. Douwes, J. and Pearce, N., Commentary: The end of the hygiene hypothesis? *Int. J. Epidemiol.*, 37, 570, 2008.
5. Gereda, J.E., Leung, D.Y.M, Liu, A.H. et al., Levels of environmental endotoxin and prevalence of atopic disease, *J. Am. Med. Assoc.*, 284, 1652, 2000.
6. Braun-Fahrlander, C., Riedler, J., Herz, U. et al., Environmental exposure to endotoxin and its relation to asthma in school-age children, *N. Engl. J. Med.*, 347, 869, 2002.
7. Gehring, U., Bischof, W., Fahlbusch, B. et al., House dust endotoxin and allergic sensitization in children, *Am. J. Respir. Crit. Care Med.*, 166, 939, 2002.
8. Bottcher, M.F., Bjorksten, B., Gustafson, S. et al., Endotoxin levels in Estonian and Swedish house dust and atopy in infancy, *Clin. Exp. Allergy*, 33, 295, 2003.
9. Douwes, J., Le Gros, G., Gibson, P. et al. Can bacterial endotoxin exposure reverse atopy and atopic disease? *J. Allergy Clin. Immunol.*, 114, 1051, 2004.
10. Douwes, J., van Strien, R., Doekes, G. et al., Does early indoor microbial exposure reduce the risk of asthma? The prevention and incidence of asthma and mite allergy birth cohort study, *J. Allergy Clin. Immunol.*, 117, 1067, 2006.
11. Liu, A.H. and Leung, D.Y., Renaissance of the hygiene hypothesis, *J. Allergy Clin. Immunol.*, 117, 1063, 2006.

12. Iossifova, Y.Y., Reponen, T., Bernstein, D.I. et al., House dust, (1–3)-beta-D-glucan and wheezing in infants, *Allergy*, 62, 504, 2007.
13. Thorne, P.S., Kulhankova, K., Yin, M. et al., Endotoxin exposure is a risk factor for asthma: The National Survey of endotoxin in U.S. housing, *Am. J. Respir. Crit. Care Med.*, 172, 1371, 2005.
14. Michel, O., Kips, J., Duchateau, J. et al., Severity of asthma is related to endotoxin in house dust, *Am. J. Respir. Crit. Care Med.*, 154, 1641, 1996.
15. Dharmage, S., Bailey, M., Raven, J. et al., Current indoor allergen levels of fungi and cats, but not house dust mites, influence allergy and asthma in adults with high dust mite exposure, *Am. J. Respir. Crit. Care Med.*, 164, 65, 2001.
16. Bolte, G., Bischof, W., Borte, M. et al., Early endotoxin exposure and atopy development in infants: Results of a birth cohort study, *Clin. Exp. Allergy*, 33, 770, 2003.
17. Osborne, M., Reponen, T., Adhikari, A. et al., Specific fungal exposures, allergic sensitization, and rhinitis in infants, *Pediatr. Allergy Immunol.*, 17, 450, 2006.
18. Liu, A.H., Hygiene theory and allergy and asthma prevention, *Paediatr. Perinat. Epidemiol.*, 21 (Suppl 3), 2, 2007.
19. von Mutius, E., Allergies, infections and the hygiene hypothesis—The epidemiological evidence, *Immunobiology*, 212, 433, 2007.
20. Hubálek, Z., Emerging human infectious diseases: Anthroponoses, zoonoses, and sapronoses, *Emerg. Infect. Dis.*, 9, 403, 2003.
21. WHO, *The Global Burden of Disease 2004 Update*, World Health Organization, Geneva, Switzerland, 2008.
22. WHO, Global tuberculosis control: A short update to the 2009 report, WHO/HTM/TB/2009.426 (www.who.int/tb/publications/global_report/2009/update/en/index.html, website accessed December 15, 2010).
23. Coultas, D.B. and Samet, J.M., Respiratory disease prevention, in *Public Health and Preventive Medicine*, 14th edn., Wallace, R.B. (Ed.), Appleton & Lange, Stamford, CT, 1998, p. 981.
24. Nair, H., Nokes, D.J., Gessner, B.D. et al., Global burden of acute lower respiratory infections due to respiratory syncytial virus in young children: A systematic review and meta-analysis, *Lancet*, 375, 1545, 2010.
25. Cashat-Cruz, M., Morales-Aguirre, J.J., and Mendoza-Azpiri, M., Respiratory tract infections in children in developing countries, *Pediatr. Infect. Dis. J.*, 16, 84, 2005.
26. Mangili, A. and Gendreau, M.A., Transmission of infectious diseases during commercial air travel, *Lancet*, 365, 989, 2005.
27. Porta, M., *A Dictionary of Epidemiology*, 5th edn., Oxford University Press, New York, 2008.
28. Jones, K.E., Patel, N.G., Levy, M.A. et al., Global trends in emerging infectious diseases, *Nature*, 451, 990, 2008.
29. Bramley, T.J., Lerner, D., and Sarnes, M., Productivity losses related to the common cold, *J. Occup. Environ. Med.*, 44, 822, 2002.
30. Hellgren, J., Cervin, A., Nordling, S. et al., Allergic rhinitis and the common cold—High cost to society, *Allergy*, 65, 776, 2009.
31. Akazawa, M., Sindelar, J.L., and Paltiel, A.D., Economic costs of influenza-related work absenteeism, *Value Health*, 6, 107, 2003.
32. Thompson, W.W., Shay, D.K., Weintraub, E. et al., Mortality associated with influenza and respiratory syncytial virus in the United States, *J. Am. Med. Assoc.*, 289, 179, 2003.
33. Molinari, N.A., Ortega-Sanchez, I.R., Messonnier, M.L. et al., The annual impact of seasonal influenza in the US: Measuring disease burden and costs, *Vaccine*, 25, 5086, 2007.
34. Editorial, On a wing and a prayer, *Nature*, 435, 385, 2005.
35. WHO, *Pandemic (H1N1) 2009, Weekly Update 101*, May 21, 2010, (www.who.int/csr/don/2010_08_06/en/index.html, website accessed August 4, 2011).
36. Morgan, O.W, Parks, S., Shim, T. et al., Household transmission of pandemic (H1N1) 2009, San Antonio, Texas, USA, April–May 2009, *Emerg. Infect. Dis.*, 16, 631, 2010.
37. Cauchemez, S., Donnelly, C.A., Reed, C. et al., Household transmission of 2009 pandemic influenza A (H1N1) virus in the United States, *N. Engl. J. Med.*, 361, 2619, 2009.
38. Cowling, B.J., Chan, K.-H., Fang, V.J. et al., Facemasks and hand hygiene to prevent influenza transmission in households, a cluster randomized trial, *Ann. Intern. Med.*, 151, 437, 2009.
39. Larson, E.L., Ferng, Y.H., Wong-McLoughlin, J. et al., Impact of non-pharmaceutical interventions on URIs and influenza in crowded, urban households, *Public Health Rep.*, 125, 178, 2010.

40. Cowling, B.J., Zhou, Y., Ip, D.K. et al., Face masks to prevent transmission of influenza virus: A systematic review, *Epidemiol. Infect.*, 138, 449, 2010.
41. Evans, D., Epidemiology and etiology of occupational infectious diseases, in *Occupational and Environmental Infectious Diseases*, Couturier, A.J. (Ed.), OEM Press, Beverly Farms, MA, 2008, Chap. 3.
42. Chen, S.C., Chio, C.P., Jou, L.J. et al., Viral kinetics and exhaled droplet size affect indoor transmission dynamics of influenza infection, *Indoor Air*, 19, 401, 2009.
43. Fabian, P., McDevitt, J.J., DeHaan W.H. et al., Influenza virus in human exhaled breath: An observational study, *PloS One*, 3, e2691, 2008.
44. Huynh, K.N., Oliver, B.G., Stelzer, S. et al., A new method for sampling and detection of exhaled respiratory virus aerosols, *Clin. Infect. Dis.*, 46, 93, 2008.
45. Jue, R., Schmalz, T., Carter, K. et al., Outbreak of cryptosporidiosis associated with a splash park—Idaho, 2007, *J. Am. Med. Assoc.*, 302, 938, 2009.
46. Jewett, D.L., Heinsohn, P., Bennett, C. et al., Blood-containing aerosols generated by surgical techniques a possible infectious hazard, *Am. Ind. Hyg. Assoc. J.*, 53, 4, 228, 1992.
47. Mauderly, J.L., Cheng, Y.S., Johnson, N.F. et al., Particles inhaled in the occupational setting, in *Particle–Lung Interactions*, Gehr, P. and Heyder, J., (Eds.), Marcel Dekker Inc., New York, 2000, Chap. 3.
48. Musher, D.M., How contagious are common respiratory tract infections? *N. Engl. J. Med.*, 348, 1256, 2003.
49. Chai, L.Y., Netea, M.G., Vonk, A.G. et al., Fungal strategies for overcoming host innate immune response, *Med. Mycol.*, 47, 227, 2009.
50. Wells, W.F., Ratcliffe, H.L., and Crumb, C., On the mechanics of droplet nuclei infection. II. Quantitative experimental airborne tuberculosis in rabbits, *Am. J. Hyg.*, 47, 11, 1948.
51. Wells, W.F., Eds., Response and reaction to inhaled droplet nuclei contagium, in *Airborne Contagion and Air Hygiene*, Harvard University Press, Cambridge, MA, 1955, Chap. 11.
52. Chen, S.C. and Liao, C.M., Probabilistic indoor transmission modeling for influenza (sub)type viruses, *J. Infect.*, 60, 26, 2010.
53. Riley, E.C., Murphy, G., and Riley, R.L., Airborne spread of measles in a suburban elementary school, *Am. J. Epidemiol.*, 107, 421, 1978.
54. Catanzaro, A., Nosocomial tuberculosis, *Am. Rev. Respir. Dis.*, 125, 559, 1982.
55. Remington, P.L., Hall, W.N., Davis, I.H. et al., Airborne transmission of measles in a physician's office, *J. Am. Med. Assoc.*, 253, 1574, 1985.
56. Nardell, E.A., Keegan, J., Cheney, S.A. et al., Airborne infection. Theoretical limits of protection achievable by building ventilation, *Am. Rev. Respir. Dis.*, 144, 302, 1991.
57. Gammaitoni, L. and Nucci, M.C., Using a mathematical model to evaluate the efficacy of TB control measures, *Emerg. Infect. Dis.*, 3, 335, 1997.
58. U.S. Department of Health and Human Services, *Biosafety in Microbiological and Biomedical Laboratories*, 5th edn., U.S. Government Printing Office, Washington, DC, 2007.
59. Watanabe, T., Bartrand, T.A., Weir, M.H. et al., Development of a dose-response model for SARS coronavirus, *Risk Anal.*, 30, 1129, 2010.
60. Bartrand, T.A., Weir, M.H., and Haas, C.N., Dose-response models for inhalation of *Bacillus anthracis* spores: Interspecies comparisons, *Risk Anal.*, 28, 1115, 2008.
61. Gutting, B.W., Channel, S.R., Berger, A.E. et al., Mathematically modeling inhalation anthrax, *Microbe*, 3, 78, 2008.
62. Noakes, C.J., Beggs, C.B., Sleigh, P.A. et al., Modelling the transmission of airborne infections in enclosed spaces, *Epidemiol. Infect.*, 134, 1082, 2006.
63. Chen, S.-C., Chang, C.-F., and Liao, C.-M., Predictive models of control strategies involved in containing indoor airborne infections, *Indoor Air*, 16, 469, 2006.
64. Li, Y., Leung, G.M., Tang, J.W. et al., Role of ventilation in airborne transmission of infectious agents in the built environment—A multidisciplinary systematic review, *Indoor Air*, 17, 2, 2007.
65. Dan, Y.Y., Tambyah, P.A., Sim, J. et al., Cost-effectiveness analysis of hospital infection control response to an epidemic respiratory virus threat, *Emerg. Infect. Dis.*, 15, 1909, 2009.
66. Li, S., Eisenberg, J.N.S., Spicknall, I.H. et al., Dynamics and control of infections transmitted from person to person through the environment, *Am. J. Epidemiol.*, 170, 257, 2009.
67. Pearce, N., Douwes, J., and Beasley, R., Is allergen exposure the major primary cause of asthma? *Thorax*, 55, 424, 2000.

68. Douwes, J., Pearce, N., and Heederik, D., Does environmental endotoxin exposure prevent asthma? *Thorax*, 57, 86, 2002.
69. Douwes, J. and Pearce, N., The end of the hygiene hypothesis? *Int. J. Epidemiol.*, 37, 570, 2008.
70. World Health Organization, Global surveillance, prevention and control of chronic respiratory diseases: A comprehensive approach, 2007 (www.who.int/gard/publications/GARD%20Book%202007.pdf, website accessed August 4, 2011).
71. American Lung Association (ALA), *Trends in Asthma Morbidity and Mortality*, ALA Epidemiology and Statistics Unit, Research and Program Services Division, February 2010 (www.lungusa.org/finding-cures/our-research/trend-reports/asthma-trend-report.pdf, website accessed December 15, 2010).
72. Johnston, S.L., Pattemore, P.K., Sanderson, G. et al., Community study of role of viral-infections in exacerbations of asthma in 9–11 year-old children, *Br. Med. J.*, 310, 1225, 1995.
73. Johnston, S.L., Innate immunity in the pathogenesis of virus-induced asthma exacerbations, *Proc. Am. Thorac. Soc.*, 4, 267, 2007.
74. Stein, R.T., Sherrill, D., Morgan, W.J. et al., Respiratory syncytial virus in early life and risk of wheeze and allergy by age 13 years, *Lancet*, 354, 541, 1999.
75. Sigurs, N., Gustafsson, P.M., Bjarnason, R. et al., Severe respiratory syncytial virus bronchiolitis in infancy and asthma and allergy at age 13, *Am. J. Respir. Crit. Care Med.*, 171, 137, 2005.
76. Lemanske, R.F., Jackson, D.J., Gangnon, R.E. et al., Rhinovirus illnesses during infancy predict subsequent childhood wheezing, *J. Allergy Clin. Immunol.*, 116, 571, 2005.
77. IOM (Institute of Medicine), Major issues in understanding asthma, Methodological considerations in evaluating the evidence, and Indoor biologic exposures, in *Clearing the Air: Asthma and Indoor Air Exposures*, Committee on the Assessment of Asthma and Indoor Air, National Academy Press, Washington, DC, 2000, Chaps. 1, 2, and 5.
78. IOM (Institute Of Medicine), *Damp Indoor Spaces and Health*, National Academy Press, Washington, DC, 2004.
79. *WHO Guidelines for Indoor Air Quality: Dampness and Mould*, Copenhagen, Denmark: World Health Organization Regional Office for Europe, 2009.
80. Fisk, W.J., Lei-Gomez, Q., and Mendell, M.J., Meta-analyses of the associations of respiratory health effects with dampness and mold in homes, *Indoor Air*, 17, 284, 2007.
81. National Center for Health Housing, Housing intervention and health: A review of the evidence, January 2009 (www.nchh.org/LinkClick.aspx?fileticket=2lvaEDNBIdU%3d&tabid=229, website accessed December 15, 2010).
82. Matheson, M.C., Abramson, M.J., Dharmage, S.C. et al., Changes in indoor allergen and fungal levels predict changes in asthma activity among young adults, *Clin. Exp. Allergy*, 35, 907, 2005.
83. Stark, P.C., Celedón, J.C., Chew, G.L. et al., Fungal levels in the home and allergic rhinitis by 5 years of age, *Environ. Health Perspect.*, 113, 1405, 2005.
84. Reponen, T., Vesper, S., Levin, L. et al., High environmental relative moldiness index during infancy as a predictor of asthma at 7 years of age, *Ann. Allergy Asthma Immunol.*, 107, 120, 2011.
85. Mudarri, D. and Fisk, W.J., Public health and economic impact of dampness and mold, *Indoor Air*, 17, 226, 2007.
86. Douwes, J., Thorne, P., Pearce, N. et al., Bioaerosol health affects and exposure assessment: Progress and prospects, *Ann. Occup. Hyg.*, 47, 187, 2003.
87. State of New Jersey, Industries and Asthma-Causing Agents (www.state.nj.us/health/eoh/survweb/wra/documents/asthmagens.pdf, website accessed December 15, 2010).
88. Tarlo, S.M., Balmes, J., Balkissoon, R. et al., Diagnosis and management of work-related asthma, American College of Chest Physicians consensus statement, *Chest*, 134, S1, 2008.
89. Chan-Yeung, M.J. and Malo, L., Epidemiology of occupational asthma, in *Asthma and Rhinitis*, Busse, W. and Holgate, S.T. (Eds.), Blackwell Scientific, Oxford, U.K., 1994, 44–57, 1994.
90. Kogevinas, M., Anto, J.M., Soriano, B. et al., The risk of asthma attributable to occupational exposures—A population-based study in Spain, *Am. J. Respir. Crit. Care Med.*, 154, 137, 1996.
91. Fishwick, D., Pearce, N., D'Souza, W. et al., Occupational asthma in New Zealanders: A population based study, *Occup. Environ. Med.*, 54, 301, 1997.
92. Meredith, S. and Nordman, H., Occupational asthma: Measures of frequency from four countries, *Thorax*, 51, 435, 1996.
93. Meredith, S.K., Taylor, V.M., and McDonald, J.C., Occupational respiratory disease in the United-Kingdom 1989—A report to the British Thoracic Society and the Society of Occupational Medicine by the Sword Project Group, *Br. J. Ind. Med.*, 48, 292, 1991.

94. Karjalainen, A., Kurppa, K., Martikainen, R. et al., Work is related to a substantial portion of adult-onset asthma incidence in the Finnish population, *Am. J. Respir. Crit. Care Med.*, 164, 565, 2001.

95. Shusterman, D., Upper respiratory tract disorders, in *Current Occupational and Environmental Medicine*, 4th edn., LaDou, J. (Ed.), McGraw Hill, New York, 2007, Chap. 19.

96. Zacharasiewicz, A., Douwes, J., and Pearce, N., What proportion of rhinitis symptoms is attributable to atopy? *J. Clin. Epidemiol.*, 56, 385, 2003.

97. Goldberg, A., Confino-Cohen, R., and Waisel, Y., Allergic responses to pollen of ornamental plants: High incidence in the general atopic population and especially among flower growers, *J. Allergy Clin. Immunol.*, 102, 210, 1998.

98. ISAAC Steering Committee (Writing Committee: Beasley, R., Keil, U., von Mutius, E. et al.), Worldwide variation in prevalence of symptoms of asthma, allergic rhinoconjunctivitis and atopic eczema: ISAAC, *Lancet*, 351, 1225, 1998.

99. Fink, J.N., Ortega, H.G., Reynolds, H.Y. et al., Needs and opportunities for research in hypersensitivity pneumonitis, *Am. J. Respir. Crit. Care Med.*, 171, 792, 798, 2005.

100. Lacasse, Y., Selman, M. Costabel. U. et al., Clinical diagnosis of hypersensitivity pneumonitis, *Am. J. Respir. Crit. Care Med.*, 168, 952, 2003.

101. Erkinjuntti-Pekkanen, R., Reiman, M., Kokkarinen, J.I. et al., IgG antibodies, chronic bronchitis, and pulmonary function values in farmer's lung patients and matched controls, *Allergy*, 54, 1181, 1999.

102. Roussel, S., Reboux, G., Dalphin, J.C. et al., Microbiological evolution of hay and relapse in patients with farmer's lung, Occup, *Environ. Med.*, 61, e3, 2004.

103. Sood, A., Sreedhar, R., Kulkarni, P. et al., Hypersensitivity pneumonitis-like granulomatous lung disease with nontuberculous mycobacteria from exposure to hot water aerosols, *Environ. Health Perspect.*, 115, 262, 2007.

104. Rose, C.S., Martyny, J.W., Newman, L.S. et al., "Lifeguard lung": Endemic granulomatous penumonitis in an indoor swimming pool, *Am. J. Pub. Health*, 88, 1795, 1998.

105. Dang, B., Chen, L., Mueller, C. et al., Ocular and respiratory symptoms among lifeguards at a hotel indoor waterpark resort, *J. Occup. Environ. Med.*, 52, 207, 2010.

106. Douwes, J., Gibson, P., Pekkanen, J. et al., Non-eosinophilic asthma: Importance and possible mechanisms, *Thorax*, 57, 643, 2002.

107. Norbäck, D., Björnsson, E., Janson, C. et al., Current asthma and biochemical signs of inflammation in relation to building dampness in dwellings, *Int. J. Tuberc. Lung Dis.*, 3, 368, 1999.

108. Centers for Disease Control and Prevention (CDC), Acute idiopathic pulmonary hemorrhage among infants, *Morb. Mortal. Wkly. Rep.*, 53, 1, 1994.

109. Centers for Disease Control and Prevention (CDC), Update: Pulmonary hemorrhage/hemosiderosis among infants—Cleveland, Ohio, 1993–1996, *Morb. Mortal. Wkly. Rep.*, 46, 33, 1997.

110. Etzel, R.A., Montaña, E., Sorenson, W.G. et al., Acute pulmonary hemorrhage in infants associated with exposure to *Stachybotrys atra* and other fungi, *Arch. Pediatr. Adolesc. Med.*, 152, 757, 1998.

111. Centers for Disease Control and Prevention (CDC), Update: Pulmonary hemorrhage/hemosiderosis among infants—Cleveland, Ohio, 1993–1996, *Morb. Mortal. Wkly. Rep.*, 49, 180, 2000.

112. Vesper, S.J. and Vesper, M.J., Stachylysin may be a cause of hemorrhaging in humans exposed to *Stachybotrys chartarum*, *Infect. Immunol.*, 70, 2065, 2002.

113. Hastings, C., Rand, T., Bergen, H.T. et al., *Stachybotrys chartarum* alters surfactant-related phospholipid synthesis and CTP-cholinephosphate cytidylytransferase activity in isolated fetal rat type II cells, *Toxicol. Sci.*, 84, 186, 2005.

114. Pestka, J.J., Yike, I. Dearborn, D.G. et al., *Stachybotrys chartarum*, trichothecene mycotoxins, and damp building-related illness: New insights into a public health enigma, *Toxicol. Sci.*, 104, 4, 2008.

115. Norbäck, D., An update on sick building syndrome, *Curr. Opin. Allergy Clin. Immunol.*, 9, 55, 2009.

116. Weschler, C.J., Wells, J.R., Poppendieck, D. et al., Workgroup report: Indoor chemistry and health, *Environ. Health Perspect.*, 114, 442, 2006.

117. Seppänen, O., Fisk, W.J., and Lei, Q.H., Ventilation and performance in office work, *Indoor Air*, 16, 28, 2006.

118. Iossifova, Y.Y., Reponen, T., Ryan, P.H. et al., Mold exposure during infancy as a strong predictor of potential asthma development, *Ann. Allergy Asthma Immunol.*, 102, 131, 2009.

119. Douwes, J. and Brooks, C.P.N., The protective effects of farming on allergies and asthma: Have we learnt anything since 1873? *Exp. Rev. Clin. Immunol.*, 5, 213, 2009.

120. Maier, R.M., Pepper, I.L., and Gerba, C.P., *Environmental Microbiology*, 2nd edn., Academic Press, San Diego, CA, 2008.

121. Sylvia, D.M., Fuhrmann, J.J., Hartel, P.G. et al., *Principles and Applications of Soil Microbiology*, 2nd edn., Prentice-Hall, Upper Saddle River, NJ, 2004.
122. Burrows, S.M., Butler, T., Jöckel, P. et al., Bacteria in the global atmosphere—Part 2: Modelling of emissions and transport between different ecosystems, *Atmos. Chem. Phys. Discuss.*, 9, 10829, 2009.
123. Schloss, P.D. and Handelsman, J., Toward a census of bacteria in soil, *PLoS Comp. Biol.*, 2, e92, 2006, doi: 10.1371/journal.pcbi.0020092.
124. Yeo, H.G. and Kim, J.H., SPM and fungal spores in the ambient air of west Korea during the Asian dust (Yellow sand) period, *Atmos. Environ.*, 36, 5437, 2002.
125. Huang, C.Y., Lee, C.C., Li, F.C. et al., The seasonal distribution of bioaerosols in municipal landfill sites: A 3-yr study, *Atmos. Environ.*, 36, 4385, 2002.
126. Rylander, R., Health effects among workers in sewage treatment plants, *Occup. Environ. Med.*, 56, 354, 1999.
127. Thorn, J. and Kerekes, E., Health effects among employees in sewage treatment plants: A literature survey, *Am. J. Ind. Med.*, 40, 170, 2001.
128. Baertsch, C., Paez-Rubio, T., Viau, E. et al., Source tracking aerosols released from land-applied class B biosolids during high-wind events, *Appl. Environ. Microbiol.*, 73, 4522, 2007.
129. Viau, E. and Peccia, J., Evaluation of the enterococci indicator in biosolids using culture-based and quantitative PCR assays, *Appl. Environ. Microbiol.*, 75, 164, 2009.
130. Fleming, L.E., Bean, J.A., Kirkpatrick, B. et al., Exposure and effect assessment of aerosolized red tide toxins (brevetoxins) and asthma, *Environ. Health Perspect.*, 117, 1095, 2009.
131. Bienfang. P.K., DeFelice, S.V., Laws, E.A. et al., Prominent human health impacts from several marine microbes: History, ecology, and public health implications, *Int. J. Microbiol.*, 2011, 152815, 2011.
132. Abraham, W.M. and Baden, D.G., Aerosolized Florida red tide toxins and human health effects, *Oceanography (Wash DC)*, 19, 107, 2006.
133. Erdner, D.L., Dyble, J., Parsons, M. et al., Centers for Oceans and Human Health: A unified approach to the challenge of harmful algal blooms, *Environ. Health*, 7(Suppl 2), S2, 2008.
134. Genitsaris, S., Kormas, K.A., and Moustaka-Gouni, M., Airborne algae and cyanobacteria: Occurrence and related health effects, *Front. Biosci.*, 3, 772, 2011.
135. Matthias-Maser, S. and Jaenicke, R., Examination of atmospheric bioaerosol particles with radii ~0.2 µm, *J. Aerosol Sci.*, 25, 1605, 1994.
136. Gruber, S., Matthias-Maser, S., Brinkmann, J. et al., Vertical distribution of biological aerosol particles above the North Sea, *J. Aerosol Sci.*, 29, S771, 1998.
137. Gruber, S., Matthias-Maser, S., and Jaenicke, R., Concentration and chemical composition of aerosol particles in marine and continental air, *J. Aerosol Sci.*, 30, S9, 1999.
138. Ebert, M., Weinbruch, S., Hoffmann, P. et al., Chemical characterization of North Sea aerosol particles, *J. Aerosol Sci.*, 31, 613, 2000.
139. Bauer, H., Kasper-Giebl, A., Loflund, M. et al., The contribution of bacteria and fungal spores to the organic content of cloud water, precipitation and aerosols, *Atmos. Res.*, 64, 109, 305, 2002.
140. Bauer, H., Schueller, E., Weinke, G. et al., Significant contributions of fungal spores to the organic carbon and to the aerosol mass balance of the atmospheric aerosol, *Atmos. Environ.*, 42, 5542, 2008.
141. Elbert, W., Taylor, P.E., Andreae, M.O. et al., Contribution of fungi to primary biogenic aerosols in the atmosphere: Wet and dry discharged spores, carbohydrates, and inorganic ions, *Atmos. Chem. Phys.*, 7, 4569, 2007.
142. Heald, C.L. and Spracklen, D.V., Atmospheric budget of primary biological aerosol particles from fungal spores, *Geophys. Res. Lett.*, 36, L09806, doi:10.1029/2009GL037493, 2009.
143. Penner, J.E., Carbonaceous aerosols influencing atmospheric radiation black and organic carbon, in *Aerosol Forcing of Climate*, Charlson, R.J. and Heintzenberg, J. (Eds.), John Wiley & Sons, New York, 1995, p. 91.
144. Jaenicke, R., Abundance of cellular material and proteins in the atmosphere, *Science*, 308, 73, 2005.
145. Burrows, S.M., Elbert, W., Lawrence, M.G. et al., Bacteria in the global atmosphere—Part 1: Review and synthesis of literature data for different ecosystems, *Atmos. Chem. Phys. Discuss.*, 9, 10777, 2009.
146. Després, V.R., Nowoisky, J.F., Klose, M. et al., Characterization of primary biogenic aerosol particles in urban, rural, and high-alpine air by DNA sequence and restriction fragment analysis of ribosomal RNA genes, *Biogeosciences*, 4, 1779, 2007.
147. Fierer, N., Liu, Z., Rodríguez-Hernández, M. et al., Short-term temporal variability in airborne bacterial and fungal populations, *Appl. Environ. Microbiol.*, 74, 200, 2008.
148. Brownstein, J.S., Wolfe, C.J., and Mandl, K.D., Empirical evidence for the effect of airline travel on inter-regional influenza spread in the United States, *PLoS Med.*, 3, e401, 2006.

149. Park, B.J., Sigel, K., Vaz, V. et al., An epidemic of coccidioidomycosis in Arizona associated with climatic changes, 1998–2001, *J. Infect. Dis.*, 191, 1981, 2005.

150. Centers for Disease Control and Prevention (CDC), Increase in coccidioidomycosis—California, 2000–2007, Bissell, S.R. and Weiss, E.C., *Morb. Mortal. Wkly Rep.*, 58, 105, 2009.

151. D'Amato, G., Cecchi, L., D'Amato, M. et al., Urban air pollution and climate change as environmental risk factors of respiratory allergy: An update, *J. Investig. Allergol. Clin. Immunol.*, 20, 95, 2010.

152. Muilenberg, M.L., Pollen in indoor air: Sources, exposures, and health effects, in *Indoor Air Quality Handbook*, Spengler, J.D., Samet, J.M., and McCarthy, J.F. (Eds.), McGraw-Hill, New York, 2001, Chap. 44.

153. Käpylä, M., Diurnal variations of non-arboreal pollen in the air in Finland, *Grana*, 20, 55, 1981.

154. Lin, W.H. and Li, C.S., Associations of fungal aerosols, air pollutants, and meteorological factors, *Aerosol Sci. Technol.*, 32, 359, 2000.

155. Liao, C.-M. and Luo, W.-C., Use of temporal/seasonal- and size-dependent bioaerosol data to characterize the contribution of outdoor fungi to residential exposures, *Sci. Total Environ.*, 347, 78, 2005.

156. Chen, Q. and Hildemann, L.M., Size-resolved concentrations of particulate matter and bioaerosols inside versus outside of homes, *Aerosol Sci. Technol.*, 43, 699, 2009.

157. Hock, N., Scneider, J., Borrmann, S. et al., Rural continental aerosol properties and processes observed during the Hohenpeissenberg Aerosol Characterization Experiment (HAZE2002), *Atmos. Chem. Phys. Discuss.*, 7, 8617, 2007.

158. Di Giorgio, C., Krempff, A., Guiraud, H. et al., Atmospheric pollution by airborne microorganisms in the City of Marseilles, *Atmos. Environ.*, 30, 155, 1996.

159. Jones, B.L. and Cookson, J.T., Natural atmospheric microbial conditions in a typical suburban area, *Appl. Environ. Microbiol.*, 45, 919, 1983.

160. Bovallius, A., Bucht, B., Roffey, R. et al., Three-year investigation of the natural airborne bacterial flora at four localities in Sweden, *Appl. Environ. Microbiol.*, 35, 847, 1978.

161. Kelly, C.D. and Pady, S.M., Microbiological studies of air masses over Montreal during 1950 and 1951, *Can. J. Bot.*, 32, 591, 1954.

162. Tong, Y. and Lighthart, B., The annual bacterial particle concentration and size distribution in the ambient atmosphere in a rural area of the Willamette Valley, Oregon, *Aerosol Sci. Technol.*, 32, 393, 2000.

163. Kujundzic, E., Hernandez, M., and Miller, S.L., Particle size distributions and concentrations of airborne endotoxin using novel collection methods in homes during the winter and summer seasons, *Indoor Air*, 16, 216, 2006.

164. Reponen, T., Nevalainen, A., Jantunen, M. et al., Normal range criteria for indoor air bacteria and fungal spores in a subarctic climate, *Indoor Air*, 2, 6, 1992.

165. Chew, G.L., Rogers, C., Burge, H.A. et al., Dustborne and airborne fungal propagules represent a different spectrum of fungi with differing relations to home characteristics, *Allergy*, 58, 13, 2003.

166. LeBouf, R., Yesse, L., and Rossner, A., Seasonal and diurnal variability in airborne mold from an indoor residential environment in northern New York, *J. Air Waste Manag. Assoc.*, 58, 684, 2008.

167. Adhikari, A., Reponen, T., Grinshpun, S.A. et al., Correlation of ambient inhalable bioaerosols with particulate matter and ozone: A two-year study, *Environ. Pollut.*, 140, 16, 2006.

168. Levetin, E., Fungi, in *Bioaerosols*, Burge, H.A. (Ed.), Lewis Publishers, Boca Raton, FL, 1995, Chap. 5.

169. Muilenberg, M.L., The outdoor aerosol, in *Bioaerosols*, Burge, H.A. (Ed.), Lewis Publishers, Boca Raton, FL, 1995, Chap. 9.

170. Bornehag, C.G., Blomquist, G., Gyntelberg, F. et al., Dampness in buildings and health. Nordic interdisciplinary review of the scientific evidence on associations between exposure to "dampness" in buildings and health effects (NORDDAMP), *Indoor Air*, 11, 72, 2001.

171. Fröhlich-Nowoisky, J., Pickersgill, D.A., Després, V.R. et al., High diversity of fungi in air particulate matter, *Proc. Natl Acad. Sci. U S A*, 106, 12814, 2009.

172. Ariano, R., Canonica, G.W., and Passalacqua, G., Possible role of climate changes in variations in pollen seasons and allergic sensitizations during 27 years, *Ann. Allergy Asthma Immunol.*, 104, 215, 2010.

173. D'Amato, G. and Cecchi, L., Effects of climate change on environmental factors in respiratory allergic diseases, *Clin. Exp. Allergy*, 38, 1264, 2008.

174. Griffin, D.W., Atmospheric movement of microorganisms in clouds of desert dust and implications for human health, *Clin. Microbiol. Rev.*, 20, 459, 2007.

175. Hyvärinen, A., Meklin, T., Vepsäläinen, A. et al., Fungi and actinobacteria in moisture-damaged building materials—Concentrations and diversity, *Int. Biodeterior. Biodegrad.*, 49, 27, 2002.

176. Rintala, H., Nevalainen, A., and Suutari, M., Diversity of streptomycetes in water-damaged building materials based on 16S rDNA sequences, *Lett. Appl. Microbiol.*, 34, 439, 2002.

177. Rintala, H., Hyvärinen, A., Paulin, L. et al., Detection of streptomycetes in house dust—Comparison of culture and PCR methods, *Indoor Air*, 14, 112, 2004.
178. Andersson, M.A., Mikkola, R., Kroppenstdt, R. et al., Mitochondrial toxin produced by *Streptomyces griseus* strains isolated from indoor environments is valinomycin, *Appl. Environ. Microbiol.*, 64, 4764, 1998.
179. Hirvonen, M.R., Suutari, M., Routsalainen, M. et al., Effect of growth medium on potential of *Streptomyces anulatus* spores to induce inflammatory responses and cytotoxicity in RAW264.7 macrophages, *Inhal. Toxicol.*, 13, 55, 2001.
180. Roponen, M., Toivola, M., Meklin, T. et al., Differences in inflammatory responses and cytotoxicity in RAW264.7 macrophages induced by *Streptomyces anulatus* grown on different building materials, *Indoor Air*, 11, 179, 2001.
181. Murtoniemi, T., Nevalainen, A., Suutari, M. et al., Induction of cytotoxicity and production of inflammatory mediators in RAW264.7 macrophages by spores grown on six different plasterboards, *Inhal. Toxicol.*, 13, 233, 2001.
182. Murtoniemi, T., Hirvonen, M.R., Nevalainen, A. et al., The relation between growth of four microbes on six different plasterboards and biological activity of spores, *Indoor Air*, 13, 65, 2003.
183. Torvinen, E., Meklin, T., Torkko, P. et al., Mycobacteria and fungi in moisture-damaged building materials, *Appl. Environ. Microbiol.*, 72, 6822, 2006.
184. Górny, R.L., Dutkiewicz, J., and Krysinska-Traczyk, E., Size distribution of bacterial and fungal bioaerosols in indoor air, *Ann. Agric. Environ. Med.*, 6, 105, 1999.
185. Mueller-Anneling, L., Avol, E., Peters, J.M. et al., Ambient endotoxin concentrations in PM10 from Southern California, *Environ. Health Perspect.*, 112, 583, 2004.
186. Chen, C.M., Mielck, A., Fahlbusch, B. et al. Social factors, allergen, endotoxin, and dust mass in mattress, *Indoor Air*, 17, 384, 2007.
187. Giovannangelo, M.E.C.A., Gehring, U., Nordling, E. et al., Levels and determinants of $\beta(1 \rightarrow 3)$-glucans and fungal extracellular polysaccharides in house dust of (pre-) schoolchildren in three European countries, *Environ. Int.*, 33, 9, 2007.
188. Solomon, G.M., Hjelmroos-Koski, M., Rotkin-Ellman, M. et al., Airborne mold and endotoxin concentrations in New Orleans, Louisiana, after flooding, October through November 2005, *Environ. Health Perspect.*, 114, 1381, 2006.
189. Reponen, T., Singh, U., Schaffer, D. et al., Visually observed mold and moldy odor versus quantitatively measured microbial exposure in homes, *Sci. Total Environ.*, 408, 5565, 2010.
190. Park, J.H., Spiegelman, D.L., Burge, H.A. et al., Longitudinal study of dust and airborne endotoxin in the home, *Environ. Health Perspect.*, 108, 1023, 2000.
191. Bouillard, L.A., Devleeschouwer, M.J., and Michel, O., Characteristics of the home bacterial contamination and endotoxin related release, *J. Pharm. Belg.*, 61, 63, 2006.
192. Adhikari, H. Lewis, J.S., Reponen, T. et al., Exposure matrices of endotoxin, (1 3)-β-D-glucan, fungi, and dust mite allergens in flood-affected homes in New Orleans, *Sci. Total Environ.*, 408, 5489, 2010.
193. Kline, J.N., Cowden, J.D., Hunninghake, G.W. et al., Variable airway responsiveness to inhaled lipopolysaccharide, *Am. J. Respir. Crit. Care Med.*, 160, 297, 1999.
194. Park, J.H., Gold, D.R., Spiegelman, D.L. et al., House dust endotoxin and wheeze in the first year of life. *Am. J. Respir. Crit. Care Med.*, 163, 322, 2001.
195. Meyer, H.W., Würtz, H., Suadicani, P. et al. Molds in floor dust and building-related symptoms in adolescent school children, *Indoor Air*, 14, 65, 2004.
196. Licorish, K., Novey, H.S., Kozak, P. et al., Role of *Alternaria* and *Penicillium* spores in the pathogenesis of asthma, *J. Allergy Clin. Immunol.*, 76, 819, 1985.
197. Hirvonen, M.-R., Ruotsalainen, M., Savolainen, K. et al., Effect of viability of actinomycete spores on their ability to stimulate production of nitric oxide and reactive oxygen species in RAW264.7 macrophages, *Toxicology*, 124, 105, 1997.
198. Green, B.J., Mitakakis, T.Z., and Tovey, E.R. Allergen detection from 11 fungal species before and after germination, *J. Allergy Clin. Immunol.*, 111, 285, 2003.
199. Rao, C.Y., Burge, H.A., and Chang, J.C.S., Review of quantitative standards and guidelines for fungi in indoor air, *J. Air Waste Manag. Assoc.*, 46, 899, 1996.
200. Husman, T.M., The Health Protection Act, national guidelines for indoor air quality and development of the national indoor air programs in Finland, *Environ. Health Perspect.*, 107 (Suppl 3), 515, 1999.
201. ACGIH, Biologically derived airborne contaminants and agents under study, in *2010 TLVs and BEIs, American Conference of Governmental Industrial Hygienists*, Cincinnati, OH, 2010, p. 181.

202. Verhoeff, A.P. and Burge, H.A., Health risk assessment of fungi in home environments, *Ann. Allergy Asthma Immunol.*, 78, 544, 1997.

203. AIHA, Report of microbial growth task force, American Industrial Hygiene Association, Fairfax, VA, 2008.

204. Centers for Disease Control and Prevention (CDC), *State of the Science on Molds and Human Health*, Statement for the record before the Subcommittees on Oversight and Investigations and Housing and Community Opportunity, Committee on Financial Services, United States House of Representatives, Washington, DC, 2002.

205. Spicer, R.C. and Gangloff, H.J., Differences in detection frequency as a bioaerosol data criterion for evaluating suspect fungal contamination, *Building Environ.*, 45, 1304, 2009.

206. AIHA, Guidance for assessment and remediation of indoor microbial growth, in *Recognition, Evaluation, and Control of Indoor Mold*, Prezant, B., Weekes, D.M., and Miller, J.D. (Eds.), American Industrial Hygiene Association, Fairfax, VA, 2008, Chap. 2.

207. WHO, *Development of WHO Guidelines for Indoor Air Quality*, World Health Organization, Geneva, Switzerland, 2007.

208. Górny, R.L., Filamentous microorganisms and their fragments in indoor air—A review, *Ann. Agric. Environ. Med.*, 11, 185, 2004.

209. Green, B.J., O'Meara, T.O., Sercombe, J.K. et al., Measurement of personal exposure to outdoor aeromycota in northern New South Wales, Australia, *Ann. Agric. Environ. Med.*, 13, 225, 2006.

210. Brasel, T.L., Martin, J.M., Carriker, C.G. et al., Detection of airborne *Stachybotrys chartarum* macrocyclic trichothecene mycotoxins in the indoor environment. *Appl. Environ. Microbiol.*, 71, 7376, 2005.

211. Douwes, J., $(1 \rightarrow 3)$-Beta-D-glucans and respiratory health: A review of the scientific evidence, *Indoor Air*, 15, 160, 2005.

212. Rylander, R., Indoor air-related effects and airborne $(1 \rightarrow 3)$-ß-D-glucan, *Environ. Health Perspect.*, 107 (Suppl 3), 501, 1999.

213. Douwes, J., Doekes, G., Montijn, R. et al., Measurement of $\beta(1 \rightarrow 3)$-glucans in occupational and home environments with an inhibition enzyme immunoassay, *Appl. Environ. Microbiol.*, 62, 3176, 1996.

214. Douwes, J., Doekes, G., Heinrich, J. et al., Endotoxin and $\beta(1 \rightarrow 3)$-glucan in house dust and the relation with home characteristics: A pilot study in 25 German houses, *Indoor Air*, 8, 255, 1998.

215. Douwes, J., Zuidhof, A., Doekes, G. et al., $(1 \rightarrow 3)$-ß-D-glucan and endotoxin in house dust and peak flow variability in children, *Am. J. Respir. Crit. Care Med.*, 162, 1348, 2000.

216. Gehring, U., Bolte, G., Borte, M. et al., Exposure to endotoxin decreases the risk of atopic eczema in infancy: A cohort study, *J. Allergy Clin. Immunol.*, 108, 847, 2001.

217. Chew, G.L., Douwes, J., Doekes, G. et al., Fungal extracellular polysaccharides, beta $(1 \rightarrow 3)$-glucans and culturable fungi in repeated sampling of house dust, *Indoor Air*, 11, 171, 2001.

218. Paterson, R.R. and Lima, N. Toxicology of mycotoxins, *EXS*, 100, 31, 2010.

219. van Egmond, H.P., Schothorst, R.C., and Jonker, M.A., Regulations relating to mycotoxins in food: Perspectives in a global and European context, *Anal. Bioanal. Chem.*, 389, 147, 2007.

220. Croft, W.A., Jarvis, B.B., and Yatawara, C.S., Airborne outbreak of trichothecene toxicosis, *Atmos. Environ.*, 20, 549, 1986.

221. Ammann, H., Hodgson, M., Nevalainen, A. et al., Indoor mold, basis for health concerns, in *Recognition, Evaluation and Control of Indoor Mold*, Prezant, B., Miller, J.D., and Weekes, D. (Eds.), American Industrial Hygiene Association, Fairfax, VA, 2008, Chap. 1.

222. Bloom, E., Bal, K., Nyman, E. et al., Mass spectrometry-based strategy for direct detection and quantification of some mycotoxins produced by *Stachybotrys* and *Aspergillus* spp. in indoor environments, *Appl. Environ. Microbiol.*, 73, 4211, 2007.

223. Brasel, T.L., Douglas, D.R., Wilson, S.C. et al., Detection of airborne *Stachybotrys chartarum* macrocyclic trichothecene mycotoxins on particulates smaller than conidia, *Appl. Environ. Microbiol.*, 71, 114, 2005.

224. Charpin-Kadouch, C., Maurel, G., Felipo, R. et al., Mycotoxin identification in moldy dwellings, *J. Appl. Toxicol.*, 26, 475, 2006.

225. Miller, J.D., Sun, M., Gilyan, A. et al., Inflammation-associated gene transcription and expression in mouse lungs induced by low molecular weight compounds from fungi from the built environment, *Chem. Biol. Interact.*, 183, 113, 2010.

226. Ahearn, D.G., Crow, S.A., Simmons, R.B. et al., Fungal colonization of air filters and insulation in a multi-story office building: Production of volatile organics, *Curr. Microbiol.*, 35, 305, 1997.

227. Korpi, A., Pasanen, A.L., and Pasanen, P., Volatile compounds originating from mixed microbial cultures on building materials under various humidity conditions, *Appl. Environ. Microbiol.*, 64, 2914, 1998.

228. Claeson, A.-S. and Sunesson, A.-L., Identification using versatile sampling and analytical methods of volatile compounds from *Streptomyces albidoflavus* grown on four humid building materials and one synthetic medium, *Indoor Air*, s9, 41, 2005.
229. Wilkins, K., Larsen, K., and Simkus, M., Volatile metabolites from mold growth on building materials and synthetic media, *Chemosphere*, 41, 437, 2000.
230. Wilkins, K., Larsen, K., and Simkus, M., Volatile metabolites from indoor molds grown on media containing wood constituents, *Environ. Sci. Pollut. Res. Int.*, 10, 206, 2003.
231. Schurchardt, S. and Kruse, H., Quantitative volatile metabolite profiling of common indoor fungi: Relevancy for indoor air analysis, *J. Basic Microbiol.*, 49, 350, 2009.
232. Zureik, M., Neukirch, C., Leynaert, B. et al., Sensitization to airborne moulds and severity of asthma: Cross sectional study from European Community respiratory health survey, *Br. Med. J.*, 325, 325, 2002.
233. Eduard, W., Fungal spores: A critical review of the toxicological and epidemiological evidence as a basis for occupational exposure limit setting, *Crit. Rev. Toxicol.*, 39, 799, 2009.
234. IOM (Institute of Medicine), Magnitude and dimensions of sensitization and disease caused by indoor allergens, in *Indoor Allergens, Assessing and Controlling Adverse Health Effects*, Pope, A.M., Patterson, R., and Burge, H. (Eds.), National Academy Press, Washington, DC, 1993, p. 44.
235. Lau, S., Nickel, R., Niggemann, B. et al., The development of childhood asthma: Lessons from the German Multicentre Allergy Study (MAS), *Paediatr. Respir. Rev.*, 3, 265, 2002.
236. California Department of Industrial Relations, Subchapter 7. General Industry Safety Orders, Group 16. Control of Hazardous Substances, Article 109. Hazardous Substances and Processes, §5199, *Aerosol Transmissible Diseases* (www.dir.ca.gov/title8/5199.html, website accessed December 15, 2010).
237. Yu, V.L. and Stout, J.E., Community-acquired legionnaires disease: Implications for underdiagnosis and laboratory testing, *Clin. Infect. Dis.*, 46, 1365, 2008.
238. Carratalà, J. and Garcia-Vidal, C., An update on Legionella, *Curr. Opin. Infect. Dis.*, 23, 152, 2010.
239. Purcell, B.K., Worsham, P.I., and Friedlander, A.M., Anthrax, in *Medical Aspects of Biological Warfare*, Dembek, A.F. (Ed.), Office of the Surgeon General United States Army, Falls Church, VA, and Borden Institute, Walter Reed Army Medical Center, Washington, DC, On U.S. Army Medical Department, Borden Institute Online Information Server, 2007, Chap. 4 (www.bordeninstitute.army.mil/published_volumes/biological_warfare/BW-ch04.pdf, website accessed December 15, 2010).
240. Dull, P.M., Wilson, K.E., Kournikakis, B. et al., *Bacillus anthracis* aerosolization associated with a contaminated mail sorting machine, *Emerg. Infect. Dis.*, 8, 1044, 2002.
241. Kournikakis, B., Ho, J., and Duncan, S., Anthrax letters: Personal exposure, building contamination, and effectiveness of immediate mitigation measures, *J. Occup. Environ. Health*, 7, 71, 2009.
242. Wald, P.H. and Stave, G.M., Eds., Biological hazards in the workplace, in *Physical and Biological Hazards of the Workplace*, 2nd edn., Van Nostrand Reinhold, New York, 2002, Part II.
243. Couturier, A.J., *Occupational and Environmental Infectious Diseases*, Wright, W.E. (Ed.), 2nd edn., OEM Press, Beverly Farms, MA, 2009.
244. Kennedy, S.M., Greaves, I.A., Kriebel, D. et al., Acute pulmonary responses among automobile workers exposed to aerosols of machining fluids, *Am. J. Ind. Med.*, 15, 627, 1989.
245. Mackerer, C.R., Health effects of oil mists: A brief review, *Toxicol. Ind. Health*, 5, 429, 1989.
246. Zacharisen, M.C., Kadambi, D.P., Schlueter, D.P. et al., The spectrum of respiratory disease associated with exposure to metal working fluids, *J. Occup. Environ. Med.*, 40, 640, 1998.
247. Fox, J., Anderson, H., Moen, T. et al., Metal working fluid-associated hypersensitivity pneumonitis: An outbreak investigation and case–control study, *Am. J. Ind. Med.*, 35, 58, 1999.
248. Kreiss, K. and Cox-Ganser, J., Metalworking fluid-associated hypersensitivity pneumonitis: A workshop summary, *Am. J. Ind. Med.*, 32, 423, 1997.
249. Rossmore, H.W., Antimicrobial agents for water-based metalworking fluids, *J. Occup. Med.*, 23, 247, 1981.
250. Elsmore, R. and Hill, E.C., The ecology of pasteurized metalworking fluids, *Int. Biodeterior.*, 22, 101, 1986.
251. Foxall-van Aken, S., Brown, J.A. Jr., Young, W. et al., Common components of industrial metal-working fluids as sources of carbon for bacterial growth, *Appl. Environ. Microbiol.*, 51, 1165, 1986.
252. Mattsby-Baltzer, I., Sandin, M., Ahlström, B. et al., Microbial growth and accumulation in industrial metal-working fluids, *Appl. Environ. Microbiol.*, 55, 2681, 1989.
253. Wallace, R.J., Jr., Zhang, Y., Wilson, R.W. et al., Presence of a single genotype of the newly described species *Mycobacterium immunogenum* in industrial metal-working fluids associated with hypersensitivity pneumonitis, *Appl. Environ. Microbiol.*, 68, 5880, 2002.

254. Centers for Disease Control and Prevention (CDC), Respiratory illness in workers exposed to metalworking fluid contaminated with nontuberculous mycobacteria—Ohio, 2001, *Morb. Mortal. Wkly Rep.*, 51, 349, 2002.
255. Throne, P.S., Adamcakova-Dodd, A., Kelly, K.M. et al., Metal working fluid with mycobacteria and endotoxin induces hypersensitivity pneumonitis in mice, *Am. J. Respir. Crit. Care Med.*, 173, 759, 2006.
256. Rylander, R., Bake, B., Fischer, J.J. et al., Pulmonary function and symptoms after inhalation of endotoxin, *Am. Rev. Respir. Dis.*, 140, 981, 1989.
257. Pillai, S.D. and Ricke, S.C., Bioaerosols from municipal and animal wastes: Background and contemporary issues, *Can. J. Microbiol.*, 48, 681, 2002.
258. Dutkiewicz, J., Krysinska-Traczyk, E., Skorksa, C. et al., Exposure to airborne microorganisms and endotoxin in a potato processing plant, *Ann. Agric. Environ. Med.*, 9, 225, 2002.
259. ASTM, *Standard Practice for Personal Sampling and Analysis of Endotoxin in Metalworking Fluid Aerosols in Workplace Atmospheres*, Practice E2144-07, American Society for Testing and Materials, West Conshohocken, PA, 2007.
260. CEN, European Committee for Standardization, *Workplace Atmospheres—Determination of Airborne Endotoxin*, CEN/TC 137 Work Programme, Project Reference EN 14031, Brussels, Belgium, 2003.
261. White, E.M., Environmental endotoxin measurement methods: Standardization issues, *Appl. Occup. Environ. Hyg.*, 17, 606, 2002.
262. DECOS, *Endotoxins*, publication 1998/03WGD, Dutch Expert Committee on Occupational Standards (DECOS), Health Council of the Netherlands, Rijswijk, the Netherlands, 1998.
263. Milton, D.K., Endotoxin and other bacterial cell-wall components, in *Bioaerosols: Assessment and Control*, Macher, J.M., Ammann, H.M., Burge, H.A. et al. (Eds.), ACGIH, Cincinnati, OH, 1999, Chap. 23.
264. Chang, C.W., Chung, H., Huang, C.F. et al., Exposure of workers to airborne microorganisms in open-air swine houses, *Appl. Environ. Microbiol.*, 67, 155, 2001.
265. Autrup, H., Human exposure to genotoxic carcinogens: Methods and their limitations, *J. Cancer Res. Clin.*, 117, 6, 1991.
266. Olsen, J.H., Dragsted, L., and Autrup, H., Cancer risk an occupational exposure to aflatoxins in Denmark, *Br. J. Cancer*, 58, 392, 1988.
267. Kristensen, P., Andersen, A., and Irgens, L.M., Hormone-dependent cancer and adverse reproductive outcomes in farmers' families—Effects of climatic conditions favoring fungal growth in grain, *Scand. J. Work Environ. Health*, 26, 331, 2000.
268. Nordby, K.C., Andersen, A., and Kristensen, P., Incidence of lip cancer in the male Norwegian agricultural population, *Cancer Causes Control.*,15, 619, 2004.
269. Halstensen, A.S., Nordby, K.C., Eduard, W. et al., Real-time PCR detection of toxigenic Fusarium in airborne and settled grain dust and associations with trichothecene mycotoxins. *J. Environ. Monit.*, 8, 1235, 2006.
270. Nordby, K.C., Halstensen, A.S., Elen, O. et al., Trichothecene mycotoxins and their determinants in settled grain dust related to grain production. *Ann. Agric. Environ. Med.*, 11, 75, 2004.
271. Lucas, D., Lucas, R., Boniface, K. et al., Occupational asthma in the commercial fishing industry: A case series and review of the literature, *Int. Marit. Health*, 61, 13, 2010.
272. Mainelis, G., Willeke, K., Baron, P. et al., Induction charging and electrostatic classification of micrometer-size particles for investigating the electrobiological properties of airborne microorganisms, *Aerosol Sci. Technol.*, 36, 479, 2002.
273. Mainelis, G., Willeke, K., Adhikari, A. et al., Design and collection efficiency of a new electrostatic precipitator for bioaerosol collection, *Aerosol Sci. Technol.*, 36, 1073, 2002.
274. Reponen, T., Willeke, K., Ulevicius, V. et al., Effect of relative humidity on the aerodynamic diameter and respiratory deposition of fungal spores, *Atmos. Environ.*, 30, 3967, 1996.
275. Madelin, T.M. and Johnson, H.E., Fungal and actinomycete spore aerosols measured at different humidities with an aerodynamic particle sizer, *J. Appl. Bacteriol.*, 72, 400, 1992.
276. Blachere, F.M., Lindsley, W.G., Pearce, T.A. et al., Measurement of airborne influenza virus in a hospital emergency department, *Clin. Infect. Dis.*, 48, 438, 2009.
277. Lindsley, W.G, Blachere, F.M., Davis, K.A. et al., Distribution of airborne influenza virus and respiratory syncytial virus in an urgent care medical clinic, *Clin. Infect. Dis.*, 50, 693, 2010.
278. Linder, H.P., Morphology and the evolution of wind pollination, in *Reproductive Biology*, Owens, S.T. and Rudall, P.J. (Eds.), Royal Botanic Gardens, Kew, London, U.K., 1998, p. 123.

279. Linder, H.P., Pollen morphology and wind pollination in Angiosperms, in *Pollen and Spores: Morphology and Biology*, Harley, M.M., Morton, C.M., and Blackmore, S. (Eds.), Royal Botanic Gardens, Kew, London, U.K., 2000, p. 73.

280. Gregory, P.H., Spores: Their properties and sedimentation in still air, in *The Microbiology of the Atmosphere*, Aylesbury, Leonard Hill Books, London, U.K., 1973, Chap. 2.

281. Reponen, T.A., Grinshpun, S.A., Conwell, K.L. et al., Aerodynamic versus physical size of spores: Measurement and implication on respiratory deposition, *Grana*, 40, 119, 2001.

282. Fuchs, N.A. and Sutugin, A.G., Generation and use of monodisperse aerosols, in *Aerosol Science*, Davies, C.N. (Ed.), Academic Press, New York, 1966, Chap. 1.

283. Reponen, T., Aerodynamic diameters and respiratory deposition estimates of viable fungal particles in problem buildings, *Aerosol Sci. Technol.*, 22, 11, 1995.

284. Macher, J.M., Chen, B.T., and Rao, C.Y., Chamber evaluation of a personal, bioaerosol, cyclone sampler, *J. Occup. Environ. Hyg.*, 5, 702, 2008.

285. Hjelmroos, M., Benyon, F.H.L., Culliver, S. et al., *Airborne Allergens. Interactive Identification of Allergenic Pollen Grains and Fungal Spores*, Institute of Respiratory Medicine, University of Sydney, New South Wales, Australia, 1999.

286. Lacey, J., Aggregation of spores and its effect on aerodynamic behaviour, *Grana*, 30, 437, 1991.

287. Anderson, M.C. and Baer, H., Allergenically active components of cat allergen extracts, *J. Immunol.*, 127, 972, 1981.

288. Platts-Mills, T.A. and Carter, M.C., Asthma and indoor exposure to allergens, *N. Engl. J. Med.*, 336, 1382, 1997.

289. Custovic, A., Green, R., Fletcher, A. et al., Aerodynamic properties of the major dog allergen *Can f* 1: Distribution in homes, concentration, and particle size of allergen in the air, *Am. J. Respir. Crit. Care Med.*, 155, 94, 1997.

290. Custovic, A., Simpson, A., Pahdi, H. et al., Distribution, aerodynamic characteristics, and removal of the major cat allergen *Fel d* I in British homes, *Thorax*, 53, 33, 1998.

291. Wood, R.A., Laheri, A.N., and Eglleston, P.A., The aerodynamic characteristics of cat allergen, *Clin. Exp. Allergy*, 23, 733, 1993.

292. Raja, S., Xu, Y., Ferro, A.R. et al., Resuspension of indoor aeroallergens and relationship to lung inflammation in asthmatic children, *Environ. Int.*, 36, 8, 2010.

293. Montoya, L.D. and Hildemann, L.M., Size distributions and height variations of airborne particulate matter and cat allergen indoors immediately following dust-disturbing activities, *J. Aerosol Sci.*, 36, 735, 2005.

294. Reponen, T.A., Gazenko, S.V., Grinshpun, S.A. et al., Characteristics of airborne *Actinomycete* spores, *Appl. Environ. Microbiol.*, 64, 3807, 1998.

295. Górny, R.L. and Dutkiewics, J., Evaluation of microorganisms and endotoxin levels of indoor air in living rooms occupied by cigarette smokers and non-smokers in Sosnowiec, upper Silesia, Poland, *Aerobiologia*, 14, 235, 1998.

296. Lighthart, B. and Shaffer, B.T., Increased airborne bacterial survival as a function of particle content and size, *Aerosol Sci. Technol.*, 27, 439, 1997.

297. Hyvarinen, A., Vahteristo, M., Meklin, T. et al., Temporal and spatial variation of fungal concentrations in indoor air, *Aerosol Sci. Technol.*, 35, 688, 2001.

298. Meklin, T., Reponen, T., Toivola, M. et al., Size distributions of airborne microbes in moisture-damaged and reference school buildings of two construction types, *Atmos. Environ.*, 36, 6031, 2002.

299. Lignell, U., Meklin, T., Putus, T. et al., Effects of moisture damage and renovation on microbial conditions and pupils' health in two schools—A longitudinal analysis of five years. *J. Environ. Monit.*, 9, 225, 2007.

300. Torsvik, V., Sorheim, R., and Goksoyr, J., Total bacterial diversity in soil and sediment communities—A review, *J. Ind. Microbiol.*, 17, 170, 1996.

301. Kujundzic, E., Matalkah, F., Howard, C.J. et al., UV air cleaners and upper-room air ultraviolet germicidal irradiation for controlling airborne bacteria and fungal spores, *J. Occup. Environ. Hyg.*, 3, 536, 2006.

302. Wang, H., Reponen, T., Lee, S.-A. et al., Submicron size airborne endotoxin—A new challenge for bioaerosol exposure assessment in metalworking fluid environments, *J. Occup. Environ. Hyg.*, 4, 157, 2007.

303. Adhikari, A., Jung, J., Reponen, T. et al., Aerosolization of fungi, $(1 \rightarrow 3)$-β-D glucan, and endotoxin from flood-affected materials collected in New Orleans homes, *Environ. Res.*, 109, 215, 2009.

304. Kanaani, H., Hargreaves, M., Ristovski, Z. et al., Fungal spore fragmentation as a function of airflow rates and fungal generation methods, *J. Aerosol Sci.*, 43, 3725, 2009.

305. Lee, J.H., Hwang, G.B., Jung, J.H. et al., Generation characteristics of fungal spore and fragment bio-aerosols by airflow control over fungal cultures, *J. Aerosol Sci.*, 41, 319, 2010.

306. Cho, S.-H., Seo, S.-C., Schmechel, D. et al., Aerodynamic characteristics and respiratory deposition of fungal fragments, *Atmos. Environ.*, 39, 5454, 2005.

307. Górny, R.L., Reponen, T., Willeke, K. et al., Fungal fragments as indoor air biocontaminants, *Appl. Environ. Microbiol.*, 68, 3522, 2002.

308. Reponen, T., Seo, S.-C., Grimsley, F. et al., Fungal fragments in moldy houses: A field study in homes in New Orleans and Southern Ohio, *Atmos. Environ.*, 41, 8140, 2007.

309. Seo, S.-C., Reponen, T., and Grinshpun, S.A., Size-fractionated $(1 \rightarrow 3)$-β-D-glucan concentrations released from different moldy materials, *Sci. Total Environ.*, 407, 806, 2008.

310. Salares, V.R., Hinde, C.A., and Miller, J.D., Analysis of settled dust in homes and fungal glucan in air particulate collected during HEPA vacuuming, *Indoor Built Environ.*, 18, 484, 2009.

311. Hjelmroos, M., Evidence of long-distance transport of *Betula* pollen, *Grana*, 30, 215, 1991.

312. Rantio-Lehtimaki, A., Viander, M., and Koivikko, A., Airborne birch pollen antigens in different particle sizes, *Clin. Exp. Allergy*, 24, 23, 1994.

313. Matikainen, E. and Rantio-Lehtimaki, A., Semiquantitative and qualitative analysis of preseasonal birch pollen allergens in different particle sizes—Background information for allergen reports, *Grana*, 37, 293, 1998.

314. Miguel, A.G., Taylor, P.E., House, J. et al., Meteorological influence on respirable fragment release from Chinese Elm pollen, *Aerosol Sci. Technol.*, 40, 690, 2006.

315. Takahashi, Y., Sasaki, K., Nakamura, S. et al., Aerodynamic size distribution of the particle emitted from the flowers of allergologically important plants, *Grana*, 34, 45, 1995.

316. Agarwal, M.K., Swanson, M.C., Reed, C.E. et al., Airborne ragweed allergens: Association with various particle sizes and short ragweed plant parts, *J. Allergy Clin. Immunol.*, 74, 687, 1984.

317. Schumacher, M.J., Griffith, R.D., and O'Rourke, M.K., Recognition of pollen and other particulate aero-antigens by immunoblot microscopy, *J. Allergy Clin. Immunol.*, 82, 608, 1988.

318. Knox, R.B., Suphioglu, C., Taylor, P. et al., Major grass pollen allergen *Lol p* 1 binds to diesel exhaust particles: Implications for asthma and air pollution, *Clin. Exp. Allergy*, 27, 246, 1997.

319. Packe, G.E. and Ayres, J.G., Asthma outbreak during a thunderstorm, *Lancet*, 2, 199, 1985.

320. Knox, R.B., Grass pollen, thunderstorms and asthma, *Clin. Exp. Allergy*, 23, 354, 1993.

321. Newson, R., Strachan, D., Archibald, E. et al., Effect of thunderstorms and airborne grass pollen on the incidence of acute asthma in England, 1990–94, *Thorax*, 52, 680, 1997.

322. Marks, G.B., Colquhoun, J.R., Girgis, S.T. et al., Thunderstorm outflows preceding epidemics of asthma during spring and summer, *Thorax*, 56, 468, 2001.

323. Taylor, P.E., Flagan, R.C., Miguel, A.G. et al., Birch pollen rupture and the release of aerosols of respirable allergens, *Clin. Exp. Allergy*, 34, 1591, 2004.

324. Lewis, S.A., Corden, J.M., Forster, G.E. et al., Combined effects of aerobiological pollutants, chemical pollutants and meteorological conditions on asthma admissions and A & E attendances in Derbyshire UK, 1993–96, *Clin. Exp. Allergy*, 30, 1724, 2000.

325. Dales, R.E., Cakmak, S., Judek, S. et al., The role of fungal spores in thunderstorm asthma, *Chest*, 123, 745, 2003.

326. Thorne, P.S., Inhalation toxicology models of endotoxin- and bioaerosol-induced inflammation, *Toxicology*, 152, 13, 2000.

327. Heine, H.S., Bassett, J., Miller, L. et al., Determination of antibiotic efficacy against *Bacillus anthracis* in a mouse aerosol challenge model, *Antimicrob. Agents Chemother.*, 51, 1373, 2007.

328. Salyers, A.A. and Whitt, D.D., Introduction to infectious diseases, and the lung, a vital but vulnerable organ, in *Microbiology: Diversity, Disease, and the Environment*, Fitzgerald Science Press, Bethesda, MD, 2001, pp. 289, 315.

329. Larsson, L., Szponar, B., Ridha, B. et al., Identification of bacterial and fungal components in tobacco and tobacco smoke, *Tob. Induc. Dis.*, 4, 4, 2008.

330. Sapkota, A.R., Berger, S., and Vogel, T.M., Human pathogens abundant in the bacterial metagenome of cigarettes, *Environ. Health Perspect.*, 118, 351, 2010.

331. Ryan, P.H., Bernstein, D.I., Lockey, J. et al., Exposure to traffic-related particles and endotoxin during infancy is associated with wheezing at age 3 years, *Am. J. Respir. Crit. Care Med.*, 180, 1068, 2009.

332. Lappalainen, S., Salonen, H., Lindroos, O. et al., Fungal species in mold-damaged and nondamaged office buildings in southern Finland, *Scand. J. Work Environ. Health*, 4 (Suppl l), 18, 2008.

333. Park, J., Cox-Ganser, J.M., Kreiss, K. et al., Hydrophilic fungi and ergosterol associated with respiratory illness in a water-damaged building, *Environ. Health Perspect.*, 116, 45, 2008.

334. Cox, C.S. and Wathes, C.M. (Eds.), *Bioaerosols Handbook*, Lewis Publishers, Boca Raton, FL, 1995.
335. AIHA, *Field Guide for the Determination of Biological Contaminants in Environmental Samples*, American Industrial Hygiene Association, Fairfax, VA, 1996.
336. Hurst, C.J. (Ed.), *Manual of Environmental Microbiology*, 2nd edn., American Society for Microbiology, Washington, DC, 2001.
337. Muilenberg, M.L., Bioaerosol sampling and analytical techniques, in *Air Sampling Technologies: Principles and Applications*, American Conference of Government Industrial Hygienists, Cincinnati, OH 2009, http://www.acgih.org/products/AS121_TOC.pdf
338. Wurtz, H., Sigsgaard, T., Valbjorn, O. et al., The dustfall collector—A simple passive tool for long-term collection of airborne dust: A project under the Danish Mould In Buildings Program (DANIB). *Indoor Air*, 15 (Suppl 9), 33, 2005.
339. Noss, I., Wouters, I.M., Visser, M. et al., Evaluation of a low-cost electrostatic dust fall collector for indoor air endotoxin exposure assessment, *Appl. Environ. Microbiol.*, 74, 5621, 2008.
340. Sivasubramani, S.K., Niemeier, R.T., Reponen, T. et al., Assessment of the aerosolization potential for fungal spores in moldy homes, *Indoor Air*, 14, 405, 2004.
341. Kildesø, J., Würtz, K.F., Nielsen, P. et al., Determination of fungal spore release from wet building materials, *Indoor Air*, 13, 148, 2003.
342. Willeke, K. and Macher, J.M., Air sampling, in *Bioaerosols, Assessment and Control*, Macher, J.M., Ammann, H.M., Burge, H.A. et al. (Eds.), American Conference of Government Industrial Hygienists, Cincinnati, OH, 1999, Chap. 11.
343. Hinds, W.C., Filtration, bioaerosols, in *Aerosol Technology: Properties, Behavior, and Measurement of Airborne Particles*, 2nd edn., John Wiley & Sons Inc., New York, 1999, Chap. 9, p. 19.
344. Stewart, S.L., Grinshpun, S.A., Willeke, K. et al., Effect of impact stress on microbial recovery on an agar surface, *Appl. Environ. Microbiol.*, 61, 1232, 1995.
345. Mainelis, G. and Tabayoyong, M., The effect of sampling time on the overall performance of portable microbial impactors, *Aerosol Sci. Technol.*, 44, 75, 2010.
346. Yao, M. and Mainelis, G., Analysis of portable impactor performance for enumeration of viable bioaerosols, *J. Occup. Environ. Hyg.*, 4, 514, 2007.
347. Willeke, K., Lin, X., and Grinshpun, S.A., Improved aerosol collection by combined impaction and centrifugal motion, *Aerosol Sci. Technol.*, 28, 439, 1998.
348. Aizenberg, V., Bidinger, E., Grinshpun, S.A. et al., Airflow and particle velocities near a personal aerosol sampler with a curved, porous aerosol sampling surface, *Aerosol Sci. Technol.*, 18, 247, 1998.
349. Liu, B.Y.H. and Lee, K.W., Efficiency of membrane and nucleopore filters for submicrometer aerosols, *Environ. Sci. Technol.*, 10, 345, 1976.
350. Burton, N., Grinshpun, S.A., and Reponen, T., Physical collection efficiency of filter materials for bacteria and viruses, *Ann. Occup. Hyg.*, 51, 143, 2007.
351. Schmechel, D., Górny, R.L., Simpson, J.P. et al., Limitations of monoclonal antibodies for monitoring of fungal aerosols using *Penicillium brevicompactum* as a model fungus, *J. Immunol. Meth.*, 283, 235, 2003.
352. Li, C.S., Hao, M.L., Lin, W.H. et al., Evaluation of microbial samplers for bacterial microorganisms, *Aerosol Sci. Technol.*, 30, 100, 1999.
353. Lin, W.H. and Li, C.S., The effect of sampling time and flow rates on the bioefficiency of three fungal spore sampling methods, *Aerosol Sci. Technol.*, 28, 511, 1998.
354. Lin, W.H. and Li, C.S., Evaluation of impingement and filtration for yeast bioaerosol sampling, *Aerosol Sci. Technol.*, 30, 119, 1999.
355. Adhikari, A., Martuzevicius, D., Reponen, T. et al., Performance of the Button Personal Inhalable Sampler for the measurement of outdoor aeroallergens, *Atmos. Environ.*, 34, 4723, 2003.
356. Douwes, J., Versloot, P., Hollander, A. et al., Influence of various dust sampling and extraction methods on the measurement of airborne endotoxin, *Appl. Environ. Microbiol.*, 61, 1763, 1995.
357. Wet Electrostatic Precipitator (WEP), portable high-throughput liquid assisted air sampler (PHTLAAS), Zaromb Research Corporation, Burr Ridge, IL, (www.zaromb.com/products.htm, website accessed December 15, 2010).
358. Han, T. and Mainelis, G., Design and development of an electrostatic sampler for bioaerosols with high concentration rate, *J. Aerosol Sci.*, 39, 1066, 2008.
359. Han, T., An, H.R., and Mainelis, G., Performance of an electrostatic precipitator with hydrophobic surface when collecting airborne bacteria, *Aerosol Sci. Technol.*, 44, 339, 2010.
360. Agranovski, V., Ristovski, Z., Hargreaves, M. et al., Performance evaluation of the UVAPS: Influence of physiological age of airborne bacteria and bacterial stress, *J. Aerosol Sci.*, 34, 1711, 2003.

361. Kanaani, H., Hargreaves, M., Ristovski, Z. et al., Performance assessment of UVAPS: Influence of fungal spore age and air exposure, *J. Aerosol Sci.*, 38, 83, 2007.

362. McFarland, A.R., Haglund, J.S., King, M.D. et al., Wetted wall cyclones for bioaerosol sampling, *Aerosol Sci. Technol.*, 44, 241, 2010.

363. Pyankov, O.V., Agranovski, I.E., Pyankova, O. et al., Using a bioaerosol personal sampler in combination with real-time PCR analysis for rapid detection of airborne viruses, *Environ. Microbiol.*, 9, 992, 2007.

364. Ligler, F.S., Sapsford, K.E., Golden, J.P. et al., The array biosensor: Portable, automated systems, *Anal. Sci.*, 23, 5, 2007.

365. Berti, G., Isocrono, D., Ropolo, L. et al., An experience of data quality evaluation in pollen monitoring activities, *J. Environ. Monit.*, 11, 788, 2009.

366. British Aerobiology Federation, *Airborne Pollens and Spores: A Guide to Trapping and Counting*, European Aeroallegen Network (UK), University of Worcester, Worcester, U.K., 1995.

367. Rogers, C. and Muilenberg, M., *Comprehensive Guidelines for the Operation of Hirst-Type Suction Bioaerosol Samplers*, Pan-American Aerobiology Association, Standardized Protocols, 2001 (www.paaa.org/StandardizedProtocols.pdf, website accessed December 15, 2010).

368. UNI 11108:2004, *Air Quality. Method for Sampling and Counting of Airborne Pollen Grains and Fungal Spores*, ed. UNI Ente Nazionale Italiano di Unificazione, Milano, Italy, 2004.

369. ASTM, *Standard Test Method for Categorization and Quantification of Airborne Fungal Structures in an Inertial Impaction Sample by Optical Microscopy*, ASTM International, West Conshohocken, PA, 2009.

370. Gottardini, E., Cristofolini, F., Cristofori, A. et al., Sampling bias and sampling errors in pollen counting in aerobiological monitoring in Italy, *J. Environ. Monit.*, 11, 751, 2009.

371. Chun, D.T., Chew, V., Bartlett, K. et al., Second inter-laboratory study comparing endotoxin assay results from cotton dust, *Ann. Agric. Environ. Med.*, 9, 49, 2002.

372. Foto, M., Plett, J., Berghout, J. et al., Modification of the *Limulus* amebocyte lysate assay for the analysis of glucan in indoor environments, *Anal. Bioanal. Chem.*, 379, 156, 2004.

373. Iossifova, Y., Reponen, T., Daines, M. et al., Comparison of two analytical methods for detecting (1–3)-β-D-glucan in pure fungal cultures and in home dust samples, *Open Allergy J.*, 1, 26, 2008.

374. Ding, J.L., Navas, M.A., and Ho., B., Molecular cloning and sequence analysis of factor C cDNA from the Singapore horseshoe crab, *Carcinoscorpius rotundicauda*, *Mol. Mar. Biol. Biotechnol.*, 4, 90, 1995.

375. Alwis, K.U. and Milton, D.K., Recombinant factor C assay for measuring endotoxin in house dust: Comparison with LAL, and (1 → 3)-beta-D-glucans, *Am. J. Ind. Med.*, 49, 296, 2006.

376. Thorne, P.S., Perry, S.S., Saito, R. et al., Evaluation of the *Limulus* amebocyte lysate and recombinant factor C assays for assessment of airborne endotoxin, *Appl. Environ. Microbiol.*, 76, 4988, 2010.

377. Saito, R., Cranmer, B.K., Tessari, J.D. et al., Recombinant factor C (rFC) assay and gas chromatography/mass spectrometry (GC/MS) analysis of endotoxin variability in four agricultural dusts. *Ann. Occup. Hyg.*, 53, 713, 2009.

378. Reeslev, M., Miller, M., and Nielsen, K.F., Quantifying mold biomass on gypsum board: Comparison of ergosterol and beta-n-acetylhexosaminidase as mold biomass parameters, *Appl. Environ. Microbiol.*, 69, 3996, 2003.

379. Lignell, U., Meklin, T., Putus, T. et al., Microbial exposure, symptoms and inflammatory mediators in nasal lavage fluid of kitchen and clerical personnel in schools, *Int. J. Occup. Med. Environ. Health*, 18, 139, 2005.

380. Seshadri, S., Han, T., Krumis, V. et al., Application of ATP bioluminescence method to characterize performance of bioaerosol sampling devices, *J. Aerosol Sci.*, 40, 113, 2009.

381. Brera, C., Caputi, R., Miraglia, M. et al., Exposure assessment to mycotoxins in workplaces: Aflatoxins and ochratoxin A occurrence in airborne dusts and human sera, *Microchem. J.*, 73, 167, 2002.

382. Andersson, M.A., Nikulin, M., Köljalg, U. et al., Bacteria, molds, and toxins in water-damaged building materials, *Appl. Environ. Microbiol.*, 63, 387, 1997.

383. Peltola, J., Andersson, M.A., Haahtela, T. et al., Toxic-metabolite-producing bacteria and fungus in an indoor environment, *Appl. Environ. Microbiol.*, 67, 3269, 2001.

384. Vesper, S., Dearborn, D.G., Yike, I. et al., Evaluation of *Stachybotrys chartarum* in the house of an infant with pulmonary hemorrhage: Quantitative assessment before, during, and after remediation, *J. Urban Health*, 77, 68, 2000.

385. Luczynska, C.M., Arruda, L.K., Platts-Mills, T.A.E. et al., A two-site monoclonal antibody ELISA for the quantification of the major *Dermatophagoides* spp. allergens, *Der p* I and *Der f* I, *J. Immunol. Methods*, 118, 227, 1989.

386. Price, J.A., Pollock, I., Little, S.A. et al., Measurement of airborne mite antigen in homes of asthmatic children, *Lancet*, 336, 895, 1990.

387. Leaderer, B.P., Belanger, K., Triche, E. et al., Dust mite, cockroach, cat, and dog allergen concentrations in homes of asthmatic children in the Northeastern United States: Impact of socioeconomic factors and population density, *Environ. Health Perspect.*, 110, 419, 2002.

388. Swanson, M.C., Agarwal, M.K., Reed, C.E., An immunochemical approach to indoor aeroallergen quantitation with a new volumetric air sampler: Studies with mite, roach, cat, mouse, and guinea pig antigens, *J. Allergy Clin. Immunol.*, 76, 724, 1985.

389. Schou, C., Fernandez-Caldas, E., Lockey, R.F. et al., Environmental assay for cockroach allergens, *J. Allergy Clin. Immunol.*, 87, 828, 1991.

390. Hollander, A., Heederik, D., and Doekes, G., Respiratory allergy to rats: Exposure-response relationships in laboratory animal workers, *Am. J. Respir. Crit. Care Med.*, 155, 562, 1997.

391. Pollart, S.M., Smith, T.F., Morris, E.C. et al., Environmental exposure to cockroach allergens: Analysis with monoclonal antibody-based enzyme immunoassays, *J. Allergy Clin. Immunol.*, 87, 505, 1991.

392. Iversen, M., Korsgaard, J., Halla, T. et al., Mite allergy and exposure to storage mites and house dust mites in farmers, *Clin. Exp. Allergy*, 20, 211, 1990.

393. Earle, C.D., King, E.M., Tsay, A. et al., High-throughput fluorescent multiplex array for indoor allergen exposure assessment, *J. Allergy Clin. Immunol.*, 119, 428, 2007.

394. Sander, I., Fleischer, C., Borowitzki, G. et al., Development of a two-site enzyme immunoassay based on monoclonal antibodies to measure airborne exposure to $(1 \rightarrow 3)$-beta-D-glucan, *J. Immunol. Methods*, 337, 55, 2008.

395. Milton, D.K., Alwis, K.U., Fisette, L. et al., Enzyme-linked immunosorbent assay specific for $(1 \rightarrow 6)$ branched, $(1 \rightarrow 3)$-beta-D-glucan detection in environmental samples, *Appl. Environ. Microbiol.*, 67, 5420, 2001.

396. Douwes, J., van der Sluis, B., Doekes, G. et al., Fungal extracellular polysaccharides in house dust as a marker for exposure to fungi: Relations with culturable fungi, reported home dampness, and respiratory symptoms, *J. Allergy Clin. Immunol.*, 103, 494, 1999.

397. Goodfellow, M. and O'Donnell, A.G. (Eds.), *Chemical Methods in Prokaryotic Systematics*, John Wiley & Sons, Chichester, U.K., 1994.

398. Spurny, K.R., Chemical analysis of bioaerosols, in *Bioaerosols Handbook*, Cox, C.S. and Wathes, C.M. (Eds.), Lewis Publishers Inc., Boca Raton, FL, 1995, Chap. 12.

399. Hyvarinen, A., Sebastian, A., Pekkanen, J. et al., Characterizing microbial exposure with ergosterol, 3-hydroxy fatty acids, and viable microbes in house dust: Determinants and association with childhood asthma, *Arch. Environ. Occup. Health*, 61, 149, 2006.

400. Saraf, A., Park, J.-H., Milton, D.K. et al., Use of quadrupole GC–MS and ion-trap GC–MSMS for determining 3-hydroxy fatty acids in settled house dust: Relation to endotoxin activity, *J. Environ. Monit.*, 2, 163, 1999.

401. Young, J.C., Microwave-assisted extraction method of the fungal metabolite ergosterol and total fatty acids, *J. Agric. Food Chem.*, 43, 2904, 1995.

402. Miller, J.D. and Young, J.C., The use of ergosterol to measure exposure to fungal propagules in indoor air, *Am. Ind. Hyg. Assoc. J.*, 58, 39, 1997.

403. Fabian, M.P., Miller, S.L., Reponen, T. et al., Ambient bioaerosol indices for indoor air quality assessments of flood reclamation, *J. Aerosol Sci.*, 36, 763, 2005.

404. Liu, W.-T. and Stahl, D.A., Molecular approaches for measurement of density, diversity, and phylogeny, in *Manual of Environmental Microbiology*, 2nd edn., Hurst, C.J. (Ed.), American Society for Microbiology, Washington, DC, 2001, Chap. 11.

405. Buttner, M.P., Cruz-Perez, P., and Stetzenbach, L.D., Enhanced detection of surface-associated bacteria in indoor environments by quantitative PCR, *Appl. Environ. Microbiol.*, 67, 2564, 2001.

406. Mukoda, T.J., Todd, L.A., and Sobsey, M.D., PCR and gene probes for detecting bioaerosols, *J. Aerosol Sci.*, 25, 1523, 1994.

407. Mastorides, S.M., Oehler, R.L., Greene, J.N. et al., The detection of airborne *Mycobacterium tuberculosis* using micropore membrane air sampling and polymerase chain reaction, *Chest*, 115, 19, 1999.

408. Schafer, M.P., Martinez, K.F., and Mathews, E.S., Rapid detection and determination of the aerodynamic size range of airborne mycobacteria associated with whirlpools, *Appl. Occup. Environ. Hyg.*, 18, 41, 2003.

409. Williams, R.H., Ward, E., and McCartney, H.A., Methods for integrated air sampling and DNA analysis for detection of airborne fungal spores, *Appl. Environ. Microbiol.*, 67, 2453, 2001.

410. Alvarez, A.J., Buttner, M.P., Toranzos, G.A. et al., Use of solid-phase PCR for enhanced detection of airborne microorganisms, *Appl. Environ. Microbiol.*, 60, 374, 1994.

411. Roy, S.R., Schlitz, A.M., Marotta, A. et al., Bacterial DNA in house and farm dust, *J. Allergy Clin. Immunol.*, 112, 571, 2003.

412. Meklin, T., Haugland, R.S., Reponen, T. et al., Quantitative analysis of house dust can reveal abnormal mold conditions, *J. Environ. Monit.*, 6, 615, 2004.

413. Vesper, S., McKinstry, C., Ashley, P. et al., Quantitative PCR analysis of molds in the dust from homes of asthmatic children in North Carolina, *J. Environ. Monit.*, 9, 826, 2007.

414. Morrison, J., Yang, C., Lin, K.T. et al., Monitoring *Aspergillus* species by quantitative PCR during construction of a multi-storey hospital building, *J. Hosp. Infect.*, 57, 85, 2004.

415. Vesper, S., McKinstry, C., Haugland, R. et al., Development of an environmental relative moldiness index for us homes, *J. Occup. Environ. Med.*, 49, 829, 2007.

416. Gilbert, Y., Veillette, M., and Duchaine, C., Metalworking fluids biodiversity characterization, *J. Appl. Microbiol.*, 108, 437, 2010.

417. Rodriguez de Evgrafov, M., Walker, J.J., and Pace, N.R., Molecular source tracking of bioaerosols in the quarantined Katrina Flood zone, *Aerosol Sci. Technol.*, 44, 230, 2010.

418. Rintala, H., Pitkäranta, M., Toivola, M. et al., Diversity and seasonal dynamics of bacterial community in indoor environment, *BMC Microbiol.*, 8, 56, 2008.

419. Tringe, S.G., Zhang, T., Liu, X. et al., The airborne metagenome in an indoor urban environment, *PloS One*, 3, e1862, 2008.

420. APHA, *Control of Communicable Diseases Manual*, 19th edn., Heymann, D.L. (Ed.), American Public Health Association, Washington, DC, 2008.

421. Hjelmroos, M., Benyon, F.H.L., Culliver, S., and Jones, A.S., *Airborne Allergens. Interactive Identification of Allergenic Pollen Grains and Fungal Spores*, Institute of Respiratory Medicine, University of Sydney, New South Wales, Australia, 1999.

422. Samson, R.A., Hoekstra, E.S., Frisvad, J.C., and Filtenborg, O., *Introduction to Food- and Airborne Fungi*, 6th edn., Centraalbureau voor Schimmelcultures, Utrecht, the Netherlands, 2000.

423. Larone, D.H., *Medically Important Fungi: A Guide to Identification*, 4th edn., ASM Press, Washington, DC, 2002.

424. Watanabe, T., *Pictorial Atlas of Soil and Seed Fungi, Morphologies of Cultured Fungi and Key to Species*, 3rd edn., CRC Press, Boca Raton, FL, 2010.

425. ASM, *Manual of Clinical Microbiology*, 9th edn., Murray, P.R. (Ed.), ASM Press, Washington, DC, 2007.

13 Aerosols, Global Climate, and the Human Health Co-Benefits of Climate Change Mitigation*,†

George D. Thurston and Michelle L. Bell

CONTENTS

13.1 OVERVIEW

Aerosols are airborne suspensions of particulate matter (PM). They can have effects on both human health and climate. Indeed, greenhouse gas (GHG) mitigation steps can also provide more immediate and localized health co-benefits to the localities that implement these mitigation measures, especially in terms of reductions in PM air pollution and their associated adverse health impacts. These co-benefits, if fully considered and quantified, can potentially be a significant motivator to encourage the near-term adoption of GHG measures. In addition, changes in aerosol concentrations can result in their own short-term effects on climate change. In this chapter, we introduce the principles of aerosol climate change and health effects, and present a framework for the estimation of the ancillary health co-benefits that can be derived from various GHG strategies that result in reductions of PM air pollution.

---

* *Competing interests*: The authors declare that they have no competing interests.
† *Authors' contributions*: Both authors made substantial contributions to the conception and design of this paper, were involved in drafting and revising the manuscript, and have approved the final version.

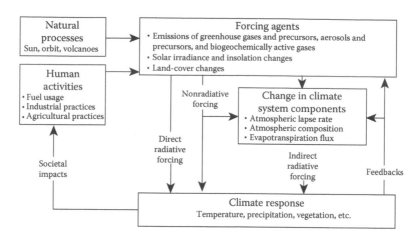

FIGURE 13.1 Framework of climate change influences. (From National Academy of Sciences (NAS), *Radiative Forcing of Climate Change: Expanding the Concept and Addressing Uncertainties*, National Research Council, Committee on Radiative Forcing Effects on Climate, Climate Research Committee, Washington, DC, 2005.)

13.2 AEROSOLS AND CLIMATE CHANGE

Most climate change efforts have focused on controlling carbon dioxide (CO_2) emissions, but there is need to control the shorter-lived greenhouse pollutants (SLGHPs) because these SLGHPs can collectively influence climate change more than CO_2, and because many of the SLGHPs, such as PM, have human health impacts as well as climate impacts. Moreover, since aerosols and other SLGHPs are shorter-lived than CO_2, aerosol atmospheric concentrations (and their associated climate effects) will respond much more quickly to reductions in PM emissions than will reductions in CO_2.

Factors that affect climate change are usefully separated into two categories: forcings and feedbacks (see Figure 13.1). A *climate forcing*, as defined by the National Academy of Sciences (2005), is an energy imbalance imposed on the climate system either externally or by human activities. Examples include changes in solar energy output, volcanic emissions, deliberate land modification, or anthropogenic emissions of GHGs, PM, and their precursors. A *climate feedback* is an internal climate process that amplifies or dampens the climate response to a specific forcing.

As noted in Figure 13.1, aerosols are among the SLGHP forcing agents that result in changes in climate system components, as well as direct radiative forcings. These SLGHP aerosol components include sulfate, organic carbon (OC), and black carbon (BC). However, as depicted in Figure 13.2, despite uncertainties about the exact size of the climate impacts, these different types of particles have very different climate implications; the first two, sulfates and OC are cooling forcings, but the third (elemental BC soot) is warming. Thus, air pollution and climate change policies that reduce elemental BC soot particles will have larger beneficial effects than many other pollutants, from a climate change perspective.

13.3 PARTICULATE MATTER HEALTH EFFECTS

Tropospheric aerosols that affect climate change also have significant human health implications. A wealth of scientific literature clearly links PM with numerous adverse health effects. Indeed, a U.S. Environmental Protection Agency (USEPA) assessment of human health effects benefits of the Clean Air Act attributed nearly 90% of the estimated monetary valuation of the human health effects benefits to be derived from the Act during 1990–2010 to reductions in PM (USEPA, 1999).

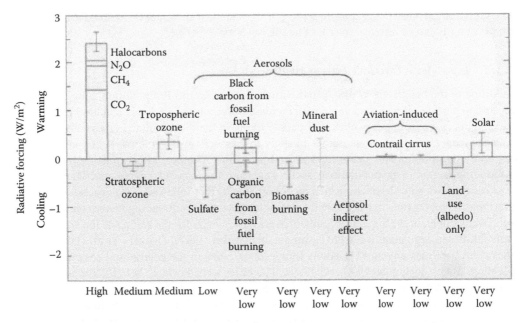

FIGURE 13.2 Estimated radiative forcings since preindustrial times for the Earth and atmosphere's troposphere system. (From National Academy of Sciences (NAS), *Radiative Forcing of Climate Change: Expanding the Concept and Addressing Uncertainties*, National Research Council, Committee on Radiative Forcing Effects on Climate, Climate Research Committee, Washington, DC, 2005.)

13.3.1 SHORT-TERM EXPOSURE EFFECTS OF PM

Acute (short-term) exposure to particulate air pollution has been found to be associated with increases in the rates of daily asthma attacks, hospital admissions, and mortality. PM is associated with increased risk of respiratory hospital admissions in many cities including New York, NY; Buffalo, NY; and Toronto, Ontario, as well as with mortality in numerous cities such as Chicago, IL, and Los Angeles, CA (e.g., Thurston et al., 1992). These effects were also observed elsewhere in the United States, and in other cities throughout the world, including national multicity studies (Schwartz, 1997; Bell et al., 2004; Dominici et al., 2006; Samoli et al., 2008).

In addition to lung damage, recent epidemiological and toxicological studies of PM air pollution have shown adverse effects on the heart, including an increased risk of heart attacks. For example, when PM stresses the lung (e.g., by inducing edema), it places extra burden on the heart, which can induce fatal complications for persons with cardiac problems. Indeed, Peters et al. (2001) found that elevated concentrations of fine particles (PM$\leq$2.5 μm in aerodynamic diameter, i.e., PM$_{2.5}$) in the air could elevate the risk of myocardial infarctions (MIs) within a few hours, and extending 1 day after PM$_{2.5}$ exposure. Others found that a 48% increase in the risk of MI was associated with an increase of 25 μg/m^3 PM$_{2.5}$ during a 2h period before the onset of MI, and a 69% increase in risk to be related to an increase of 20 μg/m^3 PM$_{2.5}$ in the 24h average 1 day before the MI onset (Peters et al., 2001).

Epidemiologic research conducted in the United States and elsewhere has indicated that acute exposure to PM air pollution is associated with increased risk of mortality. For example, a national multicity time-series statistical analysis of mortality and PM$\leq$10 μm in aerodynamic diameter (PM$_{10}$) air pollution in 90 U.S. cities indicates that a 10 μg/m^3 increase in daily PM$_{10}$ is associated with an increase of approximately 0.3% in the daily risk of death (Dominici et al., 2005). The epidemiological results—in this case a 0.3% change in the daily mortality rate—are tied to the increment of pollution, in this case a 10 μg/m^3 increase of PM$_{10}$. In other words, the studies indicate that a pollution increase larger than $10 \mu g/m^3$, or the specified increment, would be associated with a proportionally larger increase in risk of health outcome.

13.3.2 LONG-TERM EXPOSURE EFFECTS OF PM

In addition to the health effects associated with acute exposure to PM pollution, long-term chronic exposure to particles is also associated with increased lifetime risk of death, and has been estimated to take years from the life expectancy of people living in the most polluted cities, relative to those living in cleaner cities. The first studies to show this association were cross-sectional studies that compared metropolitan area death rates in high and low PM cities, after adjusting for potentially confounding factors in the populations, such as age, sex, and race (Ozkaynak and Thurston, 1987). These results have since been confirmed by cohort studies that followed large groups of individuals in various cities over time that are able to control for potential confounding factors on an individual level. For example, in the Six-Cities Study (that was a key basis for the setting of the USEPA's original health-based regulation for a $PM_{2.5}$ annual standard in 1997), Dockery et al. (1993) analyzed survival probabilities among 8111 adults living in six cities in the central and eastern portions of the United States during the 1970 and 1980s. The cities were Portage, WI (P); Topeka, KS (T); a section of St. Louis, MO (L); Steubenville, OH (S); Watertown, MA (M); and Kingston-Harriman, TN (K). Air quality was averaged over the period of study in order to study long-term (chronic) effects. It was found that the long-term risk of death, relative to the cleanest city, increased with fine particle exposure, even after controlling for potentially confounding factors such as age, sex, race, smoking, etc.

More recently, a study showed that long-term exposure to combustion-related fine particulate air pollution is an important environmental risk factor for cardiopulmonary and lung cancer mortality (see Figure 13.3). Indeed, this study indicates that the increase in risk of lung cancer from long-term exposure to $PM_{2.5}$ was of roughly the same size as the increase in lung cancer risk of a nonsmoker who breathes passive smoke while living with a smoker, or about a 20% increase in lung cancer risk (Pope et al., 2002).

Other studies indicating health risk from chronic exposure to PM include a multicity U.S. study finding than a $10 \mu g/m^3$ increase in yearly $PM_{2.5}$ is associated with approximately an 11%–21% increase in mortality (Eftim et al., 2008). A systematic review of research on long-term PM exposure found that collectively, the studies indicate a 15%–21% increase in mortality per $10 \mu g/m^3$ $PM_{2.5}$ (Chen et al., 2008).

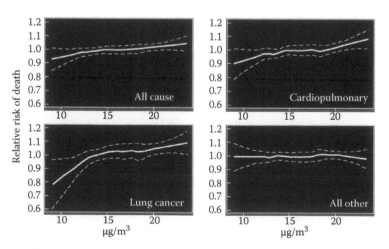

FIGURE 13.3 Cardiac, lung, and cancer mortality risks of long-term fine PM exposure increase monotonically with exposure. (Adapted from Pope, C.A. et al., *J. Am. Med. Assoc.*, 287(9), 1132, 2002.)

13.3.3 HEALTH EFFECTS OF PM CONSTITUENTS

PM is a complex mixture of a wide array of chemical constituents, and PM's chemical composition varies seasonally and regionally (Bell et al., 2007). While most past studies have investigated the effects of the PM *mass* concentration on human health effects, newer studies have begun to evaluate the mortality impacts of PM by source-specific components, including two key *aerosol components* that affect climate change: sulfates and elemental BC soot.

With regard to acute effects of PM components, Thurston et al. (2005) found that coal-burning-related sulfate-containing aerosols were among those most associated with increases in daily mortality. Bell et al. (2009) found that communities with higher $PM_{2.5}$ content of nickel (Ni), vanadium (V), and elemental carbon (EC) and/or their related sources were found to have higher risk of hospitalizations associated with short-term exposure to $PM_{2.5}$. Lall et al. (2011) similarly found that EC of traffic origins were associated with higher risk of cardiovascular disease (CVD) hospital admissions in New York than $PM_{2.5}$ mass in general. In a study of New York City mortality, Ito et al. (2011) have reported that coal-combustion-related components (e.g., selenium [Se] and sulfur) were associated with CVD mortality in summer, whereas the traffic-related EC showed associations with CVD mortality throughout the year. Zhou et al. (2011) investigated the $PM_{2.5}$ components and gaseous pollutants associated with mortality in Detroit, MI and Seattle, WA. These authors similarly found that CVD and respiratory mortality were most associated with warm season secondary aerosols (e.g., sulfates) and traffic markers (e.g., EC) in Detroit, while in Seattle, the component species most closely associated with mortality included those for cold season traffic and other combustion sources, such as residual oil and wood burning. In addition, recent evidence has implicated diesel-traffic-derived EC as a factor in increased risk of acute asthma morbidity (Spira-Cohen et al., 2011). Overall, these studies of $PM_{2.5}$ components and constituents suggest that both EC and sulfate, and their associated sources, including diesel traffic and coal burning, may be among the most explanatory of the acute adverse health effects of $PM_{2.5}$, although the health impacts of the particulate mixture are not well understood

With regard to the long-term effects of PM air pollution, Ozkaynak and Thurston (1987) conducted the first source apportionment of $PM_{2.5}$-mortality effects, finding that sulfate-related particles, largely from coal burning, were most associated with the mortality impacts of long-term exposure to $PM_{2.5}$. More recently, Ostro et al. (2007) examined daily data from 2000 to 2003 on mortality and $PM_{2.5}$ mass and components, including EC and OC, nitrates, sulfates, and various metals. The authors examined associations of $PM_{2.5}$ and its constituents with daily counts of several mortality categories: all-cause, CVD, respiratory, and mortality age >65 years, finding the strongest associations between mortality and sulfates and several metals. Ostro et al. (2010) used data from a prospective cohort of active and former female public school professionals to develop estimates of long-term exposures to $PM_{2.5}$ and several of its constituents, including EC, OC, sulfates, nitrates, iron (Fe), potassium (K), silicon (Si), and zinc (Zn), finding increased risks of all-cause and cardiopulmonary mortality from exposure to constituents derived from combustion of fossil fuel (including diesel), as well as those of crustal origin. In addition, Smith et al. (2010) undertook a meta-analysis of existing time-series studies, as well as an analysis of a cohort of 352,000 people in 66 U.S. cities during 18 years of follow-up of the American Cancer Society cohort, finding total mortality effects from long-term exposure to both the elemental BC and sulfate components of $PM_{2.5}$ aerosols.

13.4 SULFATES AND GEOENGINEERING OF CLIMATE CHANGE

One of the repeatedly raised theories regarding how to mitigate climate change is the deliberate injection of sulfates into the atmosphere because of their aforementioned climate cooling properties (Rasch et al., 2008; Ammann et al. 2010; Ban-Weiss and Caldeira, 2010). The theory behind this proposal is that the higher level of sulfates would cool the global environment much in the manner that past large volcanic eruptions have. For example, the June 1991 eruption of Mount Pinatubo sent fine ash and gases high into the stratosphere, forming a large volcanic cloud that drifted around the world. The sulfur dioxide (SO_2) in this cloud—about 22 million tons—combined with water to form droplets of sulfuric acid, a type of sulfate particles, blocking some of the sunlight from reaching the Earth, thereby cooling temperatures in some regions of the world by as much as 0.5°C (Kious and Tilling 1996). One recent analysis estimated that annual aerosol injections of 5–10 Mton of sulfate aerosol (delivering a constant 4 W m^{-2} reduction in radiative forcing, similar to a 1991 Pinatubo eruption every 18 months) could delay climate-change-related sea-level rise by 40–80 years (Moore et al., 2010). However, they note that such aerosol injections fail cost–benefit analysis unless they can be maintained indefinitely, and that, if ever stopped, the climate sea-level rise effects would then be "dramatic." In addition, as noted for the actual Pinatubo eruption itself, the resultant cooling from sulfate injection is not experienced evenly across the globe, and is unlikely to align with GHG heating effects in latitude, potentially inducing other, new unplanned weather pattern changes, such as changes in spatial precipitation patterns (Goldstein et al., 2010). Also, as discussed earlier, these sulfates have been associated with significant adverse human health effects, including increased risk of premature death. Thus, as discussed in Smith et al. (2010), any such geoengineering needs to be analyzed carefully for potential unintended consequences and uncertainties.

13.5 ANCILLARY HEALTH BENEFITS OF CLIMATE CHANGE MITIGATION

Policies designed to avert the course of climate change would eventually result in human health benefits directly associated with lessened global temperature changes and associated impacts, but many would also bring more immediate ancillary health benefits from reduced ground-level air pollution in the short term (Swart et al., 2004; Haines et al., 2007; Thurston, 2007; Smith et al., 2008; Walsh, 2008). Fossil fuel combustion processes that generate GHGs also emit other harmful air pollutants, such as toxic metals and OC. Several measures aimed at reducing GHG emissions can also improve local air quality, most notably PM air pollution. Further, whereas the benefits from climate change mitigation would materialize far in the future, these co-benefits, or ancillary benefits, would occur in the short term. Similarly, policies aimed at short-term improvements in air quality could lower GHG emissions. Much of the discussion that follows of the co-benefits of air pollution mitigation is further detailed in Bell et al. (2008).

13.5.1 FRAMEWORK OF CLIMATE MITIGATION CO-BENEFITS ASSESSMENT

Figure 13.4 describes the relationships among the health consequences of climate change and air quality policies and the general framework of how these responses can be assessed. Air quality policies are routinely evaluated in terms of the estimated health outcomes avoided and their economic impact (USEPA, 1997, 1999). However, assessment of the health impacts of GHG strategies often considers only consequences in the far future (i.e., left side of Figure 13.1), without integration of the short-term benefits of related policies (Ebi et al., 2006). Well-informed public health and environmental strategies require full consideration of consequences, including co-benefits and potential ancillary harms.

A broad array of tools to evaluate the health-related ancillary costs and benefits of climate change is currently available, and some examples are provided in italics in Figure 13.4. As described in detail in Bell et al. (2008), the general structure for most assessments involves three key steps: (1) estimating changes in air pollutant concentrations, comparing levels in response to GHG mitigation to concentrations under a baseline "business-as-usual" scenario; (2) estimating the adverse health impacts avoided from reduced air pollution; and (3) for some studies, estimating the monetary benefit from these averted health consequences, often with comparison to the cost of the climate change mitigation measure.

The first step in such a co-benefit analysis is often the development of emissions scenarios and information regarding how emissions translate into pollutant concentrations, such as with air quality modeling systems. The second step employs concentration–response functions from existing

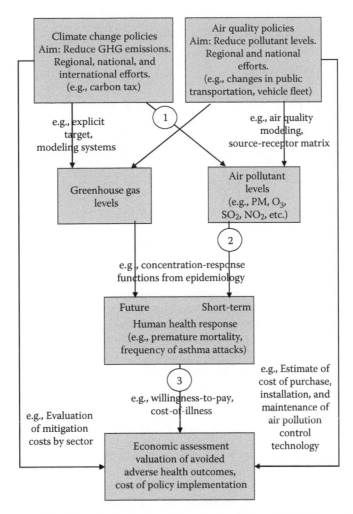

FIGURE 13.4 Framework of air pollution co-benefit estimation. (From Bell, M.L. et al., *Environ. Health*, 7, 41, 2008.)

epidemiological studies on ambient air pollution and health. The third stage utilizes a variety of techniques to translate health benefits into monetary terms. Potential additional steps include sensitivity analysis, such as applying multiple climate change scenarios or concentration–response functions for health effects.

13.5.2 Studies of Health and Air Pollution Benefits of Climate Change Mitigation

A variety of studies have been conducted to estimate the health and air pollution ancillary benefits from GHG reduction, with a wide range of methods and study areas. Energy scenarios, emission inventories, and global change and regional air quality modeling systems have been linked to estimate the short-term incremental changes in public health and the environment that could result from various GHG mitigation policies (Burtraw and Toman, 1997; McCarthy et al., 2001).

There are now numerous analyses indicating substantial health co-benefits from reductions in PM pollution, which can be induced by GHG mitigation measures that involve reductions in fossil fuel combustion emissions. As shown in Figure 13.5, a study of New York City and three Latin American cities identified significant health benefits from reducing GHG, including about 64,000 cases of avoided premature mortality over a 20 year period (Cifuentes et al., 2001a). Countrywide

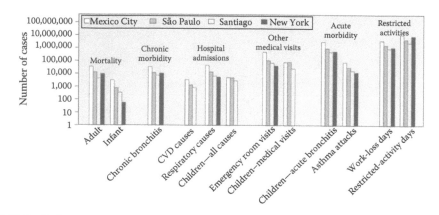

FIGURE 13.5 Estimated potential human health benefits from reductions in air pollution associated with implementing GHG mitigation measures in four cities (2001–2020). (From Cifuentes, L. et al., *Science*, 293(5533), 1257, 2001b.)

assessments of GHG mitigation policies on public health have been produced for Canada (Last et al., 1988) and selected energy sectors in China (Wang and Smith, 1999; Cao et al., 2008), under differing baseline assumptions. A synthesis of research on co-benefits and climate change policies in China concluded that China's Clean Development Mechanism potentially could save 3,000–40,000 lives annually through co-benefits of improved air pollution (Vennemo et al., 2006). Several studies investigated the links between regional air pollution and climate policy in Europe (Working Group on Public Health and Fossil Fuel Combustion, 1997; Alcamo, 2002; van Harmelen et al., 2002).

13.5.3 MONETARY VALUATIONS OF MITIGATION CO-BENEFITS

To help decision makers assess policies with a wide array of health consequences, outcomes are often converted into comparable formats. One approach is to convert health outcomes into economic terms to allow direct comparison of costs and benefits. There are several common approaches for economic valuation of averted health consequences (step 3 of Figure 13.4): cost of illness (COI); human capital; willingness to pay (WTP) methods; and quality-adjusted life year (QALY) approaches. The COI method totals medical and other out-of-pocket expenditures and has been used for acute and chronic health endpoints. For instance, separate models of cancer progression and respiratory disease were used to estimate medical costs from these diseases over one's lifetime (Hartunian et al., 1981). However, early attempts to value mortality risk reductions applied the human capital approach, which estimates the "value of life" as lost productivity. This method is generally recognized as problematic and not based on modern welfare economics, where preferences for reducing death risks are not captured. Another limitation is incorporation of racial- or gender-based discrimination in wages. This method assigns value based solely on income, without regard to social value, so unpaid positions such as homemaker and lower-paid positions such as social worker receive lower values. Because data are often available for superior alternatives, this approach is rarely used in health benefit studies. WTP generates estimates of preferences for improved health that meet the theoretical requirements of neo-classical welfare economics, by aiming to measure the monetary amount persons would willingly sacrifice to avoid negative health outcomes. Complications arise in analysis and interpretation because changes in environmental quality or health often will themselves change the real income (utility) distribution of society. A valuation procedure that sums individual WTP does not capture individual preferences about changes in income distribution. Another complication is that the value of avoided health risk may differ by type of health event and age. The QALY approach attempts to account for the quality of life lost by adjusting for time "lost" from disease or death, but these estimates may be very insensitive to different severities and types of acute morbidity (Miller, 2006).

Estimating the ancillary public health consequences of GHG policies is a challenging task, drawing upon expertise in economics, emission inventories, air pollution modeling, and public health. However, most assessments to date have focused more heavily on one aspect of the framework (i.e., a portion of Figure 13.4), whether it be estimation of changes in air pollutant concentrations, health response, or economic analysis.

Results from current ancillary benefits studies may be underestimates due to unquantified benefits, as only a subset of the health consequences from air pollution have adequate exposure–response relationships (Voorhees et al., 2001; Committee on Estimating the Health-Risk-Reduction Benefits of Proposed Air Pollution Regulations, 2002; USEPA, 2005; Voorhees, 2005). A USEPA evaluation of the Clean Air Interstate Rule (CAIR) noted numerous unquantified health impacts such as chronic respiratory damage for ozone (O_3), loss of pulmonary function for PM, and lung irritation for nitrogen oxide (NO_x) (USEPA, 2005). The nature of unquantified effects is continually evolving. Some pollution and health relationships considered unquantifiable by USEPA (1999) have since been identified, such PM air pollution's association with lung cancer (Filleul et al., 2005; Krewski et al., 2005). Furthermore, some endpoints may be included in one analysis, but regarded as too uncertain for another, perhaps due to a different study location or differences in researchers' judgment. One approach to addressing health endpoints with uncertain concentration–response functions is to include these effects qualitatively in discussion of unquantified benefits. Another is to incorporate these effects within a sensitivity analysis.

Valuations of mortality risk reductions associated with environmental policies are usually the largest category of benefits, both among health responses and compared to other attributes. For instance, a USEPA analysis of the Clean Air Act estimated a value of $100 billion annually for reduced premature mortality out of $120 billion in total benefits, compared to costs of approximately $20 billion (USEPA, 1999). European and Canadian studies similarly found that mortality risk dominates analysis of pollution reductions (Stratus Consulting, 1999; Bickel and Friedrich, 2005). Next to mortality, reductions in the probability of developing a chronic respiratory disease have been estimated to have the highest monetary value, recognizing that values for other types of diseases are sparse.

Recently, the Stern Review (Stern, 2007) addressed a wide range of global benefits and costs associated with climate change, including air pollution co-benefits. Citing a study by the European Environmental Agency, the Review notes that limiting global mean temperature increase to 2°C would lead to annual savings in the implementation of existing European air pollution control measures of €10 billion and additional avoided annual health costs of €16–€46 billion. Even larger co-benefits are estimated in developing countries, including via the substitution of modern fuels for biomass. The Stern Review also recognizes some of the trade-offs between climate change objectives and local air quality gains. For instance, switching from petrol to diesel reduces CO_2 emissions but increases pPM_{10} and NO_x emissions. Other GHG mitigating actions present fewer environmental trade-offs (e.g., reductions in aircraft weight can decrease CO_2 emissions and simultaneously improve local air quality).

Overall, although still a work in progress, the present techniques available for the analyses of the ancillary public health costs and benefits are adequate and appropriate for implementation by those comparing the relative merits and overall value of various GHG mitigation policies. Estimates of considerable benefits that remain after a variety of sensitivity analyses can alleviate some concerns regarding limitations of individual methods or assumptions. The PM air-pollution-associated public health changes associated with GHG mitigation strategies should be a key factor in the choice of GHG policies.

13.6 IMPLICATIONS

The most recent Intergovernmental Panel on Climate Change (IPCC) (2007) report has concluded, in the most definitive terms yet, that global climate change is occurring, stating that "[w]arming of the climate system is unequivocal, as is now evident from observations of increases in global average air and ocean temperatures, widespread melting of snow and ice, and rising global average sea level." The fact that the "man-made" contribution to the climate change pollutants is largely caused by the same activity that causes the air pollution health effects indicates that if a city, state, or nation acts to reduce the combustion of fossil fuels and the air pollution caused by them, it will reap not only the climate change benefits, but also the localized health benefits associated with that air pollution reduction. Thus, substantial near-term air-pollution-associated health benefits of climate control measures could go to the cities and countries that act most vigorously to control their combustion emissions of GHGs . These locally enjoyed health "co-benefits" of reductions in the PM air pollution from fossil fuel combustion should be considered a key factor in decisions relating to climate change mitigation measures.

REFERENCES

Alcamo, J., Mayerhofer, P., Gaurdans, R. et al. An integrated assessment of regional air pollution and climate change in Europe: Findings of the IAR-CLIM project. *Environmental Science and Policy* 2002, 5:257–272.

Ammann, C.M., Washington, W.M., Meehl, G.A., Buja, L., Teng, H. Climate engineering through artificial enhancement of natural forcings: Magnitudes and implied consequences. *Journal of Geophysical Research D: Atmospheres* 2010, 115(22), Article no. D22109.

Ban-Weiss, G.A., Caldeira, K. Geoengineering as an optimization problem. *Environmental Research Letters* 2010, 5(3), Article no. 034009.

Bell, M.L., Davis, D.L., Cifuentes, L.A., Krupnick, A.J., Morgenstern, R.D., Thurston, G.D. Ancillary human health benefits of improved air quality resulting from climate change mitigation. *Environmental Health* 2008, 7:41 (review).

Bell, M.L., Dominici, F., Ebisu, K., Zeger, S.L., Samet, J.M. Spatial and temporal variation in $PM_{2.5}$ chemical composition in the United States for health effects studies. *Environmental Health Perspectives* 2007, 115(7):989–995.

Bell, M.L., Dominici, F., Samet, J.M. Time-series of particulate matter. *Annual Review of Public Health* 2004, 25:247–280.

Bell, M.L., Ebisu, K., Peng, R.D., Samet, J.M., Dominici, F. Hospital admissions and chemical composition of fine particle air pollution. *American Journal of Respiratory and Critical Care Medicine* 2009, 179(12):1115–1120.

Bickel, P., Friedrich, R.E. *ExternE, Externalities of Energy, Methodology 2005 Update*. Luxembourg, Belgium: European Commission, 2005.

Burtraw, D., Toman, M. *The Benefits of Reduced Air Pollutants in the U.S. from Greenhouse Gas Mitigation Policies*. Washington, DC: Resources for the Future, 1997, Discussion Paper 98-01-REV.

Cao, J., Ho, M.S., Jorgenson, D.W. *"Co-benefits" of Greenhouse Gas Mitigation Policies in China*. Washington, DC: Resources for the Future, 2008.

Chen, H., Goldberg, M.S., Villeneuve, P.J. A systematic review of the relation between long-term exposure to ambient air pollution and chronic diseases. *Reviews on Environmental Health* 2008, 23(4):243–297.

Cifuentes, L., Borja-Aburto, V.H., Gouveia, N., Thurston, G.D., Davis, D.L. Assessing the health benefits of urban air pollution reductions associated with climate change mitigation (2000–2020): Santiago, São Paulo, Mexico City, and New York City. *Environmental Health Perspectives* 2001a, 109:S419–S425.

Cifuentes, L., Borja-Aburto, V.H., Gouveia, N., Thurston, G.D., Davis, D.L. Climate change. Hidden health benefits of greenhouse gas mitigation. *Science* 2001b, 293(5533):1257–1259.

Committee on Estimating the Health-Risk-Reduction Benefits of Proposed Air Pollution Regulations. *National Research Council: Estimating the Public Health Benefits of Proposed Air Pollution Regulations*. Washington, DC: National Academies Press, 2002.

Dockery, D.W., Pope, C.A. III, Xu, X. et al. An association between air pollution and mortality in six U.S. cities. *New England Journal of Medicine* 1993, 329(24):1753–1759.

Dominici, F., McDermott, A., Daniels, M., Zeger, S.L., Samet, J.M. Revised analyses of the National Morbidity, Mortality, and Air Pollution Study: Mortality among residents of 90 cities. *Journal of Toxicology and Environmental Health Part A* 2005, 68:1071–1092.

Dominici, F., Peng, R.D., Bell, M.L. et al. Fine particulate air pollution and hospital admissions for cardiovascular and respiratory diseases. *Journal of the American Medical Association* 2006, 295(10):1127–1134.

Ebi, K., Mills, D.M., Smith, J.B., Grambsch, A. Climate change and human health impacts in the United States: An update on the results of the U.S. National Assessment. *Environmental Health Perspectives* 2006, 114:1318–1324.

Eftim, S.E., Samet, J.M., Janes, H., McDermott, A., Dominici, F. Fine particulate matter and mortality: A comparison of the six cities and American Cancer Society cohorts with a Medicare cohort. *Epidemiology* 2008, 19(2):209–216

Filleul, L., Rondeau, V., Vandentorren, S. et al. Twenty five year mortality and air pollution: Results from the French PAARC survey. *Occupational and Environmental Medicine*, 2005, 62:453–460.

Goldstein, B., Kobos, P.H., Brady, P.V. Unintended consequences of atmospheric injection of sulphate aerosol. SAND2010-7571. Albuquerque, New Mexico: Sandia National Laboratories, 2010.

Haines, A., Smith, K.R., Anderson. D. et al. Policies for accelerating access to clean energy, improving health, advancing development, and mitigating climate change. *Lancet* 2007, 370:1264–1281.

van Harmelen, T., Bakker, J., de Vries, B., van Vuuren, D., den Elzen, J., Mayerhofer, P. Long-term reductions in costs of controlling regional air pollution in Europe due to climate policy. *Environmental Science and Policy* 2002, 5:349–365.

Hartunian, N.S., Smart, C.N., Thompson, M.S. *The Incidence and Economic Costs of Major Health Impairments: A Comparative Analysis of Cancer, Motor Vehicle Injuries, Coronary Heart Disease, and Stroke.* Lexington, MA: Lexington Books, 1981.

Intergovernmental Panel on Climate Change (IPCC). *The Physical Science Basis: Summary for Policy Makers*, Climate change 2007: The physical science basis. Summary for policymakers. Contribution of Working Group I to the Fourth Assessment Report of the Intergovernmental Panel on Climate Change. Geneva, Switzerland: IPCC Secretariat, 2007.

Ito, K., Mathes, R., Ross, Z., Nádas, A., Thurston, G.D., Matte, T. Fine particulate matter constituents associated with cardiovascular hospitalizations and mortality in New York City. *Environmental Health Perspectives* 2011, 119(4):467–473.

Kious, W.J. and Tilling, R.I. *This Dynamic Earth: The Story of Plate Tectonics.* Washington, DC: USGS, USGPO, 1996.

Krewski, D., Burnett, R., Jerrett, M., Pope, C., Rainham, D., Calle, E., Thurston, G., Thun, M. Mortality and long-term exposure to ambient air pollution: Ongoing analyses based on the American Cancer cohort. *Journal of Toxicology and Environmental Health Part A* 2005, 68:1093–1109.

Lall, R., Ito, K., Thurston, G.D. Distributed lag analyses of daily hospital admissions and source-apportioned fine particle air pollution. *Environmental Health Perspectives*, 2011, 119(4):455–460.

Last, J., Trouton, K., Pengelly, D. *Taking Our Breath Away: The Health Effects of Air Pollution and Climate Change.* Vancouver, British Columbia, Canada: David Suzuki Foundation, 1988.

McCarthy, J.J., Canziani, O.F., Leary, N.A., Dokken, D.J., White, K.S. (eds.) Climate change 2001: Impacts, adaptation & vulnerability. Contribution of Working Group II to the Third Assessment Report of the Intergovernmental Panel on Climate Change. Cambridge, U.K.: Cambridge University Press, 2001.

Miller, W., Robinson, L.A., Lawrence, R.S. (eds.) Committee to evaluate measures of health benefits for environmental, health, and safety regulation board on health care services. Valuing Health for Regulatory Cost-Effectiveness Analysis. Washington, DC: Institute of Medicine, National Academies Press, 2006.

Moore, J.C., Jevrejeva, S., Grinsted, A. Efficacy of geoengineering to limit 21st century sea-level rise. *Proceedings of National Academic Science USA* 2010, 107(36):15699–15703.

National Academy of Sciences (NAS). *Radiative Forcing of Climate Change: Expanding the Concept and Addressing Uncertainties.* Washington, DC: National Research Council. Committee on Radiative Forcing Effects on Climate, Climate Research Committee, 2005.

Ostro, B., Feng, W.Y., Broadwin, R., Green, S., Lipsett, M. The effects of components of fine particulate air pollution on mortality in California: Results from CALFINE. *Environmental Health Perspectives* 2007, 115(1):13–19.

Ostro, B., Lipsett, M., Reynolds, P. et al. Long-term exposure to constituents of fine particulate air pollution and mortality: Results from the California teachers study. *Environmental Health Perspectives* 2010, 118(3):363–369.

Ozkaynak, H., Thurston, G.D. Associations between 1980 U.S. mortality rates and alternative measures of airborne particle concentration. *Risk Analysis* 1987, 7:449–460.

Peters, A., Dockery, D.W., Muller, J.E., Mittleman, M.A. Increased particulate air pollution and the triggering of myocardial infarction. *Circulation* 2001, 103(23):2810–2815.

Pope, C.A. III, Burnett, R.T., Thun, M.J. et al. Lung cancer, cardiopulmonary mortality and long-term exposure to fine particulate air pollution. *Journal of American Medical Association* 2002, 287(9):1132–1141.

Rasch, P.J., Tilmes, S., Turco, R.P. et al. An overview of geoengineering of climate using stratospheric sulphate aerosols. *Philosophical Transactions of the Royal Society A* 2008, 366:4007–4037.

Samoli, E., Peng, R., Ramsay, T. et al. Acute effects of ambient particulate matter on mortality in Europe and North America: Results from the APHENA study. *Environmental Health Perspectives* 2008, 116(11):1480–1486.

Smith, K.R., Haigler, E. Co-benefits of climate mitigation and health protection in energy systems: Scoping methods. *Annual Review of Public Health* 2008, 29:11–25.

Smith K.R., Jerrett, M., Anderson, H.R. et al. Public health benefits of strategies to reduce greenhouse-gas emissions: Health implications of short-lived greenhouse pollutants. *Lancet* 2010, 374(9707):2091–2103.

Spira-Cohen, A., Chen, L., Kendall, M., Lall, R., Thurston, G.D. Personal exposures to traffic-related air pollution and acute respiratory health among Bronx school children with asthma. *Environmental Health Perspectives*, 2011, 119(4):559–565.

Stern, N. *The Economics of Climate Change: The Stern Review*. New York: Cambridge University Press, 2007.

Stratus Consulting. *Stratus Consulting: Air Quality Valuation Model Documentation, for Health Canada*. Boulder, CO: Stratus Consulting, Inc., 1999.

Swart, R., Amann, M., Raes, F., Tuinstra, W. A good climate for clean air: Linkages between climate change and air pollution: An editorial essay. *Climate Change* 2004, 66:263–269.

Thurston, G.D. Air pollution, human health, climate change and you. *Thorax* 2007, 62:748–749.

Thurston, G.D., Ito, K., Kinney, P.L., Lippmann, M. A multi-year study of air pollution and respiratory hospital admissions in three New York State metropolitan areas: Results for 1988 and 1989 summers. *Journal of Exposure Analysis and Environmental Epidemiology*, 1992(4):429–450.

Thurston, G.D., Ito, K., Mar, T. et al. Workgroup report: Workshop on source apportionment of particulate matter health effects—Intercomparison of results and implications. *Environmental Health Perspectives* 2005, 113(12):1768–1774.

Tuck, A.F., Donaldson, D.J., Hitchman, M.H. et al. On geoengineering with sulphate aerosols in the tropical upper troposphere and lower stratosphere. *Climatic Change* 2008, 90:315–331.

USEPA. *The Benefits and Costs of the Clean Air Act 1970 to 1990*. Washington, DC: USEPA, 1997.

USEPA. *The Benefits and Costs of the Clean Air Act 1990 to 2010*. Washington, DC: USEPA, 1999. EPA-410-R-99-001.

USEPA. *Regulatory Impact Analysis for the Final Clean Air Interstate Rule*. Washington, DC: USEPA, 2005. EPA-452/R-05-002.

Vennemo, H., Aunan, K., Jinghua., F. et al. Domestic environmental benefits of China's energy-related CDM potential. *Climate Change* 2006, 75:215–239.

Voorhees, S.A. Benefits analysis of particulate matter control programs—A case study of Tokyo. *Journal of Risk Research* 2005, 8:331–329.

Voorhees, A.S., Sakai, R., Araki, S., Sato, H., Otsu, A. Benefits analysis of nitrogen dioxide control programmes: A case-study of Chiyoda-ku, Tokyo. *Journal of Environmental Planning and Management* 2001, 44:149–165.

Walsh, M.P. Ancillary benefits for climate change mitigation and air pollution control in the world's motor vehicle fleets. *Annual Review of Public Health* 2008, 29:1–9.

Wang, X., Smith, K.R. *Near-Term Health Benefits of Greenhouse Gas Reductions: A Proposed Assessment Method and Application in Two Energy Sectors of China*. Geneva, Switzerland: World Health Organization, 1999. WHO/SDE/PHE/99.1.

Working Group on Public Health and Fossil Fuel Combustion. Short-term improvements in public health from global-climate policies on fossil-fuel combustion: An interim report. *Lancet* 1997, 350:1341–1349.

Zhou, J., Ito, K., Lall, R., Lippmann, M., Thurston, G.D. Time-series analysis of mortality effects of fine particulate matter components in Detroit and Seattle. *Environmental Health Perspectives* 2011, 119(4):461–466.

14 Health Effects of Metals in Ambient Air Particulate Matter

Morton Lippmann

CONTENTS

KEYWORDS

CAPs   Concentrated ambient air particulate matter
Cardiovascular effects   Cardiovascular system responses to inhaled particulate matter
PM Components   Chemical component masses within overall PM mass
Pulmonary effects   Pulmonary system responses to inhaled particulate matter
PM_x Particulate matter below cut-point x in aerodynamic diameter
Ultrafine PM   PM with diameters below $0.1\,\mu m$, expressed as either mass or
                                     number concentration
Fine PM   PM mass concentration with aerodynamic diameters below $2.5\,\mu m$
Accumulation mode PM   PM mass concentration with aerodynamic diameters between 0.1
                                     and $2.5\,\mu m$
Coarse thoracic PM   PM mass concentration with aerodynamic diameters between 2.5
                                     and $10\,\mu m$

14.1   INTRODUCTION

Studying the health effects of ambient air pollution has been a challenging endeavor for environmental health scientists for many reasons. Epidemiologists have documented significant associations between the routinely measured mass concentrations of particulate matter (PM) and excess mortality, morbidity, lost function, and lost time at work or school, and these associations are stronger for $PM_{2.5}$, that is, particles with aerodynamic diameters less than $2.5\,\mu m$, than those for PM in other particle size ranges, or those for routinely measured pollutant gases (USEPA 2004, 2008). Although the relative risks (RRs) for mortality and nonscheduled hospital admissions are relatively small, requiring sophisticated mathematical models for analysis in epidemiological studies, the populations studied and the populations at risk are quite large (Miller et al., 2007; Eftim et al., 2008; Schwartz et al., 2008; Pope et al., 2009), resulting in large public health impacts (e.g., thousands of cases annually in the United States). The bulk of this risk appears to be borne by the elderly, those in poor health, or both.

It seems highly unlikely that the health effects are caused by nonspecific PM mass. Rather, it is likely that some specific chemical components within the PM mixtures are more potent than other components. The situation is complicated by the fact that PM is present in the air over a broad range of chemical compositions and particle sizes. Coarse dust particles with aerodynamic diameters above $10\,\mu m$ do not normally penetrate beyond the larynx, have not been associated with health effects due to ambient air pollution exposures, and are not routinely monitored. Particles with aerodynamic diameters below $10\,\mu m$, known as PM_{10}, can deposit along the conductive airways in the thorax, and nearly all of those with aerodynamic diameters below $2.5\,\mu m$ penetrate into the gas exchange region where particle retention is much greater than for those that deposit on the conductive airways. A mucociliary blanket covering the conductive airways facilitates fairly rapid particle removal to the gastrointestinal tract. Furthermore, the smaller particles ($PM_{2.5}$) are chemically quite different from the larger ones. The larger particles are mostly of mineral origin, while the $PM_{2.5}$ is composed largely of diesel engine soot and particles formed by chemical transformations in the atmosphere from fossil fuel combustion products (both solid and gaseous) and organic vapors. Most of these particles initially form as ultrafine PM (UFP), but rapidly aggregate into the accumulation mode, that is, PM in the $0.1–2.5\,\mu m$ size range. Suspicion concerning adverse health effects has centered on fossil fuel combustion products, including elemental and organic carbon (EC/OC) and on mineral ash in the form of inorganic compounds containing metals within the $PM_{2.5}$. A focus has often been on transition metals, such as iron (Fe), vanadium (V), nickel (Ni), chromium (Cr), copper (Cu), and zinc (Zn), on the basis of their ability to generate reactive oxygen species (ROS) in biological tissues. Most of the evidence pointing to the biological effects of metals, as well as of EC and OC has come from studies involving exposures of laboratory animals *in vivo*, or of cells *in vitro*.

While no studies involving inhalation exposures of laboratory animals *in vivo* or of cells *in vitro* to pure chemicals and their compounds, at doses with environmental relevance, have produced adverse health effects, some toxicological studies using high PM mass exposures of diluted tailpipe emissions, especially whole diesel engine exhaust (WDE) or to source-related PM mixtures containing multiple metals, such as residual oil fly ash (ROFA), coal fly ash (CFA), and concentrated ambient particles (CAPs), have produced effects that appear to be related to their relatively low contents of metals and carbonaceous material. However, it has been difficult to determine the roles played by the individual components in the effects observed. Also, many laboratory-based studies have used resuspended dusts at relatively high mass concentrations, and the relevance of the effects observed to human ambient air exposures at much lower PM mass levels is, therefore, uncertain. While effects found in high-dose laboratory *in vitro* exposures have occasionally been suggested to also occur with exposures near to ambient concentrations (e.g., inflammatory indicators in the CAPs study of Maciejczyk and Chen [2005] and Maciejczyk et al. [2010]), more often those effects have not been found (e.g., no abnormal levels of cytokines in human volunteers in the CAPs exposure study of Ghio et al. [2000]).

Studies of the effects of relatively low concentrations of airborne PM components in humans have all involved complex mixtures. These include those short-term inhalation exposures to (1) CAPs in healthy human volunteers, and (2) diluted WDE, and (3) natural exposures to ambient PM, where data from simultaneous daily and/or seasonal or annual average PM compositional analyses were available for time-series and cross-sectional studies of effects in large human populations. Due to the limitations of statistical power in such natural population studies, the epidemiological analyses have focused more on identifying the contributions to the effects of factors or source-related mixtures than of individual components within the mixtures. Additional information comes from laboratory studies that have involved instillation of particle suspensions into human lungs and subsequent analyses of bronchoalveolar lavage fluid (BALF) samples for particle retention and biomarkers of effects.

Studies of the effects of relatively low concentrations of airborne PM components in laboratory animals that involve complex mixtures include: (1) short-term inhalation exposures to CAPs in mice, rats, and dogs; (2) subchronic inhalation exposures of CAPs to mice and rats; and (3) inhalation and intratracheal (IT) lung instillation of components and source-related mixtures.

A major objective of this chapter is to combine the analyses of the experimental studies with CAPs and other ambient air PM components in humans and other animals with the associations between ambient air concentrations of PM and its components to determine the nature and extent of the effects of ambient air PM and its components of major organ systems and their cross-species consistency, and to identify, as possible, the more potent PM components. Mauderly and Chow (2008) have reviewed the health effects on carbonaceous compounds in ambient air, and Chen and Lippmann (2009) have reviewed the health effects of metals in ambient air. The emphasis in this chapter will be on the metal content of the PM on the basis that they appear to be more potent, at typical ambient air concentrations, than those of EC and OC.

It is important to remember that all three particle size ranges are chemically nonspecific pollutant classes, and may originate from, or been derived from, various emission source types. Thus, PM toxicity may well vary, depending on its size distribution, source, and chemical composition. If the PM toxicity could be associated with specific source signatures, then health effects research could be better focused on specific PM components that come from those sources and specific biological mechanisms could be postulated for further consideration by toxicological studies. PM health effects research is, therefore, now being increasingly focused on source apportionment of PM using chemical speciation data, and this review of the CAPs literature emphasizes those CAPs studies that used PM compositional data to identify associations of exposures to PM source categories or to individual PM components that have been associated with health-related effects.

In addition to this introduction, this chapter contains sections discussing studies in humans (Section 14.2), studies in laboratory animals and *in vitro* (Section 14.3), and unresolved issues and conclusions (Section 14.4).

14.2 REVIEW OF STUDIES OF THE EFFECTS OF PM AND COMPONENTS IN HUMANS

14.2.1 INTRODUCTION

This review of human responses is focused on CAPs inhalation studies and their health effects, with an emphasis on studies that identify the particle size ranges and components most closely associated with the observed effects. Lung instillation studies involving PM suspensions of materials found in ambient air have also been reviewed. Finally, reported associations between ambient air PM size fractions and components and human health-related responses in natural settings have been reviewed. These include studies of limited numbers of individuals where there is information related to personal exposures and effects (panel studies) and larger-population studies that rely on central site air monitoring data and grouped responses.

14.2.2 HUMAN CAPs INHALATION STUDIES

Ghio et al. (2000) reported that the 2h CAPs exposures in Chapel Hill, NC caused neutrophilic inflammation in the lungs and increased fibrinogen levels in the blood. Of the soluble components extracted from the air-sampling filters, Fe, As, Se, and $SO_4^=$ were highly correlated with the $PM_{2.5}$ mass concentration, while Ni and Cu were least correlated. In terms of biological responses, an $Fe/Se/SO_4^=$ factor was associated with increased BALF percentage of polymorpho-neutrophils (PMNs) and a Cu/Zn/V factor with increased blood fibrinogen. The increase in plasma fibrinogen correlated with decreases in PMNs and platelets, consistent with a state of systemic inflammation and increased platelet aggregation.

Normotensive, nonsmoking, healthy volunteers in Toronto, Ontario, Canada were exposed to $PM_{2.5}$ CAPs and O_3 for which there were data on $PM_{2.5}$ composition. Brachial artery diameter (BAD), an index of cardiovascular response, decreased 0.09 mm compared to FA. There were no significant responses in endothelial-dependent flow-mediated dilatation (FMD), endothelial-independent nitroglycerin-mediated dilatation (NMD), or blood pressure (Brook et al., 2002; Urch et al., 2004). The linear regression analyses of change in BAD in relation to $PM_{2.5}$ components yielded p-values of 0.04 for OC, 0.05 for EC, 0.06 for Cd, 0.09 for K. The p-values were between 0.13 and 0.17 for Zn, Ca, and Ni, and values were even larger for all of the other measured components. The p-value for $PM_{2.5}$ as a whole was 0.40. In a follow-up study, there was a significant increase (6mm Hg) in diastolic blood pressure in those exposed to O_3 plus CAPs (p=0.013). In relation to the $PM_{2.5}$ components, there was a significant association (p=0.009) with OC, whereas the association with $PM_{2.5}$ mass was not significant (p=0.27). In this study, the EC and metals were not significantly associated with BAD constriction or blood pressure (Urch et al., 2005).

14.2.3 SHORT-TERM RESPONSES TO AMBIENT AIR PM INHALATION EXPOSURES IN HUMAN PANEL STUDIES

A panel study by Sorensen et al. (2005) included 49 students in Copenhagen. Their personal exposure to soluble V and Cr was associated with significant increases in oxidative stress and DNA damage (as measured by 8-oxodG concentrations in lymphocytes). Other soluble metals (Fe, Ni, Cu, and Pt) did not.

Lanki et al. (2006) studied the influence of ambient air $PM_{2.5}$ component exposures on exercise-induced ischemia in 45 elderly nonsmokers with stable coronary heart disease in Amsterdam (the Netherlands), Erfurt (Germany), and Helsinki (Finland). Two $PM_{2.5}$ source classes (traffic and long-range transport) were associated with ST-segment depression during submaximal exercise testing in a clinical laboratory. In a multi-pollutant model, with which the authors were able to separate effects of secondary $SO_4^=$ from effects of vehicular emissions, only the traffic emissions were significantly associated with the effect. The authors also examined whether potentially toxic transition metals

(Fe, Cu, Zn, and V) might be associated with ST-segment depression, given that both these, and OC, may have the capability to induce oxidative stress in the lung. However, when adjusted for ABS (absorbance, a measure of EC emissions from motor vehicles), none of the metals that were measured were associated with ST-segment depression, while the ABS associations remained significant with only slight variation.

Riediker et al. (2004a,b) studied a panel of nine nonsmoking healthy male highway patrol officers (ages 23–30) in North Carolina over four late-shift tours of duty. Their patrol cars had air samples that were analyzed for vapor and $PM_{2.5}$ components each day. $PM_{2.5}$ components were correlated to cardiac and blood parameters measured 10 and 15 h after the work shift. They reported that in-vehicle $PM_{2.5}$ mass was associated with changes in cardiac parameters—blood proteins associated with inflammation, hemostasis, and thrombosis—and increased red blood cell (RBC) volume. In a follow-up study, Riediker (2007) used data on $PM_{2.5}$ components. Those associated with health-related endpoints were Ca (increased uric acid and von Willebrand factor [vWF], and decreased protein C), Cr (increased WBC and IL-6), Cu (increased blood urea nitrogen, mean cycle length of normal R–R intervals), and S (increased ventricular ectopic beats). Control for the gaseous pollutants had little effect on the effect estimates.

Gent et al. (2009) studied 149 children with physician-diagnosed asthma and symptoms or medication use within the previous 12 months who were living in New Haven, CT and vicinity. Air sampling filters were collected daily and analyzed for trace elements by x-ray fluorescence (XRF) and black carbon (BC) by light reflectance. Using factor analysis/source apportionment, they identified six sources of $PM_{2.5}$. They were motor vehicle, road dust, S (for regional $PM_{2.5}$), biomass burning, oil combustion, and sea salt. They attributed 42% of the $PM_{2.5}$ to the motor vehicle source, and 12% to road dust. Increased likelihood of symptoms and inhaler use was largest for 3 day averaged exposures, with a 10% increased likelihood of wheeze per 5 µg/m³ of the motor vehicle source, and a 28% likelihood increase for shortness of breath associated with road dust. There were no associations with increased health outcome risks for $PM_{2.5}$ *per se* or the other source factors.

In summary, a broad variety of short-term cardiovascular effects have been significantly associated with peaks in ambient air concentrations of $PM_{2.5}$ and/or one or more of its chemical components in panel studies. In young, healthy highway patrol officers, $PM_{2.5}$, but not gaseous pollutants, was associated with changes in cardiac parameters such as Ca (vWF), Cr (WBC and IL-6), Cu (R-R intervals). These various cardiac-related responses, while not necessarily associated with specific $PM_{2.5}$ components, are certainly consistent with the excess cardiovascular mortality and morbidity in the ever-growing air pollution health effects literature.

In addition, peaks in ambient air PM were associated with a variety of pulmonary effects among children with physician-diagnosed asthma and symptoms or medication use within the previous 12 months living in New Haven, CT, and vicinity. The sources of $PM_{2.5}$ were motor vehicle, road dust, S for regional $PM_{2.5}$, biomass burning, oil combustion, and sea salt, with 42% of the $PM_{2.5}$ attributed to the motor vehicle source, and 12% to road dust. Increased likelihood of symptoms and inhaler use was largest for 3 day averaged exposures, with a 10% increased likelihood of wheeze per 5 µg/m³ of the motor vehicle source, and a 28% likelihood increase for shortness of breath associated with road dust. There were no associations with increased health outcome risks for $PM_{2.5}$ *per se* or the other source factors (Gent et al. 2009).

14.2.4 Large Population-Based Studies in Humans Dealing with Responses to PM Components

In a Hong Kong, China sulfur-in-fuel intervention study (Hedley et al., 2002), SO_2, Ni, and V fell promptly and substantially after the intervention, while other criteria pollutants and metals did not fall (Hedley et al., 2004). Thus, it is possible that the large changes in the three pollutants that fell may account for at least some of the changes in the intervention-related cardiovascular mortality and bronchial hyperreactivity in this study.

Another study, also of interest in this context, was that of Lippmann et al. (2006). It noted the high daily mortality associated with PM_{10} in New York City (NYC) in the 90-city National Mortality and Morbidity Air Pollution Study (NMMAPS), and showed that in those 60 cities with speciation data, only Ni and V were significantly associated with NMMAPS mortality. The NMMAPS mortality coefficient for NYC was 3.8 times higher than the average and the Ni in NYC was 9.5 times higher than the U.S. average.

Dominici et al. (2007) extended this analysis of NMMAPS data in relation to $PM_{2.5}$ speciation in terms of additional cities and years of data and confirmed the associations of daily mortality coefficients with Ni and V, but noted that with the exclusion of the NYC data, the overall association was no longer statistically significant.

Lipfert et al. (2006) found V and Ni to be significantly associated with long-term mortality, but that the traffic density variable was more robust and had larger explanatory value. Other studies also point to traffic emissions and particular metals as both having significant associations with health endpoints (Janssen et al., 2002; Grahame and Hidy, 2004).

Studies done largely in the absence of any but light traffic tend to show only the effects of particular metals (Maciejczyk and Chen, 2005). It is important to recognize that studies that have the capability of examining higher levels of both metals and of vehicular emissions tend to find both of health importance, and if the exposure to vehicular emissions is of good quality, tend to find little else of health significance (Ebelt et al., 2005; Gold et al., 2005; Schwartz et al., 2005).

Janssen et al. (2002), in a study on the influence of air conditioning as a modifier of hospital admission in relation to PM_{10} concentrations, modeled source contributions to ambient air PM using emissions data, rather than data from measured individual components. Cardiovascular admissions were significantly associated with a number of sources (highway vehicles, oil combustion, and metal processing), but there were no significant associations of the sources with COPD or pneumonia admissions.

Franklin et al. (2008) modeled EPA air quality speciation data available for every third or sixth day and daily mortality data for 25 U.S. cities between 2000 and 2005 to determine how the associations between $PM_{2.5}$ and morality were modified by $PM_{2.5}$ composition. They first determined the association between daily $PM_{2.5}$ and mortality, and then used meta-regression to examine how the pooled association was modified by community and by season-specific $PM_{2.5}$ composition. The association was increased when the $PM_{2.5}$ mass had a higher proportion of Al, As, $SO_4^=$, Si, and Ni. The extent to which the intercity heterogeneity in the PM association could be explained by was greatest for Al (45%), Ni (41%), and in a multivariate model by a combination of Al, Ni, and $SO_4^=$ or Al, Ni, and As (100%). These findings suggest that the sources of soil dust (indexed by Al, Si), residual oil combustion (indexed by Ni), and coal combustion (indexed by As and/or $SO_4^=$) are especially influential.

Zhou et al. (2011) obtained daily $PM_{2.5}$ filters from Detroit and Seattle for the years 2002–2004 and analyzed trace elements using XRF and BC using light reflectance. They used Poisson regression and distributed lag models to estimate excess hospitalization for all causes and for cardiovascular and respiratory diseases, with adjustments for time-varying covariates. The $PM_{2.5}$ components and gaseous pollutants most closely associated with cardiovascular and respiratory hospitalization in Detroit were secondary aerosols and traffic markers. In Seattle, the component species most closely associated with hospitalization were those for cold season traffic and other combustion sources, such as residual oil and wood burning.

Hsu et al. (2011) studied the influence of components of $PM_{2.5}$ and PM_{10} on heart rate and pulmonary function on panels of COPD patients in NYC and Seattle. Nickel in NYC $PM_{2.5}$ was the only component that produced a significant response, that is, an increase in heart rate.

In summary, a broad variety of cardiovascular and pulmonary health effects have been significantly associated with peaks in ambient air concentrations of $PM_{2.5}$ and/or one or more of its chemical components in a variety of large population cohort studies. These cohorts include the thorough studies of morbidity and longevity in relation to ambient air concentrations of PM_{10},

$PM_{2.5}$, and their components and associated gaseous criteria pollutants (Harvard 6-Cities, American Cancer Society, U.S. Military Veterans, NHANES III, MESA, Women's Health Initiative [WHI], and the Southern California Children's Health Study). They also include studies of the effects of air quality interventions and their health consequences (i.e., Hong Kong S-in-fuel intervention), and examination of newly available speciation data in relation to available daily mortality data from the NMMAPS study. The results of these various studies continue to implicate $PM_{2.5}$ as a useful index of excess mortality risk, and many of them, having access to speciation data, implicate transition metals within the $PM_{2.5}$ as being especially likely to be causal factors for the associations.

14.2.5 EXPOSURES OF HUMAN VOLUNTEERS VIA INTRATRACHEAL INSTILLATION

The only other studies in human volunteers for which component analyses were available involved administration of PM by IT instillation of particle suspensions.

On the basis that Fe is the most abundant of the transition metals in ambient air, Lay et al. (1998) and Ghio et al. (1998) instilled ~5 mg doses containing both soluble and insoluble 2.6 μm Fe particle agglomerates suspended in saline into the lungs of volunteer subjects to investigate oxidative stress. BALF samples were collected from 1 to 91 days later. At 1 day, Lay et al. (1998) and Ghio et al. (1998) reported inflammatory responses. Recent research has suggested that traffic-generated PM can account for pulmonary effects, with Gent et al. (2009) showing that the motor vehicle source was associated with a $10\%/5\,\mu g^3$ increase in wheeze in asthmatic children, while the road dust source was associated with a 28% increase in shortness of breath. This response is consistent with the findings reported by Gottipulo et al. (2008) based on the instillation of two kinds of tire dust into the lungs of male WKY rats in relation to the elemental composition. There were increases in BALF markers of inflammation and injury, and similar effects were seen for instilled Zn and Cu. Thus, the acute pulmonary effects of tire dust could be due to the metals (Ghio et al. 1988). reported decreased transferrin concentrations and increased concentrations of ferritin and lactoferrin. By 4 days, iron homeostasis was normal.

These kinds of tests have proved to be more informative when they were applied to real-world PM samples that were associated with adverse effects in human populations, such as those from the Utah Valley, where there was a 14 month-long strike at a steel mill complex. There were significantly lower rates of mortality and hospital admissions during the strike than in the preceding and following years (Pope, 1989, 1991; Pope et al., 1992). Analyses of the PM collected on air sampling filters during those 3 years indicated that the concentrations of many airborne metal PM components were also significantly lower during the strike interval than in the preceding and following years (Frampton et al., 1999; Dye et al., 2001; Ghio and Devlin, 2001). Extracts of metals from sampling filters were used to test whether soluble components or ionizable metals, which accounted for 20% of the PM mass, could be responsible for the adverse health effects.

14.3 REVIEW OF AMBIENT PARTICULATE MATTER STUDIES IN LABORATORY ANIMALS AND *IN VITRO*

14.3.1 INTRODUCTION

This review of laboratory animal responses is focused on CAPs inhalation studies and their health effects, with an emphasis on studies that identify the particle size ranges and components most closely associated with the observed effects. Lung instillation studies involving PM suspensions of materials found in ambient air have been covered. Associations between ambient air PM size fractions and components and human health-related responses in panel and larger-population studies were presented in Section 14.2, including those indicating responses to chronic exposures. Studies in animals have not been limited to short-term exposures and acute responses. Section 14.5 compares the effects of chronic exposure in the animal models to those associated with chronic exposure in the epidemiological studies summarized in Section 14.2.

Epidemiological studies have clearly established that $PM_{2.5}$ air pollution is associated with cardiopulmonary effects. However, because of the very low ambient air concentration levels measured in these studies, particularly of the trace metal components, the biological plausibility of these epidemiologically demonstrated associations needs to be substantiated. While we would prefer to substantiate them in human controlled exposure studies with an experimental design, such studies are not feasible and animal experimental studies can serve the purpose effectively. To date, while there are many toxicological studies that investigated the response of animals to ambient PM by inhalation or IT instillation, only a few had investigated the contributions from specific air pollution components, either as an individual compound or as part of a mixture, in producing adverse health effects. It is, therefore, critical to systematically investigate the potential cardiopulmonary effects of components of ambient PM in different regions of the United States, since PM of different composition and from different sources may vary markedly in their potency for producing adverse health effects. In this section, we first discuss short-term inhalation exposure studies (up to 1 week in duration). This discussion is followed by a review of longer-term CAPs exposure studies in animals.

14.3.2 SHORT-TERM CAPs INHALATION STUDIES

In order to determine if CAPs inhalation can induce cardiopulmonary effects, Clarke et al. (2000) investigated pulmonary inflammatory and hematological responses of canines after exposure to Boston $PM_{2.5}$ CAPs. For pulmonary inflammatory studies, normal dogs were exposed in pairs to either CAPs or filtered air (paired studies) for 6h/day on 3 consecutive days. For hematological studies, dogs were exposed for 6h/day for 3 consecutive days with one receiving CAPs while the other was simultaneously exposed to filtered air; crossover of exposure took place the following week (crossover studies). No statistical differences in biologic responses were found when all CAPs and all sham exposures were compared. However, the variability in biologic response was considerably higher with CAPs exposure. Subsequent exploratory graphical analyses and mixed linear regression analyses suggested associations between CAPs constituents and biologic responses. Factor analysis was applied to the compositional data from paired and crossover experiments to determine elements consistently associated with each other in CAPs samples. In paired experiments, four factors were identified; in crossover studies, a total of six factors (V/Ni, S, Al/Si, Br, Na/Cl, and Cr) were observed. Increased BAL PMN percentage, total peripheral WBC counts, circulating PMNs, and circulating lymphocytes (LYM) were associated with increases in the Al/Si factor. Increased PMNs and increased BAL macrophages were associated with the V/Ni factor. Increased BAL PMNs were associated with the Br/Pb factor only when the compositional data from the third day of CAPs exposure were used. Decreases in RBC counts and hemoglobin levels were correlated with the S factor. BAL or hematologic parameters were not associated with increases in total CAPs mass concentration. In terms of significant individual components, $SO_4^=$ was associated with increased WBC; BC, Al, Mn, Si, Zn, Ti, V, Ni, and Fe were associated with increased PMNs; Na was associated with increased LYM; and Al, Mn, and Si were associated with decreased LYM. These data suggest that specific components of CAPs may be responsible for biologic responses, but the lack of overall statistically significant alterations in pulmonary and systemic responses diminished the impact of this study.

Saldiva et al. (2002) studied the effects of CAPs inhalation on lung inflammation. They exposed normal and bronchitic rats to Boston $PM_{2.5}$ CAPs or filtered air for 5h/day for 3 days. The CAPs produced significant pulmonary inflammation. Some inorganic CAPs components (Si, V, Pb, $SO_4^=$, and Br) were significantly associated with increases in PMN in BALF and lung tissue.

Rhoden et al. (2004) reported that N-Acetylcysteine (NAC) could prevent lung inflammation due to Boston $PM_{2.5}$ CAPs inhalation. They also reported the results of regression analyses showing strong associations between increases in thiobarbituric reactive substances (TBARS) accumulation and the CAPs content of Al, Si, and Fe, and between BALF PMN count and Cr, Zn, and Na.

Gurgueira et al. (2002) demonstrated that oxidative stress can be induced by exposure to high levels of metals in Boston CAPs, and could be responsible for changes in cardiac parameters. *In vivo* CAPs exposures also triggered adaptive responses.

Hamada et al. (2002) exposed female BALB/c mice for 30 min to an aerosol nebulized from a solution of PM components dissolved from ROFA that had been collected at a Boston power plant, Penh, an index of airway hyperresponsiveness (AHR), was increased in a time- and dose-related manner, peaking at 48 h postexposure. PMNs in BAL peaked at 12 h postexposure. A simulated ROFA extract, containing the same concentrations of Ni, V, Zn, Co, Mn, and Cu as the ROFA, produced the same AHR response, but the summed responses to each metal separately did not.

Gurgueira et al. (2002) exposed adult SD rats exposed to Boston $PM_{2.5}$ CAPs at 300 μg/m³ for 5 h and showed significant oxidative stress in the lung and heart, but not in the liver. The increase in the lung concentrations of ROS upon exposure to CAPs was rapid, indicating an almost immediate effect of PM, or PM components, on the intracellular sources of free radicals. Furthermore, the transient nature of these increases points to a reversible interaction of PM components with cellular targets. Both observations are compatible with Fenton-type reactions catalyzed by transition metals, redox-cycling processes, or biochemical changes triggered by non-covalent binding to membrane receptors. Using single-component regression analysis, increases in chemiluminescence (an index of oxidant load) showed strong associations with the CAPs content of Fe, Mn, Cu, and Zn in the lung and with Fe, Al, Si, and Ti in the heart. The oxidant stress imposed by 5 h exposure to CAPs was associated with slight, but significant, increases in the lung and heart water content, and with increased serum levels of lactate dehydrogenase (LDH), indicating mild damage to both tissues. In addition, CAPs inhalation also led to tissue-specific increases in the activities of the antioxidant enzymes.

Morishita et al. (2004) exposed normal and allergic Brown Norway rats to $PM_{2.5}$ CAPs for 10 h at concentrations ranging from 300 to 650 μg/m³ in a mobile laboratory in Detroit. The allergic rats had, compared to the normal rats, increased pulmonary retention of La, V, Mn, and S, as well as increased lung inflammation. Using source-apportionment analyses, Morishita et al. (2006) concluded that the pattern of the airway responses was likely associated with local refineries and incinerators, and independent of $SO_4^=$ and $PM_{2.5}$ mass.

Lippmann et al. (2006) exposed ApoE$^{-/-}$ mice to $PM_{2.5}$ CAPs in Sterling Forest (Tuxedo, NY) on weekdays for 6 h/day, for 6 months at an average mass concentration of 85 μg/m³, and cardiac function was monitored continuously over the 6 months and CAPs composition was determined for each exposure day. Most of the results of this study are described in the following section on long-term animal inhalation exposures. In effect, this was a time-series study as well as a chronic effects study. Exposures to Ni, Cr, and Fe were much higher on 14 days than on the other 89 exposure days, corresponding to days with unusually high HR and unusually low HRV. In addition, V was lower than normal on the days with high Ni, since the source of high Ni was a distant Ni smelter rather than residual oil combustion, the usual source of elevated Ni concentrations. The authors attributed the acute effects on cardiac function to peaks in Ni.

Kodavanti et al. (2000a,b) performed a series of 2 and 3 day 6 h exposures to FA, $PM_{2.5}$ CAPs, and aerosolized ROFA of SD rats in Research Triangle Park, NC (RTP) using rats with and without SO_2-induced bronchitis. The CAPs concentrations ranged from 475 to 907 μg/m³, and the ROFA concentration was 1 mg/m³. The CAPs exposures produced some CAPs-concentration-related pulmonary injury in the bronchitic rats, but not in the healthy rats, and the FA and ROFA did not produce measurable effects in either group of rats. The concentrations of leachable $SO_4^=$ and metals in the CAPs were not associated with the effects.

An *in vivo* inhalation study by Kodavanti et al. (2005), utilizing two different rat strains, made a number of important findings They exposed two different strains of rats (spontaneously hypertensive [SHR] and Wistar-Kyoto [WKY]) to CAPs from ambient air in RTP, concentrated by a factor of 40–60 times. The CAPs were drawn from an area in reasonably close proximity to a major freeway near the intersection with another major road, suggesting that the effects seen might have been related to vehicular emissions. Plasma fibrinogen levels in SH rats were better correlated with the levels of water-soluble metals, particularly Zn, than mass and other components of the CAPs (Kodavanti et al., 2005). This study demonstrated that strain-specific systemic effects were not linked to high mass but appear to be dependent on CAP chemical composition.

14.3.3 LONGER-TERM CAPs INHALATION STUDIES

A series of longer-term $PM_{2.5}$ CAPs inhalation studies were conducted at NYU's laboratory in Sterling Forest. Because these studies collected both long-term ECG data as well as simultaneous data on $PM_{2.5}$ composition, such studies can have more power to identify possible causal components of ambient $PM_{2.5}$. These NYU subchronic CAPs inhalation studies involved a series of experiments that were used to study both the acute and cumulative effects of daily inhalation exposures to $PM_{2.5}$ CAPs in a mouse model of atherosclerosis. The results of the first of these studies, involving 5–6 months of warm-season daily exposures (5 day/week, 6 h/day to an average CAPs concentration = 110 μg/m³) were described n a special issue of *Inhalation Toxicology* (Chen and Hwang, 2005; Chen and Nadziejko, 2005; Gunnison and Chen, 2005; Hwang et al., 2005; Lippmann et al., 2005a,b; Maciejczyk and Chen, 2005; Veronesi et al., 2005). These papers documented CAPs exposure-associated acute and chronic effects on cardiac function, increased amounts of, and more invasive, aortic plaque, and changes in brain cell distribution and in gene expression markers, as well as data on the effects of daily CAPs exposures *in vitro* on NfkB activation.

The biological plausibility of ambient air PM contributing to changes in brain cell distribution was enhanced by a follow-up study by Sama et al. (2007) in which they conducted *in vitro* assays of the cellular and genomic responses of immortalized microglia cells (BV2) to CAPs collected during the same study described in the special issue of *Inhalation Toxicology*. Two composite samples were applied to the microglia cells; one composed of CAPs from days with high potency (HP) in their stimulation of NF*k*B release in human bronchial epithelial cells and the other from CAPs collected on days with low potency (LP). The LP composites reduced intracellular ATP at doses >250 μg/mL, and depolarized mitochondrial membranes (>6 μg/mL) within 15 min. HP and LP CAPs (>25 μg/mL) differentially affected the endogenous scavengers, glutathione and nonprotein sulfhydryl, after 1.5 h. Both HP and LP CAPs stimulated the release of proinflammatory cytokines TNF*a* and IL-6 after 6 h of exposure. Microarray analysis of both HP and LP exposed microglia (75 μg/mL) identified 3200 (HP) and 160 (LP) differentially expressed (up- and downregulated) genes relative to the media controls. The results implicate Ni and/or V in the production of these effects in that these two metals were much higher in concentration in the HP than the LP CAPs. The biological plausibility for fine and ultrafine PM in ambient air to be translocated, from the lungs to the brain, and to have neurological effects, is supported in a review paper by Peters et al. (2006).

To investigate the contributions of $PM_{2.5}$ components to cardiovascular effects, Lippmann et al. (2005b) used the 5 months of daily 6-h source apportionments of Maciejczyk and Chen (2005), the continuous HR data for exposure days (weekdays only) used in Hwang et al. (2005), and the corresponding HRV data used in Chen and Hwang (2005) to determine the source-related $PM_{2.5}$ components' associations with HR and HRV. They used HR and HRV data collected on normal (C57) mice and a murine model for atherosclerotic disease (ApoE$^{-/-}$) (Chen and Hwang, 2005; Hwang et al., 2005). Daily 6 h $PM_{2.5}$ air samples were also collected and analyzed by XRF, permitting attribution to major $PM_{2.5}$ source categories (secondary $SO_4^=$, suspended soil, residual oil combustion, and a remainder category, which was largely due to long-range transported motor vehicle traffic). Lippmann et al. (2005c) examined associations between these $PM_{2.5}$ components and both HR and HRV for three different daily time periods: (1) during exposure, (2) the afternoon following exposure, and (3) late at night. For HR, there were significant transient associations (p = <0.01) for secondary sulfate during exposure, and for residual oil combustion (predominantly V and Ni) in the afternoon. For HRV, there were comparable associations with suspended soil (predominantly Si, Al, Ca) in the afternoon and for both residual oil combustion and traffic (Br, Fe, elemental carbon) late at night. The biological bases for these various associations and their temporal lags are not known at this time, but may have something to do with the differential solubility of the $PM_{2.5}$ components at the respiratory epithelia, and their access to cells that release mediators that reach the cardiovascular system.

One important parameter that was not addressed in the above study, but that could influence metals' ability in mediating biological response is the extent of soluble metal components present in the $PM_{2.5}$ mass. In a follow-up subchronic $PM_{2.5}$ CAPs inhalation study of $ApoE^{-/-}$ mice at $85\,\mu g/m^3$ (Lippmann et al., 2006), there was a dramatic change in cardiac function in the fall months in the $ApoE^{-/-}$ mice. As previously discussed, the 14 days with northwest winds carried more Ni, Cr, and Fe, but less of the other elemental tracers than the 89 days with winds from all other directions, and were associated with significant increases in HR and significant decreases in HRV (Lippmann et al., 2006). V was lower than normal on the 14 days with unusually high levels of Ni, Cr, and Fe in this mouse study. Back-trajectory analyses from Sterling Forest for the 14 days with northwest winds led through lightly populated areas to Sudbury, Ontario, which is the location of the largest Ni smelter in N. America. At the end of the 6 months of exposure in this study, Sun et al. (2005) compared the mice in the CAPs exposed subgroup on a high fat diet (HF) with those exposed to filtered air (FA). For the CAPs-exposed mice, the plaque area in the aorta was 41.5% versus 26.2% (p=0.001), while for the subgroup on a normal diet, the CAPs-exposed difference versus FA was 19.2% versus 13.2% (p=0.15). Lipid content in the aortic arch in the HF versus NC groups exposed to CAPs was 30% versus 20% (p=0.02). Vasoconstrictor challenges in the thoracic aorta were increased in the CAPs-exposed HF mice versus the FA mice (p=0.03), and relaxation in response to acetylcholine was greater (p=0.04). In addition, HF mice exposed to CAPs had marked increases in macrophage infiltration, expression of inducible NO-synthase, ROS generation, and immunostaining for 3-nitrotyrosine (all with p<0.001. Thus, the 30h/week subchronic CAPs exposure of $ApoE^{-/-}$ mice at $85\,\mu g/m^3$ altered vasomotor tone, induced vascular inflammation, and potentiated atherosclerosis.

14.3.4 INHALATION STUDIES IN ANIMALS WITH PM COMPONENTS

This section summarizes some studies that used PM components that have been known to produce significant health-related effects at concentrations that are relevant to current or recent human exposures, but does not cover studies with complex mixtures or pure materials at much higher concentrations that are not considered to be relevant to subsequent comparisons of results from the CAPs studies.

Campen et al. (2001) examined responses to Ni and V in conscious rats by whole-body inhalation exposure. The authors tried to ensure valid dosimetric comparisons with prior instillation studies, by using concentrations of V and Ni ranging from 0.3 to $2.4\,mg/m^3$. The concentrations used incorporated estimates of total inhalation dose derived using different ventilatory parameters. HR, core temperature, and ECG data were measured continuously throughout the exposure. The rats were exposed to aerosolized Ni, V, or Ni + V for 6h/day for 4 days, after which serum and BAL samples were taken. While Ni caused delayed bradycardia, hypothermia, and arhythmogenesis at concentrations $>1.2\,mg/m^3$, V failed to induce any significant change in HR or core temperature, even at the highest concentration. When combined, Ni and V produced observable delayed bradycardia and hypothermia at $0.5\,mg/m^3$ and potentiated these responses at $1.3\,mg/m^3$ to a greater degree than were produced by the highest concentration of Ni ($2.1\,mg/m^3$) alone. The results are suggestive of a possible synergistic relationship between inhaled Ni and V, albeit these studies were performed at metal concentrations orders of magnitude greater than their typical ambient concentrations.

In a second study using dogs with preexisting cardiovascular disease, Muggenburg et al. (2003) evaluated the effects of short-term inhalation exposure (oral inhalation for 3h on each of three successive days) to aerosols of transition metals. HR and the ECG readings were studied in conscious beagle dogs (selected for having preexisting cardiovascular disease) that inhaled respirable particles of oxide and sulfate forms of transition metals (Mn, Ni, V, Fe, Cu oxides, and Ni and V sulfates) at concentrations of $0.05\,mg/m^3$. Such concentrations are 2–4 orders of magnitude higher than for typical ambient U.S. levels (usually $0.1–1.0\,\mu g/m^3$ for such metals). No significant effects of exposure to the transition metal aerosols were observed. The discrepancy between the results of Muggenburg et al. (2003) and those of Clarke et al. (2000) leave open major questions about PM effects on the cardiovascular system of the dog. The use of ROFA samples from different sources may have accounted for the differences in response that were reported.

Kodavanti et al. (2003) exposed male SD, WKY, and SH male rats to Boston ROFA, which contained bioavailable Zn at doses of 2, 5, or $10\,mg/m^3$ for 6h/day for four consecutive days. A second exposure paradigm used exposure to $10\,mg/m^3$ ROFA for 6h/day, 1 day/week, for 4 or 16 consecutive weeks. Cardiovascular effects were not seen in SD and SH rats with the acute or chronic exposure, but WKY rats from the 16 week exposure group had cardiac lesions consisting of chronic-active inflammation, multifocal myocardial degeneration, fibrosis, and decreased numbers of granulated mast cells. These results suggest that myocardial injury in sensitive rats can be caused by long-term inhalation of high concentrations of ROFA.

14.3.5 Lung Instillation Studies with PM Components

Besides using CAPs inhalation study, which can be expensive and time consuming, studies using collected urban PM for intratracheal instillation/aspiration (IT/IA) to healthy and compromised animals have also produced interesting information concerning influential PM components and their health-related effects. Although there are many issues such as extrapolation and dosimetry that need to be addressed when IT is used in a toxicological study, the results of IT of ambient PM collected from different geographical areas can be used to support the hypothesis that PM composition is one of the most relevant parameters affecting ambient PM-associated health effects. Similarly, IT delivery of well-defined components of ambient air PM, in studies of comparative toxicity, can also be informative for the identification of particularly influential components.

14.3.5.1 Ambient Air PM

Instillation in rats of Ottawa PM extracts at 2.5 mg induced pronounced biphasic hypothermia, a severe drop in HR, and increased arrhythmias (Watkinson et al., 2000a,b) that were not seen with a comparable instilled dose of Mt. St. Helens volcanic ash. The results of this study showed that urban sites with high contributions from vehicles and industry were most toxic. This study also showed that the biological effects differ as a function of site and season. The analysis based on chemical class indicated that PM containing metal oxides, transition metals (Pb, Mn, Cu, Se, Zn, and As), EC, OC, and hopanes/steranes were the most important predictors of cytotoxic and inflammatory responses. The analysis also indicated that $SO_4^=$, secondary organic aerosols, meat cooking, and vegetative detritus were not correlated with the biological responses. On the other hand, analysis based on the source apportionment, the most toxic samples were from the sites during seasons with the largest contributions of diesel and gasoline emissions, whereas wood burning was only weakly correlated with toxicity endpoints. The analysis also indicated that $SO_4^=$, secondary organic aerosols, meat cooking, and vegetative detritus were not correlated with the biological responses. This study supports the concept that specific constituents and/or sources of PM affect its toxicity.

Adamson et al. (2000) sought to determine which component(s) of the urban air PM from Toronto could account for its pulmonary toxicity. They did an aqueous extraction of the whole dust and instilled the extract and equivalent concentrations of the soluble metals within it into the lungs of mice. Three days later, in comparison to IT saline, only the whole extract and the Zn solution produced significant increases in inflammatory cells and protein in BALF. With 28 days of exposure, the Zn produced focal necrosis of Type 1 alveolar cells, and focal fibrosis was seen at 4 weeks.

Gerlofs-Nijland et al. (2007) collected $PM_{2.5}$ and $PM_{10-2.5}$ CAPs from six European cities with contrasting traffic profiles, PM composition, and *in vitro* analyses, and exposed SH rats intratracheally with 3 or 10 mg PM/kg. They assessed changes in biochemical markers, cell differentials, and histopathological changes in the lungs and blood 24 h later. They reported dose-related adverse effects with both $PM_{2.5}$ and $PM_{10-2.5}$ CAPs that were mainly related to cytotoxicity, inflammation, and blood viscosity. There was a trend toward greater toxicity with increasing traffic levels. They selected component markers for traffic-related PM sources, that is, polynuclear aromatic hydrocarbons (PAHs), Zn, Cu, Ba, and K. There was no correlation of any of the effect markers with combustion-exhaust-related PAHs except for an increase of lymphocytes associated with $PM_{2.5}$ CAPs (p=0.04). There was a significant correlation between $PM_{2.5}$ Zn and BALF protein (p=0.01) and LDH (p=0.03), and $PM_{2.5}$ K with total BALF cells and PMNs. In pathological assays, there were significant associations of $PM_{2.5}$ K with alveolar inflammatory foci, and ascorbate with $PM_{2.5}$ Cu (p=0.02) and Ba (p=0.01). For the $PM_{10-2.5}$ CAPs, there were significant correlations of BALF protein with Cu (p=0.02) and with Ba (p=0.05), and for alveolitis with Cu (p=0.04). They concluded that the effects were attributable to components derived from brake wear (Cu and Ba), tire wear (Zn), and wood smoke (K).

14.3.5.2 Utah Valley Dust

Some of the most convincing evidence to demonstrate that the lung dose of bioavailable transition metals, not just instilled PM mass, was the primary determinant of the acute inflammatory response was derived from a series of studies using ambient PM_{10} collected in the Utah valley (Frampton et al., 1999; Dye et al., 2001; Ghio and Devlin, 2001). Frampton et al. (1999) showed that the extract of PM_{10} collected during the strike (having the lowest metal content, specifically soluble Fe, Cu, Pb, and Zn), showed no apparent cytotoxicity, minimal induction of cytokines, and lowest oxidant generation ability compared to extracts from PM_{10} (collected before and after the strike) having higher metal content. These experiments indicate: (1) that instillation of ambient air particles, albeit at a very high concentration, can produce cardiovascular effects; and (2) that exposures of equal mass dose to particle mixtures of differing composition did not produce the same cardiovascular effects, suggesting that PM composition rather than just mass was responsible for the observed effects.

To investigate the dose, time course, and the roles of specific metals, Dye et al. (2001) exposed SD rats, by IT, with equivalent masses of aqueous extracts of the same Utah ambient PM_{10}, described earlier, at 0, 0.83, 3.3, 8.3, or 16 mg extract/kg body weight in 0.3 mL saline (Dreher et al., 1997). Twenty-four hours after IT, rats exposed to extracts of PM_{10} collected when the plant was open developed significant pulmonary injury and PMN inflammation. Additionally, 50% of rats exposed to these extracts had increased airway responsiveness to acetylcholine, compared to 17% and 25% of rats exposed to saline or to the extracts of PM_{10} collected when the plant was closed. By 96 h, these effects were largely resolved, except for increases in lung lavage fluid PMNs and lymphocytes in rats exposed to PM_{10} extracts from prior to the plant closing. Analogous effects were observed with lung histologic assessment. Chemical analysis of extract solutions demonstrated that extracts of PM_{10} collected when the plant was open contained more $SO_4^=$, cationic salts (e.g., Ca, K, Mg), and certain metals (e.g., Cu, Zn, Fe, Pb, As, Mn, Ni). The strong qualitative coherence among these human epidemiological, clinical, and animal toxicological studies clearly showed that soluble metals could be the most important components related to PM exposure-related health outcomes.

14.3.5.3 Residual Oil Fly Ash

SD rats were exposed IT to ROFA suspension, leachate, washed, neutralized suspension, neutralized leachate, neutralized leachate supernate, and suspension + deferoxamine (2.5 mg ROFA/rat). The leachate produced similar lung injury to that induced by the suspension, implicating the soluble components for the effects. The inflammatory effects were abrogated by depletion of metals from the ROFA leachate. A mixture of transition metal sulfate containing Fe, V, and Ni largely reproduced the lung injury induced by ROFA. Neutralization of ROFA, soluble Ni, and transition metal sulfate mixtures produced fine precipitates in the solutions, leading to the production of a more progressive acute lung injury for ROFA particles, and enhanced morbidity/mortality for Ni and the transition metal sulfate mixtures, providing direct evidence for soluble transition metals to produce pulmonary injury, at least for exposures at relatively high doses of ROFA. To further demonstrate how ROFA with differing V and Ni may differ in terms of their ability to cause *in vivo* acute pulmonary injury, male SD rats were exposed IT with either saline or saline suspension of 10 ROFA samples collected at various sites within a power plant (Kodavanti et al., 1998). After 24 h, ROFA containing the highest concentrations of water-leachable Fe, V, and Ni or V and Ni caused the largest increase in these biochemical indices of lung injury, while ROFA containing primarily soluble V caused more dramatic PMN influx. It appears that while V was responsible for the recruitment of PMNs, Ni was responsible for increase pulmonary permeability, suggesting that the potency and the mechanism of pulmonary injury differ between V and Ni.

BAL inflammatory markers in normotensive WKY and SH rats were examined after a single IT to either saline or ROFA. BAL lung injury markers were measured at 24 and 96 h post-IT (Kodavanti et al., 2001). Rats were also IT instilled with either VSO_4 or $NiSO_4 \cdot 6H_2O$ in saline and assessed at 6 and 24 h post-IT. ROFA-induced increases in BALF markers of inflammation were generally greater in SH rats than in WKY rats and had resolved by 96 h post-IT in both strains. In response to a single-metal IT exposure, both the onset and duration of the inflammatory response, were metal and strain dependent. V-induced increases in BALF protein and LDH peaked at 6 h post-IT in WKY rats. In SH rats, BALF protein and LDH were not affected by V. Ni increased BALF protein in both strains. The Ni-induced increase in LDH activity was progressive over 24 h (WKY > SH). The number of macrophages decreased following V and Ni exposure at 6 h, and this decrease was reversed by 24 h in both strains. V caused BALF PMNs to increase only in WKY rats. The Ni-induced increase in BALF PMNs was more dramatic and progressive than that of V, but was similar in both strains. Lung histology similarly revealed more severe and persistent edema, perivascular and peribronchiolar inflammation, and hemorrhage in Ni- than in V-exposed rats. This effect of Ni appeared slightly more severe in SH than in WKY rats. This study showed that inflammatory response to metallic constituents of ROFA is both strain and dose dependent, and that V caused pulmonary injury only in WKY rats, whereas Ni was toxic to both strains. In subsequent studies (Kodavanti et al., 2001; Wu et al., 2003), however, Zn was found to be the responsible component in a batch of different oil combustion emission particles.

Some, or perhaps most, of the differences in biological responses to the metals in the ROFA could be explained by aqueous solubility, and the effects of solubility on translocation. Wallenborn et al. (2007) measured the elemental content of lungs, plasma, heart, and liver of male WKY rats after IT administration of either saline or 8.3 mg/kg of ROFA from a Boston power plant, and measured tissue concentrations 4 and 24 h after the instillation. Water-soluble metals (V, Ni, Zn, and Mn) were detected at both time points, while Al and Si were not.

The effects of two ROFA samples of equivalent diameters, but having different metal and $SO_4^=$ content, on pulmonary responses in SD rats were studied (Gavett et al., 1997). One sample had higher saline-leachable $SO_4^=$, Ni, V, and Fe, whereas the other sample had higher Zn. At a dose of 2.5 mg, 4 of 24 rats exposed to high-Zn ROFA suspension or supernatant had died 4 days post-IT, while none had in high-Ni, V, and Fe groups. Pathological indices, such as alveolitis, early fibrotic changes, and perivascular edema, were greater in both high-Zn suspension and supernatant-exposed groups than the other ROFA. In surviving rats, exposures to high-Zn ROFA also worsened the baseline pulmonary function parameters and AHR to acetylcholine as well as BAL PMNs. This study confirmed the finding of an earlier study in guinea pigs that soluble forms of Zn are capable of producing a greater pulmonary response than other sulfated metals in combustion generated particles (Amdur et al., 1978).

AHR induced by ROFA and its soluble components was also observed in mice exposed to an aerosolized soluble leachate of ROFA (ROFA-s). AHR to acetylcholine challenge occurred in a time- and dose-dependent manner after exposure to ROFA-s with peak at 48 h post-IH exposure. AHR was accompanied by an earlier onset of BAL PMNs, which was maximal at 12 h after exposure. The AHR caused by ROFA-s was reproduced by a mixture of its major metal components

(Ni, V, Zn, Co, Mn, Cu), but not by any individual metal alone. Intraperitoneal pretreatment of mice with the antioxidant dimethylthiourea abrogated ROFA-s-mediated AHR, confirming the role of ROS in metal-induced inflammation. Interestingly, ROFA-s had no effect on AHR of 2 week old mice, in contrast to the AHR seen in 3 and 8 week old mice. This study also found that ROFA treatment does not initiate neurogenic inflammation because ROFA-s-mediated AHR was unchanged in neurokinin-1 receptor knockout mice and in mice treated with a neurokinin antagonist. After either IT or inhalation, of ROFA, lung injury is evident within 24 h of exposure, with a dose-dependent recruitment of PMNs, eosinophils, and monocytes into the airway. The peak of this influx occurs 18–24 h after exposure. The cellular influx persisted 96 h later, and resolution occurs slowly. Inflammatory lung injury after ROFA was accompanied by airway, an increase in susceptibility to infections, and, at high concentration, noncardiogenic pulmonary edema.

AHR induced by ROFA and its soluble components has also been observed in mice that were exposed to aerosolized ROFA-s. AHR to acetylcholine challenge occurred in a time- and dose-dependent manner after exposure to ROFA-s with a peak at 48 h post-IH exposure. AHR was accompanied by an earlier onset of BAL PMNs, (maximal at 12 h after exposure). The AHR caused by ROFA-s was reproduced by a mixture of its major metal components (Ni, V, Zn, Co, Mn, Cu) but not by any individual metal alone. Intraperitoneal pretreatment of mice with the antioxidant dimethylthiourea abrogated ROFA-s-mediated AHR, confirming the role of ROS in metal-induced inflammation. Interestingly, ROFA-s had no effect on AHR of 2 week old mice, in contrast to the AHR seen in 3 and 8 week old mice. ROFA treatment did not initiate neurogenic inflammation, because ROFA-s-mediated AHR was unchanged in neurokinin-1 receptor knockout mice and in mice treated with a neurokinin antagonist.

14.3.5.4 Tire Dust

Gottipulo et al. (2008) instilled two kinds of tire dust into the lungs of male WKY rats. TP1 was made from ground tires of recycled styrene butadiene rubber, while TP2 was from scrap tires. Elemental analyses were available for both dusts. Tests were done with administered saline, TP1, and TP2. Additional tests were done with soluble Zn or Cu, or both. For TP1 and TP2, there were increases in BAL fluid markers of inflammation and injury (TP2 > TP1) but no effects on cardiac enzymes. Instillation of Zn, Cu, and Zn + Cu decreased the activity of cardiac aconitase, isocitrate dehydrogenase, succinate dehydrogenase, cytochrome-c-oxidase, and superoxide dismutase as indicative of cardiac oxidative stress.

14.3.5.5 Metal Oxide Nanoparticles

Lu et al. (2009) compared *in vitro* assays of intrinsic free radical generation, oxidative activity in an extracellular environment, cytotoxicity to A549 lung epithelial cells, hemolysis of healthy human erythrocytes, and inflammation potency in WKY rat lungs. They used nanoparticles of carbon black (CB), and metal oxide nanoparticles (2–30 nm) composed of NiO, CeO_2, Co_3O_4, MgO, SiO_2, anatase, rutile, and three kinds of alumina; ZnO as UFP (90–210 nm); and alumina as 0.3 μm microparticles. For assays using equivalent surface areas, only NiO and alumina #2 (7 nm) caused significant lung inflammation, while four of the 13 metal oxides (NiO, CeO_2, Co_3O_4, and CB) caused significant free radical generation (of these, only NiO was inflammogenic), and 3 of 13 (NiO, CeO_2, and alumina #2) were significantly hemolytic (of these, only NiO was inflammogenic). Thus, *in vitro* assays cannot be relied on to predict lung inflammation.

14.4 SUMMARY OF UNRESOLVED ISSUES AND CONCLUSIONS

14.4.1 Where Do PM Metal Components Fit in the Larger Picture of PM-Associated Health Effects?

There is clearly emerging evidence that the inhalation of some components of ambient air PM are associated with adverse health effects at concentrations near or not much higher than current ambient

$PM_{2.5}$ mass concentrations. In addition to EC, these components include Ni, V, and Pb, and suggestive evidence exists for other components, such as Al, Zn, and OC. There is also a rapidly growing literature implicating motor-vehicle-related pollution in human health effects, as indexed by proximity to major roadways, and by measured concentrations of OC, NO_2, and UFP. However, there are also metals in motor vehicle exhaust and resuspended road dust whose role, if any, in causing traffic-related human health effects at contemporary ambient air concentrations is largely unknown. There is some evidence that the metals in resuspended road dust may be important. Recent research has suggested that traffic-generated PM can account for pulmonary effects, with Gent et al. (2009) showing that the motor vehicle source was associated with a $10\%/5\,\mu g^3$ increase in wheeze in asthmatic children, while the road dust source was associated with a 28% increase in shortness of breath. This response is consistent with the findings reported by Gottipulo et al. (2008) based on the instillation of two kinds of tire dust into the lungs of male WKY rats in relation to the elemental composition. There were increases in BALF markers of inflammation and injury, and similar effects were seen for instilled Zn and Cu. Thus, the acute pulmonary effects of tire dust could be due to the metals.

Furthermore, there is some evidence that adverse health effects are significantly associated with aerosol acidity originating from fossil fuel combustion, which could be due to its irritancy, or to its role in solubilizing metals within the particles.

14.4.2 Are There Specific Sources of Metals That Can Account, at Least in Part, for Health Effects Associated with $PM_{2.5}$?

It is known that ROFA, which is a mixture that is presumed to be similar in composition to the fly ash emitted by power plants burning residual oil, and which is notably high in the content of Ni and V, as compared to other metals, and Utah Valley dust, which is a mixture enriched in steel mill emissions, were more toxic than other source-related mixtures that have been tested in laboratory animals *in vivo*, or in cells *in vitro*. For acute pulmonary system responses, it appears, from such tests, that V and Zn may play prominent roles, and that the effects may depend on interactions among the metals. For acute cardiovascular effects, Ni appears to play a more important role. By contrast, other source-related mixtures, such as coal combustion effluents, that are notable for their content of Se, Fe, and Mn, and resuspended soil that contains more refractory metals, have been found to be less acutely toxic.

Janssen et al. (2002) modeled source contributions to ambient air PM using emissions data. Cardiovascular admissions were significantly associated with a number of sources (highway vehicles, oil combustion, and metal processing), but there were no significant associations of the sources with COPD or pneumonia admissions.

In summary, toxic metals from power plants, and possibly from resuspended road dust, are worthy of increased concern.

14.4.3 Addressing Research Needs Relating to Health Effects of Components in Ambient Air Particulate Matter

There are many reasons why past research has not resolved the roles that PM components may play in the health-related effects of ambient air PM. These are as follows: (1) concentrations of components in ambient air PM generally range from a few $\mu g/m^3$ in some refractory metals to less than $10\,ng/m^3$ for transition metals that are known to generate ROS, raising the issue of biological plausibility; (2) epidemiologic research opportunities have been limited because of the paucity of data on the concentrations of PM components—even now, when $PM_{2.5}$ speciation data have been available since 2000 for many U.S. cities, they are mostly limited to every third or sixth day; (3) few toxicologists or clinical researchers have had the resources needed to perform subchronic CAPs inhalation studies that include speciation data on the PM in the exposure samples; (4) controlled exposures to pure compounds at concentrations of environmental relevance have been uniformly negative, even when sensitive animal models were used; (5) a lack of studies defining the relationship between personal exposure and ambient air levels for most metal species; and (6) most controlled clinical and laboratory animal exposure studies have been limited to one or a few days, which may not be sufficient to elicit responses of concern.

The subchronic CAPs inhalation studies in New York, Ohio, and California suggest that such studies overall, and those with elevated concentrations of Ni and EC in particular, can yield evidence that current levels of ambient air concentrations produce health effects of interest in terms of public health. Furthermore, there are many toxicology studies of ROFA cited earlier, that are buttressed by epidemiological studies (Hedley et al., 2002, 2004; Janssen et al., 2002; Lipfert et al., 2006), suggesting a line of continuity between both types of studies with regard to damage from the combination of V and Ni. There are also some studies showing oxidative stress and DNA damage associated with V (Sorensen et al., 2005). If the inhalation of Ni, or Ni in combination with V, at current, relatively low, ambient air concentrations, does appreciably affect cardiac function and mortality in humans, one may wonder why the effects of such exposures have not previously been recognized. One reason may be that the increment in cardiovascular mortality that they may have produced is a relatively small part of the very large cardiovascular mortality. Also, it is possible that other $PM_{2.5}$ components may account for the more numerous studies, not having speciation data, showing associations of $PM_{2.5}$ mass with health effects because of the influence of other components alone or because the other components potentiate the effects of Ni.

Also, the statistically significant transient and progressive changes that Ni produced in cardiovascular function in the ApoE$^{-/-}$ mice were relatively subtle, required advanced analytical techniques for their detection, and are unlikely to be detected in the kinds of short-term exposure studies that have previously been undertaken in laboratory animals.

In addition, the exact physical and chemical characteristics of Ni-bearing ambient PM have not been determined. Based on the previous work at NYU, the fact that the potency of ultrafine Zn particles with a thin coating of sulfuric acid is much greater than uncoated acid particles (Amdur and Chen, 1989), raises the likelihood that a specific form of a metal that has not been reproduced in the laboratory could be responsible for the observed biological effects.

Many studies have used ROFA as a surrogate for ambient PM in various *in vivo* and *in vitro* experiments. ROFA contains many soluble metals, and since it is clear, from this review, that they interact with each other, chemically and biologically, it should not be surprising that there are inconsistent and confusing results. Although ROFA was useful in providing plausible evidence that metals are important in eliciting adverse cardiopulmonary effects, it should not be the focus of future studies. The experimental *in vitro* design of Maciejczyk and Chen (2005), in which CAPs were collected in a biosampler impinger simultaneously with a series of daily CAPs inhalations, provided a sound basis for parallel daily *in vitro* assays. Performing such assays in parallel with future animal inhalation studies and/or human clinical studies, could provide opportunities for gaining a better understanding of the source profile that may contribute to the adverse effects seen in animals and humans.

Much of the remaining skepticism concerning the biological plausibility of the premature mortality and increased morbidity associated with ambient air $PM_{2.5}$ has been due to the paucity of exposure–response data in laboratory studies involving $PM_{2.5}$ inhalation, and the heretofore seemingly impossible task of identifying any specific causal components. The subchronic CAPs inhalation studies that were performed in Sterling Forest (Lippmann et al., 2005b, 2006; Sun et al., 2005) helped to establish such plausibility, and have also developed a mechanistic base for the initiation and progression of effects attributable to the long-range transported aerosol in the northeastern United States (Sun et al., 2005). The consistency in these analyses and reexaminations of available data lead to the conclusions that (1) Ni is a particularly influential component of ambient $PM_{2.5}$ in terms of cardiac responses to the inhalation of ambient air $PM_{2.5}$; (2) further research is needed on the specific influences of both Ni and V, which are both generally most closely associated with residual oil combustion effluents, on both acute and chronic respiratory and cardiovascular health effects; and (3) further research is also needed on the currently unknown impacts of other toxic metals in ambient air.

14.5 FURTHER PERSPECTIVE ON BIOLOGICAL PLAUSIBILITY

For most scientists today, biological plausibility for adverse effects caused by ambient air metals needs to be based on a clear understanding of the underlying biological mechanisms that can account for the empirical observations arising from laboratory-based exposures or epidemiological studies of populations in the real world. Until recently, mechanistic understanding, based on *in vitro* studies, has been quite sparse, and the evidence for causal relationships between exposures to ambient air metals and adverse health effects has been more indirect. Of necessity, in the absence of much knowledge on underlying mechanisms, we have had to rely on the consistency of significant associations between indices of PM mass exposures and specific responses, their stability over time and space, and the coherence of each specific response with other responses that one would expect to see if the first was really reliable. In part, this situation was necessitated by the fact that we have had to rely on exposure indices based on particle mass or number concentrations, and have seldom had access to the concentrations of the causal components within the PM, and their temporal concentration variation.

We have only recently begun to have opportunities to frame the questions about underlying biological mechanisms in conjunction with observational studies of exposure–response relationships. A broad range of health effects have recently been associated with subchronic CAPs inhalation studies in rodent models, and it is reasonable to assume, based on this review, that many of these effects were due to the metals within the CAPs. The following are some examples:

1. Sun et al. (2005) demonstrated that the *in vivo* CAPs exposure increase of plaque in the aorta was associated with lipid content, responses to vasoconstrictor challenge, relaxation in response to acetylcholone, increases in macrophage infiltration, expression of inducible NO-synthase, ROS generation, and immunostaining for 3-nitrotyrosine.
2. Sun et al. (2008a) demonstrated that the CAPs exposure also increased tissue factor expression in the aorta.
3. Sama et al. (2007) demonstrated that CAPs that altered brain cell distributions *in vivo* caused changes in immortalized microglial cells *in vitro*, intracellular ATP, depolarization of mitochondrial membranes, glutathione, nonprotein sulfhydryl, TNF a, IL-6, and gene regulation. The responses were dose related, with the primary determinant being the capacity of CAPs to activate NFkB in human bronchial epithelial cells.
4. Sun et al. (2008b) demonstrated that, after CAPs exposure, infusion of angiotensis II increased mean arterial pressure, potentiated aortic vasoconstriction to phenylephrine, exaggerated relaxation a Rho-kinase inhibitor, and increased superoxide levels in the aorta.
5. Sun et al. (2009) demonstrated that the CAPs exposure in an obese mouse model had insulin-signaling abnormalities that were associated with abnormalities in vascular relaxation to insulin and acetylcholine, increased adipose tissue macrophages (F4/80+ cells) in visceral fat, and expression of higher levels of TNF-a/IL-6 and lower levels of IL-10. In coordinate *in vitro* tests, PM induced cell accumulation in visceral fat and potentiated cell adhesion in the microcirculation, supporting a link between CAPs exposure and Type II diabetes mellitus and metabolic syndrome.
6. In association with their CAPs inhalation study of the influence of CAPs exposure on nonalcoholic fatty liver disease (NAFLD), Tan et al. (2009) performed *in vitro* exposures using a reference PM sample (NIST SRM1649a). For the mice exposed by injection, particles were detected only in Kupffer cells from livers not seen in sham injected mice. Cell culture studies were done using macrophage (RAW) and stellate (LX2) cell lines. Direct exposure to $PM_{2.5}$ activated IL-6 production by Kupffer cells in a TLR4-dependent manner. Thus, exposure to ambient air PM may be a significant risk factor for NAFLD progression.

7. Xu et al. (2011) showed that ambient particulate air pollution induces oxidative stress and alterations of mitochondria and gene expression in brown and white adipose tissues, contributing to cardiac stress.
8. Xu et al. (2009) demonstrated that diesel exhaust exposure induces angiogenesis.
9. Xu et al. (2010) showed the effect of early particulate air pollution exposure on obesity in mice.
10. Ying et al. (2009a,b) reported that ambient particulates alter vascular function through induction of reactive oxygen and nitrogen species.

These examples identify some of the biological pathways that are activated by particles taken as a result of inhalation exposure and the cells that participate in the responses. Future studies can utilize pure materials as well as ambient air PM of mixed composition to obtain more information on the roles of specific PM metal components and their interactions and combined effects, and thereby identify those components most in need of control in order to reduce the health impacts of airborne PM.

GLOSSARY

ABS	absorbance (of light by BC on a sampling filter)
AHR	airway hyperresponsiveness
ApoE$^{-/-}$	apolipoprotein A-deficient (knockout) mouse
BAD	brachial artery diameter
BAL	bronchoalveolar lavage
BALF	bronchoalveolar lavage fluid
BC	black carbon, aka soot, measured as light absorbance by, or reflectance off a sampling filter
BP	blood pressure
Br	bromine
CA	California
CAPs	concentrated ambient air particles
Cd	cadmium
CFA	coal fly ash
CHD	coronary heart disease
CI	confidence interval
CO	carbon monoxide
COPD	chronic obstructive pulmonary disease
Cr	chromium
CRP	C-reactive protein
Cu	copper
CYP	cytochrome p isoenzyme
eNO	exhaled nitric oxide
eNOS	endothelial nitric oxide synthase
EC	elemental carbon
ECG	electrocardiogram or electrocardiographic
EPA	Environmental Protection Agency
FA	filtered air
Fe	iron
FEV$_1$	forced expiratory volume in 1 s
FVC	forced vital capacity
HDL	high-density lipoprotein
HEI	Health Effects Institute

HR	heart rate
HRV	heart rate variability
IL-6	interleuken-6, an interleukin that acts as both a pro-inflammatory and anti-inflammatory cytokine. It is secreted by T cells and macrophages to stimulate immune response to trauma, especially burns or other tissue damage leading to inflammation
IL-8	interleuken-8, a chemokine (ability to induce directed chemotaxis in nearby responsive cells) produced by macrophages and other cell types such as epithelial cells
IT	intratracheal
K	potassium
LDH	lactate dehydrogenase
LYM	lymphocyte
MEF	mid-expiratory flow rate
MI	myocardial infarction
MMAD	mass median aerodynamic diameter
MS	metabolic syndrome
NAC	N-Acetylcysteine
NAFLD	nonalcoholic fatty liver disease
NFkB	nuclear factor kappa B
NHANES III	Third National Health and Nutritional Examination Survey
Ni	nickel
NMMAPS	National Morbidity and Mortality Air Pollution Study
NO_2	nitrogen dioxide
O_3	ozone
OC	organic carbon
ppb	parts per billion
PAH	polynuclear aromatic hydrocarbon
PEF	peak expiratory flow rate
PM	particulate matter
$PM_{0.18}$	PM with diameters <0.18 μm, a practical pseudo-ultrafine aerosol
PM_{10}	PM with aerodynamic diameters <10 μm
$PM_{10-2.5}$	PM with aerodynamic diameters between 2.5 and 10 μm, aka coarse thoracic PM
$PM_{2.5}$	PM with aerodynamic diameters <2.5 μm, aka fine PM
PMN	polymorpho-nucleus leukocyte, neutrophil
Pt	platinum
QT interval	interval between Q and T waves of a normal ECG tracing
R-R interval	interval between two adjacent R waves of a normal ECG tracing
RBC	red blood cell
RMSSD	root mean square of the standard deviation of normal-to-normal beat
ROFA	residual oil fly ash
ROS	reactive oxygen species
RR	relative risk
SBP	systolic blood pressure
SD	Sprague–Dawley (rat)
SDNN	standard deviation of normal-to-normal heart beat
SH	spontaneously hypertensive (rat)
SHR	spontaneously hypertensive rat
SO_2	sulfur dioxide
$SO_4^=$	sulfate ion
ST-segment	in ECG tracing, ST-segment starts at the J point (junction between the QRS complex and ST segment) and ends at the beginning of the T wave

T-wave	T wave of ECG tracing. It connects the QRS complex and the T wave
TF	tissue factor expression
TNF	tumor necrosis factor
TSP	total suspended particles
uCB	ultrafine carbon black
UBM	ultrasound biomicroscopy
UFP	ultrafine particles, usually defined as PM with diameters <0.1 µm
vWF	von Willibrand factor
V	vanadium
WBC	white blood cell
WKY	Wistar-Kyoto (rat), usually used as control for the SHR
XRF	x-ray fluorescence
Zn	zinc

ACKNOWLEDGMENTS

The literature summarized in this critical review includes a considerable number of papers that were described previously in a review of metals toxicology in *Inhalation Toxicology* (Chen and Lippmann, 2009), and a critical review of inhalations studies in humans and animals with concentrated ambient PM$_{2.5}$ (CAPs) (Lippmann and Chen, 2009). The author acknowledges the support received from a Center Grant (ES 00260) from the National Institute of Environmental Health Sciences (NIEHS) and from a research grant from the Health Effects Institute.

REFERENCES

Adamson, I. Y., Prieditis, H., Hedgecock, C., and Vincent, R. (2000). Zinc is the toxic factor in the lung response to an atmospheric particulate sample. *Toxicol. Appl. Pharmacol.* 166, 111–119.

Amdur, M. O., Bayles, J., Ugro, V., and Underhill, D. W. (1978). Comparative irritant potency of sulfate salts. *Environ. Res.* 16, 1–8.

Brook, R. D., Brook, J. R., Urch, B., Vincent, R., Rajagopalan, S., and Silverman, F. (2002). Inhalation of fine particulate air pollution and ozone causes acute arterial vasoconstriction in healthy adults. *Circulation* 105, 1534–1536.

Campen, M. J., Nolan, J. P., Schladweiler, M. C. J., Kodavanti, U. P., Evansky, P. A., Costa, D. L., and Watkinson, W. P. (2001). Cardiovascular and thermoregulatory effects of inhaled PM-associated transition metals: A potential interaction between nickel and vanadium sulfate. *Toxicol. Sci.* 64, 243–252.

Chen, L.-C., and Hwang, J. S. (2005). Effects of subchronic exposures to concentrated ambient particles (CAPs) in mice. IV. Characterization of acute and chronic effects of ambient air fine particulate matter exposures on heart-rate variability. *Inhal. Toxicol.* 17, 209–216.

Chen, L. C. and Lippmann, M. (2009). Effects of metals within ambient air particulate matter on human health. *Inhal. Toxicol.* 21, 1–31.

Chen, L-C. and Nadziejko, C. (2005). Effects of subchronic exposures to concentrated ambient particles (CAPs) in mice. V. CAPs exacerbate aortic plaque development in hyperlipidemic mice. *Inhal. Toxicol.* 17, 217–224.

Clarke, R. W., Coull, B., Reinisch, U., Catalano, P., Killingsworth, C. R., Koutrakis, P., Kavouras, I. et al. (2000). Inhaled concentrated ambient particles are associated with hematologic and bronchoalveolar lavage changes in canines. *Environ. Health Perspect.* 108, 1179–1187.

Dominici, F., Peng, R. D., Ebisu, K., Zeger, S. L., Samet, J. M., and Bell, M. L. (2007). Does the effect of PM$_{10}$ on mortality depend on PM nickel and vanadium? A reanalysis of the NMMAPS data. *Environ. Health Perspect.* 115, 1701–1703.

Dreher, K. L., Jaskot, R. H., Lehmann, J. R., Richards, J. H., McGee, J. K., Ghio, A. J., and Costa, D. L. (1997). Soluble transition metals mediate residual oil fly ash induced acute lung injury. *J. Toxicol. Environ. Health* 50, 285–305.

Dye, J. A., Lehmann, J. R., McGee, J. K., Winsett, D. W., Ledbetter, A. D., Everitt, J. I., Ghio, A. J., and Costa, D. L. (2001). Acute pulmonary toxicity of particulate matter filter extracts in rats: Coherence with epidemiological studies in Utah Valley residents. *Environ. Health Perspect.* 109(Suppl. 3), 395–403.

Ebelt, S. T., Wilson, W. E., and Brauer, M. (2005). Exposure to ambient and nonambient components of particulate matter. *Epidemiology* 16, 396–405.

Eftim, S. E., Samet, J. M., Janes, H., McDermott, A., and Dominici, F. (2008). Fine particulate matter and mortality: A comparison of the six-cities and American Cancer Society cohorts with a Medicare cohort. *Epidemiology* 19, 29–216.

Frampton, M. W., Ghio, A. J., Samet, J. M., Carson, J. L., Carter, J. D., and Devlin, R. B. (1999). Effects of aqueous extracts of PM_{10} filters from the Utah Valley on human airway epithelial cells. *Am. J. Physiol.* 277, L960–L967.

Franklin, M., Koutrakis, P., and Schwartz, J. (2008). The role of particle composition on the association between $PM_{2.5}$ and mortality. *Epidemiology* 19, 680–689.

Gent, J., Koutrakis, P., Berlanger, K., Triche, E., Holford, T., Bracken, M., and Leaderer, B. (2009). Symptoms and medication use in children with asthma and traffic-related sources of fine particle pollution. *Environ. Health Perspect.* 117, 1168–1174.

Gerlofs-Nijland, M. E., Dormans, J. A., Bloemen, H. J., Leseman, D. L., John, A., Boere, F., Kelly, F. J. et al. (2007). Toxicity of coarse and fine particulate matter from sites with contrasting traffic profiles. *Inhal. Toxicol.* 19, 1055–1069.

Ghio, A. J. and Devlin, R. B. (2001). Inflammatory lung injury after bronchial instillation of air pollution particles. *Am. J. Respir. Crit. Care Med.* 164, 704–708.

Ghio, A. J., Kim, C., and Devlin, R. B. (2000). Concentrated ambient air particles induce mild pulmonary inflammation in healthy human volunteers. *Am. J. Respir. Crit. Care Med.* 162, 981–988.

Ghio, A. J., Richards, J. H., Dittrich, K. L., and Samet, J. M. (1998). Metal storage and transport proteins increase after exposure of the rat lung to an air pollution particle. *Toxicol. Pathol.* 26, 388–394.

Gold, D. R., Litonjua, A. A., Zanobetti, A., Coull, B. A., Schwartz, J., MacCallum, G., Verrier, R. L. et al. (2005). Air pollution and ST-segment depression in elderly subjects. *Environ. Health Perspect.* 113, 883–887.

Gottipulo, R. R., Landa, E. R., Schladweiler, M. C., McGee, J. K., Ledbetter, A. D., Wallenborn, G. J., and Kodavanti, U. P. (2008). Cardiopulmonary responses of intratracheally instilled tire particles and constituent metal components. *Inhal. Toxicol.* 20, 473–484.

Grahame, T. and Hidy, G. (2004). Using factor analysis to attribute health impact to particulate pollution sources. *Inhal. Toxicol.* 16(Suppl. 1), 143–152.

Gunnison, A. and Chen, L. C. (2005). Effects of subchronic exposures to concentrated ambient particles (CAPs) in mice. VI. Gene expression in heart and lung tissue. *Inhal. Toxicol.* 17, 225–233.

Gurgueira, S. A., Lawrence, J., Coull, B., Murthy, G. G., and Gonzalez-Flecha, B. (2002). Rapid increases in the steady-state concentration of reactive oxygen species in the lungs and heart after particulate air pollution inhalation. *Environ. Health Perspect.* 110, 749–755.

Hamada, K., Goldsmith, C.-A., Suzuki, Y., Goldman, A., and Kobzik, L. (2002). Airway hyperresponsiveness caused by aerosol exposure to residual oil fly ash leachate in mice. *J. Toxicol. Environ. Health, Part A.* 65, 1351–1365.

Hedley, A. J., Chau, P. Y. K., and Wong, C. M. (2004). The change in sub-species of particulate matter [PM10] before and after an intervention to restrict sulphur content of fuel in Hong Kong. Poster presented at *Better Air Quality/Asian Development Bank Meeting*, 2004, Agra, India.

Hedley, A. J., Wong, C. M., Thach, T. Q., Ma, S., Lam, T. H., and Anderson, H. R. (2002). Cardiorespiratory and all-cause mortality after restrictions on sulphur content of fuel in Hong Kong: An intervention study. *Lancet* 360, 1646–1652.

Hsu, S.-I., Ito, K., and Lippmann, M. (2011). Effects of thoracic and fine PM and their components on heart rate and pulmonary function in COPD patients. *J. Expos. Sci. Environ. Epidemiol.* 21, 464–472.

Hwang, J. S., Nadziejko, C., and Chen, L. C. (2005). Effects of subchronic exposures to concentrated ambient particles (CAPs) in mice. III. Acute and chronic effects of CAPs on heart rate, heart-rate fluctuation, and body temperature. *Inhal. Toxicol.* 17, 199–207.

Janssen, N. A. H., Schwartz, J., Zanobetti, A., and Suh, H. (2002). Air conditioning and source-specific particles as modifiers of the effect of PM_{10} on hospital admissions for heart and lung disease. *Environ. Health Perspect.* 110, 43–49.

Kodavanti, U. P., Hauser, R., Christiani, D. C., Meng, Z. H., McGee, J., Ledbetter, A., Richards, J., and Costa, D. L. (1998). Pulmonary responses to oil fly ash particles in the rat differ by virtue of their specific soluble metals. *Toxicol. Sci.* 43, 204–212.

Kodavanti, U. P., Mebane, R., Ledbetter, A., Krantz, T., McGee, J., Jackson, M. C., Walsh, L. et al. (2000a). Variable pulmonary responses from exposure to concentrated ambient air particles in a rat model of bronchitis. *Toxicol. Sci.* 54, 441–451.

Kodavanti, U. P., Moyer, C. F., Ledbetter, A. D., Schladweiler, M. C., Costa, D. L., Hauser, R., Christiani, D. C., and Nyska, A. (2003). Inhaled environmental combustion particles cause myocardial injury in the Wistar Kyoto rat. *Toxicol. Sci.* 71, 237–245.

Kodavanti, U. P., Schladweiler, M. C., Ledbetter, A. D., McGee, J. K., Walsh, L., Gilmour, P. S., Highfill, J. W. et al. (2005). Consistent pulmonary and systemic responses from inhalation of fine concentrated ambient particles: Roles of rat strains used and physiochemical properties. *Environ. Health Perspect.* 113, 1561–1568.

Kodavanti, U. P., Schladweiler, M. C., Ledbetter, A. D., Watkinson, W. P., Campen, M. J., Winsett, D. W., Richards, J. R., Crissman, K. M., Hatch, G. E., and Costa, D. L. (2000b). The spontaneously hypertensive rat as a model of human cardiovascular disease: Evidence of exacerbated cardiopulmonary injury and oxidative stress from inhaled emission particulate matter. *Toxicol. Appl. Pharmacol.* 164, 250–263.

Kodavanti, U. P., Schladweiler, M. C. J., Richards, J. R., and Costa, D. L. (2001). Acute lung injury from intratracheal exposure to fugitive residual oil fly ash and its constituent metals in normo- and spontaneously hypertensive rats. *Inhal. Toxicol.* 13, 37–54.

Lanki, T., de Hartog, J. J., Heinrich, J., Hoek, G., Janssen, N. A. H., Peters, A., Stolzel, M. et al. (2006). Can we identify sources of fine particles responsible foe exercise-induced ischemia on days with elevated air pollution? The ULTRA study. *Environ. Health Perspect.* 114, 655–660.

Lay, J. C., Bennett, W. D., Kim, C. S., Devlin, R. B., and Bromberg, P. A. (1998). Retention and intracellular distribution of instilled iron oxide particles in human alveolar macrophages. *Am. J. Respir. Cell Mol. Biol.* 18, 687–695.

Lipfert, F. W., Baty, J. D., Miller, J. P., and Wyzga, R. E. (2006). $PM_{2.5}$ constituents and related air quality variables as predictors of survival in a cohort of U.S. military veterans. *Inhal. Toxicol.* 18, 645–657.

Lippmann, M., Gordon, T., and Chen, L. C. (2005a). Effects of subchronic exposures to concentrated ambient particles (CAPs) in mice. I. Introduction, objectives, and experimental plan. *Inhal. Toxicol.* 17, 177–187.

Lippmann, M., Gordon, T., and Chen, L. C. (2005b). Effects of subchronic exposures to concentrated ambient particles in mice. IX. Integral assessment and human health implications of subchronic exposures of mice to CAPs. *Inhal. Toxicol.* 17, 255–261.

Lippmann, M., Hwang, J. S., Maciejczyk, P., and Chen, L. C. (2005c). PM source apportionment for short-term cardiac function changes in ApoE$^{-/-}$ mice. *Environ. Health Perspect.* 113, 1575–1579.

Lippmann, M., Ito, K., Hwang, J. S., Maciejczyk, P., and Chen, L. C. (2006). Cardiovascular effects of nickel in ambient air. *Environ. Health Perspect.* 114, 1662–1669.

Lu, S., Duffin, R., Poland, C., Daly, P., Murphy, F., Drost, E., MacNee, W., Stone, V., and Donaldson, K. (2009). Efficacy of simple short-term *in vitro* assays for predicting the potential of metal oxide nanoparticles to cause pulmonary inflammation. *Environ. Health Perspect.* 117, 241–247.

Maciejczyk, P. B. and Chen, L. C. (2005). Effects of subchronic exposures to concentrated ambient particles (CAPs) in mice: VIII. Source-related daily variations in *in vitro* responses to CAPs. *Inhal. Toxicol.* 17, 243–253.

Maciejczyk, P. B., Zhong, M., Lippmann, M., and Chen, L. C. (2010). Oxidant generation capacity of source-apportioned $PM_{2.5}$. *Inhal. Toxicol.* 22, 1–8.

Mauderly, J. L. and Chow, J. C. (2008). Health effects of organic aerosols. *Inhal. Toxicol.* 20, 257–288.

Miller, K. A., Siscovick, D. S., Sheppard, L., Shepherd, K., Sullivan, J. H., Anderson, G. L., and Kaufman, J. D. (2007). Long-term exposure to air pollution and incidence of cardiovascular events in women. *N. Engl. J. Med.* 356(5), 447–458.

Morishita, M., Keeler, G. J., Wagner, J. G., and Harkema, J. R. (2006). Source identification of ambient $PM_{2.5}$ during summer inhalation exposure studies in Detroit. *Atmos. Environ.* 40, 3823–3834.

Morishita, M., Wagner, J. G., Marsik, F. J., Timm, E. J., Dvonch, J. T., and Harkema, J. R. (2004). Pulmonary retention of particulate matter is associated with airway inflammation in allergic rats exposed to air pollution in urban Detroit. *Inhal. Toxicol.* 16, 663–674.

Muggenburg, B. A., Benson, J. M., Barr, E. B., Kubatko, J., and Tilley, L. P. (2003). Short-term inhalation of particulate transition metals has little effect on the electrocardiograms of dogs having preexisting cardiac abnormalities. *Inhal. Toxicol.* 15, 357–371.

Peters, A., Veronesi, B., Calderon-Garciduenas, L., Gehr, P., Chen, L.-C., Geiser, M., Reed, W., Rothen-Rutishauser, B., Schurch, S., Schulz, H. (2006). Translocation and potential neurological effects of fine and ultrafine particles: A critical update. *Part. Fibre Toxicol.* 3, 1–13.

Pope, C. A., III. (1989). Respiratory disease associated with community air pollution and a steel mill, Utah Valley. *Am. J. Public Health* 79, 623–628.

Pope, C. A., III. (1991). Respiratory hospital admissions associated with PM_{10} pollution in Utah, Salt Lake, and Cache Valleys. *Arch. Environ. Health* 46, 90–97.

Pope, C. A., III, Ezzati, M., and Dockery, D. W. (2009). Fine-particulate air pollution and life expectancy in the United States. *N. Engl. J. Med.* 360, 376–386.

Pope, C. A., III, Schwartz, J., and Ransom, M. R. (1992). Daily mortality and PM_{10} pollution in Utah Valley. *Arch. Environ. Health* 47, 211–217.

Pope, C. A., III, Thun, M. J., Namboodiri, M. M., Dockery, D. W., Evans, J. S., Speizer, F. E., and Heath, C. W. Jr. (1995). Particulate air pollution as a predictor of mortality in a prospective study of U.S. adults. *Am. J. Respir. Crit. Care Med.* 151, 669–674.

Rhoden, C. R., Lawrence, J., Godleski, J. J., and Gonzalez-Flecha, B. (2004). *N*-acetylcysteine prevents lung inflammation after short-term exposure to concentrated ambient particles. *Toxicol. Sci.* 79, 296–303.

Riediker, M. (2007). Cardiovascular effects of fine particulate matter components in highway patrol officers. *Inhal. Toxicol.* 19(Suppl. 1), 99–105.

Riediker, M., Cascio, W. E., Griggs, T. R., Herbst, M. C., Bromberg, P. A., Neas, L., Williams, R.,W., and Devlin, R. B. (2004a). Particulate matter exposure in cars is associated with cardiovascular effects in healthy, young men. *Am. J. Respir. Crit. Care Med.* 169, 934–940.

Riediker, M., Devlin, R. B., Griggs, T. R., Herbst, M. C., Bromberg, P. A., Williams, R. W., and Cascio, W. E. (2004b). Cardiovascular effects in patrol officers are associated with fine particulate matter from brake wear and engine emissions. *Part. Fibre Technol.* 1, 2 (open access).

Saldiva, P. H. N., Clarke, R. W., Coull, B. A., Stearns, R. C., Lawrence, J., Krishna Murthy, C. G., Diaz, E. et al. (2002). Ling inflammation induced by concentrated ambient air particles is related to particle composition. *Am. J. Respir. Crit. Care Med.* 165, 1610–1617.

Sama, P., Long, T. C., Hester, S., Tajuba, J., Parker, J., Chen, L.-C., and Veronesi, B. (2007). The cellular and genomic response of an immortalized microglia cell line (BV2) to concentrated ambient particulate matter. *Inhal. Toxicol.* 19, 1079–1087.

Schwartz, J., Coull, B., Laden, F., and Ryan, L. (2008). The effect of dose and timing of dose on the association between airborne particles and survival. *Environ. Health Perspect.* 116, 64–69.

Schwartz, J., Litonjua, A., Suh, H., Verrier, M., Zanobetti, A., Syring, M., Nearing, B. et al. (2005). Traffic related pollution and heart rate variability in a panel of elderly subjects. *Thorax* 60, 455–461.

Sorensen, M., Schins, R. P. F., Hertel, O., and Loft, S. (2005). Transition metals in personal samples of $PM_{2.5}$ and oxidative stress in human volunteers. *Cancer Epidemiol. Biomarkers Prev.* 14(5), 1340–1343.

Sun, Q., Wang, A., Jin, X., Natanzon, A., Duquaine, D., Brook, R. D., Auinaldo, J. G. S. et al. (2005). Long-term air pollution exposure and acceleration of atherosclerosis and vascular inflammation in an animal model. *J. Am. Med. Assoc.* 294, 3003–3010.

Sun, Q., Yue, P., Deiuliis, J. A., Lumeng, C. A., Kampfrath, T., Mikolaj, M. B., Cai, Y. et al. (2009). Ambient air pollution exaggerates adipose inflammation and insulin resistance in diet induced obesity. *Circulation* 119, 538–546.

Sun, Q., Yue, P., Kirk, R. I., Wang, A., Moatti, D., Jin, X., Lu, B. et al. (2008a). Ambient air particulate matter exposure and tissue factor expression in atherosclerosis. *Inhal. Toxicol.* 20, 127–137.

Sun, Q., Yue, P., Ying, Z., Cardounel, A. J., Brook, R. D., Devlin, R., Hwang, J. S., Zweier, J. L., Chen, L. C., and Rajagopalan, S. (2008b). Air pollution exposure potentiates hypertension through reactive oxygen species mediated activation of Rho/Rock. *Arterioscler. Thromb. Vasc. Biol.* 28, 1760–1766.

Tan, H.-H., Friel, M. I., Sun, Q., Guo, J., Gordon, R. E., Chen, L.-C., Friedman, S. I., Odin, J. A., and Alina, J. (2009). Kupffer cell activation by air particulate matter exposure may exacerbate fatty liver disease. *J. Immunotoxicol.* 6, 266–275.

Urch, B., Brook, J. R., Wasserstein, D., Brook, R. D., Rajagopalan, S., Corey, P., and Silverman, F. (2004). Relative contributions of $PM_{2.5}$ chemical constituents to acute arterial vasoconstriction in humans. *Inhal. Toxicol.* 16, 345–352.

Urch, B., Silverman, F., Corey, P., Brook, J. R., Lukic, K. Z., Rajagopalan, S, and Brook, R. D. (2005). Acute blood pressure responses in healthy adults during controlled air pollution exposures. *Environ. Health Perspect.* 113(8), 1052–1055.

U.S. Environmental Protection Agency. (2004). Air quality criteria for particulate matter. Research Triangle Park, NC: National Center for Environmental Assessment-RTP Office; report no. EPA/600/P-99/002aF, October 2004.

U.S. Environmental Protection Agency. (2008). Integrated science assessment for particulate matter. Research Triangle Park, NC: National Center for Environmental Assessment-RTP Office; report no. EPA/600/R-08/139, December 2008.

Veronesi, B., Makwana, O., Pooler, M., and Chen, L. C. (2005). Effects of subchronic exposures to concentrated ambient particles. VII. Degeneration of dopaminergic neurons in ApoE$^{-/-}$ mice. *Inhal. Toxicol.* 17, 235–241.

Wallenborn, J. G., McGee, J. K., Schadweiler, M. C., Ledbetter, A. D., and Kodavanti, U. P. (2007). Systemic translocation of particulate matter-associated metals following a single intratracheal instillation in rats. *Toxicol. Sci.* 98, 231–239.

Watkinson, W. P., Campen, M. J., Dreher, K. L., Su, W.-Y., Kodavanti, U. P., Highfill, J. W., and Costa, D. L. (2000a). Thermoregulatory effects following exposure to particulate matter in healthy and cardiopulmonary-compromised rats. *J. Therm. Biol.* 25, 131–137.

Watkinson, W. P., Campen, M. J., Nolan, J. P., and Costa, D. L. (2001). Cardiovascular and systemic responses to inhaled pollutants in rodents: Effects of ozone and particulate matter. *Environ. Health Perspect.* 109(Suppl. 4), 539–546.

Watkinson, W. P., Campen, M. J., Nolan, J. P., Kodavanti, U. P., Dreher, K. L., Su, W.-Y., Highfill, J. W., and Costa, D. L. (2000b). Cardiovascular effects following exposure to particulate matter in healthy and cardiopulmonary-compromised rats. In: Heinrich, U. and Mohr, U., eds. *Relationships between Acute and Chronic Effects of Air Pollution.* Washington, DC: ISLI Press, pp. 447–463.

Wu, W., Wang, X., Zhang, W., Reed, W., Samet, J. M., Whang, Y. E., and Ghio, A. J. (2003). Zinc-induced PTEN protein degradation through the proteasome pathway in human airway epithelial cells. *J. Biol. Chem.* 278, 28258–28263.

Xu, X., Kherada, N., Hong, X., Quan, C., Zheng, L., Wang, A., Zhong, M. et al. (2009). Diesel exhaust exposure induces angiogenesis. *Toxicol. Lett.* 191, 57–68.

Xu, Z., Xu, X., Zhong, M., Hotchkiss, I. P., Lewandowski, R. P., Wagner, J. G., Bramble, L. A. et al. (2011). Ambient particulate air pollution induces oxidative stress and alterations of mitochondria and gene expression in brown and white adipose tissues. *Part. Fiber Toxicol.* 8, 20 (p. 14).

Xu, X., Yavar, Z., Verdin, M., Ying, Z., Mihai, G., Kampfrath, T., Wang, A. et al. (2010). Effect of early particulate air pollution exposure on obesity in mice: Role of p47[phox]. *Arterioscler. Thromb. Vasc. Biol.* 30, 2518–2527.

Yue, W., Schneider, A., Stölzel, M., Rückerl, R., Cyrys, J., Pan, X., Zareba, W., Koenig, W., Wichmann, H. E., and Peters, A. (2007). Ambient source-specific particles are associated with prolonged repolarization and increased levels of inflammation in male coronary artery disease patients. *Mutat. Res.* 621, 50–60.

Ying, Z., Kampfrath, T., Thurston, G., Farrar, B., Lippmann, M., Wang, A., Sun, Q., Chen, L-C., and Rajagopalan, S. (2009a). Ambient particulate matter alters vascular function through inducing reactive oxygen and nitrogen species. *Toxicol. Sci.* 111, 80–88.

Ying, Z., Yue, P., Xu, X., Zhong, M., Sun, Q., Mikolaj, M., Wang, A., Brook, R. D., Chen, L. C., and Rajagopalan, S. (2009b). Air pollution and cardiac remodeling: A role for RhoA/Rhokinase. *Am. J. Physiol. Heart Circ. Physiol.* 296, H1540–H1550.

Zhou, J., Ito, K., Lall, R., Lippmann, M., and Thurston, G. (2011). Time-series analysis of mortality effects of fine particulate matter components in Detroit and Seattle. *Environ. Health Perspect.* (doi: 10.1289/ehp.1002613, available at http://dx.doi.org/Online December 30, 2010).

15 Radioactive Aerosols*

Lev S. Ruzer

CONTENTS

---

* I dedicate this chapter to my best friend Professor Tatjana Tolstikova.

15.1 HISTORICAL OVERVIEW

The existing lung cancer risk estimates and guidelines for exposure to radon, both occupationally and in the home, are based on the cancer experience of miners in underground mines. Here, we explain how and why we arrived at the conclusion that it is very useful to discuss radioactive aerosols, as our experience in mines in Tadjikistan relates to the validity of the existing values of risk that are adopted by many countries.

The details of the work presented show the progress made to better evaluate the actual exposure of miners to radon and its decay products. Originally, much effort was made to measure and characterize the air concentrations in mines. Many of the characterization techniques were developed at the Aerosol Laboratory in Moscow. These are described in later sections. A method was developed to measure the radioactivity content of the miners' lungs directly, and it was shown that measured air concentrations alone did not accurately describe a miner's exposure. Ultimately, the direct measurements in the lungs could be correlated with health effects, and this is the fundamental reason for this study.

15.2 AEROSOL LABORATORY OF THE ALL-UNION INSTITUTE OF PHYSICO-TECHNICAL AND RADIOTECHNICAL INSTITUTE (VNIIFTRI) IN MOSCOW (FORMER USSR)

I am a nuclear physicist by training. For most of my professional life in the former Soviet Union, nearly 20 years, I worked as the Chairman of the Laboratory of Aerosols in one of the main Soviet Metrological Centers in Moscow, the VNIIFTRI.

The primary goal of the laboratory was to develop the metrological basis for the measurement of aerosol parameters, both radioactive and nonradioactive. As a result of this work, lasting approximately 15 years, the State Standard on Aerosols, which consisted of a facility for the generation of stable aerosols and precision measurement techniques for measuring its parameters, was developed and officially approved by the Committee of Standards of the USSR. This facility allowed highly accurate measurements of

- The concentration of radioactive aerosols, both natural and artificial
- The particle size distribution in all particle size ranges
- The electrical parameters of aerosol particles

A secondary goal of this laboratory was to test the methods developed in the laboratory for measurement of the concentration of radon decay products in mine atmospheres and to develop a direct measurement of the activity (dose) levels in the lungs of miners under actual mining conditions.

The metrological part of the work demanded a careful consideration and quantitative assessment of every source of error (uncertainty) and methods to calculate the total error. By definition, the error (uncertainty) should be a small fraction of the value of interest itself.

15.3 REPORTED UNCERTAINTIES IN THE EXPOSURE OF MINERS (SOME BEIR VI REMARKS)

If we follow the typical path of metrology, that is, carefully assess every uncertainty in each step from the air concentration measured using standard procedures to the activity (dose) of individuals (or groups) of miners, we find that the total uncertainty in this case will be in hundreds of percent. This fact makes the data on dosimetry, and consequently on the risk assessment, open to question. Because the error cannot exceed 100% in the negative direction, it is better to say that the true value (of concentration, and consequently exposure) can be an order (or even more) of magnitude greater or smaller than the current estimated value. A broad discussion of some (not all) sources of uncertainties in dosimetry is presented in BEIR VI (NAS, 1999).

We were not surprised that the accuracy in the measurement of radon decay product concentrations in mines was in tens of percent (40%–50%), because in mines the instrumentation must be portable and easy to operate. Such poor accuracy is necessarily acceptable under some field conditions.

What was not acceptable from a dosimetric point of view was the fact that the air concentration of radon decay products (or exposure) measured by a standard procedure, especially in real mining conditions, was not the physical parameter that is directly responsible for the biological effect, that is, lung cancer.

Moreover, no serious attempt was made to study the correlation between radon and its decay product measurement in real mining conditions, and the damaging biological factor, the activity or dose to the lungs of miners.

According to standard metrological terminology, the uncertainty estimate in such cases is called the propagation of the uncertainties. From a metrological point of view, the worst scenario in the assessment of the physical value is not when the error (uncertainty) is large, because very often by analyzing the sources of the errors we can find a way of introducing the corresponding corrections. The worst scenario is when the analysis shows that the uncertainty is itself uncertain. This is exactly what takes place in the dosimetry of miners.

The (NRC, 1999) report "Health Effect of Exposure to Radon" (BEIR VI, 1999) discussed this problem briefly as a heterogeneity of the exposure—response trends among the various miner studies. That is, there were very different results from the 11 underground mining cohorts studied to determine the relationship between radon decay product exposure and lung cancer.

Still, it seems that the assessment of miner dosimetry appears more qualitative than quantitative because no serious attempt to simply follow the rules of metrology was practiced.

Another question arises. Are there different approaches, that is, methods in improving this situation, to diminish partial and consequently total uncertainty? It seemed to us that such an opportunity to at least diminish the uncertainty existed in using a direct measurement on miners' lungs, and this technique was subsequently proved in measurements on miners in Tadjikistan. Most of the results of these studies were not published, and exist only in a few dissertations and my *Radioactive Aerosols* (2001, in Russian).

We had access to the health effects data in these same miners on whom lung measurements were made. We decided it could clarify the risk in mines from radon and its decay product if we carried out a study on the health effects as a function of radon and especially the detailed decay product exposure.

Some (if not most) of the studies on exposure present effect results without error bars on both vertical and horizontal axes, that is, without any serious attempt to assess the reliability of the dosimetric and consequently risk data, as well as the death data.

Decreasing the statistical uncertainty in the biological effect (lung cancer mortality in the case of miners) is usually achieved by choosing a substantial number of miners with reported exposure histories. The uncertainties in exposure in most of the studies are not mentioned at all. The irony is because the uncertainty is so large and undetermined, we simply ignore it.

If someone suggested that in grouping many mines, averaging of the various exposure factors takes place, and that there is therefore no problem in risk assessment, we can respond, "Maybe." But such a point of view should be validated by statistical modeling or on-site experimentation in the mines. We should also remember that miner data are the only source of quantitative risk assessment in the epidemiology of radon and its decay product.

Our measurements in mines, and especially the direct measurement of radon decay product activity in the lungs of miners, suggest that there is no correlation between the data on measured concentration as is measured routinely and the activity (dose) in the lungs of miners.

The significance of the uncertainty in risk estimates in this case pertains not only to the abstract metrological factors but also to the economic factors. Risk estimates are directly related to the permissible concentration in mines and therefore to the atmospheric controls, ventilation rate, predicted health effects, etc.

On the other hand, it will be shown through these studies that the radon decay product concentration, together with the work load, directly correlated with the lung sickness of miners, which in turn correlated with the economic factors.

15.4 DIRECT MEASUREMENT OF ACTIVITY IN THE LUNGS: PROBLEMS WITH PRACTICAL APPLICATION

When I studied dosimetry, the idea and the resulting analyses to correlate the absorbed dose from alpha-emitting radon decay products with the external body counting of gamma-emitting decay products arose.

The importance of such a correlation was obvious. Alpha-radioactivity and the absorbed dose from radon progeny in the lungs, the physical values directly responsible for the health effect due to alpha-particles, cannot be measured directly. Alpha-particles have relatively high energy and a very short range in tissue and thus can cause significant damage to cellular DNA in the lungs because of this high linear energy transfer (LET).

On the other hand, the gamma-activity from two of the radon decay products in this decay chain is measurable, and therefore, this correlation presented an opportunity to assess the absorbed dose *in vivo*, at least in principle.

Model measurements were initiated using a human phantom and human body spectrometer in order for quantitative external measurement of the radon progeny gamma-activity to be performed. The amounts of radioactivity used in the phantom approximated the level that corresponds to that in the lung from breathing the average air concentration in the mine atmosphere. Preliminary measurements confirmed that such measurements are possible.

After the direct method was officially approved by the Committee of Radiology of the Ministry of Health of the USSR, the following statements were made:

1. The proposed method of estimating radiation effects from radon decay products contained in the lung is in contrast to the generally applied method of indirect estimation of the effect based on the measurement of concentration of radioactive aerosol in the atmosphere.
2. The proposed method is original.
3. There is no doubt that it is possible to measure the intensity of the gamma-radiation of radioactive aerosols deposited in the lungs of miners using high-sensitivity equipment designed to estimate low concentrations of radioactive substances in human organisms.

The next logical step was testing the method under actual working conditions, that is, making direct measurements on uranium miners. However, for me personally, this was to be a very difficult task.

15.5 GEOGRAPHY AND UNDERGROUND CONDITIONS OF MINE REGIONS

The mining departments in which the study was conducted were located in the territory of the former Soviet Republic of Tadjikistan and also partly in the former Soviet Republic of Uzbekistan (Tashkent province).

Tadjikistan is a mountainous country, with 93% of its territory being occupied by mountains with narrow valleys in between. The main part of Tadjikistan belongs to the Tjan-Shan mountain system. The southeastern part of the territory is located in the Pamir mountain system. All mining departments of the Republic were located on the southern spurs of Tjan-Shan and so were the mines of Uzbekistan. Only a few mines (six prospecting mines) were located in the Pamir mountain system.

Deposits of nonferrous and rare metals (lead, zinc, bismuth, tungsten, molybdenum, etc.) are concentrated mainly in North Tadjikistan. Almost all of them belonged to the Karamazar mining region. In Central Tadjikistan, deposits of Hg, W, and fluorspar were located (Takob region) and some of the uranium mines were located in North Tadjikistan.

15.6 DIVERSITY OF MINING AND WORKING CONDITIONS

The complexity of the terrain causes great diversity in the climate of underground working sites. North of the Republic, mines were located at an altitude of 1–2 km above sea level. The summer weather is hot and dry with temperatures between 25°C and 28°C. The mines in Central Tadjikistan had average temperatures of 112 and 212 in July and January, respectively (altitude 2100–2600 m). Some mines, primarily involved in geology and prospecting, are located above 3 km and activity in this region takes place only in the summer because daily temperatures are less than 15 during the day and less than 0 at night. In this zone of permafrost, the walls and roofs of the pit were covered with ice, which formed a solid monolith with the rock.

The mining departments in this area represented either the industrial complex, consisting of the mining section with a concentration factory or simply a mining section, consisting of the mines, which were independent in the sense of production and developed separate mining fields. Thus, a mine represents a distinctive "organism" with distinctive characteristics of ventilation, geological conditions, and production functions.

The mining industry, especially gold and silver mining, was primitive in this region. We were told that these were open pit mines made by the soldiers of Alexander the Great. Skeletons in the vertical position still exist in some mines.

The results of the primary and secondary studies in mines, together with measures for improving the working condition, were presented in the dissertations of A.D. Alterman and S.A. Urusov. In all mines, only one type of technological working plan (exploitation) was used: boring (drilling) of the blast hole (bore hole) with a pneumatic boring instrument (types PR-30k, TR-45) followed by blasting, ventilation of the ore face, loading the ore mass in the trunk, and exporting (carting) by electric locomotive to the vertical shaft or mouth (estuary) of the mining gallery. The most labor-intensive activities were preparation and timbering (fastening) work, which required substantial physical pressure, especially in a drift (a rise in the mine shaft).

All work is done by the "brigade" method. All members of a "brigade" (team) take part in the management of any emergency situation (derailing of the trucks, water breaks in the main line, etc.) and other hard work.

The important parameters for working conditions are the following:

- Time distribution of work with known atmospheric decay product characteristics
- Work load, which determines the breathing rate

For many job categories, it is difficult to determine the specific work site (locksmiths, electricians, mine surveyors, samplers, electric locomotive engineers, mining engineers, etc.). For them a "working site" can be all of the working sites, including "nonworking sites." A much more constant working place exists for the drillers, despite the fact that in specific situations their time at the pit face can vary within broad limits. According to the time sheets provided on six shaft sinkers (drifters) during two shifts, the working time varied from 8% to 92% of the shift time. Some of the miners were constantly situated at the same working site.

Most of the mining industry in this region used artificial ventilation. In the period of the primary study, only three mines did not have artificial ventilation. Blind (cul-de-sac) mine shafts were ventilated by ventilators (types VAM-450, BM-200). In most cases, suction-type ventilation was used together with a dispersed set of ventilators in the pipelines. A substantial disadvantage to this method is the possibility of contamination (pollution) of the clean air entering the mine working with radon and also the possibility of suctioning polluted air from nonworking sites.

In general, numerous nonworking sites connected with working sites created a negative influence on ventilation and consequently on radon and its decay product concentration in the working sites, increasing it. Due to the noise, often miners turn off ventilation during the shift and mine working takes place without air replacement. The worst air conditions, therefore, took place in blind shafts, particularly in shaft sinker (drifter) working sites. In cleaning sites, due to trough ventilation, air exchange was constant.

In the nonworking sites, air usually stagnates. These nonworking sites are visited from time to time by geologists, mine surveyors, and others, who carry out different types of work, such as dismantling, taking down the rail, pipes, cables, etc.

The main tool in the fight with dust, besides ventilation, was wet drilling. Even in mines where the average dust concentration did not exceed the APC standard level in mine shafts, the dust concentration was substantially higher and therefore, in terms of dust, the worst situation was for ore face-workers (getters).

Especially difficult working conditions arose for the geology-prospecting groups due to the following reasons:

- Remoteness (distance) of this type of work from main transport arterial roads and industrial centers
- Difficulties in energy supply
- Difficulties in alpine (mountain), arid (waterless), deserted, permafrost conditions
- Difficulties with maintenance equipment
- Absence of ventilation and mountain life-saving services

All these situations existed in Tadjikistan. The prospecting mine shaft is blind, that is, it has only one exit to the surface, which makes the ventilation of such working sites very difficult. In addition, a cross section of such sites is usually less than $5\,m^2$, making it impossible to use ventilation pipes with large diameters. A low cross section also makes it impossible to use loading mechanisms (gearing), which resulted in most work being manual with a very high physical work load. All these circumstances came to our attention after we became closely acquainted with mine working conditions.

It became clear that mining conditions are very diverse and, therefore, too complicated to use one or two measurements per month of radon and its decay product concentration (or exposure) as a characteristic of the dose of individuals or a group of miners.

15.7 DIRECT METHOD: THE TADJIKISTAN STUDY AS AN OPPORTUNITY TO REDUCE LUNG DOSIMETRIC UNCERTAINTY

Our first underground experience was a very dramatic one. We arrived in the capital of Tadjikistan, Dushanbe, after an 8 h flight from Moscow, late in the evening. The next day, early in the morning, we went to the mine located outside the city.

In order to obtain the best results, we asked the mining administration to show us the places with the highest concentrations. At first, our guide led us to the wrong site and it was only by chance that we did not perish. We arrived at the site, where the concentrations were 10^4–10^5 times higher than permissible. We wore respirators, but I was so tired after the flight and the almost sleepless night that I had difficulty breathing, and so took off the respirator. Radon decay products were not only in our lungs, they were on our faces, clothes, everywhere.

We immediately went to another place where the concentration was closer to the average levels, and continued measuring the radon decay products in our lungs. After this experience, we continued to make direct lung measurements in metal mines in North Tadjikistan (Leninabad, now the Khodgennt region). The measurements in uranium mines were performed in the mines of the Uranium Industrial Complex, located in the city of Chkalovsk near the Sir-Darjinsky reservoir. The mines of this complex were located on the huge territory of three former Soviet Republics (all now independent states)—Tadjikistan, Uzbekistan, and Kirgizstan. For many years, together with the highly qualified and enthusiastic personnel of the local dosimetric laboratory, we made measurements in the mines of the Yangiabad region. Then, similar measurements were performed in North Kazakhstan.

It was rumored among the Soviet population that prisoners were working as miners in uranium mines. People called them "smertniki" (prisoners sentenced to death). However, we did not see any prisoners underground. Most of the uranium miners were recruited from other parts of the former Soviet Union in order to make money. They became worried when we measured the radioactivity in their lungs instead of routine air measurements. From this, we understood that these types of measurements have a special psychological impact on both miners and the mining administration.

At one meeting at the Industrial Mining Complex, the minister of this former secret ministry ("Middle Machinery") told an audience of miners, "You should not worry about your health, because radon is going in and out of your body." But suddenly a voice from the hall loudly said, "But the decay products remain in the lung."

Our measurements directly proved that radon decay products are in miners' lungs in measurable amounts, and this made an impression on the mining administration. As will be shown later, the practical application of this method inaugurated some improvements in working conditions, which resulted in declines in the lung sickness of miners.

It should also be mentioned that when the dosimetric data in miner studies are used, the problem of objectivity (lack of bias) should be taken into account. We will discuss this problem later in connection with the Polish study. From our own experience, we know that each time we asked for official approval of our measurements (which was mandatory), the chief engineer would very politely and cunningly exclude the results that exceeded permissible levels. With a smile, he hinted that it would be difficult for us to come again if such "incorrect" data were reported.

Those years until the middle of the 1970s were very productive, especially in measurements in Tadjikistan. Together with the measurements of the distribution of radon and radon decay products for different working sites, groups of workers, and the direct measurement of activity in the lungs of miners, a study of lung cancer mortality and lung sickness was provided. Unfortunately, after the conclusion of the study in Tadjikistan, it was impossible to continue similar measurements in other mines, including in the Far East.

After arriving in the United States, I received an invitation from a publisher in Washington, DC, to prepare a monograph on Aerosol Research in the USSR.

When I left Moscow, I was afraid to take some of my papers, dissertations, and other materials. No one knew what the KGB would demand from me at the airport.

Based on what was available for me, I prepared a monograph in 1989 entitled *Aerosol R&D in the Soviet Union*. The book was published mostly for government organizations. I attended the Aerosol Conference in Reno in 1989 and asked the speaker at the plenary session Dr. Naomi Harley about the quality of mining dosimetric data. From her response, I understood that the data of our Tadjikistan miners' study could present useful information to this problem.

My last book *Radioactive Aerosols* was published in Russia in 2001. This book is different from previous publications, because it includes materials on developing Aerosol Standards, dose measurements in the lungs of miners, and other materials on radioactive aerosols published together with American colleagues.

15.8   RADIOACTIVE AEROSOLS AND LUNG IRRADIATION

The incidence of elevated rates of lung cancer among miners exposed to radon decay products has led to programs that control worker exposure. Furthermore, such evidence is the basis for concern about lung cancer risk among the general public due to radon in homes, which can lead to a substantial number of putative lung cancer deaths.

In conducting epidemiological studies or in controlling worker or general-public risks, human exposures to the decay products must represent or estimate, in some way, time-integrated airborne concentrations of the relevant radionuclides ("exposure") or calculated energy deposition in the lung tissues affected ("absorbed dose"). Intermediate quantities that may also be used are the amount of activity deposited in the whole lung or relevant part of the lung ("deposited activity") or intake, which was officially included together with concentration in the Norms of Radiation safety. The use of any of these quantities in either epidemiological studies or control programs involves, either explicitly or implicitly, presumptions about their relationships with one another. However, although it is the dose that is the cause of changes in tissue that lead to cancer, it is airborne concentration that is measured and corresponding estimates of exposure that are used as correlates or predictors of risk.

The absorbed dose or the activity (intake) in lung tissue is calculated to understand the actual insult leading to lung cancer, or to permit the quantitative comparison of doses arising from atmospheres having different characteristics such as in mines and homes. If exposure is used as an index of dose (activity), for example, in epidemiological studies, presumptions are made implicitly about a comparability or constancy of breathing rates and of the retention of decay products in the lung.

Since both dosimetric and epidemiological data for miners are the main source of a risk cancer mortality assessment for miners and the general population, it is important to understand to what degree correlation between the measured radon decay products concentration in the air and dose to the lung takes place in the real mining environment.

In this connection, it should be noted that, especially in mining environments, inhalation rates depend on the groups involved and on working conditions, specifically the physical load, which can vary substantially. Retention of the decay products depends on the properties of the aerosols and also on the load of work and is typically not known accurately.

It should be pointed out that measurements of the airborne concentrations are not complete enough to provide directly the value of exposure, because the concentrations to which individuals are exposed vary substantially in time and space. For example, variation of the ventilation rate even for a short period of time can lead to a substantial change in concentration. Moreover, the concentrations of radon decay products and other nuclides in the breathing zone may differ substantially from the value measured by the actually implied standard instrument (Domanski et al., 1989). Finally, the very concept of "workplace," associated concentration, and load of work are indefinite, since miners are typically at a number of places during their working day, with variable concentrations (as well as nature of work). Since the measurements of concentrations in mines may be performed only once or twice a month (not to mention estimated retrospectively), we cannot expect reliable correspondence between exposures estimated from measurement results and actual personal exposures, that is, time-integrated concentrations. Thus, the use of measurements of airborne concentration as a basis for estimating or comparing dose can lead to substantial errors, both because of the lack of correspondence to concentration measurements and personal exposures and because of uncertainty and variability in breathing rate and radionuclide retention. These errors, in aggregate, may be as much as an order of magnitude and therefore make dosimetric and consequently risk assessment data unreliable (Ruzer, 2001).

It should also be mentioned that the dose to the lung from radon itself is negligible in comparison with that from its decay products. However, until now assessment of the risk in many studies is based on the measurement of radon with an assumption on the degree of equilibrium between radon and radon progeny in mines. The degree of equilibrium varies substantially in time and space, both for underground workers and the general population.

Unfortunately, even after concentration (exposure) of decay products (not of radon) was established as a measure of the irradiation of the lung more than 40 years ago, no discussion and experimental study in mines were conducted to determine the degree of correlation between the measured concentration and the actual dose (intake) for individuals and a group of miners.

To assess the reliability of dosimetric data for miners, we should carefully consider every factor related to dose (activity) calculated or measured together with assessment of the uncertainties for every partial factor and the total error (uncertainty).

It seems that a lack of such careful consideration of the metrological aspects in the assessment of the dose to the lungs of miners from radon decay products in the past resulted in large uncertainty in the dose to the lung, and correspondingly unreliable data in estimating the risk of lung cancer.

It is obvious that because irradiation of the lung by radioactive aerosols, and particularly radon decay products, depends on physical activity, there are some biological factors that will contribute to the total uncertainty in the dose assessment. In order to simplify the task, we will put aside, at least for now, the biological factors and focus only on the physical factors, especially because the uncertainty in physical factors is very large in itself.

Let us call all these factors that contribute to the release of energy to the lung tissue as "dosimetric factors." There are two groups of dosimetric factors. The first group—radioactive dosimetric factors—determines the total energy of radiation in the air or in the lung. The second—nonradioactive group—is mainly responsible for the portion of this energy deposited in the lung.

The following are the factors in the first group—radioactive parameters:

1. Concentration measured by standard procedure
2. Breathing zone concentration
3. Concentration of unattached radon decay products
4. Exposure
5. Activity in the lung
6. Energy of the alpha and beta particles
7. Integral absorbed dose
8. Absorbed dose

The parameters of the second nonradioactive group are

1. Counting, surface area, and weighted aerosol concentration
2. Aerosol particle size distribution
3. Breathing rate (minute volume)
4. Deposition coefficient
5. Filtration ability of the lung (FAL)—combination of the volume breathing rate and deposition coefficient
6. Efficiency in using respirators (when applicable)
7. Parameters of biokinetic processes

Besides all these factors, mention of the "work itinerary" ("scenario of exposure") is also a necessary element for correct dose assessment.

In the NRC Report, the structure of the lung inner surface layers from the point of view of alpha-dosimetry was studied.

It is well known that the dose to lung tissue from the inhalation of radon progeny cannot be measured. It must be calculated by modeling the sequence of events involved in inhalation, deposition, clearance, and decay of radon progeny within the respiratory airways. We must discuss this statement because of its great importance to the dosimetry of radon decay products.

In the first step of this consideration, that is, the correlation of the concentration in the breathing zone and that in the place of sampling, it is very difficult to predict how some of the parameters of the first and second group will interact. It is difficult, or even impossible, to model the concentration in the breathing zone, especially for hard-working miners. Even the sparse data available showed that the ratio between the breathing zone concentration and that measured at some distance varies within very broad limits. This is especially true for aerosol particle size distribution and activity particle size distribution, which are responsible for the deposition in different parts of the lungs.

The second step of the distortion of particle size distribution spectra is the change of particle size distribution inside the lung pathways due to change in humidity and temperature. The modeling of this step can lead to very high uncertainty in the calculation of particle activity deposition.

It is true that the physical effect of alpha-radiation of the lung tissue cannot be measured directly. But very often, when the direct measurements of the physical factors are not possible, a different approach is used. In such cases, we often try to find other measurable physical parameters, and at the same time use a correlation (analytically or in another way) between this measurable and nonmeasurable physical value. One method is external gamma-ray counting of the lung.

Such analytical correlation of the gamma-activity of radon progeny (a measurable factor) and the alpha-dose to the lung tissue (nonmeasurable factor) was derived in (Ruzer, 1958, 1964). The possibilities of practical application of this "direct method," including an introduction of the necessary corrections and assessment of the accuracy of measurement, were presented in Ruzer (1962), Vasin et al. (1975), Ruzer and Urusov (1969), and Urusov (1972).

The practical applications of the direct measurement of the activity in the lungs of miners were carried out in uranium and nonuranium mines of the republics of Kazakhstan, Uzbekistan, and Tadjikistan on more than 500 miners. The results of these studies were published in three dissertations (Urusov, 1972; Alterman, 1974) (in Russian) and in my book *Radioactive Aerosols* published in Russia in 2001.

The radon decay products of gamma-radiation of ^{214}Pb and ^{214}Bi are natural markers, and the relatively high historic radon progeny concentration in mines makes it possible to obtain information from direct measurements of the gamma-activity of radon decay products deposited in the lungs of miners.

In principle, using more sophisticated gamma-detectors together with collimators, it is possible to measure the distribution of radon progeny in different parts of the lungs under real mining conditions.

A detailed analysis of all corrections in such measurements suggested that the accuracy in the activity assessment is satisfactory from the point of view of practical dosimetry.

In all calculations of the dose to the lung tissue provided in NRC Report (1998), it is assumed that activity of the radon decay products in the lung or in part of it is known, without taking into account uncertainties in the activity assessment.

However, here lies the main problem. There are no real data on the activity in the lung. From a practical point of view, it makes no sense to use data concerning the activity of radon progeny without mention of the errors of the calculation of the dose distribution through lung tissue, especially because the errors in the activity assessment are so large.

The purpose of practical dosimetry is to present a set of rules (algorithm) for determining the quantity responsible for a biological effect (in this case, the absorbed dose due to alpha-radiation of radon progeny) using the value—in the case of miners—measured by the standard procedure radon progeny concentration. This should be done with the assessment of associated errors.

The method of direct measurement of activity in the lungs of miners allowed the study of the transition coefficient between activity in the air and in the lungs. It was demonstrated in Ruzer and Urusov (1969), Urusov (1972), Alterman (1974), and Ruzer et al. (1995) that this coefficient—FAL—is different for different physical activities, which resulted in different doses (intake) in different groups of miners.

As a result of this nonuniformity in dose, the variability in lung sickness and lung cancer mortality among nonuranium miners in Tadjikistan was established.

This aspect of nonuniformity both for dosimetric and epidemiological data is very important. In previous epidemiological studies, a uniformity approach to dosimetry was used and this can lead to a substantial error in risk determination.

The nonuniformity problem together with other aspects mentioned in the following suggests that there is good cause for reevaluating the previous epidemiological data.

15.9 AEROSOL CONCENTRATION MEASUREMENT

15.9.1 RADON AND ITS CONTRIBUTION TO ABSORBED DOSE

In a real atmospheric environment, we have a combination of radon (inert gas) and its decay products (natural radioactive aerosols). For the assessment of the biological effect of the inhalation of radon and its decay products, it is important to estimate the contribution to the dose of radon itself and its decay products separately. Such an assessment is important because of different physicochemical properties of radon and radon progeny, and subsequently their different behaviors in the lung.

15.9.1.1 Experimental Study on Animals

In Leites and Ruzer (1959), the results of the experiments on white rats inhaling radon with its decay products in the small chamber ("emanatorium") are presented. Doses were determined separately for radon and radon progeny. Besides dosimetry, pathology—anatomical study and weighting were provided.

The animals inhaled radon for 2 h every day with radon concentrations of 4.83×10^8 Bq/m^3 (first series), 5.03×10^7 Bq/m^3 (second series), and 1.33×10^6 Bq/m^3 (third series).

As there were no available methods for the measurement of the concentration of radon decay products for the time of this study, especially for the small-volume chamber, the simplified variant of the method proposed in Ruzer (1960a,b) was used.

From these data, it is clear that the general absorbed energy due to the decay of radon in the organism is much smaller than that due to the decay of radon progeny. The main reason for this is that the deposition of radon decay products in the lung is very high and at the same time biological clearance of the inert gas radon from the organism is high. In other words, the behavior of decay products in the organism is determined completely by radioactive decay in the lungs with no observed trace of the clearance. At the same time, radon, due to its chemical nature, does not deposit in the organism for a long period and therefore does not produce a substantial amount of decay products and a subsequently absorbed dose. A difference in terms of absorbed dose will be much more substantial if we take into account the mass of irradiation tissue—lungs for decay products and whole body for radon—because with some approximations we can assume that the distribution of radon in the body is uniform. According to our calculations, the absorbed dose to the lung from decay products of radon is two to three times higher in magnitude than from radon itself.

In the first series of experiments, animals beginning from the second week of exposure were inert and sluggish, with decreasing appetite and decreasing weight in comparison with the control group. They died after the completion of the irradiation within 7–72 days in the first series of experiments and within 69–94 days in the second series. The main macro- and microscopic changes took place in the lungs, especially in the front and median parts of it. In the third series, no changes in the weight and behavior of animals were detected during the period of exposure and even 10 months later.

These biological results show that in this case typical radiation sickness of the lung took place, which was in agreement with the dosimetric data.

For directly checking the fact that the biological effect in this case was determined mainly by the radon decay products, and not from radon itself, animals took a bath in radon water in such a way as to practically exclude the entry of radon and its decay products through the lungs. The amount of radon that entered the body from the skin was close to that in the first series of experiments, that is, the highest concentration. Contrary to the case of breathing, no difference in the biological effect from the control group was found.

The comparative contribution of radon itself and its decay products is important in terms of understanding what accuracy of measuring concentration of radon itself is important for dosimetric purposes.

The measurement of radon concentration itself is practically important because it shows the potential upper bound of danger associated with radon decay products, that is, with the equilibrium concentration of radon progeny.

However, it does not represent the real qualitative assessment of the dose, because depending on the shift of equilibrium the ranges of the dose can be of 2–3 orders of magnitude. The contribution of radon will be substantial only in cases when the shift of equilibrium is small.

Still, the results of radon concentration measurements are practically important in cases when preliminary assessment of the danger associated with natural radioactivity, especially in houses, should be made.

Data on measurement of radon decay products in air by alpha and beta spectrometry are presented in Ruzer and Sextro (1997a).

Data on measurements of the unattached activity of radon progeny are presented in Ruzer and Sextro (1997b).

15.10 DOSIMETRY

15.10.1 Intake versus Exposure: Propagation of the Uncertainties in Dose Assessment in Mining Studies

Epidemiological studies of underground miners are the basis for estimating the risk of indoor radon. Although there have been a number of studies investigating the health effects of exposure to radon decay products in mines (the most recent compilation has been carried out by NRC 1998 [BEIR VI]), there are several unresolved issues in the assessment of the actual radiation dose to the miners' lungs:

- Lack of detailed spatial and temporal data on radon and radon decay product concentrations
- Variability in the ratio between concentrations as measured by the standard inspection procedure and as measured in the breathing zone
- Variability in breathing rates and deposition coefficients for radon decay products in the lungs for different types of work and among different groups of miners
- Information not known about the use of respirators by different groups of miners
- Very little data on the work itinerary (scenario of exposure) for individuals or groups of miners

The presence of such errors in the exposure estimates for miners has been widely recognized and discussed in NRC 1998 (BEIR VI). In this context, it has been noted that concentrations of radon and its decay products vary spatially and temporally within mines, although little data have been published in this regard. In New Mexico mines, for example, information presented on dosimetry documented extensive variation in the concentrations of radon progeny across various locations within mines in Ambrosia Lake, New Mexico (BEIR VI).

It has also been pointed out in BEIR VI that exposure estimates for individual miners would be ideally based on either a personal dosimeter, as used for low-LET occupational exposure, or on detailed information on concentrations at all locations where participants in the studies received significant exposure. For miners, information would be needed on the location where time was spent, the duration of time spent in the location, and the concentration in the location when miners were present. Personal dosimeters for radon and its progeny have not been developed until recently and their usage has been limited; hence, detailed information on concentrations within mines and time spent in various locations has not been available to most epidemiological studies.

In the epidemiological studies, these ideal approaches have been replaced by various pragmatically determined strategies for exposure estimates that draw on measurements made for regulatory and research purposes and extend the measurements using interpolation and extrapolation to complete gaps for miners in particular years. Additionally, missing information for mines in the earliest years of some of the studies was completed by either expert judgment or by recreation of operating conditions. It should be noted that very little has been done to quantitatively assess uncertainties related to the replacement of this ideal approach with the pragmatic approach.

This is a very important issue because, in reality, uncertainties in the assessment of the dose and even exposure are many times greater than statistical errors in the assessment of lung cancer mortality, which make risk assessment very uncertain. In Alterman (1974), Urusov (1972), Ruzer et al. (1995), and measurements of radon decay product concentrations in different working sites, the direct measurement of the activity in the lungs of individual miners and lung cancer mortality and lung sickness for different groups of miners was studied in nonuranium mines of Tadjikistan. It was shown that variations in the radon and radon decay product concentrations vary by a factor of 2–10 and that calculated radon progeny intake varies by a factor of 3–14. A similar nonuniformity was found in lung cancer mortality and lung sickness. The uniformity approach can also be the source of uncertainties much greater than the statistical errors in the assessment of mortality.

From the point of view of dose (alpha particle energy deposited in the lung) assessment of all the factors, that is, concentration of radon decay products measured by standard procedure, the ratio between this measurement and the concentration in the breathing zone, volume breathing rate, and deposition coefficient are equally important, including efficiency in using respirators, because all these factors affect the amount of radioactivity deposited in the lung.

In this section, based on the quantitative analysis of accuracy of dosimetric factors, we will show that the conception of uniformity is incorrect in terms of both dosimetry and epidemiology.

As a result, we suggest that intake, which takes into account not only radon progeny concentration in the working site but also the physical load of work for different groups of miners and the scenario of exposure, is a better measure of radon and its decay products dosimetry than exposure.

15.10.1.1 Discussion on Miner Radiation Dosimetry: Quantitative Approach

In every miner epidemiological study, two important parameters should be determined together with their errors (uncertainties):

1. Lung cancer mortality
2. Radon decay products dosimetric characteristics: concentration, exposure, activity in the lung, intake, dose of alpha-radiation

In a majority of published studies, only the assessment of concentration and exposure takes place, despite the fact that both are characteristics of air and not of tissue irradiation. It is not even a characteristic of air really breathed in by miners, as was shown in studies (Domanski et al., 1989).

It should also be mentioned that to achieve reliable data on risk assessment, uncertainties (errors) for both mortality and dosimetry must be comparable.

One of the most comprehensive studies of this kind is BEIR VI, which summarizes dosimetric and epidemiological data from 11 local studies in uranium and nonuranium mines in different countries. Similar data from the former Soviet uranium industry were not included in this report, because at the time of preparation data were yet to be classified. In the report BEIR VI, two objective criteria were established for inclusion in the study:

1. A minimum of 40 lung cancer deaths. This criterion established the level of uncertainty in lung cancer mortality in the range of 15%–20%.
2. Estimates of Rn progeny exposure in units of WLM for each member of the cohort based on historical measurements of either Rn or Rn progeny.

We will try to assess the uncertainties in every step of the calculation from the air concentration to the activity (intake, dose) of radon decay products in the lung, assuming that activity (intake) is the main physical value responsible for the biological effect in this case.

1. In some of the studies mentioned, assessment of radon progeny was made based on reconstruction and some assumptions (China, New Foundland, Sweden, Beaverlodge). In such cases, it is impossible even to determine the uncertainty (errors). The use of these data should be called into question. It seems that criterion 2 was not applied in this case because it was impossible to estimate exposure for each member of the cohort. In the Workshop on Uncertainty in Estimating Exposure to Radon Progeny Studies of Underground Miners (BEIR VI, E—Annex 2), questions were raised about removing certain cohorts, ranking of cohorts, and obtaining additional information.
2. In many of the earliest studies, only radon concentration was measured. It is well known that the contribution of radon itself to the dose is negligible in comparison to its decay products. It was shown that the equilibrium factor varies in mines and has a wide range from 0.2 to 0.9; Domanski et al. [1989], Poland). Therefore, assuming that the average equilibrium factor is 0.5 for all situations, we can have uncertainties on the order of 100%.
3. Measurements of radon decay products are directly related to irradiation of lung tissue. Errors in calibration and measurements should be taken into account. Assessment of the errors related to radon progeny measurements was the topic of discussion in the special workshop (BEIR VI, E—Annex 2). It was estimated that uncertainty in this case was about 50%. This is consistent with the standard adopted in the former Soviet Union for radon decay products measurements in mines—the errors should be on the order of 30%–40% (Antipin et al., 1980).
4. Another source of error is the correlation between concentration measured by a standard procedure (or area monitor) and that in the breathing zone. In BEIR VI (1998), the suggestion was made that in some cases there are no substantial differences in these two concentrations. But in Domanski et al. (1989), results were also presented on the correlation between these two factors. The one measured by the air sampling system (ASS) was based on the field monitoring of radon progeny in air; the second one, called the individual dosimetric system (IDS), was based on the individual dosimeter worn by miners. The ratio ASS/IDS varies from 11.0 to 0.14. that the ratio of concentrations measured by a personal aerosol sampler (PAS) and a standard aerosol sampler (SAS) depends substantially on the strategy of measurements and sampler location. In short, this means that if the concentration was measured only by standard procedure and the ratio PAS/SAS is not known, additional uncertainty in the assessment of the concentration related to lung irradiation can be on the order of hundreds of percent.
5. By definition, irradiation of the lung by radon decay products should depend on physical activity. The data on the breathing rate for different types of physical activity are presented in Alterman (1974), Ruzer et al. (1995), and Layton (1993). The problem for miners is that this parameter changes substantially within the shift from 10 to 30–40 L/min, and by using the average value for the breathing rate an error of 100% can be made. It should be mentioned that the measurement of the actual breathing rate was usually made only for low physical activity because the measurement method itself disturbs real breathing conditions, especially in the case of hard work.

6. For deposited activity in lungs, the combination of breathing rate and deposition coefficient is important, and not these values separately.

   A new value, FAL, was introduced as a combination of breathing rate v and deposition coefficient k (FAL $= vk$) as a bridge between concentration in the breathing zone and activity in the lungs. FAL can be estimated by measurement of the gamma-activity in the lungs of miners and radon progeny concentration in the breathing zone. This method permits measurements on miners without disturbing their real working conditions (noninvasion approach) and was used for the measurement of different groups of miners. The error in estimating FAL was 20%, which is many times smaller than the combined error in the assessment of breathing rate and deposition coefficient (Urusov, 1972; Alterman, 1974; Ruzer et al., 1995).

7. In Ruzer (1958, 1964) and Ruzer and Urusov (1969), the method of direct measurement of the activity in the lungs of miners was developed and used in miner measurements both for the assessment of activity in the lung and for the estimation of deposition of radon decay products. The accuracy of activity assessment in the lungs of miners, including corrections on the shift equilibrium in the air and geometric factors, was 30%–40%.

No correlation was found between direct measured activity in the lungs of miners and calculated activity based on radon progeny concentrations measured by the standard procedure. On the other hand, a relationship was found between the concentration in the breathing zone, the parameter of physical activity (FAL), and direct measured activity in the lungs.

As a result of this, conclusions should be made that the total uncertainty of the activity in the lungs or the dose for miners calculated on concentration measurements can be on the order of hundreds of percent. But because errors close to even 100% make no sense, it is better to express this uncertainty in such a way that the true value of the dose or the activity in the lung can be at least on the order of magnitude lower or higher than calculated.

Only direct measurements of activity can produce an accuracy acceptable from the dosimetric point of view and close to the statistical error that is used in the assessment of lung cancer mortality.

In Domanski et al. (1989), the results of a long-term study were described by comparing two independent systems, that is, the ASS and the IDS, which were implemented and tested for 6 years simultaneously in Polish underground metal-ore mines.

Each of these systems has certain different inherent advantages and critical weak points. The main feature of the ASS is usually the relatively high precision of each single measurement; however, the strategy of monitoring and the selection of the proper frequency of monitoring and the site of the system on the area of the mine still remains the weak point of the ASS.

On the other hand, the critical point of the IDS lies in the cost of the measuring devices or in the doubtful precision of the measurement technique. The term "dosimetry system" should also not be used, because such types of instruments measure only concentration, not the dose, which in this case depends on physical activity.

The ASS system was implemented in all Polish underground copper and zinc—lead ore mines between 1981 and 1983, and the crucial point, that is, strategy of monitoring and sampling, was thoroughly considered, discussed, and finally approved by the Institute of Occupational Medicine. It was recommended that from tens to several hundreds of potential alpha-energy concentration measurements of radon progeny in the year would be done at the sites where miners actually work and in the local air stream outlets.

The IDS, based on "individual dosimeters" worn by miners on the backs of their helmets, was introduced in all the mines under consideration in the period between 1977 and 1979. This system is based on the use of small cassettes containing a track etch detector foil sensitive to alpha-particles emitted by radon decay products.

Thousands of measurements conducted simultaneously in 11 metal-ore mines under these two technically compatible, but entirely independent, long-term systems of radiation exposure

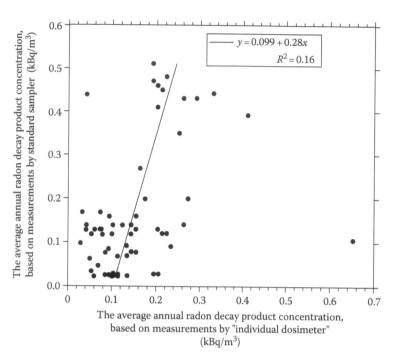

FIGURE 15.1 Average annual radon decay product concentration, based on measurements by individual dosimeter (kBq/m³).

assessment have brought results that have led to important conclusions concerning the reliability and validity of these systems.

The summary of the results of these measurements is presented in Figure 15.1. The first general conclusion is that these systems yield inconsistent results. It can be seen from Figure 15.1 that the correlation coefficient is very low—0.16. Thus, the general question arises as to which system is more reliable and practicable.

The ratio of annual concentration, that is, mean ASS/mean IDS, varies from 11 to 0.14, depending on the mine and the year of implementation of the ASS. The results show that the policy of air sampling is very important, because soon after ASS implementation the effect which the authors called "hunting for results" takes place, that is, radiation officers or dosimetrists look for places where concentration will be higher, or in other words they "hunt" for results, neglecting the strict instruction given to them during the training courses before implementing ASS. The opposite effect, revealed by results collected in two other mines, can be called the "avoidance of inconvenient results." The authors suspected that in these cases, the radiation safety officers probably tried to hide the results that would increase the mean value, or simply avoid places in which miners work at relatively higher concentration.

Both the aforementioned effects distort the real picture of both concentrations and miners' exposure. The "hunting" effect appears to be disappearing over the years, but it sometimes turns into the "avoidance" effect, and the "avoidance" effect appears to be more serious.

The conclusion was drawn that IDS is more reliable in miners' exposure assessment than the currently implemented ASS. Nevertheless, ASS should not be abandoned, because it is obviously better than nothing and plays a useful role as a tool for technical preventive actions.

Still, as a measure of real irradiation of the lungs of all miners, these standard types of measurements should be called into question, because the calculation of the activity (dose) to the lungs of individuals or groups of miners based on the results of such measurements will yield results with such uncertainty, which is unacceptable in the dose and, correspondingly, the risk assessment.

The authors mention that the results presented in Ruzer and Sextro (1997a,b) can have a possible impact on and contribution to the discussion on the value of radiation mortality risk. It was shown in a study of American miners that the mortality risk was several times lower than in an analogous study of Czechoslovakian mines. Such a discrepancy can at least be partially explained by different strategies of measurement. This would lead to an overestimation of miners' expected exposure and, consequently, to the obtaining of a much lower mortality risk for American miners than for Czechoslovakians.

Summing up, Domanski et al. (1989) declared the following opinion:

> The postirradiation lung cancer risk factor due to exposure to radon progeny should be thoroughly reconsidered in the light of discovered effects. The correctness of the miners' exposure can be provided only by the IDS.

Based on the results the following conclusions were drawn:

- Since the beginning of the study of the health effects of radon and its decay products, concentration and exposure were chosen as an adequate measure of the irradiation of the lungs of miners. No serious and systematic attempts were made to establish the individual dosimetric control for underground workers from alpha-emitters of radon series despite the fact that for low-LET radiation in occupational exposure such individual control is mandatory.
- In the majority of dosimetric and epidemiological studies on miners, no quantitative considerations were made according to the rules of metrology for accuracy in the evaluation of exposure, intake, and dose to the lungs for every dosimetric parameter, that is, concentration in the breathing zone of miners, volume breathing rate, and deposition for different degrees of the load of work.
- Routine methods of assessment of exposure (intake, dose) in the lungs of miners are very sensitive to the strategy of the measurements and consist of errors (uncertainties) in hundreds of percent, which make the correct risk assessment of miners, and consequently of the general population, very uncertain.
- Direct measurements of the activity (dose) in the lungs of miners permit one to assess dosimetric factors with an accuracy close to the statistical error in the assessment of lung cancer mortality.
- Because practically it is impossible to provide individual measurements for every miner, an intake for the group of miners is proposed, which takes into account diversities of mine environment as an alternative to exposure as a characteristic of miners' irradiation.

15.10.2 Method of Direct Measurement of the Activity (Dose) in the Lungs of Miners

The determination of the absorbed radiation dose in the lungs of miners due to the inhalation of radon and its decay products can in principle be estimated from the air concentration of radon progeny at the work sites, the rate of inhalation and the retention in the lung airways. However, the value of the rate of inhalation is indefinite, since it depends on the physical load (the nature of the work) and varies within broad limits from 10 L/min for the rest to 30–40 L/min for the hard work. The air concentration varies with time and the value of the lung deposition fraction depends on the particle size of the inhaled aerosols and the inhalation rate and was not measured under actual working conditions.

The accurate measurement of the air concentration during exposure is complex. First the concentration at the same working site is subject to substantial variations. For example, a variation of the intensity of ventilation even for a short time period of time leads to a change in the radon and its progeny concentration of several-fold. Moreover, the content of aerosols directly in the breathing zone may differ substantially from the value measured by an instrument, located in another part of the working area.

Finally, the very concept of the "work place" and "concentration at the work place" are indefinite since miners are at several work places during their work shift, each with different possible concentrations of radon decay products and physical work loads.

For a correct evaluation of the exposure under actual working conditions, we have also taken into account that in some cases miners used respirators for protection of the lung from the aerosol conditions in the mine. All these factors are important for the assessment of the irradiation of the lungs and we will call them "the exposure scenario " or "the working itinerary."

Since practically all measurements of the air concentration of radon progeny at work sites in mines are performed only once or twice a month, there cannot be a precise correspondence between the actual and measured individual (or for group of miners) breathing zone concentrations.

No systematic studies have been made for the assessment of the breathing rate and deposition in lungs for individual or group of miners in real underground conditions. Therefore, it is impossible to calculate the individual exposures correctly.

We present here a method for making direct measurements of the radioactivity in miners' lungs, including the experimental development, the assessment of error, the necessary corrections and the results of direct measurements of the activity (dose) of miners using portable instruments. The research studies, that is, the model and phantom measurements, were performed in Moscow (in the former Soviet Union) in the All-Union Institute of Physico-Technical and Radio-Technical Measurements (VNIIFTRI) and Institute of Biophysics (Ministry of Health). The practical application took place in uranium and non-uranium mines in Uzbekistan and Tadjikistan (former USSR). Some aspects of the work have been published, for example Ruzer et al. (1995). The complete experimental and theoretical details of the method along with the compilation of the results and observed health effects in these mines have not been published before.

15.10.2.1 Theory of the Method

The basic equations in the most generalized form for the radon series were derived in Ruzer (1958, 1960a,b). The derivations of the equations are based on the equations described in Bateman (1910) with some appropriate transformations. Bateman's equations for the decay chain transformations were used for the determination of the correlation between the measured air concentration and activity of each decay product on the filter (or lungs).

In order to use these equations for the buildup of activity in the lungs due to filtration, that is, breathing, we assumed that each member of this chain of decay products supplies the decay products to the lungs at a constant rate $Q_i = q_i v k / \lambda_i$. From a mathematical point of view, the constant rate of supply is equal to the equilibrium between the first and the second member of the chain of radioactive transformation. In this case, the number of atoms N of each decay product in the lungs can be found according to

$$N_i = c_1 e^{-\lambda_1 t} + c_2 e^{-\lambda_2 t} + \cdots + c_i e^{-\lambda_i t}$$

where
$c_i = N_{1,0} \lambda_1 \lambda_2 \ldots \lambda_{i-1} / (\lambda_1 - \lambda_i)(\lambda_2 - \lambda_i) \ldots (\lambda_{i-1} - \lambda_i)$
λ_i is the decay constant of the "i" progeny

The idea in Ruzer (1958) was to find a correlation between the dose from the alpha-radiation of [218]Po and [214]Po, which cannot be measured directly in the lung, and the gamma-radiation from [214]Pb and [214]Bi, which at least theoretically can be measured by external counting. The carcinogenic dose is delivered by the alpha-radiation due to its high-energy transfer to the irradiated cells.

Due to the gamma-emission from the decay products [214]Pb and [214]Bi and also the relatively high maximum permissible air concentration in comparison with other radioactive aerosols, radon decay products present a unique opportunity for the direct measurement of activity in the lungs.

The final expression for the alpha-activity of ^{218}Po and ^{214}Po on the filter (i.e., lungs) is

$$A_\alpha = vk\left\{q_a\left[F_a\left(\theta,t\right)+F_{a,c}\left(\theta,t\right)\right]+q_bF_{b,c}\left(\theta,t\right)+q_cF_{c,c}\left(\theta,t\right)\right\} \qquad (15.1)$$

Correspondingly, the beta- and gamma-activity of ^{214}Pb and ^{214}Bi is

$$A_{\beta,\gamma} = vk\left\{q_a\left[F_{a,b}\left(\theta,t\right)+F_{a,c}\left(\theta,t\right)\right]+q_b\left[F_{b,b}\left(\theta,t\right)+F_{b,c}\left(\theta,t\right)\right]+q_cF_{c,c}\left(\theta,t\right)\right\} \qquad (15.2)$$

The integral absorbed dose to the lung will be

$$D = E_A\,A_a\left(t\right)\mathrm{d}t + E_c \times A_c\left(t\right)\mathrm{d}t \qquad (15.3)$$

where E_A and E_c are the energies of the alpha-particles of ^{218}Po and ^{214}Po.
After a series of transformations, Equation 15.3 can be present in the form

$$D_a = vk\left[q_aX_a\left(\theta\right)+q_bX_b\left(\theta\right)+q_cX_c\left(\theta\right)\right] \qquad (15.4)$$

where X_i represents the contribution of each decay product in the absorbed dose. Expressions (15.1) and (15.2) are true for the dose to the whole lung. But practically measurements were made over the pulmonary region of the lung. By comparing (15.2) and (15.3), the correlation between the integral absorbed alpha-dose to the lung and the gamma-activity of ^{214}Pb and ^{214}Bi A_γ can be presented in the form

$$D_a = A_\gamma F\left(\theta,t;\eta_{ba};\eta_{ca}\right) \qquad (15.5)$$

where
  F is the ratio of the right-hand sides of Equations 15.1 and 15.2
  θ is the duration of exposure
  t is the time after exposure; $\eta_{ba}=q_b/q_a$ and $\eta_{ca}=q_c/q_a$

The direct correlation between the absorbed dose from alpha-emitters ^{218}Po and ^{214}Po and beta- and gamma-emitters ^{214}Pb and ^{214}Bi in the lungs was at first introduced in Ruzer (1958,1960a,b,1968).
As a first step, it is important to know to what extent function F depends on shift of equilibrium between ^{218}Po, ^{214}Pb, and ^{214}Bi. The results of the calculations are presented in Table 15.1.
The results in Table 15.1 suggest that by measuring directly the gamma-activity in the lungs, we could avoid the radon decay products concentration measurements. In this case, the maximum deviation from the mean value will be not more than 15%.
The final expression for the integral absorbed dose of alpha-radiation for an exposure time of more than 3 h can be written as

$$D_a = 2.1\times10^8\,A_\gamma\,\theta\,\mathrm{erg} \qquad (15.6)$$

where
  θ is the duration of inhalation in hours
  A_γ is the activity of gamma-emitters in Ci.

For the maximum permissible concentration (APC) of 30 pCi/L (1100 Bq/m^3) (or in terms of [PAEC] 3.8×10^4 MeV/L), a breathing rate of 20 L/min, and a standard lung deposition coefficient

TABLE 15.1

Relationship between the Function $F(\theta, \eta_{ba}; \eta_{ca})$ and the Shift of Equilibrium of Radon Decay Products for Various Time of Inhalation

	$F(\theta, \eta_{ba}; \eta_{ca})$	10^4 erg·min
$\eta_{ba}; \eta_{ca}$	$\theta = 3\,h$	$\theta = 6\,h$
$\eta_{ba} = \eta_{ca}$	23.8	47.8
$\eta_{ba} = \eta_{ca} = 0$	40.8	81.4
$\eta_{ba} = \eta_{ca} = 0.8$	24.2	48.2
$\eta_{ba} = \eta_{ca} = 0.6$	24.5	49.0
$\eta_{ba} = \eta_{ca} = 0.5$	24.9	49.8
$\eta_{ba} = \eta_{ca} = 0.4$	25.3	50.7
$\eta_{ba} = \eta_{ca} = 0.2$	27.2	54.5
$\eta_{ba} = \eta_{ca} = 0.1$	30.0	60.0
$\eta_{ba} = \eta_{ca} = 0.01$	38.2	75.8

of 0.25 for a duration of 3 h or more, the activity in the lungs will be about 0.02 μCi (740 Bq). This activity can be measured by external counting with good statistical accuracy. To prove this, model measurements were taken.

15.10.2.2 Correction for the Shift of Equilibrium of Radon Progeny in the Air and in the Lungs

The laboratory measurement of radon decay product gamma activity in simulated lungs was conducted using a calibrated sample of ^{226}Ra. The ^{226}Ra gamma spectrum is identical to the entire gamma-spectrum of radon progeny. However, there is a difference in measurement of miners. In the ^{226}Ra calibration source all radon decay products are in equilibrium, that is, their activities are equal. The shift in equilibrium in the lungs will depend on the shift in equilibrium in the air and this will change with time in the lungs.

This is especially important because the gamma-spectra of RaB and RaC are different, which will result in the change of gamma-detection efficiency with the change of equilibrium both in the air and in the lungs.

Let us denote e_0 as the detection efficiency of RaB and RaC in equilibrium in the lungs or in the standard source of ^{226}Ra, e_b the detection efficiency of RaB, e_c the detection efficiency of RaC, e the detection efficiency in the case of the shift of equilibrium between RaB and RaC, and $k_2(\theta,t) = e_0/e$ the correction factor on inequality of detection.

$$e_0 = \frac{(N_{b,0} + N_{c,0})}{2A_0}$$

where
  $N_{b,0}$ and $N_{c,0}$ are the count rates of RaB and RaC, respectively, from equilibrium RaB and RaC in the lungs or in the standard source of ^{226}Ra in the phantom
  A_0 is the activity in the equilibrium source

$$e = \frac{[N_b(\theta,t) + N_c(\theta,t)]}{[A_b(\theta,t) + A_c(\theta,t)]}$$

where N_b, N_c, A_b, A_c are the count rates and activities of RaB and RaC, respectively, in the lungs at time t after exposure time θ. Then

$$k_2 = \frac{e_0}{e} = \frac{(N_{b,0} + N_{c,0})/2A_0}{(N_b + N_c)/(A_b + A_c)}$$

$$N_{b,0} = e_b A_0, N_{c,0} = e_c A_c$$

$$N_b = e_b A_b, N_c = e_c A_c$$

$$k_2 = \frac{(e_b A_0 + e_c A_0)(A_c + A_b)}{2A_0(e_b A_b + e_c A_c)}$$

and finally

$$k_2(q,t) = \frac{(1 + e_c / e_b)(1 + A_c / A_b)}{2[1 + (e_c / e_b)(A_c / A_b)]}$$

When the shift of equilibrium between RaA, RaB, and RAC in the air changes from 1:1:1 to 1:0.03:0.01, the ratio of $A_c(t)/A_b(t)$ in the lungs changes from 0.31 to 2.93. But due to the mode of the function k_2 (the ratios e_c/e_b and A_c/A_b are both in the numerator and denominator), the difference in k_2 is not significant. For example, for a NaI(Tl) crystal 80×40 mm, the change in k_2 is not more than 20% (Ruzer and Urusov, 1969).

In calculating the detection efficiency for RaB and RaC, we have to take into account the number of gamma quanta per decay for RaB (0.823) and RaC (1.397) with their average energies of 0.316 and 0.717 MeV, respectively.

Relative efficiency e_c/e_b was calculated for the 80×40 mm NaI(Tl) crystal and the model of the lungs (Ruzer and Urusov, 1969) according to the formula

$$\frac{e_c}{e_b} = \frac{\sum h_i k_{l,i} k_{b,i} k_{c,i}}{\sum h_j k_{l,j} k_{b,j} k_{c,j}}$$

where
h is the yield of the correspondent gamma-line
k_l and k_b are the absorption coefficients of the lungs and the body, respectively
k_c is the detection efficiency of the crystal
Indexes i and j belong to RaC and RaB, respectively

Simplifying the expression for the average absorbed dose, assuming that the mass m of the lung is 1000 g,

$$\frac{D}{m} = 8.05 \times 10^{-7} \, k_1 k_2 k_3 N_{0,\gamma} \theta \, \text{mrad} \tag{15.7}$$

where
θ is the duration of the inhalation
$N_{0,\gamma}$ is the count rate above the "background" (the "background" is the count rate measured for the person before entering the workplace)
k_1 is the correction coefficient for decay of radon progeny in the lungs after the end of exposure
k_3 is the reciprocal of detection efficiency of the gamma-radiation of radon progeny in equilibrium

It should be pointed out that the measurement of the gamma-activity in the lungs of miners was provided in the environment with a relatively high gamma-background, especially when the measurements took place near the working sites. In such cases, the parameter "quality of measurement" $q = n^2/n_b$ (n, count rate above the background; n_b, background count rate) was chosen to achieve the best measurement conditions.

15.10.2.3 Accounting for Parametric Variations: Variations of Concentrations, Breathing Rate, and Deposition Coefficients in Real Working Conditions

All dosimetric parameters under actual working conditions are not uniform. This is important both in the case of the estimated or indirect calculation of the dose (lung activity) and in the case of direct measurement.

Let us denote Q (the activity in the lung) as a combination of all parameters (air concentration, breathing rate, lung deposition, respirator efficiency):

$$Q = qvk(1 - f)$$

where
  f is the coefficient of effectiveness of a respirator
  Q is the rate of the intake

To take into account the variation of intake, two options were used:

1. Experimental, by placing a portable gamma-counter near the work site and, for parts of the shift, with special attention to the situation when Q is high.
2. Theoretical, by calculation of corresponding corrections on the inequality of Q during the time of exposure for different types of working itinerary (scenario of exposure).

To perform such a calculation, the whole working shift time, θ, should be divided by n periods in which every dosimetric parameter will be constant. Different typical scenarios of exposure (work itinerary) in terms of the rate of intake, Q, changing during the shift were calculated.

Variants present situations during the work shift:

1. Ventilation turned off (on) resulting in increasing (decreasing) q
2. Change in physico-chemical properties of aerosols resulting in an increase (decrease) of the coefficient k
3. Change in the work load resulting in changed breathing rate and deposition coefficient
4. Miners use of respirators resulting in a change of the factor $1 - f$

For the introduction of a correction factor for Q we have to obtain additional information on each concrete working situation (variation of Q during the shift). The degree of the error in determining the correction factor depends on the proximity of the chosen variant of the function Q to the variant under actual conditions.

According to the calculations (Urusov 1972), even if the magnitude of Q_{max}/Q_{min} is close to 30 the total error in the measurement of activity in the lungs of miners by the direct method will not exceed 40%, which is acceptable for the purposes of individual dosimetry.

15.10.2.4 Model Measurements

Experiments to determine the detection limit of the gamma-activity measurement of radon progeny in the lungs of miners were provided using a Human Body Spectrometer at the Biophysics Institute in Moscow. It consisted of a measurement chamber protected by a 20 cm layer of iron.

TABLE 15.2
Human Body Spectrometer Measurements

Mode of Measurement	Impulses per 100 s	
	Single Detection	Dual Detection
Background (open door)	914	1540
Background (closed door)	427	843
Background (closed door) with a person	500	978
Person with a ^{226}Ra sample		
590 Bq.	—	1244
175.4 Bq.	557	1088
0.016 μ Ci.	704	—

TABLE 15.3
Relative Errors (%) in Human Body Spectrometer Measurements

Activity of ^{226}Ra	One Detector			Two Detectors			
t_b	10 min			100 s		10 min	
t_b+p+s	100 s	3 min	5 min	10 min	100 s	5 min	10 min
590 Bq	9.5	7.4	6.4	—	12	—	—
175 Bq	46	36	30	24	57	15	12

t_b, time of the background measurement; t_b+p+s, time of the measurement of the person+^{226}Ra source + background.

The detector was a NaI(Tl) crystal and samples of ^{226}Ra containing RaCl$_2$ solution with activity 590 Bq (a) and 175.4 Bq (b) were placed in different positions in the chest of the phantom, and a 80×50 mm NaI(Tl) detection crystal and photomultiplier was placed on the back. The activity of the sample (a) corresponded to approximately 0.75 of annual permissible concentration (APC) and (b) 0.25 APC in air.

Tables 15.2 and 15.3 show the results of these measurements concerning the assessment of the detection limit and the statistical errors with one and two detectors (minimal distance from the sample to the detector was 23 cm). The results of these model experiments suggested that at the level of concentration of radon progeny in the air of 0.75APC.

15.10.2.5 Phantom Measurements and Geometric Corrections

For calibration purpose in laboratory conditions a phantom of the human torso was used. The measurements were made with a NaI(Tl) crystal with a window diameter of 77 mm, a lead shield and cylindrical sample of ^{226}Ra activity of 3.9×10^{-5} Ci size 160×16 mm placed in different positions in the phantom.

Results of phantom measurements including positions of the sample and detector are shown in Table 15.4. Measurements were provided by scintillation detector of NaI(Tl) with lead shield (diameter of the window 77 mm).

For the field measurements a more primitive phantom was used, that is, a long-sized vessel filled with water size close to the size of lungs, in which the sources of ^{226}Ra were placed in different positions.

The systematic errors associated with the contamination of the body and work clothes in real working conditions, for example, the influence of radon accumulated in the adipose tissue of the

TABLE 15.4
Phantom Measurements

i/j	$j=1$	$j=2$	$j=3$
1	3500	450	50
2	1600	600	130
3	2200	1050	130
4	1550	1350	110
5	620	800	460
6	750	1400	470
7	600	1900	400
8	100	650	3650

i, detector position; j, source of
^{226}Ra position.

abdominal cavity, were eliminated by the introduction of corresponding corrections. It has been shown that the contribution of gamma radiation from the abdomen to the total count rate 30 min after leaving the working place does not exceed 5%.

The errors associated with contamination of the clothes and the body of miners dropped to 15% and 1% of the measured value, respectively, after work clothes have been taken off and a shower was taken.

The variability of the background in underground measurements with the same person was no more 3% (Ruzer et al. 2004).

Dependency of the gamma-background from the body thickness is shown in Ruzer et al. (2004).

The decay of radon progeny in the lungs of miners presented in Ruzer et al. (2004) shows that activity in the lung of miners decreased according to the law of radioactive decay of radon progeny, that is, no substantial clearance was found during 3 h period.

The dependence of the count rate from the distance between detector and the chest is shown in Ruzer et al. (2004).

15.10.2.6 Portable Instrument for Direct Measurement of the Activity of Radon Decay Products in the Lungs of Miners

Some variants of portable instruments were developed for measuring miners both near the working sites and in the sanitary building, where miners took showers and changed their working clothes. Instruments with one and two detectors (for the front and back of the chest) were developed. For field measurement NaI(Tl) crystals were used with standard lead blocks as a shield. The main demand on the instrument was to measure the concentration of radon progeny at a concentration of 1.3×10^5 MeV/L (1 WL) and taking into account the general tendency of decreasing the maximum permissible concentration even on the level of 3.9×10^4 MeV/L (0.3 WL). The radiometric quality of the instrumentation can be improved by increasing counting efficiency by using a crystal of larger size, higher density CsI(Tl), or semiconductor detectors or by decreasing the background. A study of the correlation between background and shield thickness have shown that after increasing the thickness above 5 cm, no decrease of the background took place. Therefore, 5 cm thickness of lead shield was chosen as optimal (Urusov, 1972).

The measurement of activity in the lungs can be provided by means of general gamma activity measurement or spectrometry. The calculation shows that the quality factor of the first group of measurement is higher.

TABLE 15.5

Radioactive Aerosols Epidemiology: Miners Studies

Miners	1	2	3 N_0	4 N	5	6
ä-v	420	20	574	671	17.0	55
V-n	420	25	574	667	17.0	57
S-v	420	25	574	638	16.0	39
ä-v	420	25	574	664	16.0	55
O-v"	420	25	574	609	3.0	21
ï-v"	420	25	574	614	3.0	24
ë-Í	420	30	574	614	4.0	26
Ñ-Ó"	420	30	574	608	4.0	22
ë-n"	420	30	574	604	13.0	20
F-v	420	30	574	613	13.0	26
ë-n	420	40	574	562	14.0	0
L-i"	420	30	574	550	14.0	0
ë-Í"	420	30	574	609	13.0	23
I-Ó"	420	30	574	615	4.0	26
V-n	420	25	574	586	3.0	0
K-n"	420	20	574	680	17.0	60
S-j	420	20	574	681	17.0	61
S-n	420	25	574	671	16.0	59

N_0, background count rate, that is, before entering radon atmosphere; N, count rate after leaving working site; 1, time of exposure (min); 2, time of measurement (min); 3 and 4, counts rates (multiplied by 16); 5, PAEC (MeV/L); 6, absorbed dose (mrad).

More than 500 measurements of the activity (dose) in the lungs of miners were provided in uranium and non-uranium mines of former Middle Asia Republic of the USSR over an 11 year period. Some of the results are presented in Table 15.5 Measurements took place in June 1969 in Tadjikistan.

15.10.3 LUNG CANCER MORTALITY AND LUNG SICKNESS AMONG NONURANIUM MINERS IN TADJIKISTAN

15.10.3.1 Lung Cancer Mortality Data

In order to study the lung cancer mortality of miners, data were collected in Alterman (1974) about mortality among all male populations older than 19 year in the cities of Tadjikistan (200,000), a similar contingent of mining settlements (11,000), and different groups of underground workers (2,400) from 1960 to 1970. In this region, 30 mines and geology-prospecting teams were located.

The primary diagnosis of cancer took place in the health department of the industrial complex. All such complexes had x-ray diagnostic equipment with special medical personnel trained in the diagnosis of the sickness of breathing organs. Patients with suspected cancer were directed to the oncology department (hospitals) in the cities of Leninabad (now Hodjent), Dushanbe, Tashkent, and Alma-Ata for final diagnosis.

The mortality of miners was studied according to registry office records. For all men older than 18 year (the minimum age for underground work) from mining settlements, who died from breathing organs illness, special files were set up. Then according to the data from the personnel department, the duration of underground work, professional itinerary, and number of underground workers per year, including miners in the drilling group, were established. Data on male populations of the mining settlement were taken from local Soviet authorities according to the 1959 and 1970 census.

A study of lung cancer mortality was provided by means of long-time observations of miners in mines, where radiation—hygienic conditions were studied. As a control, the average of three very close mortality rate data of men older than 18 year was used: the population of all cities of Tadjikistan, mining settlements, and miners who worked in mines with a very low radon progeny concentration.

Measurements of the radon progeny concentrations were provided by means of the procedure; in some of the cases, direct measurements of gamma-activity (dose) in the lungs of miners were provided.

Besides this, the gamma-ray intensity, thoron decay products concentration, and ^{210}Po in the air were measured, but it is known that from a dosimetric point of view the contribution of these factors is very small.

For every group of miners, the annual intake was calculated based on the average radon decay products concentration and the FAL. In order to diminish the influence of other factors such as dust concentration (and correspondingly, silicosis) and smoking habits of the miners as much as possible, data on these factors were collected.

According to Alterman (1974), the sickness and mortality of miners in this region from 1949 to 1959 was registered for those with a duration of work of 3 years. In 1959–1961, when the dust concentration was diminished to the level of some mg/m^3, only eight workers were diagnosed with silicosis from miners with a duration of work of more than 10 years. At the end of the 60th year, only a single occurrence of silicosis took place.

As far as the problem of smoking is concerned, there is a high probability that the distribution of smokers among underground workers is very similar as for the population of mining settlements (control group) of the same age. Moreover, it is difficult to imagine that the distribution of smokers should be different for different groups of miners, working in sites with different levels of radon decay products concentrations.

However, it was shown that the lung cancer mortality of nonuranium miners in this study (Alterman, 1974) differs substantially for different groups of miners, with the highest level for the drilling group.

Lung cancer mortality was studied in connection with the level of exposure. For this purpose, all mines were divided into two groups:

1. Mines with the median concentration of radon decay products in the working sites (q) lower than APC
2. Mines with q larger than APC

These data suggest that the mortality of miners exceeded the mortality of the control group only in the second group of mines, with concentrations larger than APC according to the ordinary index by 3.5 times and according to the standard index by 3.1 times. At the same time, the mortality of drilling workers was greater than for the control group by 5.4 and 6.5 times, respectively.

The mortality of the drilling workers of the first group of mines, and the mortality of other miners in both groups did not differ from the mortality of the control group.

Thus, the higher lung cancer mortality of miners in this mining region was caused mainly by the high level of mortality of drilling workers of the second group. Apparently, it is difficult to consider it accidental that this group also had the highest intake of radon decay products.

15.10.3.2 Lung Sickness Data

Lung cancer mortality represents a long-time effect of the irradiation of miners. The study of this effect presents many difficulties in terms of collecting information during a long period of time, when professional itinerary, working conditions, and other factors can change substantially.

On the other hand, both from the medical and economical point of view, it is important to assess the effect of radon progeny on sickness rate disability as a short-time effect.

For this purpose, a comparative study was provided in the mining department of Karamazar, Tadjikistan (Alterman, 1974), by studying the radon decay products concentration with a parallel observation of the sickness rate disability of miners. The study was carried out in two steps. In the first period (1963–1965) and in the second (1967–1971), measures were taken to reduce the concentration of radon decay products mainly by improving the efficiency of the ventilation system.

These results suggest that due to measures provided at the end of the first period, the radon decay product concentration was reduced:

- In all working sites, 4–9 times
- In drilling sites, 6.6–13 times
- Including 4–12 times in passing sites

Results with the maximum concentration decreased even more sharply: in all working sites 8–32, in drilling 14–43, and in passing 10–60 times. Similar data on aerosol concentration suggested that from period 1 to 2, a reduction in aerosol concentration took place on an average of 2.3 and 5 times in terms of the maximum and was very close to the level of a limit dust permissible concentration of $2\,mg/m^3$. Data in Figure 15.2 suggest that sickness of breathing organs in terms of days of disability for

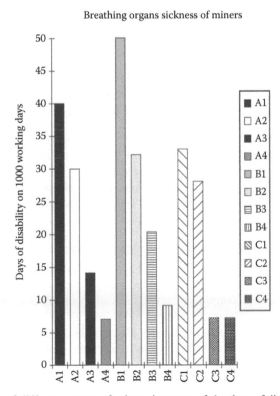

Breathing organs sickness of miners

FIGURE 15.2 Sickness of different groups of miners in terms of the days of disability for two periods of observation with the higher and lower concentration.1, high exposure (radon decay product concentration), general number of days of sickness; 2, low exposure (radon decay product concentration), general number of days of sickness; 3, high exposure, days in a hospital; A, for all miners; B, for drillers; C, for others.

all workers has decreased by 23%, mainly due to the reduction of average duration of the case by 28% and in terms of numbers of days in hospital by 44%. In the first period, sickness of drilling workers was 44% higher than among others; in the second period, it was not a reliable difference in this index.

In general, these results suggest that it is a great possibility that correlation between concentration of radon decay products in the working sites and sickness of breathing organs in terms of days of disability of miners took place especially for drilling workers, that is, for the group with the higher intake.

15.10.3.3   Comparison of Tadjikistan Data with Data from Other Epidemiological Studies

Analyses of many epidemiological studies are presented in Lubin (1994) and BEIR VI (1999) together with advantages and disadvantages in every study. It was also mentioned, especially in Ruzer (2001), that the uncertainties in dose assessment are very high and no data are available on the individual dose of miners or at least for different groups of miners.

In this sense, data on nonuranium miners in Tadjikistan present a new approach to the problem:

- A large diversity in terms of radon progeny concentration for different groups of miners, with especially high concentrations for the drilling group, was pointed out.
- The parameter of physical activity, FAL, differs substantially within groups of miners, with the highest for the drilling group.
- As a result, the drilling group had the highest intake among other miners.
- Both lung cancer mortality and lung sickness of miners were highest for the drilling group.
- For the first time in such studies, direct measurements of the activity (dose) in the lungs of miners were used.

From these results, the conclusion should be drawn that it is incorrect to consider miners as a uniform cohort in the risk assessment study, both from a dosimetric and epidemiological point of view.

15.10.4   QUALITY OF DOSIMETRY AND THE RISK ASSESSMENT FOR MINERS: SOME ASPECTS OF THE COMPARISON OF A "JOINT ANALYSIS OF 11 UNDERGROUND MINERS STUDIES" AND A STUDY OF NONURANIUM MINERS IN TADJIKISTAN

The direct measurement technique presents an opportunity to measure directly the deposition of radon decay products in the lung and, due to the correlation between alpha-dose and gamma-activity, to determine the absorbed dose to the lung tissue.

Direct measurement also provides important information about the correlation between concentration in the air and activity in the lung. This transition coefficient called FAL is a product of minute volume rate and deposition coefficient. It also plays an important role in dosimetry of non-radioactive inhaled particles because it does not depend on radioactivity but depends on physical activity and particle size distribution.

The accuracy of dose assessment by direct measurement is the highest achievable (35%–40% uncertainty), and for this reason it presents an opportunity to assess the optimal uncertainty in the dose under real conditions and, consequently, the uncertainty in risk assessment.

In Lubin (1994), a joint analysis of underground miners in China, Czechoslovakia, Colorado, Ontario, Newfoundland, Sweden, New Mexico, Beaverlodge, Port Radium, Radium Hill, and France was presented. The objective criteria for inclusion in the study of joint analysis were

- A minimum of 40 lung cancer deaths
- An estimate of Rn progeny exposure in WLM for each member of the cohort based on historical measurements of either Rn or Rn progeny

Let us look at both these criteria from the point of view of reliability (uncertainty) in the risk assessment.

Criterion (1) suggests a willingness to tolerate 15% in mortality.

From Table D-12, we can see that from the highest number of deaths (980) in the China study to the minimum by established criteria (40), the fractional statistical error in mortality σ_M will change only from 3% to 15%. And even when the number of deaths is in the range 20–30, σ_M will be in the range of 20%.

In contrast to the criteria that mortality must be known with an accuracy of about 15%, the criteria on dose (or exposure) are much more lenient.

Assessment of radon progeny concentration in mines is subject to an uncertainty of about 650%. Furthermore, in most of the studies, exposure was reconstructed from work histories and historical radon measurements rather than being directly measured.

Taking into account that the contribution to the dose from radon itself is negligible in comparison with its progeny and that, according to Lubin (1994), the range of the shift of equilibrium in mines varies from 0.2 to 0.9, we should assume that uncertainty in exposure assessment is in hundreds of percent.

It should also be pointed out that personal dosimetry even in terms of so-called personal dosimeters, that is, devices that measure concentration in the breathing zone of miners, except in a French study with a very small number of deaths, were not provided.

From this point of view, data on WLM for all studies presented in Table D-12 look completely unrealistic (accuracy up to tenths of WLM instead of at least 50%).

Because the statistical error in mortality is already much smaller than the uncertainty in exposure, it makes no sense to decrease δ_M by including it in the risk assessment data from the early years when the uncertainty in the exposure δ_E was very high.

One way of improving the accuracy of dose (exposure) assessment is by using the weighted average exposure where the weights are inversely proportional to the square of the variance

$$W \sim \frac{1}{(\delta)^2}$$

If these data are excluded, the number of deaths will decrease, but the statistical error in mortality will still be much lower than the error in dose or exposure assessment.

There is a trade-off: by choosing the largest cohort and correspondingly the greatest number of deaths, to try to increase the statistical accuracy of mortality, we include cases with great dosimetric (exposure) uncertainty, which results in decreasing the reliability of risk assessment.

A professor of epidemiology said at a conference, "Unfortunately, not so many people died from this epidemic." Paradoxically, in the case of miner studies, by improving the mortality statistic we can make the risk assessment less reliable.

In Lubin (1994), a summary of the strengths and weaknesses of the various studies is presented. Among the strengths, the authors mostly mentioned the large cohort and the long follow-up of the studies. Among the weaknesses, limited exposure data, no or limited smoking data, and in six studies limited numbers of lung cancer were cited.

In the study of nonuranium miners in Tadjikistan, the number of deaths was even more limited: 34 (14 miners and 20 men from mining settlements). We can say, following the professor of epidemiology, "Unfortunately, only 14 miners died from lung cancer in our study." In this case, the fractional statistical error in mortality will be in the range of 40% for miners.

On the other hand, the assessment of dosimetric factors for all 11 year of the Tadjikistan study was much better than in most of the studies:

- All instruments for radon and radon progeny measurements were properly calibrated with the accuracy of PAEC measurement better than 25%–30%.
- Radon decay product concentration measurements were provided in different working sites and for different groups of workers.

- For hundreds of miners, direct measurements of activity (dose) in lungs were provided.
- By making around 300 measurements of the activity of radon progeny in the lungs of miners and a similar number of measurements of the concentration in the breathing zone, FAL was established for different physical activities (group of miners).

In the Tadjikistan study by dividing all mines into two categories, with an average annual concentration $q <$ APC and $q >$ APC, another goal was achieved.

In the control group, three categories of miners were included

1. Men older than 19 year from all cities of Tadjikistan
2. The same age nonminers from miner settlements
3. Miners who worked in mines with concentrations of the first group

All these three groups had very close mortality rates. One can suggest that smoking habits of groups 1 and 2 are different from miners. But it is likely that smoking habits for miners who work in mines with high radon concentrations are similar to those of miners who work in mines with lower concentrations.

Comparison of lung cancer for miners in high and low radon concentration mines thus allows the estimation of mortality as a function of cumulative exposure.

This chapter focuses on the importance of correct dosimetry, that is, measurement or calculation of dose factors with a critical assessment of uncertainty in measurement or calculation.

It focuses on dosimetry even in a broader sense, particularly on how values measurable in real mining conditions (in the case of miners, concentration of Rn and its decay product, or calculated exposure) are related to the main physical factor that determines the radiation damage to the lung-absorbed dose. It especially focuses on what kind of propagation errors occur, and what kinds of measures and methods can be used to diminish uncertainty in the dose, which makes data of risk assessment unreliable.

It seems to us that, based on the data and ideas presented in this chapter, some review of the data in epidemiological studies, including Lubin (1994), should be made:

- Try to find some additional data on concentration and mortality among different groups of miners.
- Try to achieve more reliable data in the risk assessment (optimal value of σ_R) by choosing groups of miners with more reliable dosimetry.

NOMENCLATURE

A_i	activity in the lung, $i = \alpha, \beta, \gamma$ (Bq)
D	integral absorbed dose (erg; g rad)
D/m	absorbed dose (rad)
e	detection efficiency registration for the decay products
I	intake (rate of intake) (Bq/s)
k	retention coefficient (k)
m	mass of the lung (g)
q_i	radon decay products concentration, $i = a, b, c$ (Bq/m^3)
v	rate of inhalation (m^3/s)
η_i	degree of equilibrium of radon progeny, $i = a, b, c$

REFERENCES

Alterman, A.D., The problems of radiation hygiene in underground nonuranium mines, Candidate's dissertation, Institute of Hygiene and Professional Diseases, Medical Academy of Science, Moscow, Russia, USSR, 1974 (in Russian).

Antipin, N.I. et al., Special state standard vor the volumetric activity of radioactive aerosols, *Izmer. Tekh.*, Moscow, USSR No. 1, 5, 1980.

Bateman, H., *Proc. Cambridge Philos. Soc.*, 15, 423, 1910.

Domanski, T., Kluszczynski, D., Olszewski, J., and Chrusciewski, W., Field monitoring vs individual miner dosimetry of radon daughter products in mines, *Polish J. Occup. Med.*, 2, 147–160, 1989.

Health Effect of Exposure to Radon, BEIR VI; National Research Council, Washington DC, National Academy Press, 1999.

Layton, D.W., Methodically consistent breathing rate for use in dose assessments, *Health Phys.*, 64, 23–26, 1993.

Leites, F.L. and Ruzer, L.S., *Arkhiv Patologii*, Moscow, USSR No. 1, 20–27, 1959 (in Russian).

Ruzer, L.S., Estimate of dose in the inhalation of radon, *At. Energy*, February, 189–194, 1958 (in English).

Ruzer, L.S., Determination of absorbed doses in organisms exposed to emanations and their daughter products, *At. Energy*, 8, 542–548, 1960a.

Ruzer, L.S., Determination of degrees of equilibrium of short-lived radon in the air, *At. Energy*, 8, 557–559, 1960b.

Ruzer, L.S., Gamma-control in the inhalation of radon, *At. Energy*, 13, 384–385, 1962.

Ruzer, L.S., The method of determining the concentration of the gas, Patent No. 234746, *Bulletin of Inventions*, No. 28, 1964 (in Russian).

Ruzer, L.S., Non-emanating samples from Ra, Th, and Ac, *Radiat. Prot. Dosimetry*, 46, 127–128, 1993.

Ruzer, L.S., *Radioactive Aerosols*, Energoatomizdat, Moscow, Russia, 2001, 230 p. (in Russian).

Ruzer, L.S., Harley, N.H. eds. 2004. Aerosol Handbook: Measurement, Dosimetry, and Health Effects, New York: CRC Press.

Ruzer, L.S., Nero, A.V., and Harley, N.H., Assessment of lung deposition and breathing rate of underground miners in Tadjikistan, *Radiat. Prot. Dosimetry*, 58, 261–268, 1995.

Ruzer, L.S. and Sextro, R.G., Measurement of radon decay products in air by alpha and beta spectrometry, *Radiat. Prot. Dosimetry*, 72, 43–48, 1997a.

Ruzer, L.S. and Sextro, R.G., Assessment of very low aerosol concentrations by measuring the unattached fraction of [218]Po, *Prot. Dosimetry*, 71, 135–140, 1997b.

Ruzer, L.S. and Urusov, S.A., Determining the lung irradiation dose from radon disintegration products, *At. Energy*, 26, 301–303, 1969 (in English).

Urusov, S.A., Method and measurement technique for determination of the intake of radon decay products in the lungs of underground workers. Candidate's dissertation (in Russian). Biophysics Institute of the USSR Ministry of Health, Moscow, Russia (1972).

Vasin, V.A., Zalmanson, Yu.E., Kuznetzov, Yu.V., Lekhtmakher, S.O., Ruzer, L.S., and Sidorov, V.V., Phantom measurements for the determination of the intake of the radon decay products in the lungs by direct method, *Proc. VNIIFTRI*, 22, 46–47, 1975.

Volkova, E.A., Ziv, D.U., Labushkin, V.G., Ruzer, L.S., Stepanov, E.K., and Tyutikov, N.V., *Prib. Tek. Eksp.*, 4, 36–39, 1966.

Yu, K. et al., Geterogene Phantom of the Human Torso, Patent No. 402070, *Bulletin of Inventions* No. 41, 1973.

Zhivet'ev, V.M., Labushkin, V.G., and Ruzer, L.S., *At. Energy*, 20, 511, 1966.

16 Unattached Fraction of Radon Progeny as an Experimental Tool in the Assessment of the Risk of Nanoparticles*

Lev S. Ruzer

CONTENTS

I think it makes sense to analyze the relationship between nanoaerosol measurements and the health effect by comparing the studies on health effect associated with radon and its decay products and nanoaerosols especially because we are now at the beginning of a new global problem—the assessment of the dose and health effect of nanoparticles.

The similarity in both cases is that we have to study two groups of separate problems—dosimetry and health effect.

The difference is that in the case of radiation in general and the problem of radon and its progeny in particular, there were many studies on the correlations between radon and its progeny both

---

* I dedicate this chapter to my best friend, Professor Tatjana Tolstikova.

in terms of measurements and health effect, including humans. Unfortunately, similar data for nanoparticles are not available right now.

So, the question arises why we did not have the quantitative assessment of the correlation between measured radon and its progeny and health effect.

From my point of view there are two reasons:

1. In the majority of the radon and its progeny—effect studies not the dose itself, but its surrogate physical value was measured. Moreover, no relationship between this surrogate value and real dose as a main factor was established.
2. In some cases the neglect of elementary rules of metrology took place, especially in terms of uncertainty of assessment of the measurements and calculations.

I will try to illustrate this with the results of two mega-studies conducted by the National Research Council (1999) and Darby et al. (2004).

It should be mentioned that in BEIR VI a great work was accomplished; the distribution of radon and its progeny concentrations in the mines of 11 different countries were presented and analyzed together with the data on lung cancer mortality among miners.

In Darby et al., the similar very important data on radon concentrations in different countries in Europe were collected and analyzed. Again, the authors did a very good job in collecting and analyzing information on distribution of radon concentration in Europe.

Unfortunately they try to connect radon concentration itself with lung cancer mortality of general population, despite the fact that practically in this case only the radon progeny, not a radon itself, is responsible for biological effect. Experimental and theoretical studies suggested that only radon progeny, not the radon, deposited in the lung produces harmful effect. So, it is a classical case where a dose is not a cause of the effect, but rather the surrogate of the dose was used and the correlation between and real cause was not established. But that is not all.

In this study the authors found that "the absolute risks of lung cancer by age 75 years at usual radon concentrations of 0, 100, and 400 Bq/m^3 would be about 0.4%, 0.5%, and 0.7%, respectively, for lifelong nonsmokers, and about 25 times greater (10%, 12%, and 16%) for cigarette smokers."

It is obvious that if we take into account the uncertainty in the risk assessment in this case the numbers (0.4%, 0.5%, and 0.7%) and (10%, 12%, and 16%) will be in the range of errors, that is, the same. In other words, increasing radon concentration from 0 to 400 Bq/m^3 does not produce elevation in cancer risk. By the way, these concentrations are considerably lower than permissible concentrations for mines.

In Table D.12 of BEIR VI the results of "average exposures" in "average Working Levels" (WLM) are presented for 11 countries (China—286.0, Czechoslovakia—196.8, France—59.4, Canada, United States—578.6, etc.). We know that the uncertainty in the exposure assessment is on the order at least of tens of percent. So, we cannot trust these data with tenths of WLMs.

Our study on dosimetry and health effect of miners in Tajikistan suggested that different groups of miners got substantially different exposure (dose) and different lung cancer mortality. So averaging in this case can lead to additional uncertainty.

It seems to me that in epidemiological studies there exists some sort of tendency to present as much as possible cases of effect (mortality, morbidity, etc.) in order to get good statistical data. With such tendency we used old and questionable data on concentrations, even based on data on ventilation. So, our good statistic in this case is compromised with bad dosimetry.

As it was shown in Chapter 14 of Ruzer and Harley (2004, pp. 483–493) in order to get correct result we should establish a balance between the number of cases of health effect and correct dose assessment.

Data presented in BEIR VI are very valuable in terms of the assessment of concentration in mines of different countries. But it cannot be connected to the health effect.

As a result, we have to assume that the results of this study cannot be used for quantitative correlation of dose (and even exposure) of radon and its progeny and lung cancer mortality in mines, not to mention for general population in homes.

It is understandable that in such studies we try to simplify measurements and averaging data. But if we cannot assess the dose directly, measuring the surrogate physical value instead and using averaging data, we have to pay the price for this, that is, try to assess the loss in the reliability in the risk assessment.

As it is clear from the dosimetric road map for nanoaerosols, presented in the following, the problem of the dose and risk assessment for nanoaerosols is even more complicated than for radiation, including radon and its progeny. So, we should avoid the errors that were made in the case of radon.

16.1 NANOAEROSOLS: DOSIMETRIC ROAD MAP

16.1.1 Introduction

The nanotechnology industry is rapidly growing with promises of substantial benefits that will have significant global, economic, and scientific impacts, applicable to a whole host of areas from engineering and electronics to environmental remediation and medical healthcare. However, at present there is a growing concern over the safety of nanomaterials with respect to occupational, consumer, and environmental exposures.

Particularly, information on nanomaterial exposure, dosimetry, risk assessment, and health effect is negligible. Furthermore, the gap between these two tendencies widens at an alarming rate.

According to Kulinowski (2009), "there's a lot more data now than there was back in the early days. Between 2001 and 2008 (the last year for which complete data are available), the annual NanoEHS publication rate grew between 20%-120% per year with over 3600 individual papers."

Unfortunately, however, "it becomes equally difficult to say that all these data are conclusive. A recent analysis found that much of the "nanotoxicology" research is done *in vitro*, focusing on acute toxicity and mortality induced by native nanoparticles, with limited relevance to human health or environmental impacts and little attention to consumer products."

Up to the present time, publicly available quantitative data related to dosimetry of nanomaterials, and particularly aerosolized nanoparticles, are very difficult to find.

And what is even more important is that there is a lack of information in scientific literature on the development of strategy in the study of sequence of physical parameters, each of which is extremely important for the assessment of the main cause of the effect in the case of nanoparticles—dose.

In this chapter we present and discuss the dosimetric road map, that is, the consequence of particular steps, which we need to take in order to assess the dose—main cause of biological effect from nanoaerosols.

16.1.2 Nanoparticle

In general, a particle is defined as a small object that behaves as a whole unit in terms of its transport and properties. It is further classified according to size: In terms of diameter, fine particles cover a range between 100 and 2500 nm, while ultrafine particles, on the other hand, are sized between 1 and 100 nm. Similar to ultrafine particles, nanoparticles are sized between 1 and 100 nm, though the size limitation can be restricted to two dimensions.

Nanoparticles may or may not exhibit size-related properties that differ significantly from those observed in fine particles or bulk materials.

There is no accepted international definition of a nanoparticle, but one given in the new PAS71 (the British Standards Institution, BSI) document developed in the United Kingdom is "particle with one or more dimensions at the nanoscale."

Correspondingly, the nanoscale is defined as "having one or more dimensions of order of 100 nm or less." There is a note associated with this definition: "Novel properties that differentiate nanoparticles from the bulk material typically develop at a critical length scale of under 100 nm."

This makes the size of particles or the scale of its features the most important attribute of nanoparticles. What is different about a nanoparticle?

There is no strict dividing line between nanoparticles and non-nanoparticles. The size at which materials display different properties as compared with bulk material is material dependent and can certainly be claimed for many materials much larger in size than 100 nm.

Most experts in United Kingdom and United States define nanoparticles as particles smaller than 100 nm. But, for example, in Japan, particles between 50 and 100 nm are classified as "ultrafine" and only those below 50 nm in one dimension as nanoparticles. Even some agencies in United States use the term "ultrafine" to describe particles under 100 nm (usually in connection with natural or incidental nanoparticles).

In order to solve such discrepancies, national and international agencies such as International Organization for Standardization (ISO), International Electronically Commission (IEC), American National Standard Institute (ANSI), and American Society for Testing Materials (ASTM) are discussing the standardization of terminology, metrology, characterization, and approaches to safety and health.

Another definition from Malvern.com is as follows:

A particle having one or more dimensions on the order of 100 nm or less. There is a note associated with this definition: "novel properties that differentiate nanoparticles from the bulk material typically developed at a critical length scale of under 100 nm."

The "novel properties" mentioned are entirely dependent on the fact that at the nanoscale, the physics of nanoparticles mean that their properties are different from the properties of bulk material.

It should be noted that these nanoparticle definitions require special consideration. First, because in practice nanoparticles represent a complicated structure, it is not easy to measure its linear size.

We can say also that the second part of the definition stated that nanoparticles may or may not exhibit properties completely different from the fine particles or bulk materials, can be call in question.

In Oberdörsted et al. it was demonstrated that nanoparticles appeared to be more inflammatory in the lungs than microscale particles. But if the data were plotted against surface area instead of mass, response was identical for both nano and larger particles (Duffin et al., 2007; Oberdörster et al., 2005).

In other words, in principle there is no difference between the nanoparticles, fine particles, and bulk materials. The difference is only that in the case of nanoparticles the ratio of the number of molecules on the surface to all molecules of particle "surface to volume ratio" (SVR) is much higher than that of fine particles or bulk materials.

The importance of the surface area as a dosimetric parameter in the study of health effect of nanoparticles is shown in Duffin et al. (2007).

Of course not only the surface area is responsible for nanoparticles toxicity, but rather surface area in combination with surface reactivity, and elemental toxicity. Still the surface area is likely the best dose metric in the case of nanoparticles (Kelly, 2009).

Based on this consideration we propose another definition for nanoparticles based on the ratio of molecules on the surface to all molecules of the particle. As an appropriate number for this SVR boundary portion we propose 3% or 0.03%, which corresponds to spherical particles with a diameter of 100 nm.

Our proposed definition of the nanoparticles:

Nanoparticles are particles which have the ratio of number of molecules on the surface to all particle molecules, surface to volume ratio (SVR), of 3% (0.03) or higher.

16.1.3 NANORISK

First, we have to present the definitions of all particular parameters associated with the risk, and consequently the dose as a cause of biological effect in the case of nanoaerosols.

The term "exposure" as "the amount (c) present in one cubic meter of air expressed in milligrams and multiplied by the time (t) in minutes" came to the aerosol field from the study of the effect of gases as a "Haber Law." (Ruzer, Apte, Sextro; in Chapter 7 of Ruzer and Harley [2004].)

Strictly speaking, it was incorrect to simply transfer the term "exposure" from gases to aerosols. The definition of exposure in the case of aerosols is much more complicated than that for gases. The main difference in dosimetry between gases and aerosols is that aerosol distribution is uneven both spatially and temporally, and deposition of aerosol in the lungs is not uniform.

According to the National Academy of Science report (NAS, 1991), the definition of exposure is as follows:

An event that occurs when there is contact at a boundary between humans and the environment with a contaminant of a specific concentration for an interval of time; the units are concentration multiplied by time.

From our point of view this definition is not free from some contradiction. If exposure is an event, it cannot be expressed in physical units.

A statistical definition of exposure has been proposed (Ott, 1966):

An exposure at some instant of time is a joint occurrence of two events:

1. the pollutant of concentration C is present at a particular location in space at a particular time, and
2. the person is present at the same time and location in space.

A later definition (Duan and Ott, 1989) addresses the notion that the target remains important and also that different parts of the target can receive different exposures at the same time.

The last definition is more adequate for aerosols, because it takes into account the specific aerosol problem of nonuniformity.

In the Guidelines for Exposure Assessment (Zartarian et al., 2005) a slightly different definition was proposed:

Exposure – Contact of a chemical, physical, or biological agent with outer boundary of an organism. Exposure is quantified as the concentration of the agent in the medium in contact integrated over the time duration of that contact.

In all these definitions, the key word is contact, which means that in the case of aerosols only breathing zone measurement should be used for the exposure and particle size measurement. If concentration (and particle size distribution) is provided at a distance from the breathing zone, the correlation should be established between breathing zone and sampling site measurement.

The risk has to be defined in quantitative terms as a ratio of the number of cases of mortality, morbidity, attendance of hospitals, etc., divided by the cause of the effect, that is, dose.

Based upon this discussion, we can formulate the definition of dose as a physical value responsible for a biological effect (Ruzer and Harley, 2004):

Dose is the specific quantity of aerosols delivered to a target site that is directly responsible for a biological effect.

The term "quantity" is defined as follows:

1. In the case of radioactive aerosols, quantity is the deposited energy per unit mass for alpha, beta, or gamma radiation which is expressed in units of J/kg (Gray) or rads (100 erg/g) or the equivalent.
2. In the case of nonradioactive aerosols, quantity is the deposited number of particles, surface area, or mass of a discrete particle size.
3. The term "directly" means that dose is a quantity of the deposited amount of aerosol particles after the completion of all biokinetic processes.

According to EPA (1996) of the United States, "in epidemiological studies, an index of exposure from personal or stationary monitors of selected pollutants is analyzed for associations with health outcomes, such as morbidity or mortality. However, it is a basic tenet of toxicology that the dose delivered to the target site, not the external exposure, is the proximal cause of a response. Therefore, there is increased emphasis on understanding the exposure–dose–response relationship. Exposure is what gets measured in the typical study and what gets regulated; dose is the causative factor."

16.1.4 AIRBORNE NANOPARTICLE CONCENTRATION MEASUREMENT

Let us consider every step in our dosimetric nanoaerosols road map from the point of view of available approach, methods and measurement technique. We begin with the branch which describes the study of airborne nanoparticles.

Manufacturing and handling processes for nano-sized materials are widely variable. For example the materials may be fabricated in a fluidic system closed to the environment (e.g., colloidal suspension of metallic crystals formed in a liquid reaction vessel), or in an open-air system where they may directly mix with ambient air (e.g., manufacture of carbon black using combustion techniques). For each *nanotechnology* the potential for fugitive emissions leading to an *airborne concentration* can be different. Information on the release of the nanomaterial to the air for the manufacture of different nanomaterials is very scarce.

It should be noticed that the problem of metric in nanoaerosol concentration is complicated.

First, nanoaerosols often exist in practice as a structure not as a single particle with the size at least in one dimension in the range from 1 to 100 in diameter. So, generally speaking, it is not always possible to use diameter itself as a characteristic of the particle.

Second, from the point of view of the dose to the lung the aerosol mass concentration as a characteristic also has many disadvantages:

1. In the nanometer range when mass concentration is very small, but the number concentration, particle density at the lung tissue, and correspondingly dose can be very high.
2. By the same mass concentration particle size distribution can be different, so will be different the local particle deposition inside the lung and correspondingly the dose and biological effect.
3. In the case of aerosols for the dose assessment, the important role is played by only the "respirable" particles, that is, particles with a diameter less than 5–7 μm. Unfortunately, this limit is uncertain (Martonen et al., 1992).

According to the majority of the studies of ultrafine and nano-sized aerosols, it is not the mass concentration, but the *particle number and surface area concentrations* that should be used for the assessment of dose because they appear to be better predictors of health effects (Aitken et al., 2004; Hankin et al., 2008; Oberdörster et al., 2005; Royal Society, 2004).

As we already mentioned the use of mass concentration data alone is insufficient for the expression of dose, and the number concentration and/or surface area need to be included.

Unfortunately there is a lack of information on measurement of nanoaerosols particles, and especially the size distribution and surface area concentration in the working environment.

A review of the literature on environmental health in the new rapidly developing nanotechnology industry shows that the problem of exposure has not been adequately assessed (Oberdörster et al., 2005). Worker health and safety is of initial concern as occupational groups are likely to be among the first to be exposed to elevated concentrations of nanomaterials. A gap exists between the existing particle measurement methods and those truly appropriate for nanoaerosol exposure assessment. Until now, the primary tools available for measurement of nano-sized aerosols have been Condensation Particle Counters (CPCs), and Differential Mobility Analyzers (DMAs).

A new instrument on the market, the Nanoparticle Surface Area Monitor (TSI 3550, Nanoparticle Surface Area Monitor, 2005), is used for assessment of deposited surface area (DSA) in the lung. Lung deposition estimates from this instrument are based on correlations developed (Wilson et al., 2004) between the electrical signal and modeled DSA. The instrument is said to be capable of detecting particles with diameters down to 10 nm.

One of the important problems in the safety of people working with nanomaterials is the efficiency of respirators, discussed in many reports: (Strategic Plan for NIOSH).

16.1.5 Nanoparticle Respirators True Efficiency Measurements

The problem of respirator efficiency is discussed in many reports: Nanotechnology (2005), Aitken et al. (2004), White Paper on Nanotechnology, IRGC, Orwin Renn and Mike Roco Geneva, Switzerland, Shaffer (2008). It is well known that the determining factor which governs the effectiveness of respiratory protective equipment (RPE) is not absolute penetration through the filter, but rather face-seal leakage, which bypasses the device.

In Ruzer et al. (1995) the new idea for measuring the true efficiency of respirators was used.

Let us denote the terms as follows:

A_0—measured gamma-activity in the lung of miners before entering radon atmosphere (background measurement).

$A_1 - A_0$—measured gamma-activity in the lung of miners without wearing respirator

$A - A_0$—measured gamma-activity in the lung of miners with respirator

Then respirator efficiency will be $(A - A_0)/(A_1 - A_0)$ 100%

Penetration coefficient $\{1 - [(A - A_0)/(A_1 - A_0)]\}$ 100%

Our measurements in the mines demonstrated that true respirator efficiency varies from 67% to 95% depending on individual training, duration, and type of work.

At the present time the data on the true efficiency of respirators, including the face-leakage problem, are unavailable.

As discussed in Geraci, NIOSH (2009), in terms of nanoparticles we still do not know nature and extent of hazard, nature and extent of exposure, nature and extent of risk, what measure to use, limitation of controls, limitations of protection, and what limits are appropriate. There are no specific exposure limits.

According to Friends of the Earth (2009) there is

no consistent nomenclature, terminology and measurement standards to chacterise and describe nanoparticles and exposure. Inadequate understanding of nanotoxicity, in particular to determine whether acceptable exposure limits exist. No effective methods to measure and assess workplace exposure to nanoparticles; no data on existing or predicted workplace exposure. No effective control methods to protect workers from exposure.

Despite the hundreds of products containing nanomaterials that are already being manufactured commercially, and the emerging body of scientific literature demonstrating the serious risk associated with nanotoxicity, there are still no laws to manage workplace exposure and to ensure workers' safety. This suggested that governments have learnt little from their experiences with asbestos.

16.1.5.1 Conclusions

1. Until now it is very difficult to find studies on measurement of nanoaerosol concentration and correspondingly exposure, especially on surface area concentration in working conditions.
2. There is a lack of experimental data on the assessment of exposure in occupational settings.
3. In the majority of studies even on animals still the mass concentration, not a surface area, in contradiction to new scientific data is used as a measure of dose.

16.1.6 NANOPARTICLE DEPOSITION IN THE LUNG (RIGHT BRANCH OF THE DOSIMETRIC ROAD MAP)

There is a lack of systematic studies on nanoparticles lung deposition.

One of the most important difficulties in the assessment of the dose as a main cause of the effect of aerosols, particularly in the nanometer range, is the lack of information on local deposition in human lung.

Our approach to this fundamental problem we presented in Ruzer and Apte (2010a,b).

Operationally, the concept uses these ^{218}Po radon progeny as a radiolabel. These particles have a very high diffusion coefficient and readily attach to other particles in air. When attached, or aggregated, with the environmental aerosol, these particles are called "attached activity." Given their high diffusion coefficient, their attachment efficiency can approach 100%. Thus, almost every particle in the environmental aerosol becomes labeled with a radioactive radon progeny particle, destined to decay and emit Gamma particles (^{214}Pb and ^{214}Bi). Particle inhalation experiments may be designed in which relatively low concentrations of radon gas is mixed with a non-active study aerosol that will subsequently be inhaled by subjects. As the radon atoms decay, their progenies attach to the study aerosol particles and thus radiolabel them.

In this case every measured gamma-quantum corresponds to a nonradioactive aerosol particle in the nanometer range locally deposited in the lung. So, the measured gamma-activity will represent dose of nonradioactive nanoaerosols at the target.

According to dosimetric road map for nanoparticle dose assessment we have to have data on translocation of nanoparticles across the air-blood barrier and cell penetration.

In SCENIHR (2007), Geiser et al. (2008), and Muhlfeld et al. (2007), a review of such studies is presented.

In animal studies for titanium dioxide as well as for diesel exhaust particles (Geiser et al., 2005; Muhlfeld et al., 2007; Nemmar et al., 2002), it was demonstrated that a small fraction of the nanoparticles can be translocated to the circulation and can reach extrapulmonary organs via the bloodstream.

In human lung one study (Neemar et al., 2002) reported a rapid and significant translocation of inhaled carbonaceous nanoparticles to the systemic circulation and extrapulmonary organs. In contrast, other subsequent studies failed to confirm this finding and detected only a low degree of translocation for iridium (Kreiling et al., 2002) or carbon nanoparticles (Mills et al., 2006; Moller et al., 2008; Wiebert et al., 2006).

The studies of Nemmar et al. and Mills et al. had a very similar design and Mills et al. have argued supporting the fact that the strong translocation of ^{99m}Tc labeled particles observed by Nemmar et al. was mainly related to the translocation of soluble ^{99m}Tc.

A very specific property that has been assigned to inhaled nanoparticles refers to translocation from the nasal epithelium to the brain. Elder et al. showed that ultrafine manganese oxide particles translocate to the olfactory bulb and other regions of the central nervous system.

Additional support for translocation of ultrafine particles via olfactory axons comes from a study (Oberdörster et al., 2004) that showed that inhaled radioactive carbon particles were significantly enhanced in the olfactory bulb after exposure, in contrast to other brain regions that showed only inconsistent increases in carbon particles.

In conclusion we could say that nanoparticles can enter the blood circulation from the respiratory tract and be deposited mainly in the liver, spleen, and other organs. Even if the amount of these nanoparticles are small in terms of mass in comparison with the total amount that entered the body, we cannot say that the locally deposited number at the target, that is, dose and effect will be small and unsubstantial.

As the knowledge of long-term behavior of nanoparticles is very limited, a conservative estimate should be assumed that insoluble nanoparticles may accumulate in secondary target organs during chronic exposure with unknown consequences.

16.1.7 CELLULAR ENTERING MECHANISMS

A review of literature on existing knowledge and uncertainties regarding nanoparticles into cell penetration is presented in IOM report CBO407.

Nanoparticles into cell penetration have been observed to be size and temperature dependent (IOM, CBO407).

It was shown that after inhalation 25 nm particles of TiO_2 has been observed in greater amount in lung interstitium than 250 nm particles.

The uptake of nano- and microparticles of polylactic polyclycolic acid copolymer of 100 nm was 2.5-fold greater than that of 1,000 nm and 6-fold greater than 10,000 nm particles. The uptake rate was greater at 37°C than at 4°C (Desal et al., 1997).

In a study (Rejman et al., 2004) it was observed that the uptake of 50 nm beads was three- to fourfold greater than for 100 nm and 8–10 times greater than for 1000 nm.

There is very limited work in published literature on the interaction of nanoparticles with cell membranes and their behavior within the cells.

As it was underlined in IOM, CBO407, it is clear from the review of literature that many studies on nanoparticles into cell penetration did not investigate systematically the physicochemical factors which control these processes. Therefore, it is difficult to understand the role of each of the factors from these studies where many factors and variables are reported (size, surface coating, and charge). Moreover, such properties of nanoparticles as surface chemistry or aggregation in biological or other fluids cannot be in equilibrium or even at steady state and these properties may well change as a function of time.

In Bhabra et al. (2009) it was reported that nanoparticles of chromium, 29.5 ± 6.3 nm in diameter, can damage human fibroblast cells even without having to cross the barrier.

16.1.7.1 Conclusions

Assessment of the dose and correspondent risk associated with nanoaerosols is a very complicated process consisting of many individual steps and parameters which should be determined by measurement or calculation.

Unfortunately the strategy for this process is not developed and as a result it is not clear exactly what kind of physical values determining the dose as a cause of health effect should be measured.

In this chapter we presented the possible strategy for the dose assessment of nanoaerosols in the form of dosimetric road map that allows analyzing the gaps between available and necessary information on every dosimetric parameter in the case of nanoaerosols.

Dosimetric road map consists of two main branches:

1. Parameters related to nanoaerosols airborne concentration measurement
2. Parameters of lung deposition and lung dosimetry for nanoaerosols

A review of the literature suggested that in the case of nanoparticles not the mass concentration but rather the particle surface area and particle concentration are better metric especially for health effect characterization. Generally speaking aerosol mass concentration cannot be used as a measure of the health effect in the case of nanoaerosols.

As a result, the calculation of the exposure based on measured concentration and time cannot be considered as a measure of the health effect in the case of nanoaerosols. So, at a present time it is not possible to get a valuable quantitative assessment of the exposure in the case of nanoaerosols. In practice, it is difficult to find studies with the assessment of the exposure in occupational setting or other types of contacts between human and nanoaerosols.

In principle, there cannot be a correlation between exposure and effect in the case of nanoaerosols, especially because we can see from dosimetric road map in this case so many different factors will affect the final assessment of the dose.

In terms of the second part of dosimetric road map, it should be pointed out that the main obstacle in the assessment of the risk associated with nanoaerosols is the lack of data on the local deposition of nanoparticles in the lung.

In Ruzer and Apte (2010a,b) the new approach is proposed for this problem based on using the unattached radon progeny as a safe radioactive marker.

A review of the literature on translocation of nanoparticles in the body suggested that nanoparticles can enter blood circulation from respiratory tract. These processes are likely dependent on the physico-chemical properties such as size and physiological state of the organs of entry. The translocation seems to be rather low in terms of the mass, but it is not known what amount of particles can reach the target.

As the knowledge of long-term behavior of nanoparticles is not known we should estimate that insoluble nanoparticles may accumulate in secondary organs during chronic exposure with an unpredictable amount.

Due to the lack of information both on the air concentration determination and lung local deposition data the problem of the dose and correspondent risk assessment for nanoaerosol cannot be solved at the present time.

In order to at least diminish the potentially harmful effect of nanoparticles the respirators should be recommended. Unfortunately, as we already mentioned earlier it is not known at the present time what the real efficiency of respirators against the nanoparticles is.

16.2 RADON PROGENY AS AN EXPERIMENTAL TOOL FOR DOSIMETRY OF NANOAEROSOLS

The health effects from aerosols in air pollution are discussed widely in scientific literature. During the last 10 years the need to understand the relationships between aerosol exposure and biological effects has become especially important due to the rapid development of the new, revolutionary industry—nanotechnology. Commerce in nanoproducts could top $1 trillion by 2015 (Aitken et al., 2004).

The understanding of the relationship between aerosol concentration in the breathing air and particle deposition in different portions of the lung remains poor, in spite of several decades of research. Without direct validation through measurement of aerosol deposition and dose, uncertainty in estimates from inhalation models will persist.

According to the Strategic Plan for NIOSH Nanotechnology Research (NIOSH, 2005) there are many gaps in our knowledge of aerosols in the nanometer range, which we need to fill in order to improve risk assessment and dosimetry of nanoaerosols, including

- Measurement studies of nanoparticles in the workplace
- Particle surface area as a dose metric
- Dosimetry and risk assessment
- Evaluation of pulmonary deposition and translocation of nanomaterials

One experimental approach that addresses these problems uses safe doses of radioactive tracer particles: after particles are inhaled, their spatial distribution in the lung is determined through two- and possibly three-dimensional spectrometric imaging. In recent years the published literature on the effectiveness of risks associated with using radioactive markers in human imaging has matured to the point where reliable and safe protocols have been devised for their measurement.

As mentioned by Oberdörster et al. (2005), it is not clear if there are significant human exposures to airborne-engineered carbon nanotubes or C_{60} fullerene particles. This is of concern, because, for example, at the very low mass concentration of $10\,\mu g/m^3$ of unit density 20 nm particles, the number concentration is greater than 10^6 particles/cm^3. At this concentration, what is the particle surface dose delivered per cm^2 to different lung tissue? Animal studies using ultrafine and nanometer-sized aerosols showed that such particles induced significant pulmonary inflammatory responses as well as effects in extrapulmonary organs (Oberdörster et al., 2005).

Yet, without direct measurements of the response in humans with accurate dosimetry we cannot make conclusive estimates of the risks. Examples of the safe use of radioisotopes to label aerosol particles for human inhalation studies are presented in the literature.

A number of studies on the exposure of ultrafine and, especially, nanometer aerosols (Donaldson et al., 1998; Oberdörster et al., 2005), have indicated that health effects associated with low-solubility inhaled particles may be more appropriately associated with particle surface area than with mass concentration. Such data on the correlation between number, surface area, and mass concentration are needed for exposure investigations.

The use of particle surface area as a dose metric for nanoaerosols is discussed in many reports: NIOSH (2006), Royal Society (2004), NIOSH (2005), NIOSH (2006a), Aitken et al. (2004), DEFRA (2007), SCENIHR (2006), and ASCC (2006).

It has been mentioned that particle surface area might provide the most suitable criterion for assessing inhalation exposure. Currently however, there is a need to develop and expand methods available by which particle surface area can be assessed in the workplace. The main concern is with free nanoaerosols that are more available for absorption and distribution within the body. Investigators found that when lung burdens and clearance rates were expressed as a function of the surface area, there was a much closer correlation with biological responses, ASCC (2006).

The special importance of the surface area of nanoparticles in the study of the risk assessment is discussed in Maynard (2007) by comparing the three characteristics: particle number, surface area, and mass concentration.

Another important issue in the safety of people working with nanomaterials is the efficiency of respirators, discussed in many reports: NIOSH (2006b), Aitken et al. (2004), Renn and Roco (2006). It is well known that the determining factor governing the effectiveness of RPE is not absolute penetration through the filter, but rather face-seal leakage causing particles to bypass the device (ASCC, 2006; Brown et al., 2001).

Here we discuss a new approach to the measurement of aerosol surface area concentration based on the rate of deposition of the unattached activity of radon progeny on aerosol particles. The correlation, results of calculation, and the assessment of the sensitivity of this method will be presented.

16.2.1 Approach: Safety of Radioactive Markers in Aerosol Exposure Study

We may assume that a radioactive marker is safe in an experiment with human subjects if the radiation exposure is negligible relative to the subjects' background exposures. We already discussed in this chapter the three scenarios of exposure to radon and its decay products.

A comparison of these three cases shows that radiation exposure in the human experiment was less than 1/1000th the magnitude of the lifetime background exposure. From a radioactive exposure point of view, the PSI experiment was safe.

Nonetheless, such human experiments need radiation and environmental health and safety reviews. Also, the type of radiation, half-life, clearance, and the particle size of the markers should be taken into account. The chemical characteristics and the size distribution of the nonradioactive aerosol under study are also important in assessing the safety of human exposure experiments. The experiment at the PSI, mentioned earlier, was conducted after a human subject's internal review board (IRB) approval by the "Uberregionale ehtische Kommision fur klinische Forschung der Schweizerischen Akademie der medizinischen Wissenschaften" was granted (Butterweck et al., 2001).

The expected fast particle transport from nasal passages, larynx, and mouth to the GI tract was not found. The explanation may be that a substantial fraction of deposited activity with diameter of 1 nm is bound to lung tissue. The most important aspect of this study is that from a radiation safety point of view it is possible to provide similar human experiments in laboratory conditions, after radiation safety and IRB approval.

Similar studies (Ruzer and Harley, 2004) were conducted with miners who during their normal work activities were exposed to much lower concentrations of radon, much higher gamma-background, and simple instrumentation that quantified post-exposure gamma emissions from the chest of the subjects.

16.2.2 Characteristics of Radon Progeny

The presence of radon and its decay products in the air is due to the abundance of unstable heavy metals (radioactive elements at the end of the periodic table) in the earth. One of them, uranium, undergoes a long series of transformations to yield radium. The chain of radioactive decay continues beyond radium to generate radon, a radioactive noble gas. Due to its inert chemical properties, radon does not bind completely to surficial soils in the earth or stay in water but enters the atmosphere as a gas.

The elements following radon in the radioactive decay chain—isotopes of polonium, bismuth and lead, atom-sized radionuclides—may attach to aerosol particles to become radioactive aerosols or exist in unattached forms in the air. Eventually, they may be deposited in the lung and cause irradiation of the lung tissue. The specific biological consequences depend upon the dose of radioactive aerosols which in turn depends on physiological characteristics including human breathing rates, especially changes with physical activity as well as the amount of the radium in the soil and radon and its decay products in the air, and atmospheric conditions both in the open air, dwellings and the underground environment.

The decay products represent a very complicated system consisting of a series of radioactive elements and various types of decay (alpha, beta, and gamma). In terms of the radiation safety, the most important radionuclides are alpha-emitters because the alpha particles have the greatest ionization density (Linear Energy Transfer [LET]). Given identical absorbed energy, the biological effect of alpha-particles is thought to be 20 times greater than the corresponding effect of beta-particles and gamma-radiation (i.e., the "quality coefficient" for alpha particles is 20). However, due to low particle penetration through human tissue, it is impossible to externally measure the alpha-activity of aerosols deposited in the lung of a living subject. As a result, this alpha-radioactivity is typically measured in the air and the absorbed dose to the lungs is then calculated according to the known concentration, breathing rate, and coefficient of deposition in the lungs (which is not accurate).

16.2.3 Assessment of Particle Deposition in Lungs

Ruzer (1964a,b) and Ruzer and Harley (2004) presented another, more precise, opportunity for assessment of the alpha-dose to the lung from radon progeny. It was based on the derived correlation between the alpha-dose and gamma-activity of radon progeny measured directly from the lung. This possibility was studied first on animals, then in model experiments, and finally, after certification from the Soviet Ministry of Health, on hundreds of miners in the former Soviet Republics of Tajikistan, Uzbekistan, and Kazakhstan. It was shown that for concentrations in the range of maximum permissible in mines, gamma-activity of radon progeny in the lung can be measured directly by means of a simple technique, such as the use of NaI (Tl) crystal detectors with standard lead shielding.

This approach of direct measurement of the natural marker such as radon progeny can also be used for the assessment of deposition of nonradioactive aerosols, particularly nanometer aerosols, in the lung.

The formula for gamma-activity $A\gamma$ is

$$A\gamma = a \, v \, k \, (q_b + q_c) \qquad (16.1)$$

where
  a represents proportionality coefficient
  v represents the volumetric breathing rate
  k represents the lung deposition coefficient
  q_b and q_c represent the concentrations of the ^{214}Pb (RaB) and ^{214}Bi (RaC) correspondingly

Based on (16.1) these results yield values for the product vk, which has been termed "Filtration Ability of Lungs" (FAL) (Ruzer and Harley, 2004), where

$$FAL = vk = \frac{A\gamma}{aq} \qquad (16.2)$$

where $q = q_b + q_c$.

FAL is an important breathing parameter reflecting the gross particle removing behavior of the respiratory system. It is a "bridge" between the quantity of aerosols in the air and in the lung for different physical activities.

Measurements were performed without disturbing the working conditions for three occupational groups: drillers, auxiliary drillers, and inspection personnel, totaling approximately 100 workers. The average, standard error, and the median values for a total of 297 air samples and 391 lung measurements are shown in Ruzer et al. (1995).

The measurements were also carried out in a non-uranium mine in Tajikistan (former USSR) with a special instrument having two probes (Antipin et al., 1978), one for lung gamma-activity measurement and the other for air alpha-activity measurement.

As we mentioned before, these experiments, conducted in specific underground conditions with a high gamma-background, were provided with simple, portable instrumentation. Therefore, it was difficult to study the detailed distribution of activity in the lung. The successful measurements made in this study under rugged mining conditions illustrate the possibilities for using radon progeny as a tracer in the study of aerosol distribution in the lung. This method would be particularly suited for use in the nanometer range, under laboratory conditions, where the background is low, and instrumentation is much more sensitive, similar to the studies conducted at PSI. This approach could be used to map aerosol dose to the lung because it can provide graphic information on where these particles of different sizes are deposited in the respiratory system.

Thus, the approach is to use radon progeny as a marker at safe doses in the study of deposition of nonradioactive nanoaerosols in the human body. The proposed radiation dose during a human experiment will be negligible in comparison with the natural background exposure over time. This is consistent with the use of radiological tracers for other medical research. For human experiments we propose using a generator of unattached fraction of radon progeny. This could be a small environmental chamber such as used in the Swiss research (Butterweck et al., 2001), or using a respirator mask exposure apparatus attached to a small chamber. Both radon and monodisperse aerosols of known size and morphology will be injected into the chamber under controlled conditions.

In terms of radioactive safety, when properly handled, the gamma-activity of radium is negligible. Through radioactive decay, radium produces radon (^{222}Rn), which in turn will produce atoms of ^{218}Po. Once it is released into the air, about 10–12 molecules of air constituents naturally diffuse onto and surround the ^{218}Po atom. These clusters are about 1 nm in diameter and have a diffusion coefficient of $\sim$0.06 cm^2/s. They are called unattached activity of radon progeny. Again through natural diffusion processes, these Rn progenies deposit on particles coexisting in the air ranging in size from nanometers to micrometers.

Controlled experiments where nanometer (or larger)-sized aerosols of known diameter are radiolabeled through natural attachment to unattached activity (e.g., becoming "attached activity") can be used to trace aerosol lung or surface deposition behavior. To do this, a size-characterized monodisperse aerosol is inserted into an exposure chamber containing particle-filtered air. Rn progenies are also inserted into the chamber at a controlled or known rate. By measuring the unattached fraction in the air we can assess the particle surface area of aerosols, see Ruzer (2008) and Ruzer and Apte (2005), and for monodisperse aerosols consequently we assess the particle concentration. The ratio between particles attached to the nanoparticle activity and the nanoaerosol particle concentration itself will be known. After the nanometer aerosol (labeled with radon progeny) is inhaled into the lungs, the nanometer particles will be locally deposited according to their size depending on some breathing parameters (volume breathing rate, humidity, and temperature).

Until now, experimental data on nanoaerosol deposition in human lungs have not been available. For larger size aerosols mostly bulk deposition data are available, based on the difference in concentration in exhaled and inhaled air. However, it is well known that biological effects depend on local deposition.

After exposure, the local gamma emission distribution in the lung can be measured using a gamma-spectrometer, with the local activity being proportional to the aerosol deposition (dose). In addition, it may be possible to use SPECT scanning to provide a more precise spatial resolution of particle deposition and local dosing (Kao et al., 1997; Piai et al., 2004).

As with all such radiotracer studies, the protocol must meet the approval of an IRB and radiological screening review. In these experiments, as in other studies, when radiation is used as a tool, for example, in using radiation in the study of Alzheimer disease, we have to compare the risk with benefit. The use of such experiments will enable us to close the gaps in our knowledge. Quantitative assessment of the local deposition of aerosol is at the core of aerosol, and particularly nanoaerosol exposure and risk assessment. So, our goal will be to find the safest possible and most appropriate marker.

Radon progenies are attractive as a marker for several reasons:

1. Radon and its progeny belong to the natural background of radioactivity to which the general population is exposed during their lifetime. Therefore, it is easy to assess the additional risks due to its use by the methods proposed.
2. Part of radon progenies, called "unattached activity," are 1 nm sized particles with diffusion coefficient close to $0.06 \, cm^2/s$ (a size that attaches readily to nanoaerosols), which make it very attractive as a marker for nanoaerosols with a built-in signal.
3. Radon decay products are easy to generate.
4. Radon decay products are short-lived nuclei.

Direct measurement on humans is needed in order to validate the hollow cast, animal studies, and modeling. From our point of view, this kind of study will be strategically important in nanoaerosol dosimetry and risk assessment. And it will partially close one of the many gaps in our understanding of nanoaerosol exposure.

The first step of this study should be human experiments with monodisperse spherical nanoparticles. In the case of nonspherical particles, typically found in aerosol studies, we should use the term "equivalent diameter," that is, the diameter of a monodisperse aerosol with the same local deposition as the aerosol of interest. Study of polydisperse aerosols adds complexity that can be resolved after the monodisperse aerosol lung deposition characterization across a broad nanometer size range is completed.

16.2.4 RADON PROGENY AS A TOOL IN THE ASSESSMENT OF PARTICLE SURFACE AREA

As discussed earlier, one very important property of radon decay is that after radon decay, the newly formed atom of Po forms clusters that are useful as markers in studies of properties of

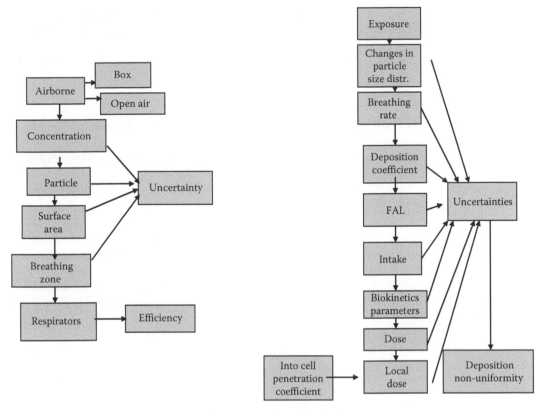

FIGURE 16.1   Road map nanoaerosols dosimetry presents a conceptual mapping of the processes starting with the manufacture of nanomaterial, through its possible release into the environment and ultimate dose to the target cell.

nonradioactive aerosols. Figure 16.1 (Ruzer, 2008) depicts the basic processes of radon decay producing "unattached" and "aerosol attached" activities. According to this figure there should be some correlation between unattached activity and aerosol concentration. In other words, the smaller the aerosol concentration, the bigger will be the unattached activity and vice versa. In quantitative terms, the unattached fraction of radon progeny can be used as a measure of aerosol particle concentration. This is the basis for the approach proposed in Ruzer (1964a,b) for measuring very small aerosol concentrations.

In the following, we present the details and calibration procedure for unattached fraction measurement.

16.2.5   METHOD AND CALIBRATION PROCEDURE

The method and calibration procedure for measurement of radon progeny unattached fraction is presented in Ruzer and Sextro (1997). The equipment for calibration and measurement of unattached and attached radon progeny consists of the following:

1. A chamber with regulated concentration of the unattached fraction of radon decay products in the range from 0, to close to 0.95–1.0 (note that a high aerosol concentration, $>10^6$ particles/cm^3, creates surface for diffusion of nearly all unattached fraction, while a very low aerosol concentration will remove almost none of the attached fraction).
2. An alpha-spectrometer.

3. A set of diffusion batteries or diffusion batteries composed of wire screen can be used for collecting aerosol particles in the ultrafine and nanometer range (Ruzer and Sextro, 1997).
4. An air Pump with regulated flow rate.
5. High collection efficiency (99.999% for particle sizes <1000 nm) gravimetric air filters.

The calibration procedure consists of the following measurements:

1. e_1, e_2—deposition coefficients of the unattached and attached radon progeny in diffusion battery, respectively
2. e_3, e_4—detection efficiencies of the unattached and attached radon progeny deposited in diffusion battery, respectively
3. e—detection efficiencies of the summary (unattached and attached radon progeny) air-borne activity on the filter

The procedure for measurement of the unattached fraction of radon progeny consists of measuring the activity of radon progeny inside the diffusion battery, in the diffusing battery backing filter, and in the open filter.

The unattached fraction of radon progeny, f, can be calculated as a ratio of the unattached (q_u) and sum [($q_u + q_a$), unattached +attached] concentration according to the formula (Ruzer and Sextro, 1997):

$$f = \frac{q_u}{[(q_u + q_a)]} = \frac{[eN_{db}(1-e_2) - e_2 e_4 N_{bf}]}{N_{of}[e_1 e_3 (1-e_2) - e_2 e_4 (1-e_1)]} \qquad (16.3)$$

where N_{db}, N_{bf}, and N_{of} are measured activity in the diffusion battery, backing filter, and open filter, respectively.

A summary of the results for this approach is presented in Figure 16.2 (Ruzer, 2008), which shows the measurement of unattached concentration of ^{218}Po relative to particle surface area concentration in the range of 0.3–2.1 μm, made with monodisperse latex aerosols of different sizes and concentrations (Dokukina and Ruzer, 1976). In each case, the measured aerosol concentration was converted to the aerosol surface area concentration. These results suggest that for aerosols in the size range covered by this calibration, 0.3–2.1 μm in diameter, the particle surface area concentration is in the corresponding range from 10^{-5} to 0.3 cm^{-1}, and is related to the unattached fraction of ^{218}Po, f. The calibration procedure used is for monodisperse spherical particles, from which the surface area can be directly calculated. In practice, for nonspherical and polydisperse aerosols an "equivalent surface area" should be used, which is the surface area of a spherical aerosol having the same diffusion deposition property as the real aerosol. Under actual measurement conditions

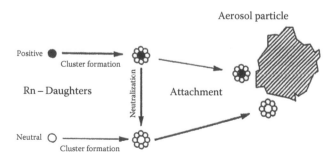

FIGURE 16.2 Basic process of Rn decay product behavior in air defining "unattached" and "aerosol-attached" activities.

(e.g., in a building or in a mine), the only measurements necessary to determine the unattached fraction (and to infer the average aerosol surface area) are those described earlier—a diffusion battery followed by a backing filter and an open face filter operated in parallel. The alpha activity from [218]Po on each of these three collectors is then measured and used with appropriate calibration factors to yield the unattached fraction (Figures 16.3 through 16.9).

Results of the calculation showed that

1. For a constant unattached fraction, f, aerosol particles surface area, practically speaking, (in the range of measurements errors) does not depend on particle size in the nanometer range from 2 to 100 nm in diameter.
2. The presented idea for assessment of the surface area will work for polydisperse nanometer-sized particles.
3. This method will not be sensitive to very small individual particles, because its contribution to the general surface area is proportional to the square of their diameter.

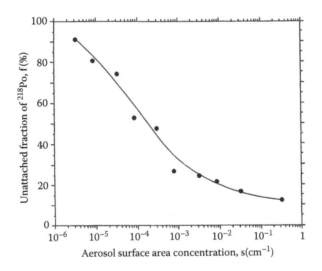

FIGURE 16.3 Unattached fraction of [218]Po, f vs. aerosol surface area concentration.

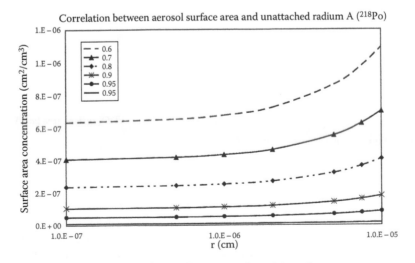

FIGURE 16.4 Relationship between particle surface area and particle radius.

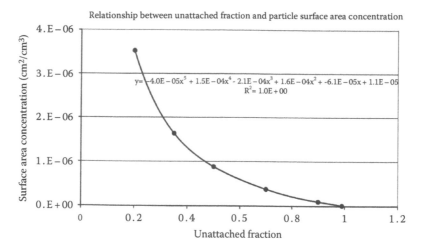

FIGURE 16.5 Relationship between particle surface area and unattached fraction.

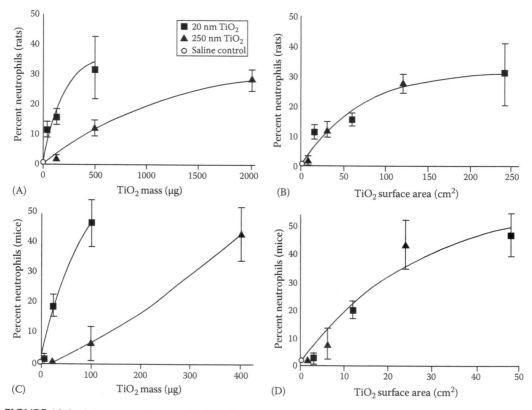

FIGURE 16.6 Mass vs. surface area health effect.

16.2.6 Discussion and Conclusions

The problem of aerosol deposition and lung dosimetry is very complicated. First, it is still very difficult to measure the particle size distribution of aerosol particles in the breathing zone of humans. But, even if such data were available, the behavior of aerosol particles according to aerosol size spectra in different parts inside the lung is difficult to predict with high accuracy due to humidity, temperature, lung morphology, and other factors. Therefore, direct measurement of lung

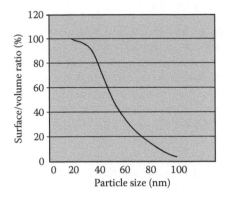

FIGURE 16.7 Surface area/volume ration (%) vs. particle size.

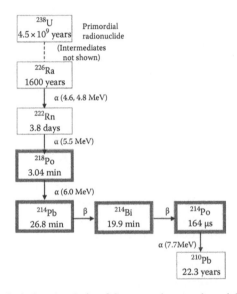

FIGURE 16.8 ^{226}Ra family depicting the chain of decay products and particle emissions.

FIGURE 16.9 Instrument for filtration ability of lung (FAL) measurement. 1—NaI detector; 2—collimators; 3—lead shield; 4—air sample; 5—electronic.

deposition parameters is still needed in order to validate the results of modeling. We have discussed the potential use of radon progeny as a radioactive marker, which can be used as an experimental tool in measurement of local deposition parameters and dosimetry of nanoaerosols.

The first problem faced in using radioactive markers is that of safety. We discuss three settings of human exposure experiments using radon decay products. It is clear from the comparison, with careful controls, that the risk of such experiments is minimal. The benefits from gaining information on lung deposition and dosimetry will be extremely valuable. Inter-species variability in respiratory system morphology and physiology is so great that similar data will be impossible to obtain in animal experiments.

A review of literature on environmental health in the new rapidly developing nanotechnology industry shows that the problem of exposure has not been adequately assessed (Oberdörster et al., 2005). Worker health and safety is of initial concern as occupational groups are likely to be among the first to be exposed to elevated concentrations of nanomaterials. A gap exists between existing particle measurement methods and those truly appropriate for nanoaerosol exposure assessment. Until now, the primary tools available for the measurement of nano-sized aerosols have been CPCs, and DMA. Results of the particle counting detection efficiency of the Condensation Particle Counter TSI CPC 3762 for different operating parameters have been presented (Banse et al., 2001). This study showed a substantial decrease in the efficiency in the range of particle diameter of 6–10 nm. DMA measurements suffer mainly due to the low probability with which in this range of sizes nanoparticles are charged (NSF, 2003). Even with improved aerosol instrumentation for nano-sized particles, the issues of respiratory tract deposition quantitation cannot be resolved without a direct localized measurement of particle dose.

The experiments at PSI and the measurements on miners serve as a model for experiments that can be performed in laboratory conditions. These experiments can provide accurate data on human breathing characteristics, deposition, lung dosimetry, and the assessment of true efficiency of respirators.

A new instrument on the market, the Nanoparticle Surface Area Monitor, is used for the assessment of DSA in the lung. Lung deposition estimates from this instrument are based on correlations developed (Wilson et al., 2004) between the electrical signal and modeled DSA. The instrument is said to be capable of detecting particles with diameters down to 10 nm. Another attempt to solve the problem of surface area assessment, previously presented (Maynard, 2003), is based on simultaneous number and mass concentration measurements. Of course, with these data, only for air concentration, it is not possible to calculate the true dose for target cells in the respiratory system in terms of the density of the number of particles per cm^2, especially because results from modeling (Balásházy and Hoffman, 2000) showed that such particles can form "hot spots," that is, extremely high local density of particles.

Clearly, all human experiments with nanometer particles labeled with radioactive markers need careful consideration from the point of view of both radioactive and nonradioactive nanometer aerosols safety. Still, in the balance, the scientific need for information from good human lung dosimetry experiments suggests that the problem should be considered.

From these observations, the following conclusions can be made:

1. Radon decay products (as a radioactive marker) can be used in a controlled manner that provides minimal risk in human studies under laboratory conditions.
2. Particle deposition, lung dosimetry, measurement of the breathing characteristics, and respirator efficiency for nanometer-sized particles may be measured by noninvasive techniques under laboratory conditions, in the range of exposures described in this chapter.
3. The unattached fraction of radon progeny can be used for quantitative assessment of a very important characteristic of nanometer-sized aerosols, that is, the aerosol particle surface area.

REFERENCES

Aitken, RJ, Creely, KS, Tran, CL. Nanoparticles: An occupational hygiene review. Prepared by the Institute of Occupational Medicine for Health and Safety Executive. Edinburgh; Research Report 174; Nanotechnology Now (July 17, 2007) www.nanotech-now.com (2004).

Antipin, NI, Kuznetzov, YuV, Ruzer, LS. Instrument for determination of the filtration ability of lungs. *At. Energy*, 40, 371–372 (1978).

ASCC. A review of the potential occupational health & safety implications of nanotechnology. Australian Government; Australian Safety and Compensation Council. http://www.ascc.gov.au/ascc/AboutUs/Publications/ResearchReports/ (2006).

Balásházy, I, Hoffman, W. Bronchial airway quantification of local deposition bifurcations. *Health Phys.*, 78(2), 147–158 (2000).

Banse, DF, Esfeld, K, Hermann, M, Sierau, B, Wiedensohler, A. Particle counting efficiency of the TSI CPC 3762 for different operating parameters. *J. Aerosol Sci.*, 32(1), 157–161 (2001).

Bhabra, G, Sood, A, Fisher, B, Cartwright, L, Saunders, M, Evans, WH, Surprenant, A et al. Roger Newson & Charles Patrick Case. Nanoparticles can cause DNA damage across a cellular barrier. *Nat. Nanotechnol.*, 4, 876–883 (2009).

Brown, DM, Wilson, MR, MacNee, W, Stone, V, Donaldson, K. Side-dependent proinflammatory effect of ultrafine polystyrene particles: A role for surface area and oxidative stress in the enhanced activity of ultrafines. *Toxicol. Appl. Pharmacol.*, 175, 191–199 (2001).

Butterweck, G, Vezzu, G, Schuler, Ch, Muller, R, Marsh, JW, Thrift, S, Birchall, A. In vivo measurement of unattached radon progeny deposited in the human respiratory tract. *Radiat. Prot. Dosimetry*, 94(3), 247–250 (2001).

Darby, S, Hill, D, Auvinen, A, Barros-Dios, JM, Baysson, H, Bochicchio, F, Deo H et al. Radon in homes and risk of lung cancer: Collaborative analysis of individual data from 13 European case-control studies. BMJ. DOI:10. 1136/bmj.38308.477650.63 (published December 21, 2004).

DEFRA. Characterising the potential risks posed by engineered nanoparticles; A Second UK Government Research Report. Department for Environment, Food and Rural Affairs, London, U.K. http://www.defra.gov.uk/environment/nanotech/research/pdf/nanoparticles-riskreport07.pdf (2007).

Dokukina, VL, Ruzer, LS. *Proceedings of All-Union Institute of Physico-Technical and Radiotechnical Measurements* (*VNIIFTRI*). Moscow (in Russian) 30(60) (1976).

Donaldson, K, Li, XY, MacNee, W. Ultrafine (nanometer) particle mediated lung injury. *J. Aerosol Sci.*, 29(5/6), 553–560 (1998).

Duan, N, Ott, WR. Comprehensive definitions of exposure and dose to environment pollution, in *Proceedings of the EPA/A&WMA Specialty Conference on Total Exposure Assessment Methodology*, Las Vegas, NV (November 1989).

Duffin, R, Tran, L, Brown, D, Stone, V, Donaldson, K. Proinflammogenic effects of low-toxicity and metal nanoparticles in vivo and in vitro: Highlighting the role of particle surface area and surface reactivity. *Inhal. Toxicol.*, 19, 849–856 (2007).

EPA. Review of the National Ambient Air Quality Standards for Particular Matter: Policy Assessment of Scientific and Technical Information, OAQPS Staff Paper, EPA/452/R-96-013, Office of Air Quality Planning and Standards, Research Triangle Park, NC (1996).

Friends of the Earth Nanotechnology Project. Workplace exposure to nanomaterials and the question of will nano be the next asbestos. AzoNano.com (2009)

Geiser, M, Rothen-Rutishauser, B, Kapp, N, Schürch, S, IOM, Project CB0407. CELL PEN: A study to identify the physico-chemical factors controlling the capacity of nanoparticles to penetrate cells (August 12, 2008).

Geraci, C. *An Update of NIOSH Nanotechnology Research Program*. Berkeley, CA: US School of Public Health. Nanoparticle Safety Symposium (2009).

Hankin, M, Tran, CL, Ross, B, Donaldson, K, Stone, V, Chaud, QM, Project CB0407, August 12, 2008, Cell pen: A study to identify the physico-chemical factors controlling the capacity of nanoparticles to penetrtate cells. Nanoparticle Surface Area Monitor, TSI Incorporated. June 2005. http://www.particle.tsi.com

Kao, CH, ChangLai, SP, Shen, YY, Lee, JK, Wang, SJ. Technetium-99m tetrofosmin SPECT imaging of lung masses: A negative study. *J. Nucl. Med.*, 38, 1015–1019 (1997).

Kelly, RJ. Occupational medicine implications of engineered nanoscale particulate matter. *ACS J. Chem. Health Saf.* 1–42 (March 2008).

Kulinowski, KM. Director, International Council on Nanotechnology Temptation, Temptation: Why easy answers about nanomaterial risk are probably wrong (2009).

Martonen, TB, Katz, I, Fults, K, Hickey, AJ. Use of analytically defined estimates of aerosol respirable fraction to predict lung deposition patterns. *Pharm. Res.*, 9(12), 1634–1639 (1992).

Maynard, AD. Estimating aerosol surface area from number and mass concentration measurements. *Ann. Occup. Hyg.*, 47(2), 123–144 (2003).

Maynard, AD. Nanotechnology: The next big thing, or much ado about nothing? *Ann. Occup. Hyg.*, 51(1), 1–12 (2007).

Mills, NL, Amin, N, Robinson, SD, Anand, A, Davies, J, Patel, D et al. Do inhaled carbon nanoparticles translocate directly into the circulation in humans? *Am. J. Respir. Crit. Care Med.*, 173(4), 426 (2006).

Möller, W, Felten, K, Sommerer, K, Scheuch, G, Meyer, G, Meyer, P et al. Deposition, retention and translocation of ultra-fine particles from the central airways and lung periphery. *Am. J. Respir. Crit. Care Med.*, 177(4), 426 (2008).

Muhlfeld, C, Geiser, M, Kapp, N, Gehr, P, Rorthen-Rutishauser, B. Re-evaluation of pulmonary titanium dioxide nanoparticle distribution using the "relative deposition index": Evidence for clearance through microvasculature. *Part Fibre Toxicol.*, 4, 7 (2007).

National Academy of Sciences. Human exposure assessment for airborne pollutants. Washington, DC: National Research Council (1991).

National Institute for Occupational Safety and Health. *Approach to Safe Nanotechnology: An Information Exchange with NIOSH*. Atlanta, GA: National Institute for Occupational Safety and Health, Centers for Disease Control and Prevention. Draft (October 2006a).

National Institute for Occupational Safety and Health. *Current Ultrafine Particle and Nanotechnology-Related Research*. Atlanta, GA: National Institute for Occupational Safety and Health, Centers for Disease Control and Prevention (October 27, 2006b). www.cdc.gov/niosh/topics/nanotech/oltrares.html

National Research Council, *Health Effect of Exposure to Radon, BEIR VI*, Washington, DC: National Academy Press (1999).

National Science Foundation. *NSF Workshop Report on Emerging Issues in Nanoparticle Aerosol Science and Technology (NAST)*. Los Angeles, CA: University of California (2003).

Nemmar, A, Hoet, PHM, Vanquickenborne, B, Dinsdale, D, Thomeer, M, Hoylaerts, MF, Vanbilloen, H et al. Passage of inhaled particles into the blood circulation in humans. *Circulation*, 105(4), 411–414 (2002).

NIOSH. Strategic Plan for NIOSH Nanotechnology Research—Filling the Knowledge Gaps. Nanotechnology Research Program, National Institute for Occupational Safety and Health, Centers for Disease Control and Prevention. Draft. http://www.cdc.gov/niosh/topics/nanotech/default.html (2005).

NIOSH. Approach to Safe Nanotechnology; An Information Exchange with NIOSH; National Institute for Occupational Safety and Health, Centers for Disease Control and Prevention. Draft (October 2006a).

NIOSH. Current Ultrafine Particle and Nanotechnology-Related Research. Atlanta, GA: National Institute for Occupational Safety and Health, Centers for Disease Control and Prevention. www.cdc.gov/niosh/topics/nanotech/oltrares.html (October 27, 2006b).

Oberdörster, G, Oberdörster, E, Oberdörster, J. Nanotoxicology: An emerging discipline evolving from studies of ultrafine particles. *Environ. Health Perspect.*, 113, 823–839 (2005).

Ott, WR. Concepts of human exposure to air pollution. *Environ. Int.*, 7, 179–196 (1966).

PAS BSI 71:2005 Nano particle III–Vs review; 18(5), 5 (June–July 2005) www.bsi-global.com/nano

Rejman, J, Oberle, V, Zuhorn, IS, Hoekstra, D. Size-dependent internalization of particles via the pathways of clathrin-and caveolae-mediated endocytosis. *Biochem. J.*, 377(1), 159–169 (2004).

Renn, O, Roco, M. White paper no. 2, nanotechnology risk governance, international risk governance council (IRGC). *With annexes*, Rocco, M, Litten, E. (eds.). Geneva, Switzerland: IRGC (2006). http://www.irgc.org/Nanotechnology.html

Rothen-Rutishauser, BM, Schurch, S, Haenni, B, Kapp, N, Gehr, P. Interaction of fine particles and nanoparticles with red blood cells visualized with advanced microscopic techniques. *Environ. Sci. Technol.*, 40, 4353–4359 (2006).

Ruzer, LS. The method of determining of the absorbed doses in organism upon inhalation of radon. Patent No. 165250. *Bull. Invent.*, 28 (1964a) (in Russian).

Ruzer, LS. The method of determining the concentration of the gas. Patent No. 234746. *Bull. Invent.*, 18 (1964b) (in Russian).

Ruzer, LS. Radioactive aerosols. Determination of the absorbed dose. Doctoral dissertation, Moscow, USSR (1970) (in Russian).

Ruzer, LS. Assessment of nanoparticles surface area by measuring of unattached fraction of radon progeny. *J. Nanopart. Res.*, 10, 761–766 (2008).

Ruzer, LS, Apte, MG. Radon progeny as a potential experimental tool in aerosol dosimetry, *Frontiers in Aerosol Research*, Irvine, CA (October 24–25, 2005).

Ruzer, LS, Apte, MG. Unattached radon progeny as an experimental tool for dosimetry of nanoaerosols primarily less than 100 nm in size: Proposed method and research strategy. *Inhal. Toxicol.*, 1–7, (2010).

Ruzer, LS, Apte, MG. Environmental Energy Technologies Division, Indoor Environment Department Lawrence Berkeley National Laboratory, Berkeley, CA 94720, US. *J. Int. Res. Publ.: Ecol. Saf.*, 5 (2010), ISSN 1313-7999.

Ruzer, LS, Harley, NH, eds. *Aerosol Handbook: Measurement, Dosimetry, and Health Effects.* New York: CRC Press (2004).

Ruzer, LS, Nero, AV, Harley, NH. Assessment of lung deposition and breathing rate of underground miners in Tadjikistan. *Radiat. Prot. Dosimetry*, 58(4), 261–268 (1995).

Ruzer, LS, Sextro, RG. Assessment of very low aerosol concentrations by measuring the unattached fraction of ^{218}Po. *Radiat. Prot. Dosimetry*, 71, 135–140 (1997).

SCENIHR. The appropriateness of existing methodologies to assess the potential risks associated with engineered and adventitious products of nanotechnologies, Health and Consumer Protection Directorate—General, Scientific Committee on Emerging and Newly Identified Health Risks, European Commission, Brussels: European Commission (2006).

Shaffer, RE. Respirator Fit: Past, Present, and Future National Institute for Occupational Safety and Health (NIOSH) National Personal Protective Technology Laboratory (NPPTL) *No Fit Test Respirator Research Workshop* (November 6, 2008).

Wiebert, P, Sanchez-Crespo, A, Seitz, J, Falk, R, Philipson, K, Kreyling, WG, Möller, W et al. Negligible clearance of ultrafine particles retained in healthy and affected human lungs. *Eur. Respir. J.*, 28(2), 286–229 (2006).

Wilson, WE, Stanek, J, Pui, DYH, Han, HS. Use of the electrical aerosol detector as an indicator for the total particle surface area deposited in the lung, in *Proceedings of Symposium on Air Quality Measurement Methods and Technology*, Research Triangle, Park, NC (2004).

Zartarian, V, Bahadori, T, Mckone, T. Adoption of an official ISEA glossary. *J. Exposure Anal. Environ. Epidemiol.*, 15, 1–5 (2005).

17 Filtration and Sampling of Aerosols by Fibrous Filters

A.K. Budyka, A.A. Kirsch, and B.I. Ogorodnikov

CONTENTS

17.1 INTRODUCTION

Air and gas purification from submicron particles is one of the prime components of modern technologies applied in high-precision mechanics, electronics, medicine, nanotechnologies, high-purity materials obtaining, etc. High requirements to air quality are declared for production departments where radioactive agents and different origin hazardous aerosols containing toxic chemical compounds are being generated. Therefore, adequate means of personal respiratory system protection against such aerosols as well as against bioaerosols containing viruses and bacteria are essential.

A great variety of filtration conditions does not let us use "universal filters," because every case demands considering requirements to the purified media. Thus, filters for aerosol sampling (analytical filters) should meet other requirements than filters for superfine dust collecting. On the analytical filter, the largest amount of substance should be sampled during the least time. Analytical filters should not be as effective in aerosols capturing as finish filters for gas purification, because the uncertainty of particle characteristic measurements for any analysis technique is usually not less than several percent. Such analytical filters should have relatively low aerodynamical resistance in order to provide high productivity for aerosol sampling. And finally, analytical filters should consist of substances that will not hamper but instead facilitate the process of sediment analysis.

Fibrous filtering materials appeared to be the most effective means for air purification of submicron aerosol particles because these materials—when flow resistance is the same—have the highest particle capturing efficiency compared with all other type materials. Fibrous filtering materials are used for manufacturing industrial filters designated for ventilating streams purification, for analytical filters and respirators production. Filters manufactured of thin-layer high-porous nanofiber-based membranes are used for superfine purification.

Thin-layer filtering materials with surface filtering mode are usually used at the finishing stage in gas purification. To prolong finish filters' working lifespan they are preceded by preliminary purification filters (pre-filters) manufactured of coarse fiber materials. Pre-filters perform volume filtering, that is, aerosol particle sediment forms on the fibers of all the filtering material.

To minimize pressure drop on the filter, the gas flow velocity does not exceed some cm/s, so the particles having come into contact with fibers are not blown off.

Langmuire was the first to describe main mechanisms of particle deposition on thin fibers and to give the method to calculate filter efficiency, so he can be ranked, as of right, as the founder of the gas filtration theory as an independent trend of research [1]. Scientists of the Karpov Institute of Physical Chemistry have contributed significantly to the development of the theory, their research performed in the middle of the twentieth century [2–10] forms the basis of the modern theory. Works on the theory of filtration [2–16] should be remarked as well. We should mention especially the experimental works of Ya.I. Kogan, who developed in the 1950s an original device KUST [11] that lets us detect nanoparticles less than 1 nm.

Nowadays thin-layer filtering materials made of or based on fiberglass are widely used as ultrafine filters. Filters containing ultrafine fiber materials are called in scientific and technical literature "HEPA filters" (high-efficiency particulate air filters). In Russia for purposes of fine purification there are used filters of micron and submicron polymer fibers obtained by electrospinning (electrostatic polymer solution spraying). These are so called Petryanov filters (FP) [17,18], their method of manufacturing developed by I.V. Petryanov, N.A. Fuchs, and N.D. Rosenblyum over 70 years ago [19]. Electrostatic charge on fibers makes these filters very efficient. A valveless low resistance and high-efficiency respirator "Lepestok" (petal) [20] is based on the FP.

In the latest decade there appeared growing interest to obtain fibers thinner than 0.1 μm by electrospinning technique [21–24]. It arises from the theory of filtration that the resisting force of a fiber decreases with the fiber diameter going down, whereas particle capture efficiency increases [25]. Accordingly, the principal trend to improve filters has always the tendency to manufacture filtering material from the thinnest possible fibers. Filtering materials from nanofibers ensure particle concentration decrease by nearly 10 orders of magnitude [26].

Amounts of purified air have been growing incessantly, so it is reasonable to reduce power consumption for air circulating through filters. Thus, the main goal of filtration is not only to achieve high purity of air at any cost, but to develop energy-conserving purifying systems, that is, systems with high particle capturing efficiency, low pressure drop, and maximal lifespan.

The first edition of this chapter was generally connected with analytical filtration. The present chapter will illustrate the main results in research of superfine air cleaning, obtained in the Karpov Institute of Physical Chemistry or in collaboration with their scientists and published mainly in the

Russian journals, the latter, apparently, being the reason for unawareness among scientists, studying filtration. Issues of aerosol sampling with fiber filters and their analyzing will be considered as well.

17.2 FILTERING CHARACTERISTICS OF FIBROUS FILTERS

The main characteristics of filters are the efficiency of capturing specific size particles at selected velocity of the air flow through the filter and the pressure drop (Figure 17.1).

The ratio $K = N/N_0$ is called penetration coefficient or simply *penetration* through the filter. The value $E = 1 - K$ is called filter *efficiency*. Here N_0 represents concentration of particles before the filter and N is the concentration of particles after the filter. The value of *pressure drop* ΔP depends on the filter thickness H and gas flow velocity U.

Fibrous filters consist of thin fibers, randomly oriented in planes perpendicular to the flow. The volume portion of the fibers or the *packing density*, α, is very small, and usually does not exceed 5%. The *porosity* of the filter is $1 - \alpha$.

It is assumed in the theory of fine filtration that the part of particles deposited on the fiber from the incoming flow is constant through the whole thickness of the filter, that is, it is independent of the filter thickness. This part is called the *capture coefficient*. The capture coefficient η depends on the filter parameters, conditions of gas flowing through the filter and particle size. Knowing the capture coefficient and filter parameters, we can calculate efficiency of the filter E:

$$E = 1 - \frac{N}{N_0} = 1 - \exp(-2aL\eta) \tag{17.1}$$

where L is the total length of fibers per unit area of the filter

$$L = \frac{\alpha H}{\pi a^2} \tag{17.2}$$

where
  a is the fiber radius
  α is the packing density

Pressure drop Δp on the filter is equal to the sum of drag forces of fibers with the length L:

$$\Delta p = U\mu LF \tag{17.3}$$

Here F is the dimensionless drag force, applied to the fiber unit length. The ratio of the logarithm of the penetration to the pressure drop, called criterion of quality, is selected to compare filters' initial qualities:

$$q_f = -\frac{\ln(N/N_0)}{(\Delta p/U\mu)} \tag{17.4}$$

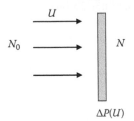

FIGURE 17.1 Parameters for defining the main characteristics of a filter.

This value is dimensional [L] and depends on the filtration conditions. The higher the value of q_f, the better the filter suits various conditions of air purification.

The major task of the theory of gas filtration is to calculate the capture coefficient for a fiber of a given radius r_p, depending on velocity U, viscosity μ, temperature T, and pressure P of air, existence of external forces F_i, electrical field, filter parameters—packing density α, mean fiber diameter $2a$ and their dispersion σ, filter internal structure:

$$\eta = \eta \ (r_p, \rho, \ U, \ \mu, \ T, \ P, \ F_i, a, \ \alpha, \ \sigma, \ \varepsilon) \qquad (17.5)$$

where
  ρ is the density of particles
  ε is the parameter of inhomogeneity, characterizing the filter structure

Besides, the capture coefficient depends both on the particle shape, its electrostatic charge q, its dielectric permittivity and on charges present on the fibers, their dielectric permittivity, and the fibers' cross-sectional shape. In nonstationary filtration (deposit accumulates on fibers) the capture coefficient depends on the number of deposited particles and the deposit packing density as well.

17.3 AEROSOL PARTICLE CAPTURE MECHANISMS IN FIBROUS FILTERS

There exist several physical mechanisms of aerosol particle capturing by filter fibers. First, it should be mentioned that the so-called "sieve" mechanism (screening particles larger than the filter cell) is not the main mechanism gas of particle capturing.

For low filtration velocity *diffusive capture* by fiber filters dominates. Aerosol particles do not move along streamlines flowing round an obstacle (in our case, the fiber) but become displaced from them due to incessant collisions with gas molecules (Figure 17.2). The smaller the size of the particle and its velocity, the higher the probability of collision. The diffusive capture coefficient can be more than 1.

If a particle moves along a streamline flowing in the immediate vicinity of a fiber, then the increase of the particle size leads to its capture probability increase. This mechanism is called the "interception effect."

Large and heavy particles can be displaced from the streamline due to its own inertia. The probability of collision of aerosol particle with fibers depends on the Stokes number (St) and the Reynolds (Re) number. There exists the critical value of St_{crit}, below which the inertial deposition does not appear [27].

For cylindrical fibers $St_{crit} \leq 0.25$ [28,29]. Particle capture due to inertia is the dominant mechanism of micron and submicron aerosols capturing during high velocity filtration. Regularities of inertia deposition are used for aerosol particle fractioning according to their dimensions (see the following).

If a particle and a filter fiber are charged, then the probability of the particle capture due to the Coulomb interaction increases. In some filtering installations, particles are intentionally charged in the corona field in order to increase filtration efficiency (in the atmosphere the charges on the fine particle are distributed according to the Boltzmann law and, as a rule, their amount does not exceed 1–2 elementary charges). For example, filtering material fibers, obtained by electrospinning, are being charged directly while producing. They are widely used, therefore, in disposable means for respiratory system personal protection [20].

Gravity affects aerosol particle deposition at low velocity of air through filter. This effect is substantial for heavy particles and can be most clearly observed during filtration from "below to up" or from "up to bottom." In the first case the integrate capture coefficient is increasing,

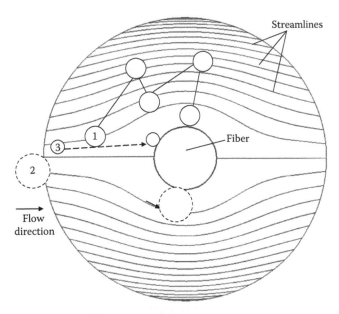

FIGURE 17.2 Main mechanisms of aerosol capture by fibers: 1, diffusion; 2, interception; 3, inertia.

and in the second it is diminishing by the value of order $\eta_G \propto G(1 + R)$, where G is the ratio of the settling velocity to flow velocity U [8]. Deposition efficiency of the aerosol particles is essentially affected by the forces of intermolecular interaction, or van der Waals forces. In some cases neglecting this mechanism can result in 15%–20% underrating of the design efficiency of the filter.

The task of calculating the filter efficiency comes down to determining the capture coefficient depending on the parameters of the filter, the particles and the medium:

$$\eta = \eta(Re,\ Kn,\ Pe, R,\ St,\ G,\ F_i,\ a,\ \alpha,\ \sigma,\ \varepsilon,\ \dots) \tag{17.6}$$

where
  $Kn = \lambda/a$ stands for the Knudsen number, that characterizes the ratio of the mean free-path length of the gas molecules to the fiber radius
  $Pe = 2Ua/D$ is the diffusive Peclet number, that characterizes convective transport predominance over the diffusive one, D is the coefficient of the particle diffusion, given by

$$D = \frac{k_B TC(Kn)}{6\pi\mu r} \tag{17.7}$$

where
  $C(Kn)$ is the Cunningham correction factor [30]
  k_B is the Boltzmann constant
  T is the temperature
  μ is the gas viscosity

Table 17.1 gives the values of the diffusion coefficient for particles smaller than 1 μm.
  $R = r_p/a$ is the interception parameter
  $St = 2C(Kn)r_p^2\rho U/9\mu a$ is the Stokes number (inertia parameter)

TABLE 17.1
Diffusion Coefficients of Aerosol Particles in
Air, Pressure 1 atm Temperature 20°C

D (µm)	0.01	0.1	0.5	1
D (cm²/s)	5.31E–04	6.84E–06	6.31E–07	2.76E–07

Source: Fuchs, N.A., *The Mechanics of Aerosols*, Pergamon Press, Oxford, U.K., 1964.

$Re = 2aU/\nu$ is the Reynolds number
ν is the gas kinematic viscosity
$G = U_G/U$ is the sedimentation parameter
U_G is the particles setting velocity
F_i is the dimensionless parameter, characterizing particle deposition due to external forces

Research of particle deposition in the area of the highest penetration (the worst particle deposition) is the major task. It is necessary to take into account the particle's own size and electrostatic and van der Waals forces. Joint action of all filtration mechanisms should be regarded in the area of the highest penetration. For nonstationary filtration calculations of particle deposition turn out to be even more complicated, the advance of the theory being significantly slowed down by the absence of experimental works containing all the necessary quantitative data.

First, theoretical works on particle deposition on filter fibers were based on the flow field in the vicinity of a separate cylinder [1]. However, quantitative correspondence between theoretical calculations and filters experimental data was out of the question. First, because the flow field in the vicinity of a separate fiber is determined by Re, whereas the flow field in the filter is independent of the Reynolds number. Second, the measured values of F and η for different filters vary several times [25]. Because of a complicated microstructure of the fibrous filtering materials and hence complicated flow in them, it was necessary to use fibrous filter models.

17.4 MODELING FILTRATION PROCESSES

17.4.1 HYDRODYNAMIC RESISTANCE OF PARALLEL FIBER MODEL FILTERS

A *system of parallel cylinders* perpendicular to the flow direction was assumed as the simplest model of a fibrous filter [5–7]. As distinct from the Lamb flow field for a separate cylinder, the flow function for a system of cylinders in Stokes approximation does not depend on the Reynolds number, but is defined by α, the packing density of the system. The flow field in such a system was obtained with the help of the Kuwabara cell model [31]. The value of the resisting force for the range of α 0.0043–0.27, experimentally obtained in Ref. [6], corresponded directly to the equation

$$F = \frac{4\pi}{k_0} = 4\pi(-0.5\ln\alpha - 0.75 + \alpha - 0.25\alpha^2)^{-1} \tag{17.8}$$

The results of measurements for the flow field and resistance for a *hexagonal* model were theoretically proved in Refs [31,32]. The simple view of the Kuwabara flow function made this model popular. It became the basis for estimating particle deposition on fibers. While studying deposition of larger particles the fiber rows were drawn apart to exclude the influence of the diffusive trace behind the fibers; in this case rows reciprocal action disappeared. Hydrodynamic resistance of rows did not depend on the Reynolds number, but was defined by the value a/h, which is the ratio of the fiber diameter and the distance between fibers in the row.

Analytical solutions for F both for dense and rarefied rows were obtained earlier. Keller was the first to determine the fiber drag force at $a/h \to 1$ using the approximation of the hydrodynamic lubrication theory [33]:

$$F = \frac{9\pi}{2\sqrt{2}} \frac{1}{(1 - a/h)^{5/2}} \tag{17.9}$$

The Miyagi equation appeared to be applicable to the rarefied row [34]:

$$F = 8\pi \left(1 - 2\ln 2t + \frac{2}{3}t^2 - \frac{1}{9}t^4 + \frac{8}{135}t^6 - \frac{53}{1350}t^8 + \cdots \right)^{-1} \tag{17.10}$$

where $t = \pi a/2h$. Figure 17.3 shows comparison of the fiber drag force $F(a/h)$ (point 1) [35] and the curves based on Equations 17.9 and 17.10.

The separate row model was used to investigate the effect of the fiber cross-sectional shape on flow resistance and particle deposition. Papers [36,37] show that band-type fibers with dumbbell-shaped cross section obtained by electrospinning (Figure 17.4) can be approximated as elliptical cross-sectional ones or as a couple of doubled fibers. As calculations showed the dumbbell jumper dimension does not affect the fiber drag force.

Figure 17.5 shows the comparison of the drag force for different fibers having the same middle cross section with the drag force of the fibers couple. It should be noted that calculation data for the range of elliptic fibers aligned with the analytical solution in another Kuwabara paper [38].

The effect of nonuniformity in fibers arrangement and their polydispersity on the flow resistance at $Re \ll 1$ and on particle deposition was investigated on a separate row of parallel fibers [25].

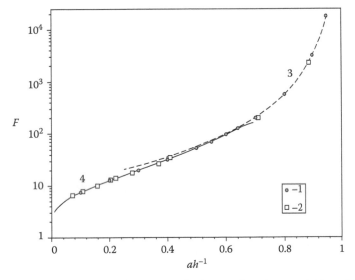

FIGURE 17.3 Dimensionless fiber drag force versus lattice spacing: 1, numerical modeling (From Kirsh, V.A., *Colloid J.*, 68(3), 261, 2006.); 2, experiment (From Kirsch, A.A. and Fuchs, N.A., *Ann. Occup. Hyg.*, 10, 23, 1967.); 3, Equation 17.9; 4, Equation 17.10.

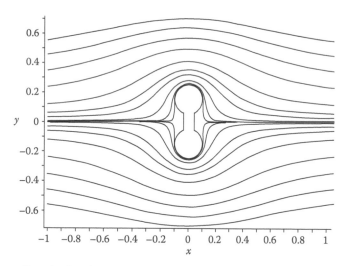

FIGURE 17.4 Streamlines in the vicinity of a dumbbell cross-sectional fiber. Radii of fibers joined with a jumper $b = 0.1$, $\varphi = 90°$, jumper length on the symmetry axis (gap) $d = 0.1$, jumper thickness $\delta = 0.08$; streamlines in the vicinity of an elliptic fiber/streamlines correspond to the values of the flow function $\Psi = \pm [5 \times 10^{-5}, 5 \times 10^{-4}, 5 \times 10^{-3}, 2.5 \times 10^{-2}, 0.05, 0.1:0.1:0.5]$. (From Kirsh, V.A. et al., *Colloid J.*, 70(5), 547, 2008.)

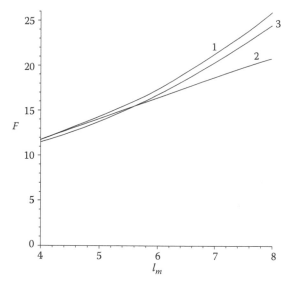

FIGURE 17.5 Drag force of fibers with different cross-sectional versus the midsection width: 1, dumbbell fiber; 2, a couple of parallel fibers (radius b is fixed, the gap varies); 3, elliptical fiber (semi minor axis equals b); $\varphi = 90°$, $b = 0.1$, dumbbell jumper thickness $\delta = 0.04$. (From Kirsh, V.A. et al., *Colloid J.*, 70(5), 547, 2008.)

In particular, it was shown both theoretically and experimentally that if fibers are coupled in a row the couple drag force equals the drag force of separate fibers. Model of a separate row was used to study confined flow in the vicinity of porous fibers [39] and fibers covered with permeable membranes [40], and while modeling flow in porous sphere particle sediments [41]. It was only the idea of separate rows that appeared to be effective in the experimental study of preliminary air purifying of large particles. A model filter made of rows of parallel equidistant wires with $2a = 8.9$ and $2h = 62\,\mu m$ was used to study (1) inertia deposition of micron and submicron drops at $St < 10$ and $Re < 1$ [42] and (2) pressure drop variation while solid [43] and liquid [44] particles accumulating.

The results of this research appeared to be in good conformity with recent theoretical research [45–47], respectively.

Model filters of parallel fibers were used to study deposition of nanoparticles [48]. No deposition mechanisms except the diffusive one are to be considered while calculating nanoparticle deposition. No demands are laid down concerning nanoparticles nature because their diffusivity does not depend on density. Apparently, this fact may be the reason for the phenomenon of particle diffusive deposition to be the most studied.

17.4.2 DIFFUSIVE DEPOSITION OF PARTICLES IN MODEL FILTERS

First works on point particle deposition in model filters were performed by Fuchs and Stechkina [49]. They obtained the diffusion capture coefficient from the solution of the equation for convective diffusion:

$$\eta_D = 2.9\, k_0^{-1/3} Pe^{-2/3} \qquad (17.11)$$

Later Stechkina [50] refined this equation:

$$\eta = 2.9 k^{-1/3} Pe^{-2/3} + 0.624 Pe^{-1} \qquad (17.12)$$

where

$$k = \frac{4\pi}{F} \qquad (17.13)$$

Calculation results appeared to be in good conformity with experimental ones on deposition of nanoparticles sized 1.5–7 nm in model hexagonal structure filters [7]. Although calculation of diffusive deposition is based on the theory of boundary layer applicable when $Pe \gg 1$, good conformity with experiment was obtained up to $Pe \sim 2$ and $\eta \sim 1$. Conformity with experiment for loose [7] and dense [51] models permitted applying the relationship $\eta \sim Pe^{-2/3}$ to solving the inverse problem—obtaining particle diffusion coefficient based on their penetration through fiber systems with known geometry and at known velocity [52–55].

Both theoretical and experimental results in papers [48,56–59] show the existence of geometrical limitation for the capture coefficient in the system of parallel cylinders at low Peclet numbers. At $Pe \ll 1$ the capture coefficient is described by the equation [59]:

$$\eta = \frac{2\pi}{Pe}\left(K_0\left(\frac{Pe}{4}\right) + 2\sum_{m=1}^{\infty} K_0\left(\frac{mPeh}{2a}\right) \right)^{-1} \qquad (17.14)$$

where $K_0(z)$ is the modified Bessel function. Calculation results as per Equations 17.11 and 17.14 aligned exactly with both the experiments for nanoparticles [48] at $a/h \ll 1$, $Re \ll 1$, $Pe \gg 1$ and $Pe \ll 1$ and direct numeric calculations [56] (Figure 17.6).

It flows out from the data in Figure 17.6 that the calculated curves for $\eta(Pe)$ for each a/h when $Pe \to 0$ transform into straight lines parallel the X-axis, the capture coefficient being constant, $\eta \to h/a$. It can be seen that the denser the grids are, the higher the Pe's are when the curves reach the plateau $\eta = h/a$. At $a/h \ll 1$ in the area of $Pe < 1$ there can be seen matching with the curve 5 calculated as per (17.14) for the limiting capture coefficient. Conformity with experiment at $a/h \ll 1$ can be seen for the wide range of the Pe numbers.

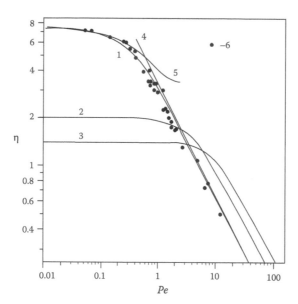

FIGURE 17.6 Capture coefficient versus the Peclet number for rows of parallel fibers with $a/h=0.133$ (1, 4, 5); $a/h=0.5$ (2); $a/h=0.714$ (3). Curves 1–3, numeric calculations (From Kirsh, V.A., *Colloid J.*, 65(6), 726, 2003.); 4, as per Equation 17.11; 5, as per 17.14; 6, experiment. (From Kirsch, A.A. and Chechuev, P.V., *Aerosol Sci. Technol.*, 4(1), 11, 1985.)

The capture coefficient being independent of Pe is peculiar to particle deposition in the dense row at low or intermediate values of Pe. Thus, the existence of the limit for $\eta = h/a$ at $Pe \to 0$ should always be considered, including solving the inverse problem—determining diffusive agility of nanoparticles by their penetration through nets [60]. Moreover, it should be noted that deviation from regular arrangement of cylinders in the model fiber results in a sharp decrease of deposition efficiency. For example, if neighboring cylinders in each row touch one another, then η nearly halves [25,36].

The problem of diffusive particle deposition considering their own size is solved numerically. It is shown that the overall capture coefficient exceeds the simple sum of separate capture coefficients for diffusion and interception. The results of computer calculations could be approximated by a simple equation [5]:

$$\eta_{D+R} = \eta_D + \eta_R + \eta_{DR} \tag{17.15}$$

where $\eta_{DR} = 1.24k^{-1/3}Pe^{-1/2} R^{2/3}$.

Where the particle capture coefficient for diffusion is calculated as per (17.11), interception capture is defined from dimensionless flow function:

$$\eta_R = \Psi\left(1 + R, \left(\frac{\pi}{2}\right)\right) \tag{17.16}$$

Analytical view of the relationship (17.15) lets find particle radius r^*—radius of particles of the most penetrating size (MPPS), conforming to the minimal deposition at the given velocity through the model filter with known parameters α, H, a.

Interesting results were obtained with model filters made of polydisperse parallel fibers [61–63]. It was shown that for practical calculations the fibers arithmetic mean radius can be used. Paper [64] offers a method to calculate the efficiency of capture for coarse nondiffusing particles by a rough fibrous filter depending on the ratio of thick to thin fibers, calculation results being in good accordance with experimental data. Paper [65] gives corresponding results for the case of gas sliding flow.

17.4.3 MODEL FILTERS OF ULTRAFINE FIBERS

The fiber diameter $2a$ for modern filtering materials made from polymer and glass fibers and for the best specimens of membrane filters is comparable to the mean free path of air molecules λ under normal conditions (approximately 70 nm). Gas flow through such filters is seriously affected by the so-called gas sliding along the surface of so thin fibers. Due to this phenomenon pressure drop on the filter decreases with decreasing a (increase in the Knudsen number, $Kn = \lambda/a$), the flow velocity being constant. Efficiency of aerosol particle deposition increases. It is usually assumed, while describing gas flow in the vicinity of thin fibers, that tangential flow velocity U on the fiber surface is not zero. At $\alpha \ll 1$ and $Kn \ll 1$ allowing gas sliding gives the equation for flow function in the linear approximation according to the Knudsen number (Figure 17.7) [25]:

$$\Psi = \frac{1}{2k_1}\left[\frac{1}{r} - r + 2r\ln r + 2\tau Kn\left(r - \frac{1}{r}\right)\right]\sin\theta$$

where

$$k_1 = k_0 + \tau Kn \tag{17.17}$$

Dimensionless drag force, applied to the fiber unit length, can be presented as $F = 4\pi/k_1$. Numerous experiments showed that the linear relationship between $1/F$ and Kn is still in force at $Kn > 1$ for filtering materials with both identical and polydisperse fibers (polydispersity degree being not high) [66]. The linear relationship between $1/F$ and Kn up to the values $Kn = 3$ was obtained in experiments with model filters (grids with wire diameter 8.9 μm). The slope ratios of this relationship let determine the value $\tau \approx 1.15$ [67], which is in conformity with the theoretical value $\tau = 1.147$ [68]. The linear relationship between $1/F$ and Kn for dilute gas flow is in force even at high temperature up to 500°C [69]. It should be noted that the linear relationship between $1/F$ and Kn for dilute gas flow through loose porous medium at $Re \ll 1$ and $Kn \sim 1$ was considered so far as an experimental

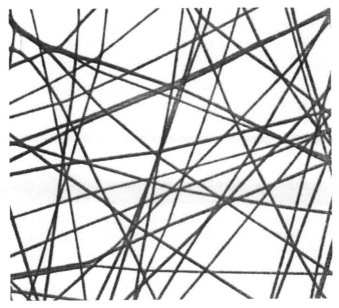

FIGURE 17.7 Photograph of a layer of fluoropolymer fibers with mean diameter 90 nm obtained by electrospinning. (From Kirsh, V.A. et al., *Colloid J.*, 70(5), 547, 2008.)

fact. It is only recently that this fact has been validated [70]. The research in this paper uses methods of gas kinetic theory, developed for analyzing transfer processes in porous media. The approximating calculation equation for the dimensionless resisting force in this range of packing densities looks as follows:

$$F^{-1} = F_0^{-1} + \frac{\tau f(\alpha) Kn}{4\pi}$$ (17.18)

where $\tau f(\alpha) = 1.27 - 3\alpha$. For a fan model filter this value is equal to $1.43(1 - \alpha)$ [66]. Deposition of nondiffusing particles noticeably increases with the increase in the Knudsen number. When $R \ll 1$ and $\alpha \ll 1$ the capture coefficient is [4]

$$\eta_R = \frac{R^2}{k_1}\left(1 + \left(\frac{\tau Kn}{R}\right)\right)$$ (17.19)

when it follows that gas sliding in the vicinity of the fiber surface effects significantly small particle deposition irrespective of the fiber diameter.

Diffusive deposition of point particles ($R=0$) out of a thin boundary layer in the vicinity of a fiber (thickness $\delta = (2k/Pe)^{1/3} \ll 1$) in a high-porous system of parallel cylinders with $\alpha \ll 1$ was studied in Ref. [25] at $Re \ll 1$ and $Kn \ll 1$. The relations obtained for the capture coefficient are as follows:

$$\eta_D = 3.2\sqrt{\frac{\tau Kn}{Pek_1}}$$ (17.20)

for $Kn > \delta$, and

$$\eta_D = 2.9\frac{1 + 0.55\tau Kn(Pe/k_1)^{1/3}}{(Pe^2 k_1)^{1/3}}$$ (17.21)

for $Kn < \delta$. If to take into account the particle finite size, as shown in Ref. [71], the numerically calculated capture coefficients at $Kn < 1$ coincide with, and at $Kn > 1$ exceed the interception and diffusion capture coefficients obtained analytically in the approximation $Kn \ll 1$. These results are confirmed in Ref. [72].

There has not been an experimental check of particle deposition in model filters with known parameters as function of Kn performed, although the calculated values of η_R and η_D are in good qualitative conformity with the data obtained for real filters (Figure 17.8).

As mentioned earlier, the deposition of particles of finite size is affected by the dispersion forces (van der Waals forces) [73]. Particle deposition is explicitly studied in Refs [74–77], depending on surface curvature of fibers and gas sliding. It is shown that the particle deposition on fibers due to the van der Waals attraction can contribute about 15% to the overall capture coefficient (Figure 17.9). Paper [78] presents the calculated capture coefficient as a function of particles size and density, angle between the vectors of gravity force and flow velocity before the filter. It is established that the radius of the most penetrating "heavy" particles for the down coming flow appears to be smaller than for the rising one.

17.4.4 Particles of the Most Penetrating Size for Nanofibrous Filters

As mentioned earlier, the most difficult task in the theory of filtration with high performance thin fiber filters is to describe deposition under the worst conditions, that is, in the area of maximal

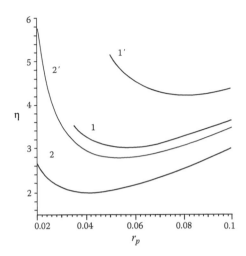

FIGURE 17.8　Capture coefficient versus particle radius r_p, μm: $U=3$ cm/s (1, 1′), 10 cm/s (2, 2′). Curves 1, 2, air; 1′, 2′, helium; $a=0.04$ μm; $\alpha=1/36$. (From Kirsh, V.A., *Colloid J.*, 66(3), 311, 2004.)

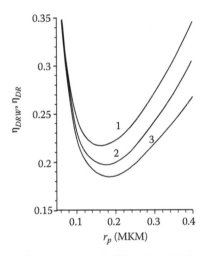

FIGURE 17.9　Diffusive capture coefficient versus particle radius: (1, 2) considering the particle finite size and van der Waals attraction η_{DRW}; (3) η_{DR} without considering van der Waals attraction. Fiber radius $a=1$ μm, flow velocity before the filter $U=1$ cm/s, constant of the retarded van der Waals attraction $A_7=10^{-18}$ erg cm (1), $A_7=10^{-19}$ erg cm (2), packing density of the filter $\alpha=1/16$. (From Kirsh, V.A., *Colloid J.*, 66(4), 444, 2004.)

penetration. In Ref. [70], capture coefficients were calculated both for nondiffusing particles interception and point particles diffusion at medium Knudsen numbers in the model filter with parallel fibers. Diffusive particle deposition in the range $Kn=0.1$–10 considering interception was studied in Ref. [71]. The problem is solved in the approximation to the diffusion boundary layer δ. The calculated capture coefficients $\eta(R/\delta)$ (*R*—interception parameter) can be approximated to straight lines up to $R/\delta=2$ at $Kn>1$ and at different values of δ. It is shown that at low Knudsen numbers and on condition that the Knudsen layer thickness is smaller than the diffusion boundary layer thickness the capture coefficient η is higher than the sum of η_R and η_D, which conforms with the calculations performed in Refs [5,25] at $Kn=0$.

At $Pe \gg 1$ and $Kn>1$ the value η is clearly less than the sum of η_R and η_D. At medium *Pe* numbers and $Kn>1$ the overall capture coefficient practically equals this sum. This result is important

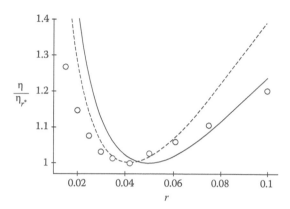

FIGURE 17.10 Function of the capture coefficients ratio in the area of maximal particles penetration. Curves, theory; dots, experiment. (From Tamaru, S. et al., Development and application of ultra-high performance fluorocarbon polymer air filter, *Proceedings of 13th International Symposium on Contamination Control*, September 16–20, 1996, The Hague, The Netherlands, ICCCS, Stouthart Communicatie, Waddinxveen, The Netherlands, pp. 236–242, 1996.)

for practical calculations of analytical filtration to determine particle size with the help of simple analytical equations according to the fraction of particles deposited on or penetrated through the filter with known parameters and at known velocity. Experimental data available on particles penetration through high performance filters consisting of ultrafine fibers for which the Knudsen number under normal pressure can achieve the value of the order of unity usually correspond to big capture coefficients and small or medium Pe numbers.

The authors of the paper [26] used monodisperse latex aerosols as test ones; they found out that the most penetrating particle radius $r_p^* \approx a \approx 0.04$ μm at the velocity 1.5 cm/s. Figure 17.10 shows curves for relationship between the capture coefficients ratio η/η^* (η^* is the capture coefficient corresponding to the minimum on the curve $\eta(r)$, that is $\eta^* = \eta(r_p^*)$) and the particle radius r_p. Filtering material fiber dimensions plotted on the grid were obtained from microphotograph. Hence, the real flow velocity could be higher than 1.5 cm/s.

In Figure 17.10, the block curve was calculated for $U = 1.5$ cm/s, the point curve—for doubled velocity. It can be seen from data in Figure 17.10 that the type of the relationship and the area of the curves' minimum corresponding to r_p^*, agree with experiment, a better agreement being seen for higher velocity. Thus, it is possible to evaluate on the basis of the advanced theory the location of r_p^* and the type of the curve describing relationship between particles penetration and their radii for ultra fine fiber filters even for high performance filters with $\eta \sim 1$.

Relatively simple models of parallel cylinders with the known flow field appeared to be useful while studying various problems of the filtration theory, such as modeling of solid particles deposit growth [79], growth of droplets on fibers [47], effect of external electric field on particles capturing efficiency [80], deposition of particles on porous fibers [81] and on fibers, covered porous shell [82], inertial deposition of particles [83], etc.

17.4.5 FAN-TYPE MODELING FILTER

The filter model with parallel fibers let us solve many problems in the theory of filtration. However, the flow field in real filters is 3D and is extremely hard to be determined. The closest in its qualities and structure to the real filters idea is a *fan model filter*, where layers of parallel fibers are turned relative each other through an arbitrary angle φ. The resisting force F^f of the fan-type model equals F of a *stagger* model whose packing density is 1.6 times lower [6]. This problem was not paid due cognizance for a long time despite the importance of determining hydrodynamic drag force for grids with 3D flow. Investigations with numeric calculations of resistance for various 3D grids [84,85] were carried out in the latter half of the 1990s.

The first step on the way of analytical description of resistance in the 3D grid was executed in Ref. [86]. Simple analytical expressions for drag force were obtained for the case when the pair layers are turned in their plane relative to the impair ones. These expressions describe qualitatively the model filters resistance which is being observed experimentally.

Hydrodynamic flow in the polydisperse 3D fibrous porous medium was analytically scrutinized for the first time in Ref. [87]. The theory of self-consistent field was applied to study averaged equations of liquid flow in isotropic medium and in anisotropic fibrous filter. One of the important practical results that follow this research is determination of the Reynolds number, when self-similarity flow mode in the filter with the given packing density is deranged. It is of particular importance to know the onset of the self-similarity derangement to evaluate the inertia particles deposit during analytical filtration, that is, at high flow velocities.

Specific character of flow in the 3D model filter was studied in Ref. [35] as well. Numerical methods were used to determine the relationships between hydrodynamic resistance and a step and distance between rows for a model filter consisting of two rows of parallel cylinders perpendicular to the Stokes flow direction, cylinders being turned through $-90°$ in their planes relative to each other. Approximation equations were obtained for the cylinder drag force in the adjoining rows:

$$F = 8\pi \left\{ 1 - 1.5\ln\left(\frac{2a}{h}\right) - 7.48\frac{a}{h} + 26\left(\frac{a}{h}\right)^2 - 37.5\left(\frac{a}{h}\right)^3 + 21\left(\frac{a}{h}\right)^4 \right\}^{-1} \tag{17.22}$$

at $a/h < 0.6$ and

$$F = \frac{9\pi}{2\sqrt{2}}\left(1 - \frac{a}{h}\right)^{-5/2} + 10.34\left(1 - \frac{a}{h}\right)^{-1} - 10.45 \tag{17.23}$$

at $a/h > 0.6$. Equations 17.22 and 17.23 can be used to calculate the resistance of thin nets.

Studies of the fan-type model resistance at tenuous air flow showed that gas sliding effect exerts in this model more notably than in the parallel one. Proportionality factor considering interaction of air molecules with the fiber surface and calculated from the angle of slope of the linear relationship between $1/F^f$ and Kn equals 1.43 instead of 1.18 for the parallel model [66].

The air flow and the particle deposition process are similar for the fan and the real filter. In particular, a very important common feature is discovered: diffusive aerosol deposition efficiency for the fan model filter does not depend on the packing density in a wide range of α from 0.01 to 0.15, the capture coefficient being determined as per the equation [7]:

$$\eta_D^f = 2.7Pe^{-2/3} \tag{17.24}$$

While collating F for real filters with F^f for the fan-type model, it was found that the force character dependent on the packing density is described correctly by the equation for the fan-type model, however, absolute magnitudes $F^f > F$, which is caused by inhomogeneity of the real filters. Paper [88] provides theoretical support for the diffusive deposition law in disorder fibers system with fibers perpendicular the flow, which is being experimentally observed. It has been shown that in the area of low packing densities α the capture coefficient dependence on α is really rather weak.

The simple view of the Equation 17.7 contributed to the wide spread of the diffusive method of sampling submicron aerosol particles by means of diffusive grid batteries [52–54].

17.5 METHOD OF CALCULATION OF PARTICLE DEPOSITION EFFICIENCY IN REAL FILTERING MATERIALS

It was assumed at the development of the method to calculate efficiency in real filters that a fan model filter is a reference standard of an ideal homogeneous fibrous filter. That enables introducing a quantity characteristic of inhomogeneity grade of a real filter structure presented as a relation of nondimensional force F^f acting on a unit of fiber in a fan model to the value F in a real filter at equal values of α and a.

It was established as the result of numerous experiments with fan-type model and real fibrous filters that structure inhomogeneity of real filters equally affects both their resistance and particle deposition:

$$\frac{F^f}{F} = \frac{\eta^f}{\eta} \tag{17.25}$$

Although no complete theoretical proof of this empiric correlation has been found until now, it is of fundamental significance for calculation of filter efficiency as it enables to calculate their efficiency from hydrodynamic resistance using equations for particle deposition in a fan model filter. Penetration of particles with the radius r_p at velocity U (at $Re < 1$) through a filter with known resistance Δp may be evaluated in the following way [25]:

$$\frac{N}{N_0} = \exp\left(-\frac{2a\eta^f \Delta p}{U\mu F^f}\right) \tag{17.26}$$

where

$$\eta^f = \eta_R + \eta_D + \eta_{DR} \tag{17.27}$$

$$\eta_D = 2.7 Pe^{-2/3}\left\{1 + 0.55(k_1)^{-1/3} Pe^{1/3} Kn\right\} \tag{17.28}$$

$$\eta_R = (2k_1)^{-1}\left\{(1+R)^{-1} - (1+R) + 2(1+R)\ln(1+R) + 2.86 Kn(2+R)R(1+R)^{-1}\right\} \tag{17.29}$$

$$\eta_{DR} = 1.24(k_1)^{-1/2} Pe^{-1/2} R^{2/3} \tag{17.30}$$

$$k_1 = \left(-0.5\ln\left(\frac{a}{(1+\sigma)}\right) - 0.52 + \frac{0.64a}{(1+\sigma)} + 1.43\left(\frac{1-a}{(1+\sigma)}\right)Kn\right)^{-1}, \quad \sigma = \frac{\left(\overline{a^2} - \overline{a}^2\right)}{\overline{a}^2}$$

A possibility to calculate the efficiency of real HEPA-filters consisting of polydisperse ultrathin fibers was demonstrated on the example of national and international fibrous filters intended for fine gas purification. Compliance of the theory with experiments is acquired for different filters with particles of different size, including that in the area of maximal penetration. The model fan filter was an analogue to FP filters made of monodisperse fibers, in which there is no effect of fibers sticking together due to the presence of static charge on fibers [89,90]. Results of extensive studies of highly efficient filters are given in Refs [91,92,140].

17.6   ON TESTING SUBMICRON AND NANOFIBER FILTERS WITH NANOPARTICLES

The researched mechanisms of aerosol particle deposition on fibers from a flow enable finding the size of the most penetrative particles for a filter with given parameters depending on the flow velocity. It leads to the possibility to choose correctly a method for testing high effective filters.

There are several reasons explaining the discrepancy between experimental and calculation data. Such discrepancy may relate to the coagulation of tested solid particles; to partial clogging of a filter; to the influence of charges on particles or fibers that are not taken into account in calculation; to incomplete drying out of particles (e.g., to the presence of a "coat" on polystyrene latex particles); to polydispersity of particles; to defects in the filter structure (e.g., thin through-holes in a filter), etc. Besides, it is important while testing filters to take into account the polydispersity of aerosols and particles shapes. There are also many unsolved problems related, for example, to calculation errors, taking into account charges. The mechanism of blowing out particles and fiber fragments at filter vibrations is not clear. The mechanism of particles rebound from fibers is not investigated when there is a local crack in the particle deposit on the filter and velocity in the local crack increases multiply. It is difficult to evaluate the influence of hidden defects of highly effective filters on the penetration of particles. One can discover filter defects with the help of nanoparticles with diameter size of order of 0.01 μm: probability of penetration of such small particles is negligibly small at a low flow velocity and their presence after a filter testifies the presence of a defect. It is necessary to carefully check the measurement technique if the experimental data sharply differ from the calculated one. It is especially important while measuring the deposition of particles smaller than MPPS.

As it was mentioned earlier, modern highly effective filtering materials are manufactured from very thin fibers; that is why the area of mostly penetrative particles moved to the range lower than 0.1 μm. At the same time according to commonly accepted standards one should measure the penetration of particles around its maximum point. Not many photoelectric counters have sensitivity sufficient to detect particles in a range of 1 μm that is why for their registration now as half a century ago one has to use method of enlargement of particles.

At present, a procedure of measuring size and concentration of nanoparticles based on application of condensational counters is widely used. Test particles are obtained while separating polydisperse particles charged with one elementary charge in a differential mobility analyzer. Particles are obtained while spraying of liquids—dioctyl phthalate or common salt solution.

It should be taken into account while measuring penetration of small particles through filters that not all particles get enlarged in condensational counters using oversaturated spirit vapors but a part thereof depending on their size and concentration. It affects the test accuracy of highly efficient filters. This explains why penetration curves may differ from the calculated ones. Penetration of particles in the area of sizes less than 0.07 μm varies very little with the variation of particle sizes, especially in the case of $K < 0.01$.

Applicability of condensational counters to testing of high-efficiency filters is easily proved by measuring the penetration of particles through several sequentially placed similar units of filtering materials. In accordance with the basic filtration law a correlation should be fulfilled $n/n_0 = (n_i/n_0)^m$, where m is the number of units of material and n_i is the concentration of particles with radius r_j after one filter layer.

Measurements that we conducted at a TSI mod 6030 installation (which incorporated a generator of monomobile particles of dioctyl phthalate and sodium chlorate as well as two condensational counters) with five layers of filtering material, demonstrated that this correlation was strictly observed only for particles with the size larger than 0.07 μm. For particles with a diameter of 0.02 μm at velocity 2 cm/s the value of n_0/n_1 for the first filter was 115 and for the fifth $n_4/n_5 = 6$ only. Regretfully, most of the authors do not take that into account. As a result, not quite true generalizations and conclusions are published. The most reliable method of testing high-efficiency filters is a comparison method with standard reference filters [92].

17.7   NONSTATIONARY FILTRATION AND OPTIMIZATION PROBLEMS

The difficulties of principal character at solving the problems of determination of a pressure drop Δp growth, efficiency growth, and estimation of time expectancy before clogging of a filter with solid particles are related to the necessity of taking into account numerous additional factors and, first of all, those related to variation of the flow field while accumulating sediment. An approach to the solution of this problem proved to be possible for the first time due to approximating a dust-covered fiber to a cylinder with a co-axial porous permeable membrane [82,93]. The distribution of deposited particles through the filter depth and the reverse influence of a deposit growing on fibers on the flow field in the proximity of the fibers is taken into account in the proposed approach. Such an approach is used for calculation of pre-filters [94,141].

A correlation of packing density of a deposit layer on the filter surface and radius of particles is obtained for calculation of surface clogging of a high-efficiency filter [95]. It is demonstrated that a 3D layer of submicron particles building at flow velocity $U = 1$ cm/s on the front surface of a filter has a high-porous structure with the maximal packing density equal to 0.15. Approaches based on filters clogging kinetic research are proposed for optimizing the filter parameters in a two-stage air purification system composed of a pre-filter and a finishing high-efficiency filter [94,96,97].

In conclusion, we note that different attempts are being continuously made to solve the problem of creating high-efficiency filters with low resistance to the flow on the basis of nanofibers with special configuration. For example, it is experimentally demonstrated in Ref. [98] that "nano-moustache" of nanotubes developed on filter fibers noticeably increase filter efficiency at a relatively low increase of its resistance. Quantitative theory of filtration with such filters was developed earlier in the work [93], where it is demonstrated that quality criterion of a loose filter may considerably increase at the initial stage if to coat fibers with a permeable porous shell consisting, for example, of ultrathin "whiskers." Such filtering materials alongside with nanofiber membranes may find wide application in industry and in analytical filtration of dust-loaded gases.

17.8   ANALYTICAL FILTRATION

The theory of air sampling by filters with a purpose to analyze aerosol particles is one of the most important directions of the filtration theory that incorporates evaluation of particle capture completeness at a filter and that of separation of particles using filters depending on filter parameters and sampling conditions. We have mentioned previously a diffusive method of particles' determination. In this section we shall consider the traditional approach to sampling description based on deposition particles on fibers of a filter due to their inertia.

The high rate filtration is poorly known theoretically until now despite the wide use of analytical filters. There are a number of experimental studies of filtering materials efficiency in inertial mode of particles capture [12,99–103], including by FP filtering materials [103–106]. Equations for calculating inertial capture factors are obtained from experiments with real and model filters but one can acknowledge full absence of universal functional connections for a wide range of structural characteristics, velocities, densities, and sizes of aerosol particles. It is worth noting that figuring out of a character of inertial deposition dependence on Stokes numbers under flow conditions at $Re \sim 1$ is required for justification of the evaluation of aerodynamic diameter of radioactive particles by the multilayer filter method (MFM) [107].

Inertia deposition of aerosol particles onto the cylindrical obstacle is one of the earliest study trends in the area of aerosols mechanics. Applied to the theory of filtration, the inertia deposition was being studied for long time on the basis of the Lamb flow field in the vicinity of a separate fiber at low Reynolds numbers. Later, the Kuwabara [31] cell model with the Stokes flow field [25] was used for considering neighboring fibers influence. Experimental and theoretical (mainly numerical) research was carried out for low flow velocity that conforms to the Stokes flow condition, $Re \ll 1$ [42]. In this case the pressure drop in the filter is strictly linearly dependent on the flow velocity.

It should be noted that the cell model which is very fruitful while solving many problems in the filtration theory appeared to be not quite applicable to studying the inertia deposition because the deposition efficiency is strongly affected by the prequel of the particle movement far off the fiber. So, taking this into account, model filters consisting of separate non-affecting each other rows of parallel equidistant fibers perpendicular the flow are chosen for experimental and theoretical study of the particles inertia deposition (Section 17.4.1).

At low air flow velocity through the filter, which conforms to the condition $Re \ll 1$, particle deposition depends strongly on the fiber system packing density and is determined by the Stokes number and the interception parameter R. With the increase of velocity U the linear relation between pressure drop and velocity is observed to break. Flow lines in the vicinity of fibers start being asymmetric, which slightly affects the fiber resisting force, but strongly influences the particle capture coefficient value. In this case the inertia particle deposition must depend on Re as well.

In the recent paper [83] the inertia particle deposition has been numerically studied for various deposition modes. Figures 17.11 and 17.12 present examples of the capture coefficients calculation

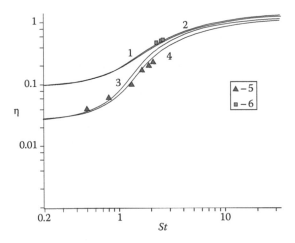

FIGURE 17.11 Capture coefficient versus St: (1, 3) considering Re; (2, 4) Stokes flow: $2a=10\,\mu m$; (1, 2) $r_p=2\,\mu m$; (3, 4) $r_p=1\,\mu m$; $a/h=0.1419$; (5, 6) experiment (From Kirsch, A.A. and Fuchs, N.A., *Ann. Occup. Hyg.*, 10, 23, 1967.): 5, $R=0.206$; 6, $R=0.382$. (From Kirsh, V.A. et al., *Colloid J.*, 72(2), 206, 2010.)

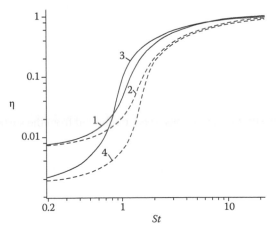

FIGURE 17.12 Capture coefficient versus St: curves (1, 3) considering Re; (2, 4) the Stokes flow: $2a=10\,\mu m$; (1, 2) $r_p=0.5\,\mu m$; (3, 4) $r_p=0.25\,\mu m$; $a/h=0.1419$. (From Kirsh, V.A. et al., *Colloid J.*, 72(2), 206, 2010.)

for particles with density $\rho = 1\,g/cm^3$ for the Stokes flow field and for the flow field at $Re \sim 1$. The solid curves 1 and 3 are calculated with due consideration of air flow inertia under normal conditions, dotted curves—for the Stokes flow field. As is clear from this data, discrepancy between block and dotted curves for high values of the interception parameter ($R = 0.2-0.4$) in the range of the Stokes number 0.7–10 is insignificant. Absolutely different picture can be observed in the case of smaller particles, that is, low interception parameters R. Discrepancy between block and dotted curves takes place at lower Stokes numbers, in the area of $St \sim 1$ the capture coefficients difference is more than 10 times particularly for the case with $R \sim 0.1$ at $Re \approx 1$. Moreover, solid curves intersect because for the identical St smaller particles have higher velocity (Re increases). Thus, it can be seen that taking into account real flow field at $Re \sim 1$ affects significantly the value of η. In this case the capture coefficient depends, as mentioned earlier, not only on St and R but also on the Reynolds number. The Reynolds number magnitude is different in every part of the curve. In the end it should be noted that *here and now empiric relations based on the performed experiments have to be used to calculate the fibrous filters efficiency in the inertia particles capture area as yet.*

17.9 REBOUND OF PARTICLES

The questions connected with nonstationary filtration are important for selecting duration and velocity of sampling while aerosol monitoring. Sampling velocity is limited by the increase in particles rebound from fibers and, connected with it, decrease in inertia capture efficiency. Duration is determined by dynamics of the deposit accumulation on filter fibers.

A particle touching a fiber is held on it due to the van der Waals forces. Their magnitudes depend on the particle and fiber composition, shape and surface as well as on a range of other factors. Force values can vary several orders of magnitude even for a uniform system.

Aerosol particle energy, which it possesses after collision, may be sufficient to overcome adhesion energy. The latter can vary at collision (e.g., due to the particle distortion). Apparently, there exists a certain velocity above which the collision of a particle with a fiber becomes ineffective.

There is no quantitative theory that allows predicting a decrease in the capture coefficient of aerosol particles due to rebound. We should note experimental works [107–109], which present studies on laws of behavior for polystyrene particles at their collision with flat surfaces. Paper [110] gives a equation for probability of adhesion versus St, Re, and other parameters for spherical particles and the case of van der Waals adhesion. The probability of adhesion decreases with the increase in the particle size, velocity, and elastic properties and with reduction in the fiber diameter. Calculations were performed for a separate cylinder and they are in conformity with the experimental data for quartz and paraffin particles with diameters 3–20 µm up to velocity 1 m/s.

Correlation between adhesion and kinetic energy of the particle is found in Ref. [111]. The calculation performed in the paper [111] with the help of the Kuwabara cell model showed the existence of a strong dependence between the collision effectiveness and the Stokes number, however, the data differ more than twice from the experiment. The similar results were obtained in the work [112].

Experiments where the particles were observed to rebound from the surface of the FP fibers were carried out in Ref. [104]. It was stated while testing material made of acetyl cellulose fibers with diameters 5–7 µm that decrease of capturing efficiency of aerosol particles with diameter 0.8 µm started at velocities higher than 5 m/s. Data on Figure 17.13 illustrate effect of rebound.

Experiments with copper dioxide and tungsten carbide aerosols showed the shift of velocity corresponding to the maximal inertia capture coefficient to lower magnitudes with the particle size increase. The results for latex particles rebound from polymer fiber of FP material and from tungsten wires with the diameter 11 µm used to make fan-type models are described in Refs [113,114].

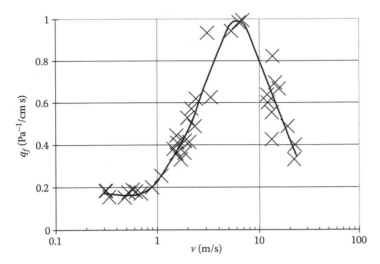

FIGURE 17.13 Effect of rebound of aerosol particles on filtration characteristics of a fibrous filter: Quality factor (q_f) versus velocity of particles (v). Mean fiber diameter is 7 μm, packing density=0.03, particle diameter=0.8 μm. (From Skitovich, V.I. et al., IPG Works, 1976, 21, 8.)

The fibers of the filtering material FPA-70 are made of cellulose diacetate. Hydrodynamic fiber diameter is 5.5 μm and packing density of the filtering material is 0.03. Fan-type models are made of cylinder tungsten wire with diameter 11.1 μm. The packing density of model #1 is 0.01 and of model #2 is 0.025.

A critical velocity V_{crit} was found. The exceedance of this velocity leads to increase of penetration through the fan-type model and the fiber filter and, thus, to reduction in their capture efficiency by the cylinder. Values of V_{crit} obtained on models #1 and #2 coincide. This means that the lattice packing density does not affect the rebound process, at least at $\alpha < 0.025$. Experimental results are presented in Table 17.2.

As it can be seen from the data presented in Table 17.2 values of V_{crit} shift to lower magnitudes with the increase of the particle size. The relationship between V_{crit} and r is well described by the equation

$$V_{kp} = Cr^{-0.7}$$

where the constant $C=1.6$, if the particle radius r is given in microns and V_{crit} in m/s.

It is easy to notice that the relationships for the cellulose diacetate fibers with hydrodynamic diameter of 5.5 μm are the same as for the tungsten wire with diameter 11.1 μm. However, the constant C equals 0.78, which is less than for the grids. Obviously, the difference in the constants is caused by different diameters and material of the cylinder grids and filter fibers.

TABLE 17.2
Critical Velocities (m/s) for Models 1 and 2 and for Filtering Material FPA-70

Particle Radius (μm)	0.36	0.51	0.77	0.98
Models I and II	3.20	2.55	1.90	1.35
Filter FPA-70	1.60	1.25	0.95	0.80

17.10   FIBROUS FILTERING MATERIALS OF FP TYPE

Fibrous FP filtering materials are statistically uniform and homogeneous [115] layers of charged ultrafine polymer fibers. The fibers are produced by drawing of polymer solution under action of an electrostatic field (electrospinning technology) [19]. At present more than 40 FP filtering materials made from different polymers are developed in the Karpov Institute of Physical Chemistry and Russian industry produces several millions square meters of filtering materials annually. Some international companies, such as Hollingsworth & Vose Company, Donaldson Company, Inc., and some others, are also active currently in the production of filtering materials by electrospinning technology.

FP filtering materials are produced as rolls or cloth sized 0.6×1.6 m. They are used for production of respirators, analytical filters, and aerosol filters of HEPA and ULPA classes. FP filtering materials for analytical filtration are used in high capacity aspirators designed for air intake of hundreds and thousands cubic meters per hour. Airplane devices for sampling of atmospheric aerosols (filtering pods) may have capacity up to 20,000 m^3/h [116]. They can accommodate up to 6 m^2 of filtering material. FP filtering materials are easily pressed after sampling into compact premolds (tabs), which is especially convenient at measuring of radioactive aerosols containing gamma-radiation sources.

Portable or stationary sampling devices are used at industrial premises. They contain analytical AFA aerosol filters having surface from 3 to 20 cm^2. Capacity of such devices is relatively small (dozens to hundreds l/min) but it is quite sufficient for dosimetric purposes.

AFA analytical filters are also produced from FP filtering materials. They constitute disks of a filtering material, whose edges have paper spacer with a fixture for installation of the filter into the filter holder. A list of basic AFA filters produced by Russian industry is given in Table 17.3 [117].

Filtering belts (ribbons) are used in devices for continuous monitoring of radioactive aerosols.

Analytical NEL-3 filtering belt has been developed for using in devices for measurement of dust weight concentration by absorption of α- or β-radiation from a standard radioactive source in captured sediment. The belt constitutes a uniform layer made of ultrathin fibers' mixture having a diameter of 5–7 and 0.5 μm and thickness of 1.5 ± 0.2 mg/cm^2.

NEL-4 filtering belt is designed for use in devices for measuring aerosol concentration (volumetric activity) by radiometric, optical, and chemical activation and other methods. The belt may also be used for aerosol sampling at enhanced mass concentration up to building of dust sediments on the filtering layer of 2.5 mg/cm^2. The belt is used at ambient air pressure that is lower than atmospheric as well as for sampling at high rates of filtration. NEL-4 filtering belt constitutes a uniform layer made of ultrathin fibers' mixture having diameter of 7 and 1 μm. Thickness of the layer is 4.0 ± 0.5 mg/cm^2.

TABLE 17.3
Russian Filters for Aerosol Analysis

Type of AFA Filter	Function	Filter Denomination by Method of Analysis
AFA-HP	Determination of concentration of chemical and	Chemical
AFA-HA	radiochemical aerosol composition	
AFA-RMP	Determination of volumetric activity of aerosols	Radiometric
AFA-RMA		
AFA-RSP	Determination of volumetric activity and isotope composition of aerosols	Radiospectrometric
AFA-RGP	Determination of aerosol particle sizes and activity	Radiographic
AFA-VP	Determination of aerosol weight concentration	Weight
AFA-DP	Determination of aerosol particle sizes under microscope	Dispersive
AFA-BA	Determination of aerosol concentration	Bacterial

TABLE 17.4
Filtering Belt Characteristics

Grade of Analytical Belt	Average Pore Diameter (μm)	Resistance at 1 cm/s (Pa)	Penetration Index by 0.15 μm Particles at 120 cm/s (%)	Penetration Index by 0.34 μm Particles at 170 cm/s (%)
Millipore FSLW	1.5	110	0.01	0.001
LFAS-2	2.0	20	10	5
LFAS-4	1.1	40	0.1	0.05

LFS-2 analytical filtering belt is developed mainly for use in radiometric and spectrometric instruments for measuring of volumetric activity and nuclide composition of radioactive aerosols. LFS-2 belt consists of two fibrous layers. The upper (front) filtering layer consists of ultrathin fibers with diameter of 0.3–0.5 μm. The lower layer fulfills the role of a substrate (space frame) and consists of glued to each other fibers having diameter of 5–7 μm. The operation layer is additionally covered with thick fibers to enable its fixing to the substrate.

Ultrathin fibers of the main filtering layer build up a structure having package density of 0.07–0.1. One can consider the fibrous structure as a porous layer having an average size of pores of 2.0 μm.

LFS-2, NEL-3, and NEL-4 belts were developed about 40 years ago but until now remain reliable and effective means of capturing and monitoring of radioactive aerosols of different origin. Nevertheless, their relatively low effectiveness in relation to particles of the size range of 100–200 nm at filtration rates of about 1 m/s does not allow reliable evaluation of volumetric activity of aerosols containing particles of that size.

Aerosol particles containing daughter products of radon fit exactly into that size range. Such radionuclides as iodine and ruthenium can be discovered at the smallest particles [118]. A new filtering material *LFAS-2* is developed for the analysis of such aerosols and upgraded filters AFA-RSP as well as pilot belts *LFAS-2* are developed on its base [117]. LFAS-2 differs from LFS-2 by the presence of thinner fibers (0.1–0.5 μm) used for manufacture of a front filtering layer. Apart from that, one managed to avoid using of thick fibers for fixing the front layer to the space frame by implementing a special process method. LFAS-4 belt contains more of thinner fibers, which brings about enhanced effectiveness of particle capturing.

As is evident from Table 17.4 the LFAS-4 belt practically equals to one of the best analytical belts FSLW of Millipore Company by its filtering characteristics [119] but its hydrodynamic resistance is essentially lower (Figure 17.14).

17.11 USING FIBROUS FILTERS FOR DETERMINATION OF PARTICLE SIZE DISTRIBUTION

The problem of determination of aerosol size distributions is solved by different methods based on the use of correlation of the measured physical property with the size of a particle. Widely known are precise but time-consuming analytical methods under optical or electronic microscopy usage. But it is difficult to determine which chemical elements or materials are carried by particles of this or that size if the aerosol was originated from several sources. The identical problem appears when using particle size distribution instruments based on light diffusion [120]. Impactors [121] mostly often used for analysis of particle size distribution of aerosols have relatively low productivity of sampling. That is why they are not so useful for the cases when concentration of the analyzed aerosols is extremely low in spite of their relatively high precision of distribution by size fractions.

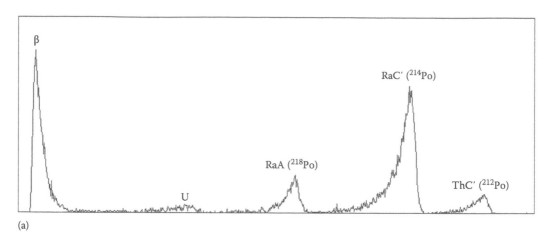

(a)

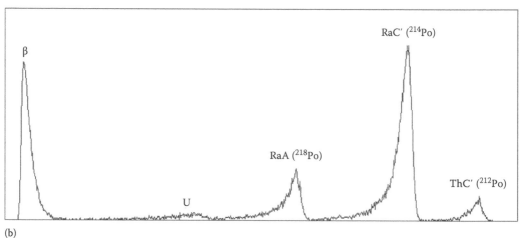

(b)

FIGURE 17.14 α-Radiation spectra of Uranium and radon daughters received by Millipore FSLW (a) and LFAS-4 (b) analytical belts and measured by installation UDA-1AB. (From Budyka, A.K. and Borisov, N.B., *Fibrous Filters for Air Pollution Control*, IzdAT, 360p., 2008.)

A task to reproduce the aerosol particles' size distribution spectrum belongs to the class of tasks not quite correctly formulated, if no presumption is made concerning unknown function. At filtration of aerosols by a package consisting of fibrous material placed sequentially such task is being presented as

$$N_i(\vec{u}) = \int_0^\infty E_i(r)f(r,\vec{u})dr, \quad I = 1,\ldots,n \tag{17.31}$$

where
  r is the radius of particles
  N_i is the portion of substance, captured by i-filter
  E_i is the efficiency of i-filter in a package considering the substance captured by previous $i - 1$
    filters $\vec{u}$ assemblage of distribution parameters
  f is the unknown function (density of distribution by sizes)
  n is the number of filters

Efficiency E_i of the I-filter in the package equals to

$$E_i = E_i^* \prod_{j=1}^{i-1} \left(1 - E_j^*\right) \tag{17.32}$$

where E_j^* is the efficiency of the j-filter.

Functions $E_i(r)$ are equation kernels in the system of n Fredholm integral equations of first kind. A system of ill-conditioned equation appears after replacing of integrals by a finite sum.

Some authors [122,123] have analyzed the size distribution of sub-micron radioactive aerosols using a package of several fibrous filters by solving a reverse task (17.31) without any additional presumptions. Naturally, such an approach turned out to be inconvenient in praxis.

In this work [124] authors managed to reduce the task (31) to a correctly posed one thanks to introduction of priory information about size distribution spectrum and to determine size distribution of alpha-active submicron aerosols in the surface air layer.

It is known that the sizes of the majority of natural and artificial aerosols can be well approximated by logarithmically normal distribution (LND), whose density equals to

$$f(r) = \frac{1}{\sqrt{2\pi}\,\ln\sigma \cdot r}\exp\left\{-\frac{(\ln r - \ln r_0)^2}{2(\ln\sigma)^2}\right\} \tag{17.33}$$

where
r is the radius of particles
r_0 and σ are the LND parameters (median radius and standard geometrical deviation)

So, if to assume that aerosol particle sizes are subject to LND then the task to determine their distribution is reduced to determination of LND parameters from a system of equations:

$$N_i = \int_{R_1}^{R_2} \frac{E_i(r)}{\sqrt{2\pi}\,\ln\sigma} \cdot \exp\left\{-\frac{(\ln r - \ln r_0)^2}{2\ln^2\sigma}\right\} \frac{dr}{r} \tag{17.34}$$

To find LND parameters it is sufficient to have three filters, at that the third one is to be highly effective (to capture practically all aerosol particles).

Task (17.34) is solved by minimizing the deficiency functional

$$Q = \sqrt{\sum_{i=1}^{2}\left(N_i - \int_{R_1}^{R_2} E_i(r)f(r)dr\right)^2} \tag{17.35}$$

As it was mentioned before filter efficiency at inertial capturing mechanism is proportional to the particle size raised to the second power. Quantitative correlations of effectiveness of FP filtering materials with Stokes numbers in the range of 0.2–5 [125] were received to solve the task (17.34), calculation algorithms and programming realization for the search of required parameters [126,127] were proposed, a range of method use was widened, optimal composition of a filter package and characteristics of filters [125] were determined. The process for production of FP fibrous filtering materials and analytical filter kits intended for aerosol particles size distribution analysis was upgraded, which lead to production of very thin layers of uniform and moldable filtering materials.

It was exactly these advantages of electrospinning technology that enabled realization in the praxis of the method known in Russia as an MFM.

17.11.1 SIZE RANGE

The range of particle sizes determined by the MFM depends on the filtration rate and on the selected composition of a filter kit. All limitations follow from the beginning and the end of action of the inertial mechanism of aerosols' capturing. So, if the Stokes number is lower than 0.2, then the capture inertial mechanism does not work, that is why this method is not effective at low filtration rates of small particles. At higher filtration rates (over 2 m/s) the capturing factor grows not as sharply as it follows from the theory and the further increase of the filtration rate leads to reduction of the capturing factor due to the rebound effect of particles from fibers.

It is evident that conducting a size distribution analysis by MFM is impossible if practically all particles are captured by the first layer of a kit. For that reason the bigger size of the particles the less effective should be the first and the second layers. To arrange that one should either select thinner filters having the same fiber diameter or select filters made of thicker fibers.

Analysis of particle sizes in the range of 0.1–2 μm is possible at a sampling rate of about 1.5 m/s and [ΔP] of the first layer of 1.2 Pa/cm/s. After reduction of the rate to 0.5 m/s and having reduced resistance of the first layer down to 0.8–1 Pa/cm/s one can determine particle size with median diameter up to 7–8 μm [128].

17.11.2 SAMPLING

A package of filters for size distribution analysis is placed into a standard filter holder or at a cage screen of an aspiration installation. Filtering layers are placed along the flow in the order of their [ΔP] increasing. We have realized in our studies a mode of aerosols' inertial capture in three-layer packages, where the first and the second filters were made of polymer fibers having diameter of about 7 μm. Standard resistance of the first filter usually is 1 Pa/cm/s. [ΔP] of the second filter was 1.73 times higher in order to reduce analysis errors of the second filter [125]. To capture by the third filter all particles that penetrated the first two filters that third filter is to consist of thin fibers and have [ΔP] not less than 20 Pa/cm/s. FPP-15-1.5 filtering material is normally used for the third layer. Surface of filtering materials, of which the package is made, can be of any value. In our studies it was in the range from 0.3 to 6×10^4 cm².

Linear flow velocity is to be kept constant during the whole time period of sampling. Its value is to be chosen from the range of 0.5–2 m/s depending on supposed size of analyzed particles and volume flow rate of sampling may reach 20,000 m³/h. Thanks to that a size distribution analysis of such aerosols is possible, whose concentration is extremely low (e.g., of aerosols containing ^{137}Cs that originated at nuclear tests in the atmosphere). The sampling time period is determined on the base of sensitivity of instruments used for measuring of the captured substance at each of the filtering layers of the package.

17.11.3 COMPARISON WITH THE IMPACTOR

1. Form of the distribution function
   Distribution density in the MFM method is assigned *a priori* and its parameters are recaptured. The distribution function is recaptured by the resulting histogram with the help of the impactor. MFM usage for the analysis of bimodal distributions is not effective but one can replace the final cascade of the impactor by a filter package [129].
2. Number of experimental points
   The MFM method uses three filters, impactor has not less than five separation stages, and some impactors (Andersen Ambient, Andersen MK III) consist of 8 or even 10 cascades

(Sierra/Marple Model 210). Therefore, when using the MFM one requires less measurements that is important from the point of view of time and cost reduction for conducting a study.

3. Productivity of sampling

Theoretically, there are no limitations on the productivity of the MFM. What is limited it is linear velocity but not the filtering surface. In reality one managed to reach productivity up to 20,000 m³/h and using standard sampling devices. Productivity of the most powerful impactors is considerably (approximately 300 times) lower.

4. Range of determined sizes

The MFM realized on the base of fiber materials having diameter of about 7 μm enables to determine d_0 in the size range of 0.04–7 μm. The error of LND parameters determination is higher at both ends of the size range than in its center and the best results are received in the range of 0.1–5 μm. None of the existing impactor marks allows for covering of such a wide range: MFM has advantages here giving way to the majority of impactor marks in the area of micron sizes, especially, at $d_0 > 5$ μm.

5. Auxiliary equipment and sampling procedures

Devices intended for sampling for fibrous filters are quite appropriate for use with the MFM method. Impactor being a rather complicated design having high aerodynamic resistance requires the use of powerful high-head air blowers to achieve a required flow rate. In that respect MFM method is simpler and cheaper in operation.

17.11.4 MFM APPLICATION

Application of the MFM method of multilayer filters for determination of size distribution of aerosols enabled us to receive a number of interesting scientific results. Namely, starting with 1986 this method is the main one for the study of radioactive aerosols produced during Chernobyl accident (see Chapter 18).

During the course of a full scale study of aerosols' transfer from the Aral Region [130] one and the same atmospheric aerosol has been analyzed using an impactor, a laser aerosol spectrometer and the MFM method. Value deviations of mass median aerodynamic diameter (MMAD) accounted for no more than 20%. The method was used to study the sizes of natural radioactive aerosols of ⁷Be, ²¹²Pb [131], atmospheric aerosol over Moscow [132], for evaluation of size distribution of some specific aerosols in process communications of industrial plants [133], nuclear reactors [134] and for complex study of other aerodispersed systems [135].

Recently [136] experiments were accomplished, which compare results of determination of sizes of NaCl and Se particles, produced by laboratory generators. The results received by the MFM practically coincide with those received by other known methods realized in the TSI installation and by a laser aerosol spectrometer (difference between calculated and measured values does not exceed 0.02 μm for particles having their size of about 0.30 μm).

17.12 FILTERS FOR CAPTURING GASEOUS SUBSTANCES

Some chemical compounds of substances (iodine, ruthenium, tellurium, cesium, mercury and others) exist in the atmosphere both in aerosol and gaseous state simultaneously. It is known that fibrous filters are effective in capturing aerosols but are "transparent" for vapors. For that reason two-component FP (SFM) sorption-filtering materials were developed that consist of fibers and sorbent particles introduced into the intra-fiber space [137]. Such materials are seldom used in the process purification systems due to their relatively low sorption capacity, but as analytical means they have a number of advantages as compared to other analytical means (carbon cartridges, sorption columns, etc.). First of all, that is sufficiently high effectiveness and low resistance to air flow. Second, such analytical filters can be accommodated in standard sampling devices operating with

TABLE 17.5

Characteristics of Some Analytical Filters for Capture of Gaseous Compounds

Analytical Filter	Sorbent	Effectiveness of Gaseous Substances Capture at Velocity up to 5 cm/s	Resistance to Air Flow (Pa), at Velocity of 1 cm/s	Application Field
AFAS-I	Carbon OU-A with silver nitrate	90–99	50	Iodine
AFAS-P	Carbon OU-A	99	60	Polonium
AFAS-R	Carbon OU-A with iodine	99	50	Mercury
AFAS-U	Carbon SKT	99	60	Low-volatile substances

Source: Budyka, A.K. and Borisov, N.B., *Fibrous Filters for Air Pollution Control*, IzdAT, 360 p., 2008.

fibrous filters. Finally, that is a simple way of analyzing captured substances. Sorption analytical AFAS filters are manufactured on the base of SFM materials that are intended for gaseous compounds of some hazardous substances. Characteristics of some analytical AFAS filters are given in Table 17.5.

It is worth noting that gaseous compounds of the same chemical elements are adsorbed in a different way. For example, radioactive iodine can be found inside both easily adsorbed (I_2) and poorly adsorbed (CH_3I and other) compounds, which is expressed in differing of their dynamic sorption factors and effectiveness of their capture. Several identical analytical AFAS filters are to be placed after the aerosol analytical filter to separate volatile iodine compounds with different sorption properties. By a distribution character of iodine in such a package one can determine not only phase composition of iodine but also correlation between easily adsorbed and poorly adsorbed gaseous phase fractions [138].

SFM layers are placed after aerosol filters. Depending on analyzed substance SFMs are used with different sorbents. SFM application considerably simplifies conducting of monitoring. SFM mat cushion enables preparing the samples for measuring in a form of compact pellets as in case of filters.

17.12.1 Correction for Volatile Substances Desorption

Volatile substances after sedimentation on an aerosol particle can get desorbed from its surface. Time period of such substances holding on a particle is determined by desorption probability (λ_{des}). If that time period is comparable to the duration of sampling then there is a secondary redistribution of a substance between the capturing stages of a sampling device. Thus experimental values of gaseous substance portion may be considerably overstated as compared to the real ones. In the final effect it leads to incorrect dose evaluations.

Unfortunately, it is not possible to calculate the value of λ_{des}, but it is possible to estimate its value judging from experimental data. The bulk of data is available for radioactive ^{131}I—product of Chernobyl Atomic Power Station disaster. Using a simple model of iodine entrainment from the aerosol particle, one succeeds in determining with the help of nonlinear regression analysis that $\lambda_{des} = 0.13/h$ [139].

It goes from the received value of desorption probability that within a characteristic time $t_{1/2des} = -\ln2/\lambda_{des}$, which equals approximately 5.3 h, half of the radioactive iodine is transferred from the aerosol filter to the second stage of a sampling device.

17.13 COMPOSITION OF FILTERING MATERIALS FOR ATMOSPHERIC MONITORING

Sampling for capturing and subsequent analysis of aerosols and substances present in the atmosphere simultaneously as aerosols and gases one should conduct using a row of sequentially placed filters and SFMs (Figure 17.15).

The first three filters are required for the determination of aerosol size distribution via multilayer filtering method. The subsequent SFM layers are intended for determination of gaseous components.

One can vary the number of filters in a package. So, filter 3 may consist of several layers to exclude aerosols penetration onto adsorbing layers, that is, aerosols are to be captured not by four filters as it is shown on Figure 17.15 but by more filters. On the other hand, if a gaseous substance is captured with high effectiveness by one SFM layer, there is no need using a greater number of layers. On the contrary, if gaseous substances are captured with insufficient effectiveness as it is in case of presence of poorly adsorbed iodine compounds, one may require greater SFM number (4–6) for separation of gaseous forms into different fractions. Sorbent content in each of the SFM is $2–10\,mg/cm^2$ depending on its grade.

Using such packages allows determining after measuring of the captured substance at each filter all basic characteristic of gaseous-aerosol system including size distribution of aerosols and correlation between aerosol and gaseous components of contaminating substances.

In a number of cases such filter package is a simple, effective, and cheap way to get authentic information about characteristics of aerosols. We stress once more that when operating this method instruments and installations are used that are intended for routine sampling of aerosols by fibrous filters.

17.14 CONCLUSION

Fibrous filtering materials are the most effective means for purification of gases from suspended micro- and nanoparticles as well as for sampling of aerosols.

The necessity to develop aerosol filtration theory is driven by its most important practical applications. Exactly from the filtration theory follows, which filtering materials and filters are required for solving this or that problem, how variation of filtration conditions influences upon the gas purification grade, life expectancy of filters, power consumption, etc.

Understanding objective laws of filtration enables to choose analytical filters that fit the purpose in each concrete case in the best way as well as to determine continuation time period and rate of sampling, to choose the best method for the analysis of the captured sediment.

Introduction of compositions from different filtering materials enables to determine not only concentration and chemical (radionuclide) composition of aerosols but also to determine size distribution of aerosol particles, correlation between aerosol and gaseous components of a substance under consideration of correction for desorption.

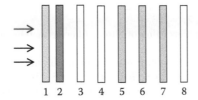

FIGURE 17.15 Arrangement of filters for monitoring. 1–3, filters for size distribution analysis of aerosols; 4, filter for prevention of aerosols penetration into adsorbing-filtering materials; 5–7, SFM; 8, protective fibrous filter.

ACKNOWLEDGMENT

Chapter translated from original Russian by Natalya A. Tchuksina.

REFERENCES

1. Langmuir I. Report on Smokes and Filters. Section I. U.S. Office of Scientific Research and Development, No. 865, Pt IV, 1942.
2. Natanson G.L. Diffusional precipitation of aerosols on a streamlined cylinder with a small capture coefficient. *Proc. Acad. Sci. USSR Phys. Chem. Sec.*, 1957, 112, 21–25.
3. Natanson G.L. Diffusional deposition of aerosols on a flow-around cylinder under impact of electrostatic attraction. *Dokl. AN USSR*, 1957, 112(4), 696–699 (in Russian).
4. Fuchs N.A. *Advances in Mechanics of Aerosols*. Moscow, Russia: AN USSR, 1961.
5. Stechkina I.B., Fuchs N.A. Studies of fibrous aerosol filters—I. Calculation of diffusional deposition of aerosols in fibrous filters. *Ann. Occup. Hyg.*, 1966, 9, 59–64.
6. Kirsch A.A., Fuchs N.A. Studies on fibrous aerosol filters—II. Pressure drops in systems of parallel cylinders. *Ann. Occup. Hyg.*, 1967, 10, 23–30.
7. Kirsch A.A., Fuchs N.A. Studies on fibrous aerosol filters—III. Diffusional deposition of aerosols in fibrous filters. *Ann. Occup. Hyg.*, 1968, 11, 299–304.
8. Stechkina I.B., Kirsch A.A., Fuchs N.A. Studies of fibrous aerosol filters—IV. Calculation of aerosol deposition in model filters in the range of the maximum penetration. *Ann. Occup. Hyg.*, 1968, 12, 1–8.
9. Kirsch A.A., Fuchs N.A. The fluid flow in a system of parallel cylinders. *J. Phys. Soc. Jpn.*, 1967, 22(5), 1251–1255.
10. Stechkina I.B., Kirsch A.A., Fuchs N.A. Studies of fibrous aerosol filters—V. Experimental determination of the efficiency of fibrous filters in the range of maximum particle penetration. *Colloid J.*, 1969, 31(2), 227–232 (in Russian).
11. Kogan Ja.I., Burnashova Z.A. Enlarging and measuring of condensation nuclei in a continuous flow. *Russ. Phys. Chem. J.*, 34, 2630, 1960.
12. Lee K.W., Liu B.Y.H. Experimental study of aerosol filtration in fibrous filters. *Aerosol Sci. Technol.*, 1982, 1(1), 35–46.
13. Lee K.W., Liu B.Y.H. Theoretical study of aerosol filtration in fibrous filters. *Aerosol Sci. Technol.*, 1982, 1(2), 147–161.
14. Davies C.N. *Air Filtration*. New York: Academic Press, 1973.
15. Chen C.Y. Filtration of aerosols by fibrous media, *Chem. Rev.*, 1955, 55, 595.
16. Brown R.C. *Air Filtration*. Oxford, U.K.: Pergamon Press, 1993, p. 269.
17. Kirsh A.A., Budyka A.K., Kirsch V.A. Aerosol filtration in FP fibrous materials. *Russ. Chem. J.*, 2008, 52(5), 97–101.
18. Petryanov-Sokolov I.V. *Selected works*. Moscow, Russia: Nauka, 2007.
19. Fuchs N.A. Highly effective filtration of gases and liquids in fibrous materials. *Russ. Chem. Ind. J.*, 1978, (11), 688.
20. Petryanov I.V., Koshtsheev V.S. et al. *"Lepestok" (Lighe Respirators)*. Moscow, Russia: Nauka, 1984.
21. Chronakis I.S. Novel nanocomposites and nanoceramics based on polymer nanofibers using electrospinning process—A review. *J. Mater. Process. Technol.*, 2005, 167(2–3), 283–293.
22. Filatov Yu., Budyka A., Kirichenko V. *Electrospinning of Micro-and Nanofibers and Their Application in Filtration and Separation Processes*. New York: Begell House Inc., 404pp., 2007.
23. Reneker D.H., Chun I. Nanometer diameter fibres of polymer, produced by electrospinning. *Nanotechnology*, 1996, 7(3), 216–223.
24. Shutov A.A., Astakhov E.Yu. Formation of fibrous filtering membranes by electrospinning. *Tech. Phys.*, 2006, 51(8), 1093–1096.
25. Kirsch A.A., Stechkina I.B. Theory of aerosol filtration with fibrous filters. In *Fundamentals of Aerosol Science*. D.T. Shaw (Ed.), John Wiley & Sons, NY, pp. 165–256, 1978.
26. Tamaru S., Aomi H., Tanaka O. Development and application of ultra-high performance fluorocarbon polymer air filter. *Proceedings of 13th International Symposium on Contamination Control*, September 16–20, 1996, The Hague, The Netherlands, ICCCS, Stouthart Communicatie, Waddinxveen, The Netherlands, pp. 236–242, 1996.
27. Levin L.M. Studies in physics of coarse-dispersed aerosols, *Dokl. AN USSR*, 1953, 91(6), 1329 (in Russian).

28. Budyka A.K., Ogorodnikov B.I., Petryanov I.V., Skitovich V.I. Hydrodynamics of fan model of fibrous filter and aerosol capture with Stokes numbers from 0.4 to 4. *Dokl. AN SSSR*, 284(5), 1161–1164.

29. Ushakova E.N., Kozlov V.I., Petryanov I.V. *Dokl. AN USSR*, 1972, 206(4), 916 (in Russian).

30. Fuchs N.A. *The Mechanics of Aerosols*. Oxford, U.K.: Pergamon Press, 1964.

31. Kuwabara S. Forces experienced by randomly distributed parallel circular cylinders or spheres in a viscous flow at small Reynolds number. *J. Phys. Soc. Jpn.*, 1959, 14(4), 527–532.

32. Golovin A.M., Lopatin V.A. Flow of a viscous fluid through a doubly periodic series of cylinders. *Pricl. Mech. Teknich. Fiz.*, 1969, 2, 99–105 (in Russian).

33. Keller J.B. Viscous flow through a grating or lattice of cylinders. *J. Fluid Mech.*, 1964, 18, 94–96.

34. Miyagi T. Viscous flow at low Reynolds numbers past an infinite row of equal circular cylinders. *J. Phys. Soc. Jpn.*, 1958, 13(5), 493–496.

35. Kirsh V.A. The viscous drag of three-dimensional model fibrous filters. *Colloid J.*, 2006, 68(3), 261–266.

36. Kirsh V.A. Budyka A.K., Kirsh A.A. Simulation of nanofibrous filters produced by electrospinning method. I—Pressure drop and deposition of nanoparticles. *Colloid J.*, 2008, 70(5), 547–583.

37. Kirsh V.A., Budyka A.K., Kirsh A.A. Simulation of nanofibrous filters produced by electrospinning method. II. The effect of gas slip on the pressure drop. *Colloid J.*, 2008, 70(5), 583–588.

38. Kuwabara S. Forces experienced by lattice of elliptic cylinders in a uniform flow at small Reynolds number. *J. Phys. Soc. Jpn.*, 1959, 14(4), 522–527.

39. Kirsh V.A. Stokes flow past periodic rows of porous cylinders. *Theor. Found. Chem. Eng.*, 2006, 40(5), 465–471.

40. Kirsh V.A. Stokes flow in periodic systems of parallel cylinders with porous permeable shells. *Colloid J.*, 2006, 68(2), 173–181.

41. Kirsh V.A. The drag of the row of parallel chains of spherical particles in the Stokes flow. *Colloid J.* 2006, 68(3), 387–389.

42. Kirsch A.A., Stechkina I.B. Inertial deposition of aerosol particles in model filters at low Reynolds numbers. *J. Aerosol Sci.*, 1977, 8(5), 301.

43. Kirsch A.A., Dvuhimyonnyi V.A. Study of particles deposition in model filters in the course of deposits' accumulation. *Theor. Found. Chem. Eng.*, 1982, T.16(5), c.711–c.714.

44. Kirsh A.A. Increase of pressure drop in a model filter during mist filtration. *J. Colloid Interface Sci.*, 1978, 64(1), 120–125.

45. Kirsh V.A. Inertial deposition of aerosol particles on fibrous filters. *Colloid J.*, 2004, 66(5), 547–552.

46. Kirsh V.A. Method for the calculation of an increase in the pressure drop in an aerosol filter on clogging with solid particles. *Colloid J.*, 1998, 60(4), 439–444.

47. Chernyakov A.L., Kirsch A.A. Growth in hydrodynamic resistance of aerosol fibrous filter during deposition of liquid dispersed phase. *Colloid J.*, 1999, 61(6), 791–796.

48. Kirsch A.A., Chechuev P.V. Diffusion deposition of aerosol in fibrous filters at intermediate Peclet numbers. *Aerosol Sci. Technol.*, 1985, 4(1), 11–16.

49. Fuchs N.A., Stechkina I.B. To the theory of fibrous aerosol filters. *Dokl. AN USSR*, 1962, 147(5), 1144–1146 (in Russian).

50. Stechkina I.B. Diffusion deposition of aerosol in fibrous filters. *Dokl. AN USSR*, 1966, 167(6), 1327–1330 (in Russian).

51. Cheng Y.S., Yamada Y., Yeh H.C. Diffusion deposition on model fibrous filters with intermediate porosity. *Aerosol Sci. Technol.*, 1990, 12(2), 286–299.

52. Kirsch A.A., Stechkina I.B. A diffusional method for the determination of the size of condensation nuclei. *Proceedings of 7th International Conference on Condensation and Ice Nuclei*, September 18–24, 1969, Prague, Czech Republic, J. Podzimek (Ed.), Academia, Prague, Czech Republic, 1969, pp. 264–287.

53. Scheibel H.G., Porstendorfer J. Penetration measurements for tube and screen-type diffusion batteries in the ultrafine particle size range. *J. Aerosol Sci.*, 1984, 15(6), 673–682.

54. Reineking A., Porstendorfer J. High-volume screen diffusion batteries and α spectroscopy for measurement of the radon daughter activity size distributions in the environment. *J. Aerosol Sci.*, 1986, 17(5), 873–879.

55. Knutson E.O. History of diffusion batteries in aerosol measurements. *Aerosol Sci. Technol.*, 2007, 31(2), 83–128.

56. Kirsh V.A. Deposition of aerosol nanoparticles in fibrous filters. *Colloid J.*, 2003, 65(6), 726–732.

57. Kirsh V.A. Deposition of nanoparticles in a model fibrous filter at low Reynolds numbers. *Russ. J. Phys. Chem.*, 2005, 79(12), 2049–2052.

58. Kirsch V.A., Kirsch A.A. Deposition of aerosol nanoparticles in model fibrous filters. In *Aerosols-Science and Technology*, I. Agranovski (Ed.), Wiley–VCH, 2010, pp. 283–314.

59. Chernyakov A.L., Kirsch A.A., Roldugin V.I., Stechkina I.B. Diffusion deposition of aerosol particles on fibrous filters at small Peclet numbers. *Colloid J.*, 2000, 62(4), 490–494.

60. Kirsh V.A., Kirsh A.A. Penetration of nanoparticles through screen-type diffusion batteries. *Colloid J.*, 2010, 72(4), 491–498.

61. Kirsch A.A., Stechkina I.B. Pressure drop and diffusional deposition of aerosol and polydisperse model filter. *J. Colloid Interface Sci.*, 1973, 43(1), 10–16.

62. Kirsch A.A., Stechkina I.B., Fuchs N.A. Gas flow in aerosol filters made of polydisperse ultrafine fibres. *J. Aerosol Sci.*, 1974, 5, 39–45.

63. Chernyakov A.L., Lebedev M.N., Stechkina I.B., Kirsh A.A. Hydrodynamic resistance of the row of parallel polydisperse fibers. *Colloid J.*, 1998, 60(1), 91–103.

64. Brown R.C., Thorpe A. Glass-fibre filters with bimodal fibre size distributions. *Powder Technol.*, 2001, 118(1), 3–9.

65. Lebedev M.N., Stechkina I.B., Chernyakov A.L. Viscous drag of periodic row of polydisperse fibers with allowance for slip effect. *Colloid J.*, 2003, 65(2), 211–221.

66. Kirsch A.A., Stechkina I.B., Fuchs N.A. Effect of gas slip on the pressure drop in fibrous filters. *J. Aerosol Sci.* 1973, 4, 287–293.

67. Kirsch A.A., Stechkina I.B., Fuchs N.A. Effect of gas slip on the pressure drop in a system of parallel cylinders at small Reynolds numbers. *J. Colloid Interface Sci.*, 1971, 37(2), 458–461.

68. Cercignani C. *Mathematical Methods in Kinetic Theory*. London, U.K.: MacMillan, 1969.

69. Kirsch A.A., Chechuev P.V. Experimental study of air filtration by fibrous filters at high temperature. *AAAR 1990 Annual Meeting*, June 18–22, Philadelphia, PA, Abstract book, p. 190, 1990.

70. Roldughin V.I., Kirsch A.A., Emel·yanenko A.M. Simulation of aerosol filters at intermediate Knudsen numbers. *Colloid J.*, 1999, 61(4), 492–504.

71. Roldughin V.I., Kirsch A.A. Diffusion deposition of finite size particles on fibrous filters at intermediate Knudsen numbers. *Colloid J.*, 2001, 63(5), 619–625.

72. Kirsh V.A. The deposition of aerosol submicron particles on ultrafine fiber filters. *Colloid J.*, 2004, 66(3), 311–315.

73. Kirsh V.A. Calculation of the van der Waals force between a spherical particle and an infinite cylinder, *Adv. Colloid Interface Sci.*, 2003, 104, 311–324.

74. Kirsh V.A. Gravitational deposition of aerosol particles in a fibrous filter with account for the effect of van der Waals forces. *Colloid J.*, 2001, 63(1), 68–74.

75. Kirsh V.A. Inertial deposition of "heavy" aerosol particles in fibrous filters. *Theor. Found. Chem. Eng.*, 2005, 39(1), 47–52.

76. Kirsh V.A. Diffusional deposition of heavy submicron aerosol particles on fibrous filters. *Colloid J.*, 2005, 67(3), 313–317.

77. Kirsh, V.A. The effect of van der Waals forces on the deposition of highly dispersed aerosol particles on ultrafine fibers. *Colloid J.*, 2004, 66(4), 444–450.

78. Kirsh V.A. Diffusional deposition of heavy submicron aerosol particles on fibrous filters. *Colloid J.*, 2005, 67(3), 313–317.

79. Kirsh V.A. Method for the calculation of an increase in the pressure drop in an aerosol filter on clogging with solid particles. *Colloid J.*, 1998, 60(4), 439–443.

80. Kirsch A.A. The influence of an external electric field on the deposition of aerosols in fibrous filters. *J. Aerosol Sci.*, 1971, 2(5), 29–33.

81. Kirsh V.A. Deposition of nanoparticles in filters composed of permeable porous fibers. *Colloid J.*, 2007, 69(5), 609–614.

82. Kirsh, V.A. Deposition of aerosol nanoparticles in filters composed of fibers with porous shells. *Colloid J.*, 2007, 69(5), 615–619.

83. Kirsh V.A. Pripachkin D.A., Budyka A.K. Inertial deposition of aerosol particles from laminar flow in fibrous filters. *Colloid J.*, 2010, 72(2), 206–210.

84. Higdon J.J.L., Ford G.D. Permeability of three-dimensional models of fibrous porous media. *J. Fluid Mech.*, 1996, 308, 341–361.

85. Clague D.S., Philips R.J. A numerical calculation of the hydraulic permeability of three-dimensional disordered fibrous media. *Phys. Fluids*, 1997, 9(6), 1562–1572.

86. Chernyakov A.L., Kirsh A.A. Resistance force in a three-dimensional grid formed by cylindrical fibers. *Colloid J.*, 1997, 59(5), 698–708.

87. Chernyakov A.L. Fluid flow through three-dimensional fibrous porous media. *J. Exp. Theor. Phys.*, 1998, 86(6), 1156–1166.

88. Chernyakov A.L., Kirsh A.A. The effect of long-range fiber space and orientation correlations on the hydrodynamic resistance and diffusion deposition in fibrous filters. *Colloid J.*, 2001, 63(4), 506–510.

89. Ushakova R.N., Kozlov V.I., Petryanov I.V. Regularity in capturing of aerosols by fibrous filtering materials in diffusional area. *Colloid J.*, 1973, 35(3), 388–391.

90. Ushakova R.N., Kozlov V.I., Petryanov I.V., Regularity in capturing of aerosols by FP filtering materials. *Colloid J.*, 1973, 35(5), 993–995.

91. Fuchs N.A., Kirsch A.A., Stechkina I.B. A contribution to the theory of fibrous aerosol filters, *Faraday Symposia of the Chemical Society*, Swansea, England, No. 7, 1973, pp. 143–156.

92. Kirsh A.A., Zhulanov Yu.V. Measurement of the penetration of aerosols through high-efficiency filters. *Colloid J.*, 1977, 39(3), 288–294.

93. Kirsh V.A. Aerosol filters made of porous fibers. *Colloid J.*, 1996, 58(6), 737–740.

94. Stechkina I.B., Kirsh V.A. Optimization of the aerosol filter parameters. *Colloid J.*, 2001, 63(4), 471–475.

95. Kirsch V.A. Stokes flow in model fibrous filters. *Sep. Purif. Technol.*. 2007, 58(2), 288–294.

96. Stechkina I.B., Kirsh V.A. Optimization of parameters of filters in a multistage system of fine gas filtration. *Theor. Found. Chem. Eng.*, 2003, 37(3), 218–227.

97. Kirsch V.A., Stechkina I.B. Kinetic of clogging and optimization of pre-filters in a two-stage air purification system. *Theor. Found. Chem. Eng.*, 2011, 44(1), 76–85.

98. Park S.J., Lee D.G. Performance improvement of micron-sized fibrous metal filters by direct growth of carbon nanotubes. *Carbon*, 2006, 44(10), 1930–1935.

99. Fan K.S., Wamsley B., Gentry J.W. The effect of stokes and Reynolds numbers on the collection efficiency of grid filters. *J. Colloid Interface Sci.*, 1978, 65(1), 162.

100. Emi H., Okuyama K., Adachi M. The effect of neighboring fibers on the single fiber inertial-interception efficiency of aerosols. *J. Chem. Eng. Jpn.*, 1977, 10(2), 148.

101. Brewer J.M., Goren S.L. Evaluation of metal oxide whiskers grown on screens for use as aerosol filtration medium. *Aerosol Sci. Technol.*, 1984, 3(4), 411.

102. Gentry J.W., Choudhary K.R. Collection efficiency and pressure drop in grid filters of high packing density at intermediate Reynolds numbers. *J. Aerosol Sci.*, 1975, 6, 277.

103. Ogorodnikov B.I., Skotnikova O.G. et al., Study of size distribution of aerosols at ground-level air layer. *At. Energy J.*, 1972, 32(6), 76.

104. Skitovich V.I., Efimenko V.S., Ogorodnikov B.I. FP filters' characteristics at high rate air filtration (up to 40 m/sec). *Trudy* IPG Works. 1976, 21, 8.

105. Ushakova E.N., Kozlov V.I., Petryanov I.V. Study of effectiveness of filtering materials of Petryanov filter type in inertial area. *Colloid J.*, 1975, 37(2), 318–322.

106. Budyka A.K., Ogorodnikov B.I., Petryanov I.V., Selection of atmospheric aerosols' filtration parameters depending on the size distribution. In *Nuclear-Physical Methods of Analyses of Environment*. L. Gidrometeoizdat, Leningrad, 1987, p. 128.

107. Dahneke B. The capture of particles by surfaces. *J. Colloid Interface Sci.*, 1971, 37(2), 342.

108. Dahneke B. Measurements of bouncing of small latex spheres. *J. Colloid Interface Sci.*, 1973, 45(3), 584.

109. Dahneke B. Further measurements of the bouncing of small latex spheres. *J. Colloid Interface Sci.*, 1975, 51(1), 58.

110. Loffler F. Problems and recent advances in aerosol filtration. *Sep. Sci. Technol.*, 1980, 15, 3, 297.

111. Stenhouse J.I.T., Freshwater D.C. Particle adhesion in fibrous air filters. *Trans. Inst. Chem. Eng.*, 1976, 54, 95.

112. Kyaw Tha Paw U. Dimensional aspects of aerosol deposition on cylinder with rebound. *J. Aerosol Sci.*, 1984, 15(6), 657.

113. Budyka A.K., Ogorodnikov B.I., Skitovich V.I. Behavior of sub-micron aerosols at their collision with a cylinder in a high velocity air flow. *Materials of Sixth International Conference Applied problems of fluid mechanics*, September 29 to October 3, 1997, Sevastopol, Ukraine, Sevastopol GTU, 1997, pp. 3–8.

114. Budyka A.K, Ogorodnikov B.I., Skitovich V.I. Interactions of submicron particles with cylindrical surfaces at high flow velocities. In *Adhesion of Dusts and Air Cleaning*, Frunze polytechn. Inst. Publ., Frunze, 1985, pp. 29–30

115. Budyka A.K. Statistical characteristics of structure of fibrous filter. *Russ. Colloid J.*, 1992, 54(5), 54.

116. Matushtshenko A.M. *Chernobyl: Catastrophe. Heroic Deed. Lessons and Conclusions*, Moscow, Russia: Inter-Vesy, 1996, p. 436.

117. Budyka A.K., Borisov N.B. *Fibrous Filters for Air Pollution Control*. IzdAT, Moscow, 2008, 360 pp.

118. Budyka A.K., Ogorodnikov B.I., Skitovich V.I. *Fission Product Transport Processes in Reactor Accidents*. New York: Hemisphere Public Corporation, 1990, p. 779.

119. Baron P.A., Willeke K. (Eds.) *Aerosol Measurements: Principle, Technique and Application*, 2nd edn., New York: Wiley, 2001, p. 1001
120. Belyaev S.P., Nikiforova N.K., Smirnov V.V., Shtsheltshkov G.I. *Optoelectronic Methods of Aerosol Studies*. Moscow, Russia: Energoizdat, 1981, p. 232.
121. Hering S.V. Size selective sampling for atmospheric aerosols, Tut. No 17 presented at *1995 Annual Meeting AAAR*, October 9, 1995, Pittsburg, PA, 69p.
122. Lockhart L.B., Patterson R.L., Sanders A.W. The size distribution of radioactive atmospheric aerosols, *J. Geophys. Res.*, 1965, 24(7), 6033–6041.
123. Lockhart L.B., Patterson R.L., Sanders A.W. *NRL Report 6164*, UN Document A/AC. 82/g/l 1138, 1964.
124. Bogolapov N.V., Kashtshenko N.I., Konstantinov I.E. Method for investigation of dispersity of radioactive aerosols. In *Problems of Dosimetry and Protection from Radiation*, Moscow, Russia: Atomizdat, 1975, Issue 14, p. 136.
125. Budyka A.K. Development of the fundamentals of method of multilayer filters for the determination of radioactive aerosol particle size distribution, PhD thesis. Moscow, Russia: Moscow Engineering Physical Institute, 1986, Moscow, p. 127.
126. Sankov Yu.A., Budyka A.K. Algorithm and their realization for determination of parameters of lognormal distribution of aerosol particles. In *Environment Protection, Ecological Problems and Production Quality Control*, Moscow, Russia: NIITECHIM, Moscow, 1992, Issue 9, p. 35
127. Budyka A.K., Konstantinov I.E., Maksimov V.Yu. *Radiation Safety and Protection of NPP*, Moscow, Russia: Energoatomizdat, 1986, Issue 11, p. 106.
128. Budyka A.K., Ogorodnikov B.I. Radioactive aerosols of Chernobyl origin. *Russ. J. Phys. Chem.*, 1999, 73(2), 310–319.
129. Budyka A.K., Pripachkin D.A., Tsovjanov A.G., Rizin A.Yu. Personal impactor. Patent RU 2290624 C1 24.06.2005.
130. Andronova A.V. 1996, Private communication.
131. Lujaniene G., Ogorodnikov B.I., Budyka A.K., Skitovich V.I., Lujanas V. An investigation of changes in radionuclide carrier properties, *J. Environ. Radioact*, 1997, 35(1), 71.
132. Ogorodnikov B.I., Budyka A.K., Skitovich V.I., Brodovoy A.V. *Izvestya AN SSSR*, Series Physic of Atmosphere and Ocean, 1996, 32(2), p. 163.
133. Skitovich V.I., Sharapov A.G., Ogorodnikov B.I. Particle size distribution of some industrial radioactive aerosols, *Materials of Scientific-Technical Seminar of the Russian Nuclear Society, Present State of Nuclear Waste Handling Problem,* Sergiev Posad, Russia, 1995, p. 23.
134. Budyka A.K., Fedorov G.A. *Isotopes in the USSR*, Moscow, Russia: Energoatomizdat, 1987, Issue 1(72), p. 113.
135. Budyka A. K. Atmosphere monitoring and aerosols diagnostics, Doctoral thesis. Moscow, Russia: Karpov Institute of Physical Chemistry, 2001, p. 219 (in Russian).
136. Budyka A.K., Pripachkin D.A., Chmelevsky V.O., Tsovjanov A.G. Modelling and experimental investigation of aerosol deposition in personal impactor. *ANRI*, 2009, (3), 27–37.
137. Borisov N.B. *Isotopes in the USSR*. Moscow, Russia: Energoatomizdat, 1978, Issue 52/53, p. 66.
138. Borisov N.B., Budyka A.K., Verbov V.V., Ogorodnikov B.I. Monitoring of radioiodine in the atmosphere, *J. Aerosol Sci.*, 1994, 25(Suppl. 1), S271.
139. Budyka A.K. Relation between aerosol and gaseous components of [131]I: Account of desorption correction, *J. Aerosol Sci.*, 2000, 31(Suppl. 1), S480.
140. Roldugin V.I., Kirsh A.A., Stechkina I.B. Diffusion and sedimentation of particles in inhomogeneous porous media. *Russ. J. Phys. Chem.*, 2005, 79(5), 767–772.
141. Kirsch V.A. Inertial deposition of aerosol particles in a model filter with dust-loaded fibres. *J. Filtr. Soc.*, 2002, 2(4), 109–113.

18 Radioactive Aerosols of Chernobyl Accident

A.K. Budyka and B.I. Ogorodnikov

CONTENTS

18.1 INTRODUCTION

The accident in the fourth block of the Chernobyl Nuclear Power Plant (NPP) resulted in radioactive contamination not only in a territory of the European part of the former USSR, but also in the entire northern hemisphere, including the United States and Japan. This accident should be classified as a global disaster.

It is very clear that the contamination of territories occurred due to atmospheric transfer and fallout of radioactive aerosol particles on the earth's surface. Therefore, understanding the physico-chemical properties of radioactive aerosols, formed as a result of reactor explosion, allows not only the correct assessment of the scale of the disaster, but also the choice of a more effective means of defense for more than 300,000 people, who participated in the liquidation of the consequences of the Chernobyl accident.

Despite the fact that a few months after the accident the radioactive aerosol concentrations did not exceed permissible concentrations in the majority of the territory that surrounds Chernobyl (30 km around Chernobyl NPP), today there are places with a dangerously high level of air contamination.

For example, there are some places inside the object "Shelter" (previously called "Sarcophagus") with high levels of contamination. "Shelter" was built in 1986 in order to localize and subsequently control the situation of nuclear fuel from the destroyed reactor of the fourth block of Chernobyl NPP.

The fallout on the earth's surface slowly migrates deep into the soil. For 25 years after the accident, the property of the primary fallout changed. Secondary airborne aerosol particles were formed due to winds, forest fires, and also technogene activity (moving transport, building, and agriculture activity).

The data in this chapter is unique, because it is impossible to repeat such a situation due to the exclusiveness of the accident. Even now, for example, when we measure the radioactive aerosol concentration inside the "Shelter," the smallest change in meteorological conditions (temperature, pressure, humidity, speed and direction of the wind) and some technogene activity results in changes in the physicochemical characteristics of aerosols in the same sampling point.

It should be understood under what circumstances the measurements in the first weeks and months after the accident were taken. Dosimetric control of inner irradiation air sampling was provided in extreme conditions. In some sites, the radiation levels were so high that dosimetric personnel who would spend just a couple of minutes or hours would endanger their health. Often it was difficult to study the samples carefully, and dosimetrists were restricted only to measuring the summary of gamma- and beta-activity of the sample.

Still we hope that these materials, the majority of which were obtained by scientists and technicians of the aerosol laboratory of Karpov Institute of Physical Chemistry, will inform readers of the main properties of radioactive aerosols and the gaseous products of the accident in Chernobyl NPP.

18.2 DYNAMICS OF THE EJECTION OF RADIONUCLIDES FROM THE REACTOR

On April 26, 1986, as a result of the explosion in the fourth block of Chernobyl NPP, the core and upper part of the reactor were completely destroyed, as were all defense barriers and systems of security. It was the largest accident in the atomic industry. Most of the radionuclides were ejected from the fourth block from April 26 up to May 6, 1986 [1]. The dynamics of the ejection are shown in Figure 18.1.

In the first day after the accident, aerosol sampling and analysis of sampling composition above the destroyed reactor already began. Due to many factors (nonstationary character of the ejection, meteorological conditions, intensive work on covering the destroyed reactor with different

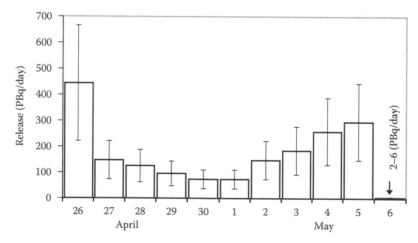

FIGURE 18.1 Daily integral ejection of radioactive products (without inert gases) from destroyed reactor block of the Chernobyl NPP in April–May 1986.

TABLE 18.1

Emission (in %) from the Total Activity of Radioactive Substances Saved up in the Core during the Operating Period of the Fourth Block of Chernobyl NPP

Radionuclide	According to Data from the USSR for the International Atomic Agency (August 1986) [1]	According to Results of the Study at the End of 2000 [2]
^{133}Xe, ^{85}Kr	~100	~100
^{131}I	20±10	55±5
^{134}Cs	10±5	33±10
^{137}Cs	13±7	33±10
Uranium and transuranium elements[a]	3±1.5	3±1

[a] Taking into account the fragments of the reactor active zone, ejected around the fourth block.

materials), the accurate determination of the dynamics of radioactive substances entering the atmosphere was difficult to assess. Two assessments of the composition of the emission rejection are shown in Table 18.1.

In the initial period after the accident, most radionuclides were ejected from the destroyed reactor in the form of dispersed fuel (mainly with a UO_2 matrix). During lava formation of the fuel materials, which took place at a temperature of about 2000°C, only volatile and light fusible substances, like Te and alkaline metals, evaporated from the fuel. More than 95% of nuclear fuel (more than 180 t) remained inside the "Shelter," which was built above the destroyed block at the end of November 1986.

The integral ejection of radionuclides with a half-time of $T_{1/2} > 20$ h, without taking into account inert gases from the destroyed reactor, was around 3×10^{18} Bq [2].

18.3 GLOBAL TRANSFER OF THE ACCIDENT PRODUCTS

At the moment of the accident, winds near the earth's surface were weak and without special direction. Still, at altitudes of more than 1500 m the wind was mostly in the southeast direction, with a speed around 8–10 m/s. Part of the radioactivity was raised to this level and moved throughout the western regions of the USSR to Finland and Sweden. There, on April 27, radioactive aerosols from the Chernobyl accident were discovered [3] for the first time outside the USSR. Over the end of April and the beginning of May, radioactive aerosols, including I and Cs, were detected in the upper troposphere and lower stratosphere (up to 15 km) above the territory of Poland [4]. The fallout of refractory elements such as Ce, Zr, Np, and Sr was detected mainly on USSR territory.

Changing meteorological conditions, especially wind direction, and also the continuous ejection of a large mass of radioactive aerosols and gases for 10 days resulted in a very complicated picture of the distribution of radionuclides in the atmosphere. Part of the nuclides was shifted on the low highs to Poland and Germany. By April 29 and 30, radioactive clouds reached other countries of Eastern and Central Europe. Radioactive substances were in the north of Italy by April 30, and in Central and South Italy the next day. In France, Belgium, and the Netherlands, radioactive contamination was detected on May 1, and in Great Britain the next day. Gases and aerosol products from the Chernobyl accident reached the north of Greece on May 2 and the south of Greece by May 3. At the beginning of May, radioactive substances were detected in Israel, Kuwait, and Turkey.

The observation of aerosol and gaseous components of radioactive iodine was provided in 19 stations in Europe, with sampling and analysis of 171 samples. Only aerosol components of radionuclides were studied on 1892 samples from 85 sites [5].

The transfer of radioactive products in the eastern direction took place very fast. Already by May 2, the first aerosol and gaseous samples of ^{131}I were found in Japan, May 4 in China, May 5 in India, and May 5 and 6 in Canada and the United States. On the North American continent, products of the Chernobyl accident came practically simultaneously from the west and east. All this confirms that the intermix of aerosols in the horizontal and vertical direction was substantial, and also that the transfer of activity took place mostly on submicrometer-sized particles. No substantial amount of radionuclides was detected in the southern hemisphere. Thus, the distribution of aerosols after the Chernobyl accident took place in the earth's atmosphere according to ideas formulated earlier based on the examples of volcano eruption and tests of nuclear weapons.

18.4 SAMPLING DEVICES

In the initial period after the accident and during the building of the "Shelter," measurements of aerosol characteristics (mainly for radiation reconnaissance and dosimetry) were taken almost continuously at different sampling points. After the "Shelter" was built, the main goal became the study of the temporary and spatial evolution of radionuclides and the disperse composition of aerosols, and also the study of secondary sources of aerosols (wind raising dust, forest fires, etc.). Sampling became more focused and related to meteorological conditions. For sampling, usually filter materials AFA-RMP and also packages from different filter materials were used.

Three-layer packages, consisting of filter materials FPA-70-0.15, FPA-70-0.25, and FPA-15-1.5, were used for the simultaneous measurement of the concentration and disperse composition of aerosols (with the help of the so-called method of multilayer filters [MMF], see Chapter 17).

For simultaneously sampling aerosol and gaseous substances with I, Te, Ru, and Cs, sorption-filter material SFM-I was used [6]. This material consists of two sorption layers, covered from above and below with FPP-70-0.2, and the frontal layer of FPP-15-1.5 is only for aerosol deposition. Each sorption layer was developed from FPP-70-0.3 material, on the fiber of which was fine-grained carbon with $AgNO_3$. The quantity of sorbent was some mg/cm^2, and the efficiency of molecular iodine catching was not less than 90%. All used filter materials and analytical filters were developed in the Karpov Institute of Physical Chemistry.

The areas of filter materials, depending on sampler devices, were in the range of $3\,cm^2$–$3\,m^2$. Such variety was needed for the operational assessment of radiation contamination of the air, taking a great number of air samples at different sites and times.

All the sampler devices used, both ground and aircraft, were variations of the measurement technique widely used in the system of routine and research aerosol monitoring in Russia. All measurements were provided by means of standard methods of spectrometry and radiometry of ionizing radiation. In some cases, radiochemical methods were used for the preliminary separation of studied substances from the samples.

18.5 AEROSOL CHARACTERISTICS IN THE FIRST HALF YEAR AFTER THE ACCIDENT

The first aerosol sample of "Chernobyl origin" was taken on April 26 above the western portion of the European part of the Soviet Union. These data were used for the assessment of the ejection of radioactive substances from the destroyed fourth block of Chernobyl NPP in the first day after the accident. Regular sampling above the breakdown reactor began at night from April 27 and 28 from aircraft An-24, which belonged to the Ministry of Defense of the USSR. In addition, for the same purpose, a helicopter was used [7].

Aerosol sampling from the aircraft above the fourth block continued until the beginning of August 1986 [8]. In Figure 18.2, the values of concentrations of beta-radioactive aerosols for this period are shown. As we can see from these data, the concentration of radioactive isotopes in the atmosphere, despite some variations, decreased by 5–6 orders of magnitude after 100 days.

This took place due to decreasing temperature in the destroyed reactor, radioactive decay of short-lived radionuclides, deactivation of the territory, measures for decreasing dust concentration, etc.

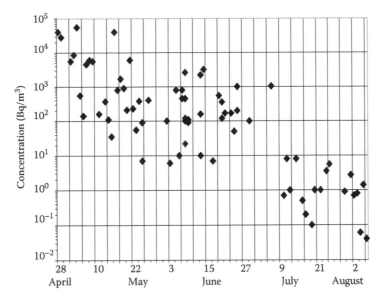

FIGURE 18.2 Concentration of the sum of beta-active aerosols over Chernobyl NPP in April–August 1986. Sampling from aircraft AN-24.

Table 18.2 presents the distribution of radionuclides in aerosols that were sampled above the destroyed reactor of the fourth block of Chernobyl NPP in April–May 1986 from helicopter Mi-8 [7], and aircraft An-30 [1], and An-24 [9].

As we can see from the Table 18.2 data, the composition of the ejection changed strongly not only day by day, but also during the same day (see the results of measurements on May 8, 14–16).

Figure 18.3 presents the dynamics of changes in aerosol concentration of ^{95}Zr, ^{95}Nb, ^{103}Ru, ^{106}Ru, ^{131}I, ^{134}Cs, ^{137}Cs, ^{140}La, ^{140}Ba, and ^{141}Ce sampling from An-24 aircraft 200–300 m above the destroyed reactor on May 8–19, 1986. This period can be described as the beginning of systematical decreasing of ejection of radioactive substances from the destroyed reactor. At this time, the burning of graphite was stopped, and the temperature in breakdown began to gradually decrease. Still, the concentrations measured on May 8 and 16 (some thousands of Bq/m^3) were comparable with the concentrations measured on May 4 and 5 (see Figure 18.2).

Usually, the radioactive composition of aerosols did not correspond to the composition of irradiated fuel at the moment of reactor explosion. For example, on May 8 and 19, volatile radioactive nuclides I, Ru, and Te were measured. On other days, the main contribution to the summary samples of gamma-radiation was made by refractory radionuclides Zr, Nb, and Ce.

Such variations of the ratio of the activities of nuclides in every sample were similar for the aerosols sampled above the territory of Belarus and Russia, which were close to Chernobyl NPP from the north and northeast, and from the research ship in the Atlantic Ocean at the beginning of May [9].

It is also possible to form an opinion about the composition of primary aerosols based on the fallout, which were sampled in a 30 km zone around the atomic station. Almost 90% of the particles were fragments of irradiated nuclear fuel. It consisted of two groups of fuel particles. The composition of one of them was identical to the composition of fuel elements at the moment of the accident. These particles were formed under heat explosion of the reactor. 85% of the activity of the second group consists of refractory ^{144}Ce. The quantity of radionuclides of ^{95}Zr and ^{95}Nb was larger than in irradiation fuel at the moment of the accident, but Cs and Ru were substantially lower. Among the particles, there were some radioactive microparticles that were displaced on a nonradioactive carrier [10]. Radioactive particles of condensation origin have, mainly, nuclides of Cs and Ru [11]. At a

TABLE 18.2
Relative Content of Radionuclides in Aerosol Particles above the Reactor of the Fourth Block of the Chernobyl NPP in April–May 1986

	April			May													
	26	28	29	1	2	3	4	8	8	14	14	15	15	16	16	17	18	19
	An-30	Mi-8	Mi-8	Mi-8	Mi-8	Mi-8	Mi-8	Mi-8	An-24	Mi-8	An-24	Mi-8	An-24	Mi-8	An-24	An-24	Mi-8	An-24
131I	39	31	11.6	81	68.4	100	17.6	19.6	35.4	—	4.2	—	8.8	1.8	1.7	3.2	—	3.7
132Te	34.7								6.9		0.92							6.2
103Ru	5.2	12.7	8.3				13.4	58.7	37.2		8.4		14	10.3	11	8	48.7	36.9
106Ru	1.7	4.6	2.5				5.1	20.6	1.3		1.1		7	4	3	2	13.5	11.4
134Cs	1.3						1.3		0.65	31.8	0.37	3.4	1	2.7	0.35	0.46	13.5	0.62
137Cs	2.6		2.5				1.9		3.3	68.2	1.3	8.2	2.5	5.8	0.87	0.93	24.3	1.5
95Zr	3.9	12.7	34.7	15.8			12.3		5.3		18.3		12.6	16.4	16.8	23.5		8.3
95Nb		16.3	28.1	19	15.8		17.6		2.4		21.6		22.3	21.2	30.2	33.2		13.5
141Ce	3.5	13.6	7.4				18.4		2.6		16.8		11.6	22.7	13.8	12.3		7.1
144Ce	3.9	9.1	5				12.3		1.5		6.2		9.1	15.2	14.2	12.3		7.4
140Ba	4.3								2.6		8.4		8.3		6.6	3.1		2.5
140La									0.9		12.5		2.8		1.5	1.3		1.2

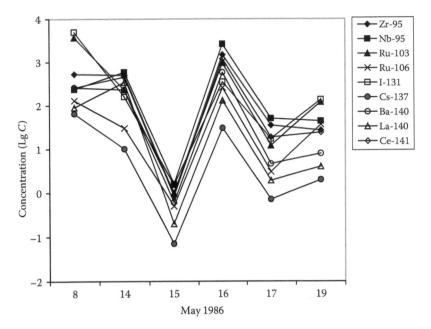

FIGURE 18.3 Concentration of the aerosols containing gamma-radioactivity above the fourth block of the Chernobyl NPP in May 1986.

distance of 15 km from the reactor, the portion of fuel particles depending on azimuth were from 50% to 100%. In the far zone, including beyond borders of the USSR, to the contrary, particles of fuel origin were measured only in a small number of cases (see, e.g., [12]).

Radioactive aerosol particle sizes directly above the destroyed reactor were measured for the first time on May 14, 1986. For this purpose, the gondola of aircraft An-24 was supplied with a package of filters for disperse analysis composition.

It was shown that the smallest size had particle carriers of ^{131}I, [132]Te, and [103,106]Ru. The activity median aerodynamic diameter (AMAD) of aerosols was 0.3–0.4 μm with geometric standard deviation $\sigma = 2.3$–2.5. The sizes of carriers of refractory radionuclides Zr, Nb, La, and Ce were larger: AMAD = 0.7 ($\sigma = 1.6$–1.8). On the same carriers, [134, 137]Cs were discovered. In some countries of the northern hemisphere, at the beginning of May [13–19] practically similar values of AMAD of radionuclides carriers of Chernobyl origin were measured. In these measurements, different types of impactors were used.

In the summer of 1986, the surface of the soil, buildings, vegetable layer, rods, etc., polluted with primary fallout became comparable to or even more important sources of contamination closest to the surface layer of the atmosphere. In this period, it was impossible even quantitatively to describe all mechanisms of forming aerosols, because in the atomic station area and around a 30 km zone very intense deactivation took place, often resulting in aerosol formation. An additional source was the intense transport movement near the Chernobyl NPP.

Figure 18.4 presents the dynamics of aerosol concentration in the summer of 1986.

This correlation was based on measurements of the total gamma-activity of samples provided by scientists of Khlopin Radium Institute from military vehicles, which traveled over Chernobyl NPP by perimeter [20]. As we can see, average aerosol concentrations near the station in the background of substantial variations were comparable with measured concentrations above the destroyed reactor in the second half of May (Figure 18.2).

Measurements provided in July–September 1986 demonstrated that practically all radionuclides were disposed on the same particle carriers. AMAD values of aerosols, averaging for the period of observation, were 2.95 ± 0.60 μm for [95]Zr, 2.89 ± 0.37 μm for [95]Nb, 2.93 ± 0.62 μm for [134]Cs and

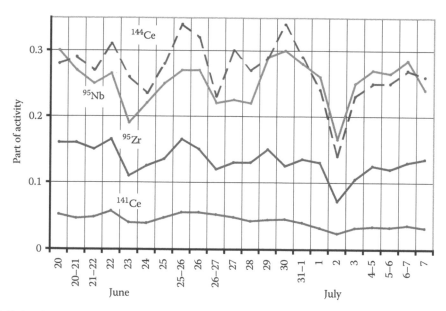

FIGURE 18.4 Concentration of the sum of gamma-radioactive nuclides in the aerosols sampled in June–July 1986 on territory of Chernobyl NPP.

[137]Cs, $2.83 \pm 0.26\,\mu m$ for [144]Ce, and $3.00 \pm 0.28\,\mu m$ for isotopes of Pu. At the same time, isotopes of [103]Ru, [106]Ru were discovered on smaller particle carriers (AMAD = $2.57 \pm 1.01\,\mu m$) with the greatest variation from average size. This can be explained by the desorption of volatile compounds Ru from particle carriers, and subsequently their deposition on aerosol particles of submicron range. On some days, the portions of small disperse fractions of Ru were near 40%.

Observation in the 1986 summer–autumnal period showed that AMAD values of aerosols near the earth's surface began to increase, which was confirmed by later studies.

Table 18.3 shows typical aerosol concentrations, based on samples that were taken in the middle of June in the machine room of the third block (neighboring the destroyed fourth block) by means of packets of fiber and sorption filter materials [21].

It is clear from these data that day concentrations were substantially higher than night concentrations due to extensive activity in the daytime, resulting in the formation of higher dust concentrations.

The sizes of aerosol particles at night (AMAD = $0.75\,\mu m$) were smaller than in the daytime (AMAD = $1\,\mu m$), and all radionuclides were on the same carriers. In the "day" sample A, Ru was found in both aerosol (70%) and gaseous form (30%). But in the "night" sample C, Ru in gas form did not exceed 2%, possibly due to differences in air temperature. At the same time, in samples A and C, the portion of gaseous ^{131}I was 98%.

TABLE 18.3
Concentrations of Radionuclides in the Third Block of the Chernobyl NPP (Bq/m³)

Sample	Sampling Date and Time	[144]Ce	[141]Ce	[131]I	[103]Ru	[106]Ru	[95]Zr	[95]Nb	[137]Cs	[134]Cs
A	July 13, 18:15–19:07	437	156	1	210	—	—	—	40.7	25.9
B	July 13, 18:15–19:07	465	175	—	124	79	334	516	46.7	24.2
C	July 13–14, 19:10–10:10	51.8	18.5	8.5	15.8	12.2	36	54	5.9	2.9
D	July 13–14, 19:10–10:10	54.1	23.3	—	17.4	14.1	46.3	70.4	6.4	2.5

In July–September of 1986, in some of the rooms of the third block, the aerosol concentration (400–$1000\,Bq/m^3$) was not lower than at the same period near the destroyed reactor.

Before building the roof of the "Shelter," it was necessary to ensure the necessity of reconstruction and connecting up the filtration station, which was common for the third and fourth blocks. For this purpose, in August and September of 1986, 11 air samples were taken from a height of 10–30 m above the reactor [22].

Analysis of samples shows that aerosol concentrations (C) consisted of gamma-radiation nuclides and were stable and their values usually did not exceed limit permissible numbers. Short-lived radionuclides were not found in the samples, which proved that no chain reaction was in the fragments of the fuel; the values of C for alpha-active aerosols were in good correlation with C for gamma-radiation (C of the last was 5,006,200 times higher). By measuring ^{242}Cm activity on aerosol particles, assessments were provided for C of ^{239}Pu. It was shown that the aerosol concentration of ^{239}Pu was in the range of 0.1–$0.9\,Bq/m^3$ and was usually higher than the limit permissible value. Despite the fact that at the end of August ^{131}I had to be decayed, in one of the samples it was found completely in the gaseous form ($0.6\,Bq/m^3$). Small amounts of Ru (2%–5%) were also in the gaseous form; AMAD values of aerosol consisting of isotopes of Ce, Cs, Zr, and Nb were the same ($1.06 \pm 0.08\,\mu m$) (September 10, 1986). There were slightly smaller sizes of particles for Ru ($0.84 \pm 0.08\,\mu m$), but according to one measurement it was difficult to judge the difference. The absence of diversity of isotopes by particle sizes showed that the temperature in the higher layer of the destroyed reactor was close to the temperature of outdoor air. The next day, particle sizes were higher ($AMAD = 1.7 \pm 0.1\,\mu m$) and sizes of Ru were the same as for other isotopes.

During the same period, samples were taken 1 m above the earth's surface around 300 m from the destroyed reactor. The aerosol concentrations were on average an order of magnitude lower than near the reactor. Assessment showed that the contribution of aerosols from the reactor to aerosol concentration near the earth's surface was not more than a few percent.

Important information about the physicochemical forms of radioactive aerosols, sampled near the surface not far from the destroyed reactor, was obtained by analysis of radionuclide composition. Radionuclides that made the larger contribution in the summary gamma-radiation of samples can be divided into two groups.

In the first group are refractory nuclides Zr, Nb, La, and Ce, which despite the high temperature during the accident remained in the destroyed fourth block. The second group—^{103}Ru, ^{106}Ru, ^{134}Cs, ^{137}Cs—includes volatile compounds, which condensed on aerosol particles in the air.

From the data in Figure 18.5, we can see that the contribution of Ru and Cs in samples of 20–22, 25–26, and 30–31 of August, 1986 did not exceed 20%; therefore, on these days, the contribution of fuel aerosol was highest.

The sample of September 2 was different. Here, the contribution of Ru and Cs was near 65%. It has been mentioned very often in the literature [23–25] that the radiation of some highly radioactive particles sampled even at a great distance from Chernobyl consisted of a large portion of Ru. On other days there was a mixture of fuel and condensation particles in the atmosphere.

Only in October–November 1986, when a substantial portion near the destroyed reactor was cleaned up from radioactive fallout, sealed by gravel, sand, and concrete, aerosol concentrations decreased by an order of magnitude and reached relative levels, which were observed in July–September of 1986.

18.6 GASEOUS COMPONENTS I, TE, AND RU IN THE ATMOSPHERE

The products of the Chernobyl accident in the atmosphere were not only included in aerosols but also in the form of gases. During the sampling in May 1986, both aerosol and gas compounds such as I, Te, and Ru were measured. For the detection of gases, multilayer filter material SFM-I, consisting of an aerosol layer, upper facing filter material, and two sorption layers, was used. The ratio

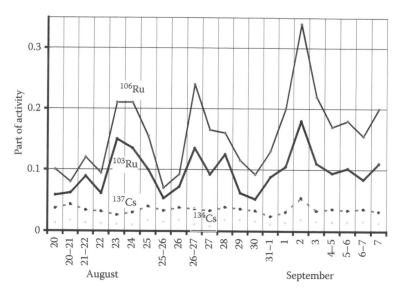

FIGURE 18.5 Relative contribution of radionuclides with volatile components in total gamma-radiation of the aerosols sampled from ground layer of atmosphere (1 m above a surface of ground) near destroyed reactor in 1986.

between aerosol and gaseous components of radionuclides was determined according to the distribution in layers of the package. It was assumed that the sorption deposition of I, Te, and Ru was near 100%.

The layer-by-layer distribution of volatile nuclides in SFM-I, which was used for air sampling above the destroyed reactor, is present in Table 18.4.

In middle of May 1986, in gaseous form, it was 0.6%–13% ^{103}Ru, 2.7%–16% ^{106}Ru, and 1%–8% ^{132}Te.

The similarity of distribution for both isotopes of Ru showed that they are present in air in the same gaseous compounds, and their absence in the second layer shows that sorption was good. Even with high filtration velocity, the efficiency of gaseous compounds of Te was also good.

Radioactive iodine was found in the second sorption filter. So the gaseous compound in which it was, sorbed worse than that with Ru and Te. Only in samples taken above the reactor on May 8 and 14 (Table 18.4) was iodine found in the upper facing and in the first sorption layer. This can be explained by the fact that iodine was in the form of vapors of I_2. These measurements point out that from May 15 to 19, new gaseous iodine compounds were in the atmosphere with smaller coefficients of dynamic sorption, that is, possible organic forms (CH_3I, etc.).

Measurements of radioactive components of I and Ru were provided also at a great distance from Chernobyl NPP [26] from aircraft at a distance of 100–300 km north of Chernobyl above Ukraine, Belarus, and Russia at a height of 1200 m, and from a research ship 4000 km in the Atlantic ocean. The results of measurements of concentrations of ^{131}I and the ratio of aerosol and gaseous components of I and Ru are shown in Table 18.5.

As we can see from Table 18.5, gaseous ^{131}I was present in all samples, and gaseous Ru in the majority of samples. Based on date and site of sampling (near 4000 km from Chernobyl), samples from May 4 to 5 belong to the first ejection of radioactive materials from the exploded reactor. Measurements provided in May 1986 showed a weak dependence of gaseous ^{131}I from the distance of the Chernobyl NPP—the lowest near the reactor and the highest in the Atlantic, North, and Baltic Sea.

Unfortunately, in the majority of countries around the world, atmosphere monitoring was provided only by aerosol filters [5]. It is, however, necessary to pay attention to the measurements

TABLE 18.4

Distribution (%) by Layers of Material SFM-I of I, Te, and Ru in Samples

Sampling Date (1986)	Component	Aerosol	Upper Facing	SFM-I Material Layers	
				First Sorption Layer	Second Sorption Layer
May 8	—		?	?	?
	^{131}I	69.4	19.4	11.2	—
	^{103}Ru	88.7	4.2	7.1	—
	^{106}Ru	84.0	5.2	10.8	—
	^{132}Te	91.7	3.3	5.5	—
May 14	—		0.25[a]	5.1[a]	2.5[a]
	^{131}I	72.5	19.8	7.7	—
	^{103}Ru	97.4	1.0	1.6	—
	^{106}Ru	96.0	2.0	2.0	—
	^{132}Te	99.0	—	1.0	—
May 15	—		0.29[a]	4.7[a]	2.7[a]
	^{131}I	42.4	19.2	18.7	19.7
	^{103}Ru	99.4	—	0.6	—
	^{106}Ru	97.3	—	2.7	—
May 16	—		0.26[a]	2.4[a]	2.7[a]
	^{131}I	41.4	34.5	20.4	3.7
	^{103}Ru	96.1	1.9	2.0	—
	^{106}Ru	97.1	—	2.9	—
May 17	—		0.28[a]	5.0[a]	4.2[a]
	^{131}I	28.0	41.0	28.0	3.0
	^{103}Ru	97.7	—	2.3	—
	^{106}Ru	100	—	—	—
May 19	—		0.24[a]	5.0[a]	3.5[a]
	^{131}I	8.9	64.5	22.2	4.4
	^{103}Ru	87.0	7.0	6.0	—
	^{106}Ru	87.8	5.7	6.5	—

[a] Quantity of sorbent in layer (mg/cm^2).

of gaseous components of the Chernobyl accident provided in Europe, United States, and Japan. Summary data of the global atmosphere monitoring phase composition of ^{131}I at different periods after the Chernobyl accident [27] are presented in Figure 18.6.

From the majority of measurements performed in the first 29 days after the accident, it is clear that gaseous I concentration was higher than aerosol concentration, and 1 month after the accident (data from Japan and United States) it was still higher than 0.7.

After deposition on the particle I compounds can leave its surface due to desorption. If desorption probability is close to the duration of sampling, the second desorption can take place between different parts of the sampler. Therefore, experimental data on aerosol–gas ratio can be higher than the actual value [28].

Taking into account the correction on desorption, it was found that the real portion of gaseous iodine was in the range of 0.33–0.63 with a mode of 0.48 [28]. In [29,30] it was found that ^{133}I was in gaseous form in the same proportion with aerosols as ^{131}I, and in the period from April 29–May 2 the gaseous fraction was 60%–80%. From this, the important conclusion can be drawn that the behavior of ^{129}I and ^{135}I in the atmosphere was similar. It was also found that ^{132}Te was in the air in gaseous form with 10% of its general concentration, which was the same

TABLE 18.5

Volume Activity of ¹³¹I, Phase Composition I and Ru in the Atmosphere above Ukraine, Belarus, and Russia and from a Research Ship in May 1986

Sampling Date (1986)	Sampling Site	Sampling Velocity (m/s)	Time of Sampling (h)	Concentration 131I (Bq/m³)	Portion of Gaseous Component (%)		
					¹³¹I	¹⁰³Ru	⁰⁶Ru
May 8	Above the reactor	0.9–1.2	1	5000	30	11.3	16
May 14				160	28	2.6	4
May 15				1.5	58	0.6	2.7
May 16				350	59	3.9	2.9
May 17				17	72	2.3	
May 19				136	91	13	12.2
May 14	Ukraine, Belarus, Russia, north of Chernobyl	0.9–1.2	1–2	0.34	57	0.7	
May 16				1.5	46	2.2	2.7
May 17				1.7	77	4	3.5
May 19				0.81	90	8	12
May 4–5	Atlantic, North, and Baltic Sea	1.0	31	0.008	79		
May 5–6			25	0.015	84	23	
May 6–7			18	0.034	90	18	
May 7–9			44	0.031	73	2	7
May 10–11			27	0.001	79		

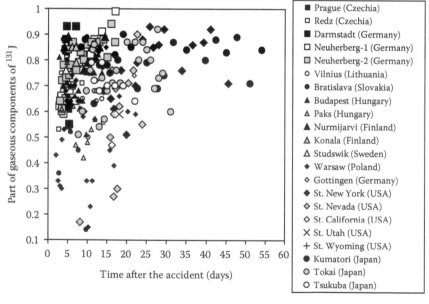

FIGURE 18.6 Results of global monitoring of a gaseous component of ¹³¹I of Chernobyl origin. (From Ogorodnikov, B.I., *At. Energy*, 93, 42, 2002.)

as the results of [26]. After the first days of the accident, both aerosol and gaseous forms of ^{137}Cs were also found [31].

So in the process of monitoring the products of the Chernobyl accident and the new previously unknown data on physicochemical forms of existence in the atmosphere, some radionuclides such as Te, Ru, Cs, and I isotopes were found.

18.7 CHARACTERISTICS OF RADIOACTIVE AEROSOLS NEAR THE EARTH'S SURFACE

Extensive territories were polluted as a result of fallout on the earth's surface of gaseous–aerosol products of the accident at Chernobyl NPP. The most polluted territories were close to the Chernobyl station, where the density of pollution by ^{137}Cs was more than 3.7 MBq/m^2 (100 Ci/km^2).

After fallout was stopped from the reactor, the main mechanism of producing aerosol was the secondary raising (resuspension) from polluted surfaces of the soil, trees, buildings, etc. Concentration of these secondary aerosols depends on many factors. Among them are meteorological situation, characteristics of the surfaces (presence of vegetable layer, buildings, surface humidity and type of soil, intensity of mechanical activity on the surface, presence of the snow layer), physicochemical properties of pollution, time after forming of primary pollution, etc.

The value that characterizes the danger of radioactive fallout, as a source of aerosols, is the resuspension coefficient:

$$K_r = \frac{C}{S}$$

where
  C is aerosol concentration (Bq/m^3)
  S is the density of pollution (Bq/m^2)

Because S depends only on the decay constant of radionuclides, the value of C is in the same way as informative as K_r (in the case of absence of aerosol transport from other regions).

As a result of extensive studies on secondary aerosols of Chernobyl origin, it was found that

- The concentration of secondary aerosols decreases three to four times approximately linearly at a height of up to 15 m above the soil [32].
- [137]Cs and [144]Ce concentrations change in the region by more than an order of magnitude, and concentration decrease takes place faster than it should according to radioactive decay.
- The intensity of fallout decreases with increasing distance from roads and sites of agricultural activity [33].
- Aerosol concentration increases linearly with the speed of wind [34].
- Values of K_r decrease with time after primary fallout according to exponential law with a time constant of 0.02–0.12/month [35].
- Strong winds resulted in aerosol transport from the zone of pollution by hundreds of kilometers, which resulted in an increase in aerosol concentration on remote territories by tens–hundreds of times [36].
- The values of K_r measured in the same periods, but at different points of Europe, differ from Chernobyl data in the range of order of magnitude, and variations are accidental in nature. In 1994, K_r was around 2.33×10^{-10}/month [35].

After completion of the "Shelter" building, measurement of aerosol disperse composition around the Chernobyl NPP continued at some points in a radius of 0.5–5 km around the reactor. All samples were taken at a height of 1.5 m from the surface. The MMF was used for sampling [37].

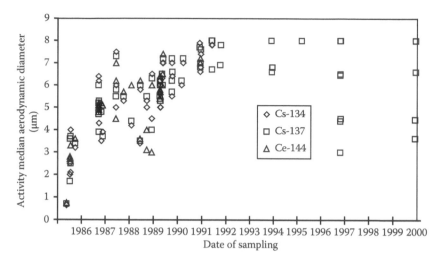

FIGURE 18.7   AMAD (in µm) of aerosols ^{134}Cs, ^{137}Cs, and ^{144}Ce in 5 km zone around the Chernobyl NPP in 1986–2002.

More extensive data on the disperse composition of aerosol carriers of radionuclides until April 2002 were obtained by analysis of ^{137}Cs. ^{144}Ce disperse composition was measured only until the end of 1992, when this radionuclide decayed practically completely.

Figure 18.7 shows the value of AMAD of ^{137}Cs and ^{144}Ce for all periods of monitoring, where AMAD is in the range of 3–8 µm. These data are comparable with the data presented in Ref. [38]. It should be noted that in the majority of cases, the sizes of particle carriers of ^{90}Sr and plutonium isotopes are the same as for ^{137}Cs.

From the presented data, it is clear that average AMAD increased from the summer–autumn of 1986 until 1991–1992, after which it became constant until now. It is difficult to find a reason for this. The sizes cannot be different from the sizes of nonactive aerosol, which were developed from resuspension from surfaces of soil, vegetable, roads, buildings, etc., because of natural causes. Such was the situation in 1992. However, the continuation of activity near the Chernobyl NPP (transportation of soil, transport movement, etc.) led to changes in aerosol concentration and disperse composition, which was observed frequently [39].

18.8   FOREST FIRES IN THE EXCLUSIVE ZONE

Forest fires taking place on territories polluted with radionuclides are an additional factor of radiation danger, because under these circumstances aerosol particles that consist of radioactive substances are formed. Their concentration can substantially exceed what was typical for this site before the fire. Physicochemical characteristics of aerosols of fire genesis can be different from "background" aerosols in the atmosphere. Smoke trains are dangerous for people who take care of the fire, as for people located at great distances from the fire. Transportation of radioactive smoke leads to a redistribution of radionuclides between "dirty" and "clean" territories.

Nearly 600 different fires were observed from 1993 to 2002 in the 30 km zone of the Chernobyl NPP. For the first time, attention was paid to the radiation effects of fires after the hot and dry summer of 1992 [40]. Then radioactive substances, moving with the smoke trains from Chernobyl, were observed as far as 500 km in Lithuania [41] and Sweden [42]. Monitoring of the radiation situation at 30 points in the 30 km zone of Chernobyl showed that even at a distance of 5–10 km from forest fires on relatively "clean" territories, the concentration of radioactive aerosols increased up to 10–100 times.

Radioactive aerosol concentration growth was caused by a sublimation of light volatile ^{137}Cs from the firing zone. It was found experimentally that the growth of concentration at a distance of some kilometers from the fire was more than 2 orders of magnitude. Assessments show that if the density of pollution of the surface is $7\,Ci/km^2$ ($2.6 \times 10^5\,Bq/m^2$) in the case of a fire close to the earth's surface, the concentration of radioactive Cs can exceed the limit permissible values [40].

A study that was provided in the framework of the experimental fire in 1993 suggested that ^{137}Cs concentration increased 1000 times, particle size distribution was bimodal (75% AMAD > $10\,\mu m$, 25% AMAD = $0.4\,\mu m$), and the water-soluble form of Cs increased by 38% [41]. It should be noted that the primary radioactive fallout of the accident contained 99% of Cs in nonsoluble chemical form. This means that fires are responsible not just for concentration growth. We have a new sub-micrometer compound of spectra size of particles much more dangerous from the point of view of dosimetry of internal irradiation. The amount of water-soluble Cs becomes more substantial. Ultrafine components of aerosol make a substantial contribution in remote transfer. For example, ^{137}Cs concentration in Vilnius (Lithuania) during the fire of 1992 increased 100 times [41].

18.9 AEROSOLS OF THE "SHELTER"

The object "Shelter" is the aggregate of structures that sealed radioactive sources outside the active zone of the reactor of the fourth block after the accident. It was completed by November 30, 1986. As a result, a unique system was created, comprising already destroyed and rebuilt constructions, which make it generally stable without any guarantee for the destruction of some of its elements. "Shelter" is a temporary construction for the localization of nuclear fuel and radioactive materials.

"Shelter" was connected with engineer communications and was equipped with devices for fuel diagnostics, neutron sorption solutions, dust suppression, contrafire lines, etc., which present a complex of the security of "Shelter" [43].

Years of operation of "Shelter" showed that the goal of the "Shelter" building was correct. Still, the difficult conditions under which it was built did not allow the construction of a truly hermetic structure. There are cracks, apertures, and technological openings for cables and pipe-lines in the structure. Natural ventilation of "Shelter" takes place through the ventilation tube in the third and fourth blocks.

For characterization of "Shelter" as a source of radioactive aerosols and assessment of its influence on environment, the following information is required: (1) mechanism of formation of aerosols inside "Shelter"; (2) physicochemical characteristics of aerosol particles in different parts of the object; and (3) values and dynamics of ejection of aerosols in the free atmosphere.

18.9.1 TYPES OF AEROSOLS

At the present time, radioactive aerosols of "Shelter" are formed from dust, present in its compartments (the concentration of uranium in dust is assessed as 5–10 t [2]), and from fuel-containing materials (FCMs). FCM is destroyed due to radioactive decay, strong fields of radiation, variations of humidity and air temperature, and also building activity and drilling work.

During the observation period, two types of radioactive aerosols were found in the "Shelter." First, aerosols of disperse origin consist of particles of irradiated fuel and, usually, with only small amounts of volatile isotopes of Cs, Ru, Te, I. Second, condensation aerosols were formed due to the absorption of these and other radionuclides on the particle carriers.

The fuel particles also consist of two groups. To the first group belong large particles (average diameter near $30\,\mu m$), consisting of grains of UO_2. To the second group belong smaller (diameter some micrometers) particles, formed after graphite burning, fuel oxidation, and its partial melting. These particles consist of small amounts of fragmental radionuclides, especially Cs.

18.9.2 AEROSOL CONCENTRATION INSIDE THE "SHELTER"

In the first years of the work of the "Shelter," the main cause of aerosol forming was drilling work inside the structure. Small radioactive particles were present in large amounts in boring liquids, on the floor, which led to high aerosol concentration.

During the existence of "Shelter," a large number of aerosol samples were taken in room 207—the corridor was divided into three sections. After the completion of drilling activity, some of the boreholes remained open and became channels of intake for air and water from sites with high concentrations of radionuclides products of the accident. According to the measurements provided in 2000, the dose rate of gamma-radiation in 207 changed in the range of 0.1–10 mSv/h.

The results of systematic aerosol sampling in room 207/5, taken every 1 h in 1988–1991, were very significant. Figure 18.8 shows variations of summary aerosol alpha-activity. After the bore equipment was cut out, their concentrations were usually at a level of 0.03 Bq/m³, and when the boring work started it increased by 1–2 or even 3–4 orders of magnitude.

A large amount of dust formed during boring activity, and its deposition on all surfaces was very substantial even for other types of activity. For example, when the metal platform was cut, the concentration of alpha-active aerosol increased by 2 orders of magnitude, up to 7 Bq/m³. The concentration decreased to normal level only 8–10 h after completion of the work.

The secondary long monitoring of radioactive aerosols in room 207/5 took place from November 28 to December 26, 2000, after Chernobyl NPP was closed. During those days, 20 samples were taken with 2–3 h exposure [44]. As can be seen from Figure 18.9, ^{137}Cs concentration varies from 0.2 to 20 Bq/m³. There was a positive correlation between ^{137}Cs concentration and long-lived beta-emitters. During monitoring, some splashes of radioactive aerosol concentration were measured, but all of them were lower than during the drilling activity.

As can be seen from Figure 18.9, the highest concentration of radioactive aerosols was measured on December 9. Besides ^{137}Cs, also ^{134}Cs, ^{154}Eu, ^{155}Eu, ^{90}Sr, ^{241}Am, and ^{60}Co were found in aerosol samples. Taking radioactive decay into account, the ratio of activities of ^{134}Cs/^{137}Cs was 0.48, which is close to the number at the time of the accident [45].

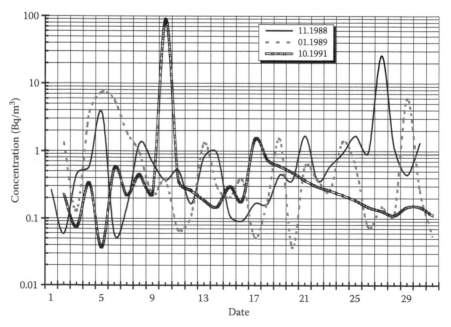

FIGURE 18.8 Concentration of alpha-active aerosols during drilling works in room 207/5 of the "Shelter" (1988–1991).

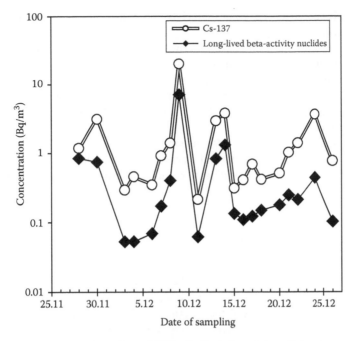

FIGURE 18.9 Contents of aerosol carriers of ^{137}Cs (in Bq/m^3) and long-living beta-active radionuclides (in pulses per second per cubic meter) in room 207/5 of the "Shelter" in 2000.

^{241}Am and ^{154}Eu concentrations were 0.05 Bq/m^3, and for ^{60}Co it was three times higher. The ratio ^{137}Cs/^{90}Sr1. 15 was also close to that calculated for the fuel of the fourth block.

Besides monitoring aerosol characteristics, simultaneous measurements of the daughter products (DPs) of radon and thoron also took place. From the data presented in Figure 18.10, it is clear that, unlike accident aerosols, the concentration of natural radioactive aerosols in room 207/5 continuously declined in the second decade of the month and increased in the third.

From a comparison of Figures 18.9 and 18.10, it is clear that the dynamics of changes of concentrations of accident products and natural radionuclides are different. The cause of the high concentrations of the first type of aerosols is connected to working operations, and for the second

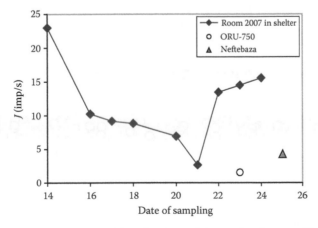

FIGURE 18.10 Rate of count of beta-particles of radon and thoron daughters on 50th min after the finish of aerosol sampling in room 207/5 of the "Shelter," and in two points near Chernobyl NPP in December 2000.

it is due to gas emissions from building materials and constructions of the destroyed block, air exchange, and atmospheric pressure.

One important cause of sharp increasing aerosol concentrations in "Shelter" is fire inside the building. Over a period of 16 years, in the room of "Shelter" seven fires were recorded. During the fire: (1) the formation of radioactive aerosols takes place, and these aerosols are transported with smoke and hot air; and (2) inside the "Shelter" a very intensive movement of air mass takes place, which overwhelms the usual ventilation flow. This results in additional ejection of radioactive aerosols into the free atmosphere.

During the 4h fire on January 14, 1993, around 33 MBq of radioactive material was ejected additionally into the atmosphere. The normal ejection for that period was approximately 1.4 MBq. The average concentration of gamma-radioactive nuclides through the ventilation stack increased 140 times.

18.9.3 Aerosol Concentration in the "Bypass" of the "Shelter" in 2002–2009

Construction of the "Shelter" had been completed in November 1986. Thereafter, monitoring radioactive aerosols emission was conducted in the "Bypass" system connecting the former central hall of the fourth power-generating unit with the high level elevated ventilation stack VT-2. The last remaining power-generating unit of the Chernobyl NPP was stopped in December 2000, and decommissioning of the station started. All that required carrying out more profound monitoring of aerosols coming out of the "Shelter" into its environment.

Some 400 aerosol samples were taken at the "Bypass" system during 2002–2009 with the help of three-layer filter packages. As a rule, concentration of total β-activity (Σβ) of aerosol carriers of Chernobyl accident products was in the range of 1–10 Bq/m³ as shown in Figure 18.11, about 30% of the activity being accounted for ^{137}Cs.

It was established that significant influence on the aerosol content in the former central hall of the fourth power-generating unit had been exercised by wind presence in the environment. So, Σβ = 165 Bq/m³ (Figure 18.11) was registered on December 8, 2003, when maximal wind gusts reached 12–13 m/s. An identical situation was observed on April 8, 2008, when Σβ = 100 Bq/m³ at wind gusts of 14 m/s.

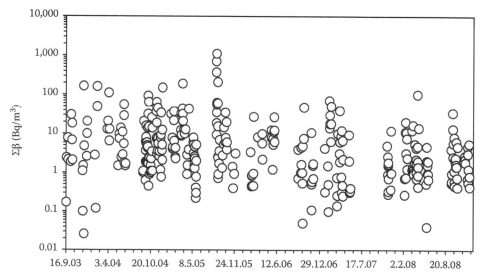

FIGURE 18.11 Concentration of total β-activity (Σβ) of aerosol carriers of Chernobyl accident products at the "Bypass" system of the "Shelter" in 2003–2008.

Considerable increase in $\Sigma\beta$ took place during intensive industrial activity in the premises of the "Shelter" object. At the end of August—beginning of September 2005—the values of $\Sigma\beta$ reaching 500–1000 Bq/m^3 during welding and slotting works were observed. Several highly active β- and γ-radiating black colored particles sized about 1 mm were discovered on a filter with the sample taken on September 1, 2005. It follows from the fact that construction works can bring about a sharp rise in total concentration of radioactive aerosols (including "hot" particles) in the area of conducting process operation, their transfer through premises and coming into the flow emitted via VT-2 stack into the atmosphere.

This fact was confirmed half a year after the sampling made on February 14, 2006. Deposit on the filter had foxy color. Nothing of the kind had been ever observed during previous monitoring (about 200 samples). Filters, as a rule, looked gray. Correlation between ^{137}Cs and $\Sigma\beta$ appeared to be unusual, as well. The value of 0.93 showed that the aerosol was enriched in ^{137}Cs. Before that the said correlation was as a rule in the range of 0.2–0.7. It turned out that intensive welding works were carried out on February 14. It brought about appearance in the air of Cesium and two of its oxides with sublimating temperature of about 650°C. Then vaporous substances were condensed on atmospheric condensation nuclei with the size of 0.2–0.4 μm and were delivered by air flow into different premises of the "Shelter" object.

18.9.4 Aerosol Transport from the "Shelter" into the Atmosphere

The assessment of the influence of the "Shelter" on the environment is complicated and many factors are problematic. Radiation, including the aerosol situation, both inside and outside of the "Shelter" is stable. Contribution of the "Shelter" as a source of radioactive aerosols is only some percents from the level permissible for the normal working of a nuclear block of 10^3 MWt.

During the experimental study provided in 1996–2000 [46], parameters of air flows in the apertures in the walls and characteristics of radioactive aerosols transported into the atmosphere were determined by means of aerosol filters.

It was found that the maximum aerosol concentrations of ^{137}Cs, ^{90}Sr, and $^{239+240}$Pu in the exits of apertures are close to permissible values. In the cold time of the year, the transportation of radioactive aerosol from the "Shelter" was substantially higher than during the warm period. In summary, the rate ejection of 290 Bq/s (7×10^9 Bq/year), part of ^{137}Cs was 220 Bq/s (76%), for ^{90}Sr—58 Bq/s (20%), and for $^{239+240}$Pu—1.1 Bq/s (0.38%). The rest of the activity was determined by ^{60}Co, ^{106}Ru, ^{125}Sb, ^{134}Cs, ^{144}Ce, ^{154}Eu, ^{155}Eu, and ^{241}Am. Data presented in Ref. [47] are smaller, but in this study a plane table and not a filter was used for sampling.

In August 2000, the concentration of aerosols of ^{137}Cs was in the range of 0.01–0.06 Bq/m^3 and exceeded the concentration of radiocesium near the earth's surface on the territory only slightly [48].

18.9.5 Aerosol Dispersity in Premises of the "Shelter"

The first study of disperse composition of aerosols inside the "Shelter" was provided in 1988 [49]. Analysis of 180 particles with about 80% of aerosol activity, collected in filters, showed that aerosol particle size distribution can satisfactorily be described by a logarithmic normal law with a median diameter of 5 μm. The results of the measurements of disperse composition of aerosols provided by the MMF in room 207/5 [44] are presented in Figure 18.12.

The dispersity of carriers of accident products was in the range of AMAD 0.5–6 μm. Such a wide range of sizes was probably because aerosols were produced from different sources, with different mechanisms of formation. Large particles were probably of disperse origin, that is, they become small, erode, and disperse some materials. Submicrometer-sized particles can be produced in the "Shelter" by dispersion of liquids, including radioactive solutions formed by condensation of moisture or sediments. After evaporation of moisture from micrometer-sized or even larger particles,

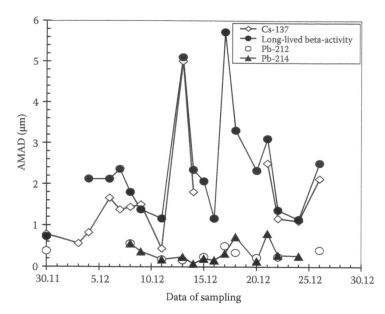

FIGURE 18.12   AMAD of aerosol carriers of ^{137}Cs, long-lived beta-active nuclides, [212]Pb, and [214]Pb in room 207/5 of the "Shelter" in December 2000.

submicrometer particles with small amounts of even dry particles can be formed. It should be noted that during welding work, sizes of radiocesium were AMAD = 1.02 μm.

Aerosols of radon and thoron daughters belong to condensation aerosols. They form due to deposition of atoms of metals (Po, Pb, Bi) that arise during radioactive decay of gaseous maternal substances on very small aerosol particles, called condensation nuclei. These aerosols can be formed both inside the "Shelter" due to emanation of radon and thoron from building constructions and from the free atmosphere. The AMAD values of such particles are in the range of 0.08–0.8 μm.

Important data were received on December 9 in room 207/5, when aerosol concentration was the highest. Three methods were used: beta-radiation radiometry, gamma-spectrometry, and beta-spectrometry. It was found that all radionuclides (^{90}Sr, [134]Cs, ^{137}Cs, [241]Am) were disposed on the same particles with AMAD values near 1.5 μm. The ratio of ^{137}Cs/^{90}Sr was near 1.15 and was practically equal to the data presented in Ref. [45] for average nuclear fuel of the fourth block at the moment of the accident.

During the study of ejection from the "Shelter" from 1996 to 2000, a five-cascade impactor of the Institute of Biophysics (Russia) and an Andersen impactor PM-10 were used.

A majority of the measurements were provided for [137]Cs (22 samples) and [241]Am (16 samples). It was found that the AMAD for aerosols of ^{137}Cs was 2.9 ± 1.5 μm, and for [241]Am it was 1.4 ± 0.4 μm. No systematically measured AMAD values of aerosols of ^{90}Sr and [239 + 240]Pu were in the range of 3.1–4.5 and 6.0–9.3 μm, respectively. It should be noted that these data are similar to the data obtained by the MMF. Unfortunately, in Ref. [46] there were no comments on the substantial difference in the size of Pu and Am, which is a DP of [241]Pu.

18.9.6   RADIOACTIVE AEROSOL DISPERSITY IN THE "BYPASS" SYSTEM OF THE "SHELTER" IN 2002–2010

Over 300 samples taken on three-layer filters at the "Bypass" system enabled us to determine the size distribution of aerosols. It was established that the radioactive products of Chernobyl accident had been leaving the "Shelter" on particles with AMAD > 1 μm and DPs of radon and thoron—on submicron particles.

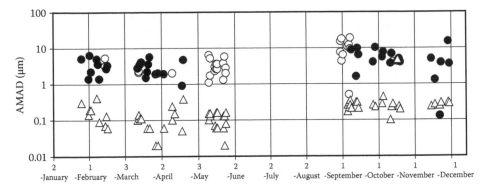

FIGURE 18.13 AMAD of aerosol carriers of $\Sigma\beta$-radiating nuclides—Chernobyl accident products (O, sampling during the dust suppression operation; ●, sampling in absence of the dust suppression operation), and of radon and thoron daughters (Δ) at the "Bypass" system of the "Shelter" in 2005.

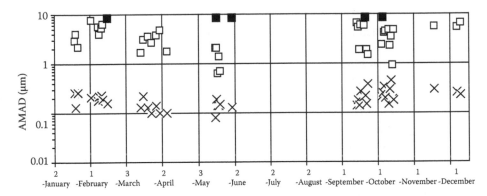

FIGURE 18.14 AMAD of aerosol carriers of Chernobyl accident products (□) and DPs of radon and thoron (×) at the "Bypass" system of the "Shelter" in 2010: samples with AMAD over 8 μm are displayed by mark.

Processing of results demonstrated that in 2002–2006 radio nuclides—products of the accident practically always had the AMAD in the range of 1–10 μm at s values from 1.1 to 3.5. For example, the AMAD values for 80 samples taken in 2003–2004 were presented in logarithmic-normal coordinates. It was established that 50% interval fell at the AMAD = 3.8 μm and 67% of all samples had the AMAD from 1.7 to 8.5 μm. Fractionating of ^{90}Sr, ^{137}Cs, ^{154}Eu, ^{241}Am by particles of different size was not observed.

At the same time DPs of radon and thoron had the AMAD in the range 0.02–0.8 μm, the 50% interval falling to the AMAD = 0.15 μm. Consequently, radio nuclide carriers of the Chernobyl genesis had appeared as a rule because of dispersion and DPs of radon and thoron as a result of condensation. The similar data were received in 2005 and 2006 (Figure 18.13).

Particles with AMAD over 2 μm continued, as a rule, being radio nuclide carriers—Chernobyl accident products in the "Bypass" system in 2010 (Figure 18.14). AMAD exceeded 5 μm in the majority of samples and in four samples exceeded even 8 μm, DPs of radon and thoron being associated, as before, with aerosol particles with AMAD in the range of 0.08–0.4 μm.

18.10 RADIOACTIVE AEROSOLS CLOSE TO THE SURFACE LAYER OF THE ATMOSPHERE NEAR THE "SHELTER"

After 1992, monitoring radioactive aerosols in the vicinity of the "Shelter" has been continued by specialists of Nuclear and Radiation Safety Department of ISTC "Shelter" of National Academy of

Science of Ukraine. Round the clock sampling was executed at all three aspiration installations AU, located in the northern (AU-1), the northwestern (AU-2) and the southern (AU-3) sectors of the area. FPP-15-1.5 filtering material with the surface $0.8\,m^2$ was used and exposed for 2 weeks. Air flow rate was 400–$500\,m^3/h$. The material was placed on a horizontal cage screen, its axis being situated $1\,m$ above the ground. A small roof made of roofing iron was erected over the installation at the height of $1.8\,m$ to protect it from rain and snow. The installation was enclosed in a metal mesh. After sampling had been over the filtering material was delivered to Chernobyl for conducting gamma-ray spectrometric and radiochemical analysis.

A convincing illustration of the influence of "Shelter" on the composition and concentration of radioactive aerosols near Chernobyl NPP was the fire in room 805/3 of January 14, 1993 [43]. During the fire, concentration of $\Sigma\beta$ in VT-2 stack exceeded the usually observed level by approximately 30 times.

Figure 18.15 presents the monthly results of monitoring close to the earth's layer of air in the area of AU-1, AU-2, and AU-3 in 2000.

In 2000, the aspiration in the area of AU-1 and AU-2 showed the same concentration values for ^{137}Cs—1–$3\,mBq/m^3$ and for ^{241}Am—0.02–$0.06\,mBq/m^3$. During wintertime, the values were a little lower than in the summer. In the south filter installation (AU-3) they are usually higher than for the north and northwest, which can be explained only by the higher density of pollution of the soil in this site.

A substantial concentration growth took place in the spring of 2000, when at a distance of 300–$500\,m$ from the "Shelter" building activity took place. Figure 18.15 reveals that in this period aerosol concentration increased substantially. It should also be mentioned that the $^{137}Cs/^{241}Am$ ratio increased at the beginning and end of 2000 (Figure 18.16)

This can be explained as a result of changing the ratio between contributions of aerosols due to dust raising in the territory and release from the "Shelter" [47–49]. In the limit situation, when dust raising is minimum, for example, in the case of a stable snow layer, the $^{137}Cs/^{241}Am$ ratio can be close to 200, which is typical for aerosols ejected from cracks in the walls [46].

The average annual concentration of aerosol carriers of total beta-radiating nuclides dynamic in 1993–2008 is given on Figure 18.17. As products of the Chernobyl accident with a

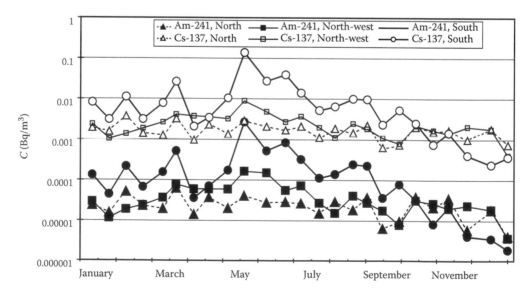

FIGURE 18.15 Concentration of radioactive aerosols of ^{137}Cs and ^{241}Am in ground layer of atmosphere near the "Shelter" in 2000. Sampling was executed in three points.

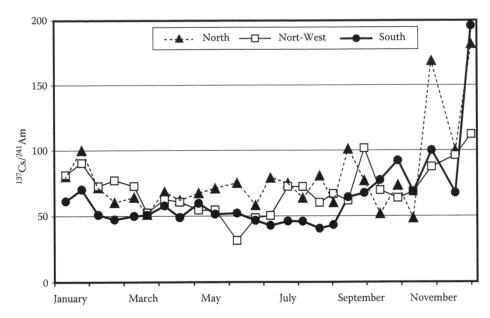

FIGURE 18.16 Ratio of activity ^{137}Cs/^{241}Am in aerosols of ground level of atmosphere near the "Shelter" in 2000.

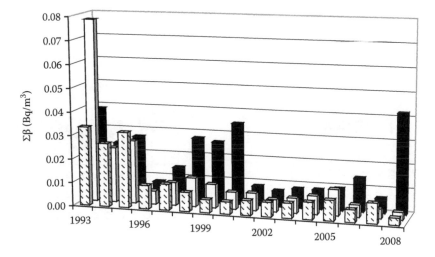

FIGURE 18.17 Average annual aerosol concentration $\Sigma\beta$ of radiating nuclides in the area of the northern (\\\\), the northwestern ($\square$) and the southern ($\blacksquare$) installations in 1993–2008.

half-life time of less than 300 days had practically disappeared by 1993, $\Sigma\beta$ comprises ^{90}Sr + ^{90}Y, ^{137}Cs and ^{241}Pu only.

As it is apparent from Figure 18.17, a systematic decrease with the course of time of beta-active aerosol carriers of Chernobyl accident products took place. The sharpest drop (approximately, three times) happened in 1993–1995. In the subsequent 13 years, the concentration in the areas AU-1 and AU-2 did not exceed 0.01 Bq/m^3 and did not descend below 0.005 Bq/m^3. A more sophisticated dynamic was observed in the area of AU-3. Here the concentration increased again to 0.03 Bq/m^3 after a drop in 1996–1997. Then in 2001 it again dropped down to 0.005–0.007 Bq/m^3 and stayed at this level indicative for area of AU-1 and AU-2 for 7 years. But in 2008 a sharp rise of the concentration up to $0.06 \, Bq/m^3$ was observed in the area of AU-3. This value proved to be the highest for 15 previous years. A higher concentration of about $0.08 \, Bq/m^3$ took place in the area of AU-2 only in 1993, as it follows from Figure 18.17.

A detailed concentration dynamic of aerosol carriers ^{137}Cs and ^{241}Am as a result of the 15 day sampling in 1993–2007 is presented in Figure 18.18. It is evident that ^{137}Cs and ^{241}Am concentrations are very close in the area of AU-1, AU-2, and AU-3 positions. But some considerable fluctuations (up to 2 orders of magnitude) are observed sometimes at a general tendency of concentration decline.

Although the nuclear-physical origins of ^{137}Cs and ^{241}Am are different (the first one is a product of uranium fission, the second appears at beta-decay of ^{241}Pu ($T_{1/2} = 15.2$ years), the dynamic of their concentrations in the air of local areas is practically identical for the period of 1993–2007 (see Figure 18.18). As it follows from the results of measurements about 900 samples, concentrations of ^{241}Am were prevalent in the range of $0.01–0.1 \, mBq/m^3$. Minimal concentration was not lower than $0.002 \, mBq/m^3$ and maximal approached $1 \, mBq/m^3$ in several cases only.

Concentration increase was caused by works conducted in the local zone or in its vicinity. Let us consider some of the reasons.

One of the highest concentrations of ^{137}Cs ($0.06 \, Bq/m^3$) was registered in the period from March 24 to April 3, 1998, in the southern installation area (AU-3). At that time a lifting crane "Demag" used at construction of the "Shelter" in 1986 was being dismantled and decontaminated.

Concentrations of ^{137}Cs and ^{241}Am increased several fold at AU-3 in the spring and summer, 2000, because of site preparation works for construction of an irradiated fuel storage facility

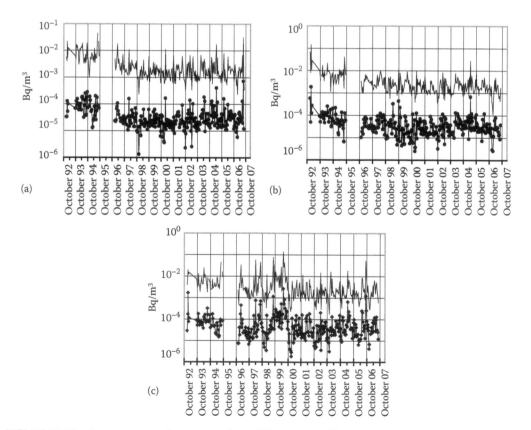

FIGURE 18.18 Concentration of aerosol carriers of ^{137}Cs (—) and ^{241}Am (–●–) by 1992–2007 samples at the northern (a), northwestern (b), and southern (c) installations in the local zone of the "Shelter."

(HOYaT-2) at a 2 km distance south-east of the "Shelter." There, the stumps were pulled, ground was leveled by scrapers, foundation ditches were excavated, and ground irradiated by radionuclides was carried out. A part of the radioactive substances came to the area of AU-3 as a result of the dust rising to air.

A road was constructed along the southern fence of the local zone for delivery of hardware and automobiles to the berm, which was to be partially destroyed in order to obtain place for the construction of a new and safer containment. In that time period the concentrations of ^{137}Cs and [241]Am increased 10-fold in the area of AU-3 (see Figure 18.18). The maximal content of ^{137}Cs reached 0.05 Bq/m^3 in samples taken from August 7–22.

$\Sigma\beta$ of 12 and 15 mBq/m^3 was registered in February and March 2007, in the area of AU-2, which are approximately 10 times higher than average for samples of that year. Most probably the surge of concentrations was caused by dismantling of the concrete protection wall and of the site for top assembly of steel trusses.

Distinctive minimums apart from peaks of radioactive substance contents in the air of the local zone can be noted in Figure 18.18. They practically always coincide with the cold season of the year, for example, in 1995, 1998, 2000, 2002, 2003, and 2006. One can conclude from the earlier discussion that dust elevation from underlying terrain was decreasing during a rainy season, moreover, in the presence of snow, and consequently quantity of radioactive aerosols in the air was decreasing, too.

18.11 CONCENTRATIONS OF RN, TN, AND THEIR DAUGHTER PRODUCTS INSIDE AND OUTSIDE OF THE "SHELTER" IN 2002–2009

Radon ([222]Rn) and thoron ([220]Rn) daughters occupy a special place among the radioactive aerosols presented in the "Shelter". They affect radiation environment at the premises of the "Shelter" and detection of radioactive aerosols of Chernobyl genesis. Both gases emanate from the concrete constructions of both the former fourth unit and the new elements of the "Shelter" (cascade and separation walls, materials of the reactor filling from helicopters in 1986 and others) containing natural radionuclides [226]Ra и [232]Th. Radon and thoron are also coming from the ground (mainly sand), into which the foundation and underneath premises of the Chernobyl NPP are embedded. Some quantity of thoron may appear from the irradiated fuel left in the ruins of the fourth unit as a result of decay of [232]U that had accumulated within 2 years of the reactor operation before the catastrophe. A chain of three subsequent α-decays of [232]U brings about appearance of thoron, out of which [216]Po and [212]Pb appear later on. As the half-life period of [232]U is 72 years, then the emanation of thoron from it will last long.

DPs of radon and thoron represent an obstructive factor while monitoring radioactive aerosols in the "Shelter" as they aggravate radiometry of samples. One should hold filters with the samples for 6 h to decrease radon DP content approximately 1000-fold to determine concentration of the aerosol products of Chernobyl accident. Around 4 days are necessary to make the same decrease of DPs for thoron. But there is also a positive moment: DPs of radon and thoron are sui generis markers of submicron aerosols. One can use them, for example, to evaluate the efficiency of filtering installations' and respirators' operation.

Long-term monitoring of radon and thoron DPs was conducted for the first time at room No. 207/5 of the "Shelter" in December 2000 [44]. Systematic sampling from ventilation flow coming to the "Bypass" system started in 2002, radon and thoron DPs being captured with a package of filters simultaneously with aerosol products of Chernobyl accident (see Chapter 17). The results of determination of volumetric activities of [212]Pb in 2002–2008 are presented in Figure 18.19. As a rule, they stayed in the range of 0.5–9 Bq/m^3. That is 50–100 times higher than at simultaneous sampling conducted by the same technique in March–May 2007, in the territory of the "Shelter" object. It followed from this fact that the thoron source was situated inside the "Shelter" and [212]Pb originated at the same place and did not come with the outside air.

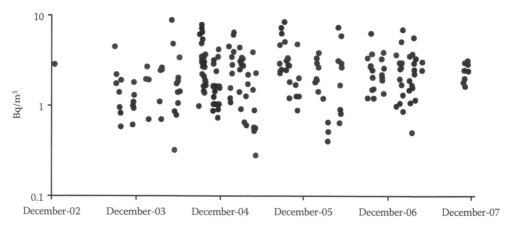

FIGURE 18.19 Concentration of aerosol carriers of ^{212}Pb at the "Bypass" system in 2002–2007.

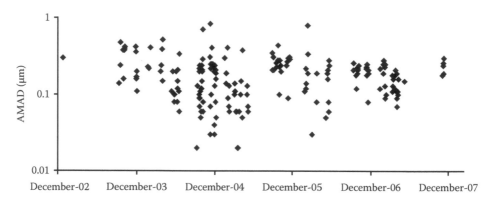

FIGURE 18.20 AMAD of aerosols carriers of ^{212}Pb at the "Bypass" system in 2002–2007.

Carriers of radon and thoron DPs both in the "Bypass" system and in the territory of the "Shelter" object are aerosol particles with AMAD of 0.05–0.4 μm (Figure 18.20). Average value of AMAD = 0.15 μm, which is characteristic for carriers of radon and thoron DPs in other regions both in the environment and inside premises [50].

Researches demonstrated that radon and thoron DPs as well as their mother substances can accumulate in a number of premises of the "Shelter" when air flow through the "Bypass" system is stopped. Usually this happens in the second half of the spring when during the daytime the air in the vicinity of the Chernobyl NPP warms up to 30°C and above and the premises inside the "Shelter" still remain cold after the winter time. Such situations were recorded, for example, in premise No. 207/5 in the third week of May 2003, and on May 23 and 24, 2007. Concentration of ^{212}Pb reached in the first case 15 Bq/m³, and in the second one 12–13 Bq/m³.

DPs of radon and thoron represent the main danger while inhaling air that contains them. First, some of them radiate α-particles. Secondly, being placed on submicron aerosols they intrude the lowest parts of lungs—bronchi and alveolus.

Discovering the radon concentration over 100 Bq/m³ and thoron over 6 Bq/m³ in the premises of the "Shelter" is a negative factor, which has not been taken into account while monitoring the radiation environment. Due to inhalation of aerosol carriers of DPs of natural noble gases the average radiation dose for personnel of the "Shelter" object may reach 20 mSv/g, that is, the value of the effective radiation dose limit.

18.12 CONCLUSION

The accident in the reactor of the fourth block of Chernobyl NPP resulted in the ejection of radioactive materials with a summary activity of 3×10^{18} Bq without taking inert gases into account. As a result of the fallout of radioactive gas–aerosol products of the accident on the earth's surface, extensive territories not only of the former USSR but also of other countries were polluted.

Ejection from the reactor was not stable both in terms of concentration and radioactive composition. Usually, the composition of aerosol particles did not correspond to the composition of irradiated nuclear fuel at the moment of the accident. Transport of activity from the reactor at long distances took place on particle carriers of submicrometer range.

Volatile compounds of I, Te, Ru, and Cs for a very long period of time were found in both aerosol and gaseous form. The largest portion of the gaseous component was for radioactive iodine.

After stopping ejection from the destroyed reactor, radioactivity at the ground level was determined by secondary raising of aerosol particles from polluted surfaces. Aerosol concentration depends on many natural and technogene factors. Variations of concentration under typical conditions were in the range of an order of magnitude. A decrease in concentration took place substantially faster than according to the law of radioactive decay. Radionuclides were disposed on the same carriers, and sizes of the particles were in the micrometer range.

Fires on the territory, polluted by radionuclides of Chernobyl origin, resulted in the sharp growth of aerosol concentrations of radiocesium both in the zone close to the source and in the far zone at a distance of hundreds of kilometers. Its transport from the fire zone takes place on submicrometer carriers and the water-soluble forms of Cs in aerosols of fire origin are increased substantially.

"Shelter" was built above the reactor in order to prevent pollution of the environment from radioactive materials. Only some percent of permissible levels for 1 GWt capacity nuclear object ejection go through cracks in the walls of the "Shelter." Inside this construction, aerosol concentration depends on the type and intensity of provided work and changes in the range of some orders of magnitude. The composition of particles usually corresponds to the composition of nuclear fuel, and long-lived α- and β-nuclides are disposed on particles of the same sizes of micrometer and sometimes submicrometer range.

Radon and thoron in premises of the "Shelter" is a negative factor that has not been taken into account before 2000. The main dangers are the DPs of radon and thoron. First, some of them radiate α-particles. Secondly, being placed on submicron aerosols they intrude the lowest parts of lungs—bronchi and alveolus. Due to the inhalation of aerosol carriers of DPs of natural noble gases the average radiation dose for personnel of the "Shelter" object may reach 20 mSv/g, that is, the value of the effective radiation dose limit.

The dispersity of carriers of accident products was in the range of AMAD 0.5–8 μm. Such a wide range of sizes was probably because the aerosols were produced from different sources, with different mechanisms of formation. DPs of radon and thoron had the AMAD in the range of 0.02–0.8 μm, the 50% interval falling to the AMAD = 0.15 μm. Consequently, radio nuclide carriers of the Chernobyl genesis had appeared as a rule because of dispersion and DPs of radon and thoron as a result of condensation.

The physicochemical characteristics of radioactive aerosols of Chernobyl origin are still being studied very extensively. At the present time, the 30 km zone around Chernobyl NPP and "Shelter" is a unique experimental testing area, where it is possible to conduct a wide spectrum of studies, particularly studies of the behavior of disperse systems, and the testing of new methods of measurements and analysis of radioactive aerosols.

ACKNOWLEDGMENT

Chapter translated from original Russian by Lev S. Ruzer and Natalya A. Tchuksina.

REFERENCES

1. USSR State Committee on the Utilization of Atomic Energy, The accident at the Chernobyl NPP and its consequences, *IAEA Post-Accident Review Meeting*, Vienna, Austria, August 25–29, 1986.
2. Borovoi, A.A. and Gagarinski, A.Yu., Chernobyl 15 years after: Radioactive release, *Nucl. Eur. Worldscan*, 1–2, 34–35, 2001.
3. Persson C., Rodhe Y., and De Geer, L.E., The Chernobyl accident—A meteorological analysis of how radionuclides reached and were deposited in Sweden, *Ambio*, 16, 1, 1987.
4. Kownacka, L. and Jaworowski, Z., Vertical distribution of ^{131}I and radiocesium in the atmosphere over Poland after Chernobyl accident, *Acta Geophys. Polonica*, 34, 405–412, 1986.
5. Raes, F., Graziani, G., Stanners, D., and Girardi, F., Radioactive measurements in air over Europe after the Chernobyl accident, *Atmos. Environ.*, 24A, 900–916, 1990.
6. Borisov, N.B., New sorption-filtering materials for the analysis of aerosols and vapors, *Isotopy in the SSSR*, 52/53, 66–67, 1978.
7. Dobrynin, Yu.L. and Khramtsov, P.B., Date verification methodology and new data for Chernobyl source term, *Radiat. Prot. Dosimetry*, 50, 307–310, 1993.
8. Gavrilin, Yu.I., The consequences of evolution of two scenarios of Chernobyl accident, *Bull. At. Energy*, 8, 20–28, 2001.
9. Borisov, N.B., Verbov, V.V., Kaurov, G.A., Ogorodnikov, B.I., Skitovich, V.I., and Churkin, S.L., Composition and concentration of gaseous and aerosols fractions above the 4-th unit of Chernobyl NPP and far from reactor in May 1986, Science Technical Report Series, Environment Protection, Questions of Ecology and Food Quality Control, Moscow, Russia, No. 1, 1992, pp. 11–17 (in Russian).
10. Ter-Saakov, A.A., Glebov, M.V., and Gordeev, S.K., Chernobyl–90, Reports of *All-Union Conference*, Chernobyl, Ukraine, Vol. 1, Part 2, 1990, p. 3 (in Russian).
11. Ter-Saakov, A.A., Kurinny, V.D., and Michaelyan, A.I., Chernobyl–90, Reports of *All-Union Conference*, Chernobyl, Ukraine, Vol. 1, Part 2, 1990, p. 9 (in Russian).
12. Dewell, L., Nuclide composition of Chernobyl hot particles, in *Hot Particles from the Chernobyl Fallout*, von Philisborn, H. and Steinhausler, F. (Eds.), Bergbau, Theuern, Germany, 1988, p. 16.
13. Ooe, H., Sirinuntavid, S., Ootsuji, M., Seki, R., and Ikeda, N., Size distribution of radionuclides in airborne dust (April 1986), *J. Radiat. Res.*, 28, 68, 1987.
14. Yanase, N., Parpyatipsakul, Y., Matsunaga, T., and Kasai, A. Concentration and particle size distribution of airborne dust at Chiba from reactor accident at Chernobyl, *J. Radiat. Res.*, 28, 67, 1987.
15. Ooe, H., Seki, R., and Ikeda, N., Particle size distribution of fission products in airborne dust collected at Tsukuba from April to June 1986, *J. Environ. Radioact.*, 6, 219, 1988.
16. Maqua, M. and Bonka, H.-G., Deposition velocity and washout coefficients of radionuclides bound to aerosol particles and elemental radioiodine, *Radiat. Prot. Dosimetry*, 21, 43, 1987.
17. Georgi, B., Helmeke, H.-J., Hietel, B., and Tschiersch, J., Particle size distribution measurements after the Chernobyl accident, in *Hot Particles from the Chernobyl Fallout*, von Philisborn, H. and Steinhausler, F. (Eds.), Bergbau, Theuern, Germany, 1988, p. 16.
18. Kauppinen, E.I., Hillemo, R.E., Aaltonen, S.H., and Sinkko, K.T.S., Radioactivity size distributions of ambient aerosols in Helsinki, Finland, during May, 1986 after the Chernobyl accident: Preliminary report, *Environ. Sci. Technol.*, 20, 1257, 1986.
19. Erlandsson, B., Askind, L., and Swietlicki, E., Detailed early measurements of the fallout in Sweden from the Chernobyl accident, *Water Air Soil Pollut.*, 35, 335, 1987.
20. Belovodsky, L.F. and Panfilov, A.P., Ensuring radiation safety during construction of the facility "Ukrytie" and restoration of unit 3 of the Chernobyl nuclear power station, one decade after Chernobyl: Summing up the consequences of the accident, Poster Presentations, Vol. 2, International Conference, Vienna, Austria, April 8–12, 1996, IAEA, Vienna, Austria, 1997, pp. 574–597.
21. Ogorodnikov, B.I., Pavluchenko, N.I., and Pazukhin, E.M., *Radioactive Aerosols of "Shelter" Object (A Review), Part 1. Aerosol Statement in Industrial Zone of the Chernobyl NPP under Building of the "Shelter" Object*, Preprint No. 02-10, National Academy of Sciences of Ukraine, Interdisciplinary Scientific and Technical Centre "Shelter," Chernobyl, Ukraine, 2002, 48 p. (in Russian).
22. Ogorodnikov, B.I., Radioactive products over the damaged block-4 of Chernobyl nuclear power plant before the completion of "Sarcophagus," *Proceedings of the International Seminar on Fission Products Transport Processes in Reactor Accidents*, Dubrovnik, Yugoslavia, May 22–26, 1989, in Rogers, J.T. (Ed.), Hemisphere Publishing Corporation, New York, 1990, pp. 799–806.
23. Devell, L., Tovedal, H., Bergstrom, U., Appelgren, A., Chyssler, J., and Andersson, L., Initial observations of fallout from the reactor accident in Chernobyl, *Nature*, 321, 192–193, 1986.

24. Devell, L., Tovedal, H., Bergstrom, U., Appelgren, A., Chyssler, J., and Andersson, L., Initial observations of fallout at Studsvik from the reactor accident at Chernobyl, Studsvik Energiteknik AB, Sweden, Report No. NP-86/56, 1986, 23 p.

25. Interim Report on Fallout Situation in Finland from April 26 to May 4 1986, Finnish Centre for Radiation and Nuclear Safety, Finland, STUK-B-YALO 44, May 1986, 38 p.

26. Borisov, N.B., Ogorodnikov, B.I., Kachanova, N.I., Kaurov, G.A., Borisova, L.I., Churkin, S.L., Skitovich, V.I., Verbov, V.V., Polevov, V.N., Naidenov, Y.A., and Budyka, A.K., Observation of gaseous-aerosols components of radioiodine and radioruthenium at first weeks after the Chernobyl NPP Accident, Science Technical Report Series, Environment Protection, Questions of Ecology and Food Quality Control, Moscow, Russia, No.1, 1992, pp. 17–24 (in Russian).

27. Ogorodnikov, B.I., Problems of environment and natural resources, Rev. Inform., 53, 1998 (in Russian).

28. Budyka, A.K. Phase transformations of iodine and other volatile radionuclides in free atmosphere, J. Aerosol Sci., 31, Suppl. 1, S478, 2000.

29. Hotzl, H., Rosner, G., and Winkler. R., Ground depositions and air concentrations of Chernobyl fallout radionuclides at Munich-Neuherberg, Radiochim. Acta, 41, 181–190, 1987.

30. Raes, F., Radioactivity Measurements in Europe after the Chernobyl Accident Part 1, Air. EUR-12269, Office for Official Publications of the European Communities, 989, pp. 35, 229.

31. Styro, B.I., Filistovich, V.I., and Nedvetskaite, T.A., Isotopes of Iodine and Radiation Safety, Gidrometeoizdat, St. Petersburg, Russia, 1992 (in Russian).

32. Garger, E.K. Air concentrations of radionuclides in the vicinity of Chernobyl and effects resuspension, J. Aerosol Sci., 25, 745, 1994.

33. Kashparov, V.A., Protsak, V.P., Ivanov, Y.A., and Nicholson, K.W., Resuspension of radionuclides and contamination of village area around Chernobyl, J. Aerosol Sci., 25, 755, 1994.

34. Hollander, W., Resuspension factors of ^{137}Cs in Hannower after the Chernobyl accident, J. Aerosol Sci., 25, 789, 1994.

35. Garland, J.A. and Pomeroy, I.R., Resuspension of fall-out material following the Chernobyl accident, J. Aerosol Sci., 25, 793, 1994.

36. Budyka, A.K. and Ogorodnikov, B.I., Radioactive aerosols generated by Chernobyl, Russ. J. Phys. Chem., 73, 310–319, 1999.

37. Skitovich, V.I., Budyka, A.K., and Ogorodnikov, B.I., Method and results of aerosol size definition in 30-km zone of Chernobyl nuclear power plant in 1986–1987, in Proceedings of the International Seminar on Fission Products Transport Processes in Reactor Accidents, Dubrovnik, Yugoslavia, May 22–26, 1989, Rogers, J.T. (Ed.), Hemisphere Publishing Corporation, New York, 1990, pp. 779–787.

38. Dorrian, M.D. and Bailey, M.R. Particle size distributions of radioactive aerosols measured in workplaces, Radiat. Prot. Dosimetry, 60, 119–133, 1995.

39. Pasukhin, E.M. and Ogorodnikov, B.I., Radionuclide Conduct upon forest fire, in International Scientific Workshop Radioecology of Chernobyl Zone, September 18–19, 2002, Slavutych, Ukraine, Abstracts of Poster Display Presentations, Slavutych, Ukraine, 2002, pp. 132–133.

40. Budyka, A.K. and Ogorodnikov, B.I., Radiazionnaya biologia, Radioekologia, 35, 102–112, 1995 (in Russian).

41. Lujaniene, G., Ogorodnikov, B.I., Budyka, A.K., Skitovich, V.I., and Lujanas, V., An investigation of changes in radionuclide carrier properties, J. Environ. Radioact., 35, 71–90, 1997.

42. Hollander, W. and Garger, E.K. (Eds.), Contamination of surfaces by resuspended material, Final Report EUR 16527 EN, 1996.

43. Gerasko, V.N., Nosovsky, A.V., Shcherbin, V.N., Oskolkov, B.Ja., Klyuchnikov, A.A., Korneev, A.A., Kupny, V.I., and Garin, Y. V., Unit "Shelter": History, State and Perspective, Slavutych, 1998.

44. Ogorodnikov, B.I., and Budyka, A.K. Monitiring of radioactive aerosols in the object "Shelter", At. Energy, 91, 470–475, 2001 (in Russian).

45. Begichev, S.N., Borovoy, A.A., and Burlakov, E.V., Radioactive release due to the Chernobyl accident, in Proceedings of the International Seminar Fission Products Transport Processes in Reactor Accidents, Dubrovnik, Yugoslavia, May 22–26, 1989, Rogers, J.T. (Ed.), Hemisphere Publishing Corporation, New York, 1990.

46. Garger, E.K., Kashpur, V.A., Korneev, A.A., and Kurochkin, A.A., Results of investigations of radioactive aerosols release from the object "Shelter", Problems of Chernobyl, #10, Part 2, Slavutych, 2002, pp. 60–71.

47. Borovoy, A.A., Monitring of unorganized releases from the object "Shelter", Problems of Chernobyl, #10, Part 2, Slavutych, 2002, pp. 192–198.

48. Ogorodnikov, B.I., Origin and components of radioactive aerosols over industrial area of the object "Shelter" of the Chernobyl NPP, *At. Energy*, 93, 42–46, 2002.
49. Bogatov, S.A., Formes and characteristics of fuel particles in release at accident on Chernobyl NPP, *At. Energy*, 69, 36–40, 1990.
50. Postendörfer, J., Properties and behaviour in air of radon and their decay products, *J. Aerosol Sci.*, 25, 219–263, 1994.

19 Classical Nucleation Theory
Account of Dependence of the Surface Tension on Curvature and Translation-Rotation Correction Factor

S.V. Vosel, A.A. Onischuk, P.A. Purtov, and T.G. Tolstikova

CONTENTS

19.1 INTRODUCTION

Classical nucleation theory (CNT) developed by Volmer and Weber, Becker and Döring, Zeldovich, and Frenkel [1] has been used extensively by specialists in different areas of science and technology to calculate the rates of homogeneous nucleation from the supersaturated vapor. The theory looks very attractive to many because it uses just the bulk physical–chemical parameters available in handbooks. However, numerous deviations of the theory predictions from the experimental measurements were observed giving the difference in several orders of magnitude for water [2] and organic substances [3,4] and tens of that for metals [5–7]. One of the sources of error in CNT comes from the contribution of the translational-rotational degrees of freedom to the free energy of critical nucleus. Due to this contribution the so-called free energy correction factor arises in the formula for the nucleation rate. Lothe and Pound [8] have estimated (within the framework of the Gibbs imaginary process of drop formation) the translational-rotational contribution to the free energy of critical nucleus which gave the correction factor for water of about 10^{17}. Reiss and coworkers [9,10] have argued that the Lothe and Pound correction factor was exaggerated too much due to the neglect of the fluctuation of the center of mass of the nucleus and a new correction was proposed to be a factor of 10^3–10^6. Recently Kusaka [11] has derived a rigorous formula for the correction factor within the framework of the Gibbs process of drop formation and calculated numerically this factor for the Lennard–Jones system. The calculated values ranged from 10^9 to 10^{13}. The numerical calculation is probably the most direct way to determine the correction factor. However, the calculations of this kind are only possible for simple systems and, therefore, an analytical expression for the correction factor applicable to a wide range of real systems is necessary. In this paper the derivation of the analytical formula for the translation-rotation correction factor will be given.

Another source of error in CNT is the so-called capillarity approximation which assumes that the drop surface tension is that of a flat interface. It is evident that the rigorous formula for the nucleation rate must take into account the dependence of surface tension on the drop radius. The problem of the surface tension of curved interfaces was first analyzed by Gibbs in his thermodynamic theory of interface [12] taking into account the dependence of surface tension on radius. One should note that the theory of Lothe–Pound and Kusaka [11] for the correction factor takes into account the dependence of surface tension on radius automatically. However, the drop surface tension appears also in the Zeldovich factor in the formula for the nucleation rate. It is hardly possible to measure directly the small drop surface tension. However, there are many theoretical contributions in the literature where the surface tension of small drops was calculated as a function of radius [13–30]. These calculations have shown that the surface tension of small drops with a size of about 1 nm may be a rather sharp function of radius and differs essentially from that of a flat interface.

In this chapter we will present a rigorous derivation of the analytical formula for the nucleation rate including both the translation-rotation correction factor and Zeldovich factor taking into account the dependence of surface tension on the radius of critical drop. Finally, using this formula we will determine the surface tension of critical drops from the nucleation rate and supersaturation measured experimentally for some metals taken as an example.

19.2   HOMOGENEOUS NUCLEATION FROM SUPERSATURATED VAPOR

In the classical theory of single component systems the nucleation is considered as a process of formation of so-called critical nuclei, that is, the drops being in an unstable equilibrium with the supersaturated vapor. The nucleation rate I (the number of critical nuclei formed per unit time per unit volume) is [1]

$$I = \frac{N_{crit}}{V}\beta 4\pi R_e^2 Z \tag{19.1}$$

where
   N_{crit} is the equilibrium number of critical nuclei in the system of volume V
   $4\pi R_e^2$ is the surface area of critical nucleus
   R_e is the radius of equimolar surface
   $\beta = (N_1/V)\sqrt{k_B T/2\pi m}$ is the collision frequency for the vapor molecules with the unit surface,
      k_B is the Boltzmann constant, T is the absolute temperature, m is the mass of molecules, N_1 is
      the number of monomeric molecules in the system

The statistical mechanical analysis by Lothe–Pound [8], Reiss et al. [10,31,32], Kusaka [11], and others in the last decade has shown that N_{crit} is to be defined by the following equation [11]:

$$N_{crit} = KN_1 \exp\left(-\frac{W_{crit}}{k_B T}\right) \tag{19.2}$$

where

K is the translation-rotation free energy correction factor (arising due to the so-called replacement free energy [8,10,32])

W_{crit} is the reversible work required to form a critical drop from vapor molecules (it is assumed as if the drop is at rest in space)

Z in Equation 19.1 is the so-called Zeldovich factor taking into account the fact that the drop is able not only to grow due to the vapor condensation but also to decrease in size due to the molecule evaporation:

$$Z = \sqrt{\frac{Y}{2\pi k_B T}} \qquad (19.3)$$

In CNT the quantity Y is identified with [1]:

$$Y = -\left(\frac{d^2 W}{dn^2}\right)_{n=n_{crit}} \qquad (19.4)$$

where

W is the reversible work of formation of a non-critical droplet

n and n_{crit} are, respectively, the numbers of molecules in the drop and the critical drop (i.e., the numbers of molecules enclosed by the equimolar dividing surfaces)

In Equation 19.4 the derivation is made for the gaseous pressure being constant.

Our modern understanding of interfacial thermodynamics has its origins in the Gibbs theory of surface tension [12]. This theory considers a fluid maternal phase with i components and another fluid phase (with the same i components within it) being in equilibrium with the maternal one. This system is compared with a hypothetical reference system composed by two homogeneous bulk phases (maternal phase α and another phase β) and a mathematical dividing surface between them, at a certain position. The key parameter in the Gibbs theory is the surface tension σ_S attributed to the so-called surface of tension. In the case of spherical symmetry the radius of this surface is denoted as R_S. The chemical potential μ_i of molecules in the bulk phases is the same as that of the real system and the difference in pressure and composition between the drop and the phase β is incorporated in the value of surface tension. The critical nucleus (i.e., the embryo which is in equilibrium with the maternal phase) is often extremely small in size so that the homogeneous bulk properties are not attained even at its center, but Gibbs interfacial thermodynamics remains valid. Thus, to understand the surface thermodynamics of a two-phase system one should only know the surface tension and the location of the surface of tension.

As follows from the Gibbs theory the surface tension is a function of curvature. In the case of spherical symmetry for the single component system this dependence on curvature is governed by the Gibbs–Tolman–Koenig–Buff (GTKB) differential equation [33,34]:

$$\frac{d[\ln \sigma_S(R_S)]}{d[\ln R_S]} = \frac{(2\delta(R_S)/R_S)[1+(\delta(R_S)/R_S)+(1/3)(\delta(R_S)/R_S)^2]}{1+(2\delta(R_S)/R_S)[1+(\delta(R_S)/R_S)+(1/3)(\delta(R_S)/R_S)^2]} \qquad (19.5)$$

where $\delta(R_S)$ is the so-called Tolman length which is equal to

$$\delta(R_S) = R_e - R_S \qquad (19.6)$$

Equation 19.5 may be written with the superficial density Γ of matter computed with respect to the surface of tension [33]:

$$\frac{d[\ln \sigma_S(R_S)]}{d[\ln R_S]} = \frac{2\Gamma m}{R_S \Delta \rho}\left(1 + \frac{2}{R_S}\frac{\Gamma m}{\Delta \rho}\right)^{-1} \tag{19.7}$$

where $\Delta \rho (g/cm^3) = \rho^\beta - \rho^\alpha$ is the difference between the densities of reference bulk phases β and α. If the function $\sigma(R_S)$ is known, it is possible to solve Equation 19.5 and determine the function $\delta(R_S)$. On the other hand, Equation 19.6 gives the relationship between R_S and the real radius R_e. The radius R_e is the radius of the dividing surface which is chosen in such a way that the superficial density Γ computed with respect to this surface is equal to zero [12].

As follows from the Gibbs theory [12] the work of formation of the critical nucleus is

$$W_{crit} = \frac{4\pi R_S^2 \sigma_S(R_S)}{3} \tag{19.8}$$

The formula used in the CNT is not the expression (19.8) but uses the surface tension σ_∞ for the flat surface instead of $\sigma(R_S)$ and R_e instead of R_S which is a rough approximation. Thus, CNT gives for the nucleation rate [1]:

$$I_{CNT} = \left(\frac{N_1}{V}\right)^2 \sqrt{\frac{2m\sigma_\infty}{\pi}}\frac{1}{\rho}\exp\left(-\frac{4\pi R_e^2 \sigma_\infty}{3k_B T}\right) \tag{19.9}$$

where $\rho(g/cm^3)$ is the density of incompressible bulk liquid. Besides, the proper account of the translational and rotational free energies of the critical nucleus is missing in CNT resulting in underestimation by orders of magnitude in the classical formula for the nucleation rate. In the following sections we will give an accurate derivation of the formula taking into account both the dependence of the drop surface tension on radius and the contribution to the free energy from the translational and rotational degrees of freedom of the critical nucleus (correction factor K).

19.3 CALCULATION OF ZELDOVICH FACTOR

Our first task is to derive the expression for the Zeldovich factor which takes into account the dependence of droplet surface tension on radius. Such a calculation was made recently [7,35] in the framework of the Nishioka theory; in this section we provide a more transparent derivation. Following Nishioka and Kusaka [36] we consider a multicomponent system which is a liquid spherical drop surrounded by a vapor phase. In general the drop is not a critical nucleus. The total number of molecules for the component i in the system is specified as N_i to be governed by the equation $N_i = N_i^\alpha + N_i^\beta = const$, where N_i^α and N_i^β are the numbers of molecules in the real system belonging to the vapor and droplet, respectively. One of the main assumptions of the publication [36] is that the molecules of interfacial region are regarded as belonging to the drop. In other words, the chemical potential for surface molecules is equal to the chemical potential μ_i^β for the volume molecules of droplet. In general the droplet and vapor chemical potentials are not equal ($\mu_i^\beta \neq \mu_i^\alpha$).

It was shown in Ref. [36] (as well as in Ref. [37]) that the variation of minimum work necessary to form a non-critical droplet in a multicomponent system under the non-equilibrium process at T, μ_i^α, and P^α being constant is governed by the following equation:

$$dW = \sum_i \left(\mu_i^\beta - \mu_i^\alpha\right)dN_i^\beta \tag{19.10}$$

Equation 19.10 will be the starting point in our further discussion. Let us consider the non-equilibrium process for the single-component system. For the single-component system Equation 19.10 will transform to the evident relation:

$$\frac{dW}{dN^\beta} = \mu^\beta(P^\beta) - \mu^\alpha(P^\alpha) \tag{19.11}$$

where P^α and P^β are pressures in the corresponding reference bulk phases α and β, respectively.

In the model of Nishioka all the molecules are divided into two types: N^β molecules have the chemical potential μ^β and N^α molecules have the chemical potential μ^α. Therefore, the number N^S of molecules attributed to the interphase dividing surface is equal to zero, and

$$N = N^\alpha + N^\beta \tag{19.12}$$

On the other hand, by definition the equality $N^S=0$ corresponds to the choice of equimolar surface of radius R_e as a dividing one. For the equimolar surface the conservative condition is

$$N = n^\alpha(V - V^\beta) + n^\beta V^\beta \tag{19.13}$$

where
  V is the total volume of system
  V^β is the volume enclosed by the dividing surface
  $n^\alpha = n^\alpha(\mu^\alpha, T)$ and $n^\beta = n^\beta(\mu^\beta, T)$ are the number densities of molecules in the reference phases α and β which by definition have the chemical potentials μ^α and μ^β, respectively

The comparison of Equations 19.12 and 19.13 shows that the number of molecules N^β in the model system of Nishioka is in correspondence with the number of molecules $n = n^\beta V^\beta$ in the hypothetical system enclosed by the equimolar dividing surface:

$$N^\beta = n = n^\beta V^\beta = n^\beta \frac{4}{3}\pi R_e^3 \tag{19.14}$$

Then, differentiating Equation 19.11 under T, P^α, and μ^α being constant we get

$$-Y \equiv \left(\frac{d^2W}{dn^2}\right)_{\mu^\alpha, n=n_{crit}} = \left(\frac{d\mu^\beta}{dn}\right)_{\mu^\alpha, n=n_{crit}} = \frac{d\mu^\beta}{dP^\beta}\left(\frac{dP^\beta}{dn}\right)_{\mu^\alpha, n=n_{crit}} \tag{19.15}$$

The first factor in the right hand side of Equation 19.15 is the molecular volume in the reference phase β. We can get this volume from the Gibbs–Duhem relation which describes the reversible process ($\mu^\alpha = \mu^\beta$) and can be written in the following way at $T = const$ [12]:

$$-d\sigma_S = \Gamma d\mu^\beta = \Gamma \frac{d\mu^\beta}{dP^\beta}\left(\frac{dP^\beta}{dn}\right)dn \tag{19.16}$$

From Equations 19.15 and 19.16 we get

$$Y = \chi\frac{1}{\Gamma}\frac{d\sigma_S}{dn} = \chi\frac{1}{\Gamma}\frac{d\sigma_S}{dR_S}\frac{dR_S}{dn} \tag{19.17}$$

where

$$\chi = \left(\frac{dP^{\beta}}{dn}\right)_{\mu^{\alpha}, n=n_{crit}} \left(\frac{dP^{\beta}}{dn}\right)^{-1} \tag{19.18}$$

Thanks to Equations 19.6 and 19.14 we can write down for the last factor in Equation 19.17:

$$\frac{dn}{dR_S} = \frac{dn}{dR_e}\frac{dR_e}{dR_S} = 4\pi\frac{\rho^{\beta}}{m}R_e^2\left(1+\frac{d\delta}{dR_S}\right) \tag{19.19}$$

where $\rho^{\beta} = mn^{\beta}$. The GTKB equation (19.7) can be rewritten as

$$\frac{1}{\Gamma}\frac{d\sigma_S}{dR_S} = \frac{\sigma_S}{R_S}\frac{2m}{R_S\Delta\rho}\left(1+\frac{2}{R_S}\frac{\Gamma m}{\Delta\rho}\right)^{-1} \tag{19.20}$$

Using Equations 19.19 and 19.20 and accounting that for the gas-liquid nucleation $\Delta\rho \approx \rho^{\beta}$ we have

$$Y = 2\left(\frac{m}{\rho^{\beta}}\right)^2\frac{\sigma_S(R_S)}{R_S^2}\frac{1}{4\pi R_e^2}\left(1+\frac{2\Gamma m}{R_S\Delta\rho}\right)^{-1}\chi\left(1+\frac{d\delta}{dR_S}\right)^{-1} \tag{19.21}$$

Then, with the help of the Gibbs–Tolman equation [33]

$$\Gamma m = \Delta\rho\delta\left[1+x+\frac{x^2}{3}\right] \tag{19.22}$$

where $x = \delta/R_S$, Equation 19.3 transforms into the sought formula for Zeldovich factor:

$$Z = [1+x^2 f(x)]^{-1/2}\chi^{1/2}\left(1+\frac{d\delta}{dR_S}\right)^{-1/2}\frac{m}{2\pi R_e^2\rho^{\beta}}\sqrt{\frac{\sigma_S(R_S)}{k_B T}} \tag{19.23}$$

where $f(x) = (1+(2/3)x)(1+x)^{-2}$. Substituting Equation 19.23 to Equation 19.1 and taking Equation 19.2 into account we have for the nucleation rate:

$$I = Kn_1^2\sqrt{\frac{2m\sigma_S(R_S)}{\pi}}\frac{1}{\rho^{\beta}}[1+x^2 f(x)]^{-1/2}\chi^{1/2}\left(1+\frac{d\delta}{dR_S}\right)^{-1/2}\exp\left(-\frac{W_{crit}}{k_B T}\right) \tag{19.24}$$

where $n_1 = N_1/V$. Equation 19.24 includes the correction factor K which will be considered in the following sections. The only undetermined factor in Equation 19.24 is χ which is (see Equation 19.18) the ratio of derivatives of pressure with respect to the number of molecules n for equilibrium and non-equilibrium processes. It is natural to assume that this ratio is close to unity: $\chi \approx 1$. The values $x = \delta/R_S$ and $d\delta/dR_S$ in Equation 19.24 are related to $\sigma_S(R_S)$ by GTKB Equation 19.5 and can be calculated if $\sigma_S(R_S)$ is known. Besides, for the large enough drops (which match the inequality $x^2 \ll 1$) the quantities $x^2 f(x)$ and $d\delta/dR_S$ can be neglected with respect to unity; that is, in this case Equation 19.24 is practically the same as classical Equation 19.9 (except for the factor K).

19.4 TRANSLATION-ROTATION CORRECTION FACTOR

Frenkel [1] was the first who argued that the contribution to the free energy of the critical nucleus from the translational and rotational degrees of freedom should be taken into account when calculating the drop (cluster) size distribution. To determine the equilibrium cluster size distribution, Frenkel has considered an ensemble of clusters as an ideal gas mixture. In this case a statistical mechanical analysis gives [1,11,32]

$$N_n = q_n e^{\mu_v n / k_B T} \tag{19.25}$$

where
   N_n is the equilibrium number of n-sized clusters (consisting of n monomeric molecules)
   q_n is the partition function within the canonical ensemble for the drop containing n molecules
   μ_v is the chemical potential for the vapor molecules

According to Frenkel's model [1],

$$q_n = \frac{Q_{tr} Q_{rot}}{(Q_v)^6} Q_{3n,v} e^{-U_n / k_B T} = \frac{Q_{tr} Q_{rot}}{(Q_v)^6} q_n^{rest} \tag{19.26}$$

where
   Q_{tr} and Q_{rot} are the partition functions for three translational and three rotational degrees of freedom of n-sized cluster, respectively
   $Q_{3n,v}$ is the partition function for $3n$ vibrational degrees of freedom
   $(Q_v)^6$ is the partition function for six vibrational degrees of freedom which are to be deactivated

All the clusters in Frenkel's model have the same structure corresponding to the minimum potential energy U_n. q_n^{rest} in Equation 19.26 is the partition function for the cluster at rest [38]:

$$q_n^{rest} = Q_{3n,v} e^{-U_n / k_B T} = e^{-f_n^{rest} / k_B T} \tag{19.27}$$

where f_n^{rest} is the Helmholtz free energy for the cluster at rest. According to the simple Frenkel's model the free energy f_n^{rest} is assumed to be given by [1]

$$f_n^{rest} = f_0 n + \gamma n^{2/3} \tag{19.28}$$

where
   f_0 is the Helmholtz free energy per one molecule in the bulk liquid phase
   γ is a constant proportional to the surface tension of the flat interface

We refer to the factor $Q_{tr} Q_{rot}/(Q_v)^6$ in Equation 19.26 as the Frenkel factor. Equation 19.28 is the basis of the CNT. In the framework of CNT the embryo of the nucleating phase is regarded as a spherical incompressible liquid drop fixed in the space; the density of the drop is considered as homogeneous and equal to that of the bulk liquid; this drop has a sharply defined interface with the surrounding metastable mother phase which, in the case of vapor, is regarded as an ideal gas; the surface tension of the critical nucleus is regarded as equal to that of the flat interface (capillarity approximation).

   Using instead of Equation 19.28 a similar formula for the Gibbs free energy $g_n = \mu_l n + \gamma n^{2/3}$ (where μ_l is the chemical potential of a molecule, as if it were part of a bulk liquid at the pressure P outside the drop) Frenkel has derived a rigorous thermodynamic formula for the cluster (drop) equilibrium size distribution. This distribution proves to be [1]

$$N_n = N_1 \exp\left\{-\frac{1}{k_B T}(g_n - \mu_v n)\right\} \tag{19.29}$$

The rate of nucleation is proportional to the number N_{crit} of the critical nuclei [1,11,32].

Frenkel has failed to find the connection between the size distributions in Equations 19.25 and 19.29. To do this, Lothe and Pound have considered an imaginary process (devised by Gibbs), in which a cluster embedded in the bulk liquid is transferred to the vapor phase.

In the Gibbs theory of interface the work W_{crit} of formation of critical nucleus is represented by the following formula [12]:

$$W_{crit} = -V_S(P^\beta - P^\alpha) + \sigma_s A_S \tag{19.30}$$

where

P^β is the pressure of the reference bulk liquid having the same temperature and chemical potential as the vapor

A_S is the area of the surface of tension, which is assumed to be spherical

V_S is the volume enclosed by this surface

To make the meaning of Equation 19.30 clearer, Gibbs has introduced an imaginary process consisting of two separate stages. Let us consider this process in a nutshell. Initially the system consists of the bulk liquid reference phase (at pressure P^β) and the bulk vapor phase (at pressure P^α). Due to the pressure difference the reference phase is surrounded by an elastic envelope. However, it is assumed that the envelope is transmittable for the gas molecules. First, some number of molecules from the vapor is transferred to the bulk liquid. The volume of the reference phase increases by V_S due to this transfer but the surface area is kept constant. During this stage the benefit of work is $(P^\beta - P^\alpha) V_S$. In the next stage an aperture in the envelope opens and then closes so that a volume V_S of the liquid phase is extruded outside and the envelope intrudes inside to decrease the volume by the same magnitude V_S. The total work at the second stage is $\sigma_s A_S$. This work includes different components. For instance, during the extrusion the drop loses the interaction with the bulk liquid; therefore, $\sigma_s A_S$ contains the potential energy of this interaction. $\sigma_s A_S$ also includes the work of structural relaxation of the extruded drop and the adsorption of some gas molecules to its surface.

In the Gibbs's thought process, the translational and rotational degrees of freedom of the cluster is not taken into account, either in the bulk liquid or in the vapor. Lothe and Pound [8] proposed the way to account the difference in free energy associated with these degrees of freedom is necessary. They added to W_{crit} the translational and rotational free energies of the cluster in the gas phase and subtracted the entropy contribution to free energy associated with the vibrational translation and rotation of the embedded cluster with the relative positions of the molecules in the cluster held fixed. The partition function corresponding to those vibrational modes of fluctuation, q_{rep}, is called the replacement partition function, since these modes are replaced by the free translation and free rotation in the vapor phase. The procedure of translation-rotation correction results in the Lothe–Pound factor [11]:

$$\Phi_{LP} = \frac{Q_{tr}Q_{rot}}{q_{rep}} = \frac{Q_{tr}}{Q_{tr}^l}\frac{Q_{rot}}{Q_{rot}^l} \tag{19.31}$$

where

$$q_{rep} = Q_{tr}^l Q_{rot}^l \tag{19.32}$$

Q_{tr}^l is the partition function of vibrational translations of the embedded cluster

Q_{rot}^l is the partition function of vibrational rotations around the center of mass of the cluster

The factor Φ_{LP} is to substitute the Frenkel factor in Equation 19.26. As to the nucleation rate, it is to be multiplied by the free energy correction factor Φ_{LP}/N_1 (the appearance of the denominator N_1 will be discussed later). One should note that it is assumed in the Lothe and Pound theory that Equation 19.32 refers to the absolutely incompressible cluster because the relative coordinates of the n molecules in the cluster are fixed.

Approximating q_{rep} by $\exp(s/k_B)$ (where $s \approx 5k_B$ is the entropy of a single molecule in the bulk liquid) Lothe and Pound have estimated the magnitude of Φ_{LP}/N_1 to be 10^{17} for the water cluster containing about 100 molecules [8]. A correction factor this large appeared excessively large to many. Thus, a serious controversy has developed since the beginning of Lothe–Pound theory. Reiss et al. introduced the concept of the so-called stationary cluster [9,10,32]. The partition function of the stationary cluster is given by

$$q_n^{st} = \frac{1}{\Lambda^{3n} n!} \int_{r^n \in V^{st}} dr^n e^{-U_n/k_B T} \tag{19.33}$$

where
   n particles are all confined to some volume V^{st}, which in turn is held fixed in space
   U_n is particle's interaction potential
   Λ is the thermal wavelength of a particle

It was assumed that the reversible work of formation of the stationary cluster is equal to W_{crit} (Equation 19.30). The partition functions q_n and q_n^{st} are linked by the factor Φ_R (which we refer to as the Reiss factor):

$$q_n = \Phi_R q_n^{si} \tag{19.34}$$

When deriving the expression for the factor Φ_R Reiss et al. deactivated the rotational motion inside the volume V^{st} and the translational motion resulting from the fluctuation of the position of the center of mass of the stationary cluster prior to activating the free rotation corresponding to Q_{rot} and the free translation corresponding to Q_{tr}. Since the rotational motion of the n particles inside the volume V^{st} is essentially a free rotation, no explicit account needs to be taken for the rotational partition function [32]. Therefore, the Reiss factor is the ratio between the partition function for the free translations in the volume V and that for the translations of the center of mass in the volume V^{st} and proves to be [10,32]

$$\Phi_R = \frac{V}{(2\pi)^{3/2} \sigma^3} \tag{19.35}$$

where
   σ is the standard deviation in any of the three Cartesian coordinates of the center of mass
   $\left(\sqrt{2\pi}\sigma\right)^3$ is the volume in which the center of mass of the drop fluctuates

The evaluation of σ in the framework of the model of rigid spheres gives [10,32]

$$\sigma \cong \frac{0.2 v_n^{1/3}}{n^{1/2}} = \frac{0.2 v_l^{1/3}}{n^{1/6}} \tag{19.36}$$

where v_n is the drop volume. Under the capillarity approximation $v_n = n v_l$, where v_l is the volume per one molecule in the bulk liquid phase. If one assumes that $V^{st} = v_n$, Equations 19.35 and 19.36 lead to the quantity $\Phi_R/N_1 \approx 10^6$ for a water cluster of 100 molecules [32]. This value is considerably less than the estimation of Lothe and Pound for the ratio of translational partition functions $Q_{tr}/N_1Q_{tr}^l \approx 10^{10}$. The difference between the evaluation of Reiss and that of Lothe and Pound is due to the fact that in deriving Equation 19.35 the assumption of the incompressibility of the liquid was not used.

Reiss notes reasonably [32] that in the theory of Lothe and Pound the translational degrees of freedom are to be related to the motion of the center of mass of the embedded (compressible) cluster with respect to the fixed spherical boundary of the cluster but not to the vibrational translation of the rigid cluster as a whole.

On the other hand, Reiss believes that in the theory of Lothe and Pound (as well as in the case of stationary cluster) the ratio Q_{rot}/Q_{rot}^l is about unity assuming that the rotation of a drop which is a part of a bulk liquid is essentially the same as the free rotation [32], that is, no explicit account needs to be taken for the rotational partition function (as well as in the case of stationary cluster). However, one can agree with the statement that Q_{rot}/Q_{rot}^l is about unity only in the case of high temperatures. But at the temperature near the melting point (typical temperatures for homogeneous nucleation experiments) the mechanical behavior of the viscous liquid is to be more similar to that of the solid in the case of quick processes (for the time shorter than the Maxwell relaxation time) [1].

19.5 THEORY OF KUSAKA AND ANALYTICAL FORMULA FOR THE CORRECTION FACTOR

As is seen from the preceding section, the theory of Lothe and Pound considers the Gibbs extrusion process but uses rather rough and ill-founded approximations. Therefore, it seems to be clear that the approximate formula of Lothe and Pound (Equation 19.31) cannot be used as a basis for the calculation of correction factor. To derive a formula suitable for such a calculation, Kusaka has developed recently a rigorous statistical-mechanical approach [11] based on the Gibbs extrusion process considered in the previous section. Let us look at the milestones of the Kusaka's theory.

The isothermal–isobaric partition function of the bulk liquid held at constant (T, P_l, N) is [11]

$$Y(T, P_l, N) = \int \frac{dV}{a} \frac{\exp(-P_l V/k_B T)}{h^{3N} N!} \int dp^N dr^N \exp\left(-\frac{H_N}{k_B T}\right) \qquad (19.37)$$

where
p^N collectively denotes the momentum of each of the N particles
h is Planck's constant
H_N is the system Hamiltonian
The constant a arises from the mechanical degrees of freedom of a piston imposing the constant pressure P_l

Then a cluster embedded in the bulk liquid phase is defined by taking a spherical region of volume V_S, which contains m particles. The phase points embraced by Equation 19.37 are partitioned according to the number m of molecules inside the spherical region:

$$Y(T, P_l, N) = \int \frac{dV}{a} \exp\left(-\frac{P_l V}{k_B T}\right) \sum_{m=3}^{N} \frac{1}{h^{3(N-m)}(N-m)!} \times \int_{r^{N-m} \in V-V_S} dp^{N-m} dr^{N-m} \exp\left(-\frac{H_{N-m}}{k_B T}\right)$$

$$\times \frac{1}{h^{3m} m!} \int_{r^m \in V_S} dp^m dr^m \exp\left(-\frac{H_m}{k_B T}\right) \exp\left(-\frac{U_{int}}{k_B T}\right) \qquad (19.38)$$

where U_{int} denotes the interaction potential between the $N - m$ particles outside the sphere and the m particles inside the sphere, and is a function of r^{N-m} and r^m. The last integral in Equation 19.38, along with the coefficient $1/(h^{3m} m!)$, is regarded as the partition function ξ_m of a cluster, which consists of m particles, all confined to V_S, and is embedded in the liquid phase.

The coordinate transformation is to be done from a laboratory system to a body coordinate system which means that a set of Euler axes is embedded in the object with the origin at the center of mass $\mathbf{R}_{c.m.}$ and the rotation refers to the rotation of these axes. Denoting by s^{m-2} and t^{m-2} the coordinates and the conjugate momenta of the remaining $3m-6$ degrees of freedom of the embedded cluster, the partition function ξ_m is written as [11]

$$\xi_m = \frac{1}{h^{3m-6} m!} \int_{r^m \in V_S} dt^{m-2} ds^{m-2} \exp\left(-\frac{K_S}{k_B T}\right) \exp\left(-\frac{U_m}{k_B T}\right)$$

$$\times \frac{\varsigma_1 \varsigma_2 \varsigma_3}{\Lambda_{c.m.}^3} \int_{r^m \in V_S} d\mathbf{R}_{c.m.} \sin\theta \, d\theta \, d\varphi \, d\psi \exp\left(-\frac{U_{int}}{k_B T}\right) \qquad (19.39)$$

where
K_S is the kinetic energy of the m particles excluding those due to rigid translation and rotation of the embedded cluster as a whole
U_m is the interaction potential among the m particles
Euler angles (θ, φ, ψ) specify the orientation of the cluster
$\Lambda_{c.m.}$ and ς_i are $\Lambda_{c.m.} = h/\sqrt{2\pi M k_B T}$ and $\varsigma_i = \sqrt{2\pi I_i k_B T}/h$ with M and I_i ($i = 1, 2, 3$) denoting the mass of the cluster and its principal moments of inertia, respectively

The second integral in Equation 19.39 is regarded as the configurational partition function Z_c of the embedded cluster due to its translational and rotational degrees of freedom when it is subjected to the external field U_{int}. Thus, the configurational entropy S_c associated with these degrees of freedom is defined by the following equation:

$$Z_c = \int_{r^m \in V_S} d\mathbf{R}_{c.m.} \sin\theta \, d\theta \, d\varphi \, d\psi \exp\left(-\frac{U_{int}}{k_B T}\right) = \delta^3 \exp\left(-\frac{\langle U_{int} \rangle_c}{k_B T}\right) \exp\left(\frac{S_c}{k_B}\right) \qquad (19.40)$$

where $\langle \, \rangle_c$ denotes the thermal average taken with the Boltzmann weight $\exp(-U_{int}/k_B T)$ while imposing the constraint $r^m \in V_S$ that the m particles are confined to the volume V_S.

One should note that the quantity δ is an arbitrary length scale in Equation 19.40 and is introduced to make explicit the dimensionality of various quantities involved. Therefore, the entropy S_c thus defined is some "formal entropy."

Since the coordinates r^{N-m} and s^{m-2} are fixed when evaluating Z_c one can say that the cluster which is governed by Equation 19.40 is a rigid one in a rigid environment. The idea of Kusaka is that during the extrusion of such a cluster the modes of fluctuation associated with the factor:

$$\frac{\varsigma_1 \varsigma_2 \varsigma_3}{\Lambda_{c.m.}^3} \delta^3 \exp\left(\frac{S_c}{k_B}\right) \qquad (19.41)$$

are to be deactivated. Then, the resulting equation for q_{rep} is [11]

$$\frac{1}{q_{rep}} = \left\langle \left(\frac{\varsigma_1 \varsigma_2 \varsigma_3}{\Lambda_{c.m.}^3} \delta^3 \exp\left(\frac{S_c}{k_B} \right) \right)^{-1} \right\rangle_l \qquad (19.42)$$

where $\langle \ \rangle_l$ indicates a thermal average taken in the bulk liquid held at constant (T, P_l, N).

On the other hand, according to Kusaka, the activation of the free translation within the volume V and the free rotation of the extruded cluster, when averaged over all possible values of m and internal configurations s^{m-2}, leads to the factor

$$Q_{tr} Q_{rot} = \left\langle \frac{\varsigma_1 \varsigma_2 \varsigma_3}{\Lambda_{c.m.}^3} \right\rangle_l 8\pi^2 V \qquad (19.43)$$

Upon the extrusion, the m-sized cluster loses its interaction U_{int} with the surroundings, acquires $n-m$ particles from the vapor phase, and then undergoes structural relaxation; as in the original Lothe–Pound prescription Kusaka assumes that the reversible work associated with these processes is included fully in σ_S. Uniting Equations 19.42 and 19.43 Kusaka arrives finally at the expression for factor Φ_K (which we refer to as the Kusaka factor) designated to substitute the Lothe–Pound factor:

$$\Phi_K = \left\langle \frac{\varsigma_1 \varsigma_2 \varsigma_3}{\Lambda_{c.m.}^3} \right\rangle_l 8\pi^2 V \left\langle \left(\frac{\varsigma_1 \varsigma_2 \varsigma_3}{\Lambda_{c.m.}^3} \delta^3 \exp\left(\frac{S_c}{k_B} \right) \right)^{-1} \right\rangle_l \qquad (19.44)$$

Kusaka [11] has numerically calculated the correction factor for the Lennard–Jones system. The calculated values ranged from 10^9 to 10^{13} which were considerably higher than the Reiss correction factor and lesser than the Lothe–Pound one. However, it is hardly possible to numerically calculate the correction factor for real systems like water, organic species, metals, etc. Therefore, in this section we propose an analytical formula applicable to real systems. Later, in Section 19.7 we will compare the estimations of this formula with Kusaka's numerical simulation results [11].

To derive such a formula we will analyze the key equation in Kusaka's theory (Equation 19.40). Actually for all the m-sized clusters that happened to be inside the volume V_S the cluster's volume $v_m < V_S$. Thus, there is some volume $V_S - v_m$ accessible for the rigid cluster motion. Hence, there is some volume $\Delta V_{c.m.}$ accessible for the motion of the cluster's center of mass. Note that $\Delta V_{c.m.}$ is the volume over which the integration is made in Equation 19.40. The estimations made in Ref. [38] have shown that the size of the integration region is small enough, that is, the variation of $R_{c.m.}$ when taking the integral in Equation 19.40 occurs in a very narrow region. Therefore, the variation of the integrand function U_{int} is rather weak during the integration over $dR_{c.m.}$. Thus, we can consider the interaction potential U_{int} as independent of $R_{c.m.}$. Then the integral in Equation 19.40 can be written as the product of two integrals and we arrive at

$$Z_c = d^3 \int_{r^m \in V_S} \exp\left(-\frac{U_{int}}{k_B T} \right) \sin\theta \, d\theta \, d\phi \, d\psi = d^3 \exp\left(\frac{S_c'}{k_B} \right) \exp\left(-\frac{\langle U_{int} \rangle_c}{k_B T} \right) \qquad (19.45)$$

where $d^3 = \Delta V_{c.m.}$.

The integral in Equation 19.45 may be regarded as the configurational partition function of the embedded cluster due to its rotational degrees of freedom only, when it is subjected to the external field $U_{int}(\theta, \phi, \psi)$. As a consequence, the configurational entropy S_c' is associated here with these degrees of freedom, that is, it has a definite physical sense of the entropy linked with the cluster's rotational motion (in contrast to the formal entropy in Equation 19.40). The entropy S_c' has a definite physical sense now, because instead of the arbitrary length scale δ in Equation 19.40 we have in Equation 19.45 a definite and natural characteristic length scale of the system d which is the size of the integration area. This quantity d substitutes δ now in Equation 19.44.

We now assume that the average values in Equation 19.44 can be substituted by the products of averages:

$$\Phi_K \approx \frac{\langle \varsigma_1 \varsigma_2 \varsigma_3 \rangle_l}{\langle \Lambda_{c.m.}^3 \rangle_l} 8\pi^2 V \frac{\langle \Lambda_{c.m.}^3 \rangle_l}{\langle d^3 \rangle_l \langle \varsigma_1 \varsigma_2 \varsigma_3 \exp(S_c'/k_B) \rangle_l} \qquad (19.46)$$

One more assumption is made that the average magnitude of the volume d^3 over which the center of mass sweeps during the integration in Equation 19.40 is approximately equal to the volume in which the center of mass of the drop of volume $V_S = 4/3\pi R_S^3$ fluctuates:

$$\langle d^3 \rangle_l = \left(\sqrt{2\pi}\sigma \right)^3 \qquad (19.47)$$

In Equation 19.47 σ must be expressed by (see Equation 19.36)

$$\sigma = \frac{(0.2)V_S^{1/3}}{\langle m \rangle_l^{1/2}} = \frac{(0.2)V_S^{1/3}}{(n^\beta V_S)^{1/2}} = \frac{(0.2)}{(n^\beta)^{1/2} V_S^{1/6}} \qquad (19.48)$$

where
$<m>_l = n^\beta V_S$
n^β is the particle number density of the reference bulk liquid

After the cancellation, using Equation 19.47 we get from Equation 19.46:

$$\Phi_K \cong \Phi_R \frac{Q_{rot}^K}{Q_{rot,l}^K} \equiv \Phi \qquad (19.49)$$

where Φ_R is the Reiss factor (Equation 19.35) in which σ is described by Equation 19.48. The quantity Q_{rot}^K in the numerator of Equation 19.49 is $Q_{rot}^K = 8\pi^2 \langle \varsigma_1 \varsigma_2 \varsigma_3 \rangle_l$, where $\langle \ \rangle_l$ denotes the thermal averaging for various values of m taken with respect to the isothermal–isobaric ensemble representing the bulk liquid. Due to this averaging Q_{rot}^K has a physical sense of the rotational partition function for a free drop in the vapor, but not for a cluster. According to Kusaka [11] this drop contains $\langle m \rangle_l = n^\beta V_S$ particles and is a ball of radius R_S with the homogeneous number density n^β. Therefore, we must write [39]

$$Q_{rot}^K = \frac{\sqrt{\pi}(8\pi^2 I k_B T)^{3/2}}{h^3} \qquad (19.50)$$

where I is the moment of inertia of the spherical drop:

$$I = \frac{8}{15}\pi R_S^5 \rho^\beta \qquad (19.51)$$

where ρ^β is the density of the reference bulk liquid.

Finally, the quantity $Q_{rot,l}^K$ is

$$Q_{rot,l}^K = \left\langle \varsigma_1 \varsigma_2 \varsigma_3 \exp\left(\frac{S_c'}{k_B}\right) \right\rangle_l \tag{19.52}$$

As was shown previously the quantity S_c' in Equation 19.46 (as well as in Equation 19.52) has a physical sense of the entropy related to the rotational movement of embedded cluster. Therefore, $Q_{rot,l}^K$ (due to the averaging in Equation 19.52) has a sense of the rotational partition function for the embedded spherical drop of volume V_S. Following Frenkel's views one can hope that the rotational movement of such a drop with respect to the environmental viscous liquid at low enough temperature near the melting point will be like the rotational vibrations of a solid body. The necessary condition for this kind of vibrations is that the period of vibrations T_v be small with respect to the Maxwell relaxation time τ_M of shear stress which occurs during the drop rotations [1]: $T_v \ll \tau_M$ or $v\tau_M \gg 1$ (where $v = 1/T_v$ is the vibrational frequency). One may expect that the last inequality is valid for the vibrational processes with the frequency v equal to the Debye frequency v_{max} (or even less) [38].

There is a series of publications devoted to the evaluation of the vibrational frequency of the solid cluster [40–42]. These evaluations have shown that the frequency of rotational vibrations v of small clusters with the number of molecules $n \approx 100$ is about the Debye frequency v_{max}. Taking into account the fact that at the ordinary temperatures the partition function of an oscillator depends weakly enough (only linearly) on frequency, one can use the simplest formula for the vibrational frequency [40] $v \cong \sqrt{3} n^{-1/6} v_{max}$. One can see this formula gives $v \cong v_{max}$ for $n = 27$. We suppose that the equation $v \cong v_{max}$ is valid for all the small clusters. Thus, we assume that the rotational partition function $Q_{rot,l}^K$ can be written as the vibrational partition function for three degrees of freedom:

$$Q_{rot,l}^K \cong (Q_v)^3 \cong \left\{ \exp\left(\frac{h v_{max}}{2 k_B T}\right) \left[1 - \exp\left(-\frac{h v_{max}}{k_B T}\right) \right] \right\}^{-3} \tag{19.53}$$

19.6 ASSEMBLY OF DROPS: THE CORRECTION FACTOR FOR THE NUCLEATION RATE

The Kusaka factor considered in the previous section is to give a connection between the partition function (q_n^{rest}) of the drop at rest and the partition function (q_n) of the drop in motion:

$$q_n = \Phi_K q_n^{rest} = \Phi_K \exp\left(-\frac{f_n^{rest}}{k_B T}\right) \tag{19.54}$$

Note that the drop at rest is implied here to be such a drop the work of formation for which W is governed by the expression Equation 19.8 derived by Gibbs; f_n^{rest} is the Helmholtz free energy for this drop. Strictly speaking, Equation 19.8 is valid only for the critical nucleus. However, the rate of nucleation is proportional to the number N_{crit} of the critical nuclei [1,11,32]. Therefore, to evaluate the nucleation rate one should know just the work of formation of these critical nuclei.

To know the number of critical nuclei, one should derive the equilibrium drop size distribution. We will use Reiss's approach [31] to obtain this distribution. To determine the equilibrium size distribution, Reiss has considered an ensemble of physical clusters as an ideal gas mixture [32]. It was assumed in this model that these clusters did not interact with the molecules of the environmental ideal vapor. Therefore, the full partition function for the system of volume V consisting of $\sum_{n=2} N_n$ drops and N_1 ideal vapor monomeric molecules can be written in the following form [31]:

$$Q(V) = \sum_{\{N_n\}} Q^{id}_{N_1, N_n} \left(V - \sum_{n=2} N_n v_n \right) \prod_{n=2} \frac{q_n^{N_n}}{N_n!} \tag{19.55}$$

where $Q^{id}_{N_1, N_n}$ is the partition function of ideal vapor. The first sum in Equation 19.55 is over all distributions N_n such that

$$N_1 + \sum_{n=2} n N_n = N \tag{19.56}$$

where
   N is the total number of molecules in the system

The equilibrium distribution is found in the usual manner by finding the maximum term in the sum of Equation 19.55 subject to the conservation condition, Equation 19.56. The result is [31]

$$N_n = q_n \exp\left\{ -\frac{1}{k_B T} (P v_n - n \mu_v) \right\} \tag{19.57}$$

where P is the ambient ideal gas pressure. The substitution of Equation 19.54 into Equation 19.57 yields

$$N_n = \frac{\Phi_K}{N_1} N_1 \exp\left\{ -\frac{f_n^{rest} + P v_n - n \mu_v}{k_B T} \right\} \tag{19.58}$$

where $f_n^{rest} + P v_n = g_n^{rest}$ is the Gibbs free energy for the drop at rest. Equation 19.58 differs from the CNT expression for the drop size distribution [1] (Equation 19.29) by the factor Φ_K/N_1. Thus, the classical Equation 19.29 is to be modified by the correction factor Φ_K/N_1 and, as the nucleation rate is proportional to the number N_{crit} of critical nuclei, the same free energy correction factor will appear in the classical expression for the nucleation rate.

19.7   COMPARISON WITH THE KUSAKA'S NUMERICAL SIMULATION RESULTS

To distinguish the Kusaka factor Φ_K calculated numerically [11] from the correction factor evaluated analytically we denote the quantity $\Phi_R(Q^K_{rot}/Q^K_{rot,l})$ by Φ (see Equation 19.49). Combining Equations 19.35, 19.49 through 19.50, and 19.53 we propose the following expression for the correction factor Φ/N_1:

$$\frac{\Phi}{N_1} = \frac{\Phi_R}{N_1} \frac{Q^K_{rot}}{Q^K_{rot,l}} = \frac{1}{S n_1^{sat} (2\pi)^{3/2} \sigma^3} \times \pi^5 \left(\frac{64 R_s^5 \rho^\beta k_B T}{15 h^2} \right)^{3/2} \exp\left(\frac{3h\nu_{max}}{2 k_B T} \right) \left\{ 1 - \exp\left(-\frac{h\nu_{max}}{k_B T} \right) \right\}^3 \tag{19.59}$$

where
   $S n_1^{sat} = n_1$
   S is the supersaturation ratio
   n_1^{sat} is the saturated vapor number density

σ is to be evaluated by Equation 19.48. To estimate the Debye frequency one may use the following formula [1]: $\nu_{max} \approx u/2r$, $2r$ is the molecule diameter and u is the sound velocity in the bulk liquid reference phase.

Let us compare our correction factor Φ/N_1 as estimated by Equation 19.59 for Ar with that as determined by Kusaka (Φ_K/N_1) numerically by Equation 19.44 [11].

Figure 19.1 shows the correction factor Φ/N_1 as calculated [38] by Equation 19.59 for $R_S=2.0$ (in Lennard–Jones units) and supersaturation ratio $S=2.0$ as well as the contributions to this factor from Φ_R/N_1, Q_{rot}^K, and $(Q_\nu)^{-3}$. The important observation is that the contribution from the translational degrees of freedom Φ_R/N_1 is in the range $3 \times 10^2 - 10^4$, while the main contribution comes from the rotational degrees of freedom Q_{rot}^K which is about 3×10^7 in the temperature range 80–110 K. The vibrational contribution $(Q_\nu)^{-3}$ is small enough being in the range 1–8, which corresponds to the range of Q_ν from 1 to 0.5. Such a small magnitude of the vibrational partition function corresponds to the condition $k_B T < h\nu_{max}$, that is, to the low-temperature limit where the quantum nature of the oscillator is to be accounted; that is why the Equation 19.53 was written in the quantum form.

Figure 19.2 compares the correction factor Φ/N_1 and that of Kusaka Φ_K/N_1 for $R_S=1.8$, 2.0, and 2.5. One can see that these two factors are in a good agreement with each other. The temperature

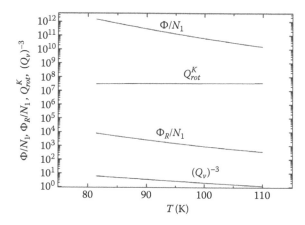

FIGURE 19.1 Correction factor Φ/N_1 (Equation 19.59) and contributions from Φ_R/N_1 (Equations 19.35 and 19.48), Q_{rot}^K (Equation 19.50), and $(Q_\nu)^{-3}$ (Equation 19.53) versus temperature for $R_S=2.0$ (in Lennard–Jones units), $S=2.0$.

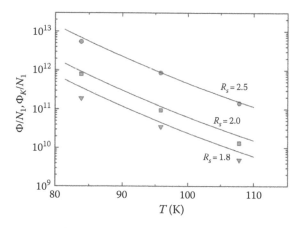

FIGURE 19.2 Free energy correction factor for different values of the drop radius R_S versus temperature. Symbols, Φ_K/N_1 (Numerical calculations by Kusaka, I., *Phys. Rev. E.*, 73, 031607, 2006.); lines, Φ/N_1 (as evaluated by Equation 19.59). The magnitudes of R_S are shown in Lennard–Jones units.

dependence of Φ/N_1 is in excellent agreement with that of Φ_K/N_1. As one can see from Figure 19.1 the dominant contribution to the temperature dependence comes from the Reiss factor Φ_R/N_1. However, the last factor $(Q_v)^{-3}$ in Equation 19.59 also depends significantly on temperature. The temperature dependence for $(Q_v)^{-3}$ is mainly governed by the first factor in Equation 19.53 which takes into account the zero-point vibrations. Without considering this factor one will lose the agreement between the temperature dependencies of Φ/N_1 and Φ_K/N_1.

The excellent agreement between the temperature dependencies for Φ/N_1 and Φ_K/N_1 supports our assumption that the rotational degrees of freedom of embedded cluster which must to be deactivated are, actually, the torsion vibrations.

19.8 APPROXIMATE ANALYTICAL FORMULA FOR THE NUCLEATION RATE: ITS APPLICATION TO THE ESTIMATION OF SURFACE TENSION OF CRITICAL NUCLEUS FROM THE EXPERIMENTAL SUPERSATURATION RATIO AND NUCLEATION RATE

The thermodynamic properties of the interphase surface are fully characterized by a state function which is called surface tension σ_S. Therefore, the knowledge of the surface tension is necessary to understand the mechanism of nanoparticle nucleation, and stability and evolution of nanosystems. In this section we will outline an application of the analytical formula for the nucleation rate to determine the surface tension of the critical drop. The metal vapor nucleation will be considered as an example.

There are experimental contributions studying the nucleation of Li, Na, Ag, Hg, Mg, Zn where the supersaturation versus temperature was measured for the nucleation rate considered as constant and being evaluated rather approximately [5–7,43–45]. However, the surface tension is a function of both radius and temperature, and to understand the interfacial thermodynamics one must determine first the dependence of surface tension on the drop radius (at constant temperature) and then the temperature dependence of surface tension. Therefore, the experimental measurements of the nucleation rate at different supersaturation ratios and fixed temperature are important. To our knowledge there is only one such investigation in the field of nucleation of metals [46]. In Ref. [46] the rate of nucleation from the vapor of Cs was measured as a function of supersaturation for three temperatures 508, 530, and 554 K. Therefore, we will start from the consideration of the experimental data on Cs vapor nucleation which are plotted in Figure 19.3. The original experimental data are shown as circles. For the convenience of presentation we have averaged the neighboring points (diamonds).

Substituting the factor K in Equation 19.24 by Φ/N_1 (Equation 19.59) and W_{crit} by Equation 19.8 and assuming $\chi \approx 1$ we get the final formula for the nucleation rate:

$$I \approx \frac{1}{Sn_1^{sat}(2\pi)^{3/2}\sigma^3} \times \pi^5 \left(\frac{64R_S^5\rho^\beta k_B T}{15h^2}\right)^{3/2} \exp\left(\frac{3h\nu_{max}}{2k_B T}\right) \left\{1 - \exp\left(-\frac{h\nu_{max}}{k_B T}\right)\right\}^3$$

$$\times \left(Sn_1^{sat}\right)^2 \sqrt{\frac{2m\sigma_S(R_S)}{\pi}} \frac{1}{\rho^\beta}\left[1 + \left(\frac{\delta}{R_S}\right)^2 f\left(\frac{\delta}{R_S}\right)\right]^{-1/2} \left(1 + \frac{d\delta}{dR_S}\right)^{-1/2} \exp\left(-\frac{4\pi R_S^2\sigma_S(R_S)}{3k_B T}\right) \quad (19.60)$$

It will be shown in the following that the factor $[1 + (\delta/R_S)^2 f(\delta/R_S)]^{-1/2} (1 + (d\delta/dR_S))^{-1/2}$ can be considered as equal to unity with adequate accuracy, and the quantum form of the vibrational partition function $Q_{rot,l}^K$ (see Equations 19.53 and 19.59) can be substituted by the classical form: $Q_{rot,l}^K \approx (k_B T/h\nu_{max})^3$ at high temperatures. Then, accounting Equation 19.48 one can rewrite Equation 19.60 as

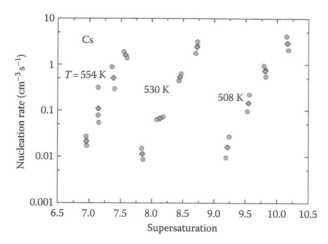

FIGURE 19.3 Rate of homogeneous nucleation from the supersaturated vapor of Cs versus supersaturation ratio for different temperatures (shown in the plot) as measured in Ref. [46]. Circles, original points; diamonds, after averaging over neighboring points.

$$I \approx \pi^{7/2} Sn_1^{sat} \frac{8^3 5^{3/2}}{9} \frac{(\rho^\beta)^2 R_S^9 v_{max}^3}{m(k_B T)^{3/2}} \sqrt{\sigma_S(R_S)} \exp\left(-\frac{4\pi R_S^2 \sigma_S(R_S)}{3 k_B T} \right)$$

$$\approx 6 \times 10^2 \pi^{7/2} Sn_1^{sat} \frac{(\rho^\beta)^2 R_S^9 v_{max}^3}{m(k_B T)^{3/2}} \sqrt{\sigma_S(R_S)} \exp\left(-\frac{4\pi R_S^2 \sigma_S(R_S)}{3 k_B T} \right) \qquad (19.61)$$

The Debye frequency in Equations 19.60 and 19.61 can be evaluated in our case as

$$v_{max}^3 \approx \left(\frac{u}{2r} \right)^3 \approx \frac{u^3}{v_l} = \frac{u^3}{m} \rho^\beta \qquad (19.62)$$

The sound velocity can be estimated as [47]

$$u \approx \frac{1}{\sqrt{\kappa \rho^\beta}} \qquad (19.63)$$

where κ is the isothermal compressibility of the bulk liquid reference phase.

As a first iteration it is possible to solve Equation 19.61 together with the Kelvin equation [36]:

$$\ln S = \frac{2\sigma_S(R_S)m}{k_B T \rho^\beta R_S} \qquad (19.64)$$

to obtain σ_S and R_S for each point shown in Figure 19.3 (at fixed I, S, and T).

v_{max} is the Debye frequency for the liquid bulk phase at pressure $P^\beta = P^\alpha + \Delta P$, where ΔP is given by the formula [12]:

$$\Delta P = \frac{2\sigma_S(R_S)}{R_S} \qquad (19.65)$$

Considering that P^α in the nucleation experiments is normally not far from 1 atm and ΔP for the critical drop is higher essentially we can put $P^\beta \cong \Delta P$. To know the Debye frequency as a function of pressure it is enough to know the dependence of sound velocity on pressure (see Equation 19.62) (the density ρ^β is a very weak function of pressure). In Ref. [48] the sound velocity of liquid Cs was measured in the range of temperatures 300–1500 K and pressure $2 \times 10^6 – 6 \times 10^7$ Pa. The experimental measurements can be approximated by the following empirical function:

$$u(\text{m/s}) \approx 1.475 \times 10^2 (P(\text{Pa}) + P_0(T))^{0.1} \tag{19.66}$$

where $P_0 = (7.35, 6.78, \text{ and } 6.10) \times 10^7$ Pa for temperatures 508, 530, and 554 K, respectively. For the first iteration we set the pressure to be 1 atm (1.013×10^5 Pa), then, using the density of liquid Cs from Table 19.1 and Equations 19.66 and 19.62 we get $u = 903$, 895, and 886 m/s and $\nu_{max} = (1.80, 1.79, \text{ and } 1.77) \times 10^{12}$ s^{-1} for $T = 508$, 530, and 554 K, respectively.

The joint solution of Equations 19.61, 19.62, and 19.64 is shown in Figure 19.4 as $\ln(\sigma_S/\sigma_\infty)$ versus $\ln R_S$. The Tolman length δ can be determined solving the GTKB differential equation (19.5) which contains $d \ln (\sigma_S(R_S)/\sigma_\infty)/d \ln(R_S)$ in the LHS and the RHS contains only δ/R_S. One can see from Figure 19.4 that within the experimental accuracy the dependence $\ln(\sigma_S(R_S)/\sigma_\infty)$ on $\ln(R_S)$ can be approximated by a linear function with a constant slope $d \ln(\sigma_S(R_S)/\sigma_\infty)/d \ln(R_S) = -0.240$,

TABLE 19.1

Important Parameters for Liquid Metals Used in Calculations

Metal	T_0 (K)	ψ (Dyne/cm/K)	C	D (K)	ρ (g/cm³)	κ (cm²/Dyne)
Li	3300	0.140	8.00	8120	0.47	1.0×10^{-11}
Na	2350	0.100	7.70	5460	0.87	2.3×10^{-11}
Cs	1790	0.048	7.25	3920	1.75	7.0×10^{-11}
Ag	6980	0.161	8.92	14464	9.35	1.4×10^{-12}
Hg	2540	0.210	8.04	3215	12.8	3.8×10^{-12}
Mg	3128	0.254	8.80	7674	1.59	5.1×10^{-12}
Zn	5430	0.167	8.35	6400	6.75	1.9×10^{-12}

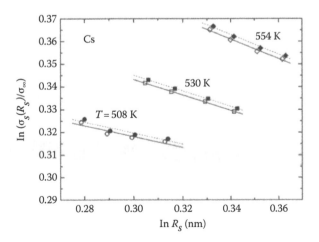

FIGURE 19.4 $\ln(\sigma_S/\sigma_\infty)$ versus $\ln R_S$ as calculated by solution of Equations 19.61 and 19.64 using the experimental measurements of the nucleation rate from the vapor of Cs (Figure 19.3). Open and filled symbols are results of the first and second iterations, respectively. Lines are linear fittings.

−0.34, and −0.43 for $T = 508$, 530, and 554 K, respectively. In other words, the RHS of Equation 19.5 is independent of R_S, that is, $\delta/R_S = const$, or $\delta = const \cdot R_S$. Then, the numerical solution of the GTKB differential equation (19.5) gives $\delta/R_S = d\delta/dR_S = -0.11$, −0.15, −0.18 for $T = 508$, 530, and 554 K, respectively. Substituting the values of $\sigma_S(R_S)$ and R_S from Figure 19.4 to Equation 19.65 we got the pressure $\Delta P \cong P^\beta$ in the range $(1.16–1.27) \times 10^8$ Pa for $508 < T < 554$ K. This pressure gives the sound velocity $u = 999 – 986$ m/s and the Debye frequency $v_{max} = (1.99 – 1.97) \times 10^{12}$ s^{-1}, respectively.

It is possible now to calculate the factor $[1 + (\delta/R_S)^2 f(\delta/R_S)]^{-1/2}(1 + (d\delta/dR_S))^{-1/2}$, and determine $\sigma_S(R_S)$ using the exact Equation 19.60 and the newly determined values of v_{max} (second iteration). As shown in Figure 19.4 the magnitudes of $\sigma_S(R_S)$ as determined in the first and second iterations differ by 0.2% but the derivatives $d \ln(\sigma_S(R_S)/\sigma_\infty)/d \ln(R_S)$ coincide absolutely which means that both the iterations give the same quantity δ/R_S. Thus, the approximate Equation 19.61 allows determining the surface tension of the critical nucleus with high accuracy. One should note that the substitution of v_{max} for the pressure of 1 atm instead of that for P^β gives negligible error for the drop surface tension but will result in the error of about 30% for the nucleation rate. The accuracy of the Equation 19.61 for the nucleation rate is given by the factor $[1 + (\delta/R_S)^2 f(\delta/R_S)]^{-1/2}(1 + (d\delta/dR_S))^{-1/2}$ and the ratio between the classical and quantum forms of the partition function $Q_{rot,l}^K$ which are in the ranges 1.05–1.08 and 1.0012–1.0015, respectively, for the experimental range of Ref. [46]. Thus, the total cost of simplifications for the Equation 19.61 for the rate of nucleation does not exceed 8%. Such a good accuracy in Equation 19.61 is due to the small value of δ/R_S. It must be noted that both the surface of tension and the equimolar dividing surface lie somewhere in the interfacial region [12] and, hence, the value $|\delta/R_S| = |(R_e/R_S) - 1|$ is expected to be less than unity in most cases. Therefore, one can hope that the approximate Equation 19.61 is suitable to evaluate the drop surface tension for a wide range of systems and nucleation conditions.

In the following section of the article, we will evaluate the surface tension of critical nucleus for different monovalent and bivalent metals using the Equations 19.61 through 19.64 and the experimentally measured nucleation rate and supersaturation for different temperatures (see Figure 19.5). The important parameters used for evaluation of σ_S/σ_∞ are summarized in Table 19.1. The temperature dependence of the planar interface surface tension is presented by

$$\sigma_\infty = \psi(T_0 - T) \tag{19.67}$$

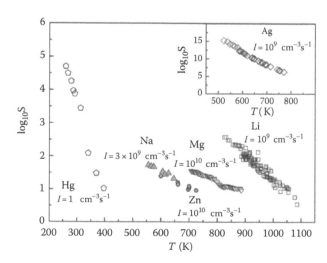

FIGURE 19.5 $\log_{10}S$ versus nucleation temperature for Li[43], Na[44], Ag[45], Mg[6], Zn[7], Hg[5] for invariant nucleation rates (shown in the plot for each metal).

where ψ and T_0 are parameters determined by fitting Equation 19.67 to the dependencies of surface tension on temperature available from reference books (see, e.g., [49,50]). The temperature dependence of saturated vapor pressure is given by the formula:

$$\log_{10}(P_{Sat}(Torr)) = C - \frac{D}{T} \tag{19.68}$$

where C and D are parameters determined by fitting Equation 19.68 to the reference books data on saturated vapor pressure. The sound velocity was evaluated using Equation 19.63. The values of σ_S/σ_∞ as determined from the solution of Equations 19.61 through 19.64 are plotted in Figure 19.6 for different temperatures. To avoid confusion we presented three points for each metal, corresponding to the ends and middle of the experimental temperature range. The points in Figure 19.6 are not dependences σ_S/σ_∞ versus R_S, but there are three independent values σ_S/σ_∞ for each metal corresponding to three different temperatures and three values of R_S. One can see from Figures 19.4 and 19.6 that all the metals considered can be divided into two groups Li, Na, Cs, Ag (monovalent metals) and Mg, Zn, Hg (bivalent metals). The monovalent metals are characterized by the ratio $\sigma_S/\sigma_\infty > 1$. The bivalent metals Zn, Hg, Mg show $\sigma_S/\sigma_\infty < 1$. It is important to note that in all the cases the surface tension for drops of radius about 1 nm differs essentially from that of the flat surface, that is, the surface tension is a strong function of size.

Table 19.2 summarizes important parameters which were determined from the experimental nucleation rate, supersaturation, and temperature by solution Equations 19.60/19.61 and 19.64. One can see that normally N_S is about a few tens of atoms, but for Ag it is $4 \leq N_S \leq 7$. Nevertheless, the Gibbs theory is rigorous even for such a small value N_S (because it is rigorous for the radius R_S whatever small). Nevertheless, Equations 19.60 and 19.61 can be considered as of less accuracy in the case of Ag with respect to those in the case of other metals, because Equations 19.3 and 19.4 for the Zeldovitch factor were derived under the assumption that the critical nucleus contains not less than 10–30 molecules [1]. However, even if the accuracy of Equations 19.60 and 19.61 in the case of Ag would be 50% it gives the error in the drop surface tension as low as 0.4%.

As seen from Table 19.2 the translational correction factor Φ_R/N_1 is about 10^6–10^7 for the majority of metals presented in Table 19.2 and only for Ag it is as high as 10^{11}. The reason in such a high value of this factor for Ag is in low concentration n_1 for the supersaturated vapor which is three to five orders of magnitude less than that for the other metals. The rotational correction factor Q_{rot}^K is

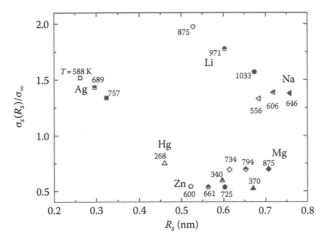

FIGURE 19.6 σ_S/σ_∞ and R_S as calculated by solution of Equations 19.61 and 19.64 using the experimental measurements as shown in Figure 19.5. Three points were selected for each metal corresponding to the ends and middle of the experimental temperature range. Temperature is shown for each point in the plot.

TABLE 19.2

Important Parameters That Were Determined from the Experimental Nucleation Rate, Supersaturation, and Temperature (Figures 19.3 and 19.5) by Solution of Equations 19.61 and 19.64 (for Li, Na, Ag, Mg, Zn, Hg) and Equations 19.60 and 19.64 (for Cs) (Second Iteration)

Metal	T (K)	S	n_1 (cm⁻³)	N_S	R_S (nm)	$\dfrac{\Phi_R}{N_1}$	Q_{nd}^K	$Q_{nd,l}^K$ Quantum Form	$Q_{nd,l}^K$ Classical Form	$\dfrac{\Phi}{N_1}$	$\sigma_S (R_S)$ (Dyne/cm)	$\dfrac{I_{exp}}{I_{CNT}}$
Li	875	173.0	1.0×10^{17}	27	0.54	1.6×10^{7}	3.8×10^{7}	1.02	1.06	4.3×10^{14}	683.0	4.3×10^{-12}
	971	33.6	1.5×10^{17}	40	0.62	1.3×10^{7}	1.2×10^{8}	1.14	1.17	8.4×10^{14}	591.0	5.7×10^{-11}
	1033	12.7	1.7×10^{17}	56	0.69	1.4×10^{7}	3.1×10^{8}	1.21	1.25	1.8×10^{15}	508.9	1.1×10^{-8}
Na	556	53.9	7.2×10^{15}	32	0.70	1.4×10^{8}	3.4×10^{8}	1.80	1.83	8.3×10^{15}	243.3	1.4×10^{-1}
	606	34.0	2.7×10^{16}	38	0.73	4.1×10^{7}	5.6×10^{8}	1.97	1.99	3.0×10^{15}	246.4	9.1×10^{-4}
	646	21.1	5.6×10^{16}	44	0.77	2.1×10^{7}	9.3×10^{8}	2.10	2.12	2.1×10^{15}	239.7	4.3×10^{-4}
Cs	508	9.2	6.0×10^{16}	85	1.37	9.0×10^{6}	1.3×10^{11}	5.32	5.33	8.4×10^{15}	84.5	6.0×10^{-10}
		9.6	6.2×10^{16}	82	1.35	8.6×10^{6}	1.2×10^{11}	5.31	5.32	7.3×10^{15}	84.7	1.7×10^{-9}
		9.8	6.4×10^{16}	79	1.34	8.3×10^{6}	1.1×10^{11}	5.31	5.31	6.5×10^{15}	84.8	3.7×10^{-9}
		10.2	6.6×10^{16}	77	1.32	7.8×10^{6}	1.0×10^{11}	5.30	5.31	5.7×10^{15}	85.2	4.6×10^{-9}
	530	7.9	1.0×10^{17}	93	1.41	5.6×10^{6}	1.8×10^{11}	5.58	5.58	6.0×10^{15}	84.2	5.3×10^{-11}
		8.1	1.1×10^{17}	90	1.39	5.3×10^{6}	1.6×10^{11}	5.57	5.58	5.2×10^{15}	84.5	9.4×10^{-11}
		8.5	1.1×10^{17}	86	1.37	5.0×10^{6}	1.4×10^{11}	5.56	5.57	4.5×10^{15}	84.9	2.0×10^{-10}
		8.7	1.1×10^{17}	83	1.36	4.8×10^{6}	1.3×10^{11}	5.56	5.56	4.0×10^{15}	85.2	3.3×10^{-10}
	554	7.0	1.8×10^{17}	99	1.44	3.2×10^{6}	2.2×10^{11}	5.86	5.86	3.8×10^{15}	84.5	3×10^{-12}
		7.1	1.9×10^{17}	96	1.42	3.1×10^{6}	2.0×10^{11}	5.85	5.86	3.4×10^{15}	84.8	5.8×10^{-12}
		7.4	1.9×10^{17}	92	1.41	2.9×10^{6}	1.9×10^{11}	5.85	5.85	3.0×10^{15}	85.2	9.0×10^{-12}
		7.6	2.0×10^{17}	90	1.39	2.8×10^{6}	1.7×10^{11}	5.84	5.85	2.7×10^{15}	85.6	1.2×10^{-11}
Ag	588	1.6×10^{12}	5.7×10^{12}	4	0.26	1.4×10^{11}	8.4×10^{6}	1.15	1.19	7.5×10^{7}	1.56×10^{3}	6.5×10^{0}
	689	3.7×10^{8}	4.5×10^{12}	6	0.30	2.1×10^{11}	2.7×10^{7}	1.36	1.39	2.2×10^{18}	1.45×10^{3}	2.6×10^{2}
	757	4.0×10^{6}	3.3×10^{12}	7	0.32	3.2×10^{11}	6.0×10^{7}	1.50	1.53	5.6×10^{18}	1.34×10^{3}	4.0×10^{4}
Mg	734	30.9	9.1×10^{15}	38	0.61	2.1×10^{8}	4.9×10^{8}	1.24	1.27	5.4×10^{16}	420.5	1.7×10^{74}
	794	18.6	3.1×10^{16}	46	0.65	6.8×10^{7}	8.7×10^{8}	1.35	1.38	2.4×10^{16}	411.9	1.3×10^{74}
	875	10.7	1.3×10^{17}	58	0.71	1.9×10^{7}	1.8×10^{9}	1.49	1.52	1.0×10^{16}	398.5	6.6×10^{74}
Zn	600	25.0	2.0×10^{15}	40	0.54	1.6×10^{9}	1.2×10^{9}	1.09	1.12	1.4×10^{18}	436.7	2.2×10^{167}
	661	14.0	9.6×10^{15}	51	0.59	3.6×10^{8}	2.4×10^{9}	1.21	1.24	5.0×10^{17}	425.9	3.7×10^{178}
	725	9.2	4.1×10^{16}	62	0.63	9.2×10^{7}	4.6×10^{9}	1.33	1.36	1.8×10^{17}	420.1	7.0×10^{182}
Hg	268	3.1×10^{4}	1.2×10^{17}	17	0.48	9.9×10^{6}	3.6×10^{8}	1.12	1.15	2.6×10^{15}	349.2	5.4×10^{75}
	340	1.2×10^{2}	1.4×10^{17}	38	0.62	1.3×10^{7}	3.7×10^{9}	1.44	1.46	1.6×10^{16}	268.4	9.8×10^{178}
	370	3.0×10^{1}	1.8×10^{17}	54	0.70	1.2×10^{7}	1.0×10^{10}	1.57	1.59	3.1×10^{16}	233.2	3.0×10^{276}

in the range 10^7–10^9 for all metals except for Cs for which it is about 10^{11}. The high values of Q_{rot}^K for Cs are due to the high radius R_S of the critical nucleus which is a few times larger for Cs than that for other metals. The vibrational partition function $Q_{rot,l}^K$ is not less than 0.91 for all the cases which gives the difference between the quantum and classical forms of the partition function not higher than 4% and it looks quite reasonable to substitute the quantum form to the classical one. Thus, in the case of metals the total correction factor is very large being 10^{15}–10^{19}. It is necessary to note once again that this correction factor takes into account the translational and rotational degrees of freedom of the critical nucleus and, therefore, the quantity $\sigma_S(R_S)$ presented in Table 19.2 is referred to the critical nucleus at rest and, hence, has the same nature as the capillary meniscus surface tension. The last column in Table 19.2 gives the ratio between the nucleation rate measured experimentally (I_{exp}) and that predicted by CNT (I_{CNT}). The CNT prediction was calculated by Equation 19.9 with R_e estimated by the classical form of the Kelvin formula:

$$R_e = \frac{2\sigma_\infty m}{k_B T \rho \ln S} \tag{19.69}$$

One should note that in general case Equation 19.69 is applicable only for large drops ($R_e > 10$–$100\,nm$); however, for the Lennard–Jones systems it was found to be correct even for 1 nm size drops [24,25]. As seen from Table 19.2 the monovalent metals demonstrate the ratio between the experimental measurements and CNT predictions of the nucleation rate in the range 10^4–10^{-12}. On the other hand, for bivalent metals the CNT nucleation rate is many tens of orders of magnitude lesser than the experimental rate. This difference between mono- and bivalent metals is related to the fact that the CNT formula neglects both the correction factor (underestimating the nucleation rate) and the dependence of σ_S on the drop radius (overestimating the nucleation rate in the case of monovalent metals and underestimating it in the case of bivalent metals). Thus, the neglect of Φ/N_1 and the size dependence of σ_S in the case of monovalent metals gives a partial compensation resulting in decrease of the total error in the CNT formula, but in the case of bivalent metals both these factors act in concert increasing the total error.

Thus, the considerations of this section show that the formula for the nucleation rate is a useful tool to study the thermodynamic properties of the nano-sized systems. The approximate formula Equation 19.61 can be used in calculations of the nucleation rate giving high enough accuracy.

19.9   CONCLUSION

An analytical formula for the translation-rotation correction factor is proposed. The formula is based on the Reiss approach considering the contribution from the clusters translational degrees of freedom, the Frenkel's kinetic theory of liquids, and Kusaka's theory. The formula for Zeldovitch factor that takes into account the size dependence of the critical nucleus surface tension is also derived. With the help of these two formulas the rigorous expression Equation 19.60 for the rate of vapor-to-liquid homogeneous nucleation is obtained. However, despite being rigorous the expression cannot be used for direct calculation of the nucleation rate as it includes the unknown quantity: the surface tension of critical nucleus σ_S. At the same time the expression 19.60 is a useful tool in the solution of reverse problem, that is, the investigation of the thermodynamic properties of interface surface which are fully characterized by the state function σ_S. Such an investigation was carried out in this paper for some mono- and bivalent metals. Using Equation 19.60 (or simplified Equation 19.61) we determined the critical drops' surface tension and radius of the surface of tension R_S. It was found that the metals considered can be divided in accordance with their valency: for monovalent metals $\sigma_S > \sigma_\infty$ and for bivalent ones $\sigma_S < \sigma_\infty$. It is important to note that for all the metals σ_S differs essentially from the flat interface surface tension σ_∞. In advance this large difference between σ_S and σ_∞ could be expected only for bivalent metals as the ratio between the experimental and CNT magnitudes of the nucleation rate is about hundred orders of magnitude for this group.

In the case of Cs, a more detailed investigation was carried out which showed that the approximate Equation 19.61 can be used instead of the rigorous Equation 19.60 to calculate the nucleation rate from critical nucleus surface tension σ_S and supersaturation ratio S or the surface tension and radius R_S of critical drop from the nucleation rate and S with adequately high accuracy.

REFERENCES

1. J. Frenkel, *Kinetic Theory of Liquids*. Dover, New York, 1946.
2. J. Wö, R. Strey, C.H. Heath, and B.E. Wyslouzil, Empirical function for homogeneous water nucleation rates, *J. Chem. Phys.*, 2002, 117, 4954–4960.
3. M. Rusyniak, V. Abdelsayed, J. Campbell, and M.S. El-Shall, Vapor phase homogeneous nucleation of higher alkanes: Dodecane, hexadecane, and octadecane. 1. Critical supersaturation and nucleation rate measurements, *J. Phys. Chem. B*, 2001, 105, 11866–11872.
4. K. Iland, J. Wedekind, J. Wö, P.E. Wagner, and R. Strey, Homogeneous nucleation rates of 1-pentanol, *J. Chem. Phys.*, 2004, 121, 12259–12264.
5. J. Martens, H. Uchtmann, and F. Hensel, Homogeneous nucleation of mercury vapor, *J. Phys. Chem.*, 1987, 91, 2489–2492.
6. F.T. Ferguson, J.A. Nuth III, and L.U. Lilleleht, Experimental studies of the vapor phase nucleation of refractory compounds. IV. The condensation of magnesium, *J. Chem. Phys.*, 1996, 104, 3205–3210.
7. A.A. Onischuk, P.A. Purtov, A.M. Baklanov, V.V. Karasev, and S.V. Vosel, Evaluation of surface tension and Tolman length as a function of droplet radius from experimental nucleation rate and supersaturation ratio: Metal vapor homogeneous nucleation, *J. Chem. Phys.*, 2006, 124, 014506 (1–13).
8. J. Lothe and G.M. Pound, Reconsiderations of the nucleation theory, *J. Chem. Phys.*, 1962, 36, 2080–2085.
9. H. Reiss and J.L. Katz, Resolution of the translation-rotation paradox in the theory of irreversible condensation, *J. Chem. Phys.*, 1967, 46, 2496–2499.
10. H. Reiss, J.L. Katz, and E.R. Cohen, Translation-rotation paradox in the theory of nucleation, *J. Chem. Phys.*, 1968, 48, 5553–5560.
11. I. Kusaka, Statistical mechanics of nucleation: Incorporating translational and rotational free energy into thermodynamics of a microdroplet, *Phys. Rev. E.*, 2006, 73, 031607 (1–10).
12. J.W. Gibbs, *Thermodynamics and Statistical Mechanics*, Nauka, Moscow, Russia, 1982.
13. V.G. Baidakov and G.Sh. Boltachev, Curvature dependence of the surface tension of liquid and vapor nuclei, *Phys. Rev. E*, 1999, 59, 469–475.
14. I. Hadjiagapiou, Density functional theory for spherical drops, *J. Phys. Condens. Matter.*, 1994, 6, 5303–5322.
15. D.J. Lee, M.M. Telo da Gama, and K.E. Gubbins, A microscopic theory for spherical interfaces: Liquid drops in the canonical ensemble, *J. Chem. Phys.*, 1986, 85, 490–499.
16. M.A. Hooper and S. Nordholm, Generalized van der Waals theory. XII. Curved interfaces in simple fluids, *J. Chem. Phys.*, 1984, 81, 2432–2438.
17. R. Guermeur, F. Biquard, and C. Jacolin, Density profiles and surface tension of spherical interfaces. Numerical results for nitrogen drops and bubbles. *J. Chem. Phys.*, 1985, 82, 2040–2051.
18. T.V. Bykov and A.K. Shchekin, A surface tension, the Tolman length and effective rigidity constant of the droplet with large radius of curvature, *Inorg. Mater.*, 1999, 35, 641–645.
19. T.V. Bykov and A.K. Shchekin, Thermodynamical characteristics of the small droplet in terms of density functional method, *Colloid J.*, 1999, 1, 144–151.
20. A.H. Falls, L.E. Scriven, and H.T. Davis, Structure and stress in microstructures, *J. Chem. Phys.*, 1981, 75, 3986–4002.
21. A.H. Falls, L.E. Scriven, and H.T. Davis, Adsorption, structure and stress in binary interfaces, *J. Chem. Phys.*, 1983, 78, 7300–7317.
22. K. Koga, X.C. Zeng, and A.K. Shchekin, Validity of Tolman's equation: How large should a droplet be? *J. Chem. Phys.*, 1998, 109, 4063–4070.
23. M. Iwamatsu, A double-parabola model for the non-classical Cahn-Hilliard theory of homogeneous nucleation, *J. Phys. Condens. Matter.*, 1993, 5, 7537–7550.
24. L.S. Bartell, Tolman's δ, surface curvature, compressibility effects, and the free energy of drops, *J. Phys. Chem. B*, 2001, 105, 11615–11615.

25. P.R. ten Wolde and D. Frenkel, Computer simulation study of gas–liquid nucleation in a Lennard-Jones system, *J. Chem. Phys.*, 1998, 109, 9901–9918.
26. J. Julin, I. Napari, J. Merikanto, and H. Vehkamaki, Equilibrium sizes and formation energies of small and large Lennard-Jones clusters from molecular dynamics: A consistent comparison to Monte Carlo simulations and density functional theories, *J. Chem. Phys.*, 2008, 129, 234506 (1–8).
27. S.M. Thompson, K.E. Gubbins, J.P.R.B. Walton, R.A.R. Chantry, and J.S. Rowlingson, A molecular dynamic study of liquid drops, *J. Chem. Phys.*, 1994, 81, 530–542.
28. J. Vrabec, G.K. Kedia, G. Fuchs, and H. Hasse, Comprehensive study of the vapour-liquid coexistence of the truncated and shifted Lennard-Jones fluid including planar and spherical interface properties, *Mol. Phys.*, 2006, 104, 1509–1527.
29. R. Bahadur and L.M. Russell, Effect of surface tension from MD simulations on size-dependent deliquescence of NaCl nanoparticles, *Aerosol Sci. Technol.*, 2008, 42, 369–376.
30. H. El Bardouni, M. Mareschal, R. Lovett, and M. Baus, Computer simulation study of the local pressure in a spherical liquid–vapor interface, *J. Chem. Phys.*, 2000, 113, 9804–9809.
31. H. Reiss, W.K. Kegel, and J.L. Katz, Role of the model dependent translational volume scale in the classical theory of nucleation, *J. Phys. Chem.*, 1998, 102, 8548–8555.
32. H. Reiss, The replacement free energy in nucleation theory, *Adv. Colloid Interface Sci.*, 1977, 7, 1–66.
33. R.C. Tolman, The effect of droplet size on surface tension, *J. Chem. Phys.*, 1949, 17, 333–337.
34. K. Nishioka, H. Tomino, I. Kusaka, and T. Takai, Curvature dependence of the interfacial tension in binary nucleation, *Phys. Rev. A*, 1989, 39, 772–782.
35. S.V. Vosel, A.A. Onischuk, and P.A. Purtov, Response to comment on evaluation of surface tension and tolman length as a function of droplet radius from experimental nucleation rate and supersaturation ratio: Metal vapor homogeneous nucleation [*J. Chem. Phys.* 133, 047101 (2010)], *J. Chem. Phys.*, 2010, 133, 047102 (1–3).
36. K. Nishioka and I. Kusaka, Thermodynamic formulas of liquid phase nucleation from vapor in multicomponent systems, *J. Chem. Phys.*, 1992, 96, 5370–5376.
37. P.G. Debenedetti and H. Reiss, Reversible work of formation of an embryo of a new phase within a uniform macroscopic mother phase, *J. Chem. Phys.*, 1998, 108, 5498–5505.
38. S.V. Vosel, A.A. Onischuk, and P.A. Purtov, Translation-rotation correction factor in the theory of homogeneous nucleation, *J. Chem. Phys.*, 2009, 131, 204508 (1–11).
39. L.D. Landau and E.M. Lifshitz, *Statistical Physics*. Nauka, Moscow, Russia, 1976.
40. J. Lothe and G.M. Pound, On the statistical mechanics of nucleation theory, *J. Chem. Phys.*, 1966, 45, 630–634.
41. J. Lothe and G.M. Pound, Concentration of clusters in nucleation and the classical phase integral, *J. Chem. Phys.*, 1968, 48, 1849–1852.
42. K. Nishioka and G.M. Pound, Theory of the replacement partition function for crystals in homogeneous nucleation, *Acta Metall.*, 1974, 22, 1015–1021.
43. F.T. Ferguson and J.A. Nuth III, Experimental studies of the vapor phase nucleation of refractory compounds. V. The condensation of lithium, *J. Chem. Phys.*, 2000, 113, 4093–4102.
44. D.M. Martinez, F.T. Ferguson, R.H. Heist, and J.A. Nuth III, Experimental studies of the vapor phase nucleation of refractory compounds. VI. The condensation of sodium, *J. Chem. Phys.*, 2005, 123, 054323-1–054323-5.
45. J.A. Nuth, K.A. Donnelly, B. Donn, and L.U. Lilleleht, Experimental studies of the vapor phase nucleation of refractory compounds. III. The condensation of silver, *J. Chem. Phys.*, 1986, 85, 1116.
46. F.A. Fisk, M.M. Rudek, J.L. Katz, D. Beiersdorf, and H. Uchtmann, The homogeneous nucleation of cesium vapor, *Atmos. Res.*, 1998, 46, 211–222.
47. L.D. Landau and E.M. Lifshitz, *Theory of Elasticity* (Volume 7 of Course of Theoretical Physics), Reed Educational and Professional Publishing Ltd., Oxford, U.K., 01999.
48. N.D. Vargaftik, V.F. Kozhevnikov, A.M. Gordeenko, D.I. Arnold, and S.P. Naurzakov, *Int. J. Thermodyn.*, 1986, 7, 821–828.
49. I.S. Grigoriev and E.Z. Meylikhov (Eds.). *Physical Magnitudes* (reference-book). Energoatomizdat, Moscow, Russia, 1991.
50. D.R. Lide (Editor-in-Chief). *Handbook of Chemistry and Physics*, 85th edn. CRD Press, Boca Raton, FL, 2004.

20 Radioactive Aerosol Standards

Lev S. Ruzer, Yu.V. Kuznetzov, V.L. Kustova,
D.E. Fertman, and A.I. Rizin

CONTENTS

20.1 USSR SPECIAL STATE STANDARD FOR THE VOLUMETRIC ACTIVITY OF RADIOACTIVE AEROSOLS

Particular attention has been paid in recent years to the pollution of ambient air and production premises by radioactive and other substances owing to rapid development rates of atomic power generation, electronic industry, usage aerosols in medicine, pharmaceutical industry, etc. A number of organizational measures have been adopted in many countries for solving this problem, that is, work is in hand for developing and mass producing instruments for testing aerosols of all types and for raising the measurement precision. A most important problem is that of improving the existing metrological-provision means and, above all, those of highest precision.

In the previous chapters the different methods and measuring techniques were presented for measuring the aerosol concentration of different types of radioactive and nonradioactive aerosols together with both theoretical and experimental studies focused on improving the accuracy and quality of the measurements of aerosol parameters.

All this information was used in the development of the Aerosol Standard in the former USSR.

The set of installations (Antipin et al., 1980) approved in 1973 and then in 1978 as a special state standard for the unit of artificial (man-made) and natural radioactive aerosol concentration had no equivalent in USSR or other countries' metrological practice. Experience gained in its operation indicates that its composition and design were selected correctly and that it was reliable in determining radioactive pollution; and also that it confirms the validity of the principles used as a basis for determining the precision levels and number of steps in the transition of units, which were established by the scheme for testing the radioactive aerosol concentration-measuring equipment.

At the same time it was found necessary to reduce the physical standard random error S by a factor of two, that is, to the value of 5×10^{-2}. Moreover, the considerable amount of work carried out in the state testing, certification, and checking of aerosol radiometers by means of physical-standard installations revealed the possibility of reducing the maximum errors of the references and working measuring equipment (ME) to 15%–30% and 30%–60%.

This was promoted to a considerable extent by the new aerosol-radiometer design.

It also became evident that the physical-standard installations should be supplemented by generators of thoron and actinon daughter products and that it is necessary to modernize the electronic equipment used in measuring the activity of aerosol samples and to extend considerable the potentialities of installations for measuring the aerosols, disperse composition and electrical parameters.

The improved special state standard for the volumetric activity (VA) unit of radioactive aerosols incorporates generators type GERA-1 of artificial aerosols; generators GERA-2 of natural radioactive aerosols; generators GERA-3 of vapors ^{131}I; spectrometric equipment type GERA-4 for measuring activity of alpha-, beta, and gamma-radiating aerosol samples; installation GERA-5 for generating and measuring the disperse composition and counted concentration of inactive aerosols; installation GERA-6 for generating and measuring the parameters of aeroions and electroaerosols.

The generators type GERA-1 are based on the method of bubbling pure air through a radioactive solution of the appropriate radionuclide salt (^{90}Sr + ^{90}Y) and ^{239}P producing beta radiation; ^{239}Pu and ^{210}Po producing alpha radiation. The aerosols are formed in the following manner: under the effect of surface tension the films of bubbles which are formed on the surface of the solution are disintegrated into drops. These drops are subsequently dried at temperatures of 140°C–150°C and converted into solid crystals. The artificial radioactive aerosols with a solid disperse phase thus obtained are diluted with pure air. The VA of artificial radioactive aerosols can be varied by changing the activity of the solution which is poured into the generation unit as well as the rate of bubbling and pure air dilution. The generators have a high reproducibility of radioactive aerosol VA in the range of (7×10^{-2}–10^{-3}) s^{-1} m^{-3} and a mean-square deviation (MSD) for measurement results not exceeding 5×10^{-2}.

The GERA-2 set comprises generators of the radon, thoron, and actinon daughter products. The radon daughter generator is based on the method of accumulation these products in an enclosed volume and it suitable for reproducing their VA in the range of (2×10^{2}–4×10^{5}) s^{-1} m^{-3} with a measurement-results MSD not exceeding 5×10^{-2}. Investigations of the generator's radon atmosphere carried out by means of the GERA-6 installation have shown that the fraction of the RaA charged aerosols amounts to 89% ± 8% for a radon daughter products VA of 259 Bq L^{-1} and a median aerosol particle radius of R$_m$ = 0.11 μm ± 15%.

The thoron and actinon daughter product generators are also based on the method of cumulating emanation daughter products in an enclosed volume. Since their half-lives are insufficient time for penetrating from the bubblier to the atmosphere of the container, the bubblier is replaced by solid unsealed sources with thin layers of a substance based on ^{227}Ac and ^{228}Th and possessing an adequate emanation capacity. The basic raw material for obtaining the ^{227}Ac and ^{228}Th preparations consisted of ^{226}Ra irradiated with neutrons which form from it the required preparations as a result of the following reaction:

$$^{236}Ra + nr - {}^{237}Ra - \beta\, {}^{227}Ac + nr - {}^{228}Ac - \beta\, {}^{228}Th$$

The ^{227}Ac and ^{228}Th sources were prepared at the V.G. Khlopin Radium Institute and placed in 1 m^3 containers type 2BP2-OS.

The thoron and actinon daughter product generator's reproducibility expressed in MSD term does not exceed 35%. This value is due to the small generator volume, but for practical cases it is completely satisfactory.

The natural radioactive-aerosols' VA can be varied by means of diluting filters. Aerosol samples are obtained with spectrometric filters types AFA-RSP-3 and AFA-RSP-10.

The Gcra-3 generator of vaporous ^{131}I is based on the method of distilling by oxidizing KI potassium dichromate solution. It possesses a high VA reproducibility in the range of (7×10^{-4}–10^{56}) s^{-1} m^{-3} with a measurement-result MSD not exceeding 5×10^{-2}.

The Gera-4 spectrometric equipment for measuring the activity of aerosol samples includes a spectrometric installation for measuring the activity of the disperse phase deposited on the filter and of artificial and natural alpha-, and beta-radiating aerosols, and an instrument for measuring the ^{131}I activity in the sample by means of betas and gamma radiation.

The spectrometric installation comprises three detection units (DUs), a pulse amplifier, one integral and one differential discriminators, a switching unit with three inputs, two count rate meters, a conversion device, a stable-amplitude generator, and a power pack.

The samples' activity measurements by means of the alpha-spectrometric method are carried out with two DUs; the first one includes a semiconductor detector with a working area of $2\,cm^2$ whereas the second one is a scintillation detector consisting of a $63*0.35\,mm$ CsI(Tl) crystal and a spectrometric photoelectron amplifier type FEU-82. The first DU is suitable for measuring under atmospheric pressure, an atmospheric rarefied down to ~0.7 Pa, and at different distances between the detector and filter. The alpha particles' registration efficiency can attain 0.7 for a resolution of ~150 keV with respect to ^{238}Pu.

The second DU comprises a collimator with a 2 mm aperture diameter and height, thus making it possible to record only the alpha particles which leave the filter almost perpendicularly to the detector. Its registration efficiency is substantially lower than that of the first DU and amounts to 0.10 for a resolution of ~250 keV with respect to RaC'. Both DUs are suitable for measuring either by means of the alpha-spectrometric or counting method. The discriminator levels are set in the first DU by means of spectrometric alpha source (RSAS) with ^{226}Ra radionuclides.

The activity of the samples with respect to beta radiation is measured with DU based on a halogen counter type SBT-11. The registration efficiency of the radionuclide $^{90}Sr+^{90}Y$ beta particles amounts to 0.22.

The instrument for measuring the VA of the vapors ^{131}I is simultaneously a beta and gamma radiator and consists of two DUs, a conversion device, and a power pack. One of the DUs, made on the basis of a $40*40\,mm$ NaI(Tl) monocrystal and a type FEU-82 photomultiplier, is intended for recording gamma count; the second DU, based on a halogen counter type SBT-11, is intended for recording beta particles. The recording efficiency of gamma-quants amounts to 0.25, and of beta particles to 0.18. Measurements are made by means of the counting method.

For computing the VA from the number of recorded pulses N the following formulae are used:

For counting method

$$q = \frac{N}{\varepsilon \upsilon \eta \theta}$$

For the alpha-spectrometric method of Radon decay products

$$q_a = \frac{N_a(\theta,t)C_1}{\varepsilon \upsilon \eta}$$

$$q_b = \frac{[C_2 N_c(\theta,t_2) - C_3 N_c(\theta,t_2) - C_4 N_a(\theta,t_2)]}{\varepsilon \upsilon \eta}$$

$$q_c = \frac{[C_5 N_c(\theta,t_1) - C_6 N_c(\theta,t_2) + C_7 N_a(\theta,t_1)]}{\varepsilon \upsilon \eta}$$

where θ is the pumping through the filter. The GERA-5 installation is intended for generating and investigating the disperse composition of aerosols virtually in the entire range of their particle dimensions that is, for reproducing the "aerosol-particle dimension scale" (Kravchenko and Ruzer, 1978) which includes the highly dispersed aerosols with particle sizes smaller than $0.1\,\mu m$, the medium-dispersed aerosols with particle sizes of ~0.1–1, and coarsely dispersed aerosols with particle sizes exceeding $1\,\mu m$.

The highly dispersed aerosols are generated by means of two methods. The first one consists of obtaining a primary NaCl aerosol with the particle size of ~0.005 μm by means of the spontaneous condensation with a subsequent consolidation of particles up to the required size in the instrument, whose principle of operation is based on condensing the vapors of given substances on nuclei. The aerosol-particle sizes are adjusted in this method by varying the instrument mixture temperature or changing the ration of the air flows with the NaCl particles and the consolidating-substance vapors.

The interaction of the hydrochloric acid vapors with the ammonia vapor flow is used in the second method. As a result of this reaction ammonium chloride aerosol particles are produced. The size of the generated particles is adjusted by varying the concentration of ammonia vapors in an excessive concentration of hydrochloric acid vapors, or vice versa. This method is suitable for obtaining aerosols with ~0.002–0.2 μm particles.

Aerosols with 0.2–10 μm particles are generated by the method of diffusing monodisperse suspensions. Coarsely dispersed aerosols with particle sizes of 5–200 μm are generated by means of a disk pulverizer whose principle of operation consists of the following. The original liquid is fed to the center of a rapidly rotating disk. Under the effect of the centrifugal force the liquid is displaced toward the disk edge and is ejected from it in the form of monodispersed drops. The particle size is adjusted by changing the disk rotation speed. The disk pulverizer serves to obtain aerosols with a liquid dispersed phase from the type VM-4 oil and with a solid dispersed phase from a rosin solution in alcohol from steric acid melts. The described generators are suitable for obtaining virtually monodispersed aerosols with particle sizes of 0.002–200 μm.

The disperse composition of highly dispersed aerosols as well as medium-dispersed aerosols up to 0.5 μm. is determined by means of the diffusion method. For this purpose 13 diffusion batteries are used with a total channel length of 0.5–5600 m in each battery. The error in measuring the average size of particles by means of the diffusion method does not exceed 20%.

For certifying suspensions with particles smaller than 1 μm. and also for studying the dispersed composition of ammonium chloride aerosols the electron microscope type EM-9 was used. Suspensions with particles exceeding 1 μm are certified and the dispersed composition of coarsely dispersed aerosols is studied by means of optical microscope type MBI-11. Medium dispersed aerosol particles in the range of 0.25–10 μm are also determined by means of photoelectric aerosol-particle counters type AZ-4.

The counted concentration of aerosols is measured after preliminary consolidation is diluted (if required) with AZ-4 instrument, whose measurement-result MSD does not exceed 20%. Counted concentrations of particles with diameter 2–4 μm. are measured by means of the sedimentation method with an error not exceeding 7%. Aerosols with particles sizes exceeding 10 μm are measured (owing to their rapid sedimentation) by a newly developed technique for determining the number of particles entrapped in instrument sampler with an error not exceeding 10%.

The GERA-5 installation was used for studying the dispersed composition of aerosols generated by the GERA-1 and GERA-2 installations and also for metrological investigation of instruments used in measuring the parameters of inactive aerosols.

The GERA-6 installation serves to generate aerosols and electroaerosols and for measuring electrical characteristics of aerosols. Its generating unit consists of light-ion and electroaerosol generators. The light-ion generator is provided with air ionization by means of ^{239}Pu radionuclides.

The volumetric density of the charge is adjusted by varying the voltage across the generator electrodes. Pneumatic pulverization of liquid forced through a nozzle and the charging of particles in the field of the indexing electrode is used in electroaerosol generator. The volumetric density of the charge is measured by means of aspiration condenser method based on depositing the charged aerosol particles in the electric field of the condenser.

The GERA-6 installation serves to reproduce the volumetric density units of positive and negative aerosol and aeroion electric charges with a random error characterized by the measurement-result MSD not exceeding 10% and a nonexcluded systematic error also not exceeding 10%. The range of reproduced charge volumetric densities amounts to 2×10^{-11}–2×10^{-7} C m^{-3} for light ions and to $1 \times 10^{-9}–1 \times 10^{-6}$ C m^{-3} for liquid electroaerosols. This installation was used for certifying and testing ion counters, for measuring the electric characteristics of aerosol generated on other physical-standard installations, and also in research work.

It was found in practice that the transmission of the VA unit of radioactive aerosols in three stages—from the physical standard to the reference equipment and then to the working measuring equipment—is optimal and can be used as a basis for developing test schemes for equipment used in measuring the counted and mass concentrations of nonradioactive aerosols, as well as for equipment used in measuring the aerosol's electrical parameters.

Thus, as a result of improving the special state standard for the VA unit of radioactive aerosols, its metrological characteristics were raised and the reference and working VA measuring-equipment errors reduced; the nomenclature of natural radioactive aerosols measured with standard was extended and the range of generated monodisperse inactive aerosols increased.

20.2 CURRENTLY APPLICABLE RADIOACTIVE AEROSOL STANDARDS*

At the present time the State Standards in the form, which includes technique for generating and measuring of parameters for both radioactive and nonradioactive aerosols in a wide range of sizes and activities, do not exist in the world. Instead, there are, in some countries, local standards for different groups of aerosols and radioactive gases.

20.2.1 RADON AND ITS DECAY PRODUCTS

In (Cotrappa et al., 1994) the application of National Institute of Standards and Technology (NIST) of United States ^{222}Rn emanation standards for calibration ^{222}Rn is described. The NIST certified parameters include the ^{222}Rn strength and emanation coefficient. When a source of ^{222}Rn is loaded into a leak tight jar of a known volume, ^{222}Rn will accumulate over time. It is possible to calculate the time integrated average radon concentration after any given accumulation time. Radon detector in the jar should be non-radon absorbing and a true integrator. In this case radon detector must yield the theoretically predicted results. In case of consistent difference, the NIST traceable correction can be derived.

The study NIST involves 34 randomly chosen electret ion chamber system (E-PERM) and 17 NIST sources. The procedures of calibration are enabled with simple equipment. E-Perm detectors were found to give predicted measurement results with an accuracy of about 5%. Commercially available continuous radon monitors also gave satisfactory performance. With the availability of this technology, ^{222}Rn measurement instruments can be made NIST traceable—a great step forward in radon metrology.

In (Budd et al., 1998) the study, within the framework of the International Atomic Energy Agency (IAEA) and European Union (EU) International Radon Metrology Program (IRMP), the results of the international intercomparison were presented in order to evaluate radon and radon decay product measurement techniques. The work was organized jointly with U.S. Environment Protection Agency Radiation and Indoor Environment National Laboratory (EPA) in Las Vegas, Nevada, and the former U.S. Bureau of Mines (BOM).

The primary goal of this project was to compare the performance of radon and radon decay product measurement instruments from around the world under both laboratory and field exposure conditions.

Nineteen organizations from 7 countries participated in this project with 32 types of radon and radon decay product measurement instruments. The laboratory exposures were conducted in an environmental radon chamber at EPA's Radon Laboratory in Las Vegas under very stable, controlled environmental conditions at relatively low concentration of radon and radon decay products.

---

* This part of the section is based partly on information presented by D.E. Fertman and A.I. Rizin (JSC "SNIIP") and Yu.V. Kuznetzov and V.L. Kustova (GP VNIIFTRI) all Moscow, Russia.

This part of the study was conducted in order to compare the instruments under such controlled environmental conditions.

The field exposure was provided at a former underground uranium mine in Colorado maintained by BOM in order to compare the instruments under fluctuating and uncontrolled conditions.

The National Institute of Standards and Technology (NIST) provided the EPA Radon Laboratory with multiple spherical glass sample ampules each containing an activity of ^{222}Rn gas known only to NIST. The primary radon measurement system used by the EPA Radon Laboratory to determine radon concentration within the chambers is 0.36 L scintillation cells (also termed Lucas cells).

The BOM facility is a previously operated uranium–vanadium mine, which was used for the purpose of conducting research. Radon concentrations were in the range of 5,110–15,535 Bq m^{-3} and radon decay products concentration varied from about 3.54 to 74.32 μJ m^{-3}. The mean equilibrium factor was 14.2%.

Of the six participants who measured radon concentration in the EPA laboratory with charcoal collectors, four produced results within 8% of the unity for performance ratio. Only one of the four participants who used alpha-track detectors in EPA laboratory produced results within 11% of unity, and two participants who used electret ion chamber in EPA laboratory gave results of 10% of unity. Of the four organizations that measured radon decay product concentrations in EPA laboratory using grab methods, two produced 10% of unity, and the remaining two within 30%.

The overall conclusion can be made that more international intercomparisons are needed as there are differences even within the same organizations. This is especially important when making interpretations about the health effect, which is the main goal of radon and its decay product measurements.

In Nezval et al. (1997), intercomparison measurement of soil-gas radon concentration is presented.

The soil-gas ^{222}Rn concentration defined as an average radon concentration in the air-filled part of soil pores in a given volume of soil. This value has a wide range of practical applications:

1. When soil-gas radon is used as indicator for uranium, indoor radon, seismic activity, location of sub-service faults, etc.
2. In studies where the focus is on radon itself

From a metrological point of view there are many problems with organizing field intercomparison measurement in natural geological environment. Such intercomparisons were previously organized in 1991 and 1995. This project was organized during *the third International Workshop on the Geological Aspect of Radon Risk Mapping* (Praque, Czech Republic, September 1996) measurements. Participants representing 10 organizations from eight countries took part in the intercomparison. The ratio of the standard deviation to the arithmetic mean (SD/mean) was used as a measure of a spread on intercomparison. For the soil-gas radon concentration the agreement among participants was very good. If all single values that were obtained over the whole area of the test site were taken into account, the intercomparison difference, expressed as a ratio SD/mean, was 24%. A more detailed assessment shows an even smaller number—about 20%. This result has been previously considered as a reliable target for intercomparison measurements of soil-gas-radon concentration.

20.2.2 Artificial Radioactive Aerosols

In Grivaud et al., 1998, the installation for testing radioactive aerosol measurement instruments is described. The EPICEA laboratory (Laboratoire d'Essais Physiques des Instruments de Mesure de la Contamination de l'Eau de del'Air), which belongs to the Institute de Protection et de Surete Nucleaire (IPSN), France, was established to carry out various types of tests on atmospheric contamination monitors under the conditions recommended by the International Electrotechnical Commission (IEC).

These tests are carried out at the request of users, scientific or industrial manufacturers, either French or foreign, to define the performance of a given aerosol radioactive contamination monitor in order to obtain type approval for the monitor by the IPSN Centre Technique d'Homologation de l'Instrumentation de Radioprotection (CTHIR). Tests can also be used for defining prototypes manufactured by industry. There are two types of tests: static tests, performed with solid standard radioactive sources; dynamic tests, performed on the ICARE bench. This bench, continuously generating natural and artificial radioactive aerosols, calibrated for size and activity, enables the true performances of the radioactive aerosols monitors to be defined under normal operating conditions. The true measurement efficiency is obtained by sampling and measuring, in real time, the activity of aerosols labeled with ^{239}Pu and/or ^{137}Sr. The influence of the natural activity upon the artificial activity measurement channels is determined by aerosols bearing radon decay products whose concentration and attached fraction can be adjusted. Knowledge of the factor of influence of the natural activity and the type of treatment (algorithm) used on this monitor makes it possible to calculate the monitor detection threshold under normal operating conditions. The dynamic test procedure described in the document has been adopted as an international standard by IEC in 1996.

Installation for performing testing of continuous air monitors (CAMs) for alpha-radioactive nuclides is described by Grivaud et al. (1998). These instruments must have adequate sensitivity to alert potentially exposed individuals that their immediate action is necessary to minimize or terminate an inhalation exposure.

The air monitor test facility at the Lovelace Respiratory Research Institute (formerly the Inhalation Toxicology Research Institute) has been developed for U.S. Department of Energy and used to test performance of prototype and commercially available alpha CAMs, personal air samplers, and fixed area filter samplers. Test conditions for these instruments are consistent with 1995 recommendations of IEC (Hoover et al., 1998).

The facility includes a station for instrument receipt and inspection, a test bench for determining detection efficiency and energy response for alpha-radioactive radio nuclides using point-type and area-type electroplated sources and ambient radon progeny; an inline aerosol delivery for testing the internal collection efficiency of sampling heads with fluorescent and other inert aerosols; an aerosol wind tunnel in which inert aerosols can be used to evaluate the inlet and transport efficiency of sampling probes and aerosol collection devices; systems for testing the normal response of monitors to ambient radon progeny aerosols or providing aerosols of plutonium or uranium with or without radon progeny aerosols and interfering dusts to air monitor under different conditions of concentration and time. The Lovelace Air Monitor Test Facility is similar, but not identical, to the EPICEA laboratory Institute which belongs to ISPN.

Many different types of tests on air monitors were provided at this facility.

Static tests are performed using clean collection substrates with no sampling flow to verify proper reports of background in the absence of radioactive source. For this purpose the standard radioactive source with traceability to the National Institute of Standard and Technology was used to determine the overall detection efficiency for uniformity distributed sources.

Dynamic tests of collection efficiency of in-line sampling heads were provided by connecting them to an aerosol generation system with parallel sampling ports for head and in-line reference filter.

Dynamic tests with radon progeny were performed to evaluate the influence of natural radioactivity on the ability of CAMs to report correctly the absence of plutonium in the presence of a low concentration of radon decay products when no plutonium is present.

Dynamic tests with radon progeny and artificial radioactivity were provided with higher concentrations of radon progeny (up to $370\,Bq\,m^{-3}$ in the current system) in the special radioactive aerosol generation system located in the Lovelace plutonium test facility.

Tests for system performance in the presence of interfering dusts were provided because the influence of salt dusts with proper detection of actinide aerosols that may be released from the storage of transuranium wastes in underground salt formation is of special concern. To evaluate these interferences, the aerosol generation system also provides dust aerosols to CAMs, either as a homogenous mixture with actinide aerosols and radon progeny or as separate aerosols.

Tests with computer-generated pulses were provided in order to understand which aspect of CAM performance can be evaluated with sources and ambient radon progeny and which responses must be evaluated with actual aerosols, including effect of alpha energy degradation from deposition of actinide particles in a filter substrate. The Alpha Energy Spectrum Simulator has been used to provide realistic electrical pulses to the instrument preamplifier to simulate different temporal concentration and combinations of actinide aerosols and radon progeny.

The minimum detectable activity and false alarm rates were also studied in the dynamic tests for a range of radon progeny concentrations, and for conditions involving both the absence and presence of plutonium.

Performance of different types of filters was also studied.

It should be pointed out that in the future the comparison results of facilities such as ISPN and Lovelace system should provide confidence that the results of the two systems do not contain unexpected bias.

The results of development and practical application of new radioactive source types (Special Aerosol Sources [SAS]) are presented in Belkina et al. (1991). SAS are manufactured for certain type of radiometers using model aerosols of plutonoium-239, strontium-ittrium-90 or uranium of natural isotopes composition. The original technology for source production allows taking into account the features of sampling, as well as geometry and conditions of measurement of the activity in the sample.

The highest accuracy in calibration, certification, and verification of the radiometers can be provided by using national standard precision equipment (Antipin et al., 1980). But its direct usage for metrological provision of measurement technique is too expensive. At each stage of development and usage, the measurement technique for metrological purposes should be performed with an optimal cost-effective ratio.

The proposed method of manufacturing SAS includes radioactive substance depositing onto a substrate, its fixation and substrate mounting into the frame-holder intended for measurement device. The check of the radioactive substance fixation quality showed that relative activity variation of the sources with plutonium-239 after 1–4 days by the dry cotton tampon is within the limits of registration instability (less than 5%). The authors have developed and tested the procedure of certifying and verifying the nuclear power plant radiation security monitoring system aerosol channels by using Sr-90–Y-90 SAS.

In Fertman et al. (1998), the problem of the Russian secondary standard for volume activity of long-lived radio nuclides aerosols and System of international comparison was discussed. The national standard design was practically reproduced as Russian secondary standard for volume activity of long-lived radioactive aerosols (VET39-1-2005) in the metrological center of the State Atomic Energy Corporation "Rosatom" in JSC "SNIIP" Moscow. It is used for the calibration, certification and verification of the radiometers, and other metrological provision.

In addition, the important problems that arise in radioactive aerosols' metrology in the last decades were resolved:

1. It was added to the generation of natural uranium aerosols.
2. Developing model aerosol sources SAS in order to test aerosol radiometers directly on the consumer place.

Model aerosols are prepared in special generators mounted in the box. Parameters of model radioactive aerosols' particle sizes were estimated by six-cascade impactor and radiometer type MS-01P. Measurement showed that model radioactive aerosols are polydisperse with a maximum diameter close to 1 micron. Thus, radionuclides and characteristics of model aerosols are close to similar parameters of national standard (Antipin et al., 1980).

The international comparison was provided between JSC "SNIIP" (Russia) and IPSN-CEA (France). As a result of cooperation between these two installations, the number of aerosol samples with certain radionuclides ingredients was prepared. In Russia aerosol samples were prepared according to technology of SAS manufacturing, in France source of Pu-239 and Cs-137.

The review of air monitoring standards not as installations for testing and verification of air monitors, but as requirements, instructions for measurement and calibration, is presented in An International Review of Currently Applicable Standards for Measuring Airborne Activity (International Electrotechnical Commission [IEC], TC News, 2003. Report can be found electronically at: http://www.iec.ch/support/tcnews.

Here is a list of some of standards related to radioactive aerosol measurements.

IEC standards

1. IEC61172 (1992-09): Radiation protection instrumentation—Monitoring equipment—Radioactive aerosols environment.
2. IEC 60761-1 (2002-01): Equipment for continuous monitoring of radioactivity in gaseous effluents—Part 1: General requirements.
3. IEC 60761-2 (2002-01): Equipment for continuous monitoring of radioactivity in gaseous effluents—Part 2: Specific requirements for radioactive aerosol monitors including transuranic aerosols.
4. IEC 61172 (1992-09): Radiation protection instrumentation—Monitoring equipment—Radioactive aerosols in the environment.
5. IEC 61577-1 (2006-07): Radiation protection instrumentation—Radon and radon decay product measuring instruments—Part 1: General principles.
6. IEC 61577-2 (2000-10): Radon and radon decay product measuring instruments—Part 2: Specific requirements for radon measuring instruments.
7. IEC 61577-3 (2002-04): Radon and radon decay product measuring instruments—Part 3: Specific requirements for radon decay measuring instruments.
8. IEC 61577-4 (2009-02): Radon and radon decay product measuring instruments—Part 4: Equipment for the production of reference atmospheres containing radon isotopes and their decay products (STAR).
9. IEC 61263 (1994-06): Radiation protection instrumentation—Portable potential alpha energy meter for rapid measurements in mines.

International Organization for Standardization (ISO) Standards

ISO 2889 (19750): General principles for sampling airborne radioactive materials.

European Standards

EN 481 (1993): Workplace atmospheres—Size fraction definitions for measurement of airborne particles.

The American National Standards Institute (ANSI).

ANSI N323C (in preparation): Radiation Protection Instrumentation Test and Calibration Air Monitoring Instruments.

Applicable French National standards

NF X 43022 (1985): Air quality—Ambient air—Concepts relating to the sampling of particular matter.

NF X 43257 (1998): Air quality—Air in workplaces individual sampling of inspirable fraction of particulate pollution.

NF M 60-763 (1998): Nuclear Energy—Measurement of radioactivity in the environment—Air-radon and short-lived decay products in the atmospheric environment.

NF M 60-767 (1999): Nuclear Energy—Measurement of environmental radioactivity Air-Radon 222 measurement methods of the volumic activity of radon in the atmospheric environment.

REFERENCES

Antipin, N.A. et al., Special state for the volumetric activitiy of radioactive aerosols, *Izmer. Techn.*, No. 2 (1973), 5–7, January 1980.

Antipin, N.A. et al., Special State Standard for the volumetric activity of radioactive aerosols, *Izmer. Techn.* (1), 5–7, January 1980.

Belkina, S.K., Zalmanson Y.E., Kuznetsov Y.V., Rizin A.I., Fertman D.E., Special aerosol sources for certification and test of aerosol radiometers, *J. Aerosol Sci.* 22(Suppl. 1), 801, 1991.

Budd, G. et al., Intercomparison of radon and decay product measurements in an underground mine and EPA radon laboratory: A study organized by the IAEA Iradon Metrology Programme, *Health Phys.* 75(5), 465–474, 1998.

Cotrappa, P. et al., Application of NIST ^{222}Rn emanation standards for calibration ^{222}Rn monitors, *Radiat. Prot. Dosim.* 55(3), 211–218, 1994.

Fertman, D., Rizin A.I., Using polydisperse aerosols and special aerosol sources for calibration of aerosol radiometers, *Materials of International Congress for Particle Technology (PARTEDC-98)*, Nurnberg, Germany, 1998.

Grivaud, L. et al., Measurement of performances of aerosol type radioactive contamination monitors, *Radiat. Prot. Dosim.* 79(1–4), 495–497, 1998.

Hoover, M.D. et al., Performance testing of continuous air monitors for alpha-emitting radionuclides, *Radiat. Prot. Dosim.* 79(1–4), 499–504, 1998.

Kravchenko, I.I. and L.S. Ruzer, Reproducing the scale of aerosol-particle dimensions, *Izmer. Techn.*, No. 4, 66–67, 1978.

Neznal, M. et al., Intercomparison measurement of soil-gas radon concentration, *Radiat. Prot. Dosim.* 72(2), 139–144, 1997.

21 Radon and Thoron in the Environment

Concentrations and Lung Cancer Risk

Naomi H. Harley

CONTENTS

21.1 INTRODUCTION

Our planet is mainly rock and metal and its rock is remarkably radioactive. The sun supplies most of our energy but the measured radioactivity in the earth contributes up to 25% of the total heat balance of the earth (Araki et al., 2005). The primordial decay series beginning with ^{238}U (4.5×10^9 year half-life), present in all terrestrial materials, supports a chain of 13 alpha-, beta-, and gamma-emitting radionuclides that includes the gas radon (^{222}Rn, 3.8 day half-life). The primordial decay series beginning with ^{232}Th (1.4×10^{10} year half-life) is also present in all basic earth materials, and supports a chain of 11 radionuclides that includes another isotope of radon—common name thoron (^{220}Rn, 55 second half-life). Both radon isotopes are produced in all soil or rock from their parent radium isotopes (^{226}Ra or ^{224}Ra). A fraction of both gases is released from all terrestrial substances and can be measured in any dwelling, outdoors, and in the case of ^{222}Rn, even at stratospheric altitudes. Thoron, however, should not exist in the stratosphere because of its short half-life and the time required for transit to altitude.

Epidemiologic follow-up studies of 11 underground mining groups show a lung cancer dose response increasing with increasing radon concentration. Some homes can attain the concentrations that existed in historic mines, and, thus, country radon surveys and residential risk estimates are now common. The global residential measurements to date show that the current residential concentration averages 55 Bq m^{-3} (about 1.5 pCi L^{-1}). The U.S. average is estimated as 46 Bq m^{-3} (about 1.2 pCi L^{-1}). There are now over 80 worldwide residential epidemiologic studies, which include 23 case control studies, initiated to estimate residential risk directly and not rely on risk projection models from the underground studies. The presence of ^{220}Rn (thoron), often unaccounted for in the gas measurements, may affect the calculated risk estimates for homes.

The residential case control studies support lung cancer risk estimates similar to the underground miner studies. The higher exposures in mines permit features of the lung cancer response to be examined. For example, miners first exposed at older ages have a lower lung cancer response per unit exposure, and lung cancer risk diminishes with time since end of exposure.

21.2 SOURCES OF RESIDENTIAL AND OCCUPATIONAL RADON EXPOSURE

The concentration of radon in the volume into which it diffuses depends on the source strength (parent ^{226}Ra concentration), the radon emanation fraction from the source, the volume, and the ventilation rate. Underground miners work in a relatively enclosed space close to the rock source; therefore, radon concentrations can be very high. In uranium mines, the rock (ore) concentration of uranium and ^{226}Ra is much higher than typical soil concentrations and radon levels in mines can be extraordinarily high, but depend upon the local air exchange rate. Radon gas is quite soluble in water and in some non-uranium mines water-bearing ^{222}Rn can produce high concentrations. An example of this are the Canadian Fluorspar mines (deVilliars, 1966; deVilliars et al., 1971; Morrison et al., 1988, 1998; Villeneuve et al., 2007). High radon concentrations in drinking water in homes with private wells can have elevated indoor air concentrations from the gas release during water use.

Average residential concentrations depend on typical ^{226}Ra concentrations in the underlying soil and radon flow into the dwelling. Residential concentrations are generally about a factor of 2 higher than outdoor concentrations. Radon concentrations in multistory apartments or offices are generally lower than single-family residences because of smaller source terms due to more rugged foundations or higher ventilation rates.

Most countries with public health programs have conducted comprehensive indoor radon surveys in homes. In epidemiologic studies, large numbers of radon measurements have been made in homes of lung cancer cases and controls to estimate lung cancer risk for the populations involved.

21.3 UNITS FOR RADON EXPOSURE

The miner studies rely on historic radon decay product measurements made by collecting a short-term (minutes) filtered air sample and alpha counting the filter at specified times following collection (NCRP, 1988). In the Czech mines only radon gas measurements were made. The decay product measurements were reported in working levels (WLs). Cumulative exposure was calculated in working level months (WLM) and is equal to WL times the total number of hours worked in mines divided by 168 h in a work month. The WL was originally defined as the potential energy released from a radon concentration of 100 pCi L^{-1} (3700 Bq m^{-3}) in 1 L of air in equilibrium with its short-lived decay products. This potential alpha energy release is equal to 1.28×10^5 MeV L^{-1} and usually rounded to 1.3×10^5 MeV L^{-1}. This potential energy quickly became the de facto definition of the WL.

In practice, the decay products are never in equilibrium with the parent radon and UNSCEAR (2006) adopted an equilibrium factor of 40% for indoor environments. The radon gas concentration times 0.4 is the equilibrium equivalent concentration (EEC) and is the equilibrium gas concentration that gives the same potential energy as the nonequilibrium mixture.

To equate units for residential and mining exposure, 1 WL is equal to 6.29×10^5 Bq m^{-3} h ^{222}Rn (EEC). Calculated cumulative exposure in WLM in a residence with a radon gas concentration of 100 Bq m^{-3} over 30 years is 100 Bq m$^{-3} \times 0.4 \times 8760 \times 30/6.29 \times 10^5 = 17$ WLM. Thirty years could be considered the relevant exposure period for residential study risk estimates.

21.4 INDOOR CONCENTRATIONS

More measurements have been made of indoor radon than for any other natural radionuclide. Some measurements of radon in the late 1970s were comparable to underground mining concentrations. The emergence of statistically significant lung cancer risks associated with underground mine exposure spurred global measurements. Countries rapidly began indoor radon surveys to assess potential residential risk (NCRP, 1984a,b; NAS, 1986; NAS/NRC, 1999a,b; WHO, 2009; UNSCEAR, 2006).

The data from 60 country indoor radon surveys are reported in UNSCEAR (2006) and are reproduced in Figure 21.1. The global indoor radon average from the survey data reported so far is 55 Bq m^{-3} with a range of from 7 to 180 Bq m^{-3}. The estimate of the U.S. average from a survey of about 5000 homes by the U.S. EPA (1992) is 46 Bq m^{-3} (1.25 pCi L^{-1}), a geometric mean of 25 Bq m^{-3} a GSD of 3.1, with 6% of homes exceeding 148 Bq m^{-3}.

21.5 OUTDOOR CONCENTRATIONS

The U.S. Environmental Protection Agency (EPA) performed an outdoor radon survey in the United States that included each major city in all 50 states. Given that the ventilation rate is 0.2–0.5 air exchanges with outdoor air per hour in most homes, the measurement of outdoor radon establishes the baseline for indoor radon. Many countries, especially those in tropical climates, have indoor concentrations that are essentially outdoor concentrations. For example, indoor and outdoor measurements over a 4 year period in Bangkok and Chiang Mai, Thailand averaged 14 ± 1 and 15 ± 1 Bq m^{-3}, and 44 ± 1 and 39 ± 1 Bq m^{-3}, respectively (Chittaporn and Harley, 2000).

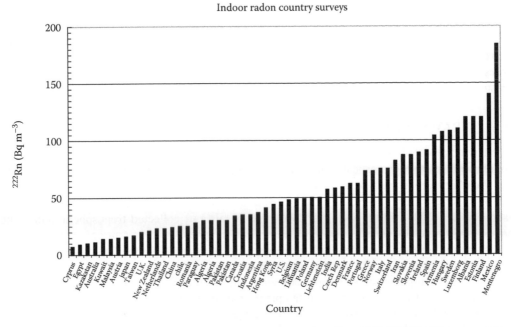

FIGURE 21.1 Global indoor radon measurements in 60 countries. (From UNSCEAR, *Sources-to-Effects Assessment for Radon in Homes and Workplaces*, United Nations Scientific Committee on the Effects of Radiation, New York, 2006.)

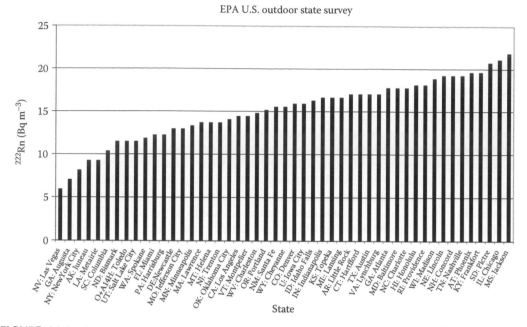

FIGURE 21.2 Average outdoor radon measurements made by U.S. EPA in all 50 U.S. states. (From NAS/NRC, *Health Effects of Exposure to Radon*, Committee on Health Risks of Exposure to Radon, Board on Radiation Effects Research, Commission on Life Sciences National Research Council, BEIR VI, National Academy Press, Washington, DC, 1999a; NAS/NRC, *Risk Assessment of Radon in Drinking Water*, National Research Council, National Academy Press, Washington, DC, 1999b.)

Outdoor radon measurements are not common. However, the U.S. average appears to represent the sparse published results. A summary of published global outdoor concentrations is given in Harley (1990). EPA has suggested target values for radon in drinking water using a multimedia approach, which depend on the knowledge of outdoor air concentrations. Any risk from radon in drinking water is associated with inhalation exposure from radon released during water use. The report on the risk from radon in drinking water (NAS/NRC, 1999a,b) includes the EPA outdoor survey. The graph of the U.S. state data is shown in Figure 21.2. The average outdoor concentration measured in the United States is 14.8 Bq m^{-3} (median 14.6 Bq m^{-3}, maximum 35.9 Bq m^{-3}).

21.6 STRATOSPHERIC CONCENTRATIONS

Radon is transported from ground release to the stratosphere by eddy turbulent diffusion and provides information on vertical transport rates. Stratospheric air samples were collected by the U.S. Weather Bureau (now NOAA) in 1962 to explore the possibility of using radon profiles as an atmospheric tracer. These data were available only in an internal report until reevaluated by Fisenne et al. (2005). In the spring of 1962, WB-57 aircraft collected tropospheric and stratospheric air samples by pressurizing steel gas collection spheres to a pressure of about 21 MPa at each sampling altitude. The sampling locations were Alaska, southwest United States, and the Panama Canal Zone at 8°, 32°, and 70° north, respectively.

Each collection sphere contained approximately 2 m^3 of air at standard temperature and pressure. The samples were transferred through a sample train consisting of two gas washers to remove carbon dioxide and water vapor and then onto low background activated charcoal traps. The adsorbed radon on the charcoal was transferred to low background (six counts per hour) 2 L ionization chambers for alpha counting. The locations and trajectories of the flights were selected in order to investigate the influences of tropospheric height, the underlying land mass, and thermal gradients on radon

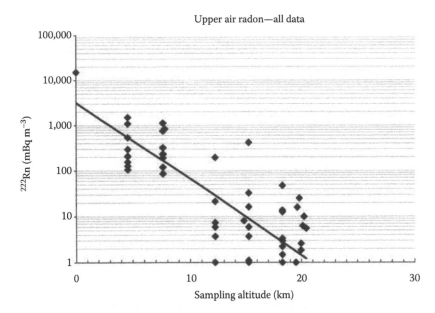

FIGURE 21.3 Radon measurements in the troposphere and stratosphere made over Alaska, the Canal Zone, and southwest United States by WB-57 high-flying aircraft. (From Fisenne, I.M., Machta, L., and Harley, N.H. Stratospheric radon measurements in three North American locations, in Radioactivity in the Environment, McLaughlin, J.P., Simopolis, S.E., and Steinhausler, F. Eds. The Natural Radiation Environment, Rhodes, Greece, 2002, Elsevier, pp. 715–721, 2005.)

concentrations in the atmosphere. An effective vertical transport rate was calculated at six different altitudes to determine an average of 0.5 ± 0.1 cm s[-1].

These data were supported by similar tropospheric concentrations reported by Machta and Lucas (1962) for air samples collected over Hawaii and later by measurements of Moore et al. (1973).

The entire data set of 54 samples is shown in Figure 21.3.

21.7 RADON IN DRINKING WATER

The major determinant of internal dose and, thus, risk from radon in drinking water is not from the ingestion of water but the lung (bronchial) dose from radon decay products in air following release of radon from the water during use (NAS/NRC, 1999a,b). An internal dose to the stomach was calculated, but is very small compared with the lung dose from inhaled decay products. For a given radon concentration in drinking water, the risk ratio from both pathways (stomach cancer versus lung cancer) is calculated to be from 1% (Harley and Robbins, 1994) to 11% (NAS/NRC, 1999a,b), and depends upon the model used to transport the gas through the stomach wall to the cells identified as targets for stomach cancer.

Radon released from water during showering and other use combines with and is indistinguishable from normal residential sources. The best estimate for the transfer factor from water to air is 10,000/1, and is based on an analysis of all published data (NAS/NRC, 1999a,b). That is, 10,000 Bq m[-3] of ^{222}Rn in water will, on average, add 1 Bq m[-3] of ^{222}Rn to the indoor air. Hess et al. (1987) measured ^{222}Rn concentrations of 3.7×10^4 Bq L[-1] (10^6 pCi L[-1]) in the state of Maine. Surface water (i.e., from reservoirs) is very low in ^{222}Rn concentration because the gas is readily removed by surface agitation.

EPA has suggested a value for water of 11 Bq L[-1] ^{222}Rn (300 pCi L[-1]). This value can be technically difficult to obtain. A multimedia approach suggested by EPA is to obtain an equivalent risk by reducing the water concentration to a value that will yield upon release the standard outdoor concentration, and reducing overall risk by other means. Thus, the average U.S. outdoor concentration of $15\,Bq\,m^{-3}$ ($0.4\,pCi\,L^{-1}$) would permit a water concentration of $150\,Bq\,L^{-1}$ to be delivered if alternative means of risk reduction are put in place. The risk reduction needed to reduce the total risk could be, for example, home radon remediation that attained the same calculated risk reduction in the population affected, similar to reducing the water to the $11\,Bq\,L^{-1}$ value.

A policy perspective was prepared to investigate the various issues regarding remediation of radon in drinking water (Hopke et al., 2000). The EPA and state agencies are responsible for water quality under the 1996 amendment to the Safe Drinking Water Act, originally passed in 1974. Radon is a known carcinogen and the maximum contaminant level was automatically set at zero. The upper limit concentration of $11\,Bq\,L^{-1}$ is considered because zero concentration could only be measured with an uncertainty of 30%. These involve technical and social decisions concerning the methods of risk reduction. There are varying opinions concerning the reduction in risk to a population exposed versus risk reduction for a few individuals in the population. These and other issues have yet to be resolved.

Some countries have strict ^{222}Rn in public water supply regulations. Beginning October 1, 1998, the Swedish government required ^{222}Rn in public water supplies greater than $100\,Bq\,L^{-1}$ be reduced, and stated that water exceeding a concentration of $1000\,Bq\,L^{-1}$ is unsafe and cannot be supplied. Finland has set a recommended maximum limit of $300\,Bq\,L^{-1}$. The World Health Organization (WHO) states that controls should be implemented if radon in public drinking water supplies exceeds $100\,Bq\,L^{-1}$. The USEPA is required to set a regulation for ^{222}Rn in drinking water in the United States. The standard they have proposed is $11\,Bq\,L^{-1}$ but to date (2003) no regulation has been set officially.

21.8 OCCUPATIONAL RADON EXPOSURE IN UNDERGROUND MINES

The first evaluation of lung cancer risk from radon decay products emerged from the high exposures in underground uranium mines (UNSCEAR, 2006). Mine operators knew little of the hazards involved with exposure to radioactive materials. In the United States, no Federal Government agency had authority to regulate the health and safety of miners. Beginning in 1954, the Atomic Energy Commission (AEC) had regulatory authority over the uranium industry after the material was mined but had no authority to regulate the mining industry. The states where uranium was mined had varied regulatory authority over the safety of miners but the agencies having such responsibility had no experience with radiation problems. In 1949, the U.S. Public Health Service (Holaday and Doyle, 1964) became concerned that the industry was exposing miners to a potential health hazard based on the experience in the Czech mines. They made some measurements and confirmed that airborne radioactivity concentrations were alarmingly high. The concentration in 24 Colorado mines ranged from 5 to $800\,kBq\,m^{-3}$ (135–$22{,}300\,pCi\,L^{-1}$) (Lundin et al., 1971). There are now 12 follow-up cohort studies for lung cancer risk in underground miners (UNSCEAR, 2006). The data for nine studies are presented in the Section 21.12 on risk.

21.9 BRONCHIAL LUNG DOSE

Risk can be estimated from the bronchial dose delivered by the decay products and use of risk factors based on the dose to A-bomb survivors (ICRP, 2007). The bronchial lung dose cannot be measured in humans and must be calculated.

The relevant bronchial dose from radon is airway deposition of the solid aerosol particles of short-lived alpha emitting decay products (^{218}Po, ^{214}Po). The short-lived decay products have an effective half-life of 30 min for the chain. As the decay product atoms form in air from ^{222}Rn decay, they rapidly attach to the ambient aerosol particles. Bronchial and pulmonary deposition is determined by the ambient particle size distribution. There are published short-term data on the size distribution of radon decay products. The median diameter indoors is about 200–300 nm (Tu and Knudson, 1988a,b; Reineking and Porstendorfer, 1986; Li and Hopke, 1993; NAS/NRC 1999a,b).

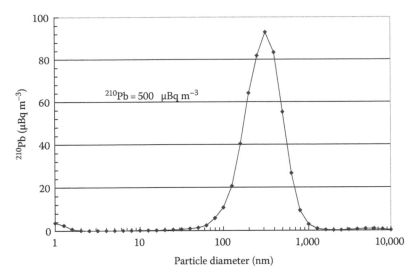

FIGURE 21.4 Aerosol particle size distribution measured indoors in a single-family home over a 3 month interval using 210Pb as a tracer.

UNSCEAR (2006) adopted a bronchial dose conversion coefficient of $9 \, nSv \, Bq^{-3} \, m^{-3} \, h$ (EEC), and included a table of 13 published dose conversion coefficients derived from bronchial dose models.

Harley et al. (2005a,b) developed an integrating particle size sampler that operates for up to 3 months indoors or outdoors. The short-lived radon decay products ultimately form long-lived ^{210}Pb and ^{210}Po (22 years, 138 days) and these radionuclides are used as tracers. Figure 21.4, shows a size distribution indoors in a typical residence.

A few percent of the ^{218}Po formed from radon decay associates with water vapor or other gases and remains as a small cluster of atoms with a diameter of a few nanometers (nm). All atmospheres contain this nanometer or unattached fraction. Because of their small diameter, the unattached fraction deposits efficiently in the bronchial airways. A small percentage of the total activity inhaled accounts for a disproportionate fraction of the bronchial dose with up to about 25% due to the unattached fraction. The unattached fraction is evident in Figure 21.4 in the 1–2 nm size region.

Saccomanno et al. (1996) evaluated the lung cancer histology in 467 Colorado Plateau uranium miners and 311 nonminers to determine the localization of lung tumors. They showed that 84% of the lung cancer in uranium miners was located in the bronchial airways and 77% in the bronchial airways of nonminers who were mostly smokers. The tumors in the pulmonary or gas exchange region were 16% and 23% in miners and nonminers, respectively. Thus, lung tumors arise mainly in the bronchial airways whether from ^{222}Rn exposure or tobacco smoke. Most of the radon decay products, that is, about 20% of what is inhaled, deposit in the lower lung while only a few percent deposit in the bronchial region. However, the very large surface area (square meters) for deposition in the lower or pulmonary lung compared with a much smaller area (a few hundred square centimeters) in the bronchial region accounts for the much larger bronchial dose, in spite of the lower actual radioactivity deposition of the decay products.

Ruzer et al. (1995) measured the deposited radon decay product gamma ray activity in the chest of metal miners in Tadjikistan by external counting. Measurements on about 100 miners were performed along with filtered air samples to assess the decay product inhalation exposure. Using these data, actual measured breathing rates were obtained for drillers, assistant drillers, and inspection personnel. The group average breathing rates were 0.0079, 00067, and $0.0052 \, m^3 \, min^{-1}$, respectively, with upper bound values of 0.023 ± 0.004, 0.020 ± 0.004, and $0.015 \pm 0.003 \, m^3 \, min^{-1}$.

The bronchial dose coefficients for a range of median particle diameters from 1 to 1000 nm (0.001–1 μm) are shown in Figures 21.5 and 21.6 from UNSCEAR (2006). Figure 21.5 calculates the dose coefficient as a function of breathing rate and Figure 21.6 calculates the dose coefficient as a function of the unattached (or nanometer size) decay products present in the decay product mixture. The unattached fraction, fpot, is specified as the fraction they contribute to the potential energy of

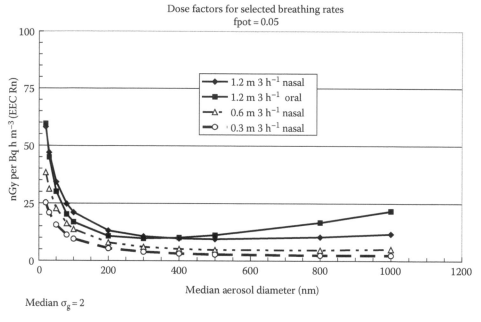

FIGURE 21.5 ^{222}Rn bronchial alpha dose coefficients as a function of breathing rate. (From UNSCEAR, *Sources-to-Effects Assessment for Radon in Homes and Workplaces*, United Nations Scientific Committee on the Effects of Radiation, New York, 2006.)

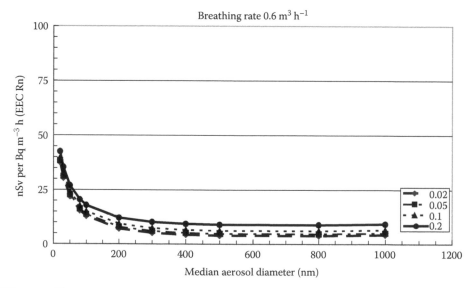

FIGURE 21.6 ^{222}Rn bronchial alpha dose coefficients as a function of unattached or nanometer fraction. (From UNSCEAR, *Sources-to-Effects Assessment for Radon in Homes and Workplaces*, United Nations Scientific Committee on the Effects of Radiation, New York, 2006.)

the mixture. UNSCEAR (2006) adopts 40% as an equilibrium factor for indoor exposure and 60% for outdoor exposure. Thus, the radon gas concentration times 0.4 or 0.6 is accepted as the numerical value of the EEC for indoor and outdoor environments.

Figures 21.5 and 21.6 show that as the inhaled particle size changes with the particular environment, the dose could change by factors of 2–3. The median particle size of inhaled particulates present in various indoor and outdoor environments can change significantly, but usually over short time periods such as during cooking when the particle size decreases. The unattached fraction can change with the total aerosol particle loading, decreasing with higher aerosol concentration.

21.10   DOSE TO OTHER ORGANS

Radon is quite soluble in body tissues and there is always some dose to organs other than the lung. The annual bronchial dose from ^{222}Rn and that to other organs such as soft tissues, female breast, and skin (also from atmospheric plate out of decay products) is small compared with the bronchial dose. For an average radon exposure of 40 Bq m^{-3} and decay product equilibrium of 40%, the dose factor from Figure 21.5 for a breathing rate of 1.2 m^3 h^{-1} and a median particle size of 0.2 μm, the dose is 10 nGy Bq^{-1} m^{-3} h. Assuming this exposure full-time each year, the calculated dose is 40×0.4×10×8760 h, or about 1.4 mGy year^{-1}. The dose to female breast and soft tissues would be 0.003 and 0.001 mGy year^{-1}, respectively (Harley and Robbins, 1992), or less than 1% of the bronchial dose. UNSCEAR (2006) adopted an annual per capita effective bronchial dose of 1.15 mSv for the global population.

21.10.1   DOSE TO THE FETUS FROM RADON IN DRINKING WATER

Robbins and Harley (2002) calculated the dose to the developing fetus from ingested water by the mother. The model indicates an increase from 9 weeks to about 14 weeks and then a decrease. This is due to the assumed changing blood flow rates. The dose at 1 week is zero. This is a consequence of the very small ^{222}Rn concentration in maternal blood in which the embryo floats.

The equivalent dose values can be compared with the average dose to any developing fetus from the natural external gamma ray and cosmic ray radiation of approximately 1 mSv during pregnancy (NCRP, 1988). The maximum equivalent dose to the fetus is about 0.4% of the fetal-life external gamma-ray and cosmic-ray dose for each 100 Bq ingested in water by the mother.

If we assume an average consumption per day of 0.6 L of raw tap water at a concentration of 100 Bq L^{-1}, the calculated total dose to the fetus over the term of the pregnancy is 0.25 mSv or 25% of the normal background radiation dose.

21.10.2   RADON AND CHILDHOOD LEUKEMIA

An epidemiologic study of childhood leukemia in Denmark (2400 cases, 6697 controls) from 1968 to 1994, suggested a weak, but statistically significant, association of residential radon exposure and acute lymphoblastic leukemia (ALL). The Danish study estimated a relative risk (RR)=1.56 (95% CI, 1.05–2.30) for a cumulative exposure of 1000 Bq m^{-3} years. For an exposure duration of 10 years, their RR corresponds to a radon concentration of 100 Bq m^{-3}. There are two dose pathways of interest for alpha particles to damage potential stem cells for ALL. One is the alpha dose to bone marrow, and two, the dose to bronchial mucosa where an abundance of circulating lymphocytes is found. Compared with an exposure of about 1 mSv year^{-1} from natural external background, radon and decay products contribute an additional 10%–60% to the bone marrow equivalent dose. The other pathway for exposure of T (or B) lymphocytes is within the tracheobronchial epithelium. Inhaled radon decay products deposit on the relatively small area of airway surfaces and deliver a significant dose to the nearby basal or mucous cells implicated in human lung cancer. Lymphocytes are collocated with basal cells and are half as abundant. Using a 10 year exposure to 100 Bq m^{-3}, dose estimates suggest that the equivalent dose to these lymphocytes could approach

1 Sv (Harley and Robbins, 2009). The relatively high dose estimate to lymphocytes circulating through the BE, potential precursor cells for ALL, may provide a dose pathway for an association.

21.11 ^{220}RN (THORON) CONTRIBUTION TO ^{222}RN (RADON) MEASUREMENTS

Historically, thoron bronchial dose was assessed through the measurement of its decay products, ^{212}Pb and ^{212}Po, and not thoron gas itself (Schery, 1985, 1990; Schery and Zarcony, 1985). Thoron gas itself was rarely measured, because of the difficulty in measuring an alpha-particle-emitting gas with a very short half-life ($t_{1/2} = 55$ s). The measurement of the two gases required real-time instrumentation with various types of decay chambers to permit a difference in signal with and without the ^{220}Rn (Israel, 1964; NCRP, 1988). Interest in thoron spurred the development of discriminative detectors to measure both gases.

Measurements of thoron gas or its decay products are now common. Several discriminative detectors that measure both radon and thoron gas or plate out of the decay products onto surfaces have been developed (Tokonami et al., 2005; Harley et al., 2010; Janik et al., 2010; Mishra et al., 2010).

Risk assessments are undoubtedly hindered by the presence of thoron in the measured radon gas signal unless measures were applied to exclude its presence. UNSCEAR (2006) provides central dose factors for radon and thoron EECs:

$$\text{Radon (EEC)} = 9 \text{ nSv per } (\text{Bq m}^{-3}\text{h})$$

$$\text{Thoron (EEC)} = 40 \text{ nSv per } (\text{Bq m}^{-3}\text{h})$$

$$\text{The EEC for radon or thoron is} = F_{eq} (\text{Equilibrium ratio}) \times (\text{gas concentration})$$

The accepted value of F_{eq} for radon (UNSCEAR, 2006) is 0.4 for indoor environments and 0.6 for outdoor environments, that is, 40% or 60% equilibrium with the decay products. Harley et al. (2010) have shown from long-term measurements of thoron gas and the thoron decay product ^{212}Pb that the average F_{eq} for thoron is 0.04 indoors and 0.004 outdoors, that is, 4% and 0.4%. Thus, thoron bronchial dose can be estimated from gas measurement similar to radon dose estimates.

Although the dose factor per unit gas concentration for thoron is larger than that for radon, this is offset by the much smaller thoron equilibrium factor, F_{eq}. Therefore, the dose from thoron decay products is usually less than that for radon decay products. Because the measurement of total gas has been used to identify radon, the historic dose and risk assessments may need to be revisited in the future.

It is unlikely that the historic measurements in uranium and other underground mines are affected by thoron, because the ore was primarily ^{238}U, the parent of ^{226}Ra and ^{222}Rn.

21.12 LUNG CANCER RISK PROJECTIONS

More published information exists concerning the lung cancer risk from radon than for any other internal radioactive emitter. Eleven underground cohorts have been studied extensively to estimate lung cancer risk and to develop risk models for the prediction of lung cancer risk in other populations (NAS/NRC, 1999a,b). The risk estimates for nine cohorts are shown in Figure 21.7 (UNSCEAR, 2006). The combined excess relative risk (ERR) from these miner studies is 0.006 WLM^{-1}. A very large study of 58,987 German miners published subsequently determines the ERR for the German cohort as 0.016 (95% CI: 0.0069–0.014) WLM^{-1} (Walsh et al., 2010).

Prior to 2003, the estimate of lung cancer risk in residences was derived from models developed from the underground miner studies. The results of the 23 case control studies and six pooled or

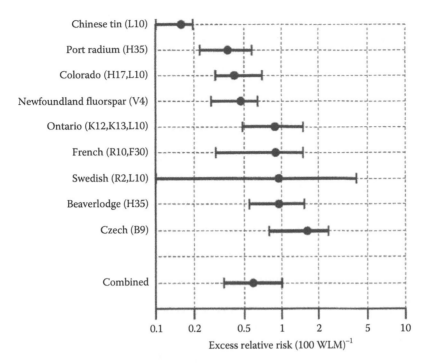

FIGURE 21.7 Underground miner lung cancer risk estimates from 11 follow-up cohort studies. (From UNSCEAR, *Sources-to-Effects Assessment for Radon in Homes and Workplaces*, United Nations Scientific Committee on the Effects of Radiation, New York, 2006.)

meta-analyses are shown in Figure 21.8. Residential risk can be calculated without reliance on underground miner risk projections.

The residential risk from pooled studies is similar to the mine studies with an ERR of 0.16 for lifetime exposure to 100 Bq m^{-3}. This is equal to an ERR of about 0.01 WLM^{-1} but depends on the years assumed for exposure in a residence. An important conclusion from both the residential (Darby et al., 2005) and miner studies (Schnelzer et al., 2010) is that the ERR is the same for both smokers and nonsmokers. The much higher lifetime lung cancer risk in smokers is due to the higher baseline lung cancer rate, the ERR multiplying the baseline.

21.13 GUIDELINES FOR INDOOR ^{222}RN

Based on the domestic radon surveys conducted in many countries, both NCRP and ICRP have set guidelines for indoor radon concentrations (NCRP, 1984a,b, 2003; ICRP, 2007). NCRP recommended that lifetime exposure to individuals above an annual rate of 2 WLM, including background, be avoided. Using the average value for the equilibrium factor, Feq, of 40%, 2 WLM equals exposure to 370 Bq m^{-3} (10 pCi L^{-1}) from all sources. ICRP (2007) recommends an upper reference level of 600 Bq m^{-3} for domestic dwellings and 1500 Bq m^{-3} for workplaces. The ICRP (2007) reference levels are undergoing review and may be reduced. The U.S. EPA recommends a domestic concentration 150 Bq m^{-3} (4 pCi L^{-1}) and Canada recommends 750 Bq m^{-3} (20 pCi L^{-1}).

21.14 SUMMARY

All atmospheres contain radon. Based on 60 countries, the global indoor residential average is 55 ± 30 Bq m^{-3} (1.5 ± 0.8 pCi L^{-1}). The estimate for the U.S. residential average is 46 Bq m^{-3}. An EPA survey of outdoor radon in all 50 U.S. states determined an average of 15 Bq m^{-3} at ground level.

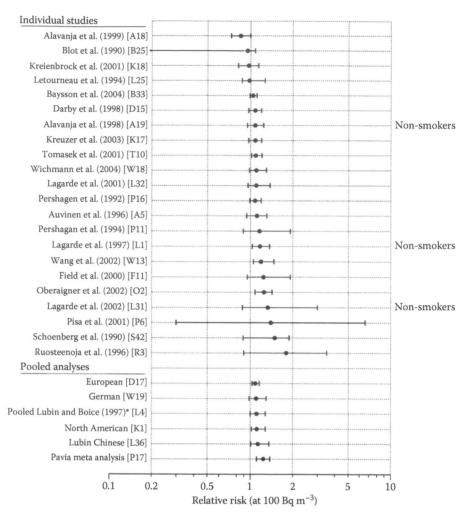

FIGURE 21.8 Residential lung cancer risk estimates from 23 case control studies and six pooled or meta-analyses. *at 150 Bq m^{-3} (From UNSCEAR, *Sources-to-Effects Assessment for Radon in Homes and Workplaces*, United Nations Scientific Committee on the Effects of Radiation, New York, 2006.)

Even at 10 km into the troposphere, near stratospheric heights, the concentration has been measured at 0.1 Bq m^{-3}.

Along with their parent radon, the alpha-emitting short-lived decay products of ^{222}Rn are also present in any atmosphere. The decay products are never in steady state or equilibrium with the parent radon. Indoors, estimated equilibrium ratios are 40% and outdoors 60%. The decay products are atoms of lead and polonium and upon formation rapidly attach to the local aerosol particles. A fraction of these aerosol particles deposit on the bronchial airway surfaces and deliver the lung dose of relevance. The majority of lung cancers associated with ^{222}Rn exposure are located in the upper airways.

Radon in water can add measurably to indoor air concentrations. The release of the gas during water use adds on average 1 unit to air for 10,000 units in water. Air concentrations derived from water can in some cases require remediation of the water. The dose to other organs, such as the stomach, is small compared with the lung dose from the gas release. A few countries have set concentration limits for ^{222}Rn in drinking water.

A fraction of ^{222}Rn gas is soluble and retained in blood and body tissues. A dose to a fetus from ^{222}Rn in maternal drinking water must be considered because of this solubility. Estimated dose to the fetus could be 25% of normal background radiation but only for relatively high concentrations of radon in drinking water.

Lymphocytes in bronchial airways are collocated with basal cells—decay product targets for radon-related lung cancer. There is a suggestion that irradiation of these lymphocytes may be responsible for a fraction of childhood leukemia. Because radon is soluble in body fluids, all organs have a small concentration of the gas and its decay products. However, the dose to other organs is very small compared with the dose to bronchial airways.

There is a clear dose response for lung cancer in underground miners exposed to high concentration of ^{222}Rn. A dose response and risk from residential radon exposure has been demonstrated in pooled studies of the 23 case control epidemiological studies. The lung cancer risk estimated from either residential or underground exposure is similar with ERR at about 0.01 WLM^{-1} for both smokers and nonsmokers. The magnitude of lifetime risk thus depends on the baseline risk. Because the lung cancer baseline is much higher for smokers, by about a factor of 10, smokers' lifetime risk is consequently higher than nonsmokers'.

REFERENCES

Araki, T, Enomoto, S, Furuno, K, Gando, Y, Ichimura, K, Ikeda, H, Inoue, K et al. 2005. Experimental investigation of geologically produced antineutrinos with KamLAND. *Nature* 436(7050):499–503.

Chittaporn, P, Harley, NH. 2000. Indoor and outdoor ^{222}Rn measurements in Bangkok and Chiang Mai, Thailand. *Technology* 7:491–495.

Darby, S, Hill, D, Auvinen, A, Barros-Dios, JM, Baysson, H, Bochicchio, F, Deo, H et al. 2005. Radon in homes and risk of lung cancer: Collaborative analysis of individual data from 13 European case-control studies. *BMJ* 330:223–228.

de Villiers, AJ. 1966. Cancer of the lung in a group of fluorspar miners. *Proceedings/Canadian Cancer Conference* 6:460–474.

de Villiers, AJ, Windish, JP, Brent Fde, N, Hollywood, B, Walsh, C, Fisher, JW, Parsons, WD. 1971. Mortality experience of the community and of the fluorspar mining employees at St. Lawrence, Newfoundland. *Occupational Health Review* 22(1):1–15.

Fisenne, IM, Machta, L, Harley, NH. 2005. Stratospheric radon measurements in three North American locations, in *Radioactivity in the Environment*, Vol. 7, McLaughlin, JP, Simopolis, SE, Steinhausler, F, Eds. The Natural Radiation Environment. Elsevier, Rhodes, Greece, pp. 715–721.

Harley, JH. 1990. Radon is out, in *Indoor Radon and Lung Cancer: Reality or Myth*, Cross, FT, Eds. *Twenty Ninth Hanford Symposium on Health and the Environment*. Battelle Press, Richland, WA.

Harley, NH, Chittaporn, P, Heikkinen, MSA, Medora, R, Merrill, R. 2005a. Airborne particle size distribution measurements at USDOE Fernald, *American Chemical Society Monograph*, Vol. 904, Berkey, E, Zachry, T, Eds. ACS Symposium Series, pp. 342–350.

Harley, NH, Chittaporn, P, Medora, R, Merrill, R. 2010. Measurement of the indoor and outdoor ^{220}Rn (thoron) equilibrium factor: Application to lung dose. *Radiation Protection Dosimetry* 141:357–362.

Harley, NH, Chittaporn, P, Merrill, R, Medora, R, Wanitsooksumbut, W. 2005b. Thoron versus radon: Measurement and dosimetry, in *High Levels of Natural Radiation and Radon Areas: Radiation Dose and Health Effects*, Vol. 1276, Sugahara, T et al., Eds. International Congress Series. Elsevier, Amsterdam, the Netherlands, pp. 72–75.

Harley, NH, Robbins, ES. 1992. ^{222}Rn alpha dose to organs other than the lung. *Radiation Protection Dosimetry* 45:619–622.

Harley, NH, Robbins, ES. 1994. A biokinetic model for ^{222}Rn gas distribution and alpha dose in humans following ingestion. *Environment International* 20(5):605–610.

Harley, NH, Robbins, ES. 2009. Radon and leukemia in the Danish study 2009: Another source of dose. *Health Physics* 97:343–347.

Hess, CT, Vietti, MA, Mager, DT. 1987. Radon from drinking water: Evaluation of water borne transfer into house air. *Environmental Geochemistry and Health* 8:68.

Holaday, DA, Doyle, HN. 1964. Environmental studies in the uranium mines, in *Radiological Health and Safety in Mining and Milling of Nuclear Materials*, Vol. 1, pp. 9–20. International Atomic Energy Agency, Vienna, Austria.

Hopke, PK, Borak, TB, Doull, J, Cleaver, JE, Eckerman, KF, Gunderson, LCS, Harley, NH et al. 2000. Health risks due to radon in drinking water. *Environmental Science and Technology* 34:921–926.

ICRP. 2007. The 2007 Recommendations of the International Commission on Radiological Protection. *Annals of the ICRP*, ICRP Publication 103, Vol. 37(2–4), Elsevier, Amsterdam, the Netherlands.

Israel, H. 1964. The radon-220 content of the atmosphere, in *The Natural Radiation Environment*, Adams, JAS, Lowder, WM, Eds. University of Chicago Press, Chicago, IL, p. 313.

Janik, M, Tokonami, S, Kranrod, C, Sorimachi, A, Ishikawa, T, Hassan, NM. 2010. International intercomparisons of integrating radon/thoron detectors with the NIRS radon/thoron chambers. *Radiation Protection Dosimetry* 141:436–439.

Li, CS, Hopke, PK. 1993. Initial size and distributions and hygroscopicity of indoor combustion aerosol particles. *Aerosol Science and Technology* 19:305–316.

Lundin, FE Jr, Wagnoer, JK, Archer, VE. 1971. *Radon Decay Product Exposure and Respiratory Cancer: Quantitative and Temporal Aspects*. National Institute for Occupational Safety and Health/National Institute of Environmental Health Sciences, Joint Monograph No. 1, National Technical Information Service, Springfield, VA.

Machta, L, Lucas, HF, Jr. 1962. Radon in the upper atmosphere. *Science* 135:296.

Mishra, R, Prajith, R, Sapra, BK, Mayya, YS. 2010. Response of direct thoron progeny sensors (DTPS) to various aerosol concentrations and ventilation rates. *Nuclear Instruments and Methods in Physics Research B* 268:671–675.

Moore, HE, Poet, SE, Martell, EA. 1973. 222Rn, 210Pb, 210Bi and 210Po profiles and aerosol residence times versus altitudes. *Journal of Geophysical Research* 78:7065.

Morrison, HI, Semenciw, RM, Mao, Y, Wigle, DT. 1988. Cancer mortality among a group of fluorspar miners exposed to radon progeny. *American Journal of Epidemiology* 128(6):1266–1275.

Morrison, HI, Villeneuve, PJ, Lubin, JH, Schaubel, DE. 1998. Radon-progeny exposure and lung cancer risk in a cohort of Newfoundland fluorspar miners. *Radiation Research* 150:58–65.

NAS/NRC. 1999a. *Health Effects of Exposure to Radon*. Committee on Health Risks of Exposure to Radon, Board on Radiation Effects Research, Commission on Life Sciences National Research Council, BEIR VI, National Academy Press, Washington, DC.

NAS/NRC. 1999b. *Risk Assessment of Radon in Drinking Water*. National Research Council. National Academy Press, Washington, DC.

National Academy of Sciences. 1986. National Academy of Sciences/National Research Council. 1988. Health risks of radon and other internally deposited alpha-emitters, committee on biological effects of ionizing radiation, IV, National Academy Press, Washington, DC.

NCRP. 1984a. Exposures form the uranium series with emphasis on radon and its daughters. Report 77, National Council on Radiation Protection and Measurements, Bethesda, MD.

NCRP. 1984b. Evaluation of occupational and environmental exposures to radon and radon daughters in the United States. Report 78, National Council on Radiation Protection and Measurements, Bethesda, MD.

NCRP. 1988. Measurement of radon and radon daughters in air. Report 97, National Council on Radiation Protection and Measurements, Bethesda, MD.

Reineking, A, Porstendorfer, J. 1986. High volume screen diffusion batteries and alpha spectroscopy for measurement of the radon daughter activity size distributions in the environment. *Journal of Aerosol Science* 17:873–879.

Robbins, ES, Harley, NH. 2002. Dose to the fetus from 222Rn in maternal drinking water, in *Radioactivity in the Environment*, Vol. 7, McLaughlin, JP, Simopolis, SE, Steinhausler, F, Eds. The Natural Radiation Environment NRE VII, Rhodes, Greece. Elsevier, Amsterdam, the Netherlands, pp. 749–755.

Ruzer, LS et al. 1995. Assessment of lung deposition and breathing rate of underground miners in Tadjikistan. *Radiation Protection Dosimetry* 58:261–268.

Saccomanno, G, Auerbach, O, Kuschner, M, Harley, NH, Michaels, RY, Anderson, MW, Bechtel, JJ. 1996. A comparison between the localization of lung tumors in uranium miners and in nonminers from 1947 to 1991. *Cancer* 77:1278–1283.

Schery, S. 1985. Measurement of airborne 212Pb and 220Rn at various indoor locations within the United States. *Health Physics* 49:1061.

Schery, S. 1990. Thoron in the environment. *Journal of the Air and Waste Management Association* 40:493.

Schery, S, Zarcony, MJ. 1985. Thoron and thoron daughters in the indoor environment. *Proceedings of the 18th Midyear Topical Symposium of the Health Physics Society*, Colorado Springs, CO, p. 443.

Schnelzer, M, Hammer, GP, Kreuzer, M, Tschense, A, Grosche, B. 2010. Accounting for smoking in the radon-related lung cancer risk among German uranium miners: Results of a nested case-control study. *Health Physics* 98:20–28.

Tokonami, S, McLaughlin, J, Tommasino, L, Harley, NH. 2010. Editorial, international workshop on environmental thoron and related issues. *Radiation Protection Dosimetry* 141:315.

Tokonami, S, Takahashi, H, Kobayashi, Y, Zhou, W, Hulber, E. 2005. Up-to-date radon-thoron discriminative detector for large scale surveys. *Review of Scientific Instruments* 76:113505-1–113505-5.

Tu, KW, Knudson, EO. 1988a. Indoor radon progeny particle size measurements made with two different methods. *Radiation Protection Dosimetry* 24:251–255.

Tu, KW, Knudson, EO. 1988b. Indoor outdoor aerosol measurements for two residential buildings. *Aerosol Science and Technology* 9:71–82.

UNSCEAR. 2006. *Sources-to-Effects Assessment for Radon in Homes and Workplaces*. United Nations Scientific Committee on the Effects of Radiation, New York.

U.S. Environmental Protection Agency (EPA). 1992. National Residential Radon Survey Summary Report, EPA 402-R-92-011. U.S. Environmental Protection Agency, Washington, DC.

Villeneuve, PJ, Morrison, HI, Lane, R. 2007. Radon and lung cancer risk: An extension of the mortality follow-up of the Newfoundland fluorspar cohort. *Health Physics* 92:57–169.

Walsh, L, Dufey, F, Tschense, A, Schnelzer, M, Grosche, B, Kreuzer, M. 2010. Radon and the risk of cancer mortality—Internal Poisson models for the German uranium miners cohort. *Health Physics* 99:292–300.

WHO. 2004. *Guidelines for Drinking Water Quality*, 3rd edn., Zeeb, H, Shannoun, F. Eds. Vol. 1, Recommendations. WHO, Geneva, Switzerland.

22 Risk from Inhalation of the Long-Lived Radionuclides Uranium, Thorium, and Fallout Plutonium in the Atmosphere

Isabel M. Fisenne

CONTENTS

22.1 INTRODUCTION

The global population is chronically exposed to naturally occurring and man-made radionuclides by both inhalation and ingestion. The inhalation pathway is the principal focus of this chapter and the ingestion pathway will only be touched upon as necessary.

All soil and rocks contain the uranium and thorium series, headed by ^{238}U ($t_{1/2} = 4.468 \times 10^9$ year) and ^{232}Th ($t_{1/2} = 1.405 \times 10^{10}$ year), respectively. The natural forces of erosion and weathering reduce the particle size of the host rock or soil and surface winds suspend the small particles. These are removed from the atmosphere by the usual scavenging processes. It is known that "dust storms" resulting from desertification carry material for thousands of kilometers and in some instances around the world. The distribution of uranium and thorium in the terrestrial environment is relatively constant at 2–3 μg U g⁻¹ (25–36 mBq ^{238}U g⁻¹) of soil and 10 μg Th g⁻¹ (4 mBq ^{232}Th g⁻¹) of soil (NCRP 1987; UNSCEAR 2000a). There are geographical areas that do have higher concentrations and have had the deposits exploited for commercial purposes. The bulk of the atmospheric inventory of uranium is soil derived from the earth's surface. Additional sources of uranium are emissions from energy-generating plants (coal, oil, nuclear), fallout resuspension from atmospheric nuclear weapons tests, satellite failures, and nuclear-related accidents (UNSCEAR 1988). Atmospheric thorium also is derived from soil resuspension and to a lesser degree energy-related emissions.

Plutonium is a man-made element and like U and Th is an actinide element. The principal source of terrestrial plutonium was atmospheric nuclear weapons testing. Plutonium is produced by an (n,γ) reaction on ^{238}U and is separated for use in nuclear weapons. Fallout plutonium consists of ^{239}Pu ($t_{1/2} = 2.411 \times 10^4$ year) and ^{240}Pu ($t_{1/2} = 6.563 \times 10^3$ year) in a 240/239 atom ratio of 0.18. For convenience, the pair will be denoted as Pu. A third Pu isotope, ^{238}Pu ($t_{1/2} = 87.7$ year), was introduced into the atmosphere in the Southern Hemisphere stratosphere when a satellite which included a Systems for Nuclear Auxiliary Power Generator (SNAP 9A) failed and reentered the atmosphere. Harley estimated that in 1970, in the Northern Hemisphere <4% of atmospheric Pu was due to ^{238}Pu from weapons plus SNAP-9A (Harley 1980). His ^{238}Pu estimate for the Southern Hemisphere was 18% of the total Pu in that hemisphere. The Pu injection into the atmosphere was about 400 kCi (1.5×10^{16} Bq), some of which remained on or close to the test sites. Because the majority of the weapons tests were conducted in the Northern Hemisphere, ~80% of the Pu fallout occurred there.

22.2 SIMILARITIES AND DIFFERENCES

As noted earlier, the principle source of U and Th in the atmosphere is derived from soil and rock. The base chemical composition is usually silicate or carbonate. The U in atmospheric aerosols exists primarily in the +6 valence state, while Th resides in the +4 state. Fallout Pu deposits on the earth's surface as the oxide and is principally in the +4 valence state.

Once aerosolized these actinides become of interest as part of the total human exposure to radiation. Unlike U, the principle exposure route for Th and Pu is inhalation rather than ingestion. The International Commission of Radiological Protection (ICRP) suggests that for insoluble compounds of Th and Pu, the gastrointestional uptake (f_1) is on the order of 5×10^{-4} (ICRP 1993). The ICRP has adopted an f_1 value of 2×10^{-2} for U, assuming equal absorption from diet and water (ICRP 1993). However, Spencer et al (1990), in the only controlled study of uptake of U in humans, showed that the principal source of uptake of U was from drinking water to the extent of about 5%.

The air concentrations of U and Th are influenced by climate and land mass while fallout Pu is dependent on the latitude of the initial injection, primarily the Northern Hemisphere.

22.3 SAMPLING

The collection of atmospheric aerosols is almost exclusively that of total suspended particulates (TSP), that is, particles <500 μm in diameter. More important are inhalable particulates, <10 μm in diameter, which may enter the nose and throat. Samplers which are designed to collect this class of particulates are referred to as PM10 samplers. Respirable particulates, <2.5 μm in diameter ($PM_{2.5}$), may reach the functional areas of the lungs.

For fallout radionuclides, high volume pumps were and are used to draw air through filters made of material such as polystyrene. The large air volumes, usually in excess of $25,000\,m^3$, were necessary to collect sufficient samples for wet radiochemical analyses. The naturally occurring radionuclides were also present in the total aerosol but from terrestrial sources.

There are remarkably few radionuclide measurements of environmental respirable particulates. Two such studies will be described later. Resuspension of deposited material has attracted little attention except for studies in the environs of nuclear weapons production facilities. There is one study of the resuspension of uranium from environmental sources. This will also be discussed later.

The sites for air collection to determine the concentration of these long-lived α emitters are almost exclusively land based in the mid-latitudes of the Northern Hemisphere. A few measurements of U have been made from collections at remote sites, including the Atlantic Ocean and Antarctica. Measurements of the Th concentration in air are rare. The U.S. Department of Energy Environmental Measurements Laboratory's (EML) Surface Air Sampling Program (SASP) was the most geographically comprehensive monitoring program for fallout Pu. Air samples were collected and analyzed from stations at Nord, Greenland (80°N, 17°W) to South Pole Station, Antarctica (90°S, 0°W), roughly along the 80th meridian. The EML database for SASP can be accessed at their website: http://www.nustl.st.dhs.gov/databases/

22.4 URANIUM IN TOTAL SUSPENDED PARTICULATES

The largest number of measurements of U in air has been performed under a U.S. Environmental Protection Agency (EPA) program named Environmental Radiation Ambient Monitoring System (ERAMS). The ERAMS air program consists of 50 stations collecting air particulates on filters. The ERAMS filters were analyzed for U starting in the late 1970s. Presently annual composites of the air particulate filters are analyzed for U and Pu.

The longest running program for the collection and measurement of air particulates was conducted at the Argonne National Laboratory (ANL), Argonne, IL. Begun in 1973, their air sampling program includes an off-site station to assess the background concentrations of nuclides of interest, including U, Th, and Pu.

Published data from the EPA and ANL sampling programs were evaluated to assess the U air concentration in the continental United States (Stevenson and Pan 1996). The U concentrations were examined from 25 ERAMS sites and an anomaly was found for stations in the northern and mid-section of the United States. This was confirmed by the independent air filter collections and measurements of U performed at ANL. The 40% decrease in the air concentrations of U at these sites was attributed to "regulatory compliance in reducing emissions from fossil burning facilities."

The data from 22 ERAMS locations averaged $2.1 \pm 0.7 \,\mu$Bq ^{238}U m^{-3} (170 pg U m^{-3}). Only the ANL data were useful to investigate seasonal variations as the ERAMS compost samples overlapped seasons. At ANL the samples showed a slight rise in April (Spring rise), a low point in August and another small rise in November. It was thought that the November rise was due to emissions from a local coal burning electrical generating plant. The non-carbonaceous residues from the TSP collections have been analyzed for U and the concentrations were found to reflect the U concentration in soil. The ^{234}U/^{238}U ratio for 22 ERAMS sites was 1.14 ± 0.24, indicative of the U source, soil.

The U concentrations in surface air are shown in Table 22.1. The impact of land mass, climate, and industrialization is apparent. Basically, there is no soil contribution at the remote sites, the ground being frozen year round or at oceanic locations. Even the remote sites in Antarctica show a difference between the pristine environment and human encampments. The global average concentration of U in air as adopted by UNSCEAR (2000a) of 1 μBq m^{-3} is based almost exclusively on data obtained in the mid-latitudes of the Northern Hemisphere.

22.5 THORIUM IN TOTAL SUSPENDED PARTICULATES

Measurements of the Th concentration in air are sparse. This was partially due to the difficulties encountered with radiochemical separation, specifically unacceptable reagent and material blanks. Again the Argonne National Laboratory's Environmental Monitoring Program has the longest running record of Th in air Measurements. As stated earlier for their U in air measurements, the impact of regional sources causes a marked decrease with time. The ANL data are included in Table 22.2 as they are unremarkable when compared with other measurements. Again, measurements of the ashed filter residues indicated that the Th was derived from soil. The activity concentrations of Th in air are in the same general range as those for U.

22.6 PLUTONIUM IN TOTAL SUSPENDED PARTICULATES

Measurements of a long-lived alpha emitter Pu were necessitated by widespread public concern over the potential hazard from global fallout, the entry of the SNAP 9 satellite to the atmosphere (1963), and the nuclear accident at Chernobyl (1986). The measurements of Pu in TSP on a global basis began in 1966. By the use of the known Pu to ^{90}Sr ratio (0.017), estimations of atmospheric Pu in TSP were extended back to 1963 for SASP samples. The USDOE EML SASP network results for Pu were part of an extensive review by Harley (1980). He estimated that some 15 PBq (400 kCi) had been produced and dispersed globally. Hardy et al. (1973) accounted for over 12 PBq (330 kCi) without estimating the deposition around test sites. The measurements of Pu on a global scale

TABLE 22.1

Uranium Concentrations in Total Suspended Particulates

	pg U m⁻³		µBq ²³⁸U m⁻³		
Remote Sites	Mean	SD	Mean	SD	References
N. Atlantic Ocean	4.1	1.2	0.05	0.015	Hamilton (1970)
S. Atlantic Ocean	2.2	0.5	0.027	0.006	Hamilton (1970)
Antarctic Ice Pack	1.2	0.3	0.015	0.004	Hamilton (1970)
Antarctic Base Camp	3.0	1.0	0.037	0.012	Hamilton (1970)
Skibotn, Norway	3.0	1.3	0.037	0.016	Kolb (1989)
Vardo	5.5	0.9	0.068	0.012	Kolb (1995)
Asia					
Tokyo, Japan	25	15	0.31	0.19	Hirose and Sugimura (1981)
Tsukba Science City	14	10	0.17	0.12	Hirose and Sugimura (1981)
Kamisaibama	234	98	2.9	1.2	Yunoki et al. (1995)
Europe					
Mol, Belgium	115	95	1.4	1.2	Janssens et al. (1975)
Olen	114	36	1.4	0.5	Janssens et al. (1975)
Berlin, Germany	133	20	1.6	0.2	Kolb (1989)
Braunschweig	85	40	1.1	0.5	Kolb (1989)
Sutton, UK	62	78	0.77	0.97	Hamilton (1970)
North America					
Ontario, Canada	100		1.2		Tracy and Prandl (1985)
United States, 23 States	170	54	2.1	0.7	USEPA (1993)
Global average	80		1		UNSCEAR (2000b)

TABLE 22.2

Thorium Concentrations in Total Suspended Particulates

	pg Th m⁻³		µBq ²³²Th m⁻³		
Remote Sites	Mean	SD	Mean	SD	References
Skibotn, Norway	100	50	0.4	0.2	Kolb (1989)
Vardo	15	7	0.065	0.003	Kolb (1995)
Asia					
Bombay, India	250–2500		1–10		Sunta et al. (1987)
Europe					
Berlin, Germany	270	20	1.1	0.1	Kolb (1989)
Braunschweig	150	70	0.6	0.3	Kolb (1989)
North America					
United States, Argonne, IL	50	20	0.2	0.1	Golchert et al. (1973)
New York, NY	100	50	0.4	0.2	Fisenne et al. (1987)
Global average	250		1		UNSCEAR (2000b)

continued until mid-1985 when the Pu concentrations in monthly composited samples approached the detection limit for the measurement. The USEPA ERAMS measurements included Pu since 1978. The collections and measurements continue on an annual basis as the Pu concentrations in the TSP are at the detection limit for the measurement. The radionuclide concentrations from the ERAMS sites are available on their website, http://www.epa.gov/enviro/html/erams. The off-site

TABLE 22.3
Plutonium in TSP

Location	Latitude	Years	Total (µBq Pu m^{-3})
Thule, Greenland	76°N	1963–1974	65.1
Moosonee, Ontario	51°N	1963–1985	71.0
New York, NY	40°N	1963–1985	98.9
Mauna Loa, HA	29°N	1963–1985	95.0
Miami, FL	25°N	1963–1985	82.5
Guayaquil, Eucador	2°S	1963–1976	10.3
Lima, Peru	12°S	1968–1985	11.6
Chalcaltaya, Chile	16°S	1963–1985	22.6
Santiago, Chile	33°S	1963–1985	30.3
Punta Arenas, Chile	53°S	1963–1985	6.5
Antarctica	64°S	1966–1975	3.4
Krakow, Poland	51°N	1990–2002	2.2×10^{-4}–1.7×10^{-2}
Krakow/Bialystok, Poland	53°N	1990–2002	2.2×10^{-4}–1.7×10^{-2}
Lublin	51°N	1993–1994	6.1×10^{-2}
		1998–1999	9.1×10^{-3}

aerosol collections at ANL begun in 1973 and continuing to the present time are still analyzed for Pu on a monthly basis, but the concentrations are nearly at the detection limit for the measurement.

In order to summarize this vast database, 11 sites were selected and the average annual concentration of Pu for the years the site was operational was summed. The site, latitude, years in operation, and the sum of the Pu concentrations measured in the TSP samples are shown in Table 22.3. The distribution pattern of fallout Pu shows the highest concentrations in the Northern Hemisphere mid-latitudes, a minimum in the equatorial region and an increase in the Southern Hemisphere mid-latitudes. Most of the atomic weapons testing was carried out in the Northern Hemisphere but a few test sites were located in the Southern Hemisphere. A summary of atmospheric weapons testing was prepared by UNSCEAR (2000b).

Harley (1980) stated that the Pu inventory for the Northern Hemisphere decreased with a half-time of 10 months until 1968 when the People's Republic of China began a series of atmospheric weapons tests. The concentration of Pu in TSP collected in New York City is depicted in Figure 22.1. With no

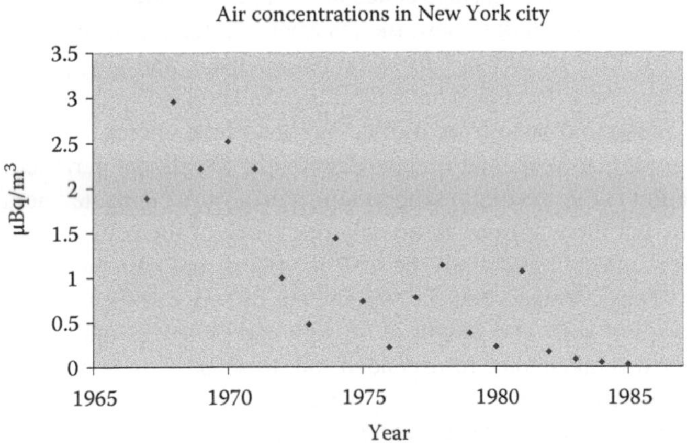

FIGURE 22.1 Plutonium in New York city aerosal samples.

substantial atmospheric injection since 1981, the TSP Pu concentrations have decreased to <0.05 μBq m^{-3} in 1985. The annual average Pu in air concentration at ANL was <0.01 μBq m^{-3} in 2001.

Lee et al. (1986) estimated the stratospheric mean residence time of 1.2 year while Holloway and Hayes (1982) estimated a 71 day mean residence time for Pu aerosol fallout in the troposphere. Both these pieces of information would suggest that all the fallout Pu was deposited on the surface of the earth by 1985. The fact that Pu is still measurable in TSP, albeit with error terms of 100% or more, leads to the conclusion that resuspension of material from the surface is now the controlling mechanism for dispersion, as it is with U and Th.

Komosa and Chibowski (2002) reported on a weekly ground level aerosol data base in Lublin, Poland initiated in 1993. Measurement of ^{239,240}Pu was performed radiochemically on 4000–8000 m^3 filtered air samples taken in 1993 to 1994 and 1998 to 1999. The ^{239,240}Pu was better correlated with sample mass than volume indicating resuspension as a source. Their measurements are reported in Table 22.3.

Kierepko et al. (2009) reported on the measurements of ^{239,240}Pu made in Krakow, Poland from 1990 to 2002 and Bialystok from 1996 to 2001. They reported a seasonal variation in the measured air concentration and that the main source appeared to be from the North Sea. Their measurements are shown in Table 22.3.

22.7 RESUSPENSION

Particle resuspension, while of importance, is an elusive process to quantify. Sehmel (1980) published a definitive review of the subject that gives an appreciation for the uncertainties associated with this topic. Resuspension is a form of large-scale erosion describing the continual movement of particles as a function of surface stresses. The stresses are saltation, surface creep and suspension, while the transport means depends on the particle diameter wind speed and turbulence. Newman et al. (1976) defined saltation as a process by which particles with diameters of 100–500 μm rise or bounce in a layer close to the surface-air interface. Surface creep particles with diameters of 500–1000 μm slide or roll, pushed along the surface by wind stresses and the impact of saltation particles. The smallest particles, <100 μm in diameter, move by suspension, following air motion. The interplay of the stresses causes suspension particles to leave a surface when saltation particles impact the surface. Although particles <50 μm and particularly <10 μm in diameter are almost impervious to wind erosion, when mixed with saltation particles, they become transportable by suspension.

Modelers have not been able to predict resuspension factors for general situations. Langham (1971) defined the resuspension factor (RF) as the ratio of the airborne pollutant concentration per unit volume of air to the pollutant surface concentration per unit area on the surface. Thus, RF has units of m^{-1}. From Sehmel's (1980) review, the resuspension factors developed under experimental conditions range from 10^{-12} to 10^{-2} m^{-1} for wind resuspension and 10^{-10} to 10^{-2} m^{-1} from human activities. These factors were developed not only from wind erosion situations but also from agricultural practices, vehicular and pedestrian traffic, and household chores. Similarly the resuspension half-life ranges from days to years and is dependent on the situational parameters.

Golchert and Sedlet (1978) several resuspension studies were conducted at nuclear facilities and the Nevada test site, but there appears to be only one study of the resuspension of U, Th, and Pu. Golchert and Sedlet collected and analyzed TSP air filters and soil samples at the off-site ANL monitoring station. They assumed that the top one cm of soil was available for resuspension and contained 925 μBq g^{-1} of soil. The weight of the non-carbonaceous material remaining after dry ashing the TSP filter was taken to be resuspended soil. Resuspension factors for U, Th isotopes and fallout Pu were calculated and are shown in Table 22.4. The RFs estimated in this study are internally consistent, despite the different sources, fallout and naturally occurring radionuclides. The Pu was "aged" deposition, assuming most of the Pu was deposited in the 1960s. This would suggest that

TABLE 22.4

Resuspension Factors for U, Th Isotopes and Fallout Pu at Argonne, IL

Nuclide	Air Concentration (μBq m^{-1})	Ground Deposition (Bq m^{-2})	Resuspension Factor (m^{-1})
^{228}Th	0.32	629	5.1×10^{-10}
^{230}Th	0.54	1036	5.3×10^{-10}
^{232}Th	0.30	555	5.5×10^{-10}
U	1.55	2516	6.1×10^{-10}
Pu	0.026	37	7.0×10^{-10}

Source: Adapted from Golchert, N.A. and Sedlet, J., Resuspension studies on fallout level plutonium, in Selected Environmental Plutonium Research Reports of the Nevada Applied Ecology Group, Report NVO-192, 1978.

the Pu in soil was in a relatively steady-state condition. Anspaugh et al. (1975) estimated RFs for 20 year old Pu deposition at the Nevada Test Site to be 3×10^{-10} and 2×10^{-9} m^{-1}. Golchert and Sedlet (1978) considered the U and Th representative of the "ultimate aged source and their resuspension factors, the equilibrium condition." However, these RF estimates are site specific and cannot be taken as the general case. It does suggest that resuspension will be the source of fallout Pu (in the absence of any atmospheric weapons testing or nuclear accidents) for a long period of time.

22.8 RESPIRABLE CONCENTRATIONS OF U, TH, AND PU

The term TSP refers to particles $<500\,\mu$m in diameter. The next smaller fraction is called the inhalable particulates with sizes of $<10\,\mu$m in diameter, sometimes designated as PM$_{10}$. Their size permits penetration into the nose and throat. The respirable particulates have diameters of $<2.5\,\mu$m (PM$_{2.5}$) and penetrate into the lung. The NCRP has defined the respirable fraction as "the fraction of airborne material that can be inhaled and possibly deposited in the lung" (NCRP 1988). "Respirable dust" was defined as the portion of inhaled dust which is deposited in the non-ciliated portions of the lung (Lippmann and Harris 1962).

Golchert and Sedlet (1978) determined the particle size distribution of U and Pu at the ANL off-site location. A commercial high-volume cascade impactor was operated for one month to collect a total air volume of 2.28×10^4 m^3. The summary of their results is shown in Table 22.5. It is worth noting that the total particulate concentration of 55 μg m^{-3} of air is in excellent agreement with the UNSCEAR (2000a) adopted value of 55 μg m^{-3} of air. As expected the Pu-bearing particulates are very small, 87% of the total activity associated with particle sizes of $\leq 2\,\mu$m. The soil-derived U particles tend to be much larger with $<20\%$ of the total activity in the $\leq 2\,\mu$m fractions. Volchok et al. (1974) collected air samples with a horizontal elutriator to determine the respirable fraction of Pu in an urban and a rural setting. The results were the same for the two locations (84% respirable) and in excellent agreement with the ANL data. With the uncertainties associated with the collections and measurements of these samples, fallout Pu will be considered 100% respirable. Although no measurements were reported for respirable fraction of Th, it is assumed that it is the same as for U, that is, $\leq 20\%$.

22.9 VALUE OF LONG-TERM MONITORING

The work performed and published by Golchert et al. (1973) demonstrates the scientific significance of long-term monitoring efforts. The measurements of two naturally occurring actinide elements,

TABLE 22.5

Particle Size Distribution of U and Pu in Air at ANL

Particle Size Range (µm)	Particulate Weight (g)	Particulate Concentration (µg m⁻³)	Air Concentration (µBq m⁻³)		Mass Concentration (mBq g⁻¹)	
			U	Pu	U	Pu
>7	0.440	19.3	1.07 (37.2%)	0.036 (2.5%)	55.5 (14.5%)	1.81 (1.3%)
3.3–7	0.169	7.4	0.67 (23.3%)	0.044 (3.1%)	89.9 (23.5%)	5.92 (4.1%)
2.0–3.3	0.097	4.3	0.59 (36.4%)	0.11 (7.8%)	139.1 (36.4%)	25.9 (18.1%)
1.1–2.0	0.102	4.5	0.41 (14.3%)	0.28 (19.7%)	91.0 (23.8%)	61.4 (42.9%)
<1.1	0.445	19.5	0.13 (4.5%)	0.96 (67.0%)	6.7 (1.8%)	48.1 (33.6%)
Total	1.253	55.0	2.87	1.43		

Source: Adapted from Golchert, N.A. and Sedlet, J., Resuspension studies on fallout level plutonium, in Selected Environmental Plutonium Research Reports of the Nevada Applied Ecology Group, Report NVO-192, 1978.

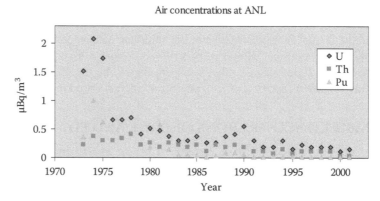

FIGURE 22.2 U, Th, and Pu in ANL aerosol samples.

U and Th, and the fallout actinide, Pu, at the same location collected on a monthly basis is a rare database. Figure 22.2 displays the annual average TSP in air concentrations for U, Th, and Pu at the ANL off-site location. Even with this gross depiction it is evident that, in the absence of atmospheric weapons testing, fallout Pu rapidly decreased below that level of the natural emitters. The graph also shows the impact of local sources (coal-fired power plants) on the U air concentration and the subsequent decline due to the enforcement of clean air act requirements. The annual means mask the features of the monthly and seasonal variations measured at this site. Fallout derived Pu concentrations followed the well-known spring rise pattern, while the U and Th air concentrations rose in the winter, a direct result of the heating pattern in the area. The data also showed that the combined air concentrations of U and Th were greater than Pu from 1973 through 2002. The record is incomplete at ANL, but Pu in air measurements in New York City, the same latitude band as Argonne, IL, from 1963 through 1973 shows that >92% of the Pu was deposited during this period. The data obtained for the New York City site and the ANL off-site location were in good agreement for the years of mutual collections. Another interesting thing concerning the ANL off-site data and the USEPA ERAMS data is that the Chicago, IL, U in air concentration of 2.4 µBq m⁻³ is close to the national average of 2.1 µBq m⁻³ of air but a factor of 5–10 greater than the ANL off-site data. This comparison of a single site with measurements of naturally occurring and a fallout radionuclide with other studies points out the difficulties in deriving reasonable estimates of air concentrations based on sampling site location.

22.10   EXPOSURE ESTIMATION OF U, TH, AND PU

The measurements of U, Th, and Pu in TSP show the vast differences in concentrations with geographical location. To generalize the inhalation estimate of these radionuclides, Th (1 Bq m^{-3} of air) are adopted, and for Pu, a single location, New York, is selected for a 23 year period from 1963 to 1985. The estimations are for the adult male with a daily breathing rate of 22.2 m^{-3} of air. The total exposure to U and Th for a 23 year period is 186 mBq for each or 372 mBq for U and Th. The exposure estimate based on New York Pu measurements from samples collected in the latitude with the highest fallout air concentration is 802 mBq or a factor of >2.2 than the sum of the U and Th exposures.

These estimates are made only for illustrative purposes and are not representative of specific geographical areas or particular time periods, except for Pu. The ANL TSP measurements for the calendar year 2001 yield air concentrations of 0.15 µBq U m^{-3}, 0.04 µBq Th m^{-3}, and 0.01 Pu µBq m^{-3}. The sum of U and Th annual exposures at site for 2001 is a factor of 15 greater than the Pu exposure.

As noted earlier, all three environmental aerosols are refractory and, thus, only very slowly transferred from the alveolar region of the lung to the bloodstream. Some of the deposited aerosol is sequestered for long periods of time in the pulmonary lymph nodes, in effect reducing the systemic body burden.

The alpha-dosimetric consequences from the inhalation of these environmental aerosols are minor compared to the dose from radon progeny.

REFERENCES

Anspaugh, LR, Shinn, JH, Phelps, PL, Kennedy, NC. Resuspension and redistribution of plutonium in soils. *Health Phys.* 29, 571–582, 1975.

Fisenne, IM, Perry, PM, Decker, KM, Keller, HK. The daily intake of ^{234,235,238}U, ^{228,230,232}Th, and ^{226,228}Ra by New York City Residents. *Health Phys.* 53, 357–363, 1987.

Golchert, NA, Sedlet, J. Resuspension studies on fallout level plutonium. In Selected Environmental Plutonium Research Reports of the Nevada Applied Ecology Group, Report NVO-192, 1978.

Golchert, NW et al. Argonne National Laboratory-East Site Environmental Report for Calendar Years 1973–2001. Argonne National Laboratory, Argonne, IL.

Hamilton, EI. The concentration of uranium in air from contrasted natural environments. *Health Phys.*, 19, 511–520, 1970.

Hardy, EP, Krey, PW, Volchok, HL. Global inventory and distribution of fallout plutonium. *Nature* 241, 444–445, 1973.

Harley, JH. Plutonium in the environment—A review. *Jpn. J. Radiat. Res.* 23, 83, 1980.

Hirose, K, Sugimura, Y. Concentration of uranium and the activity ratio of ^{234}U/^{238}U in surface air: Effect of atmospheric burn-up of cosmos-954. *Meteorol. Geophys.* 32, 317–322, 1981.

Holloway, RW, Hayes, DW. Mean residence time of plutonium in the troposphere. *Environ. Sci. Technol.* 16, 127–129 (1982).

International Commission on Radiological Protection (ICRP). Age–dependent doses to members of the public from intake of radionuclides. ICRP Publication 67, *Part 2 Ingestion Dose Coefficients. Annals of the ICRP*, Vol. 23(3/4) Elsevier Science Ltd., Oxford, U.K., 1993.

Janssens, M, Desmet, B, Dams, R, Hoste, J. Determination of uranium, antimony, indium, bromine and cobalt in atmospheric aerosols using epithermal neutron activation and a low-energy photon detector. *J. Radioanal. Chem.* 26, 305–315, 1975.

Kierepko, R, Mietelski, JW, Borowiec, W, Tarasiewicz, S, Blazej, S, Kapala, JE. Plutoniumtraces in atmospheric precipitation and in aerosols from Knakow and Bialystok. *Radiochim Acta* 97, Special issue 4–5, 253–255, 2009.

Kolb, W. Seasonal fluctuations of the uranium and thorium contents of aerosols in ground-level air. *J. Environ. Radioact.* 9, 61–75, 1989.

Kolb, W. Thorium, uranium and plutonium in surface air at Vardo. *J. Environ. Radioact.* 31, 1–6, 1995.

Komosa, A, Chibowski St. Determination of plutonium in ground-level air aerosols collected on Petrianov filters. *J. Radioanal. Nucl. Chem.* 251, 113–117, 2002.

Langham, WH. Plutonium distribution as a problem in environmental science. In *Proceedings of Environmental Plutonium Symposium*, Report LA-4756, p. 9, 1971, National Technical Information Service, Springfield, VA.

Lee, SC, Rao, HSC, Sakuragi, Y, Bakhtiar, N, Jiang, FS, Kuroda, PK. The origin of plutonium in the atmosphere. *Geochem. J.* 19, 283–288, 1986.

Lippmann, M, Harris, WB. Size-selective samplers for estimating "respirable" dust concentrations. *Health Phys.* 8, 155–163, 1962.

National Council on Radiation Protection and Measurements Report No. 94, Exposure of the Population in the United States and Canada from Natural Background Radiation, Bethesda, MD, 1987.

National Council on Radiation Protection and Measurements Report No.97, Measurement of radon and radon daughters in air, Bethesda, MD, 1988.

Newman, JE, Abel, MD, Harrison, PR, Yost, KJ. Wind as related to critical flushing speed versus reflotation speed by high-volume sampler particulate loading. In *Proceedings of the Atmosphere-Surface Exchange of Particulate and Gaseous Pollutants-1974 Symposium*, Richland, WA, September 4–6, 1974, U.S. Energy Research and Development Administration Symposium Series, CONF-740921, pp. 466–496, 1976, National Technical Information Service, Springfield, VA.

Sehmel, GA. Particle resuspension: A review. *Environ. Int.* 4, 107–127, 1980.

Spencer, H, Osis, D, Fisenne, IM, Perry, PM, and Harley, NH. Measured intake and excretion patterns of naturally occurring ^{234}U, ^{238}U, and calcium in humans. *Radiat. Res.* 124, 90–95, 1990.

Stevenson, KA, Pan, V. An assessment of uranium in surface air within the continental US. *J. Environ. Radioact.* 31, 223, 1996.

Sunta, CM, Dang, HS, Jaiswal, DD. Thorium in man and environment: Uptake and clearance. *J. Radioanal. Nucl. Chem.* 115, 149–158, 1987.

Tracy, BL, Prandl, FA. Radiological impact of coal fired power generation. *J. Environ. Radioact.* 2, 145–160, 1985.

United Nations Scientific Committee on the Effects of Atomic Radiation (UNSCEAR), Annex B, Exposures from Natural Radiation Sources, Vienna, 1988.

United Nations Scientific Committee on the Effects of Atomic Radiation (UNSCEAR), Annex B, Exposures from Natural Radiation Sources, Vienna, 2000a.

United Nations Scientific Committee on the Effects of Atomic Radiation (UNSCEAR). Sources and effects of ionizing radiation: United Nations Scientific Committee on the Effects of Atomic Radiation 2000 Report to the General Assembly, with annexes, New York, United Nations, 2000b.

USEPA. Environmental Radiation Data Report Series 402-R-93, National Air and Radiation Environmental Laboratory, Montgomery, AL, 1978–1993.

Volchok, HL, Knuth, R, Kleinman, MT. Respirable fraction of Sr-90, Pu-239 and Pb in surface air. U.S. Atomic Energy Commission Report HASL-278, I-36-39, 1974, National Technical Information Service, Springfield, VA.

Yunoki, E, Kataoka, T, Michihiro, K, Sugiyama, H, Shimizu, M, Mori, T. Background levels of ^{238}U and ^{226}Ra in atmospheric aerosols. *J. Radioanal. Nucl. Chem.* 189, 157–164, 1995.

23 Health Effects of Aerosols
Mechanisms and Epidemiology

Ira B. Tager

CONTENTS

23.1 INTRODUCTION

Health effects have been associated with a variety of aerosol components commonly encountered in the environments of people. The body of published data on the characterization (sources, composition, environmental transformations, and fate) of health-relevant aerosols is extensive as is the literature related to aerosol-specific health effects. Therefore, this chapter will discuss only selected, broad categories of health-relevant aerosols: aerosols derived from combustion processes and aerosols composed of biological materials that are encountered in ambient and/or indoor environments

(Table 23.1).* Specifically omitted are radioactive aerosols, aerosols encountered in industrial environments, and medical and pharmaceutical aerosols.

Broadly, the health-relevant aerosols to be considered can be classified as those derived from the activities of man (e.g., combustion of fuels) and those that occur in natural environments (e.g., bioaerosols) or as the result of natural processes (e.g., windblown dusts that carry man-made products such as pesticides or bioaerosols such as soil-resident fungal spores).

23.2 CHARACTERISTICS OF OUTDOOR (AMBIENT) AND INDOOR AEROSOL SOURCES OF IMPORTANCE TO HUMAN HEALTH

Clearly, the activities of people can modify the distribution and the concentrations of aerosol components that would exist in nature even in the absence of human activity. Humans come into contact with these aerosols both indoors and outdoors. Components of outdoor aerosols infiltrate into indoor environments to varying degrees, but, with the exception of products of indoor combustion (e.g., fireplaces, biomass burning stoves), indoor sources are not usually important contributors to the outdoor aerosol.

23.2.1 Outdoor Aerosol (Table 23.1, Figure 23.1)

The typical outdoor aerosol can be described conveniently in terms of three components. Combustion-generated particulate matter (PM) is the source for the vast majority of PM $\leq 1\,\mu m$ (aerodynamic diameter), and these particles are the target of federal and state clean air criteria. With regard to human health, the particle size distribution observed for PM $\leq 1\,\mu m$ has implications for the deposition distribution within the respiratory tract and toxicological properties that are related to surface area characteristics independent of the toxicology of the specific composition of the particle (e.g., see Ref. [1]). The relationship between particle size, deposition, and surface area is shown in Figure 23.2, and the implications for toxicology will be discussed in a later section. The chemical composition of the combustion-related aerosol is complex, and its exact composition depends on the mix of sources that contribute to it (Table 23.1; primary source contributions and long-range transport), meteorological conditions, and large-scale regional variations.

The so-called coarse component of the outdoor aerosol is defined in various ways. In the context of U.S. National Ambient Air Quality Standards (NAAQS), the classification of "coarse" is reserved for PM between 2.5 and 10 μm aerodynamic diameter (see Vol. I of Ref. [2]). However, given the substantial differences in sources and deposition properties and contribution to overall surface area (Figure 23.2), the classification based on a cut point of 1 μm seems more useful for health assessment. As noted in Figure 23.2, there is a small overlap of sources and particle modes, which is related to cut-point characteristics of the sampling devices and conditions during sampling particle agglomeration and to meteorological factors such as high winds (see Vol. I of Ref. [2]). Among the components of coarse PM, the bioaerosol has considerable implications for human health. Plant pollens have long been known to be important triggers of allergic reactions. However, more recently, the importance of fungal spores and toxins (e.g., see Refs [3–5]) and bacterial products, especially endotoxins (e.g., see Refs [6,7]), has been recognized. Recent studies have suggested that a major contributor of the toxicological and inflammatory potential of PM is related to bacterial endotoxins, and this activity is found almost exclusively in the $PM_{10-2.5}$ of mass distribution [6–9].

---

* Chapter 20 presents detailed information about the relevant aspects of the bioaerosol and some data on health effects. This chapter will focus on disease-specific health effects based on epidemiological and clinical data.

TABLE 23.1
Aerosols Discussed in this Chapter

	Formation (F)/Primary Sources(S)	Major Components[a] [265]
Outdoor aerosol (Figure 23.1)		
Combustion-related PM		
Ultrafine (nuclei) mode (<100 nm)[b] (includes nanoparticles (<50 nm))	F. Gas-to-particle conversion or primary during incomplete fuel combustion S. Motor vehicles, power plants, industrial sources	Organic and elemental carbon metals (especially transition metals)/elements, sulfates, nitrates, sulfuric acid
Accumulation (fine) mode (≤1 μm) (includes ultrafine PM)	F. Primary particles, gas-to-particle conversion during fuel combustion: condensation and coagulation of ultrafines; secondary transformation in atmosphere S. Motor vehicles, power plants, industrial sources	Sulfate, nitrate, ammonium Elemental carbon Organic compounds Metals (especially transition metals)
Coarse PM (>1 μm)[c]	F. Grinding, crushing, and abrasion of surfaces with suspension by wind or anthropogenic activities S. Suspended dust from industry, soil tracked onto roads and streets, mining, construction, farming Ocean spray Biological materials	Crustal metals Tire fragments Sulfates (ocean) Pollens[d] Fungal spores Pesticides Bacteria and bacterial products (e.g., endotoxins, bacterial spores)
Indoor aerosol[e] (Figure 23.2)	F. See outdoor aerosol; heterogeneous chemical reactions [14]	See outdoor aerosol
Combustion-related PM	S. Tobacco smoke; fireplaces/wood stoves, kerosene heaters, tobacco smoke, outdoor aerosol, cooking	Tobacco products (sample of tar): nicotine fluoranthenes, acrolein, benzene, pyrenes, *N*-nitrosamines, 1,3 butadiene, arsenic, chromium IV, lead, cadmium
Noncombustion PM		
Dust/lint	F. Disturbance of settled dust during cleaning and other human/pet activities S. Dirt, human	Soil and road dust tracked indoors Skin scales
Biological	F. Detritus composed of insect excrement and body parts, animal shedding, water damage; penetration of outdoor aerosol into buildings S. (1) Mites, (2) cockroaches, (3) pets, (4) water-damaged building materials/high humidity, (5) bacteria/bacterial products, and (6) outdoor aerosol	1. Fecal allergen 2. Multiple allergen-specific sources not known 3. Dander (cat, dog, etc.) 4. *Bacteria*: whole bacteria spores, peptidoglycans, endotoxins, *Molds and fungal* spores, toxins, $1 \rightarrow 3$-β-D glucans 5. Endotoxins, peptidoglycans 6. Bacteria, bacterial products, fungal spores, toxins, pollens

[a] Components of specific aerosols are spatially heterogeneous. Entries represent components found typically.

[b] Definitions taken from Ref. [266]. These terms are used because most important combustion-related components of $PM_{2.5}$ are located in the fine fraction.

[c] For regulatory purposes, coarse-mode particles usually refer to particles between 2.5 and 10 μm [2]. However, since most particles larger than 1 μm are formed by mechanical processes (see Figure 23.1), this definition seems more useful in terms of possible health effects. Some combustion-related PM may be found at the lower end of the distribution of the coarse mode (see Figure 23.1).

[d] See Chapter 20 for a more detailed presentation of the composition of bioaerosols.

[e] Refs [267–269].

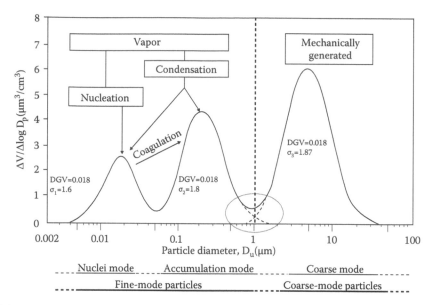

FIGURE 23.1 Distribution of particle mean diameter by particle volume for typical outdoor PM aerosol. DGV: geometric mean diameter by volume (volume mean diameter); σ_g: geometric standard deviation. The circle points to overlap area between particle diameter and source contribution. (From U.S. Environmental Protection Agency (EPA), Air quality criteria for particulate matter—Third external review draft, EPA 600/P-99/002aC, Report No.: EPA 600/P-99/002aC, Research Triangle Park, NC, 2001.)

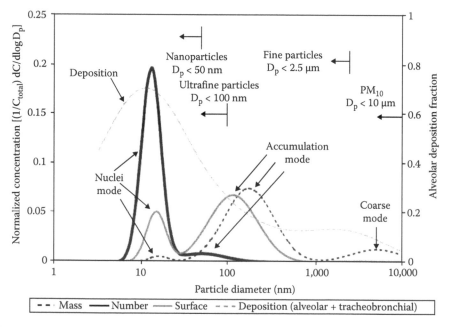

FIGURE 23.2 Normalized, mass-weighted, and number-weighted particle size distributions and alveolar deposition from typical diesel exhaust. Note that in this figure "coarse mode" is defined as $PM_{10-2.5}$. (From Kittelson, D. et al., Measurement of diesel aerosol exposure: A feasibility study, in Health Effects Institute Special Report-Research Directions to Improve Estimates of Human Exposure and Risk from Diesel Exhaust, pp. 153–179, April, 2002.)

23.2.2 Indoor Aerosol (Table 23.1 and Figure 23.3)

Depending on the characteristics of a given structure and the mode of ventilation, the outdoor aerosol can be an important contributor to the indoor aerosol. Indoor penetration of fine and coarse ambient particulate has been estimated to be close to one [10]. Source apportionment studies have indicated that the contribution of outdoor PM to the total indoor aerosol ranges from 60% or more for PM $\leq$1 µm and declines with increasing PM to about 20% for PM between 6 and 10 µm [10,11]. Estimates based on studies in the eastern United States indicate that 75% of the fine indoor aerosol during summer months is derived from the outdoors [12]. In contrast, studies carried out in the western United States indicate a much lower penetration of fine PM (~27% for sulfate and 12% for $PM_{2.5}$ [13]). In addition, chamber studies indicate that ozone, which in most residences has no indoor sources, can react with volatile organic compounds in indoor environments to produce fine particulates (<0.5 µm) that are not generated originally in the outdoor environment [14].

In homes with smokers, cigarette smoke is, by far, the most important contributor to indoor aerosols. Figure 23.4 shows the effect of the presence of a smoker in an indoor environment on the concentration of PM in the air [15]. Nicotine is found primarily in the tar (particulate) fraction of tobacco smoke [16,17] and indoor residential concentrations range from 2 to 14 µg/m^3 in homes with a single smoker [17]. Concentrations over 1000 µg/m^3 have been recorded in vehicles with ventilation systems turned off [16,17].

Cooking is an important source of indoor PM, which is often not considered in terms of potential contribution to the total indoor particle burden [18]. Frying or use of an oven transiently can be the dominant source of indoor aerosol, and a single episode of cooking on a stove can lead to

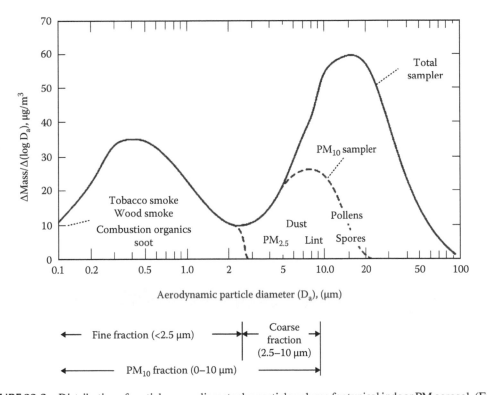

FIGURE 23.3 Distribution of particle mean diameter by particle volume for typical indoor PM aerosol. (From McDonald, B. and Ouyang, M., Air cleaning—Particles, in Spengler, J.D., Samet, J.M., and McCarthy, J.F., Eds., *Indoor Air Quality Handbook*, McGraw-Hill, New York, pp. 9.1–9.28, 2000.)

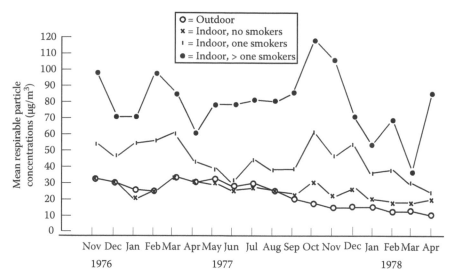

FIGURE 23.4 Effect on cigarette smoking on indoor concentrations of respirable particles (approximately PM$_{10}$). (From Spengler, J.D., *Atmos. Environ.*, 15, 23, 1981.)

increases in indoor PM that are between 1 and 2 orders of magnitude of the background PM concentrations (Figure 23.5).

The noncombustion component of indoor aerosol makes the largest contribution to total mass (Figure 23.3). Of greatest interest are the biologically active components of this aerosol. The details related to the components of this aerosol are covered in Chapter 10.

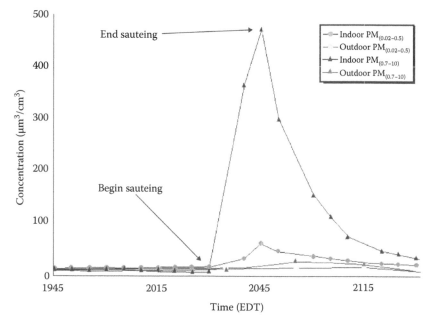

FIGURE 23.5 PM$_{10-2.5}$ and PM$_{0.7-1}$ concentrations from a sautéing event in one home. EDT: eastern daylight savings time. (From Abt, E. et al., *Environ. Health Perspect.*, 108, 35, 2002.)

23.3 DEPOSITION AND CLEARANCE OF AEROSOLS FROM THE HUMAN RESPIRATORY TRACT*

23.3.1 DEPOSITION AND RETENTION

For most people during quiet breathing, air enters the respiratory tract primarily through the nose at high velocity (Figure 23.6). Particles 0.5 μm are removed efficiently at this level. Particles <100 nm are also removed at this level, largely by impaction. Inertial impaction is proportional to velocity, the square of the particle diameter, and the sharpness of the angle of the airway [21,22]. Therefore, impaction occurs at places where there is a sudden change in the direction of the airstream, such as airway branch points, and involves a similar size range as that for gravitational impaction [23]. As air moves through the trachea and bronchi, impaction and sedimentation both play a role in the deposition of particles. Sedimentation due to gravity occurs in airways, except those that are vertical and for particles >0.5 μm aerodynamic diameter (Vol. 1 of Ref. [19]). As airway diameters narrow, particles whose distance to the airway wall surface is less than the particles' size are removed by interception [21]. As the small particles (<1 μm down to the nanometer range) move down the airways in an ever-slowing airstream, diffusion plays a greater role in deposition. Diffusion is an important mechanism for particles <0.5 μm [22]. Small particles that carry a surface charge (e.g., freshly generated combustion particles) may also be deposited as a result of electrostatic forces. Electrostatic forces are not important for particles >4 μm [22].

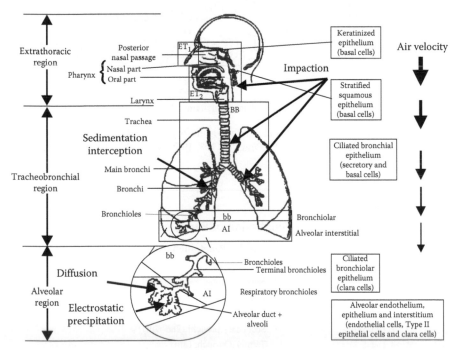

FIGURE 23.6 Schematic of human respiratory tract. (Adapted from U.S. Environmental Protection Agency (EPA), Air quality criteria for particulate matter—Third external review draft, EPA 600/P-99/002aC, Report No.: EPA 600/P-99/002aC, Research Triangle Park, NC, 2001; Casarett, L.J., Toxicolology of the respiratory system, in Casarett, L.J. and Doull, J., Eds., *Toxicology: The Basic Science of Poisons.*, MacMillan Publishing Co., New York, pp. 201–224, 1975; Lippmann, M. and Schlesinger, R.B., *J. Toxicol. Environ. Health*, 13, 441, 1984.)

---

* An exhaustive discussion of issues related to deposition of particles in the human respiratory tract can be found in Ref. [19].

The nasal passages are effective in removing particles larger than 5 μm and remove virtually all particles larger than 10 μm [22]. In addition, condensation nuclei are effectively removed in the nasal passages by diffusion. Figure 23.7 and Table 23.2 summarize the estimated deposition fractions for particles of various mass median aerodynamic diameters (MMAD). Particles <10 μm MMAD are deposited at all levels of the respiratory tract, with particles <1 μm being deposited primarily in the tracheobronchial tree and the alveoli ("Pulmonary" in Figure 23.7). A number of factors affect deposition (Table 23.3). Patterns of ventilation and underlying lung disease are of particular importance for health considerations. Exercise is associated with a switch from primarily nasal breath to oral breathing with a resultant increase in the alveolar deposition of particles >1 μm (see Section 10.5.1.4 in Ref. [19]).

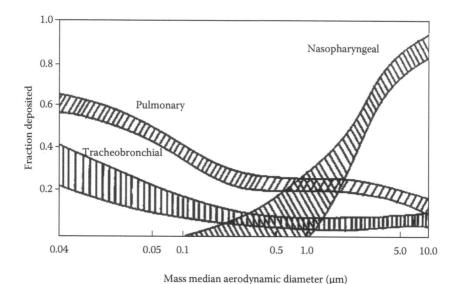

FIGURE 23.7 Fractional deposition of particles in the respiratory tract of humans. The x-axis is plotted on a logarithmic scale. (From Morrow, P.E., *Health Phys.*, 12, 173, 1966; as cited by Brain, J.D. and Valberg, P.A., *Am. Rev. Respir. Dis.*, 120, 1325, 1979.)

TABLE 23.2
Respiratory Tract Penetration of Particles of Various Sizes

Particle Size Range (μm)	Level of Penetration (Generation Number)[a]
≥11	Do not penetrate
7–11	Nasal passages
4.7–7	Pharynx
3.3–4.7	Trachea and primary bronchi (1st)
2.1–3.3	Secondary bronchi (2nd–7th)
1.1–2.1	Terminal bronchi (8th)
0.65–1.1	Bronchioles (9th–23rd)
<0.65	Alveolar ducts (24th–27th) and alveoli

Source: Adapted from Wilson, S.R. and Spengler, J.D., Emissions, dispersion, and concentration of particles, in Wilson, S.R. and Spengler, J.D., Eds., *Particles in Our Air: Concentrations and Health Effects*, Harvard University Press, Cambridge, MA, pp. 41–62, 1996.

[a] Generation numbers are from [271].

TABLE 23.3
Factors That Affect Particle Deposition Exclusive of Particle Characteristics

Respiratory tract geometry
Airway caliber
Airway branching pattern
Airway path length to terminal airways and alveoli

Ventilation
Pattern of breathing: oral, nasal, oronasal
Ventilation rate [272]
Duration of pauses between breaths
Tidal volume (volume of each breathing during quiet breathing)
Ventilation distribution

Other factors
Respiratory tract disease [273]
   Altered airway geometry, branching patterns, and path lengths
   Altered ventilation patterns
   Altered distribution of ventilation
   Changing patterns of pattern of breathing with age
   Infants are preferential nasal breathers [274]
Sex
   Females with greater deposition of nanoparticles [272]

Source: Adapted from Schlesinger, R.B., Deposition and clearance of inhaled particles, in McClellan, R.O. and Henderson, R.F., Eds., *Concepts in Inhalation Toxicology*, Taylor & Francis, Washington, DC, pp. 191–224, 1995.

Retention of PM in human lungs has been studied by a number of investigators. The number of particles retained is related to the ambient concentrations (Figure 23.8; [24]). Particle retention is also a function of the level of the airways in which the particle is deposited, the type of particle, and the functional integrity clearance mechanisms (for an extensive discussion of specific mechanisms, see Ref. [25]). In 42 left lungs obtained from the Coroner's Office in Fresno, CA (19 from cigarette smokers; all Hispanic males), carbonaceous and birefringent silica particles were rarely found in the walls of larger airways [26]. In contrast, at the level of respiratory bronchioles (beyond the 12th generation of airways), such particles were frequently found in the airway walls [26]. Similar observations were made by Churg and Vidal [23] in the lungs of nonsmokers from Vancouver, Canada (Figure 23.9). These investigators also noted that ultrafine PM constituted <15% of the retained PM (expressed as millions of particles/gram of tissue), virtually all of which were crustal minerals. This was in contrast to their findings in lungs from Mexico City, where lungs contained an average of 25% of chained aggregates of carbonaceous particles [24]. As noted previously, airway branch points are also sites of increased deposition [23]. For example, in the fourth generation of airways, Churg and Vidal [23] observed that the geometric mean particle number concentrations (mean of ten lungs) were two logs greater (per 10^{-6}/g dry weight) at bifurcation points (4.0 ± 1.2) compared to tubular segments (6.1 ± 1.4) both in the upper and lower lobes.

Deposition of particles is influenced by sex (greater deposition in females in the extrathoracic and tracheobronchial tree), age (increased deposition in the tracheobronchial region in children and young adults based on modeling), and underlying respiratory tract disease [27,28]. Figure 23.10 presents simulated deposition data presented by the U.S. EPA [28] and illustrates the variability in mass deposition (μg/day) as a function of age and sex. The largest predicted mass depositions for the

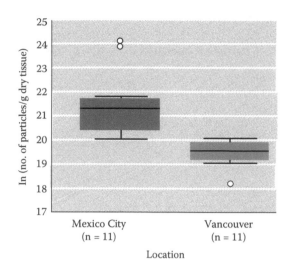

FIGURE 23.8 Number of particles per gram of lung tissue from a high PM area (Mexico City: mean $PM_{2.5} = 29.5\,\mu m$; Vancouver, Canada: mean $PM_{2.5} = 10.5\,\mu m$). The y-axis is plotted on a natural logarithm scale. Lungs from females were obtained from a general autopsy population from a referral hospital in Mexico City and a general hospital in Vancouver. (From Brauer, M. et al., *Environ. Health Perspect.*, 109, 1039, 2001.)

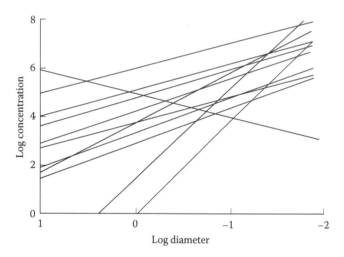

FIGURE 23.9 Relationship between airway concentration of particles and airway diameter both plotted on logarithmic scales. Lungs are from 11 nonsmoking residents of Vancouver, Canada, whose lungs were obtained from a general hospital. Vancouver is an area of low PM concentration (see text). The figure shows wide interindividual variation but fairly similar retention patterns across individuals. In these specimens, ultrafine particles constituted <15% of the retained particles at any airway site. (From Churg, A. and Vidal, S., *Occup. Environ. Med.*, 53, 553, 1996.)

tracheobronchial and alveolar regions are seen for ultrafine particles and particularly for children aged 14–18 years, a group that is likely to be physically active in outdoor activities. Figure 23.11 shows simulated data for the fraction of the total number of particles inhaled that is deposited in various regions of the respiratory tract as a function of particle diameter. The predicted deposition fraction mode falls between particles in the range of 0.1–1.0 μm for all lung regions and drops as particles increase and decrease in size.

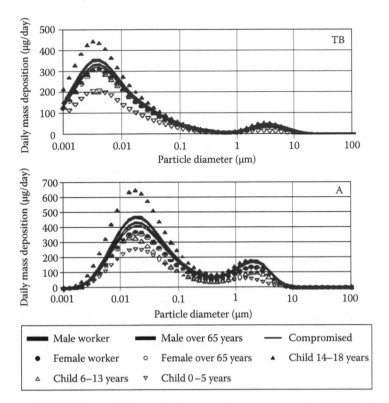

FIGURE 23.10 Daily mass particle deposition rate for 24 h exposure at 50 μg/m³ in tracheobronchial and alveolar regions as predicted by the Commission on Radiological Protection Publication 66 model [322]. Simulations used daily minute volume patterns for different demographic groups. (Adapted from U.S. Environmental Protection Agency (EPA), Air quality criteria for particulate matter Volume II, in Air Quality Criteria for Particulate Matter, Volumes I–III, EPA/600/P-95-001bF, National Center for Environmental Assessment Office of Research and Development, U.S. EPA, Washington, DC, 1996, Figure 10.43.)

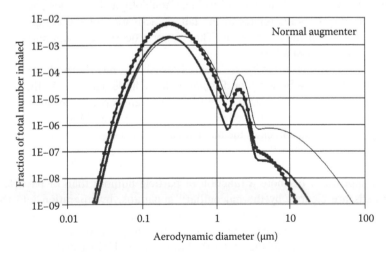

FIGURE 23.11 Fractional number deposition for a male with normal nose/mouth breathing with a general population activity pattern as predicted by the Commission on Radiological Protection Publication 66 model [322] for exposure to Philadelphia aerosol. (Adapted from U.S. Environmental Protection Agency (EPA), Air quality criteria for particulate matter Volume II, in Air Quality Criteria for Particulate Matter, Volumes I–III, EPA/600/P-95-001bF, National Center for Environmental Assessment Office of Research and Development, U.S. EPA, Washington, DC, 1996, Figure 10.50.)

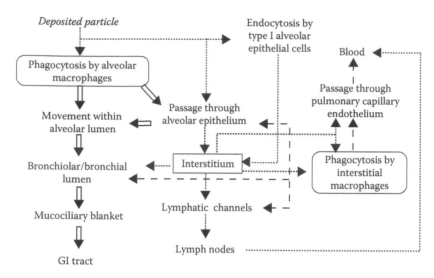

FIGURE 23.12 Known and suspected pathways for clearance of insoluble particles. This diagram does not include all of the pathways for the clearance of bioaerosol particles such as fungal spores and viable bacteria which may involve primary phagocytosis by polymorphonuclear leukocytes as well as macrophages and other immunologically mediated effector mechanisms. Mucociliary blanket (transport) refers to the combined effects of ciliary action of epithelial cells to the level of terminal bronchioles and mucus secretion by mucus glands (only found in airways with cartilage [seven generations] and goblet cells [found to the level of respiratory bronchioles]). (From Schlesinger, R.B., Deposition and clearance of inhaled particles, in McClellan, R.O. and Henderson, R.F., Eds., *Concepts in Inhalation Toxicology*, Taylor & Francis, Washington, DC, pp. 191–224, 1995; Wanner, A. et al., *Am. J. Respir. Crit. Care Med.*, 154(6 Part 1), 1868, 1996.)

23.3.2 CLEARANCE*

Clearances of nonbiological and biological aerosol particles have many common elements, and these are discussed in this section. These general pathways are depicted in Figure 23.12. The one major difference between the clearance of insoluble nonbiological particles and particles that are intact bacterial or fungal spores relates to the latter's activation of the complex immunological pathways responsible for killing these agents. These latter processes are summarized very briefly in Table 23.4 and are not discussed further.

Particles that enter the nasal cavity are cleared largely by muciliary clearance (Figure 23.11 and Ref. [29]), sneezing, nose blowing, or dissolution for soluble particles. In the tracheobronchial tree, poorly soluble particles are cleared by mucociliary transport (the net movement of which is toward the oropharynx) and are swallowed or removed by coughing. More soluble particles may be absorbed through the mucosa and enter the bloodstream.

Insoluble particles that reach the alveolar region of the lung are ingested by alveolar macrophages. Increasing particle burden results in increased numbers of cells, the maximum accumulation of which appears to be more a function of particle number than of particle mass. These particle-filled cells are cleared by the mucociliary apparatus, by migration into the connective tissue that separates the alveoli (air sacs) and surrounding airways, and into the bloodstream via lymphatic channels. Some of these latter particles may then migrate to lymph nodes that are located along the tracheobronchial tree and then enter the bloodstream. Uningested, ultrafine particles may enter the bloodstream directly across the alveolar and capillary epithelium, and this clearance into the blood can be very rapid. Within 1 min after human inhalation of [99m]Technetium-labeled carbon particles (<100 nm), labeled particles can be detected in the blood, with peak concentrations

---

* This section is a synthesis of extensive discussions presented in Refs. [25,28].

TABLE 23.4
Summary of Immunological Mechanisms for the Clearance of Intact Biological Particles[a]

Innate immune (nonspecific) response[b]
Cytokines/chemokines (pro-inflammatory, anti-inflammatory, activating)
Specific ligands (e.g., CD14 receptor for endotoxin)
Natural antibodies
  Microbe-specific antibodies found in healthy people in the absence of overt infection
Opsonin-independent phagocytosis
  Engulfment of biological particles in the absence of specific antibodies by macrophages and polymorphonuclear
  leukocytes
Acute-phase proteins

Specific immune responses[c]
Pathogen-specific antibodies
Cell-mediated immunity

Complement system
System of more than 30 proteins that acts with or without antibody to initiate inflammatory reactions and to kill viable
  biological particles

[a] This table is created from material in Chapters 4–9 of Ref. [275].
[b] The innate immune system is the part of the immune response that does not require specific response to pathogens and
  is the first response to pathogens not previously encountered by a host. The innate immune response is rapid compared
  to the delayed response of specific immunity.
[c] Specific immune responses are classified broadly as those related to antibodies and those related to direct cellular
  effects. This system is activated by a complex process of recognition, processing, and presentation of foreign antigens,
  which leads to activation of specific limbs of the specific immune response system.

achieved between 10 and 20 min [30]. Once in the bloodstream, free particles and particles contained in macrophages can be deposited in any extrapulmonary organ. Particles that are ingested by macrophages may dissolve in certain environments within the cells (e.g., the acidic environment of phagolysosomes).

Soluble particles are removed by absorption. Absorption is described as a two-stage process: dissolution (dissociation of particles into material that can be absorbed) and uptake of the material. Each of these steps is time dependent, for which surface properties, chemical structure, and surface-to-volume ratio are important determinants.

Whether a particle is cleared or retained for some period of time is dependent on the physicochemical properties of the particle, site of deposition, presence of underlying diseases, and occurrence of tobacco smoking [22]. Explanations for the increased deposition and possible retention in persons with underlying lung disease are not well understood but, in part, may be related both to structural alterations and functional changes in the lung (e.g., altered epithelial permeability) [31,32]. In healthy individuals, poorly soluble particles are estimated to be cleared over a 2.5–20 h period. Persons with chronic lung disease and certain infections (e.g., influenza) may have impaired clearance of particles. The effect of physical activity on deposition and clearance is unresolved. However, a recent study that used a mouthpiece exposure system showed that deposition of ultrafine carbon particles (count median diameter = 26 nm, GSD = 1.6) was increased in 12 healthy subjects who exercised moderately, with a 4.5-fold increase in the total number of deposited particles [33].

Figure 23.13 illustrates the effects of duration of exposure on retention of particles for three particle modes and total lung burden as a function of age for a specific aerosol composition. Of interest for health considerations is the different time course and magnitude of lung burden for the three

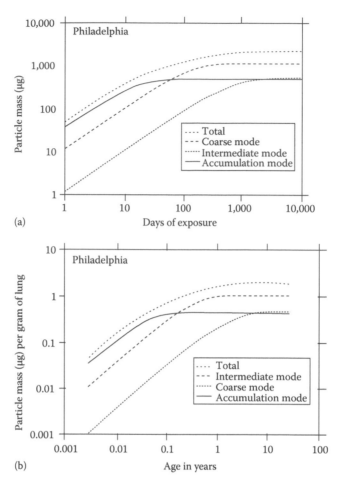

FIGURE 23.13 (a) Particle mass retained in the lung versus days of exposure, predicted by [322] for exposure to Philadelphia aerosol, under the assumption of dissolution absorption half-times of 10, 100, 1000 days for three modes of particles. (b) Same data converted to specific lung burden, based on continuous exposure to Philadelphia aerosol at $50\,\mu g/m^3$. (Adapted from U.S. Environmental Protection Agency (EPA), Air quality criteria for particulate matter Volume II, in Air Quality Criteria for Particulate Matter, Volumes I–III, EPA/600/P-95-001bF, National Center for Environmental Assessment Office of Research and Development, U.S. EPA, Washington, DC, 1996, Figures 10.55, 10.56.)

different modes. Given the importance of exposures at critical points of human development, this type of heterogeneity of accumulation may have important implications for health effects related to chronic exposures to indoor and outdoor aerosols.

23.4   MECHANISMS OF TOXICITY OF AEROSOL COMPONENTS

The mechanisms that underlie the effects of the nonbiological ambient and indoor aerosols and the biological components of the aerosols have a number of features in common. Both components directly trigger inflammatory responses and/or trigger (directly or indirectly) immunological responses that lead to specific inflammatory responses that characterize various allergic phenotypes (e.g., asthma allergic rhinitis/conjunctivitis). Reactive oxygen and nitrogen species are major products of inflammatory reactions [34,35] and can be generated by neutrophils, macrophages, and eosinophils in response to contact (ingestion) with biological materials [36] or ambient particles [37]. Moreover, the organic carbon and transition metal components of ambient and indoor aerosol components can lead directly to the production of reactive oxygen species (metabolites—ROM) [38,39]. ROM (superoxide radical [$O_{22}\cdot$], hydrogen peroxide [H_2O_2], hydroxyl radical [$OH\cdot$]; sometimes included are singlet O_2 and hypochlorous acid), in turn, are potent inducers of inflammation [40] and initiators of lipid peroxidation and DNA strand breaks [41], which may be important in carcinogenesis (summarized in Figure 12 of Ref. [42]). Since a large body of research suggests that oxidant stress may be one of the principal (if not the principal) final common pathways for the effects of ambient aerosols on human health, this section begins with a consideration of the evidence for this view. The section then considers pathways less proximate: inflammation, immunological modulation, and particle overloading effects.

23.4.1 Particle Aerosol Induction of Oxidative Damage

As epidemiological evidence of health effects related to particulate air pollution mounted, animal studies were undertaken to try to identify the components of PM and mechanisms that might explain the epidemiological observations. Some of the most important of these were animal exposure studies undertaken with residual oil fly ash (ROFA), which focused on the role of transition metals. ROFA is generated from the burning of fossil fuels and has a high concentration of metals, particularly transition metals ([43]; reviewed in Ref. [44]). These metals, which can exist in more than one valence state, can participate in electron cycling that can lead to the production of highly reactive OH through the Haber–Weiss reaction (Table 23.5) [34]. This generation can take place in acellular and cellular systems [44,45]. First-row transition metals (titanium, vanadium, chromium, manganese, iron, cobalt, nickel, and copper) are found in highest concentrations in crustal material and in the atmosphere as a consequence of human-generated pollutants [45]. A series of studies by U.S. EPA investigators and collaborators (selected references include [45–49]) established the role of the oxidative potential of the transition metal content of ROFA to generate ROM and the specificity of these metals as causes of lung injury in rodents. The metal contents of dusts from a variety of sources other than ROFA were also shown to produce lung injury that was specific to the generation of ROM [46,47,50]. A recent, comprehensive evaluation of the role of metals in the toxicology of PM generated from coals from three different locations in the United States and from PM derived from several formulations of gasoline and diesel fuel (Utah, Illinois, North Dakota) provides strong support for the toxicological importance of the oxidant properties of transition metals [51]. These investigators documented that bioavailable iron (nmol/mg of coal fly ash as measured after ferric ammonium citrate extraction) was greatest for PM $\leq 1\,\mu m$, but bioavailable iron could also be found in $PM_{10-2.5}$. Chelation of iron significantly reduced evidence of oxidative damage as measured by malondialdehyde [51].

The ability of PM_{10} to generate ROM unrelated to transition metals has been well documented [52,53]. Investigators at the University of California, Los Angeles, CA, have conducted an extensive series of studies that establish that organic components of PM lead to the production of ROM (reviewed in Ref. [54]). In particular, polycyclic aromatic hydrocarbons (PAHs) and PAH-derived quinones found in diesel exhaust particles (DEP) have been shown to produce *in vitro* evidence of oxidant stress [55].

TABLE 23.5

Haber–Weiss Reaction: Ferrous Iron–Catalyzed Generation of Hydroxyl Radicals

$O_2^{\cdot-} + Fe^{3+} \rightarrow O_2 + Fe^{2+}$

$Fe^{2+} + H_2O_2 \rightarrow Fe^{3+} + HO^{\cdot} + OH^-$

$O_2^{\cdot-} + H_2O_2 \rightarrow O_2 + HO^{\cdot} + OH^-$

All first-row transition metals can participate in this reaction.

These investigators have developed a "stratified oxidative stress model" to characterize both the production of ROM by DEP and the mechanisms through which these ROM exert their effects on inflammation and allergic response ([56]—the reader is referred to the extensive bibliography in this review—selected references relevant to this section include [38,52,55,57–61]). The model is summarized in Figure 23.14. These investigators demonstrated that low concentrations of redox-active PAHs

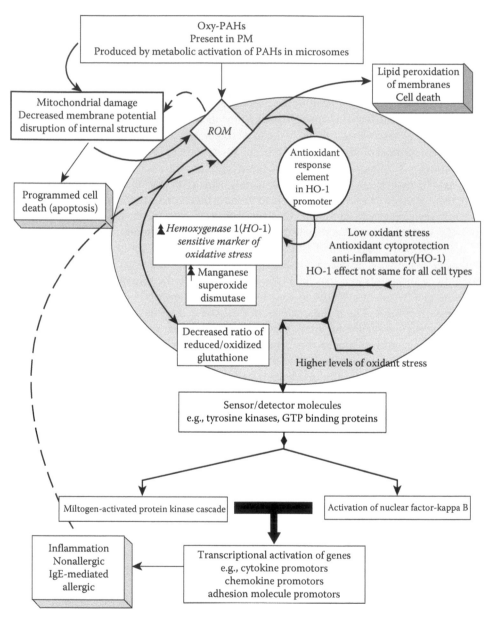

FIGURE 23.14 Basic elements of "stratified oxidative stress model" of Li et al. [56] (area enclosed in the gray circle) and consequences of ROM that are not controlled by antioxidant defenses. Solid arrows represent the pathways; dashed arrows identify feedback loops that perpetuate the adverse outcomes. ROM: reactive oxygen metabolites. (Adapted from Li, N. et al., *Environ. Health Perspect.*, 111, 455, 2003; Li, N. et al., *J. Immunol.*, 169, 4531, 2002; Li, N. et al., *J. Immunol.*, 165, 3393, 2000; Li, N. et al., *Inhal. Toxicol.*, 14, 459, 2002; Nel, A.E. et al., *J. Allergy Clin. Immunol.*, 102, 539, 1998; Hiura, T.S. et al., *J. Immunol.*, 163, 5582, 1999; Hiura, T.S. et al., *J. Immunol.*, 165, 2703, 2000; Kumagai, Y. et al., *Free Radic. Biol. Med.*, 22, 479, 1997.)

and their oxy-derivatives initially activate genes that lead to the induction of hemoxygenase-1 (HO-1) and superoxide dismutases. "Low-level" oxidative stress is defined as the normal ratio of reduced glutathione to oxidized glutathione (GSH:GSSG). The oxidation of GSH is mediated by several enzymes that convert reactive oxygen species precursors to less reactive species and is one of the major cellular pathways for the control of reactive oxygen species [62]. HO-1 is involved in the metabolism of hemoglobin and leads to the production of antioxidants (bilirubin) and anti-inflammatory (carbon monoxide) end products [62]. Superoxide dismutases catalyze the conversion of $O_2 \cdot^2$ to H_2O_2, which is converted to water through a catalase-mediated reaction. A decreasing GSH:GSSG ratio indicates a failure of these and other host defenses against oxidant stress (e.g., antioxidants in lung lining fluid and blood [vitamin C, urea] or in cell membranes [vitamin E]) to "neutralize" the oxidative stress [62,63]. Persistence of ROM leads to the activation of cell signaling cascades that ultimately lead to the activation of nonallergic and allergic inflammation, peroxidation of cell membranes, programmed cell death, and cell necrosis [57,62,64]. The inflammatory process itself generates additional ROM that are part of normal cellular defense mechanisms ([36,65] and Section V in Ref. [66]) and downregulates enzymes involved in the metabolism of ROM [67], thereby creating an ongoing cycle of cellular injury (see the discussion in Ref. [68]). Ultrafine particles derived from urban formation and receptor sites have been found to be important sources of the oxidative stress induced by PM *in vitro* [38] and in animal toxicology studies [69,70]. Components of fine and coarse PM also lead to oxidant stress either directly [38] or indirectly through bioaerosol components such as endotoxins that are part of the coarse fraction of PM and are potent inducers of inflammation [71] (see the following). Several studies of ambient PM have indicated that endotoxin may be as important a source for inflammation as are metals and organic components [6,9,72]. Through interactions with specific lymphocyte receptors, endotoxin activates many of the same signaling pathways that are activated by organic components of PM created by the activities of man [73]. This activation is the source of the potent inflammatory potential of endotoxin.

Oxidative stress provides a link between the immunomodulatory effects of DEP with regard to the immunological mechanisms that underlie asthma and allergic conjunctivitis/rhinitis (so-called hay fever) [57,74–79]. The ability of the metal content of ROFA to enhance sensitization to house dust mite in rats indicates the likelihood that a wide variety of redox-active PM sources also contribute to effects of allergen response enhancement through the generation of ROM and the pathways summarized in Figure 23.14 [80,81]. Oxidative stress plays an important role in the carcinogenic potential attributed to diesel exhaust through its capacity to induce mutagenesis, the production of DNA strand breaks and DNA adducts, and the induction or inhibition of lung metabolic enzymes [67,82–84].

23.4.2 Particle Aerosol Induction of Inflammation

The production of ROM may be the final common pathway through which anthropogenic and biological components of particle aerosols lead to the production of inflammation. Nonetheless, the ability of PM aerosol to produce inflammation is of sufficient importance to justify some expansion of this component of PM aerosol effects.

Exposure of pulmonary macrophages [6,9,72,85,86], lung epithelial cells [50,87–89], and peripheral blood monocytes [8] leads to the activation of various molecular signaling pathways for the induction of inflammation (see Figure 23.14). Transition metals, endotoxins, and organic carbon fractions of PM have all been found to trigger inflammation *in vitro* (see references in previous sections and [88]). The relevance of these *in vitro* findings to human health is supported by findings of increased levels of the pro-inflammatory cytokine interleukin 8 (IL-8) in the airways [90] and increased neutrophils and myeloperoxidase in the sputum [91] of healthy volunteers exposed to diesel exhaust in an exposure chamber under controlled conditions. Inhaled PM of anthropogenic origin has systemic effects evidenced by the stimulation of bone marrow (increased production and release of polymorphonuclear leukocytes by supernatants from human alveolar macrophages stimulated with PM_{10} [85,92,93]). Intratracheal exposure of rats to ROFA has been shown to increase the blood levels of fibrinogen, a procoagulant protein that is associated with acute inflammation and blood clotting [94]. Humans exposed to the forest fires in Southeast Asia in 1997 [85,95] showed evidence of increased pro-inflammatory cytokines and circulating neutrophils that were related to the level of exposure. A recent study of controlled exposure of healthy and asthmatic volunteers to concentrated ambient PM (CAP) showed increases in markers of systemic inflammation [96]. Other data suggest that PM stimulation of sensory nerves in the epithelium of lung airways (nonadrenergic/noncholinergic fibers and C-fibers) leads to airway inflammation through peptidenergic transmitters (substance P, neurokinin A, and calcitonin-related gene product) and that neural sensitivity is greater than that of airway epithelium to the effects of PM on inflammation [97] (see Table 23.6).

A considerable body of data has accumulated on the mechanism of the effects of diesel PM on allergic inflammation. These have been summarized in detail by Nel et al. [57]. Table 23.7 summarizes specific findings that are related to the enhancement of IgE-mediated immune responses in humans by diesel PM. The driving of the immune system toward IgE-mediated responses could be reproduced with PAHs [98]. Of particular interest for the pathogenesis of new onset allergic respiratory disease is the observation that diesel PM can induce sensitization to a neo-antigen in persons already sensitized to an allergen that triggers IgE-induced allergic response (Figure 23.15) [75]. In this study, the levels of the cytokine IL-4, which is a potent inducer of IgE synthesis, were increased in the presence of antigen and diesel PM above those seen with antigen alone. In contrast, the levels of INF-γ (interferon-γ), a Th-1 cytokine that reflects the part of the T-helper lymphocyte system that downregulates IL-4 production and stimulates cell-mediated immunity, were not increased above those observed with antigen alone. While most of the studies related to PM immune enhancement have focused on diesel PM, other sources of PM, such as ROFA, are also capable of the same type of immune enhancement [80,81,99]. These findings are not surprising, if the underlying mechanism for this enhancement ultimately relates to the production of ROM as discussed earlier.

TABLE 23.6
Summary of Effects of PM-Associated Transition Metals Derived from Combustion of Coals from Three Different Sources in the United States and from Gasoline and Diesel Exhaust

Oxidant generating capacity
$Cu(II) > Fe(II) > Va(III) > Ni(II) > Co(II) \cong Zn(II)$
Each with different time kinetics and dose–response curves

Distribution of bioavailable Fe
Greatest in $PM_{2.5}$
$PM_{10-2.5}$ about 50% of $PM_{2.5}$

Oxidant potential
Coals: $PM_1 > PM_{2.5} > PM_{10-2.5}$
  $PM_{10-2.5}$ approximately 20%–100% as potent, dependent on source
Gasoline and diesel exhaust PM
  Most oxidant production associated with transition metals (inhibited by desoxferamine (45%–97%)

Biological activity
Capacity of PM to produce IL-8 directly related to Fe content
Endotoxin content nondetectable in PM from three coal sources

Source: Aust, A.E. et al., *Res. Rep. Health Eff. Inst.*, 110, 1, 2002.

TABLE 23.7

Summary of Adjuvant Effects of Diesel Exhaust Particles on *In Vivo* Allergic Responses in Humans[a]

Nasal allergic responses

Increased production of IgE and IgE-secreting cells [76]

Qualitative difference in IgE isoforms [76,276]

Increased production of allergen-induced antigen-specific IgE [77]

Induction of broad cytokine profile in the absence of antigen [277]

Induction of Th-2-like cytokine profile with coadministration of intranasal allergen and diesel PM [77]

Allergen-specific isotype switching of B cells to IgE synthesis

Increased production of chemokines [278]

Source: Modified from Nel, A.E. et al., *J. Allergy Clin. Immunol.*, 102, 539, 1998, Table III.

[a] References cited are those cited by Nel et al.

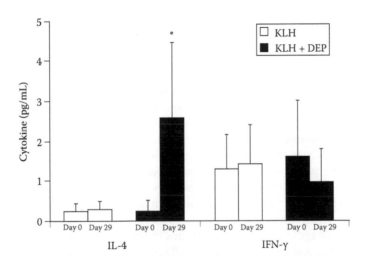

FIGURE 23.15 Concentration of IL-4 following nasal challenge of atopic subjects with keyhole limpet hemocyanin (KHO-antigen to which humans are not exposed) alone and in combination with diesel exhaust particles (DEP). IL-4 is a Th-2 lymphocyte cytokine and a potent inducer of IgE antibody whose levels are increased in persons with "hay fever" and asthma. IFN-γ is a Th-lymphocyte cytokine that is not thought to participate in IgE-mediated immune response. The figure shows a lack of IL-4 response to KLH in the absence of DEP and a lack of any IFN-γ response above that seen with KLH alone. Day 29 is one day after the last of three challenge days. (From Diaz-Sanchez, D. et al., *J. Allergy Clin. Immunol.*, 104, 1183, 1999.)

In contrast to PM enhancement of IgE-mediated immune pathways, a number of studies have demonstrated that PM can inhibit cell-mediated immune responses that represent important host defenses against a variety of microbial pathogens. Studies with *Listeria monocytogenes*, a bacterium whose clearance is dependent on an intact cell-mediated immune system, have demonstrated that inhalation and intratracheal instillation of diesel particles lead to decreased alveolar macrophage clearance of the bacteria and decreased production of cytokines (IL-1β, IL-12, tumor necrosis factor-α [TNF-α]), which are essential for the initiation and maintenance of cellular immune responses and the production of nitric oxide (NO) that is part of the oxidant antibacterial defense response [100–102]. Responses to endotoxin, a potent stimulator of IL-12 and TNF-α, were also inhibited [101,103]. Carbon black particles did not demonstrate this inhibitory effect, which suggests that the organic carbon and other components carried on the carbon core (e.g., metals) are responsible for these effects. The suppression of IL-12 may play a role in the adjuvant effect of diesel PM on IgE-mediated immune responses, since IL-12 is a potent inhibitor of Th-2 cytokine production [104]. *In vitro* studies have demonstrated that human alveolar macrophages exposed to urban air PM have impaired production of cytokines important for the control of respiratory syncytial virus and reduced ability to phagocytose the virus [105]. This virus is an important cause of lower respiratory illness in infants and upper respiratory infection in people of all ages. These same investigators also observed inhibition of phagocytosis and the oxidative burst to yeast particles in human alveolar macrophages exposed *in vitro* to ambient PM [6]. This effect was limited to the insoluble fraction (high concentration of organics) of PM_{10} and was not inhibited by chelation of iron or inhibition of endotoxin.

Products derived from bacteria, fungi, and plants may be carried by anthropogenic PM or, as is the case for plant allergens, may exist as intact particles in the air. In this section, discussion is limited to endotoxins and plant allergens. The mechanisms that related allergic responses to indoor allergens (see Chapter 14) such as dust mite and cat allergy are similar to those described for plant allergens.

Endotoxins are lipopolysaccharides (LPS) that make up the outer cell wall of Gram-negative bacteria, which are ubiquitous in the environment (see the review in Ref. [106]). Endotoxins are bound by an LPS-binding protein, which then binds to specific cell receptors (CD14 [107] and toll-like receptor [108]) and ultimately activate many of the same pathways noted for ROM (Figure 23.14) that lead to inflammation [73]. *In vitro* exposure of rat alveolar macrophages to urban air PM demonstrated that stimulation of inflammatory cytokines could be blocked by specific inhibitors of endotoxin but not by chelation of iron [72]. Further work by the same investigators showed that endotoxin, in the $PM_{2.5-10}$ fraction of ambient PM, was largely responsible for the production of pro-inflammatory cytokines (IL-6, IL-8) after *in vitro* exposure of human peripheral blood monocytes [8]. In these later experiments, chelation of iron inhibited cytotoxicity but not cytokine production. Similar results were observed with human alveolar macrophages; however, endotoxin did not appear to be responsible for the increased programmed cell death (apoptosis, which does not lead to an inflammatory stimulus) observed in the insoluble fraction of PM_{10} in which the endotoxin-mediated effects were observed [6]. Finally, exposure of rat alveolar macrophages to PM_{10} from four sites in Switzerland demonstrated that cytotoxicity, generation of ROM, and production of pro-inflammatory cytokines could be inhibited when extracts of PM were treated with an endotoxin-neutralizing protein [9]. In aggregate, these studies point to endotoxins as important contributors to the inflammatory potential of PM aerosol and the involvement of ROM in this effect.

Aeroallergens derived from plants, animals, and insects produce their effects through the stimulation of the IgE-mediated pathways (Figure 23.16). As noted earlier, organic components of PM are thought to enhance response mechanisms along this pathway and inhibit negative feedback controls by specific cytokines. Due to the current concerns related to the contribution of PM aerosol to the rise in asthma prevalence over the last several decades [109], a brief discussion of these pathways is useful. Although it undoubtedly represents an oversimplification [110,111], the immunobiology of asthma is thought to be due to a skewing of the immune response to certain agents toward what is described as the Th-2 phenotype of helper (CD41) T-lymphocytes [104]. Th-2 helper lymphocytes are defined by the secretion of IL-4, IL-5, IL-13, and granulocyte macrophage colony stimulating factor (GM-CSF), in particular and in contrast to Th-1 helper cells, which secrete IL-2, IL-2, and INF-γ. Each phenotype results in negative feedback of the alternate phenotype [112,113]. Similar phenotypes have been described for CD81 (suppressor) T-lymphocytes. The skewing of the immune response begins early in life [112] in response to as yet incompletely defined genetic predisposition and encounters with appropriate

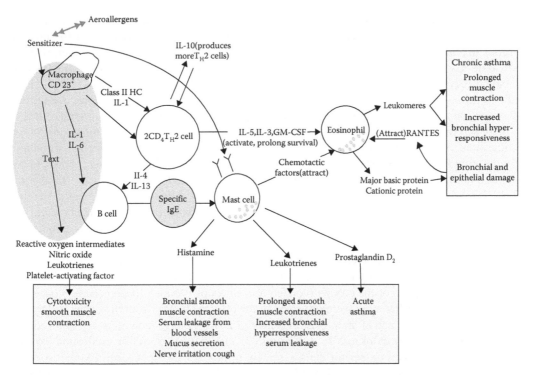

FIGURE 23.16 Major immunologic pathways related to IgE-mediated immune responses. Same pathways apply to allergens deposited in the nose. Pink-shaded areas define (approximately) pathways influenced by diesel PM. See reference for detailed explanation and references [323,324] for an additional summary related to details of cytokine network and allergic effects. (Adapted from Pandya, R.J. et al., *Environ. Health Perspect.*, 110(Suppl. 1), 103, 2002.)

environmental antigens [112]. The details of the molecular pathways that regulate Th-1/Th-2 balance have been summarized in detail by Nel [114,115].

23.5 HUMAN HEALTH EFFECTS ASSOCIATED WITH AMBIENT PM

This section summarizes an enormous body of data on the health effects that have been associated with ambient aerosols. Of necessity, the section is selective and limits itself to health effects related to PM that results from the activities of man. The emphasis is on data derived from epidemiological studies, but clinical or controlled exposures studies are cited when they serve to buttress or refute data from larger epidemiological studies. A few issues that transcend a specific consideration of specific health effects are presented prior to a more detailed discussion of health effects.

23.5.1 GENERAL INTRODUCTION

Relationship between studies of mechanisms and studies of human health effects: Although many of the studies on the molecular and cellular mechanisms relate specific components of the ambient aerosol to specific mechanistic pathways, such clarity is not possible in observational studies of humans that are the principal sources of data on human health effects. In these studies, subjects are exposed to a "chemical soup" that arises from multiple indoor and outdoor sources of anthropogenic PM aerosol that have changing source contributions over short and long timescales and is made more complex by the aging of the aerosol through atmospheric chemical processes as it moves away from its sources. In addition, the PM aerosol is experienced in the presence of other oxidant pollutants (oxides of nitrogen, ozone) and stress (e.g., infections, diet) that may contribute to and/or "account for" effects that are attributable to the PM aerosol. Animal studies provide direct evidence for the occurrence of such complex interactions. Intranasal instillation of ambient urban PM exposed to 100 ppb of ozone (O_3) prior to instillation in rats produced greater inflammation and biochemical evidence of cellular injury than a similar concentration of PM not pre-exposed to O_3 [116]. Inhalation exposure of rats to urban PM and ozone demonstrated synergistic effects of O_3 and PM on lung inflammation, an effect that was not seen for the stimulation of endothelin (potent vasoconstrictor produced by vascular endothelial cells and thought to contribute to ambient PM aerosol cardiovascular events) that was associated only with PM exposure [117]. This latter study points to the additional complexity that ambient PM likely has differential effects on different organ systems and metabolic processes that depend upon the chemical context in which the aerosol is presented to the host. A further complexity relates the qualitative and quantitative spatial heterogeneity of the aerosol constituents between areas in close geographical proximity. The effect of heterogeneity of the biological effects of PM aerosol collected from different parts of the same city is illustrated in a study conducted in Mexico City in which PM_{10} was collected from parts of the city with different sources of PM and different concentrations of O_3 [118]. Significant differences were observed in the effects of PM_{10} from the different locations on cell viability, occurrence of apoptosis, evidence of DNA damage (comet assay), and secretion of the pro-inflammatory cytokines IL-6 (Figure 23.17) and TNF-α. While controlled exposure studies of human subjects do permit the assessment of biological responses to PM aerosol exposure, they cannot provide direct evidence for acute health effects, since PM aerosol is presented

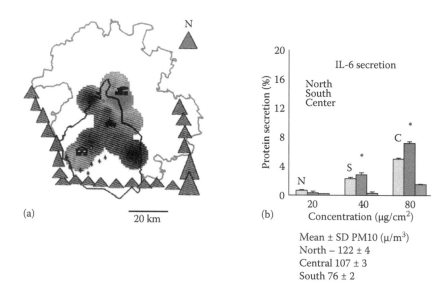

FIGURE 23.17   Effects of PM_{10} collected from three different locations in Mexico City with different levels of PM_{10} and different concentrations of ozone. IL-6 secretion was measured in mouse monocytes. (a) Map of Mexico City; the metropolitan area is composed of two regions: the Federal District (black outline) and the Estade de Mexico (☐ outline). Air quality monitoring stations cover the area shown in ■. The most polluted areas for particles and ozone are in the northeast (■) and the southwest (■), respectively. We collected particles for this study in northern (industrial), central (business), and southern (residential) zones; sites of collection are marked with ☐ squares. Mountains (▨ triangles) surround the city, mainly in the south. (Adapted from Gobierno del Distrito Federal [15]; from Alfaro-Moreno, E. et al., *Environ. Health Perspect.*, 110, 715, 2002.)

in isolation or in higher than ambient concentrations and the exposure is finite. Moreover, such controlled studies cannot offer even indirect evidence for health effects related to exposure over long periods of time.

Exposure assessment in studies of human health: Since the actual distribution of source-specific dosages of aerosol received by a population cannot be known, epidemiological studies have had to focus on measures of exposure to estimate health effects related to exposure to the toxic components of the ambient PM aerosol. Two strategies have been used to try to ascribe health effects either to the PM aerosol itself or specific components of it. By far the most common approach is the comparison of the effect estimates from statistical models that include only PM aerosol or the mass concentration of a specific component with results from the same models with other elements of the aerosol or ambient gas concentrations included. This approach is hampered by the relatively high correlations that often exist between components of the aerosol or other ambient gases and can result in unstable or biased effect estimates (e.g., see Refs [119,120]). A less common approach is to group components of the aerosol by sources without attempting to attribute health effects to any specific component from the source [121]. A variation of this approach is to identify marker components of the aggregate PM mix to narrow the source contribution to health effects. The use of black smoke (correlated with elemental carbon) and nitrogen dioxide (NO_2) as a marker for exposure to diesel exhaust PM [122,123] and traffic-related pollution [124], respectively, are examples.

Most studies of health effects of PM aerosol have based their assessment of health effects on mass concentrations of total suspended particulates (TSP; older studies) PM_{10}, $PM_{2.5}$, and $PM_{10-2.5}$ or components such as element carbon (often measured as black smoke) (e.g., see Ref. [2]). Attempts to attribute health effects to one mass fraction or another remain somewhat clouded. Some investigations in areas in which $PM_{2.5}$ dominates the PM_{10} mass have suggested that most health effects related to the anthropogenic component of the PM aerosol are confined to the $PM_{2.5}$ component [125–128]. Studies that have focused on the sulfate and acid component of PM_{10} (e.g., Refs [129–131]), by implication, emphasize the contribution of fine PM (see Figures 23.1 and 23.2) in which most of the sulfate and acid sulfates (and salts of acids from nitric acid) are found [132]. However, studies from environments in which $PM_{10-2.5}$ makes up a larger fraction of the PM_{10} mass have found that a variety of health effects are more closely associated with this coarse fraction [133,134]. Similar findings have been reported from environments whose PM aerosol mass is not dominated by $PM_{10-2.5}$ [135,136]. Further, claims have been made for a unique role of the ultrafine component of the PM aerosol in causing health effects [137–139]. The specificity of the results has been questioned in a discussion of one study (see the panel discussion in Ref. [138]) and questioned by the findings in other studies [140,141]. Thus, despite the toxicological properties attributed to ultrafine particles (refer to earlier text), it has been difficult to quantify what proportion of health effects attributed to the anthropogenic PM aerosol are related specifically to ultrafine particles or to the number of particles rather than their mass.

23.5.2 SOURCES OF PM AEROSOL AND HUMAN HEALTH EFFECTS

Mobile sources: Much of the early epidemiological research on the human health effects of ambient aerosol focused on fixed industrial sources and used home heating fuels (e.g., soft coal) that made major contributions to the ambient aerosol. The rapid growth of motor vehicle use has led to a research focus on mobile sources of the anthropogenic PM aerosol and its health effects (e.g., see Ref. [142]). This focus has fostered the extensive research on the immunotoxicology of DEP discussed previously. This emphasis derives from a large number of epidemiological studies that have observed increased risks of a variety of health outcomes related to exposure to traffic and traffic-related PM aerosol. Table 23.8 summarizes a representative sample of these studies. In general, most of these studies support an association between one of several different metrics of exposure to mobile sources

TABLE 23.8

Findings from Selected Studies on the Relationship between Exposure to Traffic Sources of PM Aerosol and Health Outcomes

Location of Study/Year Subjects (References)	Marker for Exposure to Traffic Source	Health Outcomes	Results	Comments
Munich/1993/fourth grade children [279]	Traffic counts per 24 h in school districts/distance ≤2 km	Lung function Respiratory symptoms	(<1% decrease in peak flow and mid-expiratory flow)/25,000 vehicles per 24 h	Semiindividual study [280]; effect size very small—questionable meaning for health
Birmingham, United Kingdom/1994/children <5 years [281]	Nearness (200 m cutoff) to roads with >24,000 vehicles/24 h	Hospitalization for asthma	4%–13% increase in odds of hospitalization	Case-control study with hospital and community controls; effects most closely related to density than distance from roadway
Netherlands/1996/adults and children [282]	Air pollution model based on type-specific number of vehicles, fuel source, mean traffic density, emission rates for specific engine types, local topography (e.g., street canyons), regional meteorology	Respiratory symptoms by mailed questionnaire	4–15-fold increased odds of wheeze, shortness of breath only for girls 0–15 years. Weak association with shortness of breath for adults	Comparison of subjects living along busy roads versus quiet roads. Adjusted for confounders that included SHS[1], indoor heating, moisture. Data consistent with studies of SHS that show a female predominant effect best traffic-related exposure data of any study
Austria/1997/first and second grade children [124]	Eight communities without industrial sources of PM aerosol and differing levels of diesel traffic	Respiratory symptoms; NO_2 from fixed monitor as marker	Respiratory symptoms associated only with community level NO_2 (no O_3 or SO_2) Association of "ever asthma" with community-level diesel traffic and tire dust	Semiindividual study Effects adjusted for indoor sources and SHS[a] Correlation between ambient NO_2 and diesel
Netherlands/1997/ schools and children 7–12 years [122][b]	Six areas with different traffic density Distance from roadway Density of truck/auto traffic Indoor black smoke; NO_2; wind direction at schools	Lung function	1%–6% decrease in flows at low lung volumes Effects largest for truck density and black smoke effects larger for children living <300 m from roadway	Effects adjusted for indoor sources, parental respiratory history, SHS, pets, home moisture effects: girls > boys effects on small airways similar to that seen for SHS
San Diego County/1999/ childhood (<14 years) asthma from Medical[c] database [283]	Average daily traffic flow near home by GIS	Number of claims for asthma-related medical visit	Increased occurrence of >2 visits for care for children with asthma for those living near high traffic flow (41,000 cars/day)	Case (asthma)-control (any other diagnosis) study Health implications of outcome not clear

Study	Exposure	Outcome	Results	Comments
London/1999/children 5–14 years [284]	Traffic at centroid of post code of residence (GIS); Simple Euclidean distance to main road with modeled peak/hour of >1000 vehicles computed (vehicle-meters/hour) on road within radius of 150 m	Emergency hospital visits for asthma and respiratory illness	No association between residence within 150 m of main road or traffic volume and respiratory admission	Case-control (nonrespiratory, noninjury acute admission). All studies based on hospital admission have potential bias related to criteria for admission
Harvard Six Cities/2000/all ages [121]	Factor analysis of 15 daily ambient PM metal concentrations; selenium-coal marker; lead-mobile source marker	Mortality, 1979–1988	"Mobile source" associated with 3.4% increased mortality; "coal" 1.1%; no increase with crustal factor	Time series analysis with exposure expressed as mobile source, crustal, coal factors
Nottingham, United Kingdom/2000/primary and secondary schools [285][d]	One km² grid around each school; traffic activity index (TAI—vehicle meters/day/km²)	Current wheezing by questionnaire	No association in ecological or cross-sectional analyses. Weak association (5% increased odds) for persistence of wheeze 7–8 years later	Ecological, cross-section, and longitudinal data. Results inconclusive
Amsterdam/2001/all ages [287]	Residence near road with >10⁴ vehicles/day. Day NO_2, NO, black smoke, O_3, CO, SO_2	Mortality 1987–1998	Black smoke >NO_2 associated with increased risk (BS 25% increase; NO_2 5%) for residences with >10⁴ vehicles/day	Daily time series, black smoke, CO, and NO highest at "traffic-influenced" fixed monitors
Netherlands/2002/adults 55–69 years [197]	Long-term average exposure to ambient pollutants based on regional background and pollutants from local sources (nearby streets). Distance of homes from streets (50, 100 m) with estimated exposure to black smoke and NO_2	Mortality 1986–1994	Black smoke and NO_2 associated with all-cause and cardiopulmonary mortality. Larger effect seen with indicator of distance lived from major road	Cohort study started in 1986; imbedded case cohort (with no loss to follow-up in sub-cohort). Detailed exposure assessment with source apportionment. Individual covariate data (smoking, diet, etc.). Analyses raise point that measurement error in exposure assignment offsets benefits of specificity of exposure assignment; simple distance marker more efficient[e]

(continued)

TABLE 23.8 (continued)
Findings from Selected Studies on the Relationship between Exposure to Traffic Sources of PM Aerosol and Health Outcomes

Location of Study/Year Subjects (References)	Marker for Exposure to Traffic Source	Health Outcomes	Results	Comments
Netherlands/2002/birth cohort [289]	Pollutant concentrations at residences modeled based on GIS traffic data and 40 local monitoring sites, black smoke, NO_2, $PM_{2.5}$	Respiratory symptoms by questionnaires at age 2	Weak associations with all measures of local ambient PM aerosol	Birth cohort study of asthma and mite allergy
California/2003/births [290]	1994–1996 traffic counts to calculate distance–weight traffic density (DWTD) based on dispersion of motor vehicle exhaust from roadways	Birth weight; preterm delivery	Association of preterm births with quintile of DWTD (8% increase in odds from lowest to highest)	Control for multiple factors that affect birth outcomes
			Largest effect seen for exposure during the third trimester (39% increase); no effect for spring/ summer third trimester exposure	Third trimester effect most consistent with effects on somatic growth rather than organogenesis
				Fall/winter increased risk matches "PM season" for California
			Same results for zip code with highest CO, NO_2, and PM_{10}	Major limitation is lack of data on maternal smoking during pregnancy and other indoor sources; adjustment for previous low-birth-weight delivery could have biased results

a Second-hand (passive) smoke exposure.

b Results reported for respiratory symptoms from this study in Ref. [123]. Truck density and black smoke associated with respiratory symptoms.

c California's Medicaid program.

d A later study of the same population by the same authors [286] showed association of increased wheezing and living 150 m from the roadway for children aged 4–16 years. Most of the increase was observed for children who lived within 90 m of a roadway.

e Study in Erie County, NY [288], showed association with hospitalization for asthma in children and living within 200 m of a state highway.

(residential or school distance from roadway and markers of vehicle combustion [black smoke as a surrogate for elemental carbon, NO_2 in the absence of fixed sources]) and the increased occurrence of respiratory symptoms in children, decreased lung function in children, and increased mortality across all ages. In many of the studies, the methods of exposure assessment are somewhat crude (many do not even take location of school or residence in relation to predominant wind direction) and the choice of health outcomes is of variable quality; nonetheless, the fact that distance from roadways has the most consistent association with health outcomes adds some coherence to a causal association [143]. Figure 23.18 shows the decay of black carbon as a marker of diesel exhaust from two freeways with different diesel vehicle loads in southern California [144]. At a distance between 80 and 150 m from the roadway (dependent on vehicle road), the levels of black carbon return to background. Given the diversity of methods of exposure assignment and types of health outcomes, only a qualitative conclusion is reasonable: that is, there is a strong association between exposure to traffic sources of ambient PM and a variety of health outcomes.

Indoor combustion sources: As noted earlier, indoor environments of a variety of types (home, work, vehicles) are important sources of anthropogenic PM aerosol in both developed [11,18,145] and developing countries [146]. No further comment is made with regard to human health effects that are attributable to indoor combustion sources, since in general they do not differ qualitatively from those that would be related to outdoor sources. The major issue in this regard is the relative contribution of indoor and outdoor sources to the overall burden of PM aerosol experienced by humans. A full discussion of this is beyond the scope of this chapter (see Refs [10,146,147]. The health effects related to secondhand exposure to cigarette smoke have been reviewed recently by the National Cancer Institute [148] and an update will appear in a new report of the Surgeon General of the United States that has not yet been released.

Bioaerosols: Human health effects related to bioaerosols have been discussed briefly in Chapter 14 and the mechanisms for the noninfectious components of PM aerosols have been discussed earlier.

Endotoxins are of particular interest, since they are ubiquitous in the environment either adsorbed onto the surfaces of particles generated through combustion processes (refer to earlier text) or as part of indoor dust created by human activity and tracking of soil into homes and the presence of

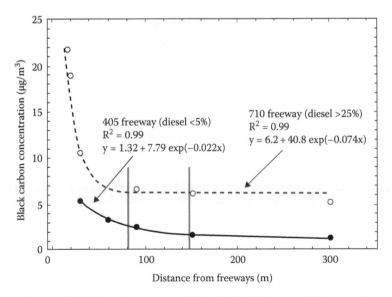

FIGURE 23.18 Decay of black carbon concentrations as a function of distance from two freeways in southern California with different loads of diesel vehicle traffic. (From Zhu, Y. et al., *Atmos. Environ.*, 36, 4375, 2002.)

animal pets [149–151]. Inhalation of endotoxins leads to activation of pro-inflammatory cytokines, which leads to marked inflammation, increases in epithelial permeability, and evidence of activation of systemic inflammation [71]. Although aerosolized endotoxin has long been known as a source of lung disease in cotton (textile) workers [152] and swine handlers [153], more recent interest has focused on the complex role of endotoxin in the occurrence of IgE-mediated allergy and asthma [107]. As noted previously, the biological responses to endotoxin, in theory, could lead both to suppression of IgE-mediated responses as well as the worsening of the lung airway inflammation, which is a hallmark of asthma.

A number of studies have associated levels of house dust endotoxin with increased respiratory symptoms in infants [154] and the worsening of asthma that is independent of the levels of other common indoor allergens (e.g., house dust mite) [155,156]. In contrast, studies have shown that house dust levels of endotoxin are inversely related to the occurrence of IgE-mediated skin test positivity in infants [151] and to the occurrence of hay fever and skin test positivity in children 6–13 years of age [157]. However, the reported cytokine responses related to other exposures appear to conflict. A study of asthma-prone infants [151] reported increased INF-γ production by CD4 and CD8 T cells with increased house dust endotoxin concentrations (expressed as EU/mL of extract), which is consistent with the finding of decreased prick skin test positivity in relation to house concentrations. The study of school-aged children [157] found decreased IL-12 and INF-γ with increased house dust (expressed as EU/m^2 of surface area sampled). These investigators invoked immune tolerance to endotoxin, which would explain the cytokine results but not necessarily the decreased prevalence of atopic disease. Several investigations have suggested that a specific polymorphism in the promotor region of the CD14 gene (refer to earlier text for the role of CD14 in endotoxin biology) may increase the severity of allergy through decreased expression of CD14 and subsequent increase in IgE concentrations [107,158], although a more recent study did not confirm this observation [159]. Given the complex associations between CD14 levels and allergic manifestations at different developmental periods [160,161], it is difficult to reconcile these studies and provide a clearer picture of how the combination of genetic predisposition and developmental timing and intensity of exposure to endotoxin that is part of the indoor and ambient PM aerosol affects the risk of atopic allergy and asthma. The picture is further complicated by the findings from studies that have evaluated the effects of other microbial surface chemicals that can be found in PM aerosol. One study that evaluated the levels of endotoxin, (1→3)-β-D-glucan (extracellular polysaccharide of fungal origin), and allergens from mites and pets found that variability in peak expiratory flow (a measure of lung function) in children aged 7–11 years was no longer associated with endotoxin after adjustment for mite allergy and the presence of dogs/cats, but the association for (1→3)-β-D-glucan persisted [162]. This result may have been influenced partly by overadjustment of pets in the endotoxin analysis, but it nonetheless points to the complexity of the study of endotoxin-related health effects in human populations.

Pollens are well-known sources of respiratory allergy (see Chapter 14) that exert their effects via IgE-mediated mechanisms similar to those described for asthma (Figure 23.16). Pollens exist in the air as pollen grains, which can penetrate into the respiratory tract or contact mucous membranes of the eye. However, it is the protein content of the grains that is responsible for the allergic responses, and pollen exists in the air as pollen grains and allergenic aerosols [163]. In addition to any direct effects of ambient PM on the modulation of immunological responses, there is good evidence that pollen and allergenic aerosol agglomerate onto ambient PM [164], and the degree of agglomeration is proportional to the aromatic component of PM [165,166]. Pollen aerosol that is carried on PM may have enhanced access to immunologically competent cells as a consequence of the inflammation that is triggered by the ambient PM itself [166]. To the extent that aromatic compounds also upregulate IgE-mediated immune responses, it is possible that the allergenic potential of pollens could be enhanced. In addition, air pollutants can alter the allergenic content of plant pollen granules [167–169] as well as the antigenicity of proteins [165]. There are limited human data that have evaluated this possibility and they fail to find evidence of interactions with pollen counts and particles [170–173]. However, all of these studies assessed pollen concentrations in terms of standard counts (and not in terms of allergenic aerosol), which the studies cited earlier indicate may not be the relevant measurement. Moreover, the presence of other pollutants may modify these effects by decreasing the release of antigens [163], which further complicates studies of this problem in human populations. Probably the strongest evidence to suggest that there are human health consequences to interactions between pollen and PM comes from studies of cedar pollinosis in Japan, which suggest that the marked increase in allergy to cedar trees in Japan after 1964 can be traced to increases in mobile sources of air pollution [174]. A study reported by Dutch investigators observed that daily all-cause mortality and cardiovascular mortality were associated with average weekly concentrations of common pollens, an effect that was not confounded by air pollutants (low correlation between pollens and ambient air pollutants, which included black smoke), day of the week, or meteorological factors [175]. The explanation for these findings is not clear, as there was no evidence of interaction with ambient pollutants. The authors offer several speculations, some of which are supported by other data [175]. In contrast to these studies, a cross-sectional study in Switzerland found an association between traffic density near residences and sensitization to pollen allergen (Timothy grass) but not with symptoms of hay fever or the number of allergens to which residents (ages 18–60 years) were sensitized [176]. The association was strongest for those who lived at their residences for 10 or more years.

23.5.3 HEALTH EFFECTS ASSOCIATED WITH CHRONIC EXPOSURE TO PM

There is no universally agreed-upon duration that separates chronic from acute or subacute exposures to PM aerosol. This section defines chronic exposures as those to which humans are exposed over months to years or during defined periods of life events such as pregnancy. There is a vast literature based on cross-sectional, population-based studies that have associated chronic exposures to elevated pollutants and the occurrence of human disease. Due to the difficulty of the evaluation of such data with respect to inferences about causality, this section is restricted to more recent cohort studies of mortality and studies of effects on birth outcomes (these latter studies can be considered as retrospective cohort studies in which the cohort is defined by onset of pregnancy and birth and exposure histories are reconstructed after the fact). The focus is on data related to outdoor PM aerosol. Indoor sources are not discussed specifically—a complete discussion of health effects related to secondhand tobacco smoke is beyond the scope of this chapter. This latter subject is reviewed comprehensively in two recent publications from the National Cancer Institute [17,148] and is the subject of a soon-to-be-released Report of the Surgeon General.

23.5.3.1 Effects of Exposure to Ambient PM and Birth Outcomes (Table 23.9)

In 1992, investigators from the Czech Republic reported that annual ambient levels of TSP, measured in 46 of 85 administrative districts in the Czech Republic, were associated with an increased risk of infant mortality [177]. Studies in the United States [178] and a subsequent study in the Czech Republic confirmed these findings [179]. Two studies of the relationship between daily changes in TSP [180] and $PM_{2.5}$ [181] also observed increased mortality risk with increased concentrations. The extent to which PM aerosol contributed uniquely to the increased risks could not be determined by any of these studies, since associations were also found for gaseous pollutants (Table 23.9). Studies from many parts of the world have identified associations between various birth outcomes (see footnote b of Table 23.9) and exposure to PM aerosol, measured most frequently as TSP but also based on PM_{10} and $PM_{2.5}$ and personal exposure to airborne PAHs (Table 23.9). While there is consistency with regard to the fact that PM aerosol affects birth outcomes, there is less consistency with regard to the specific parameters and the critical developmental periods. Figure 23.19 presents data from three representative studies from three different countries. The studies from the Czech Republic (Figure 23.19a) [182] and southern California [183] found effects of exposure during the first trimester, but the former study failed to find effects of exposure later in pregnancy. A study from Beijing, China [184], observed that exposure to TSP during the third trimester increased the

TABLE 23.9

Summary of Effects of Ambient PM Aerosol on Birth and Neonatal Outcomes

	Non-Acute Exposure		
Outcome	PM Species Measured[a]	Location	References
Neonatal (within first year) mortality	1. Total suspended particulates (TSP) (SO_2, NO_x)—cross-sectional, no personal exposure estimates	Czech Republic 1986–1988	[177]
	2. PM_{10} first 2 months of life	United States (NCHS 1989–1991)	[178]
	3. TSP (SO_2, NO_2)—monthly exposure for each month of pregnancy	Czech Republic 1989–1991	[179]
	4. Distance from a coke works within 7.5 km (no effect observed)	England, Wales, Scotland 1981–1992	[291]
Birth outcomes[b]			
1. LBW (+)	1. TSP and SO_2 exposure during each trimester and weekly prenatal moving averages	Beijing, China 1988–1991	[184]
	2. TSP (SO_2, NO_x)—ecological study (SO_2 effect only)	Czech Republic 1986–1988	[292]
2. SB (−), LBW (+)	3. PM_{10}, $PM_{2.5}$ (monthly exposure estimates)		
	4. PM_{10}, CO, NO_2, O_3 (monthly exposure estimates)		
3. IUGR (+)	5. Modeled TSP (SO_2)	Czech Republic 1994–1996	[293]
4. PTD (+)	6. PM_{10}, $PM_{2.5}$, $PM_{2.5}$ PAH (monthly average exposures)	Southern California 1989–1993	[183]
5. VLBW (+)	7. TSP (SO_2, NO_x) (trimester-specific exposure)	Georgia, United States 1988	[294]
6. IUGR (+)	8. TSP (SO_2, NO_2, CO, O_3) (first and third trimester- specific exposure)	Czech Republic 1994–1998	[182]
7. LBW (+), PTD (+) IUGR (−)	9. PM_{10} (CO, NO_2, O_3) month-specific exposure estimate—association only with CO	Czech Republic 1990–1991	[295]
8. LBW (+)	10. Distance-weighted traffic density in relation to residence (based on dispersion of vehicle exhaust within 750 m radius of residence)	Seoul, Korea, 1996–1997	[296]
9. Birth defects: Cardiac (+), orofacial (±)	11. Personal PAH during 2 days of the third trimester (nonsmoking subjects)	Southern California 1987–1993	[297]
10. PTD (+), LBW (±)		Southern California, 1994–1996	[290]
11. LBW (+)		New York City	[186]
	Daily (Acute) Exposure		
Neonatal mortality	1. $PM_{2.5}$ (O_3, NO_2, NO, SO_2)	Mexico City 1993–1995	[181]
Intrauterine mortality	1. Daily PM_{10} (NO_2, SO_2, CO, O_3) last 28 weeks of pregnancy	São Palo, Brazil, 1991–1992	[180]
Birth outcomes			
1. GA (+), PTD (+)	1. TSP, SO_2	Beijing, China, 1988	[298]

IUGR, intrauterine growth retardation; LBW, low birth weight; VLBW, very low evaluated but no or an imprecisely estimated association was observed; SGA, small for gestational age; PTD, preterm delivery; GA, gestational age; DP, duration of pregnancy; SB, stillbirth.

[a] Pollutants in parentheses are the non-PM pollutants also evaluated.

[b] Studies often assess more than one outcome. A "+" after the outcome indicates that an adverse association was seen. A "−" indicates that the outcome was evaluated but no or an imprecisely estimated association was observed.

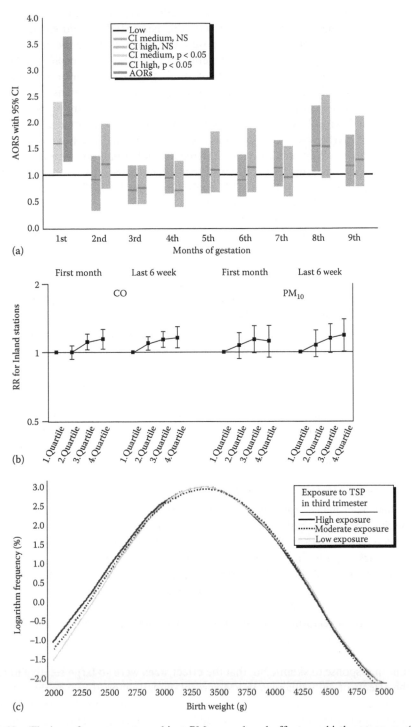

FIGURE 23.19 Timing of exposure to ambient PM aerosol and effects on birth outcomes. (a) Adjusted odds ratios (AOR) for associations of trimester-specific estimates of exposure to TSP on IUGR in the Czech Republic. (From Dejmek, J. et al., *Environ. Health Perspect.*, 108, 1159, 2000.) (b) RRs for PTD based on quintiles of exposure to CO and PM_{10} during the first month and last 6 weeks of pregnancy by quintile of exposure in southern California, United States. (From Ritz, B. et al., *Epidemiology*, 11, 502, 2000.) (c) Effect of exposure to TSP in the third trimester on the distribution of BW in Beijing, China. (From Wang, X. et al., *Environ. Health Perspect.*, 105, 514, 1997.)

risk for low-birth-weight babies without a major shift in the overall population mean. Since the third trimester is the time of maximum somatic growth, this latter study would suggest that there is a susceptible subset of pregnancies for which exposure to ambient PM may be particularly harmful. A study from the Czech Republic evaluated the effects of ambient levels of SO_2 on fecundity [185] and observed that exposure to increasing levels of SO_2 in the 2 months prior to attempts at getting pregnant was associated with decreased conception the first unprotected menstrual cycle. Given the role of SO_2 in the formation of particle sulfates, these findings are relevant in the context of health effects of ambient PM aerosol. A recent study of nonsmoking pregnant women, which included personal monitoring of PAHs, observed associations between personal PAH exposure and decreased birth weight, and birth weight and head circumference [186]. Analyses were adjusted for dietary PAHs and cotinine as a measure of exposure to secondhand tobacco smoke. A very recent study evaluated the effects on the sharp fall in TSP levels during the 1981–1982 recession in the United States on birth outcomes [187]. These investigators estimated that over 70% of the overall reduction in infant mortality during the first year of life could be attributed to the average $15\,\mu g/m^3$ decline in TSP levels that occurred during the recession. This estimate was robust to many different model specifications. Reductions were observed for deaths in the first 24 h, first 28 days, and the first year of life [187].

The fact that studies show inconsistencies in the findings related to specific outcomes does not detract from the striking extent to which the findings agree on the general effects. Differences in PM metrics, use of central monitors for exposure monitoring, inherent population differences, and individual differences in susceptibility likely all contribute to the differences observed. The associations observed with PAHs are consistent with oxidative stress as well as direct toxic effects on pregnancies. The potential implications for these findings are highlighted by the observations that asthma appears to be more common in children of low birth weight and the observation of one study that found that asthmatic children who were born either before 37 weeks gestation or of low birth weight (<2500 g) had a substantially increased risk of symptoms and reduced lung function in responses to increased levels of summertime air pollutants in the eastern half of the United States [188]. Summertime air pollution in this part of the United States is characterized by high PM (especially sulfates) and ozone.

23.5.3.2 Increased Mortality Related to Long-Term Exposure to PM Aerosol

While there had been many cross-sectional studies that suggested that air pollution might be associated with increased mortality, the seminal study in 1993 by Dockery et al. [189] gave new credence to such an association. This prospective cohort study, based on 14–16 years of follow-up in six U.S. cities, reported increased risk of death from a variety of causes that were associated with mean levels of PM_{10} and $PM_{2.5}$ (Table 23.10, rows labeled "original"). The results were not confounded by smoking history, body mass index, or education (see Table 23.2; Ref. [189]). In 1995, Pope et al. [190] report on the mortality experience of the American Cancer Society Study (ACS) cohort in relation to air pollution. This study used 151 metropolitan statistical areas and observed results broadly similar to those of the Harvard Six Cities study reported by Dockery (see Table 23.11, rows labeled "original"). Both studies pointed to $PM_{2.5}$ as the relevant mass fraction. In response to skepticism that the effect sizes were so large relative to results from studies of the effects of daily changes in PM aerosol on daily mortality (see the subsequent section), an extensive reanalysis of these data was undertaken by investigators from Health Canada and the University of Ottawa [191]. Extensive additional modeling (Tables 23.10 and 23.11) and extensive spatial analyses confirmed the results reported by the original investigators. These investigators also evaluated effect modification and observed a consistent increased risk associated with lower levels of education in both studies. A subsequent analysis of the ACS data that included a doubling of the follow-up time to more than 16 years, threefold more deaths, and more extensive $PM_{2.5}$ data confirmed the earlier findings (Figure 23.20) [192]. Effect modification by education again was observed, and the effects were largest and most consistent for

TABLE 23.10

RRs (95% CI) of Mortality by Cause of Death Associated with an Increase in Fine Particles in Risk Models with Alternative Time Axes in the Reanalysis of the Six Cities Study[a]

Risk Model	Calendar Year	Age
All causes (100%)[b]		
Base[c]	1.33 (1.14–1.54)	1.33 (1.15–1.55)
Original	1.29 (1.11–1.50)	1.29 (1.11–1.50)
Full	1.27 (1.09–1.49)	1.27 (1.09–1.48)
Extended	1.28 (1.09–1.49)	1.27 (1.09–1.48)
Cardiopulmonary disease (54%)		
Base	1.39 (1.13–1.70)	1.39 (1.14–1.71)
Original	1.35 (1.10–1.66)	1.34 (1.09–1.65)
Full	1.31 (1.06–1.62)	1.30 (1.05–1.60)
Extended	1.32 (1.07–1.63)	1.31 (1.06–1.61)
Cardiovascular disease (47%)		
Base	1.43 (1.15–1.78)	1.44 (1.16–1.79)
Original	1.41 (1.13–1.76)	1.40 (1.12–1.74)
Full	1.38 (1.10–1.72)	1.35 (1.08–1.69)
Extended	1.39 (1.11–1.73)	1.37 (1.09–1.70)
Respiratory disease (7%)		
Base	1.11 (0.62–1.97)	1.10 (0.63–1.95)
Original	0.93 (0.51–1.71)	0.95 (0.53–1.72)
Full	0.89 (0.47–1.67)	0.94 (0.51–1.73)
Extended	0.88 (0.47–1.64)	0.93 (0.51–1.69)
Lung cancer (8%)		
Base	1.53 (0.91–2.55)	1.64 (0.99–2.72)
Original	1.31 (0.76–2.25)	1.53 (0.90–2.60)
Full	1.30 (0.76–2.23)	1.42 (0.84–2.42)
Extended	1.29 (0.75–2.22)	1.45 (0.85–2.47)
Other cancers (20%)		
Base	1.05 (0.74–1.48)	1.04 (0.73–1.47)
Original	1.04 (0.73–1.47)	1.02 (0.72–1.45)
Full	1.11 (0.78–1.59)	1.09 (0.77–1.55)
Extended	1.10 (0.77–1.57)	1.08 (0.76–1.54)

Source: Adapted from Health Effects Institute, *Special Report: Reanalysis of the Harvard Six Cities Study and the American Cancer Society of Particulate Air Pollution and Mortality*, Health Effects Institute, Boston, MA, July 2000.

[a] RRs were calculated for a change in the pollutant of interest equal to the difference in mean concentrations between the most polluted city and the least polluted city; in the Six Cities study, this difference for fine particles was $18.6 \mu g/m^3$.

[b] Percentages are percentages of all deaths.

[c] Base: air pollution, no covariates; original: used by original investigators; full: many covariates; extended: most parsimonious model.

TABLE 23.11

RRs (95% CI) of Mortality by Cause of Death Associated with an Increase in Fine Particles or Sulfate in Risk Models with Alternative Time Axes in the Reanalysis of the ACS Study[a]

| | Time Axis | | | |
| Alternative Risk Model | Calendar Year | | Age | |
	Fine Particles	Sulfate	Fine Particles	Sulfate
All causes (100%)				
Base	1.27 (1.18–1.37)	1.26 (1.19–1.33)	1.26 (1.17–1.35)	1.25 (1.18–1.32)
Original	1.18 (1.10–1.27)	1.16 (1.10–1.23)	1.18 (1.10–1.27)	1.16 (1.10–1.22)
Full	1.17 (1.09–1.26)	1.15 (1.08–1.21)	1.16 (1.08–1.25)	1.14 (1.07–1.20)
Extended	1.18 (1.09–1.26)	1.15 (1.09–1.21)	1.17 (1.09–1.25)	1.14 (1.07–1.20)
Cardiopulmonary disease (50%)				
Base	1.41 (1.27–1.56)	1.39 (1.28–1.50)	1.41 (1.27–1.56)	1.38 (1.27–1.49)
Original	1.30 (1.18–1.45)	1.27 (1.17–1.38)	1.30 (1.18–1.45)	1.27 (1.17–1.37)
Full	1.28 (1.15–1.42)	1.25 (1.15–1.35)	1.28 (1.15–1.42)	1.24 (1.14–1.34)
Extended	1.30 (1.17–1.44)	1.25 (1.16–1.36)	1.29 (1.17–1.43)	1.25 (1.15–1.35)
Cardiovascular disease (43%)				
Base	1.47 (1.32–1.65)	1.47 (1.35–1.60)	1.46 (1.31–1.63)	1.46 (1.34–1.59)
Original	1.36 (1.22–1.52)	1.36 (1.25–1.48)	1.36 (1.22–1.52)	1.35 (1.24–1.47)
Full	1.34 (1.20–1.49)	1.33 (1.22–1.45)	1.33 (1.19–1.48)	1.32 (1.21–1.43)
Extended	1.35 (1.21–1.51)	1.34 (1.23–1.46)	1.34 (1.20–1.50)	1.33 (1.22–1.44)
Respiratory disease (7%)				
Base	1.07 (0.80–1.42)	0.94 (0.76–1.17)	1.09 (0.82–1.45)	0.95 (0.76–1.18)
Original	1.00 (0.76–1.33)	0.83 (0.67–1.04)	1.01 (0.76–1.34)	0.85 (0.68–1.05)
Full	0.96 (0.72–1.27)	0.81 (0.65–1.01)	0.99 (0.74–1.31)	0.82 (0.66–1.03)
Extended	0.98 (0.74–1.30)	0.82 (0.65–1.02)	1.00 (0.76–1.33)	0.83 (0.66–1.03)
Lung cancer (8%)				
Base	1.23 (0.96–1.57)	1.63 (1.35–1.97)	1.21 (0.95–1.54)	1.62 (1.34–1.95)
Original	1.02 (0.80–1.29)	1.36 (1.13–1.65)	1.02 (0.80–1.30)	1.36 (1.12–1.64)
Full	0.99 (0.78–1.26)	1.32 (1.09–1.60)	0.98 (0.77–1.25)	1.31 (1.09–1.59)
Extended	1.00 (0.79–1.28)	1.33 (1.10–1.61)	0.99 (0.78–1.26)	1.32 (1.09–1.60)
Other cancers (27%)				
Base	1.18 (1.03–1.36)	1.15 (1.03–1.28)	1.17 (1.02–1.34)	1.14 (1.02–1.26)
Original	1.14 (0.99–1.30)	1.10 (0.99–1.23)	1.13 (0.98–1.29)	1.10 (0.99–1.22)
Full	1.14 (1.00–1.31)	1.10 (0.99–1.23)	1.13 (0.98–1.29)	1.09 (0.98–1.21)
Extended	1.14 (0.99–1.31)	1.10 (0.99–1.22)	1.12 (0.98–1.29)	1.08 (0.97–1.21)

Source: Adapted from Health Effects Institute, *Special Report: Reanalysis of the Harvard Six Cities Study and the American Cancer Society of Particulate Air Pollution and Mortality*, Health Effects Institute, Boston, MA, July 2000.

[a] RRs were calculated for a change in the pollutant of interest equal to the difference in mean concentrations between the most polluted city and the least polluted city; in the ACS study, this difference for fine particles was $24.5 \mu g/m^3$, and for sulfate it was $19.9 \mu g/m^3$.

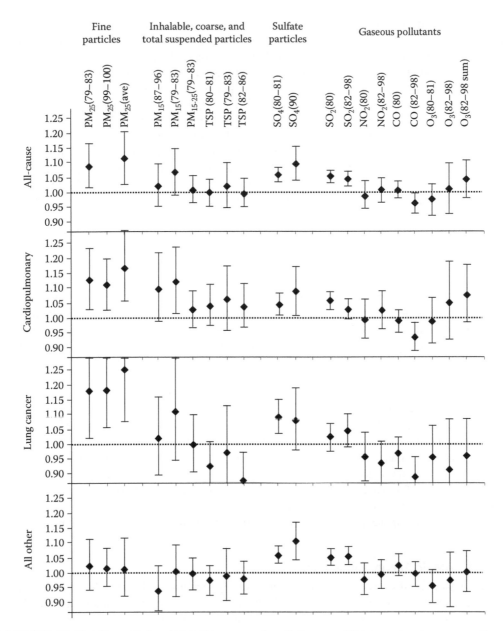

FIGURE 23.20 Summary of associations between various indicators of PM aerosol and gaseous pollutants for the ACS study over different averaging periods. (From Pope, C.A. 3rd et al., *JAMA*, 287, 1132, 2002 as presented by U.S. Environmental Protection Agency (U.S. EPA), Fourth external review draft of air quality criteria for particulate matter (June, 2003), Volume II, U.S. EPA, Report No.: EPA/600/P-99/002aD, Research Triangle Park, NC, 2003.)

PM$_{2.5}$ (similar when expressed as sulfate). Over the 21 year of follow-up each, 10 µg/m^3 increase in average PM$_{2.5}$ across the metropolitan statistical areas was associated with the following increases in mortality: all-cause 5% (95% CI: 2%–11%), cardiopulmonary 9% (3%–16%), and lung cancer 14% (4%–23%) (see Table 23.2 of Ref. [192]). The effects were approximately of the same magnitude when based on levels over the years 1979–1983 and 1999–2000. A similar finding with respect to period of exposure was reported for the Harvard Six Cities data [193].

Another cohort study that has contributed important data on the effects of long-term exposure to ambient PM aerosol is the Adventist Health Study of Smog (AHSMOG), which recruited subjects from the Adventist community in California [194,195]. This study had the advantage that cigarette smoking was virtually nonexistent as a result of religious prohibition and the fact that investigators had developed algorithms for the estimation of personal exposure of subjects to ambient PM [196]. Unlike the Harvard and ACS studies, AHSMOG reported exposure as the number of days on which PM_{10} was above $100 \mu g/m^3$ as well as in terms of mean PM_{10}; results were comparable for both metrics (see Table 3 of Ref. [194]). The initial publication from AHSMOG reported increased risk of death (1977–1992) from "any mention of non-malignant respiratory disease" for both sexes (1.28, 1.10, males and females, respectively, per interquartile range of PM_{10} of days [42.6 IQR] above 100 and $24.1 \mu g/m^3$ for mean PM_{10}). Lung cancer risks were associated with PM only for males. Similar effects were observed for nonmalignant respiratory deaths in both sexes and lung cancer in males. Mean nitrogen dioxide and sulfur dioxide were associated with lung cancer in females. Thus, this study gives a less clear picture than the Harvard and ACS studies with regard to PM effects [189,190,192]. A later publication tried to partition the PM effects between $PM_{2.5}$ (partly estimated from airport visibility data) and PM_{10-25} [195]. Effects were largely confined to the fine fraction, with risk ratios similar to those for PM_{10}.

Subjects from the prospective study of diet in the Netherlands have been evaluated to determine the effect of living near roads with heavy traffic and mortality [197]. The general results from this study were summarized in Table 23.8. Mortality risk increased for those living near major roads (100 m from a freeway or 50 m from a major nonfreeway road) compared to those who did not (e.g., RR for cardiovascular death per $10 \mu g/m^3$ increase in black smoke: near, 1.95 (1.09–3.51); not near, 1.34 (0.68–2.64) [197].

Thus, there is agreement among the prospective cohort studies with regard to associations between PM and mortality. The differences in the results in relation to gaseous pollutants remain to be explained. Given the difference in the summertime pollutant mixtures between California and more eastern portions of the United States [19], it is not likely that ozone is acting as a surrogate for PM-related pollutants as has been suggested for the eastern United States [198]. In addition, although both the Harvard study and the ACS studies pointed to sulfate as the potentially relevant component of the PM aerosol that is associated with mortality, neither of these studies, nor the AHSMOG study, conducted a comprehensive evaluation of this issue. Thus, the relevant physical and chemical components of the PM aerosol that are related to these effects remain virtually unexplained.

23.5.3.3 Atopic Allergy and Asthma

Over the past several decades, there has been a steady increase in the prevalence across all age groups, especially in the very young [109]. While there is a large body of data to support the fact that asthma symptoms are worsening on days with increased PM aerosol (see the section on acute effects), there are few prospective studies that have provided data on the extent to which PM aerosol contributes to the onset of new asthma or the onset of related atopic allergy as might be expected from data on PM mechanisms discussed previously.

In 1992, the Medical Research Council (MRC) of Great Britain reported a 36-year follow-up of a 1946 national birth cohort [199] (Table 23.12). Exposure was based on annual coal consumption during the first 11 years of life. Although not a direct measure of PM aerosol, domestic coal consumption in the United Kingdom was a major contributor to PM through the time of the great London fog of 1954 [200]. "Air pollution attributable risk" for asthma/wheeze at age 36 was 6.7% and 8.8% for exposure between birth and ages 2–11, respectively. This study is unique in that it tries to partition risk between air pollution, smoking, and history of childhood respiratory illness in a single population in which all of these exposures were measured repeatedly over a long period of time. The major limitation of this study was the lack of data on maternal smoking during pregnancy and childhood, both of which have been shown to be important risk factors

TABLE 23.12

Association between Ambient PM Aerosol and Onset of Asthma

Cohort	Follow-Up Time	Exposure Metric	Findings	Reference
Medical Research Council, National Survey of Health and Development 1946 birth cohort—national sample	Thirty-six-year multiple contacts	Annual coal consumption between birth and age 11	Self-reported wheeze and asthma at age 36 Attributable risk[a] for "air pollution" exposure between ages 0 and 2 = 6.7% Attributable risk[b] for "air pollution" exposure between ages 2 and 11 = 8.8%	[199]
Adventist Health Study of Smog Seventh Day Adventists who from various locations in California (90% southern California) enrolled in 1977	Ten years—2 contacts (20-year residential histories)	Estimated personal exposure to TSP, ozone, SO_4	Self-report of new physician diagnosis of asthma RR 2.9 (95% CI: 1.0–7.6) per $7\,\mu g/m^3$—unit increase in mean annual SO_4 exposure[c] Effects not seen for new nonasthmatic onset obstructive airways or chronic bronchitis Unable to separate SO_4 association from that with O_3	[203]
Adventist Health Study of Smog	Fifteen years—3 contacts	Estimated person exposure to SO_4, PM_{10}	Self-report of new physician diagnosis of asthma Only association observed for 8 h average ozone for females	[204]
Children's Health Study Children up to 16 from 12 southern California communities	Up to 5 years—annual contact	PM_{10}, $PM_{2.5}$, NO_2, ozone; classified as high/low O_3 and PM based on 4-year means	New onset self-report of doctor diagnosed asthma Associations limited to children who reported participation in three or more outdoor sports High-PM communities:[d] RR = 2.0 (95% CI: 1.1–3.6); low-PM communities: 1.7 (0.9–3.2) High-O_3 communities: RR = 3.3 (1.9–5.8); low-O_3 communities: RR = 0.8 (0.4–1.6)	[206]

[a] Estimate adjusted for history of childhood lower respiratory illness, parental history of bronchitis, low socioeconomic status. No data given related to maternal smoking during pregnancy or during postnatal and childhood years.

[b] Adjusted for same risk factors as for birth to 2 year exposure as well as cigarette smoking at age 36. Attributable risk for smoking = 39.7% for comparison.

[c] Adjusted for years lived with a smoker, history of past lower respiratory illness before age 16, sex, age, and education.

[d] The same communities were ranked as high for PM_{10}, $PM_{2.5}$, and NO_2. 4-year medians: low—PM_{10} $(\mu g/m^3)$ = 25.1, $PM_{2.5}$ = 7.6; high—PM_{10} = 39.7, $PM_{2.5}$ = 21.4. All models adjusted for SED, family history of allergy and/or asthma, maternal smoking, body mass index.

for asthma [201,202]. An early report from the Adventist Health Study reported an association between estimated personal exposure to sulfate over a 20-year period and the new report of a doctor diagnosis of asthma that was specific for asthma [203] (Table 23.12). However, a subsequent report, based on an additional 5 years of observation failed to confirm these findings [204]. In this latter study, new onset of asthma was associated only with mean ozone concentrations in females. Earlier reports from this cohort noted that correlations between ozone and PM aerosol (measured as TSP early on the cohort) made it difficult to distinguish PM and ozone-related health effects [205]. A recent report from the Children's Health Study based on up to 5 years of follow-up of children 9–16 years from 12 communities in southern California failed to find a clear association between PM aerosol and the onset of new asthma [206] (Table 23.12). An association with ozone was observed for children who participated in three or more outdoor sports (Table 23.12). Other reports based on a follow-up of this cohort identified important indoor and familial factors as additional determinants of new-onset asthma [207,208], but no attribution of risk was attempted between these latter exposures and ambient PM or other ambient pollutants. Although the mechanisms by which PM aerosol could contribute to the onset of asthma in susceptible individuals are compelling, the data in Table 23.12 are not so compelling. As the MRC study indicates, if PM aerosol and/or other ambient or indoor pollutants do contribute to the risk of the asthma etiology, the effect is likely to be small and will be difficult to sort out from other important exposures such as allergens and fungi. A further complication relates to the observation that the "epidemic" of asthma may have peaked, and asthma prevalence is declining (Figure 23.21). If this is the case, and the mechanisms related to PM aerosol discussed earlier are operative, then the focus of epidemiological studies will have to shift more intensely to the identification of the susceptible profiles with large populations.

A series of studies from Germany indicates that the connection between exposure to PM aerosol and the occurrence of asthma and atopic allergy may involve different mechanisms or a different time course in populations. With the fall of the East German Republic (1989) and the unification of Germany (1990), there was a sharp decline in levels of SO_2 (precursor of secondary sulfate aerosol) [209]. Early studies indicated that the prevalence of atopic diseases and asthma was *less* frequent in the old East German Republic and that symptoms of bronchitis were more frequent [210]. Surveys carried out several years after the reunification observed that prevalences of atopic allergy and positive prick skin tests were now similar in former East (Dresden) and West German (Munich) cities, but that the prevalence of asthma and reactive

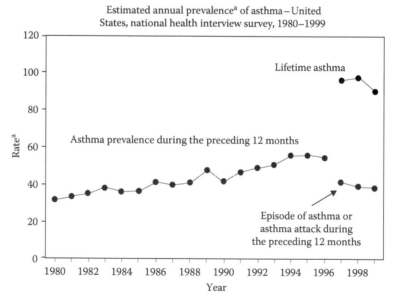

FIGURE 23.21   Annual self-reported prevalence of current asthma based on the National Health Interview Survey. Note: Format of questions changed after 1996. Data demonstrate plateau in reported prevalence of asthma in the preceding 12 months (current asthma) and report of asthma at any time during life. (From Centers for Disease Control and Prevention, *MMWR*, 51, 1, 2002.) [a] Per 1000 population; age-adjusted to the 2000 U.S. population.

airways (an asthma-associated phenotype) was still more common in the western city [209]. However, at least one study has raised the question that diagnostic bias, in part, may be playing a role in the increased reported prevalence of atopic disease in the areas of former East Germany [211]. Cross-sectional surveys of children were conducted in two formerly highly polluted areas (combustion of brown coal with high sulfur content and heavy metal containing dust from smelters) and one area of minimal pollution in areas of former East Germany in 1992–1993 and 1995–1996. No changes were noted in the self-reported prevalence of asthma or atopic disease. However, physician-diagnosed allergy increased without a concomitant change in the prevalence of specific IgE to a variety of common aeroallergens. A decrease in physician-diagnosed bronchitis was also observed. These results conflict with a study of similar design carried out in Leipzig in 1991–1992 and 1995–1996 in which the prevalence of sensitization followed that of reported symptoms [212]. Whether explanation of the differences is due to technical problems with the skin testing device in Leipzig, as claimed by one set of investigators [211], or is due to differences in air pollutant composition, population differences or chance cannot be determined from the data. In aggregate, these studies point to the difficulty in linking human health outcomes with the PM aerosol as would be expected from the data on mechanisms.

23.5.3.4 Cancer

As noted in Figure 23.20, PM aerosol has been associated with an increased risk of cancer. By and large, this excess risk has been noted for cancer of the lung [189,192]. The specific components of PM that are responsible for this excess risk have not been elaborated fully. However, several exhaustive reviews have been conducted related to the role of PM aerosol from diesel engines in the induction of cancer [213,214]. EPA provided a model for the hypothetical pathways that may be involved in carcinogenesis (Figure 23.22) and is based on concepts presented in Section 23.3. Most of the evidence to a carcinogenic potential in humans is derived from occupational studies. EPA reviewed 22 such studies (Section 7.2 of Ref. [214]) and estimated the pool relative risk at between 1.33 and 1.47. Given the much higher levels of exposure that occur in occupational environments relative to ambient environments, it is difficult to assess the risk beyond that which has been presented for lung cancer in the epidemiological studies cited previously. EPA has attempted to assess general population risk based on the conversion of exposures observed in occupational settings, for which risk estimates are available, to those experienced by the general population (Table 23.13). Based on a variety of assumptions related to the data in Table 23.13 (see pp. 8–13 to 816 of Ref. [214]), EPA estimated that the lifetime risk of lung cancer from exposure to ambient diesel PM could vary between 10^{-6} and 10^{-4}, although a zero risk could occur due to an unmeasured threshold.

23.5.4 Health Effects Associated with Acute and Subacute Exposures to PM

There is an enormous body of research that documents the association between short-term changes, usually measured over the week before a defined health outcome, and health effects. Effects on mortality, worsening of asthma and respiratory symptoms, and effects on lung function have been studied most widely. Perforce, this review is limited to a representative sample of studies that have demonstrated health effects and some of the critiques that have been levied against the validity of the associations. In addition, this section covers only the data related to the associations between daily changes in air pollution and daily mortality for several reasons: (1) mortality and years of life lost is an endpoint of major public health significance; and (2) the most extensive analyses on the effects of study design and statistical analysis techniques on estimates of effect have been carried out in relation to the mortality data. Excellent summaries of the large database related to worsening of asthma and other acute respiratory illnesses can be found in a number of the U.S. EPA references.

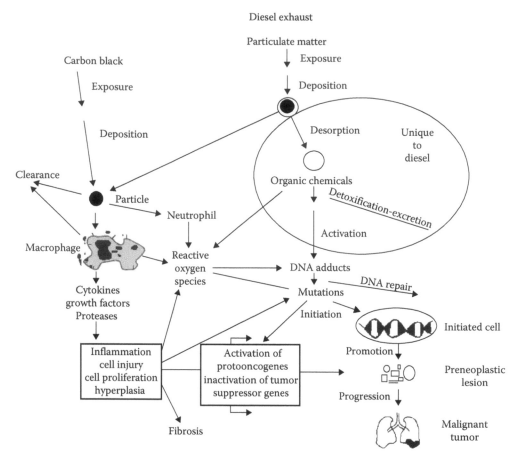

FIGURE 23.22 Postulated mechanism for carcinogenesis related to exposure to high-level diesel exhaust concentrations in rats. (From U.S. Environmental Protection Agency (EPA), Health assessment document for diesel engine exhaust, Report No.: EPA/600/8-90/057F, National Center for Environmental Assessment, Washington, DC, 2002, Figure 7.3.)

23.5.4.1 Mortality

For obvious reasons, the effects of short-term increases in ambient PM have received more attention than any other PM-aerosol-related health effect. The most famous example of the mortality associated with acute increases in PM aerosol is the London fog episode in December, 1952, which was responsible for both acute and subacute mortality that accounted for approximately 13,000 excess deaths from the start of the fog through March 1953 [215]. Bell and Davis [215] estimated that a 0.10 ppm increase in weekly SO_2 (surrogate for the PM aerosol) was associated with a relative risk (RR) of mortality of 1.31 (95% CI 1.11–1.36) (Figure 23.23). Subsequent studies of mortality in London found associations between British smoke levels [216] and acid aerosol levels [217] and mortality. In 1994, Schwartz [218] summarized mortality TSP/British smoke/coefficient of haze associations from 13 studies, all of which used the same statistical methodology and reported an RR of death per $100\,\mu g/m^3$ increase in daily TSP that ranged from 1.06 to 1.08. A detailed analysis of the causes of death in Philadelphia, PA, reveals increased RRs for deaths from chronic obstructive lung disease (COPD) (1.25 for high TSP vs. low TSP days) and pneumonia (1.13 for high vs. low days). Risk of death from cardiovascular disease was increased, but these deaths were often accompanied by underlying respiratory conditions [219]. Other studies have confirmed the increased susceptibility of patients with COPD [220] and other underlying respiratory diseases to nonrespiratory

TABLE 23.13
DPM Exposure Margins (Ratio of Occupational ÷ Environmental Exposures)

Occupational Group	Estimated Occupational Exposure/Concentration Environmental Equivalent[a]	Exposure Margin Ratio — Average Environmental Exposure for 0.8 µg/m³ of Environmental Exposure[b]	Exposure Margin Ratio — Average Environmental Exposure for 4.0 µg/m³ of Environmental Exposure[b]
public transit workers	15–98 µg/m³ 3–21 µg/m³	4–26	0.8–5
U.S. railroad workers	39–191 µg/m³ 8–40 µg/m³	10–50	2–10
Fork lift operators	7–403 µg/m³ 1–85 µg/m³	2–106	0.37–21
High-end boundary estimate	1200 µg/m³ 252 µg/m³	315	63

Source: U.S. Environmental Protection Agency (EPA), Health assessment document for diesel engine exhaust, Report No.: EPA/600/8 – 90/057F, National center for Environment Assessment, Washington, DC, 2002, Table 8.1

[a] Equivalent environmental exposure = occupational exposure × 0.21 (See Chapter 2, Section 2.4.3.1); some values are rounded.

[b] 0.8 µg/m³ = average 1990 nationwide on-road exposure estimate from HAPEM model; the companion rural estimate is 0.5 µg/m³, and 0.4 µg/m³ is a high-end estimate. The 1996 nationwide average is 0.7 µg/m³, the companion rural estimate is 0.2 µg/m³, however, a high-end estimate is not available for 1996. See Chapter 2, Sections 2.4.3.2.1 and 2.4.3.2.2.

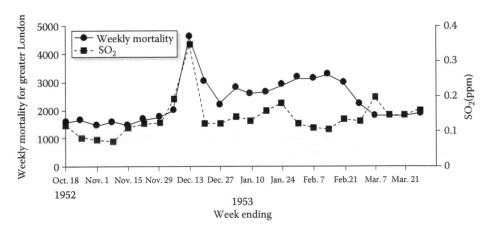

FIGURE 23.23 Relationship between London fog pollution measured as SO_2 between December 5 and 9, 1952, and mortality. Note that mortality remained elevated through February 1953. Average SO_2 during the time of the fog was 0.57 ppm (24 h U.S. NAAQS for SO_2 = 0.03 ppm). (From Bell, M.L. and Davis, D.L., *Environ. Health Perspect.*, 109(Suppl. 3), 389, 2001.)

mortality from daily changes in air pollution [221]. A recent study that evaluated deaths in New York City from 1985 to 1994 found that risk of death per interquartile range of PM_{10} was greater for persons whose death certificates indicated an underlying respiratory disease [221]. This was particularly true for those aged 75 years and older, in whom a similar increased risk was also observed for cancer deaths. Numerous other studies have supported these findings. Table 23.14 summarizes representative examples of studies based on investigations of single cities or areas (a more detailed discussion of the multicity studies in the United States and Europe follows). Studies from around the world (based on time series and case crossover) have generally reported associations between increased daily PM aerosol (or surrogates such as SO_2 and NO_2) and daily nonaccidental mortality, both from all causes and particularly from cardiovascular and respiratory causes. Although Schwartz and colleagues, in time series studies, reported that out-of-hospital deaths had a stronger association than in-hospital deaths, a case-crossover study from the Seattle area did not find any association between out-of-hospital deaths due to cardiac arrest and daily changes in any measure of PM aerosol [222] (Table 23.14). Two of the studies in Table 23.14 are of particular note. Clancy et al. [223] reported 10% and 15% decreases in daily cardiovascular and respiratory deaths in relation to daily changes in air pollution in the years that followed the 1990 ban on the sale of soft coal in Dublin, Ireland. These decreases in daily mortality were associated with decreases in average black smoke levels in 1984–1990 of 50–15 $\mu g/m^3$ in the years 1990–1996 (36% reduction) [223]. Vidal and colleagues studied daily changes in mortality in relation to ambient pollutants in an environment with low average PM_{10} levels (see footnote in Table 23.14). These investigators found an association between respiratory mortality and NO_2 and SO_2; PM_{10} was only associated with all-cause mortality at a 2-day lag. The explanation for why pollutants strongly associated with the PM aerosol seemed to have more specific effects on mortality was not explained by the author. A study from Seoul, Korea, points to the difficulties of ascribing effects to PM aerosol [224] (Table 23.14). In this study, the PM_{10} association was a 1.5% increase/interquartile interval; however, when O_3 was included with PM in the regression model, the PM effect was observed (2.7% increase) when ambient O_3 concentrations were above 13 ppb (Table 3 of Ref. [224], Table 23.14). This suggests that the overall chemical milieu is relevant to the interpretation of the magnitude of PM–mortality associations. Finally, a study in Mexico City found that over all lags and averages, the RR of death per 10 $\mu g/m^3$ of PM_{10} was greater for persons outside of hospitals at the time of death compared to deaths for persons in hospitals at the time of death [225]. Since hospitals usually have controlled environments, this supports a role for the ambient environment.

Subsequent to the publication of the time series studies noted in Table 23.16 and many other time series not cited in the table or this chapter, a problem was noted with the software that had been used by many of the studies to control the confounding effects of long-term trends in ambient PM aerosol, meteorological, and other time-dependent confounding factors [226] (see footnote in Table 23.15 for details). A number of these studies were reanalyzed, and although some of the effect estimates were reduced, the overall conclusion of an association between daily changes in PM aerosol and daily mortality was not altered (the same was observed for associations with hospital admission, which are not covered in this chapter) [227].

Given the heterogeneity of risk estimates that have been derived across studies carried out in individual cities, two large studies were undertaken in the United States and in Europe (Table 23.15), which attempted to study multiple cities with a broad range of PM aerosol environments using uniform statistical methodological approaches and to explore sources of heterogeneity between the estimates derived from individual cities. Since both studies were reanalyzed to account for the software problem noted earlier, only the reanalyzed data are summarized. The basic structure of each study is summarized in Table 23.16.

NMMAPS estimated that for each 10 $\mu g/m^3$ increase in daily PM_{10}, daily mortality increased 0.7% (posterior SE 0.06), 0.21% (0.06), and 0.10% (0.06) for lags of 0, 1, and 2 days, respectively (Figure 23.24, top panel). At lag 1, the effects were largest for cardiorespiratory diseases (Figure 23.24, middle panel), and the lag 1 PM_{10} effect was found to be robust to the inclusion of other

TABLE 23.14

Representative Sample of Time Series Studies that Relate Daily Changes in Measures of PM Aerosol to Daily Changes in Non-Accidental Mortality

General Population-Based Time Series Studies

Study Population	Years	Pollutants[a]	Methods	Results	References
Santa Clara County, CA	1980–1986	Winter COH	Multiple polynomial models; PR[2]	Results: % increase per unit change in COH (2-day lag) Respiratory 11.1 Cancer 1.0 Circulatory 9.4	
Philadelphia, PA	1972–1982	TSP, SO_2	PR, GEE	RR[2] for $100\,\mu g/m^3$ (mean of days 0 and 1) <65 year = 1.03 65+ years = 1.1 associations for COPD[1], CVD,CVD, less clear for pneumonia	[299]
Buffalo/ Rochester, NY	1988–1990	H^+, $SO_4{}^{2-}$, SO_2, CO, NO_2, O_3, PM_{10} (predicted from COH and SO_4 for 5/6 days)	PR	RR/IQR[2] for respiratory/CVD mortality PM_{10} (unfilled) 1.048 1.034 COH 1.016 1.014 $SO_4{}^2$ 1.024 1.011 O_3 1.037 1.009	[300]
Coachella Valley, CA[b]	1989–1998	PM_{10}, CO, NO_2, O_3, $PM_{2.5}$ (measured directly last 2.5 years), $PM_{10-2.5}$(predicted)	PR	RR/IQR for respiratory/CVD mortality PM_{10} (lag 0) 1.03 (1.01–1.05[c]) $PM_{10-2.5}$ (lag 0) 1.02 (1.01–1.04) $PM_{2.5}$ (lag 4) 1.03 (0.98–1.09) All $PM_{2.5}$ RR include 1 in sensitivity analyses; other PM exclude 1	[133]
Santiago, Chile	1989–1991	PM_{10}, NO_2, SO_2, O_3	OLS[2] PR	RR/$115\,\mu g/m^3$ change in PM_{10} Respiratory 1.15 (1.08–1.23) CVD 1.09 (1.04–1.14) Age 65+ years 1.11 (1.07–1.14) Male 1.11 (1.07–1.14) Female 1.06 (1.02–1.10) PM association robust to inclusion of other pollutants	[133]
Mexico City	1990–1992	TSP, PM_{10} SO_2, CO, O_3	PR	RR for $100\,\mu g/m^3$ change in TSP in regressions with SO_2, O_3 Age >65 years 1.06 (1.03–1.09) Respiratory 1.10 (1.01–1.18) Cardiovascular 1.05 (1.01–1.10) 95% CI for estimates for SO_2 and O_3 include 1 in all models	[119]

(continued)

TABLE 23.14 (continued)
Representative Sample of Time Series Studies that Relate Daily Changes in Measures of PM Aerosol to Daily Changes in Non-Accidental Mortality

General Population-Based Time Series Studies

Study Population	Years	Pollutants[a]	Methods	Results	References
Sydney, Australia	1989–1993	$PM_{0.01-2}$, NO_2, O_3	PR, GEE	Percent increase for 10th–90th percentile change in PM_{10} in models with O_3 and NO_2 Cardiovascular 2.1% (0.3–5.0) Respiratory 0.7% (−5.5, 7.5) No associations with O_3 and NO_2	[302]
Dublin, Ireland	1984–1997		PR with interrupted time series	Percent decrease in deaths/10^3 person years before and after the 1990 ban on soft coal Cardiovascular 10.3 (12.6–8.0) Respiratory 15.5 (19.1, 11.6) Age <60 years 7.9 (12.0, 3.6) Age 75+ years 4.5 (6.7–2.3)	[223]
Vancouver, Canada[d]	1994–1996	PM_{10}, CO, NO_2, SO_2, O_3	PR, GAM	Largest increases were for O_3 for respiratory mortality in summer, SO_2 for respiratory mortality in winter, and NO_2 for CVD deaths in winter. PM_{10} at lag 2 days showed association with total mortality	[303]
Seoul, Korea	1995–1998	PM_{10}, NO_2, CO, O_3	PR, GAM	Percentage change in mortality from stroke/interquartile increase in PM_{10} 1.5% (1.3–1.8) in single-pollutant model −1.2%: PM_{10} effect for O_3 concentrations < median for O_3 (13 ppb) and 2.7% for O_3 above median O_3 concentration (correlation between O_3 and PM_{10} = −0.3)	[224]
Case-crossover studies[e]					
Philadelphia, PA	1973–1980	TSP, no other pollutants evaluated	CLR^2	Adjusted odds ratio (OR)/100 μg/ m^3 increment in 48 h TSP = 1.06 >65 years 1.07 (1.04–1.11) CVD 1.06 (1.02–1.11) Pneumonia 1.08 (0.92–1.26)	[305]
Seoul, Korea	1991–1995	TSP, SO_2, O_3	CLR	RR/100 μg/m^3 increase in 3-day moving average TSP = 1.01 (0.99–1.03) SO_2 (50 ppb) 1.05 (1.02–1.08) O_3 1.02 (0.99–1.05) RR for SO_2 from PR = 1.08 (1.06–1.10)	[306]

TABLE 23.14 (continued)
Representative Sample of Time Series Studies that Relate Daily Changes in Measures of PM Aerosol to Daily Changes in Non-Accidental Mortality

General Population-Based Time Series Studies

Study Population	Years	Pollutants[a]	Methods	Results	References
Time series studies of association between exposure to mobile sources and daily mortality					
Amsterdam, the Netherlands	1987–1998	Traffic counts, black smoke (BS), PM_{10}, SO_2, CO, NO_2, NO, O_3	PR, GAM	Higher pollutant levels at traffic-influenced monitoring sites. Effects greatest when traffic-influenced site data are applied to persons who lived on roads with $>10^4$ vehicles/day; most consistent for BS $RR/100\,\mu g/m^3$ BS (lag 1), 1.89 (1.20, 2.95)	[287]

TS, time series; PR, Poisson regression; GEE, generalized estimating equations; GAM, generalized additive models; OLS, ordinary least-squares regression; CLR, conditional logistic regression; COH, coefficient of haze; RR, relative risk; IQR, interquartile (25th–75th) range; COPD, chronic obstructive lung disease; CVD, cardiovascular diseases.

[a] All studies include some variables to control for meteorological effects, season, and, for some, day of the week effects. SO_2, CO, and NO_2 are noted since they often are surrogates for fixed (SO_2) or mobile (CO, NO_2) source PM aerosol. O_3 is included, since in some areas (e.g., eastern United States) photochemical smog is highly correlated with ambient PM aerosol levels.

[b] This is a follow-up study to that in Ref. [301] and uses a longer time series and adds data on $PM_{2.5}$ to those of PM_{10}.

[c] Ninety-five percent CI.

[d] This study is included since pollutant levels are quite low. Maximum $PM_{10} = 33.9\,\mu g/m^3$, 90th percentile $= 22.8\,\mu g/m^3$.

[e] Case-crossover designs are matched designs in which each subject serves as his or her own control. Unlike the population-based time series studies, case-crossover designs provide control for individual covariates and, with proper sampling, control of temporal confounding [222,304].

pollutants in the statistical models (Figure 23.24, bottom panel). There was evidence for geographic heterogeneity, with effect estimates being greatest in the northeast and least in the northwest (Figure 23.25a). City-specific estimates for all-cause and cardiorespiratory deaths showed heterogeneity; however, within regions, there seems to be relative homogeneity (Figure 23.25b—note circled areas for southern California and the northeast). This heterogeneity likely represents a combination of differences in the ambient pollutant mixtures between regions as well as demographic differences. In the original analyses, the percentage of adults without a high school diploma in each city had an independent effect on daily mortality. This was not included in the reanalysis. The APHEA project reported percentage increases in total daily mortality of approximately 0.7% per $10\,\mu g/m^3$ increase in PM_{10} (average of lag 0, 1—see footnote a in Table 23.16) for 21 cities where such data were available (Table 23.16) [228]. The results were relatively robust to the types of modeling strategy chosen. These results are somewhat lower than those expected with a similar average of lags for NMMAPS (Figure 23.24, top panel).

A number of studies have attempted to identify specific components of mass or particle number as the source of increased risk for daily mortality (Table 23.17). Two separate reports from the Harvard Six Cities study found that only fine PM was associated with increases in daily mortality; no association was observed for coarse PM or a "crustal" component based on source attribution

TABLE 23.15
Multicity Time Series Studies of the Effect of Daily Changes in PM Aerosol on Daily Mortality

Study and References	Locations and Pollutants	Methods
The National Morbidity, Morbidity and Air Pollution Study (NMMAPS) *Initial analyses:* [307–309]	Ninety largest cities in the United States for 1987–1994 Subanalysis of 20 of the largest in this group [309] PM_{10}, O_3, CO, SO_2, NO	Three-stage regional model 1. Within-city variability estimated with log-linear semi-parametric model
Reanalyzes after discovery of the GAM problem:[a] [238,310]	Meteorological data	2. Within-region variability of the true regression coefficient with weighted second-stage regression 3. Between-region variability in the true regional regression coefficients Heterogeneity across cities and regions explored
Air Pollution and Health: a European Approach (APHEA) [228, pp. 311–315]	European cities APHEA1—15 cities, black smoke, PM_{10}, SO_2, NO_2, O_3 APHEA2—29 cities, black smoke, PM_{10}, SO_2, NO_2, O_3	Individual city analysis 1. Standard procedures for Poisson regression Between-city analyses 2. Assume city-specific means normally distributed around overall mean 3. Evaluate effect modifiers

[a] The initial analyses used the default convergence criteria in the GAM algorithm in S-Plus. These default criteria did not lead to convergence and resulted in standard errors of parameter estimates that were too small. Subsequent reanalyzes were carried out with more strict convergence criteria in S-Plus and with a generalized linear model approach with natural and penalized splines. These issues are explored in detail in the references in the table and are summarized in Ref. [226].

[121,128]. A similar conclusion was reported in a Spokane, WA, study in which the contribution of coarse PM was evaluated with an indicator for days with dust storms [126]. A meta-analysis based on 19 U.S.-based studies came to the same conclusions [229]. In contrast, studies from California, Mexico City, and Santiago, Chile, all reported that daily changes in coarse PM aerosol had as large or larger effects on daily mortality than did fine PM aerosol (Table 23.17). The extent of differences in the composition of coarse PM (concentrations of iron, bioaerosol components, and a "tail" of combustion product PM that extends into the coarse range) were not evaluated by any of these studies. Based on the available data, it does not appear appropriate to extrapolate the relative contributions of fine and coarse PM to the associations with mortality from one location to another.

A large study was undertaken in Germany to evaluate the relative strength of associations between changes in daily mortality and changes in particle number in the ultrafine range and mass fractions [138]. The authors concluded that the associations were driven by particle number effects. However, inspection of the data (Figure 23.26) does not provide a difference in the exposure–response relationships between particle number, fine mass, and total PM_{10} mass. A commentary by the reviewers of the report concluded that the evidence did not favor one component over another (begins on p. 93 of Ref. [138]).

Among the studies that have reported associations between daily changes in PM and daily mortality, there has been a general consistency that the effects are greater in the elderly (usually defined as age 65 years and older) and among individuals with selected underlying disease

TABLE 23.16

Association between a 10 μg/m³ Increment in PM$_{10}$ (Average of Lag 0.1)[a] and Percentage Increase in Total Daily Mortality

GAM[b] with Default Convergence Criterion	GAM with Strict Convergence Criterion	GLM with Natural Spline	GLM with Penalized Spline
0.7% (0.6%–0.8%)[c]	0.7% (0.5%–0.7%)	0.4% (0.3%–0.6%)	0.6% A(0.4%–0.7%)

Source: Katsouyanni, K. et al., Sensitivity analysis of various models of short-term effects of ambient particles on total mortality in 29 cities in APHEA2, in *Special Report: Revised Analyses of Time-Series Studies of Air Pollution and Health*, Health Effects Institute, Ed, Health Effects Institute, Boston, MA, pp. 157–164, 2003.

GAM, generalized additive model; GLM, generalized linear model.

[a] Not directly stated but surmised from Table 2 in Ref. [314].

[b] All results based on fixed-effects models. Results similar to random-effects models.

[c] Author's calculations from coefficients in Table 1 of Ref. [228].

(e.g., diabetes [study of hospital admissions; [230]] congestive heart failure [231]). As noted previously, persons with underlying respiratory diseases seem to have an increased risk of cardiovascular deaths.

The form of the exposure–response relationship between daily changes in PM aerosol and daily mortality has been investigated extensively (e.g., see Refs [232–237]). In general, the data have been more consistent with a no-threshold model than models with thresholds, although alternative models with thresholds still remain a consideration (Figure 13.6 from Ref. [28]). The NMMAPS study evaluated data from the 20 largest U.S. cities to assess the most likely exposure–response relationships between daily changes in PM aerosol and daily mortality [238]. A summary analysis for total mortality suggests that a no-threshold model is most consistent with the data for PM$_{10}$ (Figure 23.27). A more detailed analysis, based on the distribution of posterior probability of a threshold, suggested that a threshold was not likely for cardiovascular mortality, but was consistent for the data for "other" causes of death (Figure 23.28). In the case of all-cause and cardiovascular mortality, any likely threshold was below the federal annual PM$_{10}$ annual 24 h standard of 50 μg/m³ in force at the time over which the PM data were evaluated (Figure 23.28). In a commentary that accompanied the publication of the NMMAPS analysis, Pope [237] reviewed the various factors that could influence the detection of threshold (statistical methods, publication bias, measurement error) and presented a graphical summary of some of the more important time series studies that have addressed exposure–response relationships (Figure 23.29). He concluded that the weight of evidence "further indicates that assumptions or scientific priors of no-effects threshold levels for PM are not well supported by the empiric evidence" [237]. Other investigators, although in the minority, raise the argument that measurement errors preclude accurate specification of the exposure–response relationship [239].

In the end, the overall health impact of the associations between daily changes in PM aerosol and daily mortality depends upon the extent to which life expectancy is shortened by exposure [240]. If the people who are dying were those whose deaths were advanced by only a few days (a phenomenon that has been termed "harvesting" or mortality displacement [241]), there is general agreement (see Ref. [240] for an alternative view) that the daily increases in mortality with increases in daily PM are not due simply to harvesting [241–245]. Schwartz carried out analyses based on smoothing windows of 15, 30, 45, and 60 days and observed that the percentage increases in mortality (1979–1986) per 10 μg/m³ increase in PM$_{2.5}$ in Boston peaked at 15, 60, 60 days for COPD, pneumonia, and ischemic heart disease (IHD) deaths, respectively (Figure 23.30) [242]. COPD showed

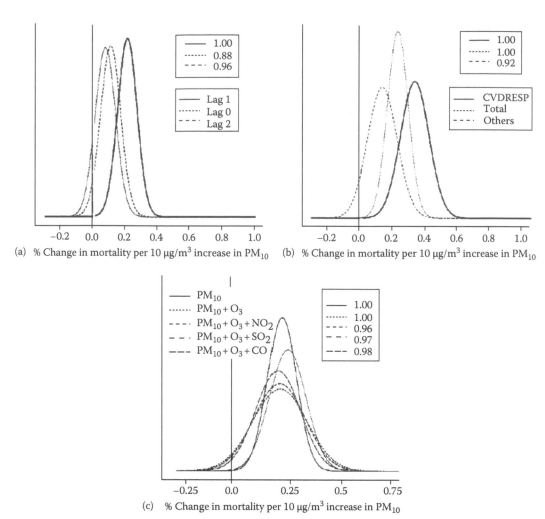

(a) % Change in mortality per 10 μg/m³ increase in PM₁₀   (b) % Change in mortality per 10 μg/m³ increase in PM₁₀

(c)   % Change in mortality per 10 μg/m³ increase in PM₁₀

FIGURE 23.24   Results from the revised NMMAPS analyses [310]. (a) Marginal posterior distribution of the national average effect of $10\,\mu g/m^3$ increase in PM_{10} on total, nonaccidental mortality based on different lags. (b) Comparative marginal posterior distribution of the national average effect of $10\,\mu g/m^3$ increase in PM_{10} at lag 1 on total, cardiorespiratory (CVDRESP), and "other" mortality. (c) Marginal posterior distribution of effect of $10\,\mu g/m^3$ increase in PM_{10} at lag 1 in models with other pollutants. Box at upper right includes posterior probabilities that PM_{10} effects are greater than 0, based on pooled city-specific estimates in a two-stage hierarchical model. ªRevised pooled estimates. (From Dominici, F. et al., Mortality among residents of 90 cities, Special Report: Revised Analyses of Time-Series Studies of Air Pollution and Health, Health Effects Institute, Ed., Health Effects Institute, Boston, MA, pp. 9–24, 2003.)

evidence of harvesting at the 60-day averaging window, and pneumonia showed evidence of short-term harvesting, followed by increased percentage increases (Figure 23.31). These data imply that COPD deaths are being brought forward by about 2 months, while deaths from pneumonia and IHD are not due to harvesting and may reflect "enrichment" of the at-risk pool as a consequence of persistent exposure to increased average levels of PM [242]. An analysis of the APHEA project, based on distributed lag models applied to ten cities, failed to find strong evidence for mortality displacement for total daily mortality (Figure 23.32) [244]. Moreover, the effect estimate for mortality for exposures 11–60 days before death was more than two-thirds as large as that for 10 days just prior to death (0.688 ± 0.261 vs. 0.922 ± 0.184, respectively—both estimates are $\times 10^3$). Similar results

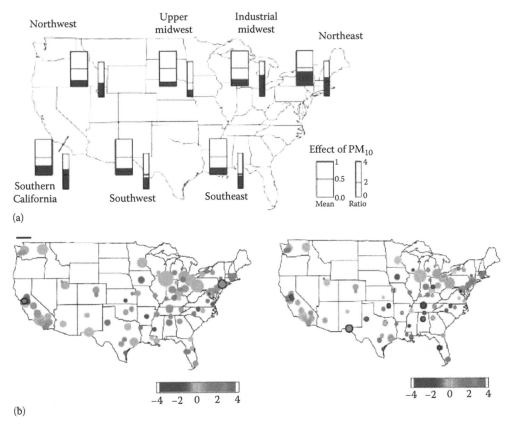

FIGURE 23.25 (a) Regional differences in estimated % change in daily total mortality per $10\,\mu g/m^3$ increase in PM_{10}. The height of the shaded area in the box on the left is the region-specific estimate. Shade area in the box on the right is the t-ratio (posterior mean divided by posterior standard deviation). (From Dominici, F. et al., Mortality among residents of 90 cities, *Special Report: Revised Analyses of Time-Series Studies of Air Pollution and Health*, Health Effects Institute, Ed., Health Effects Institute, Boston, MA, pp. 9–24, 2003.) (b) Within city total (left panel) and cardiovascular (right panel) mortality from NMMAPS. Color bars define ranges of % changes, and the sizes of the circles are proportional to the precision of the effect estimates. (From Dominici, F. et al., *Environ. Health Perspect.*, 111, 39, 2003.)

were observed by Dominici et al. [245] in a study of daily mortality in four U.S. cities between 1987 and 1994 (Figure 23.33). Results were similar for frequency domain and timescale estimates. In response to a critique of their approach [246], these authors evaluated 12 mortality rates (four cities and three categories of mortality) at timescales less than and greater than 5 days (Figure 23.34) [247]. In most cases, the estimates at greater than 5 days are greater than those at less than 5 days (Figure 23.34). While most of the data cited indicate that effects of ambient PM on daily mortality are largest for longer timescales, they do not negate the fact that shorter timescales are also associated with increased mortality. Nonetheless, the short-term associations do not appear to be the result of mortality displacement.

Although this chapter does not cover the extensive body of data related to the associations between PM and hospital admissions for respiratory and cardiovascular disease, these data have been used to support the causal connection between changes in ambient PM and daily mortality. Bates [143] argued that if the association was likely to be causal, then there should be increased risks for hospitalization for cardiorespiratory diseases. Such data have been compiled and were used to bolster the argument for the setting of the PM standard in 1996 (Figure 23.35).

TABLE 23.17

Selected Time Series Studies That Evaluate the Relationship between Changes in Specific Mass Components of PM Aerosol and Daily Mortality

Study Population	Years	Pollutants and Components	Methods	Results	References
Studies that report predominant effects for fine component of mass fraction					
Spokane, WA	1989–1996	Days with dust storms as surrogate for increased coarse PM, PM_{10}	PR^a with indicator for days with dust storms	RR for days after dust storms 1.01 (0.87–1.17)	[126]
Wasatch Front, UT	1985–1995	PM_{10} ($PM_{2.5}$ ~70%–90% of PM_{10}, $PM_{1.0}$), windblown dust episodes surrogate for $PM_{10-2.5}$	PR, GAM	Stagnant air episodes characterized by high concentrations of primary and secondary combustion source PM more associated with mortality than windblown dust episodes characterized by PM with larger contribution from crustal elements	[316]
Harvard Six Cities	1976–1987	PM_{10},[b] coarse PM, $PM_{2.5}$, H^+, SO_4^{2-}	PR, GAM	Combined % increase mortality over all cities/$10\,\mu g/m^3$ increase on same day PM metric $PM_{2.5}$ 1.3% (0.9–1.7) Coarse PM 0.4% (20.2, 0.9) (Results for two cities for course PM show comparable effects with $PM_{2.5}$)	[128]
Harvard Six Cities	1979–1988	Sources based on 15 elements, converted to five "factors"	PR, GAM	Percent increase in mortality/$10\,\mu g/m^3$ increase in mass concentration from specific source Mobile sources (fine PM) 3.4% (1.7–5.2) Crustal sources 22.3% (25.8, 1.2)	[121]
Meta-analysis based on 19 studies from United States	Studies span 1973–1990	PM_{10}, TSP, SO_2, CO, NO_2, O_3, $PM_{2.5}/PM_{10}$ ratio) for 14 U.S. cities	Random effects Bayes summaries	Percent change in mortality/$10\,\mu g/m^3$ increase in PM_{10} (ratio used directly) PM_{10} 0.67 (0.46–0.88) $PM_{2.5}/PM_{10}$ ratio as grouped variable: 0.31% (ratio <0.57), 0.68 (ratio 0.57–0.64), 0.81 (ratio >0.65) (95% CI excludes 0 for the highest two groups)	[229]

TABLE 23.17 (continued)
Selected Time Series Studies That Evaluate the Relationship between Changes in Specific Mass Components of PM Aerosol and Daily Mortality

Study Population	Years	Pollutants and Components	Methods	Results	References
Mexico City, Mexico	1993–1995	PM_{10}, SO_2, CO, NO_2, O_3	PR	Percent increase in mortality/10 µg/m³ in 5-day mean with and without changes ranges from 1.25% to 1.49%—all 95% CI includes 0 change $PM_{10-2.5}$% changes range from 4.07% to 4.28%—lower bounds of 95% CI >2%	[317]
Coachella Valley, CA[c]	1989–1998	PM_{10}, CO, NO_2, O_3, $PM_{2.5}$ (measured directly last 2.5 years) $PM_{10-2.5}$ (predicted)	PR	RR/IQR for respiratory/CVD mortality PM_{10} (lag 0) 1.03 (1.01–1.05[d]) $PM_{10-2.5}$ (lag 0) 1.02 (1.01–1.04) $PM_{2.5}$ (lag 4) 1.03 (0.98–1.09) All $PM_{2.5}$ RR include 1 in sensitivity analyses; other PM exclude 1	[133]
Santiago, Chile	1988–1996	$PM_{10-2.5}$, $PM_{2.5}$, SO_2, NO_2, CO, O_3	PR, GLM, and GAM	RR of death for increase in 2-day mean $PM_{2.5}$ 1.06 winter, 1.06 winter $PM_{10-2.5}$ 0.99 winter, 1.07 summer 95% CI for all RR >1 excludes one in single- and all two-pollutant models	[318]

Studies that report predominant effects for ultrafine fraction of PM aerosol

Erfurt, Germany	1995–1998	Particle number, PM mass fractions	PR, GLM, and GAM	See Figure 23.25. Authors interpret results as showing particle number effects. Commentary by funder raises questions about whether the number and mass effects can be separated	[138]

[a] See footnotes in Table 23.14.
[b] PM_{15} was measured until 1984. Coarse PM until 1985 was $PM_{15-2.5}$.
[c] This is a follow-up study to that in Ref. [301] and uses a longer time series and adds data on $PM_{2.5}$ to those of PM_{10}.
[d] 95% CI.

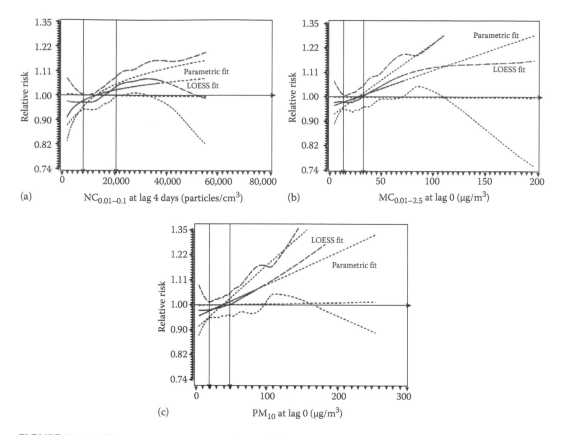

FIGURE 23.26 Exposure response curves for particle number count for particles between 0.1 and 1 μm (a), mass for particles between 0.1 and 2.5 μm (b), and PM$_{10}$ (c). (From Wichmann, H.E. et al., *Health Eff. Inst. Res. Rep.*, 98, 5, 2000.)

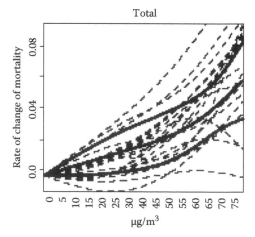

FIGURE 23.27 PM$_{10}$ total mortality exposure–response curve for the mean of the current and previous day's PM$_{10}$ concentrations for the 20 largest U.S. cities. Thick lines are the mean curve and 95% credible interval. Dashed lines are Bayesian estimates of the city-specific exposure–response curves. (From Dominici, F. et al., Shape of the exposure-response relation and mortality displacement in the NMMAPS study, *Special Report: Revised Analyses of Time-Series Studies of Air Pollution and Health*, Health Effects Institute, Ed., Health Effects Institute, Boston, MA, pp. 91–96, 2003.)

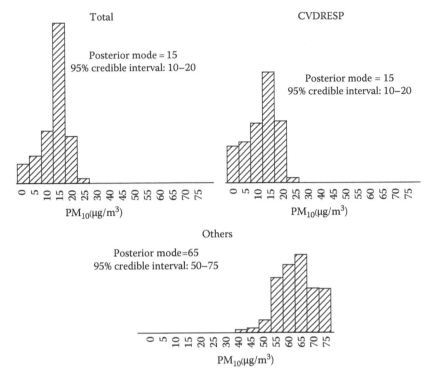

FIGURE 23.28 Posterior probabilities of the thresholds for groups of mortality based on the mean of the current and previous day's PM_{10} concentration for the 20 largest U.S. cities. (Adapted from Dominici, F. et al., Shape of the exposure-response relation and mortality displacement in the NMMAPS study, *Special Report: Revised Analyses of Time-Series Studies of Air Pollution and Health*, Health Effects Institute, Ed., Health Effects Institute, Boston, MA, pp. 91–96, 2003, Figure 4 and Table 3.)

23.5.4.2 Physiological Mechanisms Related to the Association of Daily Changes in PM and Daily Mortality

Given that mortality from cardiovascular and respiratory diseases seems most closely associated with daily changes in ambient PM, a number of studies have been undertaken to define the physiological pathways through which the pro-inflammatory, pro-oxidant effects of PM aerosol could result in the observed excess mortality.

23.5.4.2.1 Cardiovascular Mortality

Inflammation is thought to be an important component of the pathophysiology of atherosclerotic cardiovascular diseases [248]. The pro-inflammatory properties of the PM aerosol have been discussed in previous sections. In addition to any role played by inflammation in the sudden worsening of cardiovascular disease that leads to death, several other general mechanisms have emerged as being possibly relevant to the excess daily deaths from cardiovascular disease*: alterations in heart rate variability and increased cardiac arrhythmias, induction of a procoagulant state, and alteration of endothelial cell function.

The normal heart rate is quite variable as a result of control by the sympathetic and parasympathetic nervous systems [249]. Numerous studies have demonstrated that reduced heart rate variability is a risk factor for mortality after myocardial infarction (heart attack) and a general risk factor for mortality in the elderly [250,251]. A number of studies have been published that

---

* Mechanisms that could underlie cancer or disease related to inflammation have been discussed in earlier sections.

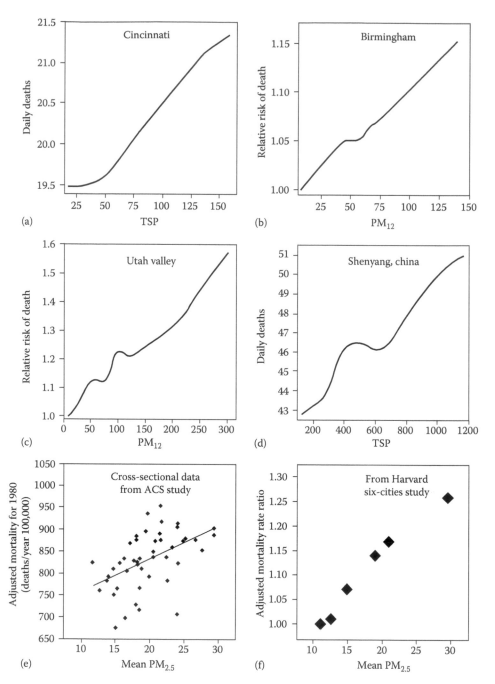

FIGURE 23.29 Plots of exposure–response relations from selected studies. (a–d) Nonparametric smooth curves of adjusted daily deaths or adjusted relative risk of death for selected daily time series mortality studies. (From Pope, C.A., *Am. J. Epidemiol.*, 152, 407, 2000, Figure 2.)

have found associations between increased levels of daily PM and reduced high-frequency heart rate variability (due to vagus nerve control) [252–257]. Gold et al. [254] evaluated changes in heart rate variability with 5 min of controlled outdoor exercise in subjects aged 53–87 years. These investigators observed an association between increased 4 h mean pre-exercise $PM_{2.5}$ concentration and decreased heart rate variability. A 221-day study in an elderly population in a

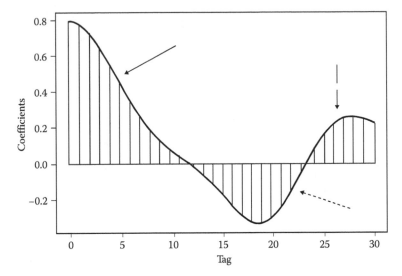

FIGURE 23.30 Schematic of mortality displacement. If frail individuals have the time of their deaths advanced by a few days due to increases in air pollution (solid arrow), the reduction of the at-risk pool will lead to a decrease in the number of deaths over ensuing days (dotted arrow). In the absence of any further increases, the hazard of death will return to its usual level of daily variability (dashed arrow). (Adapted from Zanobetti, A. et al., *Epidemiology*, 13, 87, 2002.)

retirement center observed increased odds (3.1, 95% CI: 1.4–6.6) of decreased heart rate variability on days during indoor $PM_{2.5}$ >15 $\mu g/m^3$ compared to days with PM levels ≤15 $\mu g/m^3$ [253]. A study in 20 healthy workers (nine smokers) from an occupational cohort (ages 21–58 years) who wore personal PM monitors for 24 h showed a 1.4% in heart rate variability per 100 $\mu g/m^3$ increase in the 3 h moving average $PM_{2.5}$ mean [256]. A study by the same investigators in 39 boilermakers who wore personal PM monitors observed that decreases in heart variability were associated with the lead and vanadium content of the $PM_{2.5}$ aerosol (vanadium, nickel, copper, chromium, lead, and manganese were tested) [257]. $PM_{2.5}$ mass did not show any association at any of the lags that were evaluated (see Table 4 of Ref. [257]). Either through decreases in heart rate variability or direct toxic effects on the conduction system of the heart, PM aerosol could also trigger cardiac arrhythmias that could increase the risk of death [258]. Support for this possibility in humans comes from a study of patients with implanted defibrillators [258]. The odds of increases in defibrillator discharges were associated with increases in $PM_{2.5}$, black carbon, NO_2, and O_3. The largest and most precisely estimated effects were seen for NO_2 (see Table 4 in Ref. [258]).

Acute heart attacks are thought to be initiated by the formation of a clot that obstructs the flow of blood in a blood vessel damaged by the atherosclerotic process, and anticoagulants are used routinely to reduce the risk of heart attacks [259]. Alterations in the coagulable state of blood have been suggested as one mechanism through which PM aerosol could increase the risk of death [137]. At least two studies have provided support for this hypothesis. Following a 1985 4-day episode of increased TSP and SO_2 in southern Germany, an increased risk of extreme values of plasma viscosity was observed in 3256 participants in a prospective randomized study of therapy for cardiac disease [260]. The results from this study were supported by those from a study of 112 subjects aged 60 years and older who provided repeated blood specimens over a period of 18 months (maximum of 12 specimens) [261]. Estimated personal exposure to PM_{10} over the 3 days prior to blood sampling produced changes in a set of hematological parameters that suggested sequestration of red blood cells in the circulation [261]. Such a phenomenon could increase the risk of clot formation in the presence of a hypercoagulable state.

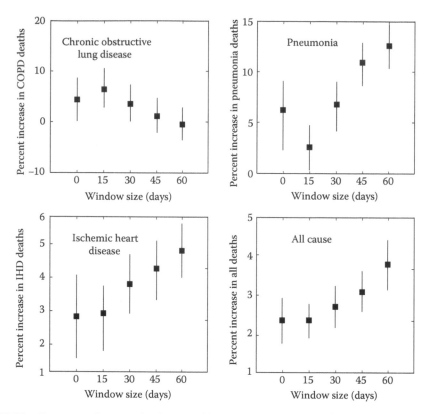

FIGURE 23.31 Percentage increase in class-specific mortality per $10\,\mu g/m^3$ increase in $PM_{2.5}$ in Boston, MA, for different smoothing windows of air $PM_{2.5}$. The figure shows increased percentage changes with longer $PM_{2.5}$ averaging times. Chronic obstructive lung disease shows some evidence of mortality displacement, with the percent change in death falling below 0 at a 60-day cycle length (see Figure 23.30). Pneumonia shows a trend to short-term harvesting, but larger effects with longer averaging times. Ischemic heart disease shows no change with the 15-day window and increasing percentage changes with progressively longer averaging times. (Adapted from Schwartz, J., *Am. J. Epidemiol.*, 151, 440, 2000.)

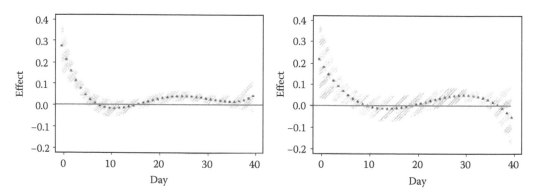

FIGURE 23.32 Shape of the association between PM_{10} and daily deaths fourth-degree distributed lag model (left panel) and a cubic-degree distributed lag model (right panel) in 10 cities of the APHEA-2 project, 1990–1997. Both fit with a random effect for city. In neither case is there strong evidence for mortality displacement (see Figure 23.30), since deaths fall to near 0 but then rise to a second peak of longer duration than the first peak. (Adapted from Zanobetti, A. et al., *Epidemiology*, 13, 87, 2002.)

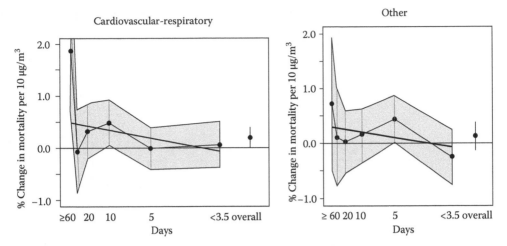

FIGURE 23.33 Pooled estimates of log relative rates of mortality at different timescales for cardiovascular-respiratory (left panel) and other causes of mortality (right panel) for four U.S. cities (Pittsburgh, PA; Minneapolis, MN; Seattle, WA; Chicago, IL), 1987–1994. Shaded regions are ±2 standard errors of the estimates. Percentage increases in mortality were greater at longer than shorter exposure intervals before death. (From Dominici, F. et al., *Am. J. Epidemiol.*, 157, 1055, 2003.)

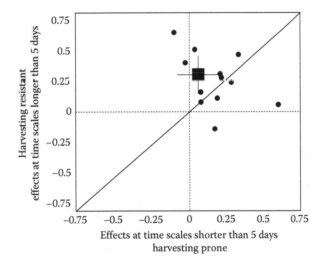

FIGURE 23.34 Estimated relative rates of mortality due to PM at timescales shorter than 5 days (harvesting-prone) versus relative rate estimates obtained at timescales longer than 5 days (harvesting-resistant) from four U.S. cities (Pittsburgh, PA; Minneapolis, MN; Seattle, WA; Chicago, Il) and three mortality outcomes (total, cardiovascular-respiratory, and other causes), 1987–1994. A large square is placed at the averages of the 12 estimates. (From Dominici, F. et al., *Am. J. Epidemiol.*, 157, 1055, 2003.)

The lining cells of blood vessels (endothelial cells) are known to play an important role in the regulation of blood flow and vascular tone. Studies in rats have observed alterations in the production of endothelin 1, a potent constrictor of blood vessels that is produced by endothelial cells after exposure to concentrated ambient PM (CAPS) and O_3 in combination [117] and vasoconstriction of small pulmonary arteries with CAPS alone [262]. The relevance of these observations to humans was demonstrated with controlled 2 h exposure of 25 healthy adults to a mixture of CAPS and O_3 compared to exposure to filtered air [263]. Brachial artery vasoconstriction was observed after the CAPS1O$_3$ exposure but not with filtered air.

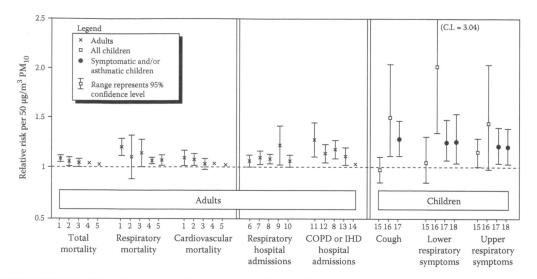

FIGURE 23.35 RRs of various health outcomes per 50 μg/m³ increase in PM₁₀ for adults and children. Each data point and error bar represents a separate study. (From Federal Register, National ambient air quality standard for particulate matter, Report No.: 40 CRF Part 50, National Archives and Records Administration, Washington, DC, December 1996.)

Thus, while the definitive physiological mechanisms that might underlie the association between increases in daily cardiovascular mortality with daily increase in PM aerosol are not known definitively, what is known is supportive of the fact that the association is likely to be causal and not due to some unmeasured confounding factors or some subtle problem with the statistical analysis approaches that have been employed by epidemiological studies. A similar case can be made for deaths from chronic lung disease, based on the expected results of the pro-inflammatory properties of the ambient PM aerosol.

23.6   CONCLUSIONS

This chapter has focused largely on the effects of the nonradioactive, nonoccupational, ambient PM aerosol and its likely effect on human health. The chapter has been representative, rather than exhaustive, in its synthesis of current data. Only selected areas related to human health have been discussed. The most exhaustive review of the material presented herein can be found in the descriptive sections of the most recent public release of the draft document for PM standard being circulated for review by the U.S. EPA [264].

While there is little controversy about the effects of the bioaerosol component of the ambient PM aerosol on human health, there has been more skepticism about the effects of the anthropogenic component of the PM aerosol on human health, as clearly indicated by epidemiological studies. However, there now appears a sufficient body of basic biological and physiological mechanistic research data that makes it far more likely that the observed associations between the anthropogenic component of the PM aerosol and human health are causal than are the methodological issues related to epidemiological study designs, statistical analysis, and issues related to exposure measurement error.

REFERENCES

1. Donaldson, K., Beswick, P.H., and Gilmour, P.S., Free radical activity associated with the surface of particles: A unifying factor in determining biological activity? *Toxicol. Lett.*, 88, 293–298, 1996.
2. U.S. Environmental Protection Agency (EPA), Air quality criteria for particulate matter—Third external review draft, EPA 600/P-99/002aC, Report No.: EPA 600/P-99/002aC, Research Triangle Park, NC, 2001.

3. Neas, L.M., Dockery, D.W., Burge, H., Koutrakis, P., and Speizer, F.E., Fungus spores, air pollutants and other determinants of peak expiratory flow, *Am. J. Epidemiol.*, 143, 797–807, 1996.

4. Delfino, R.J., Zeiger, R.S., Seltzer, J.M., Street, D.H., Matteucci, R.M., Anderson, P.R. et al., The effect of outdoor fungal spore concentrations on daily asthma severity, *Environ. Health Perspect.*, 105, 622–635, 1997.

5. Downs, S.H., Mitakakis, T.Z., Marks, G.B., Car, N.G., Belousova, E.G., Leuppi, J.D. et al., Clinical importance of Alternaria exposure in children, *Am. J. Respir. Crit. Care Med.*, 164, 455–459, 2001.

6. Soukup, J.M. and Becker, S., Human alveolar macrophage responses to air pollution particulates are associated with insoluble components of coarse material, including particulate endotoxin, *Toxicol. Appl. Pharmacol.*, 171, 20–26, 2001.

7. Schwartz, D.A., Does inhalation of endotoxin cause asthma? *Am. J. Respir. Crit. Care Med.*, 163, 305–306, 2001.

8. Monn, C. and Becker S., Cytotoxicity and induction of proinflammatory cytokines from human monocytes exposed to fine ($PM_{2.5}$) and course particles ($PM_{10-2.5}$) in outdoor and indoor air, *Toxicol. Appl. Pharmacol.*, 155, 245–252, 1999.

9. Monn, C., Naef, R., and Koller, T., Reactions of macrophages exposed to particles <10 microm., *Environ. Res.*, 91, 35–44, 2003.

10. Wallace, L., Indoor particles: A review, *J. Air Waste Manage. Assoc.*, 46, 98–126, 1999.

11. Abt, E., Suh, H.H., Catalano, P., and Koutrakis, P., Relative contribution of outdoor and indoor particles sources to indoor concentrations, *Environ. Sci. Technol.*, 34, 3579–3587, 2002.

12. Leaderer, B.P., Naeher, L., Jankun, T., Balenger, K., Holford, T.R., Toth, C. et al., Indoor, outdoor, and regional summer and winter concentrations of PM_{10}, $PM_{2.5}$, $SO_4{}^{2}$, H^1, $NH_4{}^1$, $NO_3{}^2$, $NH_3{}^2$, and nitrous acid in homes with and without kerosene space heaters, *Environ. Health Perspect.*, 107, 223–231, 1999.

13. Patterson, E. and Eatough, D.J., Indoor/outdoor relationships for ambient $PM_{2.5}$ and associated pollutants: Epidemiological implications in Lindon, Utah, *J. Air Waste Manage. Assoc.*, 50, 103–110, 2000.

14. Wainman, T., Zhang, J., Weschler C.J., and Lioy, P.J., Ozone and limonene in indoor air: A source of submicron particle exposure, *Environ. Health Perspect.*, 108, 1139–1145, 2000.

15. Spengler, J.D., Long-term measurements of respirable sulfates and particles inside and outside homes, *Atmos. Environ.*, 15, 23–30, 1981.

16. National Research Council, *Environmental Tobacco Smoke: Measuring Exposures and Assessing Health Effects*, National Academy Press, Washington, DC, 1986.

17. National Cancer Institute, Monograph 10: Health effects of exposure to environmental tobacco smoke, National Institutes of Health-National Cancer Institute, 2003, http:/cancercontrol.cancer.gov/tcrb/monographs/10/m10_complete.pdf

18. Abt, E., Suh, H.H., Allen, G., and Koutrakis, P., Characterization of indoor particle sources: A study conducted in the metropolitan Boston area, *Environ. Health Perspect.*, 108, 35–44, 2002.

19. U.S. Environmental Protection Agency (EPA), Air quality criteria for particulate matter, Report No: EPA/600/P-95-001bF, Washington, DC, April 1996.

20. Miller, F.J., Dosimetry of particles in laboratory animals and humans in relationship to issues surrounding lung overload and human health risk assessment, *Inhal. Toxicol.*, 12, 19–57, 2000.

21. Brain, J.D. and Valberg, P.A., Deposition of aerosol in the respiratory tract, *Am. Rev. Respir. Dis.*, 120, 1325–1373, 1979.

22. Yeates, D.B. and Mortensen, J., Deposition and clearance, in Murray, J.F., Nadel, J.A., Mason, R.J., and Boushey, H.A., Eds., *Textbook of Respiratory Medicine*, W.B. Saunders Co., New York, 2000, pp. 349–384.

23. Churg, A. and Vidal, S., Carinal and tubular airway particle concentrations in the large airways of nonsmokers in the general population: Evidence of high particle concentration at airway carinas, *Occup. Environ. Med.*, 53, 553–558, 1996.

24. Brauer, M., Avila-Casado, C., Fortoul, T.I., Vedal, S., Stevens, B., and Churg, A., Air pollution and retained particles in the lung, *Environ. Health Perspect.*, 109, 1039–1043, 2001.

25. Schlesinger, R.B., Deposition and clearance of inhaled particles, in McClellan, R.O. and Henderson, R.F., Eds., *Concepts in Inhalation Toxicology*, Taylor & Francis, Washington, DC, 1995, pp. 191–224.

26. Pinkerton, K.E., Green, F.H., Saiki, C., Vallyathan, V., Plopper, C.G., Gopal, V. et al., Distribution of particulate matter and tissue remodeling in the human lung, *Environ. Health Perspect.*, 108, 1063–1069, 2000.

27. Kim, C.S., Lewars, G.A., and Sackner, M.A., Measurement of total lung aerosol deposition as an index of lung abnormality, *J. Appl. Physiol.*, 64, 1527–1536, 1988.

28. U.S. Environmental Protection Agency (EPA), Air quality criteria for particulate matter Volume II, in Air Quality Criteria for Particulate Matter, Volumes I–III, EPA/600/P-95-001bF, National Center for Environmental Assessment Office of Research and Development, U.S. EPA, Washington, DC, 1996.

29. Wanner, A., Salathe, M., and O'Riordan, T.G., Mucociliary clearance in the airways, *Am. J. Respir. Crit. Care Med.*, 154(6 Part 1), 1868–1902, 1996.

30. Nemmar, A., Hoet, P.H., Vanquickenborne, B., Dinsdale, D., Thomeer, M., Hoylaerts, M.F. et al., Passage of inhaled particles into the blood circulation in humans, *Circulation* 105, 411–414, 2002.

31. Ilowite, J.S., Bennett, W.D., Sheetz, M.S., Groth, M.L., and Nierman, D.M., Permeability of the bronchial mucosa to 99mTc-DTPA in asthma, *Am. Rev. Respir. Dis.*, 139, 1139–1143, 1989.

32. Koenig, J.Q., Larson, T.V., Hanley, Q.S., Robelledo, V., Dumler, K., Checkoway, H. et al., Pulmonary function changes in children associated with fine particulate matter, *Environ Res.*, 63, 26–38, 1993.

33. Daigle, C.C., Chalupa, D.C., Gibb, F.R., Morrow, P.E., Oberdorster, G., Utell, M.J. et al., Ultrafine particle deposition in humans during rest and exercise, *Inhal. Toxicol.*, 15, 539–552, 2003.

34. Ward, P.A., Warren, J.S., and Johnson, K.J., Oxygen radicals, inflammation, and tissue injury, *Free Radic. Biol. Med.*, 5, 403–408, 1988.

35. Poli, G., Introduction—Serial review: Reactive oxygen and nitrogen in inflammation(1,2), *Free Radic. Biol. Med.*, 33, 301–302, 2002.

36. Nauseef, W.M. and Clark, R.A., Granulocyte phagocytosis, in Mandell, G.L., Bennett, J.E., and Dolin, R., Eds., *Principles and Practices of Infectious Diseases*, 5th edn., Churchill Livingston, New York, 2000, pp. 89–112.

37. Goldsmith, C.A., Frevert, C., Imrich, A., Sioutas, C., and Kobzik, L., Alveolar macrophage interaction with air pollution particulates, *Environ. Health Perspect.*, 105(Suppl. 5), 1191–1195, 1997.

38. Li, N., Sioutas, C., Cho, A., Schmitz, D., Misra, C., Sempf, J. et al., Ultrafine particulate pollutants induce oxidative stress and mitochondrial damage, *Environ. Health Perspect.*, 111, 455–460, 2003.

39. Gilmour, P.S., Brown, D.M., Lindsay, T.G., Beswick, P.H., MacNee, W., and Donaldson, K., Adverse health-effects of PM(10) particles—Involvement of iron in the generation of hydroxyl radicals, *Occup. Environ. Med.*, 53, 817–822, 1996.

40. Chang, L.-Y. and Crapo, J.D., Inhibition of airway inflammation and hyperreactivity by an antioxidant mimetic, *Free Radic. Biol. Med.*, 33, 379–386, 2002.

41. McCord, J.M., Human disease, free radicals, and the oxidant/antioxidant balance, *Clin. Biochem.*, 26, 351–357, 1993.

42. Nauss, K.M. and the HEI Diesel Working Group, Critical issues in assessing the carcinogenicity of diesel exhaust: A synthesis of current knowledge, in Institute H.E., Ed., *Diesel Exhaust: A Critical Analysis of Emissions, Exposure and Health Effects*, Health Effects Institute, Boston, MA, 1995, pp. 13–61.

43. Schroeder, W.H., Dobson, M., Kane, D.M., and Johnson, N.D., Toxic trace elements associated with airborne particulate matter: A review, *J. Air Pollut. Control Assoc.*, 37, 1267–1285, 1987.

44. Ghio, A.J. and Samet, J.M., Metals and air pollution particles, in Holgate, S.T., Samet, J.M., Koren, H.S., and Maynard, R.L., Eds., *Air Pollution and Health*, Academic Press, New York, 1999, pp. 635–651.

45. Ghio, A.J., Meng, H.H., Hatch, G.E., and Costa, D.L., Luminol-enhanced chemiluminescence after *in vitro* exposures of rat alveolar macrophages to oil fly ash is metal dependent, *Inhal. Toxicol.*, 9, 255–271, 1997.

46. Ghio, A.J., Stonehuerner, J., Pritchard, R.J., Piantodosi, C.A., Dreher, K.L., and Costa, D.L., Humic-like substances in air pollution particulates correlate with concentrations of transition metals and oxidant generation, *Inhal. Toxicol.*, 8, 479–494, 1996.

47. Pritchard, R.J., Ghio, A.J., Lehmann, J.R., Winsett, D.W., Tepper, J.S., Park, P. et al., Oxidant generation and lung injury after particulate air pollutant exposure increase with concentration of associated metals, *Inhal. Toxicol.*, 8, 457–477, 1996.

48. Dreher, K.L., Jaskot, R.H., Lehmann, J.R., Richards, J.H., McGee, J.K., Ghio, A.J. et al., Soluble transition metals mediate residual oil fly ash induced acute lung injury, *J. Toxicol. Environ. Health.*, 50, 285–305, 1997.

49. Dye, J.A., Lehmann, J.R., McGee, J.K., Winsett, D.W., Ledbetter, A.D., Everitt, J.I. et al., Acute pulmonary toxicity of particulate matter filter extracts in rats: Coherence with epidemiologic studies in Utah Valley residents. *Environ. Health Perspect.*, 109(Suppl. 3), 395–403, 2001.

50. Frampton, M.W., Ghio, A.J., Samet, J.M., Carson, J.L., Carter, J.D., and Devlin, R.B., Effects of aqueous extracts of PM(10) filters from the Utah valley on human airway epithelial cells. *Am. J. Physiol.*, 277(5 Part 1), L960–L967, 1999.

51. Aust, A.E., Ball, J.C., Hu, A.A., Lighty, J.S., Smith, K.R., Straccia, A.M. et al., Particle characteristics responsible for effects on human lung epithelial cells, *Res. Rep. Health Eff. Inst.*, 110, 1–65, 2002, discussion 67–76.

52. Li, N., Wang, M., Oberley, T.D., Sempf, J.M., and Nel, A.E., Comparison of the pro-oxidative and pro-inflammatory effects of organic diesel particle chemical in bronchial epithelial cells and macrophages, *J. Immunol.*, 169, 4531–4541, 2002.

53. Shi, T., Knaapen, A.M., Begerow, J., Birmili, W., Borm, P.J., and Schins, R.P., Temporal variation of hydroxyl radical generation and 8-hydroxy-2'-deoxyguanosine formation by coarse and fine particulate matter, *Occup. Environ. Med.*, 60, 315–321, 2003.

54. Nel, A.E., Diaz-Sanchez, D., and Li, N., The role of particulate pollutants in pulmonary inflammation and asthma: Evidence for the involvement of organic chemicals and oxidative stress, *Curr. Opin. Pulm. Med.*, 7, 20–26, 2001.

55. Li, N., Venkatesan, M.I., Miguel, A., Kaplan, R., Gujuluva, C, Alam, J. et al., Induction of heme oxygenase-1 expression in macrophages by diesel exhaust particle chemicals and quinones via the antioxidant-responsive element, *J. Immunol.*, 165, 3393–3401, 2000.

56. Li, N., Kim, S., Wang, M., Froines, J., Sioutas, C., and Nel, A., Use of a stratified oxidative stress model to study the biological effects of ambient concentrated and diesel exhaust particulate matter, *Inhal. Toxicol.*, 14, 459–486, 2002.

57. Nel, A.E., Diaz-Sanchez, D., Ng, D., Hiura, T., and Saxon, A., Enhancement of allergic inflammation by the interaction between diesel exhaust particles and the immune system, *J. Allergy Clin. Immunol.*, 102, 539–554, 1998.

58. Hiura, T.S., Kaszubowski, M.P., Li, N., and Nel, A.E., Chemicals in diesel exhaust particles generate reactive oxygen radicals and induce apoptosis in macrophages, *J. Immunol.*, 163, 5582–5591, 1999.

59. Hiura, T.S., Li, N., Kaplan, R., Horwitz, M., Seagrave J.C., and Nel, A.E., The role of a mitochondrial pathway in the induction of apoptosis by chemicals extracted from diesel exhaust particles, *J. Immunol.*, 165, 2703–2711, 2000.

60. Hashimoto, S., Gon, Y., Takeshita, I., Matsumoto, K., Jibiki, I., Takizawa, H. et al., Diesel exhaust particles activate p38 MAP kinase to produce interleukin 8 and RANTES by human bronchial epithelial cells and N-acetylcysteine attenuates p38 kinase activation, *Am. J. Respir. Crit. Care Med.*, 161, 280–285, 2000.

61. Baulig, A., Garlatti, M., Bonvallot, V., Marchand, A., Barouki, R., Marano, F. et al., Involvement of reactive oxygen species in the metabolic pathways triggered by diesel exhaust particles in human airway epithelial cells, *Am. J. Physiol. Lung Cell Mol. Physiol.*, 285, L671–L679, 2003.

62. Gilmour, P.S., Donaldson, K., and MacNee, W., Overview of antioxidant pathways in relation to the effects of air pollution, *Eur. Respir. Monogr.*, 21, 241–261, 2002.

63. van der Vliet, A. and Cross, C.E., Oxidants, nitrosants, and the lung, *Am. J. Med.*, 109, 398–421, 2000.

64. Hoidal, J.R., Reactive oxygen species and cell signaling, *Am. J. Respir. Cell Mol. Biol.*, 25, 661–663, 2001.

65. Vargas, L., Patino, P.J., Montoya, F., Vanegas, A.C., Echavarria, A., and Garcia, de Olarte, D., A study of granulocyte respiratory burst in patients with allergic bronchial asthma, *Inflammation*, 22, 45–54, 1998.

66. Barnes, P.J., Chung, K.F., and Page, C.P., Inflammatory mediators of asthma: Update, *Pharmacol. Rev.*, 50, 515–596, 1998.

67. Ma, J.Y. and Ma, J.K., The dual effect of the particulate and organic components of diesel exhaust particles on the alteration of pulmonary immune/inflammatory responses and metabolic enzymes, *J. Environ. Sci. Health C Environ. Carcinog. Ecotoxicol. Rev.*, 20, 117–147, 2002.

68. Shukla, A., Timblin, C., BeruBe, K., Gordon, T., McKinney, W., Driscoll, K. et al., Inhaled particulate matter causes expression of nuclear factor (NF)-kappaB-related genes and oxidant-dependent NF-kappaB activation *in vitro*, *Am. J. Respir. Cell Mol. Biol.*, 23, 182–187, 2000.

69. Brown, D.M., Stone, V., Findlay, P., MacNee, W., and Donaldson, K., Increased inflammation and intracellular calcium caused by ultrafine carbon black is independent of transition metals or other soluble components, *Occup. Environ. Med.*, 57, 685–691, 2000.

70. Donaldson, K., Brown, D., Clouter, A., Duffin, R., MacNee, W., Renwick, L. et al., The pulmonary toxicology of ultrafine particles, *J. Aerosol Med.*, 15, 213–520, 2002.

71. O'Grady, N.P., Preas, H.L., Pugin, J., Fiuza, C., Tropea, M., Reda, D. et al., Local inflammatory responses following bronchial endotoxin instillation in humans, *Am. J. Respir. Crit. Care Med.*, 163, 1591–1598, 2001.

72. Becker, S., Soukup, J.M., Gilmour, M.I., and Devlin, R.B., Stimulation of human and rat alveolar macrophages by urban air particulates: Effects of oxidant radical generation and cytokine production, *Toxicol. Appl. Pharmacol.*, 141, 637–648, 1996.

73. Monick, M.M. and Hunninghake, G.W., Activation of second messenger pathways in alveolar macrophages by endotoxin, *Eur. Respir. J.*, 20, 210–222, 2002.

74. Whitekus, M.J., Li, N., Zhang, M., Wang, M., Horwitz, M.A., Nelson, S.K. et al., Thiol antioxidants inhibit the adjuvant effects of aerosolized diesel exhaust particles in a murine model for ovalbumin sensitization, *J. Immunol.*, 168, 2560–2567, 2002.

75. Diaz-Sanchez, D., Garcia, M.P., Wang, M., Jyrala, M., and Saxon, A., Nasal challenge with diesel exhaust particles can induce sensitization to a neoallergen in the human mucosa, *J. Allergy Clin. Immunol.*, 104, 1183–1188, 1999.

76. Diaz-Sanchez, D., Dotson, A.R., Takenaka, H., and Saxon, A., Diesel exhaust particles induce local IgE production *in vivo* and alter the pattern of IgE messenger RNA isoforms, *J. Clin. Invest.*, 94, 1417–1425, 1994.

77. Diaz-Sanchez, D., Tsien, A., Fleming J., and Saxon, A., Combined diesel exhaust particulate and ragweed allergen challenge markedly enhances human *in vivo* nasal ragweed-specific IgE and skews cytokine production to a T helper cell 2-type pattern, *J. Immunol.*, 158, 2406–2413, 1997.

78. Diaz-Sanchez, D., The role of diesel exhaust particles and their associated polyaromatic hydrocarbons in the induction of allergic airway disease, *Allergy*, 52, 52–56, 1997.

79. Pandya, R.J., Solomon, G., Kinner, A., and Balmes, J.R., Diesel exhaust and asthma: Hypotheses and molecular mechanisms of action, *Environ. Health Perspect.*, 110(Suppl. 1), 103–112, 2002.

80. Lambert, A.L., Dong, W., Winsett, D.W., Selgrade, M.K., and Gilmour, M.I., Residual oil fly ash exposure enhances allergic sensitization to house dust mite, *Toxicol. Appl. Pharmacol.*, 158, 269–277, 1999.

81. Lambert, A.L., Dong, W., Selgrade, M.K., and Gilmour, M.I., Enhanced allergic sensitization by residual oil fly ash particles is mediated by soluble metal constituents, *Toxicol. Appl. Pharmacol.*, 165, 84–93, 2000.

82. Kumagai, Y., Arimoto, T., Shinyashiki, M., Shimojo, N., Nakai Y., Yoshikawa, T. et al., Generation of reactive oxygen species during interaction of diesel exhaust particle components with NADPH-cytochrome P450 reductase and involvement of the bioactivation in the DNA damage, *Free Radic. Biol. Med.*, 22, 479–487, 1997.

83. Seagrave, J., McDonald, J.D., Gigliotti, A.P., Nikula, K.J., Seilkop, S.K., Gurevich, M. et al., Mutagenicity and *in vivo* toxicity of combined particulate and semivolatile organic fractions of gasoline and diesel engine emissions, *Toxicol. Sci.*, 70, 212–226, 2002.

84. Moller, P., Daneshvar, B., Loft, S., Wallin, H., Poulsen, H.E., Autrup, H. et al., Oxidative DNA damage in vitamin C-supplemented guinea pigs after intratracheal instillation of diesel exhaust particles, *Toxicol. Appl. Pharmacol.*, 189, 39–44, 2003.

85. van Eeden, S.F., Tan, W.C., Suwa, T., Mukae, H., Terashima, T., Fujii, T. et al., Cytokines involved in the systemic inflammatory response induced by exposure to particulate matter air pollutants (PM10), *Am. J. Respir. Crit. Care Med.*, 164, 826–830, 2001.

86. Jimenez, L.A., Drost, E.M., Gilmour, P.S., Rahman, I., Antonicelli, F., Ritchie, H. et al., PM(10)-exposed macrophages stimulate a proinflammatory response in lung epithelial cells via TNF-alpha, *Am. J. Physiol. Lung Cell. Mol. Physiol.*, 282, L237–L248, 2002.

87. Kennedy, T., Ghio, A.J., Reed, W., Samet, J., Zagorski, J., Quay, J. et al., Copper-dependent inflammation and nuclear factor-kappaB activation by particulate air pollution, *Am. J. Respir. Cell Mol. Biol.*, 19, 366–378, 1998.

88. Quay, J.L., Reed, W., Samet, J., and Devlin, R.B., Air pollution particles induce IL-6 gene expression in human airway epithelial cells via NF-kappaB activation, *Am. J. Respir. Cell Mol. Biol.*, 19, 98–106, 1998.

89. Fujii, T., Hayashi, S., Hogg, J.C., Vincent, R., and van Eeden, S.F., Particulate matter induces cytokine expression in human bronchila epithelial cells, *Am. J. Respir. Cell Mol. Biol.*, 25, 265–271, 2001.

90. Salvi, S.S., Nordenhall, C., Blomberg, A., Rudell, B., Pourazar, J., Kelly, F.J. et al., Acute exposure to diesel exhaust increases IL-8 and GRO-alpha production in healthy human airways, *Am. J. Respir. Crit. Care Med.*, 161(2 Part 1), 550–557, 2000.

91. Nightingale, J.A., Maggs, R., Cullinan, P., Donnelly, L.E., Rogers, D.F., Kinnersley, R. et al., Airway inflammation after controlled exposure to diesel exhaust particulates, *Am. J. Respir. Crit. Care Med.*, 162, 161–166, 2000.

92. Mukae, H., Hogg, J.C., English, D., Vincent, R., and van Eeden, S.F., Phagocytosis of particulate air pollutants by human alveolar macrophages stimulates the bone marrow, *Am. J. Physiol. Lung Cell Mol. Physiol.*, 279, L924–L931, 2000.

93. Mukae, H., Vincent, R., Quinlan, K., English, D., Hards, J., Hogg, J.C. et al., The effect of repeated exposure to particulate air pollution (PM10) on the bone marrow. *Am. J. Respir. Crit. Care Med.*, 163, 201–209, 2001.

94. Gardner, S.Y., Lehmann, J.R., and Costa, D.L., Oil fly ash-induced elevation of plasma fibrinogen levels in rats, *Toxicol Sci.*, 56, 175–180, 2000.

95. Tan, W.C., Qiu, D., Liam, B.L., Ng, T.P., Lee, S.H., van Eeden, S.F. et al., The human bone marrow response to acute air pollution caused by forest fires, *Am. J. Respir. Crit. Care Med.*, 161(4 Part 1), 1213–1217, 2000.

96. Gong, H., Jr., Linn, W.S., Sioutas, C., Terrell, S.L., Clark, K.W., Anderson, K.R. et al., Controlled exposures of healthy and asthmatic volunteers to concentrated ambient fine particles in Los Angeles. *Inhal. Toxicol.*, 15, 305–325, 2003.

97. Veronesi, B. and Oortgiesen, M., Neurogenic inflammation and particulate matter (PM) airway pollutants, *NeuroToxicology*, 22, 795–810, 2001.

98. Tsien, A., Diaz-Sanchez, D., Ma, J., and Saxon, A., The organic component of diesel exhaust particles and phenanthrene, a major polyaromatic hydrocarbon constituent, enhances IgE production by IgE-secreting EBV-transformed human B cells *in vitro*, *Toxicol. Appl. Pharmacol.*, 142, 256–263, 1997.

99. Gavett, S.H., Madison, S.L., Stevens, M.A., and Costa, D.L., Residual oil fly ash amplifies allergic cytokines, airway responsiveness, and inflammation in mice, *Am. J. Respir. Crit. Care Med.*, 160, 1897–1904, 1999.

100. Yang, H.M., Antonini, J.M., Barger, M.W., Butterworth, L., Roberts, B.R., Ma, J.K. et al., Diesel exhaust particles suppress macrophage function and slow the pulmonary clearance of *Listeria monocytogenes* in rats, *Environ. Health Perspect.*, 109, 515–521, 2001.

101. Yin, X.J., Schafer, R., Ma, J.Y., Antonini, J.M., Weissman, D.D., Siegel, P.D. et al., Alteration of pulmonary immunity to *Listeria monocytogenes* by diesel exhaust particles (DEPs). I. Effects of DEPs on early pulmonary responses, *Environ. Health Perspect.*, 110, 1105–1111, 2002.

102. Yin, X.J., Schafer, R., Ma, J.Y., Antonini, J.M., Roberts, J.R., Weissman, D.N. et al., Alteration of pulmonary immunity to *Listeria monocytogenes* by diesel exhaust particles (DEPs). II. Effects of DEPs on T-cell-mediated immune responses in rats, *Environ. Health Perspect.*, 111, 524–530, 2003.

103. Yang, H.M., Barger, M.W., Castranova, V., Ma, J.K., Yang, J.J., and Ma, J.Y., Effects of diesel exhaust particles (DEP), carbon black, and silica on macrophage responses to lipopolysaccharide: Evidence of DEP suppression of macrophage activity, *J. Toxicol. Environ. Health A.*, 58, 261–278, 1999.

104. Kay, A.B., T cells as orchestrators of the asthmatic response, *Ciba Found. Symp.*, 206, 56–67, 1997, discussion 67–70, 106–110.

105. Becker, S. and Soukup, J.M., Exposure to urban air particulates alters the macrophage-mediated inflammatory response to respiratory viral infection, *J. Toxicol. Environ. Health A*, 57, 445–457, 1999.

106. Myatt, T.A. and Milton, D.K., Endotoxins, in Spencer, J.P., Samet, J.M., and McCarthy, J.F., Eds., *Indoor Air Quality Handbook*, McGraw-Hill, New York, 2000, pp. 42.1–42.14.

107. Koppelman, G.H., Reijmerink, N.E., Colin Stine, O., Howard, T.D., Whittaker, P.A., Meyers, D.A. et al., Association of a promoter polymorphism of the CD14 gene and atopy, *Am. J. Respir. Crit. Care Med.*, 163, 965–969, 2001.

108. Barton, G.M. and Medzhitov, R., Toll-like receptor signaling pathways, *Science*, 300, 1524–1525, 2003.

109. Centers for Disease Control and Prevention, Surveillance for asthma—United States, 1980–1999, *MMWR*, 51, 1–14, 2002.

110. Salvi, S.S., Babu, K.S., and Holgate, S.T., Is asthma really due to a polarized T cell response toward a helper T cell type 2 phenotype? *Am. J. Respir. Crit. Care Med.*, 164(8 Part 1), 1343–1346, 2001.

111. Magnan, A.O., Mely, L.G., Camilla, C.A., Badier, M.M., Montero-Julian, F.A., Guillot, C.M. et al., Assessment of the Th1/Th2 paradigm in whole blood in atopy and asthma. Increased IFN-gamma-producing CD8(1) T cells in asthma, *Am. J. Respir. Crit. Care Med.*, 161, 1790–1796, 2000.

112. Holt, P.G., OK, P., Holt, B.J., Upham, J.W., Baron-Hay, M.J., Suphioglu, C. et al., T-cell "priming" against environmental allergens in human neonates: Sequential deletion of food antigen reactivity during infancy with concomitant expansion of responses to ubiquitous inhalant allergens, *Pediatr. Allergy Immunol.*, 6, 85–90, 1995.

113. Holt, P.G., Yabuhara, A., Prescott, S., Venaille, T., Macaubas, C., Holt, B.J. et al., Allergen recognition in the origin of asthma, in Chadwick, D.J. and Cardew, G., Eds., *The Rising Trends in Asthma: Ciba Foundation Symposium 206*, John Wiley & Sons, New York, 1997, pp. 35–55.

114. Nel, A.E., T-cell activation through the antigen receptor. Part 1: Signaling components, signaling pathways, and signal integration at the T-cell antigen receptor synapse, *J. Allergy Clin. Immunol.*, 109, 758–770, 2002.

115. Nel, A.E. and Slaughter, N., T-cell activation through the antigen receptor. Part 2: Role of signaling cascades in T-cell differentiation, anergy, immune senescence, and development of immunotherapy, *J. Allergy Clin. Immunol.*, 109, 901–915, 2002.

116. Madden, M.C., Richards, J.H., Dailey, L.A., Hatch, G.E., and Ghio, A.J., Effect of ozone on diesel exhaust particle toxicity in rat lung, *Toxicol. Appl. Pharmacol.*, 168, 140–148, 2000.

117. Bouthillier, L., Vincent, R., Goegan, P., Adamson, I.Y., Bjarnason, S., Stewart, M. et al., Acute effects of inhaled urban particles and ozone: Lung morphology, macrophage activity, and plasma endothelin-1, *Am. J. Pathol.*, 153, 1873–1884, 1998.

118. Alfaro-Moreno, E., Martinez, L., Garcia-Cuellar, C., Bonner, J.C., Murray, J.C., Rosas, I. et al., Biologic effects induced *in vitro* by PM_{10} from three different zones of Mexico City, *Environ. Health Perspect.*, 110, 715–720, 2002.

119. Borja-Aburto, V.H., Loomis, D.P., Bangdiwala, S.I., Shy, C.M., Rascon-Pacheco, RA., Ozone, suspended particulates, and daily mortality in Mexico City, *Am. J. Epidemiol.*, 145, 258–268, 1997.

120. Burnett, R.T., Cakmak, S., Brook, J.R., and Krewski, D., The role of particulate size and chemistry in the association between summertime ambient air pollution and hospitalization for cardiorespiratory diseases, *Environ. Health Perspect.*, 105, 614–620, 1997.

121. Laden, F., Neas, L.M., Dockery, D.W., and Schwartz, J., Association of fine particulate matter from different sources with daily mortality in six U.S. cities, *Environ. Health Perspect.*, 108, 941–947, 2000.

122. Brunekreef, B., Janssen, A.H., de Hartog, J., Harssema, H., Knape, M., and van Vliet, P., Air pollution from truck traffic and lung function in children living near motorways, *Epidemiology*, 8, 298–303, 1997.

123. van Vliet, P., Knape, M., de Hartog, J., Janssen, N., Harssema, H., and Brunekreef, B., Motor vehicle exhaust and chronic respiratory symptoms in children living near freeways, *Environ. Res.*, 74, 122–132, 1997.

124. Studnicka, M., Hackl, E., Pischinger, J., Fangmeyer, C., Haschke, N., Kuhr, J. et al., Traffic-related NO_2 and the prevalence of asthma and respiratory symptoms in seven year olds, *Eur. Respir. J.*, 10, 2275–2278, 1997.

125. Schwartz, J., Dockery, D.W., and Neas, L.M., Is daily mortality associated specifically with fine particles, *J. Air Waste Manage. Assoc.*, 46, 927–939, 1996.

126. Schwartz, J., Norris, G., Larson, T., Sheppard, L., Claiborne, C., and Koenig, J., Episodes of high coarse particle concentrations are not associated with increased mortality, *Environ. Health Perspect.*, 107, 339–342, 1999.

127. Schwartz, J. and Neas, L.M., Fine particles are more strongly associated than coarse particles with acute respiratory health effects in schoolchildren, *Epidemiology*, 11, 6–10, 2000.

128. Klemm, R.J., Mason, R.M., Jr., Heilig, C.M., Neas, L.M., and Dockery, D.W., Is daily mortality associated specifically with fine particles? Data reconstruction and replication of analyses, *J. Air Waste Manage. Assoc.*, 50, 1215–1222, 2000.

129. Ostro, B.D., Lipsett, M.J., Wiener, M.B., and Selner, J.C., Asthmatic responses to airborne acid aerosols, *Am. J. Public Health*, 81, 694–702, 1991.

130. Thurston, G.D., Lippmann, M., Scott, M.B., and Fine, J.M., Summertime haze air pollution and children with asthma, *Am. J. Respir. Crit. Care Med.*, 155, 654–660, 1997.

131. Neas, L.M., Dockery, D.W., Koutrakis, P., and Speizer, F.E., Fine particles and peak flow in children: Acidity versus mass, *Epidemiology*, 10, 550–553, 1999.

132. Spengler, J.D. and Wilson, R., Emissions, dispersion, and concentrations of particles, in Wilson, R. and Spengler, J.D., Eds., *Particles in Our Air*, Harvard University Press, Cambridge, MA, 1996, pp. 41–62.

133. Ostro, B.D., Broadwin, R., and Lipsctt, M.J., Coarse and fine particles and daily mortality in the Coachella Valley, California: A follow-up study, *J. Expo. Anal. Environ. Epidemiol.*, 10, 412–419, 2000.

134. Mar, T.F., Norris, G.A., Koenig, J.Q., and Larson, T.V., Associations between air pollution and mortality in Phoenix, 1995–1997, *Environ. Health Perspect.*, 108, 347–353, 2000.

135. Loomis, D., Sizing up air pollution research, *Epidemiology*, 11, 2–4, 2000.

136. Lin, M., Chen, Y., Burnett, R.T., Villeneuve, P.J., and Krewski, D., The influence of ambient coarse particulate matter on asthma hospitalization in children: Case-crossover and time-series analyses, *Environ. Health Perspect.*, 110, 575–581, 2002.

137. Seaton S., MacNee, W., Donaldson, K., and Godden, D., Particulate air pollution and acute health effects, *Lancet*, 345, 176–178, 1995.

138. Wichmann, H.E., Spix, C., Tuch, T., Wolke, G., Peters, A., Heinrich, J. et al., Daily mortality and fine and ultrafine particles in Erfurt, Germany Part I: Role of particle number and particle mass, *Health Eff. Inst. Res. Rep.*, 98, 5–86, 2000; discussion 87–94.

139. Peters, A., Wichmann, H.E., Tuch, T., Heinrich, J., and Heyder, J., Respiratory health effects are associated with the number of ultrafine particles, *Am. J. Respir. Crit. Care Med.*, 155, 1376–1383, 1997.

140. Pekkanen, J., Timonen, K.L., Ruuskanen, J., Reponen, A., and Mirme, A., Effects of ultrafine and fine particles in urban air on peak expiratory flow among children with asthmatic symptoms, *Environ. Res.*, 74, 24–33, 1997.

141. de Hartog, J.J., Hoek, G., Peters, A., Timonen, K.L., Ibald-Mulli, A., Brunekreef, B. et al., Effects of fine and ultrafine particles on cardiorespiratory symptoms in elderly subjects with coronary heart disease, *Am. J. Epidemiol.*, 157, 613–623, 2003.

142. Kunzli, N., Kaiser, R., Medina, S., Studnicka, M., Chanel, O., Filliger, P. et al., Public-health impact of outdoor and traffic-related air pollution: A European assessment, *Lancet*, 356, 795–801, 2000.

143. Bates, D., Health indices of the adverse effects of air pollution: The question of coherence, *Environ. Res.*, 59, 336–349, 1992.

144. Zhu, Y., Hinds, W.C., Kim, S., Shen, S., and Sioutas, C., Study of ultrafine particles near a major highway with heavy-duty diesel traffic, *Atmos. Environ.*, 36, 4375–4383, 2002.

145. Levy, J.I., Houseman, E.A., Ryan, L., Richardson, D., and Spengler, J.D., Particle concentrations in urban microenvironments, *Environ. Health Perspect.*, 108, 1051–1057, 2000.

146. Ezzati, M. and Kammen, D.M., The health impacts of exposure to indoor air pollution from solid fuels in developing countries: Knowledge, gaps, and data needs, *Environ. Health Perspect.*, 110, 1057–1068, 2002.

147. Wallace, L., Correlations of personal exposure to particles with outdoor measurements: A review of recent studies, *Aerosol Sci. Technol.*, 32, 15–25, 2000.

148. National Cancer Institute, Health effects of environmental tobacco smoke: The Report of the California Environmental Protection Agency, NIH 99-4645, Report No.: NIH 99–4645, National Cancer Institute, Bethesda, MD, 1999.

149. Park, J.H., Spiegelman, D.L., Burge, H.A., Gold, D.R., Chew, G.L., and Milton, D.K., Longitudinal study of dust and airborne endotoxin in the home, *Environ. Health Perspect.*, 108, 1023–1028, 2000.

150. Park, J.H., Spiegelman, D.L., Gold, D.R., Burge, H.A., and Milton, D.K., Predictors of airborne endotoxin in the home, *Environ. Health Perspect.*, 109, 859–864, 2001.

151. Gereda, J.E., Leung, D.Y., Thatayatikom, A., Streib, J.E., Price, M.R., Klinnert, M.D. et al., Relation between house-dust endotoxin exposure, type 1 T-cell development, and allergen sensitisation in infants at high risk of asthma, *Lancet*, 355, 1680–1683, 2000.

152. Castellan, R.M., Cotton dust, in Harber, P., Schenker, M.B., and Balmes, J.R., Eds., *Occupational and Environmental Respiratory Disease*, Mosby, New York, 1995, pp. 401–419.

153. Vogelzang, P.F., van der Gulden, J.W., Folgering, H., Kolk, J.J., Heederik, D., Preller, L. et al., Endotoxin exposure as a major determinant of lung function decline in pig farmers, *Am. J. Respir. Crit. Care Med.*, 157, 15–18, 1998.

154. Park, J.H., Gold, D.R., Spiegelman, D.L., Burge, H.A., and Milton, D.K., House dust endotoxin and wheeze in the first year of life, *Am. J. Respir. Crit. Care Med.*, 163, 322–328, 2001.

155. Michel, O., Ginanni, R., Duchateau, J., Vertongen, F., Le Bon, B., and Sergysels, R., Domestic endotoxin exposure and clinical severity of asthma, *Clin. Exp. Allergy*, 21, 441–448, 1991.

156. Michel, O., Kips, J., Duchateau, J., Vertongen, F., Robert, L., Collet, H. et al., Severity of asthma is related to endotoxin in house dust, *Am. J. Respir. Crit. Care Med.*, 154, 1641–1646, 1996.

157. Braun-Fahrlander, C., Riedler, J., Herz, U., Eder, W., Waser, M., Grize, L. et al., Environmental exposure to endotoxin and its relation to asthma in school-age children, *N. Engl. J. Med.*, 347, 869–877, 2002.

158. Baldini, M., Lohman, I.C., Halonen, M., Erickson, R.P., Holt, P.G., and Martinez, F.D., A polymorphism* in the 5' flanking region of the CD14 gene is associated with circulating soluble CD14 levels and with total serum immunoglobulin E, *Am. J. Respir. Cell Mol. Biol.*, 20, 976–983, 1999.

159. Haider, S.C., Sommerfeld, A., Baldini, M., Martinez, F., Wahn, U., and Nickel, R., Evaluation of the CD14 C-159T polymorphism in the German Multicenter Allergy Study cohort, *Clin. Exp. Allergy*, 33, 166–169, 2003.

160. Jones, C.A., Holloway, J.A., Popplewell, E.J., Diaper, N.D., Holloway, J.W., Vance, G.H. et al., Reduced soluble CD14 levels in amniotic fluid and breast milk are associated with the subsequent development of atopy, eczema, or both, *J. Allergy Clin. Immunol.*, 109, 858–866, 2002.

161. Holmlund, U., Hoglind, A., Larsson, A.K., Nilsson, C., and Sverremark Ekstrom, E., CD14 and development of atopic disease at 2 years of age in children with atopic or non-atopic mothers, *Clin. Exp. Allergy*, 33, 455–463, 2003.

162. Downs, J., Zuidhof, A., Doekes, G., van den Zee, A., Wouters, I., Boezen, H.M. et al. (1–>3)-β-D-glucan and endotoxin in house dust and peak flow variability in children, *Am. J. Respir. Crit. Care Med.*, 162, 1348–1354, 2000.

163. Behrendt, H., Becker, W.M., Fritzsch, C., Silwa-Tomczok, W., Friedrichs, K.H., and Ring, J., Air pollution and allergy: Experimental studies on modulation of allergen release from pollen by air pollutants, *Int. Arch. Allergy Immunol.*, 113, 69–74, 1997.

164. Knox, R.B., Suphioglu, C., Taylor, P., Desai, R., Watson, H.C., Peng, J.L. et al., Major grass pollen allergen Lol p 1 binds to diesel exhaust particle: Implications for asthma and air pollution, *Clin. Exp. Allergy*, 27, 246–251, 1997.

165. Behrendt, H., Becker, W.M., Friedrichs, K.H., Darsow, U., and Tomingas, R., Interaction between aeroallergens and airborne particulate matter, *Int. Arch. Allergy Immunol.*, 99, 425–428, 1992.

166. D'Amato, G., Liccardi, G., D'Amato, M., and Cazzola, M., Outdoor air pollution, climatic changes and allergic bronchial asthma, *Eur. Respir. J.*, 20, 763–776, 2002.

167. Emberlin, J., Interaction between air pollution and aeroallergens, *Clin. Exp. Allergy*, 25, 33–39, 1995.

168. Emberlin, J., The effects of air pollution on allergic plants, *Eur. Resp. Rev.*, 53, 164–167, 1995.

169. D'Amato, G., Liccardi, G., D'Amato, M., and Cazzola, M., The role of outdoor air pollution and climatic changes on the rising trends in respiratory allergy, *Respir. Med.*, 95, 606–611, 2001.

170. Rossi, O.V.J., Kinula, V.L., Tienari, J., and Huhti, E., Association of severe asthma attacks with weather, pollen and air pollutants, *Thorax*, 48, 244–248, 1993.

171. Garty, B.Z., Kosman, E., Ganori, E., Berger, V., Garty, L., Wietzen, T. et al., Emergency room visits of asthmatic children, relation to air pollution, weather and airborne allergens, *Ann. Allergy Asthma Immunol.*, 81, 563–570, 1998.

172. Anderson, H.R., Ponce de Leon, A., Bland, J.M., Bower, J.S., Emberlin, J., and Strachan, D.P., Air pollution, pollens, and daily admissions for asthma in London 1987–92, *Thorax*, 53, 842–848, 1998.

173. Lewis, S.A., Corden, J.M., Forster, G.E., and Newland, M., Combined effects of aerobiological pollutants, chemical pollutants and meteorological conditions on asthma admission and A & E attendances in Derbyshire, U.K., 1993–96, *Clin. Exp. Allergy*, 30, 1724–1732, 2000.

174. Miyamoto, T., Epidemiology of pollutant-induced airway disease in Japan, *Allergy*, 52(Suppl. 38), 30–34, 1997.

175. Brunekreef, B., Hoek, G., Fischer, P., and Spieksma, F.T.M., Relation between airborne pollen concentrations and daily cardiovascular and respiratory disease mortality, *Lancet*, 355, 1517–1518, 2000.

176. Wyler, C., Braun-Fahrlander, C., Kunzli, N., Schindler, C., Ackermann-Liebrich, U., Perruchoud, A.P. et al., Exposure to motor vehicle traffic and allergic sensitization, The Swiss Study on Air Pollution and Lung Diseases in Adults (SAPALDIA) Team, *Epidemiology*, 11, 450–456, 2000.

177. Bobak, M. and Leon, D.A., Air pollution and infant mortality in the Czech Republic, 1986–88, *Lancet*, 340, 1010–1014, 1992.

178. Woodruff, T., Grillo, J., and Schoendorf, K., The relationship between selected causes of postneonatal infant mortality and particulate air pollution in the United States, *Environ. Health Perspect.*, 105, 608–612, 1997.

179. Bobak, M. and Leon, D.A., The effect of air pollution on infant mortality appears specific for respiratory causes. *Epidemiology*, 10, 666–670, 1999.

180. Pereira, L.A.A., Loomis, D., Conceicoa, G.M.S., Braga, A.L.F., Arcas, R.M., Kishi, J.S. et al., Association between air pollution and intrauterine mortality in Sao Paulo, Brazil, *Environ. Health Perspect.*, 106, 325–329, 1998.

181. Loomis, D., Castillejos, M., Gold, D.R., McDonnell, W., and Borja-Aburto, V.H., Air pollution and infant mortality in Mexico City, *Epidemiology*, 10, 118–123, 1999.

182. Dejmek, J., Solansky, I., Benes, I., Lenicck, J., and Sram, R.J., The impact of polycyclic aromatic hydrocarbons and fine particles on pregnancy outcome, *Environ. Health Perspect.*, 108, 1159–1164, 2000.

183. Ritz, B., Yu, F., Chapa, G., and Fruin, S., Effect of air pollution on preterm birth among children born in Southern California between 1989 and 1993, *Epidemiology*, 11, 502–511, 2000.

184. Wang, X., Ding, J., Ryan, L., and Xu, X., Association between air pollution and low birth weight: A community-based study, *Environ. Health Perspect.*, 105, 514–520, 1997.

185. Dejmek, J., Jelinek, R., Solansky, I., Benes, I., and Sram, R.J., Fecundability and parental exposure to ambient sulfur dioxide, *Environ. Health Perspect.*, 108, 647–654, 2000.

186. Perera, F.P., Rauh, V., Tsai, W.Y., Kinney, P., Camann, D., Barr, D. et al., Effects of transplacental exposure to environmental pollutants on birth outcomes in a multiethnic population, *Environ. Health Perspect.*, 111, 201–205, 2003.

187. Chay, K.Y. and Greenstone, M., The impact of air pollution on infant mortality: Evidence from geographic variation in pollution shocks induced by a recession, *Q. J. Econ.*, 118, 1121–1167, 2003.

188. Mortimer, K.M., Tager, I.B., Dockery, D.W., Neas, L.M., and Redline, S., The effect of ozone on inner-city children with asthma: Identification of susceptible subgroups, *Am. J. Respir. Crit. Care Med.*, 162, 1838–1845, 2000.

189. Dockery, D.W., Pope, C.A., Xu, X., Spengler, J.D., Ware, J.H., Fay, M.E. et al., An association between air pollution and mortality in six U.S. cities [see comments], *N. Engl. J. Med.*, 329, 1753–1759, 1993.

190. Pope, C.A., Thun, M.J., Namboodiri, M.M., Dockery, D.W., Evans, J.S., Speizer, F.E. et al. Particle air pollution as a predictor of mortality in a prospective study of U.S. adults, *Am. J. Respir. Crit. Care Med.*, 151, 669–674, 1995.

191. Health Effects Institute, *Special Report: Reanalysis of the Harvard Six Cities Study and the American Cancer Society of Particulate Air Pollution and Mortality*, Health Effects Institute, Boston, MA, July 2000.

192. Pope, C.A., 3rd, Burnett, R.T., Thun, M.J., Calle, E.E., Krewski, D., Ito, K. et al., Lung cancer, cardiopulmonary mortality, and long-term exposure to fine particulate air pollution, *JAMA*, 287, 1132–1141, 2002.

193. Villeneuve, P.J., Goldberg, M.S., Krewski, D., Burnett, R.T., and Chen, Y., Fine particulate air pollution and all-cause mortality within the Harvard Six-Cities Study: Variations in risk by period of exposure, *Ann. Epidemiol.*, 12, 568–576, 2002.

194. Abbey, D.E., Nishino, N., McDonnell, W.F., Burchette, R.J., Knutsen, S.F., Lawrence Beeson, W. et al., Long-term inhalable particles and other air pollutants related to mortality in nonsmokers, *Am. J. Respir. Crit. Care Med.*, 159, 373–382, 1999.

195. McDonnell, W.F., Nishino-Ishikawa, N., Petersen, F.F., Chen, L.H., and Abbey, D.E., Relationships of mortality with the fine and coarse fractions of long-term ambient PM_{10} concentrations in nonsmokers, *J. Expo. Anal. Environ. Epidemiol.*, 10, 427–436, 2000.

196. Abbey, D.E., Moore, J., Petersen, F., and Beeson, L., Estimating cumulative ambient concentrations of air pollutants: Description and precision of methods used for an epidemiological study, *Arch. Environ. Health*, 46, 281–287, 1991.

197. Hoek, G., Brunekreef, B., Goldbohm, S., Fischer, P., and van den Brandt, P.A., Association between mortality and indicators of traffic-related air pollution in the Netherlands: A cohort study, *Lancet*, 360, 1203–1209, 2002.

198. Sarnat, J.A., Schwartz, J., Catalano, P.J., and Suh, H.H., Gaseous pollutants in particulate matter epidemiology: Confounders or surrogates? *Environ. Health Perspect.*, 109, 1053–1061, 2001.

199. Mann, S.L., Wadsworth, M.E.J., and Colley, J.R.T., Accumulation of factors influencing respiratory illness in members of a national birth cohort and their offspring, *J. Epidemiol. Commun. Health*, 46, 286–292, 1992.

200. Bates, D.V., A half century later: Recollections of the London fog, *Environ. Health Perspect.*, 110, A735, 2002.

201. Gilliland, F.D., Li, Y.F., and Peters, J.M., Effects of maternal smoking during pregnancy and environmental tobacco smoke on asthma and wheezing in children, *Am. J. Respir. Crit. Care Med.*, 163, 429–436, 2001.

202. Kunzli, N., Schwartz, J., Stutz, E.Z., Ackermann-Liebrich, U., and Leuenberger, P., Association of environmental tobacco smoke at work and forced expiratory lung function among never smoking asthmatics and non-asthmatics, The SAPALDIA-Team. Swiss Study on Air Pollution and Lung Disease in Adults, *Soz. Praventivmed.*, 45, 208–217, 2000.

203. Abbey, D.E., Petersen, F.F., Mills, P.K., and Kittle, L., Chronic respiratory disease associated with long term ambient concentrations of sulfates and other air pollutants, *J. Expo. Anal. Environ. Epidemiol.*, 3, 99–115, 1993.

204. McDonnell, W.F., Abbey, D.E., Nishino, N., and Lebowitz, M.D., Long-term ambient ozone concentration and the incidence of asthma in non-smoking adults: The Ahsmog Study, *Environ. Res.*, 80, 110–121, 1999.

205. Abbey, D., Petersen, F., Mills, P., and Beeson, W., Long-term ambient concentrations of total suspended particulates, ozone, and sulfur dioxide and respiratory symptoms in a nonsmoking population, *Arch. Environ. Health*, 48, 33–46, 1993.

206. McConnell, R., Berhane, K., Gilliland, F., London, S.J., Islam, T., Gauderman, W.J. et al., Asthma in exercising children exposed to ozone: A cohort study, *Lancet*, 359, 386–391, 2002.

207. London, S.J., James Gauderman, W., Avol, E., Rappaport, E.B., and Peters, J.M., Family history and the risk of early-onset persistent, early-onset transient, and late-onset asthma, *Epidemiology*, 12, 577–283, 2001.

208. McConnell, R., Berhane, K., Gilliland, F., Islam, T., Gauderman, W.J., London, S.J. et al., Indoor risk factors for asthma in a prospective study of adolescents, *Epidemiology*, 13, 288–295, 2002.

209. Weiland, S.K., von Mutius, E., Hirsch, T., Duhme, H., Fritzsch, C., Werner, B. et al., Prevalence of respiratory and atopic disorders among children in the East and West of Germany five years after unification, *Eur. Respir. J.*, 14, 862–870, 1999.

210. Von Mutius, E., Martinez, F.D., Fritzsch, C., Nicolai, T., Roell, G., and Thiemann, H.-H., Prevalence of asthma and atopy in two areas of West and East Germany, *Am. J. Respir. Crit. Care Med.*, 149, 358–364, 1994.

211. Heinrich, J., Hoelscher, B., Jacob, B., Wjst, M., and Wichmann, H.E., Trends in allergies among children in a region of former East Germany between 1992–1993 and 1995–1996. *Eur. J. Med. Res.*, 4, 107–113, 1999.

212. von Mutius, E., Weiland, S.K., Fritzsch, C., Duhme, H., and Keil, U., Increasing prevalence of hay fever and atopy among children in Leipzig, East Germany, *Lancet*, 351, 862–826, 1998.

213. Health Effects Institute, *Diesel Emissions and Lung Cancer: Epidemiology and Quantitative Risk Assessment—A Special Report of the Institute's Diesel Epidemiology Expert Panel*, Health Effects Institute, Boston, MA, 1999.

214. U.S. Environmental Protection Agency (EPA), Health assessment document for diesel engine exhaust, Report No.: EPA/600/8-90/057F, National Center for Environmental Assessment, Washington, DC, 2002.

215. Bell, M.L. and Davis, D.L., Reassessment of the lethal London fog of 1952: Novel indicators of acute and chronic consequences of acute exposure to air pollution, *Environ. Health Perspect.*, 109(Suppl. 3), 389–394, 2001.

216. Schwartz, J. and Marcus, A., Mortality and air pollution in London: A time series analysis, *Am. J. Epidemiol.*, 131, 185–194, 1990.

217. Thurston, G.D., Ito, K., Lippmann, M., and Hayes, C., Reexamination of London, England, mortality in relation to exposure to acidic aerosols during the 1963–1972 winters, *Environ. Health Perspect.*, 79, 73–82, 1989.

218. Schwartz, J., Air pollution and daily mortality: A review and meta analysis, *Environ. Res.*, 64, 36–52, 1994.

219. Schwartz, J., What are people dying of on high air pollution days? *Environ. Res.*, 64, 26–35, 1994.

220. Sunyer, J., Schwartz, J., Tobias, A., Macfarlane, D., Garcia, J., and Anto, J.M., Patients with chronic obstructive pulmonary disease are at increased risk of death associated with urban particle air pollution: A case-crossover analysis, *Am. J. Epidemiol.*, 151, 50–56, 2000.

221. De Leon, S.F., Thurston, G.D., and Ito, K., Contribution of respiratory disease to nonrespiratory mortality associations with air pollution, *Am. J. Respir. Crit. Care Med.*, 167, 1117–1123, 2003.

222. Levy, D., Sheppard, L., Checkoway, H., Kaufman, J., Lumley, T., Koenig, J. et al., A case-crossover analysis of particulate matter air pollution and out-of-hospital primary cardiac arrest, *Epidemiology*, 12, 193–199, 2001.

223. Clancy, L., Goodman, P., Sinclair, H., and Dockery, D.W., Effect of air pollution control on death rates in Dublin, Ireland: An intervention study, *Lancet*, 360, 1210–1214, 2002.

224. Hong, Y.C., Lee, J.T., Kim, H., Ha, E.-H., Schwartz, J., and Christiani, D.C., Effects of air pollutants on acute stroke mortality, *Environ. Health Perspect.*, 110, 187–191, 2002.

225. Tellez-Rojo, M.M., Romieu, I., Velasco, S.R., Lezana, M.-A., and Hernandez, M.M., Daily respiratory mortality and PM_{10} pollution in Mexico City: Importance of considering place of death, *Eur. Respir. J.*, 16, 391–396, 2000.

226. Colburn, K.A. and Johnson, P.R.S., Air pollution concerns not changed by S-PLUS, *Science*, 299, 665–666, 2003.

227. A Special Panel of the Health Review Committee of the Health Effects Institute, Commentary on revised analyses of selected studies, *Special Report: Revised Analyses of Time-Series Studies of Air Pollution and Health*, Health Effects Institute, Ed., Health Effects Institute, Boston, MA, 2003, pp. 255–291.

228. Katsouyanni, K., Touloumi, G., Samoli, E., Petasakis, Y., Analitis, A., Le Tertre, A. et al., Sensitivity analysis of various models of short-term effects of ambient particles on total mortality in 29 cities in APHEA2, *Special Report: Revised Analyses of Time-Series Studies of Air Pollution and Health*, Health Effects Institute, Ed., Health Effects Institute, Boston, MA, 2003, pp. 157–164.

229. Levy, J.I., Hammitt, J.K., and Spengler, J.D., Estimating the mortality impacts from particulate matter: What can be learned from between study variability, *Environ. Health Perspect.*, 108, 109–117, 2000.

230. Zanobetti, A. and Schwartz, J., Are diabetics more susceptible to the health effects of airborne particles? *Am. J. Respir. Crit. Care Med.*, 164, 831–833, 2001.

231. Goldberg, M.S., Burnett, R.T., Bailar, J.C., 3rd, Tamblyn, R., Ernst, P., Flegel, K. et al., Identification of persons with cardiorespiratory conditions who are at risk of dying from the acute effects of ambient air particles, *Environ. Health Perspect.*, 109(Suppl 4), 487–494, 2001.

232. Schwartz, J. and Zanobetti, A., Using meta-smoothing to estimate dose-response relationship trends across multiple studies, with application to air pollution and daily death, *Epidemiology*, 11, 666–672, 2000.

233. Schwartz, J., Assessing confounding, effect modification, and threshold in the association between ambient particles and daily deaths, *Environ. Health Perspect.*, 108, 563–568, 2000.

234. Schwartz, J., Ballester, F., Saez, M., Perez-Hoyos, S., Bellido, J., Cambra, K. et al., The concentration-response relation between air pollution and daily deaths, *Environ. Health Perspect.*, 109, 1001–1006, 2001.

235. Schwartz, J., Laden, F., and Zanobetti, A., The concentration-response relation between $PM_{2.5}$ and daily deaths, *Environ. Health Perspect.*, 110, 1025–1029, 2002.

236. Daniels, M., Dominici, F., Samet, J.M., and Zeger, S.L., Estimating particulate matter-mortality dose-response curves and threshold levels: An analysis of daily time-series for the 20 largest US cities, *Am. J. Epidemiol.*, 152, 397–406, 2000.

237. Pope, C.A., Invited commentary: Particulate matter-mortality exposure-response relations and threshold, *Am. J. Epidemiol.*, 152, 407–412, 2000.

238. Dominici, F., Daniels, M., McDermott, A., Zeger, S.L., and Samet, J.L., Shape of the exposure-response relation and mortality displacement in the NMMAPS study, *Special Report: Revised Analyses of Time-Series Studies of Air Pollution and Health*, Health Effects Institute, Ed., Health Effects Institute, Boston, MA, 2003, pp. 91–96.

239. Lipfert, F.W. and Wyzga, R.E., Air pollution and mortality: The implications of uncertainties in regression modeling and exposure measurement, *J. Air Waste Manage. Assoc.*, 47, 517–523, 1997.

240. Murray, C.J. and Nelson, R.N., State-space modeling of the relationship between air quality and mortality, *J. Air Waste Manage. Assoc.*, 50, 1075–1080, 2000.

241. Zeger, S.L., Dominici, F., and Samet, J.M., Harvesting-resistant estimates of air pollution effects on mortality, *Ann. Epidemiol.*, 10, 171–175, 1999.

242. Schwartz, J., Harvesting and long term exposure effects in the relation between air pollution and mortality, *Am. J. Epidemiol.*, 151, 440–448, 2000.

243. Schwartz, J., Is there harvesting in the association of airborne particles with daily deaths and hospital admissions, *Epidemiology*, 12, 55–61, 2001.

244. Zanobetti, A., Schwartz, J., Samoli, E., Gryparis, A., Touloumi, G., Atkinson, R. et al., The temporal pattern of mortality responses to air pollution: A multicity assessment of mortality displacement, *Epidemiology*, 13, 87–93, 2002.

245. Dominici, F., McDermott, A., Zeger, S.L., and Samet, J.M., Airborne particulate matter and mortality: Timescale effects in four US cities, *Am. J. Epidemiol.*, 157, 1055–1065, 2003.

246. Smith, R.L., Invited commentary: Timescale-dependent mortality effects of air pollution, *Am. J. Epidemiol.*, 157, 1066–1070, 2003.

247. Dominici, F., McDermott, A., Zeger, S.L., and Samet, J.M., Response to Dr. Smith: Timescale-dependent mortality effects of air pollution, *Am. J. Epidemiol.*, 157, 1071–1073, 2003.

248. Fan, J. and Watanabe, T., Inflammatory reactions in the pathogenesis of atherosclerosis, *J. Atheroscler. Thromb.*, 10, 63–71, 2003.

249. Task Force of the European Society of Cardiology, Heart rate variability: Standards of measurement, physiological interpretation and clinical use, *Circulation*, 93, 1043–1065, 1996.

250. Tsuji, H., Venditti, F.J., Jr., Manders, E.S., Evans, J.C., Larson, M.G., Feldman, C.L. et al., Reduced heart rate variability and mortality risk in an elderly cohort: The Framingham Heart Study, *Circulation*, 90, 878–883, 1994.

251. Tsuji, H., Larson, M.G., Venditti, F.J., Jr., Manders, E.S., Evans, J.C., Feldman, C.L. et al., Impact of reduced heart rate variability on risk for cardiac events: The Framingham Heart Study, *Circulation*, 94, 2850–2855, 1996.

252. Pope, C.A., 3rd, Verrier, R.L., Lovett, E.G., Larson, A.C., Raizenne, M.E., Kanner, R.E. et al., Heart rate variability associated with particulate air pollution, *Am. Heart J.*, 138, 890–899, 1999.

253. Liao, D., Creason, J., Shy, C.M., Williams, R., Watts, R., and Zweidiner, R., Daily variation of particulate air pollution and poor cardiac autonomic function in the elderly, *Environ. Health Perspect.*, 107, 521–525, 1999.

254. Gold, D.R., Litonjua, A., Schwartz, J., Lovett, E., Larson, A., Nearing, B. et al., Ambient pollution and heart rate variability, *Circulation*, 101, 1267–1273, 2000.

255. Magari, S.R., Hauser, R., Schwartz, J., Williams, P.L., Smith, T.J., and Christiani, D.C., Association of heart rate variability with occupational and environmental exposure to particulate air pollution, *Circulation*, 104, 986–991, 2001.

256. Magari, S.R., Schwartz, J., Williams, P.L., Hauser, R., Smith, T.J., and Christiani, D.C., The association between personal measurements of environmental exposure to particulates and heart rate variability, *Epidemiology*, 13, 305–310, 2002.

257. Magari, S.R., Schwartz, J., Williams, P.L., Hauser, R., Smith, T.J., and Christiani, D.C., The association of particulate air metal concentrations with heart rate variability, *Environ. Health Perspect.*, 110, 875–880, 2002.

258. Peters, A., Liu, E., Verrier, R.L., Schwartz, J., Gold, D.R., Milttleman, M. et al., Air pollution and incidence of cardiac arrhythmias, *Epidemiology*, 11, 11–17, 2000.

259. Gaspoz, J.M., Coxson, P.G., Goldman, P.A., Williams, L.W., Kuntz, K.M., Hunink, M.G. et al., Cost effectiveness of aspirin, clopidogrel, or both for secondary prevention of coronary heart disease, *N. Engl. J. Med.*, 346, 1800–1806, 2002.

260. Peters, A., Doring, A., Wichmann, H.-E., and Koenig, W., Increased plasma viscosity during an air pollution episode: A link to mortality, *Lancet*, 349, 1582–1587, 1997.

261. Seaton, A., Soutar, A., Crawford, V., Elton, R., McNerlan, S., Cherrie, J. et al., Particulate air pollution and the blood, *Thorax*, 54, 1027–1032, 1999.

262. Batalha, J.R., Saldiva, P.H., Clarke, R.W., Coull, B.A., Stearns, R.C., Lawrence, J. et al. Concentrated ambient air particles induce vasoconstriction of small pulmonary arteries in rats, *Environ. Health Perspect.*, 110, 1191–1197, 2002.

263. Brook, R.D., Brook, J.R., Urch, B., Vincent, R., Rajagopalan, S., and Silverman, F., Inhalation of fine particulate air pollution and ozone causes acute arterial vasoconstriction in healthy adults, *Circulation*, 105, 1534–1536, 2002.

264. U.S. Environmental Protection Agency (U.S. EPA), Fourth external review draft of air quality criteria for particulate matter (June, 2003), Volume II, U.S. EPA, Report No.: EPA/600/P-99/002aD, Research Triangle Park, NC, 2003.

265. Wilson, W.E. and Suh, H.H., Fine particles and coarse particles: Concentration relationships relevant to epidemiologic studies, *J. Air Waste Manage. Assoc.*, 47, 1238–1249, 1997.

266. Health Effects Institute, *Understanding the Health Effects of the Particulate Matter Mix: Progress and Next Steps*, Perspectives, Health Effects Institute, Boston, MA, April 2002.

267. McDonald, B. and Ouyang, M., Air cleaning—Particles, in Spengler, J.D., Samet, J.M., and McCarthy, J.F., Eds., *Indoor Air Quality Handbook*, McGraw-Hill, New York, 2000, pp. 9.1–9.28.

268. Institute of Medicine Committee on the Assessment of Asthma and Indoor Air, *Clearing the Air: Asthma and Indoor Air Exposures*, National Academy Press, Washington, DC, 2000.

269. Huttunen, K., Hyvarinen, A., Nevalainen, A., Komulainen, H., and Hirvonen, M.R., Production of proinflammatory mediators by indoor air bacteria and fungal spores in mouse and human cell lines, *Environ. Health Perspect.*, 111, 85–92, 2003.

270. Wilson, S.R. and Spengler, J.D., Emissions, dispersion, and concentration of particles, in Wilson, S.R. and Spengler, J.D., Eds., *Particles in Our Air: Concentrations and Health Effects*, Harvard University Press, Cambridge, MA, 1996, pp. 41–62.

271. Murray, J.F., *The Normal Lung*, W.B. Saunders Co, Philadelphia, PA, 1986.

272. Jaques, P.A. and Kim, C.S., Measurement of total lung deposition of inhaled ultrafine particles in healthy men and women, *Inhal. Toxicol.*, 12, 715–731, 2000.

273. Kim, C.S. and Kang, T.C., Comparative measurement of lung deposition of inhaled fine particles in normal subjects and patients with obstructive airway disease, *Am. J. Respir. Crit. Care Med.*, 155, 899–905, 1997.

274. Mathew, O.P. and Sant'Ambrogio, G., Development of upper airway reflexes, in Cherniack, V. and Mellins, R.B., Eds., *Basic Mechanisms of Pediatric Respiratory Disease: Cellular and Integrative*, BC Decker, Philadelphia, PA, 1991, pp. 55–71.

275. Mandell, G.L., Bennett, J.E., and Dolin, R., Eds., *Principles and Practice of Infectious Diseases*, Churchill Livingstone, New York, 2000.

276. Diaz-Sanchez, D., Zhang, K., Nutman, T.B., and Saxon, A., Differential regulation of alternative 3′ splicing of epsilon messenger RNA variants, *J. Immunol.*, 155, 1930–1941, 1995.

277. Diaz-Sanchez, D., Tsien, A., Casillas, A., Dotson, A.R., and Saxton, A., Enhanced nasal cytokine production in human beings after *in vivo* challenge with diesel exhaust particles, *J. Allergy. Clin. Immunol.*, 98, 114–123, 1996.

278. Diaz-Sanchez, D., Jyrala, M., Ng, D., Nel, A., and Saxon, A., *In vivo* nasal challenge with diesel exhaust particles enhances expression of the CC chemokines rantes, MIP-1alpha, and MCP-3 in humans, *Clin. Immunol.*, 97, 140–145, 2000.

279. Wjst, M., Reitmeir, P., Dold, S., Wulff, A., Nicolai, T., von Loeffelholz-Colberg, E.F. et al., Road traffic and adverse effects on respiratory health in children, *Br. Med. J.*, 307, 596–600, 1993.

280. Kunzli, N. and Tager, I.B., The semi-individual study in air pollution epidemiology: A valid design as compared to ecologic studies, *Environ. Health Perspect.*, 105, 1078–1083, 1997.

281. Edwards, J., Walters, S., and Griffiths, R.K., Hospital admissions for asthma in preschool children: Relationship to major roads in Birmingham, United Kingdom, *Arch. Environ. Health*, 49, 223–227, 1994.

282. Oosterlee, A., Drijver, M., Lebret, E., and Brunekreef, B., Chronic respiratory symptoms in children and adults living along streets with high traffic density, *Occup. Environ. Med.*, 53, 241–247, 1996.

283. English, P., Neutra, R., Scalf, R., Sullivan, M., Waller, L., and Zhu, L., Examining associations between childhood asthma and traffic flow using a geographic information system, *Environ. Health Perspect.*, 107, 761–767, 1999.

284. Wilkinson, P., Elliott, P., Grundy, C., Shaddick, G., Thakrar, B., Walls, P. et al., Case-control study of hospital admission with asthma in children aged 5–14 years: Relation with road traffic in north west London, *Thorax*, 54, 1070–1074, 1999.

285. Venn, A., Lewis, S., Cooper, M., Hubbard, R., Hill, I., Boddy, R. et al., Local road traffic activity and the prevalence, severity, and persistence of wheeze in school children: Combined cross sectional and longitudinal study, *Occup. Environ. Med.*, 57, 152–158, 2000.

286. Venn, A.J., Lewis, S.A., Cooper, M., Hubbard, R., and Britton, J., Living near a main road and the risk of wheezing illness in children, *Am. J. Respir. Crit. Care Med.*, 164, 2177–2180, 2001.

287. Roemer, W.H. and van Wijnen, J.H., Daily mortality and air pollution along busy streets in Amsterdam, 1987–1998, *Epidemiology*, 12, 649–653, 2001.

288. Lin, S., Munsie, J.P., Hwang, S.A., Fitzgerald, E., and Cayo, M.R., Childhood asthma hospitalization and residential exposure to state route traffic, *Environ. Res.*, 88, 73–81, 2002.

289. Brauer, M., Hoek, G., Van Vliet, P., Meliefste, K., Fischer, P.H., Wijga, A. et al., Air pollution from traffic and the development of respiratory infections and asthmatic and allergic symptoms in children, *Am. J. Respir. Crit. Care Med.*, 166, 1092–1098, 2002.

290. Wilhelm, M. and Ritz, B., Residential proximity to traffic and adverse birth outcomes in Los Angeles county, California, 1994–1996, *Environ. Health Perspect.*, 111, 207–216, 2003.

291. Dolk, H., Pattenden, S., Vrijheid, M., Thakrar, B., and Armstrong, B., Perinatal and infant mortality and low birth weight among residents near cokeworks in Great Britain, *Arch. Environ. Health*, 55, 26–30, 2000.

292. Bobak, M. and Leon, D.A., Pregnancy outcomes and outdoor air pollution: An ecological study in districts of the Czech Republic 1986–8, *Occup. Environ. Med.*, 56, 539–543, 1999.

293. Dejmek, J., Selevan, S.G., Benes, I., Solansky, I., and Sram, R.J., Fetal growth and maternal exposure to particulate matter during pregnancy, *Environ. Health Perspect.*, 107, 475–480, 1999.

294. Rogers, J.F., Thompson, S.J., Addy, C.L., McKeown, R.E., Cowen, D.J., and Decoufle, P., Association of very low birth weight with exposures to environmental sulfur dioxide and total suspended particulates, *Am. J. Epidemiol.*, 151, 602–613, 2000.

295. Bobak, M., Outdoor air pollution, low birth weight, and prematurity, *Environ. Health Perspect.*, 108, 173–176, 2000.

296. Ha, E.H., Hong, Y.C., Lee, B.E., Woo, B.H., Schwartz, J., and Christiani, D.C., Is air pollution a risk factor for low birth weight in Seoul? *Epidemiology*, 12, 643–648, 2001.

297. Ritz, B., Yu, F., Fruin, S., Chapa, G., Shaw, G.M., and Harris, J.A., Ambient air pollution and risk of birth defects in Southern California, *Am. J. Epidemiol.*, 155, 17–25, 2002.

298. Xu, X., Ding, H., and Wang, X., Acute effects of total suspended particles and sulfur dioxides on preterm delivery: A community-based cohort study, *Arch. Environ. Health*, 50, 407–415, 1995.

299. Schwartz, J. and Dockery, D.W., Increased mortality in Philadelphia associated with daily air pollution concentrations, *Am. Rev. Respir. Dis.*, 145, 600–604, 1992.

300. Gwynn, R.C., Burnett, R.T., and Thurston, G.D., A time-series analysis of acidic particulate matter and daily mortality and morbidity in the Buffalo, New York, region, *Environ. Health Perspect.*, 108, 125–133, 2000.

301. Ostro, B.D., Hurley, S., and Lipsett, M.J., Air pollution and daily mortality in the Coachella Valley, California: A study of PM_{10} dominated by coarse particles, *Environ. Res.*, 81, 231–238, 1999.

302. Morgan, G., Corbett, S., Wlodarczyk, J., and Lewis, P., Air pollution and daily mortality in Sydney, Australia, 1989 through 1993, *Am. J. Public Health*, 88, 759–764, 1988.

303. Vedal, S., Brauer, M., White, R., and Petkau, J., Air pollution and daily mortality in a city with low levels of pollution, *Environ. Health Perspect.*, 111, 45–51, 2003.

304. Bateson, T.F. and Schwartz, J., Control of seasonal variation and time trends in case-crossover studies of the acute effects of environmental exposures, *Epidemiology*, 10, 539–544, 1999.

305. Neas, L.M., Schwartz, J., and Dockery, D., A case-crossover analysis of air pollution and mortality in Philadelphia, *Environ. Health Perspect.*, 107, 629–631, 1999.

306. Lee, J.T., Shin, D., and Chung, Y., Air pollution and daily mortality in Seoul and Ulsan, Korea, *Environ. Health Perspect.*, 107, 149–154, 1999.

307. Samet, J.M., Dominici, F., Zeger, S.L., Schwartz, J., and Dockery, D.W., The National morbidity, Mortality, And Air Pollution Study. Part I: Methods and methodologic issues, *Res. Rep. Health Eff. Inst.*, 94(Part 1), 5–14, 2000.

308. Samet, J.M., Zeger, S.L., Dominici, F., Curriero, F., Coursac, I., Dockery, D.W. et al., The National Morbidity, Mortality, And Air Pollution Study. Part II: Morbidity and mortality from air pollution in the United States, *Res. Rep. Health Eff. Inst.*, 94(Part 2), 5–70, 2000.

309. Samet, J.M., Dominici, F., Curriero, F.C., Coursac, I., and Zeger, S.L., Fine particulate air pollution and mortality in 20 U.S. cities, 1987–1994, *N. Engl. J. Med.*, 343, 1742–1749, 2000.

310. Dominici, F., McDermott, A., Daniels, D., Zeger, S.L., and Samet, J.L., Mortality among residents of 90 cities, *Special Report: Revised Analyses of Time-Series Studies of Air Pollution and Health*, Health Effects Institute, Ed., Health Effects Institute, Boston, MA, 2003, pp. 9–24.

311. Touloumi, G., Katsouyanni, K., Zmirou, D., Schwartz, J., Spix, C., Ponce de Leon, A. et al., Short-term effects of ambient oxidant exposure on mortality: A combined analysis of the APHEA project, *Am. J. Epidemiol.*, 146, 177–185, 1997.

312. Katsouyanni, K., Touloumi, G., Spix, C., Schwartz, J., Baldacci, F., Medina, S. et al., Short term effects of ambient sulphur dioxide and particulate matter on mortality in 12 European cities: Results from time series data from the APHEA project, *Br. Med. J.*, 314, 1658–1663, 1997.

313. Zmirou, D., Schwartz, J., Saez, M., Zanobetti, A., Wojtyniak, B., Touloumi, G. et al., Time-series analysis of air pollution and cause-specific mortality, *Epidemiology*, 9, 495–503, 1998.

314. Katsouyanni, K., Touloumi, G., Samoli, E., Gryparis, A., Le Tertre, A., Monopolis Y. et al., Confounding and effect modification in the short-term-effects of ambient particles on total mortality: Results from 29 European cities within the APHEA2 project, *Epidemiology*, 12, 521–531, 2000.

315. Samoli, E., Schwartz, J., Wojtyniak, B., Touloumi, G., Spix, C., Baldacci, F. et al., Investigation regional differences in short-term effects of air pollution on daily mortality in the APHEA project: A sensitivity analysis for controlling long-term trends and seasonality, *Environ. Health Perspect.*, 109, 349–353, 2001.

316. Pope, C.A., 3rd, Hill, R.W., and Villegas, G.M., Particulate air pollution and daily mortality on Utah's Wasatch Front, *Environ. Health Perspect.*, 107, 567–573, 1999.

317. Loomis, D., Castillejos, M., Borja-Aburto, V.H., and Dockery, D.W., Stronger effects of coarse particles in Mexico City, in Phalen R. and Bell, Y., Eds., *Proceedings of the Third Colloquium Particulate Air Pollution and Human Health*, Durham, NC, June 6–8, 1999.

318. Cifuentes, L.A., Vega, J., Kopfer, K., and Lave, L.B., Effect of the fine fraction of particulate matter versus the coarse mass and other pollutants on daily mortality in Santiago, Chile, *J. Air Waste Manage. Assoc.*, 50, 1287–1298, 2000.

319. Casarett, L.J., Toxicololgy of the respiratory system, in Casarett, L.J. and Doull, J., Eds., *Toxicology: The Basic Science of Poisons.*, MacMillan Publishing Co., New York, 1975, pp. 201–224.

320. Lippmann, M. and Schlesinger, R.B., Interspecies comparisons of particle deposition and mucociliary clearance in tracheobronchial airways, *J. Toxicol. Environ. Health*, 13, 441–469, 1984.

321. Morrow, P.E., Task Group on Lung Dynamics: Deposition and retention models for internal dosimetry of the human respiratory tract, *Health Phys.*, 12, 173–207, 1966.

322. International Commission on Radiological Protection, Human respiratory tract model for radiological protection, A report of a Task Group of the International Commission on Radiological Protection, *Ann. ICRP*, 24, 1–482, 1994.

323. Kay, A.B., Allergy and allergic diseases, *N. Engl. J. Med.*, 344, 30–37, 2001.

324. Busse, W.W. and Lemanske, R.F., Jr., Asthma, *N. Engl. J. Med.*, 344, 350–362, 2001.

325. Dominici, F., McDermott, A., Zeger, S.L., and Samet, J.M., National maps of the effects of particulate matter on mortality: Exploring geographic variation, *Environ. Health Perspect.*, 111, 39–43, 2003.

326. Federal Register, National ambient air quality standard for particulate matter, Report No.: 40 CRF Part 50, National Archives and Records Administration, Washington, DC, December 1996.

Index